Best Practices for Environmental Health

Environmental Pollution, Protection, Quality and Sustainability

Best Practices for Public Health

https://www.crcpress.com/Best-Practices-for-Public-Health/book-series/BPPH

Edited by Herman Koren, Indiana State University, USA.

This series of books offers public health students, academics and practitioners invaluable practical guides. It adopts a global perspective and aims to drive practical action in an increasingly globalized world.

Best Practices for Environmental Health

Environmental Pollution, Protection, Quality and Sustainability
Herman Koren

Best Practices for Environmental Health

Environmental Pollution, Protection, Quality and Sustainability

Herman Koren

Routledge
Taylor & Francis Group
New York London

Books by Herman Koren Published by CRC Press/Routledge

Best Practices for Environmental Health: Environmental Pollution, Protection, Quality and Sustainability

Illustrated Dictionary and Resource Directory of Environmental and Occupational Health 1st, 2nd editions

Handbook of Environmental Health (with Michael Bisesi), volumes 1 and 2, 3rd, 4th editions

Handbook of Environmental Health and Safety, volumes 1 and 2, 2nd edition

Management and Supervision for Working Professionals, volumes 1 and 2, 3rd edition

Other books by Herman Koren

Histories of the Jewish People of Pinellas County, Florida 1883–2005

150th Anniversary of the Jewish Community in Terre Haute, Indiana 1849–1999

Handbook of Environmental Health and Safety, volumes 1 and 2, 1st edition

Basic Supervision and Basic Management, volumes 1 and 2, 1st, 2nd editions

Environmental Health and Safety

First published 2017
by Routledge
711 Third Avenue, New York, NY 10017

and by Routledge
2 Park Square, Milton Park, Abingdon, Oxon, OX14 4RN

Routledge is an imprint of the Taylor & Francis Group, an informa business

© 2017 Taylor & Francis

Library of Congress Cataloging-in-Publication Data

Names: Koren, Herman, author.
Title: Best practices for environmental health : environmental pollution,
protection, quality and sustainability / Herman Koren.
Description: Boca Raton, FL : CRC Press, [2017] | Includes bibliographical references and index.
Identifiers: LCCN 2016037982| ISBN 9781498700221 (hbk) | ISBN 9781138196407 (pbk.) |
ISBN 9781498700238 (ebk)
Subjects: LCSH: Environmental health. | Environmental protection. |
Environmental quality. | Environmental sustainability.
Classification: LCC RA565 .K659 2017 | DDC 362.1--dc23
LC record available at https://lccn.loc.gov/2016037982

ISBN: 978-1-4987-0022-1 (hbk)
ISBN: 978-1-138-19640-7 (pbk)
ISBN: 978-1-315-11927-4 (ebk)

Typeset in Times LT Std
by Datapage International Ltd.

Printed and bound in the United States of America by
Edwards Brothers Malloy on sustainably sourced paper

Dedication

To my darling wife, Donna, my very best friend ever, who has edited and critiqued much of my work over the last 45 years, has kept me grounded, has provided the sweetest life that any human being could ever have, and has allowed me to achieve such riches in my writings that I can forever help teach my younger colleagues not only about environmental health, environmental issues, and public health but also about the joys of life.

Contents

Foreword xxxi
Preface...xxxv
Acknowledgments...xlvii
Author xlix

Chapter 1 Necessity for Effective Programming Techniques.......................................1

Author's Thoughts...1
Introduction 1
Major Concerns in Our Society Today..2
    Financial Stress 2
    Regulatory Reform in Environmental Health and Protection Programs 3
    Environmental Justice 4
    Our Changing Planet 5
Environmental Pollutants and Their Interactions 6
    Environmental Cancer..7
Emerging and Re-Emerging Infectious and Zoonotic Diseases 8
Where We Live, Work, Play, and Are Educated 9
Mobile, Growing, and Aging Population 9
    Mobile and Growing Population 9
    Aging Population..9
Sustainability...10
History of Sustainability in the United States..10
    Introduction 10
    Laws...11
    President's Council on Sustainable Development 12
    Recent Executive Orders Concerning Environmental Sustainability 13
History of Global Sustainability 13
    Introduction 13
    United Nations...13
        Agenda 21 Report, June 1992 14
Global Climate Change...14
Programs Which Achieve the Goal of Preventing Disease and Injury and
Protecting the Environment 15
    The Past 15
        How Programs Were Started 15
    The Present 16
Components of Successful Environmental Health, Safety, Protection, and
Sustainability Programs 16
    Example of a Successful Program: Community Rodent Control,
    Philadelphia, 1959–1963..17
        Statement of the Problem...17
        Solutions to the Problem...17

Stakeholders..17
The Actual Program Content..18
Funding...18
Evaluation and Results...18
Redirection 18
Summary 18
Conclusion 19
Endnotes 19
Bibliography 20

Chapter 2 Air Quality (Outdoor [Ambient] and Indoor)....................................21

Statement of Problem and Special Information for Outdoor Air Quality 21
Constituents of Air Pollution..21
Gases 21
Particulate Matter 21
Liquids 21
Origins of Air Pollutants 21
Health Effects 22
Environmental Effects..22
Types of Air Pollutants..23
Criteria Air Pollutants 23
Toxic Air Pollutants 23
Greenhouse Gases 23
Stratospheric Ozone Layer Depletion 23
Current Status...23
Climate Change 24
Measures of Air Quality...25
Standards for Criteria Pollutants 25
National Air Toxics Assessments and Toxics Release Inventory 25
Ozone Action Day..26
Air Quality Index...26
General Best Practices for All Sources of Air Pollution..................................26
Pollution Prevention as Part of Best Practices...27
Sub-Problems Including Leading to Impairment and Best Practices for
Ambient Air Quality 27
Criteria Pollutants..27
The Six Criteria Air Pollutants..27
Other Major Specific Sources of Air Pollution (Including Toxic
Air Pollutants) 33
Fracking...43
Glycol Dehydration Unit...44
Natural Gas Flaring 44
Oil and Natural Gas Production Facilities...44
Sub-Problems Including Leading to Impairment and Best Practices for
Toxic Air Pollution 48
Sources of Toxic Air Pollutants...51
Major Point Sources..51
Area and Other Sources (Smaller Sources) 51
The Aerospace Manufacturing and Rework Industry 51
Accidental Chemical Releases...51

Auto Body and Repair Shops 52
Drycleaners 53
Gas Stations and Gasoline Distribution Facilities 53
Hospital Medical and Infectious Waste Incinerators 53
Lawnmowers and Other Garden Equipment 53
Off-Site Waste Operations 54
Printing and Publishing 54
Shipyards 54
Small Chemical Plants 55
Surface Coating and Painting of Materials 55
Wood Furniture Manufacturing 55
Sub-Problems Including Leading to Impairment and Best Practices for
Greenhouse Gases 56
Sources of Greenhouse Gases 57
Electricity Production and Transmission 57
Mobile 57
Industries 57
Commercial and Residential 58
Agriculture 58
Indoor Air Quality 59
Statement of Problem and Special Information for Indoor Air Quality 59
Health Effects 59
Sub-Problems Including Leading to Impairment and Best Practices for
Indoor Air Quality 60
Laws, Rules, and Regulations 66
Clean Air Act and the Economy 68
Retrospective Study, 1970–1990 68
Prospective Study, 1990–2020 69
Resources 69
Successful Programs 71
General Control Strategies for Developing Programs for All Sources of
Air Pollution 71
Global Methane Initiative 71
Portland, Oregon, Air Toxics Solutions 72
Endnotes 73
Bibliography 77

Chapter 3   Built Environment—Healthy Homes and Healthy Communities 79

Statement of Problem and Special Information 79
Sub-Problems Including Leading to Impairment and Best Practices for the
Built Environment in Urban Areas 80
History of the Built Environment and Housing in the United States 80
Poverty and Homelessness 82
Sub-Problems Including Leading to Impairment for the Built Environment
in Urban Areas 82
Sub-Problems Including Leading to Impairment for the Built Environment
in Suburban Areas 85
Sub-Problems Including Leading to Impairment for the Built Environment
in Rural Areas 88
Rural Areas with Poverty, Lack of Resources, and Jobs 89

Sub-Problems Including Leading to Impairment for Housing..............................91
    Site and Grounds 91
    Substandard Housing...92
    Basic Sanitation 92
    Temperature Extremes...92
Specific Environmental Problems 93
    Air Pollution 93
    Asbestos...93
    Brownfield Sites...93
    Chromated Copper Arsenic..94
    Combustion Pollutants...94
    Drywall (from China) 95
    Emergencies and Disasters 95
    Environmental Injuries...95
    Food Protection 95
    Formaldehyde 95
    Hazardous Household Waste 96
    Indoor Air Pollution 96
    Insects and Rodents 96
    Lead 97
    Mold...97
    Noise 98
    On-Site Sewage Disposal 99
    On-Site Water Source 100
    Pesticides 100
    Poisoning 100
    Solid Waste Disposal 100
    Swimming Areas (Residential) 100
Laws, Rules, and Regulations 101
    Zoning Ordinances...101
    Building and Housing Codes..101
Special Resources...101
Successful Programs 101
    Dublin, Ohio, and the Bridge Street Corridor 102
    Aurora Corridor, Shoreline, Washington 102
    Belmar, Lakewood, Colorado...102
    City Centre, Houston, Texas 102
    Indianapolis, Indiana 102
    Worchester, Massachusetts 103
    Greenville, South Carolina...103
    Baltimore, Maryland, Housing Authority 103
Endnotes 104
Bibliography 105

Chapter 4   Children's Environmental Health Issues 107

Statement of Problem and Special Information 107
    Introduction 107
    Past 107
    Present 108

Uniqueness and Stages of Development of the Child Affected
by Environmental Stresses 110
    Preconception 111
    Placenta 111
    Fetus 112
    Early Childhood (0–2 Years) 113
    Young Child (2–6 Years) 114
    Younger School-Aged Child (6–12 Years) 114
    Adolescent (12–18 Years) 114
    Special Circumstances and Risk Factors 114
Societal Risk 115
    The Economy 115
    Homelessness 115
    Hunger 115
    Poverty 116
    Poor Medical Care 116
    Language Barriers 116
    Climate Change 117
Economic Effects of Pollutants 117
Special Problems of Where We Reside and Risk Factors 118
    Urban Environment 118
    Suburban Environment 119
    Agricultural Environment 119
Factors Leading to Impairment and Best Practices to Reduce Hazards
for Special Environments 121
    Home Environment 121
Factors Leading to Impairment and Best Practices for Childhood Asthma 121
Factors Leading to Impairment and Best Practices for Schools 122
    Preschool Environment 122
    School Environment 123
    Special Equipment and Facilities in the School 126
        Cooling and Ventilation Systems 126
        Drinking Water 126
        Plumbing System 126
        Fume Hoods/Spray Booths 126
        Electrical Shop 126
    Key Potential Violation Areas in Schools 126
Factors Leading to Illness and Injuries and Best Practices for the
Recreational Environment for Children 127
Factors Leading to Impairment and Best Practices for the
Occupational Environment 128
Factors Leading to Impairment and Best Practices to Reduce Hazards for
Specific Community and Home Environmental Risks 130
    Toxic Chemicals 130
    Chemical and Biological Releases Including Terrorism 131
Factors Leading to Impairment and Best Practices for Childhood Lead
Poisoning 132
    Childhood Poisoning 133
    Environmental (Secondhand) Tobacco Smoke 133
    Food 134

Indoor Air Pollution 134
Injuries.. 135
Methamphetamine-Contaminated Houses 136
Outdoor Air Pollution 137
Pesticides 137
Consumption of Unpasteurized Milk 138
Soil 138
Solid and Hazardous Waste Disposal 139
Unsafe Drinking Water 140
Insects, Rodents, and Pesticides 140
Vectors of Disease 141
Factors Leading to Impairment and Best Practices for Emergency
Environmental Risks 141
Floods, Tornadoes, and Hurricanes.. 141
Extreme Heat 141
Aftermath of Wildfires 141
Lack of Consistent Approach to Children's Environmental Health Issues 142
Inadequate Training of Professionals in Children's Environmental
Health Issues... 143
Pediatricians 143
Nurses 144
Registered Environmental Health Specialists/Registered Sanitarians........... 144
Environmental Health Aides 144
Critiques of Children's Environmental Health Efforts 145
Federal Level 145
Government Accountability Office 145
Environment and Public Health Committee of the United States Senate 145
Laws, Rules, and Regulations 146
Special Resources... 146
Children's Environmental Health Centers 146
Other Resources 149
Governmental 149
World 149
Federal 150
State and Local 153
Professional Associations and Other Organizations 153
American Academy of Pediatrics Council on Environmental Health 153
American Public Health Association—Public Health Nursing Section 153
Canadian Association of Physicians for the Environment 154
Canadian Partnership for Children's Health and Environment 154
Children's Environmental Health Network 154
Healthy Schools Network 154
National Association of City and County Health Officials 154
National Association of School Nurses 154
National Children's Center for Rural and Agricultural Health and Safety 154
National Environmental Health Association—Children's Environmental
Health Technical Section 155
National Safety Council—Environmental Health Center............................ 155
Natural Resources Defense Council, Children's Environmental
Health Initiative 155
The Coalition to End Childhood Lead Poisoning 155

The Collaborative on Health and the Environment 155
Pediatric Environmental Health Specialty Units 156
Programs 156
State and Local Level.. 156
State Programs... 156
Local Programs 157
Other Programs 157
Agricultural Environment 157
Migrant Clinicians Network 157
Urban Environment 157
Community-based Intervention Research Project—University
of Michigan 157
Healthy Homes Project Pilot, Baltimore 158
Healthy Children, Healthy Homes—Northern Miami, Florida 158
EPA Rules on Contractors Making Renovations.................................... 159
Integrated Pest Management 159
Lead Poisoning 159
Preschool Environment 159
Children's Environmental Health Network 159
Environmental Protection Agency 160
Canadian Partnership for Children's Health and Environment.............. 160
School Environment 161
Proposed Children's Environmental Health Program 161
General Issues... 161
Content of Educational Programs... 162
Homes and Communities 162
Schools and Preschools 163
Indoor Air Quality Program for Schools.. 163
Urban Schools.. 163
Site Assessment for Schools and Preschools 163
Chemicals 163
Injury Control Programs... 163
School Environmental Health and Safety Inspection Program.............. 163
Endnotes... 164
Bibliography 166

Chapter 5   Environmental Health Emergencies, Disasters, and Terrorism 169

Statement of Problem and Special Information 169
Sub-Problems Including Leading to Impairment and Best Practices
for Environmental Health and Protection Emergencies, Disasters, and
Special Issues... 171
Prevention, Control, and Mitigation of Injuries and Illness while
Protecting the Environment... 171
Topic Areas... 171
Air Quality (Outdoor and Indoor) 171
Asbestos.. 172
Biological Hazards.. 172
Biologic and Infectious Wastes.. 173
Carbon Monoxide 173
Chemical Hazards... 173

Debris.. 173
Drinking Water (On-Site Water Supply and Public Water Supply)............... 173
Electrical Systems and Appliances... 174
Food 175
Household Items and Clothing Potentially Contaminated with
Flood Waters... 175
Insects and Rodents 176
Lead-Based Paint.. 176
Molds and Other Fungi... 176
Septic Tank Systems and Public Wastewater Systems 177
Shelters... 177
Solid and Hazardous Waste Removal... 179
Structural Safety 180
Underground Storage Tanks 180
Sub-Problems Including Leading to Impairment and Best Practices for
Natural and Weather-Related Disasters 181
Introduction 181
Incidents of Natural Disasters and Weather-Related Impairment 181
Drought.. 181
Earthquakes 182
Extreme Temperatures (Heat and Cold) 185
Floods 187
Landslides and Mudslides 188
Tsunamis... 189
Volcanoes.. 189
Wildfires 190
Wind-Related Storms (Tropical Storms, Hurricanes, Tornadoes, Severe
Winter Storms, etc.) 192
Sub-Problems Including Leading to Impairment and Best Practices for Terrorism...... 194
Bioterrorism.. 195
Chemical Terrorism.. 196
Radiological Terrorism... 197
Explosives... 198
Food Terrorism 198
Water Terrorism.. 199
Disaster Site Management, Command-and-Control Problems for Emergency,
and Other Personnel, Supplies, Equipment, and Enforcement 200
Occupational Health and Safety Practices for Personnel Involved in
Emergency/Disaster Cleanups and Repairs 203
Occupational Health and Safety Practices for First Responders, Construction
Workers, and Volunteers 205
Laws, Rules, and Regulations 209
Proposed Rule—Medicare and Medicaid Programs; Emergency
Preparedness Requirements for Medicare- and Medicaid-Participating
Providers and Suppliers.. 210
Resources 210
Programs 213
City of Los Angeles Hazard Mitigation Plan 213
State of California Multi-Hazard Mitigation Plan 213
State of New York Comprehensive Emergency Management and
Continuity of Operations Plan 214

Federal Emergency Management Agency Comprehensive Preparedness
Guide on Developing and Maintaining Emergency Operations Plans............... 214
Federal Emergency Management Agency... 215
The Office of Public Health Preparedness and Response................................. 215
Texas Colorado River Floodplain Coalition Multi-Jurisdictional Hazard
Mitigation Plan Update 2011–2016... 216
Public Health Emergency Response Guide.. 216
Centers for Disease Control and Prevention—Bioterrorism
Response Documents.. 216
Radiological Emergency Response-Related Links 216
Centers for Disease Control and Prevention Health Alert Network (HAN)....... 217
Interim Planning Guide for State and Local Governments for Managing
Terrorist Incidents... 217
Federal Communications Commission's Emergency
Communications Guide.. 217
Endnotes... 217
Bibliography 221

Chapter 6   Environmental and Occupational Injury Control...223

Statement of Problem and Special Information 223
Unintentional Injuries..224
Sub-Problems Including Leading to Impairment and Best Practices for the
Built Environment (Home and Community)..224
Falls 225
Poisonings..227
Fires 228
Dog Bites 228
Drowning..228
Sub-Problems Including Leading to Impairment and Best Practices for
Hospitals and Other Healthcare Environments..229
Sub-Problems Including Leading to Impairment and Best Practices for
Transportation 229
Sub-Problems Including Leading to Impairment and Best Practices for
School Facilities and People 230
Sub-Problems Including Leading to Impairment and Best Practices for
Sports and Recreation Facilities and People 230
Sub-Problems Including Leading to Impairment and Best Practices for
Occupational Facilities and People 231
Sub-Problems Including Leading to Impairment and Best Practices for
Violence-Related Injuries (Intentional Injuries) 232
Laws, Rules, and Regulations 234
Resources 234
Programs 235
National Action Plan for Child Injury Prevention—Centers for Disease
Control and Prevention, National Center for Injury Prevention and Control...... 235
Occupational Health Surveillance Program—New York Department
of Health...236
Workplace Injury and Illness Prevention Program—California OSHA.............236
Endnotes...236
Bibliography 237

Chapter 7   Food Security and Protection...239

Statement of Problem and Special Information 239
Best Practices in Controlling Time/Temperature Events and Other
Food Problems through the Use of Hazard Analysis Critical Control
Points (HACCP) Principles 241
Determining the Scope of the Problem and Specific Best Practices through
Use of the Systems (Process) Approach 244
    Elements 244
Sub-Problems Including Leading to Impairment and Best Practices for
Production of Food 244
    Processing of Food 245
        Bottled Water..245
        Eggs 245
        Fish and Shellfish...246
        Fruits, Vegetables, Juice, and Nuts 248
        Grains, Soybeans, and Hay...251
        Meat 252
        Milk and Dairy 254
        Poultry 254
Sub-Problems Including Leading to Impairment and Best Practices for
Packing of Food 255
Sub-Problems Including Leading to Impairment and Best Practices for
Transportation of Food 256
Sub-Problems Including Leading to Impairment and Best Practices
for Storage of Food 257
Sub-Problems Including Leading to Impairment and Best Practices for
Preparation of Food 257
Types of Processes and Methods of Intervention....................................258
    Processes 258
        Interventions 258
Use of Systems Approach with Critical Control Points to Prevent and
Mitigate Potential Foodborne Disease 258
Use of the Regulatory Process to Prevent and Mitigate Potential
Foodborne Disease 259
Use of Verification of Safety Standards for Imported Food to Prevent and
Mitigate Potential Foodborne Disease 260
Function of Industry in the Prevention and Mitigation of Potential
Foodborne Disease 261
Function of Professional Associations in the Prevention and Mitigation of
Potential Foodborne Disease 263
The Problems Related to Adequate Personnel and Budgets to Prevent and
Mitigate Foodborne Disease 265
The Problems Related to Politics in the Prevention and Mitigation of
Foodborne Disease 266
The Need for Training for Environmental Health Professionals 267
A Global Response in the Prevention and Mitigation of Foodborne Disease 269
Laws, Rules, and Regulations 270
    Food Codes..270
        The FDA Food Code 270
        FDA Food Safety Modernization Act...270

Egg Products Inspection Act 271
Poultry Products Inspection Act 271
Federal Meat Inspection Act 271
Other Significant Regulations and Laws 271
Resources 272
Governmental 272
Federal Level 272
Governmental Manuals 273
Other Resources 273
State Level 273
Local Level 273
Industries, Industry Associations, Professional Associations, and
Professional Journals 274
Council to Improve Foodborne Outbreak Response 274
Professional Associations—Non-Governmental 274
Professional Journals 275
Peer-Reviewed Journals 275
Educational Procedures 275
Enforcement Procedures 275
Presidents and Food Safety 275
Programs 276
Voluntary National Retail Food Regulatory Program Standards 276
Endnotes 282
Bibliography 284

Chapter 8   Healthcare Environment and Infection Control 287

Statement of Problem and Special Information for Hospitals 287
Sub-Problems Including Leading to Impairment and Best Practices for Acute
Care Hospitals 288
Healthcare-Associated Infections 288
Emergency Medical Services as a Source for Healthcare-Associated Infections 289
Healthcare-Associated Infections Caused by Medical/Nursing Techniques 290
Healthcare-Associated Infections Caused by Environmental Hazards Internal
to the Facility 291
Building Construction or Renovation 291
Air and Airborne Infections 293
Equipment (Medical and Surgical) 294
Food 297
Hand Washing 297
Insect and Rodent Control 298
Laundry 298
Medical Waste and Other Hazardous Substances Including
Cleaning Materials 299
Noise 301
Surfaces 301
Water 303
Sub-Problems Including Leading to Impairment and Best Practices for
Employee and Patient Safety (Occupational Health and Safety) 305
Patient Safety 305
Employee Safety (Occupational Health and Safety) 308

Sub-Problems Including Leading to Impairment and Best Practices for
Potential Hazards External to the Facility 313
    Air Pollution 313
    Hazardous Waste 313
    Water Pollution 314
Sub-Problems Including Leading to Impairment and Best Practices for
Emerging and Re-Emerging Diseases.. 315
    Special Highly Infectious Pathogenic Organisms.................................... 315
    Special Problems Related to Some of the Emerging Pathogens and
    Best Practices.. 317
        Antibiotic-Resistant Bacteria.. 317
        Clostridium difficile... 318
        Respiratory and Enteric Viruses in Pediatric Care Settings 319
        Severe Acute Respiratory Syndrome (SARS) Virus 319
        Ebola Virus.. 319
        Tuberculosis 320
Sub-Problems Including Leading to Impairment and Best Practices for
Bioterrorism and Potential Agents 321
    Categories of Potential Organisms Used as Bioweapons........................ 322
        Category a Microorganisms 322
Sub-Problems Including Leading to Impairment and Best Practices
for Use of Mail as a Means of Dispersing Biological Weapons and
Other Potential Threats.. 326
Sub-Problems Including Leading to Impairment and Best Practices for
Toxicological, Environmental, and Occupational Concerns for Cleaning
Materials and Disinfectants 328
Statement of Problems and Special Information for Other Healthcare Options 329
Sub-Problems Including Leading to Impairment and Best Practices for
Home Healthcare 329
Sub-Problems Including Leading to Impairment and Best Practices for
Physicians' Offices and Medical Clinics... 330
Sub-Problems Including Leading to Impairment and Best Practices for
Stand-Alone Surgical Units... 331
Sub-Problems Including Leading to Impairment and Best Practices for
Skilled Nursing Facilities 331
    Occupational Safety and Health Problems.. 332
Laws, Rules, and Regulations 334
    Needlestick Safety and Prevention Act 334
    Clean Water Act 334
    Safe Water Drinking Act.. 334
    Resource Conservation and Recovery Act 334
    Emergency Planning and Community Right to Know Act....................... 334
    Medical Waste Tracking Act 334
    Clean Air Act.. 334
    Toxic Substances Control Act.. 335
    Federal Insecticide, Fungicide, and Rodenticide Act.............................. 335
    Federal Hazardous Materials Transportation Law................................... 335
    Nuclear Regulatory Commission 335
    Occupational Safety and Health Administration 335

Resources 335
Agency for Healthcare Research and Quality, US Department of Health
and Human Services... 335
Centers for Disease Control and Prevention.. 336
Programs 338
Successful Hospital Health and Safety Programs 338
University Medical Center at Brackenridge, Austin, Texas............. 338
St. Thomas Midtown Hospital, Nashville, Tennessee 338
Blake Medical Center, Bradenton, Florida...................................... 339
Lima Memorial Health System, Lima, Ohio.................................... 339
St. Vincent's Medical Center, Bridgeport, Connecticut 339
Cincinnati Children's Hospital, Cincinnati, Ohio 340
Tampa General Hospital, Tampa, Florida 340
Washington State Department of Social and Health Services,
Olympia, Washington 341
Endnotes.. 341
Bibliography 344

Chapter 9   Insect Control, Rodent Control, and Pesticides.. 345

Statement of Problem and Special Information 345
Sub-Problems Including Leading to Impairment and Best Practices for
Insects and Other Arthropods.. 345
Bedbugs 345
Fleas.. 346
Flies 347
Food Product Insects 347
Lice 348
Mites 348
Mosquitoes.. 349
Roaches... 350
Ticks 351
Sub-Problems Including Leading to Impairment and Best Practices for Rodents 352
Statement of Problem and Special Information for Pesticides and Antimicrobials 353
Sub-Problems Including Leading to Impairment and Best Practices for
Pesticides and Antimicrobials 353
Statement of Problem and Special Information for Integrated Pest Management 356
Laws, Rules, and Regulations 357
Federal Insecticide, Fungicide, and Rodenticide Act.......................... 357
Federal Food, Drug, and Cosmetic Act.. 358
Food Quality Protection Act of 1996 358
Pesticide Registration Improvement Act of 2003 358
Freedom of Information Act... 358
Federal Advisory Committee Act 358
Safe Drinking Water Act.. 359
Resources 359
Centers for Disease Control and Prevention.. 359
Pesticide-Related Illness and Injury Surveillance: A How-To Guide for
State-Based Programs... 359

Sentinel Event Notification System for Occupational Risk (SENSOR) 359
National Poison Data System 359
Toxic Exposure Surveillance System 359
US Environmental Protection Agency 360
    Integrated Pest Management in Buildings 360
    Pesticide Fact Sheets 360
    National Pesticide Information Center 360
US Department of Agriculture 360
    EXTOXNET 360
    Pesticide Data Program 360
Programs 361
    National Pesticide Program 361
    Endocrine Disruptor Screening Program 361
    City of Berkeley, California, Rat Control Program 361
    California Mosquito Control Programs 362
    City of Columbus Public Health Division of Environmental
    Health—Mosquito Control Program 362
Endnotes 363
Bibliography 364

Chapter 10 Recreational Environment and Swimming Areas 365

Statement of Problem and Special Information 365
Sub-Problems Including Leading to Impairment and Best Practices for
Airline Transportation 366
    Occupational Health and Safety Concerns 369
    Environmental Concerns 370
        Air Pollutants Released by Aircraft 370
Sub-Problems Including Leading to Impairment and Best Practices for
Cruise Ships 371
    Health Risks 371
        Food and Waterborne Diseases 371
        Mosquito-Borne Diseases 372
    Safety Risks 373
    Environmental Concerns 374
        Air Pollution from Cruise Ships 374
    Child Activity Centers 375
    Food Safety 375
        Hand Washing and Toilet Facilities 375
    Hazardous Materials Use, Waste Storage, and Disposal 376
    Heating, Ventilation, Air Conditioning, Fountains, Misting Systems,
    Humidifiers, and Showers 376
    Integrated Pest Management 377
    Liquid Waste (Sewage, Black Water) Disposal 377
    Liquid Waste (Gray Water) Disposal 378
    Medical Facilities 378
    Oily Bilge Water 378
    Plumbing and Cross-Connections 379
    Potable Water 380

Recreational Water Facilities on Board the Ship.....................................382
Solid Waste Storage and Disposal 382
Sub-Problems Including Leading to Impairment and Best Practices for
Occupational Health and Safety Problems on Board Ship and in Ports 383
Sub-Problems Including Leading to Impairment and Best Practices for
Children's Camps..386
Sub-Problems Including Leading to Impairment and Best Practices for
Recreational Swimming Areas 389
Disease Outbreaks..390
Physical Problems Leading to Injuries..390
Chemical Hazards 391
Aesthetic Problems..391
Swimming Pool Codes and Inspections...391
Rules and Regulations..393
Public Health Service Act of 1944 with Amendments and Executive Orders
Updated to January 4, 2012 393
Cruise Vessel Security and Safety Act of 2010....................................393
International Convention for the Prevention of Pollution from
Ships (MARPOL) 394
SOLAS 394
Act to Prevent Pollution from Ships 394
Clean Water Act 394
National Marine Sanctuaries Act 394
Resource Conservation and Recovery Act 395
Comprehensive Environmental Response, Compensation, and Liability Act.....395
Marine Protection, Research, and Sanctuaries Act (Ocean Dumping Act) 395
Oil Pollution Act..395
Virginia Graeme Baker Pool and Spa Safety Act 395
Coast Guard Regulations...395
Resources 395
Centers for Disease Control and Prevention...395
CDC Health Alert Network 395
CDC's Traveler's Health 395
National Center for Environmental Health's Division of Emergency and
Environmental Health Services..396
United States Coast Guard 396
United States Coast Guard Cruise Ship National Center of Expertise...........396
World Health Organization 396
Global Outbreak Alert and Response Network 396
Public Health Agency of Canada Global Public Health Intelligence Network 396
ILO Maritime Labor Convention and Its Guidance Concerning Health and
Safety On-Board Ships 396
Model Aquatic Health Code 397
Programs 397
CDC Vessel Sanitation Program 2011 Operations Manual...................397
CDC Vessel Sanitation Program 2011 Construction Guidelines 398
United States Coast Guard Cruise Ship National Center of Expertise 398
Interagency Marine Debris Coordinating Committee 398
Endnotes..398
Bibliography 401

Chapter 11 Sewage Disposal Systems 403

Statement of Problem and Special Information for Sewage 403
Pollutants Found in Sewage...403
Sub-Problems Including Leading to Impairment and Best Practices for
Public Sewage Systems 404
Infiltration and Inflow..404
Sewer Laterals 404
Collection Systems 405
Ancillary Structures Including Pump/Lift Stations 406
Drop Shaft..406
Valve, Diversion, and Overflow Structures 406
Manholes 406
Pump/Lift Station 407
Drop Shafts 407
Valve, Diversion, and Overflow Structures 407
Manholes 407
Combined Sewer Systems 408
Sewer Cleaning and Inspection 409
Wastewater Treatment Systems 411
Primary Treatment.. 411
Secondary Treatment.. 411
Advanced Wastewater Treatment (Tertiary Treatment)................................. 412
Land Treatment of Wastewater 413
Wetlands 413
Constructed Free Water Surface Wetlands 413
Subsurface Flow 413
Disposal of Biosolids and Other Wastewater Residuals................................... 414
Biosolids 414
Septage Treatment/Disposal: Land Application 415
Wet Weather Concerns at Combined Sewer Wastewater
Treatment Facilities 416
Flooding a Severe Danger to Water and Wastewater Treatment Facilities 417
Package Treatment Plants..418
Disinfection of Effluent 419
Chlorine 419
Chlorine Dioxide 420
Ozonation..420
Ultraviolet Radiation..420
Use of Energy in Water and Wastewater Treatment Plants 420
Statement of Problem and Special Information for Private Sewage Systems
(On-Site Sewage Disposal) 421
A Survey of Household Sewage Treatment System Failures in Ohio 423
Sub-Problems Including Leading to Impairment and Best Practices for
Private Sewage Systems (On-Site Sewage Disposal) 423
Size and Nature of Land for Houses...423
Soils and Soil Testing 424
Soil Tests..425
On-Site Wastewater Systems..426
Septic Tanks and Distribution Boxes 426

Septic Tank Systems for Large Flow Applications 428
Septic Tank Effluents, Treatment, and Disposal Systems 428
Other Advanced Treatment Systems 433
Special Toilets to Reduce Water Flow...436
High-Efficiency Toilets...436
Incineration Toilets 436
Composting Toilets 436
Personal Habits in Use of Water and the Disposal of Waste............................436
Holding Tanks 437
Laws, Rules, and Regulations 437
Stormwater Phase II Final Rule..437
Planning a New or Expansion of a Wastewater Treatment Plant 438
Resources 438
Programs 438
US Environmental Protection Agency 438
Fairfax County, Virginia, Sewer Maintenance Program 440
City of Los Angeles, California, Sewer Maintenance Program 440
Endnotes...440
Bibliography 442

Chapter 12 Solid Waste, Hazardous Materials, and Hazardous Waste Management 445

Statement of Problem and Special Information for Solid Waste 445
Global Warming 445
Sub-Problems Including Leading to Impairment and Best Practices for
Solid Waste Collection, Storage, and Transportation 445
Non-Hazardous Solid Waste Streams...445
Municipal Solid Waste...445
Agricultural and Animal Wastes 449
Industrial Wastes 452
Construction and Demolition Wastes 454
Sewage Treatment Wastes 455
Special Wastes 455
Cement Kiln Dust Waste 455
Crude Oil, Coal, and Natural Gas Waste...456
Mineral Processing Waste 457
Mining Waste...458
Medical Waste 460
Reduction of Non-Hazardous Solid Waste Prior to Generation.........................462
Source Reduction/Waste Minimization 462
Recycling 463
Sub-Problems Including Leading to Impairment and Best Practices for Solid
Waste Disposal for Municipal Landfills 464
Municipal Landfills...464
Construction and Demolition Waste Landfills 467
Sub-Problems Including Leading to Impairment and Best Practices for Solid
Waste Disposal Using Municipal Incinerators...469
Sub-Problems Including Leading to Impairment and Best Practices for Solid
Waste Disposal Using Composting 471
Best Practices in Composting...471

Sub-Problems Including Leading to Impairment and Best Practices for
Non-Hazardous Waste Disposal and Hazardous Waste Disposal Using
Underground Injection Wells 472
   Description of Underground Injection Wells... 473
      Class I Underground Injection Wells.. 473
      Class II Underground Injection Wells 474
      Class III Underground Injection Wells... 475
      Class IV Underground Injection Wells... 476
      Class V Underground Injection Wells 476
      Class VI Underground Injection Wells... 477
   In Situ Bioremediation of Groundwater 478
Statement of Problem and Special Information for Hazardous Materials.............. 479
Sub-Problems Including Leading to Impairment and Best Practices for
Production of Hazardous Materials 480
   Basic Chemicals 480
   Specialty Chemicals 482
   Agricultural Chemicals 482
   Pharmaceuticals.. 484
   Consumer Products.. 485
   Other Concerns about Chemical Plants 485
      Noise 485
      Odors... 486
      Decommissioning a Chemical Manufacturing Facility 486
Sub-Problems Including Leading to Impairment and Best Practices for
Hazardous Materials Use, Collection, Storage, and Transportation 486
   Transportation 491
Statement of Problem and Special Information for Hazardous Waste Disposal 493
Sub-Problems Including Leading to Impairment and Best Practices for
Household Hazardous Waste Management... 495
   Electronic Wastes 496
Sub-Problems Including Leading to Impairment and Best Practices for
School Hazardous Waste Management 496
Sub-Problems Including Leading to Impairment and Best Practices for
Medical Waste Management 497
   Low-Level Radioactive Waste 498
Statement of Problem and Special Information for Liners for New Surface
Impoundments and Landfills 499
Statement of Problem and Special Information for Leachate Collection Systems.....500
   Leak Detection System.. 500
Sub-Problems Including Leading to Impairment and Best Practices for
Hazardous Waste Tank Systems 501
   Sub-Problems Including Leading to Impairment and Best Practices for
   Surface Impoundments Used for Managing Hazardous Wastes.......... 502
   Sub-Problems Including Leading to Impairment and Best Practices for
   Hazardous Wastes... 502
   Land Treatment Units.. 502
Statement of Problem and Special Information for Hazardous Waste Landfills 503
Sub-Problems Including Leading to Impairment and Best Practices for
Hazardous Waste Landfills 503
Sub-Problems Including Leading to Impairment and Best Practices for
Hazardous Waste Incinerators 504

Sub-Problems Including Leading to Impairment and
Best Practices for Environmental Impacts of Hazardous Materials
and Hazardous Waste 505
   Air..505
   Land and Superfund Sites 507
   Water..507
Statement of Problem and Special Information for Occupational Health
and Safety..508
Sub-Problems Including Leading to Impairment and Best Practices for
Occupational Health and Safety Issues 509
Laws, Rules, and Regulations 513
   National Environmental Policy Act...513
   Resource Conservation and Recovery Act 513
   Comprehensive Environmental Response, Compensation, and Liability Act.....513
   Oil Pollution Act..513
   Superfund Amendments and Reauthorization Act...................................513
   Emergency Planning and Community Right-to-Know Act 514
   Hazardous Materials Transportation Uniform Safety Act of 1990.....................514
   Clean Water Act 514
   Clean Air Act..515
   Safe Drinking Water Act..515
   Toxic Substances Control Act...515
   Migratory Bird Treaty Act..515
   Federal Hazardous Substances Act 515
   Fish and Wildlife Coordination Act 515
   Federal Insecticide, Fungicide, and Rodenticide Act...............................516
   Endangered Species Act 516
   Robert T Stafford Disaster Relief and Emergency Assistance Act....................516
Resources 516
Programs 517
   Occupational Health and Safety Program for Hazardous Waste Sites................517
     Planning and Operation 517
     Hazard Assessment..518
     Training 518
     Medical Program 519
     Evaluation 519
     Integrated Solid Waste Management..519
   US Program on Recycling of Tires 520
   Battery Recycling Program 520
   Model Solid Waste Management Program..520
   Waste Management in the Oil and Gas Exploration and Production Process.....521
   US Environmental Protection Agency Hotline and Service Line Program........521
Endnotes...522
Bibliography 526

Chapter 13 Water Systems (Drinking Water Quality)...529

Statement of Problem and Special Information 529
Sub-Problems Including Leading to Impairment and Best Practices
for General and Specific Sources of Water Contamination, Prevention,
Mitigation, and Control 530

Sub-Problems Including Leading to Impairment and Best Practices for
Private Drinking Water Systems 533
Sub-Problems Including Leading to Impairment and Best Practices
for Public Drinking Water Supplies 536
    Livestock and Poultry Manure Contaminants and Raw Water Quality 537
Sub-Problems Including Leading to Impairment and Best Practices for
Public Drinking Water Treatment Plants 538
    Location, Source, and Preliminary Pretreatment for Raw Water 538
    Nitrification 539
    Drinking Water Treatment Facility 540
        Cryptosporidium 541
        Cyanotoxins 541
        Turbidity 541
    Disinfection 543
        Disinfectants 543
        Disinfectant Byproducts 544
        Nitrification 544
    National Primary Drinking Water Regulations 544
    Finished Potable Water Storage Facilities 545
Sub-Problems Including Leading to Impairment and Best Practices for
Public Drinking Water Distribution Systems 546
    Effects of Age of Water 546
    Deteriorating Buried Infrastructure 546
    Permeation and Leaching 547
    New or Repaired Water Mains 547
    Service Lines 547
    Health Risks Associated with Pressure Transients in Water
    Distribution Systems 548
    Health Risks Related to Microbial Growth and Biofilms in Water
    Distribution Systems 548
Sub-Problems Including Leading to Impairment and Best Practices for
Cross-Connections and Submerged Inlets 550
Laws, Rules, and Regulations 551
    Safe Drinking Water Act 551
    Clean Water Act 551
Resources 552
    American Water Works Association 552
    American National Standards Institute 552
    NSF International 552
    Centers for Disease Control and Prevention 552
    University Extension Services 552
    Local and State Health Departments 552
    American Ground Water Trust 553
    Water Systems Council 553
    State Governmental Agencies 553
    The US Environmental Protection Agency 553
Programs 553
    Federal Support for State and Local Response Operations in Water
    Contamination Crises 553
    Joaquin River Delta 553

      Cyanotoxin Management Plan 554
      Texas Commission on Environmental Quality—Public Water Supply
      Supervision Program 555
   Endnotes 556
   Bibliography 557

Chapter 14 Water Quality and Water Pollution 559

   Introduction 559
   Statement of Problem and Special Information 559
      Measures of Water Quality... 561
   Sub-Problems Including Leading to Impairment and Best Practices for
   Potential Sources of Water Contamination 561
      Acid Mine Drainage 561
      Aesthetics 562
      Agriculture 562
      Air Pollution Deposits 562
      Alteration of Habitats 563
      Climate Change.. 564
      Combined Storm and Sanitary Sewer 564
      Erosion 564
      Extraction of Resources 565
      Governmental Agencies 566
      Illicit Discharges 566
      Improper Disposal of Hazardous Waste 567
      Improperly Constructed or Maintained Wells 567
      Invasive Species 567
      Land Development and Building Construction 567
      Landfills 568
      Metals 568
      Low Dissolved Oxygen 568
      Non-Point Sources of Contamination 568
      Nutrients 570
      Oxygen-Demanding Substances 571
      Pesticides 571
      Sand and Salt Storage 571
      Sediment 571
      Septic Systems 572
      Sewage Discharge from Municipal Sources 572
      Spills and Runoff from Stored Chemicals and Storage Tanks 572
      Surface Impoundments for Farm Waste or Other Wastes................ 573
      Temperature 573
      Total Suspended Solids 573
      Waste from Pets 574
      Wildlife or Other Natural Sources 574
   Laws, Rules, and Regulations 574
      Rivers and Harbors Appropriation Act of 1899 574
      Federal Water Pollution Control Act (Clean Water Act) 575
   Resources 575
      Farm*A*Syst 575

US Department of Commerce—Transportation Research Board......................576
US Environmental Protection Agency 576
Programs 577
Great Lakes Restoration Initiative 577
Grand Calumet River—Great Lakes Areas of Concern 578
US Environmental Protection Agency 579
Minnesota Pollution Control Agency 579
State of Washington Department of Ecology....................................580
Philadelphia Suburban Water Company...580
US Department of Agriculture Forest Service 581
US Geological Service 581
Endnotes...582
Bibliography 583

Index...585

Foreword

In their 2010 article on a framework for public health in the United States, authors Fielding, Teutsch, and Breslow[1] noted that "in the United States as well as in the global community, public health continues to evolve to ensure healthy families and communities as well as individuals. Great achievements in the control of infectious and chronic disease and injuries will need to be sustained while we face new challenges, including providing universal access to high quality healthcare as well as addressing the underlying behavioral risk factors and the social, physical and environmental determinants of health. Meeting these challenges will require strengthening the governmental and non-governmental public health systems."

Having spent over 60 years addressing environmental health issues as an environmental researcher and practitioner at the local, state, and federal level including decades as an academic administrator and professor, it has become quite evident that we should pay more heed to the old axiom that "what is past is prologue to the future."

From a pragmatic standpoint, perhaps environmental personnel should re-craft a new axiom based on the old one to read: "what in the past has been successful should not be rejected, rather if successful should be used as a *model for the future*." That is what—at least in part—is suggested by Dr. Herman "Hank" Koren, author of this thoughtful compendium of Best Practices, resources, and successful programs in environmental health and safety.

To start with, as public health professionals we are cognizant of the great significance of environmental issues as part of that incredible mix of factors that are critical in assuring a healthier world community.

During the 20th century, the health and life expectancy of persons residing in the United States improved dramatically. Since 1900, the average lifespan of persons in the United States has lengthened by more than 30 years; 25 years of this gain are attributable to advances in public health.[2]

The Centers for Disease Control and Prevention (CDC) has noted that many great public health achievements occurred during the period 1900–1999. Most notably, 10 of these as listed below are most significant, based on the opportunity for prevention and the impact on death, illness, and disability in the United States.[3] They are as follows:

- Vaccination
- Motor vehicle safety
- Safer workplaces
- Control of infectious diseases
- Decline in deaths from coronary heart disease and stroke
- Safer and healthier foods
- Healthier mothers and babies
- Family planning
- Fluoridation of drinking water
- Recognition of tobacco use as a health hazard

What is remarkable to some, but not to the environmental health practitioner, is that more than half of the improvements noted are due in some major way to enhanced environmental health efforts and thus in some significant part due to the work of environmental health professionals.

The CDC points out the following environmental linkages of six of the ten great public health achievements of the 20th century[3]:

- Improvements in *motor vehicle safety* have resulted from engineering efforts to make both vehicles and highways safer and from successful efforts to change personal behavior (e.g., increased use of safety belts, child safety seats, and motorcycle helmets, and decreased drinking and driving). These efforts have contributed to significant reductions in motor vehicle–related deaths.[4]

- *Work-related health problems*, such as coal workers' pneumoconiosis (black lung) and silicosis—common at the beginning of the century—have come under better control. Severe injuries and deaths related to mining, manufacturing, construction, and transportation also have decreased; since 1980, safer workplaces have resulted in a reduction of approximately 40% in the rate of fatal occupational injuries.[5]
- *Control of infectious diseases* has resulted from clean water and improved sanitation. Infections such as typhoid and cholera transmitted by contaminated water, a major cause of illness and death early in the 20th century, have been reduced dramatically by improved sanitation. In addition, the discovery of antimicrobial therapy has been critical to successful public health efforts to control infections such as tuberculosis and sexually transmitted diseases (STDs).
- Since 1900, *safer and healthier foods* have resulted from decreases in microbial contamination and increases in nutritional content. Identifying essential micronutrients and establishing food fortification programs have almost eliminated major nutritional deficiency diseases such as rickets, goiter, and pellagra in the United States.
- *Healthier mothers and babies* have resulted from better hygiene and nutrition, availability of antibiotics, greater access to healthcare, and technologic advances in maternal and neonatal medicine. Since 1900, infant mortality has decreased by 90%, and maternal mortality has decreased by 99%.
- *Fluoridation of drinking water* began in 1945 and in 1999 reached an estimated 144 million persons in the United States. Fluoridation safely and inexpensively benefits both children and adults by effectively preventing tooth decay, regardless of socioeconomic status or access to care. Fluoridation has played an important role in the reductions in tooth decay (40%–70% in children) and of tooth loss in adults (40%–60%).[6]

What will the future of environmental health problems and opportunities look like as professional and political leaders consider the development of program plans in the broad field of public health?

One of the most respected environmental health leaders of the 20th century is Dr. Larry J. Gordon, a former New Mexico Cabinet Secretary for Health and Environment and long-time practitioner and chronicler of environmental health. In his address to the National Environmental Public Health Conference in Atlanta, Georgia, in December 2006, he noted the following:

> Environmental health will continue to increase in complexity, and the public will increasingly deserve, expect and demand problem/prevention and amelioration.
>
> Demographic changes, resource development and consumption, product and materials manufacturing and utilization, wastes, global environmental deterioration, technological development, international terrorism, evolving disease patterns, changing patterns of land use, transportation methodologies, resource development and utilization, and continuing organizational diversification of environmental health services will create unanticipated challenges.
>
> Environmental heath will continue to be basic to the health of the public and the quality of our environment. Environmental health problems, programs, service delivery organizations, and educational needs will evolve in ways that are unforeseen. Ensuring an adequate supply of environmental health practitioners qualified to handle the policy, leadership, managerial and scientific issues of the future should be of the highest priority.[7]

With that as background to the impact of environmental health as a key determinate in the future of the public's health as well as in the past, the process of guiding the policy process into professional "boots on the ground" practice should be fairly simple. However, the facts are anything but that. In practice, environmental programs are often generated without a clear notion of the most practical, reasoned, or even cost-effective way of proceeding. The history of past accomplishment is too often lost as new leadership assumes responsibility for public health program planning.

A key purpose of this book is to assist supervisors and managers in all areas of public health, preventive medicine, and environmental health practice who need an informed understanding of the complexity and inter-relationships of environmental health issues and their relationship to the public's health. It is also intended to assist environmental health program planners in their policy and planning efforts by bringing together in one reference document well-documented information about addressing specific environmental health problems, including Best Practices.

The suggested practices are a composite of the experience of a great many successful practitioners of environmental enhancement, gathered with the assistance of the American Academy of Sanitarians.

This book is also intended to serve the needs of those professionals who work in environmental health, occupational health, public health, and environmental protection programs and want to have the means to compare their program efforts with those which have been deemed "successful." It should also be particularly useful to those academics whose instructional responsibility deals with environmental health and of course their students who are learning about practical applications of their enhanced knowledge. Finally, it also becomes a useful reference document for professional associations and civic groups who need this kind of professional guidance to help them focus on how to deal with environmental concerns and how to assist them in the education of the general public.

Dr. Herman Koren, Professor Emeritus of Environmental Health and Safety at Indiana State University, who is the author of this tome on best management practices, resources, and successful programs in environmental health and safety to prevent and control disease and injury, is well qualified to guide colleagues in a merited effort to shape rather than await the future of this field. In addition to his 29-year tenure at Indiana State University, Professor Koren has been actively involved in environmental health and public health at every level of practice since 1955 and is a prolific contributor to the literature of this important field of public health practice. A registered environmental health specialist, he is the recipient of the prestigious Walter S. Mangold Award of the National Environmental Health Association and is a Diplomate Laureate of the American Academy of Sanitarians.

Dr. Jerrold M. Michael
Emeritus Professor and Dean
School of Public Health
University of Hawaii
Manoa, Hawaii

Adjunct Faculty
School of Public Health and Health Services
George Washington University
Washington, DC

Assistant Surgeon General (Retired)
U.S. Public Health Service
Rockville, MD

REFERENCES

1. Fielding JE, Teutsch S, Breslow L. A framework for public health in the United States. *Public Health Reviews* 2010;32:174–189.
2. Bunker JP, Frazier HS, Mosteller F. Improving health: Measuring effects of medical care. *Milbank Quarterly* 1994;72:225–258.
3. CDC. Ten great public health achievement: United States 1900–1999. *MMWR* 1999;48:241–243.
4. Bolen JR, Sleet DA, Chorba T, et al. Overview of efforts to prevent motor vehicle-related injury. In: *Prevention of motor vehicle-related injuries: A compendium of articles from the Morbidity and Mortality Weekly Report, 1985–1996*. Atlanta, GA: US Department of Health and Human Services, Centers for Disease Control and Prevention, National Center for Injury Prevention and Control, 1997.
5. CDC. Fatal occupational injuries: United States, 1980–1994. *MMWR* 1998;47:297–302.
6. Burt BA, Eklund SA. *Dentistry, dental practice, and the community*. Philadelphia, PA: WB Saunders, 1999;204–220.
7. Gordon LJ. The future of the environmental health revisited: Past recommendations and future challenges. 2006 *National Environmental Health Conference*. December 5, 2006.

Preface

INTRODUCTION

Healthy people and healthy and safe communities promote the strength and security of urban, suburban, and rural areas; states; provinces; regions; and nations. This is a programmatic book which helps explain the complexity of various environments and ecosystems and the interaction of various environmental factors as they relate to people and the environment. This book provides background information in each of the environmental areas and promotes prevention of environmental factors that lead to disease, injury, and improper modifications and/or destruction of ecosystems.

This book provides information on techniques of mitigation of contaminants, as well as containment of substances in situations that endanger the community and its citizens. Prevention, mitigation, and containment are based on accurate available data, sound scientific principles, excellent environmental practices, and education. The necessary laws, rules, and regulations to control environmental hazards which are substantial in size or when mitigation or containment cannot be achieved on a voluntary basis are part of this book.

Actual programs of prevention and control, the scope of the problem, and the sub-problems in air, water, solid and hazardous waste, infection control, food, and so on are explored and the practical learned experiences of experts in their specialized fields are shared with the reader. Often Best Practices and underlying data which have been presented may be studied in greater depth by selecting the title of a specific endnote, placing it in a search box, and then bringing up the document for review. This may work for published documents but not necessarily for information which is published solely on the Internet because updates are frequent and prior material typically is discarded. Also it may be very useful to review the material from the Special Resources section as well as the Successful Programs section and explore the biographical references at the end of each chapter. An in-depth index is provided for the individuals who want to rapidly access information about a given item.

This book brings together the concerns of all professionals in environmental health as well as environmental protection and others interested in a vast variety of biological, physical, and chemical contaminants. This book also helps professionals, students, and professors, civic and professional associations, and others in environmental protection, environmental health, safety, and sustainability understand what are the Best Practices and resources needed to have successful environmental programs to counteract environmental degradation, disease, and injury. This information may be enhanced qualitatively and quantitatively by reading the sections on endnotes, special resources, and the bibliography.

This book will save an enormous amount of time and money for the reader by providing a single source for the study and understanding of these contaminants and how to prevent, control, and mitigate their effects.

The size of this preface is unusually large because the scope of the problems and programs in the various topic areas of the environment are numerous, highly diversified, and often inter-related and complex. There may be many potential solutions that are workable. This book will present those solutions that seem to be most effective.

HISTORY OF ENVIRONMENTAL PROTECTION, SUSTAINABILITY, ENVIRONMENTAL HEALTH, AND OCCUPATIONAL HEALTH

The need for the brief explanation that follows concerning the history of environmental protection, sustainability, environmental health, and occupational health is quite evident when we explore the problems which have been created by the dividing of the environmental field into health and protection. These problems include the creation of additional bureaucracies in the same localities, where professionals in health as well as professionals in environmental protection may not be working together to resolve mutual problems and thereby causing duplication of work and waste of scarce resources. However, the division has considerable merits because it has provided considerably more resources for the professionals to try to resolve problems of the environment and people.

THE PAST

Throughout recorded history when severe problems of environmental circumstances and/or disease and injury occurred, stakeholders demanded changes. Governing bodies typically selected an individual or group who represented both environmental health and environmental protection to conduct studies which led to new laws based on the current best knowledge available, and this led to the creation of Best Practices to resolve the problems. Some examples of this are described below.

The Greeks understood that human waste and solid waste could spread disease and therefore invented toilets, sewers, and dumps outside of the city. The first dumps were developed by the Greeks outside of Athens around 500 B.C. The Greek physician Hippocrates in his book, *On Airs, Waters, and Places*, explored the potential for disease from environmental sources.

Plato and Aristotle said that no city could exist without health officers. These health officers were probably some of the first combination of environmental health and environmental protection people in the world.

The Age of Enlightenment (1750–1830, though scholars differ widely on this time frame) refers to the period of time in Europe and America when people explored the idea of what was the proper relationship of the citizen to the chief of state. The idea that society was a contract between the individual and some larger entity continued to grow, and the king or chief executive should be responsible for the health and welfare of the governed individuals as part of this contract.

The Sanitary Movement (early and mid-1800s) was brought about by a convergence of societal, environmental, and disease-related factors causing substantial loss of life and injury to people because of extreme overcrowding, highly unsanitary conditions, overwork, lack of food, and dangerous occupational and environmental conditions. A group of social reformers and health people worked together to bring this major issue to the attention of the public and to arouse them to demand the passage of appropriate laws and provide the necessary budget and people to make the laws meaningful. Edwin Chadwick (father of the sanitary movement) instigated public sanitary reform. He wrote the "Report on the Sanitary Condition of the Labouring Population of Great Britain" in 1842.

The Public Health Act of 1848 in England was a consequence of these very serious conditions and resulted in £5 million being set aside for sanitary research. It was the beginning of the commitment to a proactive rather than reactive public health philosophy. The state guaranteed certain standards of health and environmental quality and provided the necessary resources to local units of government to make the changes in the environment necessary to meet the standards.

The Sanitary Report of 1850, entitled "Report of a General Plan for the Promotion of Public and Personal Health, Devised, Prepared and Recommended by the Commissioners Appointed under a Resolve of the Legislature of Massachusetts, Relating to a Sanitary Survey of the State," was prepared by Lemuel Shattuck. His report indicated that much of the ill health and debility of people in large American cities could be traced to unsanitary conditions and recommended that local studies be carried out and programs put into effect to change these conditions. He established a prototype for other states to follow in forming a board of health.

THE MODERN ERA

As in previous times, the public was demanding action. The turbulence of the 1960s created an awareness of numerous public issues. This resulted in the questioning of the destruction of the environment at the local, state, national, and international level by industries, government, and people. This was the beginning of the modern environmental movement which ultimately raised environmental awareness across the United States and the rest of the world.

The teach-in, a means of protest by younger people, became the model for the first Earth Day on the campus of the University of Michigan in Ann Arbor on March 11–14, 1970, drawing some 50,000 participants. On April 22, some 20 million people across the country demanded healthy sustainable environments.

Earth Day 1970 achieved a rare political alignment and had support from Republicans and Democrats, labor and business, and young and old. Moreover, it stimulated interest by a large number of young people in the environment, its protection, and sustainability.

NATIONAL ENVIRONMENTAL POLICY ACT

From the perspective of the conservationist, preservationist, and environmentalist, a new era started on January 1, 1970, when President Richard M. Nixon signed the National Environmental Policy Act of 1969. In the view of the general public, this was an appropriate response to their demand that the chief executive officer of the country do something constructive to minimize or alleviate the problems causing their environmental concerns. This was an act that provided for the establishment of a national policy for the environment and the establishment of a Council on Environmental Quality, and had other purposes. The national policy was to encourage productive and enjoyable harmony between people and their environment; to promote efforts to prevent or eliminate damage to the environment and biosphere; to stimulate the health and welfare of people; and to understand the relationship of ecological systems and natural resources with the health of the nation.

Congress recognized the profound impact of the activities of people on the natural environment, the significance of population growth, the problems of high-density urbanization, industrial expansion, resource exploitation, and new technologies, and the relationship of the environment to the health and welfare of people. Congress stated that it was the continuing policy of the federal government to work with state and local governments, public and private organizations, and business and industry to preserve existing environmental quality and improve areas that had been overused while protecting the health of people.

US ENVIRONMENTAL PROTECTION AGENCY

The US Environmental Protection Agency (EPA) was established on July 9, 1970, as a separate independent agency of the executive branch of the federal government because of the public demand for cleaner water, air, and land. On December 2, 1970, the EPA, with the help of the US Public Health Service, was formed from 15 components which were parts of five other agencies. The charge of the US EPA was the improvement and preservation of the quality of the environment at the national and global level while protecting both human health and the natural resources which all people depended upon. It was a simple step to move from the EPA charge to the concept of sustainability, which in fact encompassed virtually everything related to environmental health and environmental protection.

SUSTAINABILITY

The US EPA stated, "Sustainability is based on a simple principle: everything that we need for our survival and well-being depends, either directly or indirectly, on our natural environment. Sustainability creates and maintains the conditions under which humans and nature can exist in productive harmony which permits fulfilling the social, economic, and other requirements of present and future generations."

In the rest of the world, members of the United Nations in preparation for the Stockholm, Sweden Conference on the Human Environment held in 1972, met in a preparatory conference in 1971 and expressed deep concern about the environmental consequences of the increasing development of the Earth and the associated rise in pollutants, while simultaneously recognizing the need for economic development in poorer parts of the world. This helped crystallize the concept of environmental sustainability. Out of the Stockholm conference grew the United Nations Environmental Programme, which was founded in 1972.

The modern concept of environmental sustainability is science-based and is a combination of the use of risk assessment and risk communication; prevention, minimization, and control of pollutants; preservation and proper use of resources; and avoidance of damage to air, water, land, food, ecosystems, and people. This helps prevent disease and injury while protecting the natural environment. Environmental sustainability is accomplished by use of Best Practices in all inter-related environmental areas discussed in this book.

CURRENT ENVIRONMENTAL AND OCCUPATIONAL HEALTH CONCERNS

While the environmental movement was rapidly expanding, the public health, environmental health, and occupational health and safety professionals continued to work toward preventing disease and injury and promoting good health through the prevention, minimization, and control of factors in the environment that affected people. Also, in 1970 the Occupational Safety and Health Act had been signed into law by President Nixon thereby creating the National Institute of Occupational Safety and Health (NIOSH) and the Occupational Safety and Health Administration (OSHA). NIOSH, a branch of the CDC, had three charges: providing research to reduce illnesses and injuries from work situations; using new techniques to promote safety and health in workplaces; and working with international groups to enhance safety and health on a global basis.

These health professionals embraced the recommendations of the CDC, the American Public Health Association, the National Association of Local Boards of Health, the Association of State and Territorial Health Officials, the National Association of County and City Health Officials, and the Public Health Foundation. The steering committee of this group devised the National Public Health Performance Standards Program. In 1994, they developed a list of the 10 essential public health services that should be performed in all communities. They stated that the local and state health agencies are extremely important and need to strengthen their capacity to identify and manage health problems.

HEALTHY PEOPLE 2020 AND THE FUTURE: FRAMEWORK

This book is not only a response to the environmental movement but also a partial response to the vision, mission, and goals of Healthy People 2020 and into the future.

The *vision* is of a society in which all people have the opportunity to live long and healthy lives. The *mission statement* identifies the need to establish health priorities; help the public better understand the underlying factors which contribute to health, disease, and disability; provide measurable objectives and goals; involve various groups in improving programs through Best Practices; and establish the specific needs for research, data collection, and evaluation.

The *overarching goals* are:

• Eliminate preventable disease, disability, injury, and premature death.
• Achieve health equity, eliminate disparities, and improve the health of all groups.

- Create social and physical environments that promote good health for all.
- Promote healthy development and healthy behaviors across every stage of life. (See endnote 1.)

The overall concept of Healthy People 2020 which flows through the overarching goals is that important actions can be taken much of the time to promote and preserve health and to minimize the amount and consequences of disease and injury.

PROFESSIONAL COMPETENCIES OF ENVIRONMENTAL HEALTH PROFESSIONALS

To achieve the overarching goal of Healthy People 2020, and into the future, for the environmental health field, it is necessary to ensure that environmental health practitioners throughout the country have a uniform set of competencies which they have gained through their education, continuing education, and professional experience and have been measured and certified by achieving the status of Registered Environmental Health Specialist/Registered Sanitarian as tested and then granted by the National Environmental Health Association or their state registration board. These competencies can then be modified to meet the specific problems which are encountered in various parts of the country and the world.

These competencies are found in 6 general categories and 16 specific categories. The general categories are as follows: general science including biology, microbiology, inorganic chemistry, organic chemistry, algebra, trigonometry, statistics, physics both mechanics and fluids, epidemiology, risk assessment techniques, and toxicology; communications and education including verbal, written, and computer science; planning and management including how to conduct inspections, surveys, and studies, take a variety of samples of potential contaminants using essential instrumentation, determine results, and establish programs based on facts; general technical skills in learning, teaching adults, and evaluation of data; administrative and supervisory skills in working with people and knowledge of public health laws, regulations, ordinances, and codes and their application; and professional attitudes and resolving environmental health problems while working with colleagues, supervisors, those being supervised, and the general public.

Specific competencies involve the knowledge, comprehension, and ability to implement necessary measures to prevent, mitigate, and control various environmental contaminants in the inter-related media of air, water, and land while protecting and promoting the health of the community and all its special populations. The specific areas include environmental chemical agents; environmental biological agents; environmental physical agents; ambient air; water, sewage, and liquid waste; food; solid wastes; hazardous materials and hazardous waste; built environment including healthy homes and healthy communities; indoor environment including indoor air; environmental and occupational injuries; special populations such as children, the elderly, and the immunocompromised; environmental health emergencies, disasters, and terrorism; healthcare environment and infection control; insect control, rodent control, and pesticides; and recreational environment and swimming areas. (See endnote 5.)

These competencies were established in 1972 at a special workshop at Indiana State University attended by some 75 or 80 of the top environmental health professionals in the United States from all levels of government, professional associations, and industry. These professionals were broken up into working groups and their reports as amended by the conference were compiled and published by the author. On the basis of these data, the author developed an instrument for the evaluation of the competencies of environmental health interns in 1972 and then utilized it again in 1976 for evaluating environmental health professionals. In both cases, the instrument was found to be valid and highly reliable. (See endnotes 2, 3.) These competencies have been updated in various editions of the *Handbook of Environmental Health* and will be updated again in the fifth edition. (See endnote 5.)

QUALIFIED WORKFORCE

A qualified workforce to carry out specific tasks and programs is based on the competencies shown above and the determination of the additional special competencies needed by individuals to carry out Best Practices for successful local, state, and national programs. The workforce can be improved by increasing the skill levels of existing individuals and/or increasing the necessary skills of individuals who will be graduating from collegiate programs. Employers who may be governmental agencies or business and industry should be involved in developing a curriculum which meets their needs but still prepares them to be well-qualified generalists. Accredited paid internship programs should be used to teach students practical skills.

The existing employees should have their knowledge and skills measured against the recommended competencies needed as shown above and those added to by the specific program areas. Then appropriate in-service training can be developed and utilized to enhance the existing competencies and/or to create new ones.

INTERNSHIP PROGRAM

A paid internship is an integral part of the learning experience of the student. It enables the individual to integrate the art and science of the profession taught at the college or university with the practical existing problems in the community or industry to produce a person who is current, knowledgeable, and ready to work at many different tasks. The intern is under the close supervision of a licensed environmental health practitioner at the worksite as well as the university coordinator. The worksite meets the professional standards of environmental health practice based on tested principles and practices. The student is provided a planned and rigorous experience, where he or she works 40 hours a week for 10–15 weeks per internship experience and is expected to complete all tasks in a timely and professional manner as any other member of the full-time staff. In addition, the student is required to keep a daily log of activities and to complete a college-level paper on a special project which is of value to the employer. The student should present orally the major findings and recommendations to the staff of the employing agency in a special meeting. He or she should also learn to respond to questions concerning the study, and when the individual does not have an answer, tell the questioner that he or she will investigate the issue and send a proper response in writing.

The student learns the practical problem-solving approach to resolving issues while performing an effective service for the employer. The student gains practical experience, refines his or her basic environmental health skills, refines his or her communication skills, learns to work with people, learns to identify the sources of problems and make reasonable recommendations, develops techniques for studying major problems in depth, learns to prepare comprehensive reports, and learns the fine points of professionalism.

The student receives an appropriate remuneration for the work, which allows for housing, food, and other personal expenses as well as some assistance with tuition and books. The student should receive appropriate reimbursement for the use of a vehicle while at work.

On December 15, 1969, under United States Public Health Service contract number CPS-69-002, Indiana State University was awarded a grant to develop the modern concept of the environmental health internship. A 100-page set of guidelines was developed by the author, presented to the US Public Health Service, and then used at Indiana State University for the next 25 years as well as many other universities with environmental health internship programs. Indiana State University placed and supervised 1100 paid interns in 70 different health departments and industries in local, state, and federal employment in 28 different states. The 500 graduates from the accredited program served a minimum of two distinctly different internships including one at a local level that was general in nature and the other which could be specialized. Some students served three different internships.

TWO SIDES OF THE SAME COIN

In order to use our personnel and other resources most effectively, we need to recognize that the environmental health practitioner and the environmentalist are two sides of the same coin. Before the formation of the US EPA and of the subsequent state environmental protection agencies or pollution control agencies, typically the same practitioner was responsible for most environmental issues including air, water, and land pollution and the resulting potential for disease and injury. Today, environmental health practitioners, in government and industry, including environmental health specialists, environmental engineers, occupational health and safety specialists, and public health professionals, work toward preventing disease and injury and promoting health while protecting the environment. Environmentalists including environmental protection specialists, pollution control specialists, environmental sustainability specialists, energy specialists involved with carbon-based fuels, natural resources and conservation specialists, environmental science specialists, industrial water and waste specialists, renewable energy specialists, outdoor environmental specialists, environmental law and policy specialists, environmental advocacy and communications specialists, earth science specialists, environmental conservation specialists, and others protect and preserve the environment while also helping to prevent disease and injury to people.

It is essential that all environmental professionals have a broad understanding of the potential of collateral damage which may occur when a change is made to one environmental area. Since most environmental health or environmental protection people are specialists rather than generalists, it is necessary to have a broad understanding of all environmental areas and their potential interactions. This is one of the major reasons for writing a book of this nature.

USE OF KNOWLEDGE

In an ever-changing society, where knowledge is expanding at the most rapid pace in history, the amount of accessible content is overwhelming, and thoughts can be transmitted throughout the world virtually instantaneously, we have also become a society with limited vision. If we cannot find it by the use of a search engine, it does not exist. Therefore, past programs, including Best Practices, in environmental health, environmental protection, pollution control, and environmental sustainability, which have been highly successful, if not found on the Internet, may not become the basis for present and future actions. And if on the Internet, trying to sort out these programs from the voluminous material available is very difficult and very time-consuming. In addition, there is considerable information on the Internet which is either biased or totally inaccurate and yet references to the information appear on our Internet searches. This is also a major reason for writing a book of this nature which now presents to the various environmental professionals a single-source book for all areas of the environment and helps explain the complexity of the field as well as the inter-relationships of all environmental media.

BEST PRACTICES

When we examine current and past environmental health and protection programs and the Best Practices used in communities and industries, and how well they worked, we can create viable, economically sound responses to the needs of our own communities.

The discussion and use of Best Practices (evidence-based practice) is currently being stressed by the CDC, educational institutions, and various medically based professional organizations such as the American Psychological Association, the American Nurses Association, and the American Physical Therapy Association, as well as industry. This concept will accomplish what the author was taught while earning a Master of Public Health degree at the University of Michigan in 1959, that public health as well as environmental health will be at its best when it adopts techniques used by business and industry because of the great accomplishments of the American economy and industry which has been using Best Practices for many years.

There are two basic types of Best Practices in the various environmental and public health fields, and these are discussed below.

EVIDENCE-BASED PUBLIC HEALTH PRACTICES

Evidence-based public health practices are an outgrowth of the 1970s and 1980s when expert panels accumulated information and then made recommendations to resolve problems. The evidence-based approach as discussed in connection with Developing Healthy People 2020 consists of the use of data and information systems, behavioral science theory, and program planning models to develop, implement, and evaluate effective programs and policies in public health in clinical areas. An example of evidence-based practice would be examining treatment decisions in certain disease processes and utilizing a scientific basis for the approach used, while recognizing that the art of medicine also includes individual factors such as values or quality of life. This technique may also be applied to a variety of public health practices.

Evidence-based public health practices may be a more rigorous method of securing Best Practices than the practical approach and certainly would be more effective in resolving clinical problems. However, the practical approach is more readily available and can be utilized more quickly and then adapted if and when necessary.

PRACTICAL APPROACH TO BEST PRACTICES DEVELOPED AND UTILIZED BY EXPERTS IN THEIR FIELDS

The practical approach, which is a less rigorous form of evidence-based practices, is most frequently used in environmental health programs. Typically, the Best Practices have already been established by various federal, state, and local agencies; professional associations; and industry.

Best Practices are science-based principles and practical applications of techniques (art of the field) which may help prevent, resolve, or mitigate hazardous situations and existing or potential problems of disease and injury. They may consist of planning procedures, supervision and management procedures, educational approaches, operational and/or maintenance procedures, and/or legal controls. They may exist individually or as part of a system. In any case, priority should be given to those Best Practices which most readily and economically solve, reduce, or prevent a problem and avoid collateral damage from the impact of their use. Typically, Best Practices may be found in several different categories including:

- Gathering and analyzing data in a scientific manner to determine the nature and scope of the problem
- Information and education
- Ordinances and regulations
- Elimination of dangerous and hazardous situations
- Preventing and controlling the sources of potential outbreaks of acute disease, injury, and contamination of the environment through the efforts of the appropriate environmental systems and/or the enforcement of the appropriate rules, regulations, and laws
- Preventing and controlling the underlying factors which contribute to the occurrence of chronic disease which may have been of environmental origin or is enhanced by environmental contaminants
- Making necessary program changes to control the problem
- Using various prevention and treatment measures
- Developing continued maintenance programs
- Re-evaluating program content and results and redirecting efforts as necessary

Best Practices and programs described include but are not limited to food protection, food technology, insect control, rodent control, pesticides, lead poisoning, housing and communities, the indoor environment, hospital, school, and nursing home environments, recreational environment, occupational environment, air quality, solid and hazardous waste, private and public water supplies, swimming areas, plumbing, private and public sewage disposal and soils, water pollution and water quality, terrorism, disasters, environmental health emergencies, children's environmental health issues, injury control, and so on.

RESOURCES

This book is not a technical text. For basic technical and scientific information which supports the various Best Practices, see:

1. *Handbook of Environmental Health: Biological, Chemical, and Physical Agents of Environmentally Related Disease*, Volume 1, Fourth Edition, by Dr. Herman Koren and Dr. Michael Bisesi, CRC Press, Boca Raton, FL (2003).
2. *Handbook of Environmental Health: Pollutant Interactions in Air, Water, and Soil*, Volume 2, Fourth Edition, by Dr. Herman Koren and Dr. Michael Bisesi, CRC Press, Boca Raton, FL (2003).
3. *Illustrated Dictionary and Resource Directory of Environmental and Occupational Health*, Second Edition, by Dr. Herman Koren, CRC Press, Boca Raton, FL (2005).
4. *Industrial Hygiene Evaluation Methods*, Second Edition, by Dr. Michael Bisesi, CRC Press, Boca Raton, FL (2003).
5. A list of special resources found in each chapter beginning with Chapter 2.
6. Excellent up-to-date endnotes and bibliography used by the author in recording the Best Practices and the discussion of successful programs for each of the environmental health areas of concern.
7. Names and locations of the actual local, state, and federal programs that have been successful are cited within the text.

HOW TO MAKE IT WORK

The individual reader who utilizes information from this book can work efficiently by incorporating his or her education and professional experience into the mix of information; adding the input from stakeholders; and utilizing the knowledge and experience from local and state agencies including university cooperative extension services, state environmental protection agencies, state pollution control agencies, state health departments, state departments of agriculture, and so on to develop necessary plans based on comprehensive data collected by evaluating all problems and conditions related to the subject area. He or she should then prepare a request for a new program or program review, which can be presented to the appropriate authorities for approval. This proposal should include information on appropriate expenditures, amount of people and types of people needed to successfully resolve the problems in a given size community, tested techniques of supervision and management of personnel, as well as successful techniques of communications which have been used with all individuals involved.

One of the values of this book is that all chapters other than Chapter 1 and Chapter 4 are written in a consistent manner. They provide a statement of the nature of the problem and special information, immediately followed by the division of the problem into sub-problems including factors leading to impairment and Best Practices used to resolve each of these factors and sub-problems. This is followed by sections on laws, rules, and regulations; resources; and programs. Finally, each chapter lists comprehensive endnotes and a bibliography. A very comprehensive index allows the reader to move quickly from one topic to another. This makes the information more easily retrievable and therefore more usable. It makes planning for a new or improved program more cost-effective and saves individuals a substantial amount of time in researching and gathering accurate data on problems and solutions.

UNIQUE FEATURES OF THIS BOOK

1. A one-of-a-kind programmatic book incorporating Best Practices devised by working professionals who specialize in all environmental areas.
2. Easily accessible resources for environmental health, safety, protection, and sustainability.
3. An explanation of the complexity and interaction of people, various environments, and ecosystems based on the underlying science found in other books identified in this book, as well as in the endnotes.
4. A useful problem-solving tool to plan and help evaluate programs for professionals, students, professors, governmental agencies, industry, and civic leaders in environmental health, safety, protection, and sustainability.
5. A single source of tested principles and Best Practices in all environmental areas developed by other professionals based on their experience using the appropriate science and the best and most economical available technology, thereby saving an enormous amount of time and money for the reader.
6. A rapid means of gathering additional information on a given topic by inserting the title of an endnote in the search box of the computer and bringing up the document for immediate review (this works for most published documents but not necessarily for those which are published solely on the Internet).
7. An understanding that science-based environmental sustainability is achieved through the use of Best Practices identified in each of the environmental areas.
8. An understanding of the nature of Best Practices as evidence-based practice in environmental areas and how they are achieved.
9. A recognition by all people in the environmental field that despite their specialty, they all work for the ultimate goal of protecting the environment and preventing disease and injury to people.
10. An excellent means of refreshing or upgrading the knowledge and skills of the reader in all environmental areas, as well as the potential interaction between these areas.
11. A book which depends mainly on practical principles, excellent field experience, and Best Practices in the prevention, protection, and sustainability of the environment as well as preventing disease and injury in people and promoting good health as identified by actual programs in the United States and Canada. These Best Practices can easily be used by practitioners, students, professors, governmental agencies, industry, and civic leaders worldwide.

WHY THIS BOOK NOW?

Many people are inefficient because they spend time and money developing new procedures and programs to solve existing problems, and what they think are new problems which may have already been solved in a given community and/or in other places in past times. Also, we are inundated by information from the numerous environmental health and protection programs at the local, state, federal, and international level, by the myriad of industry groups working on specific environmental pollutants, by the large number of professional associations and civic organizations, and by the substantial amount of research completed because of the enormous interest that people have in protecting their environment and their health.

Local and state communities, regional areas, and nations have been and will continue to face budget restrictions. This book, which is a single-source reference, is an extremely useful tool for planning how best to use resources to create procedures and programs to prevent disease and injury and promote good health, while protecting the environment.

This book is a reference and textbook for students, professors, governmental agencies, industries, and existing professionals who are specialists in a given field seeking to increase their overall knowledge and improve their existing skills. It provides the best opportunity to understand how to establish and manage successful programs using Best Practices in the quickest time frame and achieve the best results at the lowest cost.

ENDNOTES

1. US Department of Health and Human Services. 2008. *Developing Healthy People 2020: Objectives for 2020, Phase 1 Report: Recommendations for the Framework and Format of Healthy People 2020.* The Secretary's Advisory Committee on National Health Promotion and Disease Prevention. Washington, DC.
2. Koren, Herman. 1972. *An Instrument for Evaluating Competencies of Environmental Health Interns.* Doctoral dissertation, Indiana University, Bloomington, IN.
3. Koren, Herman. 1976. *Delineation of Responsibilities and Performance Levels of Professional Environmentalists.* Indiana State University, Terre Haute, IN.
4. Koren, Herman. 1969. *Demonstration of an Approach to the Internship Concept in Environmental Health*, CPS-69-002. Indiana State University, Terre Haute, IN.
5. Koren, Herman, Bisesi, Michael. 2003. *Handbook of Environmental Health: Biological, Chemical, and Physical Agents of Environmentally Related Diseases*, Fourth Edition, Volume 1, pp. 67–73. CRC Press, Boca Raton, FL.

BIBLIOGRAPHY

Robert J. Brulle. 2008. The US Environmental Movement. In: *20 Lessons in Environmental Sociology*, 211–227. Gould, K., Lewis, T. (editors) Roxbury Press. Sweet Springs, MO.

Stanley Johnson. 2012. *UNEP: The First 40 Years: A Narrative.* United Nations Environmental Programme. UNON/Publishing Section Services, Nairobi.

Presidential Documents, Executive Order 13423. 2007. Strengthening Federal Environmental, Energy, and Transportation Management. *Federal Register*, 72:17.

Presidential Documents, Executive Order 13514. 2009. Federal Leadership in Environmental, Energy, and Economic Performance. *Federal Register*, 74:194.

United Nations. 2012. *Report of the United Nations Conference on Sustainable Development.* A/CONF. 216/16. Rio de Janeiro, Brazil.

United Nations, Sustainable Development. 1992. *United Nations Conference on Environment and Development.* Rio de Janeiro, Brazil.

United Nations Department of Economic and Social Affairs. 2013. *Water Quality: International Decade for Action "Water for Life" 2005–2015.* United Nations Department of Economic and Social Affairs, New York, NY.

US Department of Health and Human Services. 2010. *Developing Healthy People 2020: Evidence-Based Clinical and Public Health: Generating and Applying the Evidence.* The Secretary's Advisory Committee on National Health Promotion and Disease Prevention, Washington, DC.

US Environmental Protection Agency. 2013. *Plan EJ 2014: Progress Report.* Washington, DC.

US Environmental Protection Agency, Region 10: The Pacific Northwest. 2013. *History of Sustainability: Creation of EPA and NEPA.* Washington, DC.

Acknowledgments

SPECIAL ADVISORS

Consumer advisor and proofreader who took my ideas that I had been working on for 15 years concerning preserving the best in programs of environmental health and turned them into the business concept of Best Practices—Gene DeLucia, BS (Management and Computer Science), MBA (Retired).

Graphic designer and final proofreader and dear friend, who has designed and edited all of my books since 1988—Sr. Alma Mary Anderson, CSC, M.F.A., Professor of Graphic Design, Indiana State University.

Technical advisor on content and accuracy who was one of my excellent students and is now helping guide his former professor—Keith L. Krinn, R.S., M.A., DAAS, CPHA, Environmental Health Administrator, Columbus Public Health, Columbus, Ohio, Past President National Environmental Health Association, Past Chairperson of the National Conference of Local Environmental Health Administrators.

MENTORS

Dr. Larry Gordon, Retired New Mexico Cabinet Secretary for Health and Environment, Founding Director New Mexico Environmental Improvement Agency, Rank of Naval Capt., US Public Health Service Inactive Reserve, Past President of the American Public Health Association, Chairperson of the National Conference of Local Environmental Health Administrators, Founder of the Council on Education for Public Health (the national accrediting agency), a Founder of the American Academy of Sanitarians, and recipient of innumerable environmental health and public health awards of the highest level.

Dr. Jerrold M. Michael, Emeritus Professor and Dean, School of Public Health, University of Hawaii, Adjunct Faculty, School of Public Health and Health Services, George Washington University, Assistant Surgeon General US Public Health Service (Retired), and recipient of innumerable highest level honors, many from foreign countries for his work in environmental health and public health.

Dr. Richard Spear, Professor Emeritus Health and Safety, Indiana State University, Past President of the Emergency Medical Systems Commission of the State of Indiana, who would not let me quit working on my doctorate in a time of unbelievable stress and gave me wonderful advice that I have followed throughout my life.

ACTIVE SUPPORTERS OF THE CONCEPT OF BEST PRACTICES IN ENVIRONMENTAL HEALTH, SAFETY, PROTECTION, AND SUSTAINABILITY

Gary Noonan, Retired Rank of Naval Capt., US Public Health Service, Diplomate Laureate-6, American Academy of Sanitarians, Executive Secretary-Treasurer, American Academy of Sanitarians for his support and encouragement as well as the support and encouragement of the American Academy of Sanitarians to move forward with an original innovative book on Best Practices in environmental health, safety, protection, and sustainability. Also to my peers at the Academy who gave me the great honor of being chosen as a Diplomate Laureate-7.

NATIONAL ENVIRONMENTAL HEALTH ASSOCIATION

Outstanding assistance from the many staff members over the past 60 years to me as a member and consultant to various technical sections, and to my peers who granted me their highest consideration when they named me in 2005, the Walter S. Mangold Award recipient. I have had the great fortune and honor to serve this organization and to have received four distinct Presidential Citations at different times in my continuing active career.

COLLEAGUES

Gary Becker, MA, LMFT (Florida), Certified School Psychologist (Florida, Illinois) for his work in helping me organize my thoughts as the Editor of *Best Practices Series for Public Health Programs*.

Carolyn Becker, MS, NCSP (National Certified School Psychologist), Licensed School Psychologist for her work in helping me organize my thoughts as the Editor of *Best Practices Series for Public Health Programs*.

Laura Hickerson, MS, MLS, Pesticide Specialist at the National Pesticide Information Center located at Oregon State University for her work and her colleagues' assistance in providing for me vital information and resources concerning the current status of Pesticides of Public Health Significance in order for me to accurately write Chapter 9, entitled "Insect Control, Rodent Control, and Pesticides."

Mildred Lee Tanner, BS, Environmental Health Manager, South Carolina Department of Health and Environmental Control, and for many years Technical Chairperson and Advisor on Children's Environmental Health Issues to the National Environmental Health Association, for her work in clarifying my ideas on Chapter 4, "Children's Environmental Health Issues" (a highly significant chapter in this book), and providing for me numerous sources of information.

Melissa Wenzel, BS, Natural Resources Environmental Studies, Industrial Storm Water Program Coordinator, Minnesota Pollution Control Agency for helping to guide me and providing necessary resources for Chapter 14, "Water Quality and Water Pollution," which although last in this book, was written first so that I could understand better how to write the entire book by writing a complex chapter.

SPECIAL RECOGNITION

Suzanne Zsiga, MS, Cardiac Pulmonary Rehab Department Supervisor, Judy Ciaccia, RN, Cardiac Nurse, Cindy Chilton, RN, Cardiac Nurse, Sandy Dasch, RN, Cardiac Nurse, and Christine Sostak, RRT, Respiratory Therapist at Largo Medical Center, Largo, Florida, for helping keep me alive and well for many years.

Travis Cornett, Sheena Hogan, and Jonathan Garland, Managers, Kevin Moseley, Margaret (Maggie) Zemzicki, and Summer Brooks, Assistant Managers, Panera Bread Company, Belleair Bluffs, Florida, and all of the fine young people working for them who have helped me over the years and allowed me to assimilate many reams of data while being in a lovely and homey atmosphere.

TEAM THAT MADE A 15-YEAR DREAM A REALITY

Richard O. Hanley, Publisher, CRC Press, Taylor & Francis Group, who signed me at age 81 to an immediate contract and convinced me to be the Editor of the unique series *Best Practices for Public Health*; Grace McInnes, Senior Editor, Health and Social Care, Routledge Taylor & Francis, who accepted the book and *Best Practices for Public Health* series as part of her Health and Social Care division; Carolina Antunes, Editorial Assistant, Health and Social Care, Routledge Taylor & Francis for solving numerous problems; Robert Sims, Project Editor, Taylor & Francis Group, for guiding the manuscript to excellent book form; Viswanath Prasanna, Senior Project Manager, Datapage and his team for a wonderful production; Gay O'Casey, the Copyeditor, for challenging me at every step to make sure we were totally accurate; and Alma Mary Anderson, Professor Graphic Design, Indiana State University who carried me through the intricacies of modern publishing, major proofreader, index provider, formatter, and graphic designer of the cover.

Author

Herman Koren, HSD, MPH, DLAAS, REHS is Professor Emeritus of Health and Safety, former Coordinator Environmental Health Science Program, which he founded in 1967, former Director Supervision and Management Program I and II for continuing education, which he founded in 1980, and former Coordinator Environmental Health Internship Program, which he founded in 1969, at Indiana State University at Terre Haute. He is currently the Series Editor for the Routledge, Taylor & Francis Group, for *Best Practices Series for Public Health Programs.*

He has been and continues to be an outstanding researcher, teacher, consultant, and practitioner and has served with great distinction for the past 62 years in environmental health and public health. He has been an active Registered Environmental Health Specialist (REHS)/ Registered Sanitarian, since 1963 in Pennsylvania, 1969 in Indiana, and 2010 in National Environmental Health Association. In addition to his many oral presentations to his colleagues and his numerous articles in journals, this book, part of a series of 20 books, includes in-depth, updated, and vastly expanded new editions of his outstanding works which have been recognized nationally and internationally. His *Handbook of Environmental Health: Biological, Chemical, and Physical Agents of Environmentally Related Disease*, Volume 1, Fourth Edition, Lewis Publishers–CRC Press, Boca Raton, FL, 2003, and *Handbook of Environmental Health: Pollutant Interactions in Air, Water, and Soil*, Volume 2, Fourth Edition, Lewis Publishers– CRC Press, Boca Raton, FL, 2003, are recommended to individuals who are seeking to become REHSs. He has also written *Environmental Health and Safety*, Pergamon Press, Elmsford, NY, 1974; *Handbook of Environmental and Safety Principles and Practices*, Volumes 1 and 2, Second Edition. Lewis Publishers, Chelsea, MI, 1991; two editions of the *Illustrated Dictionary of Environmental Health and Occupational Safety*, CRC Press, Boca Raton, FL, 1996 and 2004; third edition of a two-part series entitled *Management and Supervision for Working Professionals*, Lewis Publishers–CRC Press, Boca Raton, FL, 1996, and two history books. He has been asked to work on a fifth edition which would be a comprehensive, in-depth rewrite of the *Handbook of Environmental Health*, Routledge. His field experience is vast. He served as a rural practitioner, large metropolitan area practitioner in Philadelphia, District Environmental Health Supervisor of a 250,000-person district, administrator as Chief of Environmental Health and Safety at Philadelphia General Hospital (a 2000-bed institution), and coordinator of the Environmental Health Internship Program at Indiana State University, where he supervised in the field 1100 interns over 25 years in 28 different states and 70 different environmental health programs in government and industry, and has given thousands of hours without pay to numerous health departments and hospitals in all areas of the environment plus hospital infection control problems for many years.

He is an extremely innovative thinker. He has developed new program concepts, new programs, tools of evaluation, and continuing education for working professionals, which have been utilized by many others in the field, throughout his career, while involving many different types of stakeholders outside of the field of environmental health science. Some of his innovative thinking covers a community home accident control program; a community rodent control program; special self-inspection programs for the hospital housekeeping department, food service department, laundry department, physical plant and maintenance, and medical and nursing departments including surgical suites for hospital infection control; environmental health professional competencies; environmental health internship program; supervision and management of people; 10-week high school environmental health program; and now Best Practices and resources in environmental health programs; and so on.

He has been recognized by his peers many times over his career. Some recognitions are as follows:

DIPLOMATE LAUREATE OF THE AMERICAN ACADEMY OF SANITARIANS

In April 2012, Dr. Herman Koren received the distinct honor of being selected by his peers for the extremely distinguished recognition as a Diplomate Laureate of the American Academy of Sanitarians: "The attainment of this honor is based upon the demonstration of outstanding knowledge, skills and abilities and distinguished competence in their professional field." He is only the seventh person to receive this honor.

WALTER S. MANGOLD AWARD

In June 2005, Dr. Herman Koren received the highest honor in environmental health in the United States and globally, when he became the recipient of the 2005 Walter S. Mangold Award for a lifetime dedicated to environmental health practice at the field, supervisory, and administrative level, for research, teaching, and public service: "He has left a true mark upon the United States and the world from the more than five hundred environmental health professionals he graduated from the accredited Indiana State University program, with a Bachelor of Science in Environmental Health Science degree plus seven months of professional field experience served in two different types of Environmental Health Internships. He was the founder of the modern environmental health internship concept in 1969. He was the founder of the Student National Environmental Health Association and the alpha chapter at Indiana State University. He has been a mentor to thousands of young people, who went on to highly successful careers in environmental health, occupational health and public health at the governmental level and in private industry" (Walter S. Mangold Award presentation). Subsequently, he served as the chairperson of the Mangold Awards Committee for a 2-year period and rewrote both manuals that are used by applicants and the committee to determine the next Mangold award recipient.

PRESIDENTIAL CITATIONS FROM THE NATIONAL ENVIRONMENTAL HEALTH ASSOCIATION

Dr. Koren has been the recipient of four distinct Presidential Citations from the National Environmental Health Association over his career, the latest one being June 2011.

SOME OTHER MAJOR HONORS

- Keynote Speaker and Consultant to the Canadian Institute of Public Health Inspectors
- Blue Key Honor Society Award for Outstanding Teaching—Indiana State University
- Alumni and Student Plaque and Citations for Outstanding Teaching, Research, and Service—Indiana State University

1 Necessity for Effective Programming Techniques

AUTHOR'S THOUGHTS

The major thrust of this book is the provision of Best Practices in all environmental areas for students, practitioners, and all other stakeholders so they can use them to effectively and economically improve existing programs and in the development of new programs. Best Practices are the practical applications of knowledge and successful experiences in program areas used to prevent disease and injury as well as promote good health, while protecting and sustaining the natural environment.

Although it is not necessary to read the first two chapters of the *Handbook of Environmental Health-Biological, Chemical and Physical Agents of Environmentally Related Disease* by Dr. Herman Koren and Dr. Michael Bisesi (Lewis Publishers, CRC Press, Boca Raton, FL, 2003) to understand the present book and how to utilize the material it contains, it would enhance the learning experience of the reader if their knowledge was refreshed or they were exposed for the first time to the background knowledge found in that book. Chapter 1, Environment and Humans, includes a brief discussion on ecosystems, energy, health and environmental problems, concepts of chemistry, transport and alteration of chemicals in the environment, environmental health problems and the economy, risk–benefit analyses, environmental health problems and the law, creating federal laws and regulations, and environmental impact statements. Similarly, it would also be helpful to read Chapter 2, Environmental Problems and Human Health, which includes information on human systems, toxicological principles, applicable concepts of microbiology, principles of communicable disease, epidemiological principles, and risk assessment and risk management.

INTRODUCTION

The health, safety, environmental degradation, and environmental sustainability problems of our society, including global warming, continue to grow, while we have less money than ever to resolve them. At the same time, the previous concerns related to environmental degradation, environmental sustainability, injuries, and spread of disease have not subsided, but rather have increased in scope and complexity.

We need to use all of our existing resources in an effective manner including integrating the knowledge and practices of professionally trained personnel, generalists and specialists, both in environmental health and in environmental protection, while exposing them to a broad understanding of the inter-relationships between the various environmental media and the affect that a single action in one area may have on other environmental problems. We have to recognize the significance of the knowledge, experience, and enthusiasm of the large numbers of people and organizations in the environmental movement, the conservationists, the preservationists, the ecologists, the environmental justice groups, the sustainability groups, the public health and occupational health oriented groups, etc. We have to incorporate and utilize the knowledge and practical experience of the various professional associations, the high school students and teachers interested in protecting the environment, the college and university students and teachers who are working toward degrees in environmental health and environmental protection through a multitude of Bachelors and Masters programs ranging from environmental health to ecology to pollution control to environmental quality to environmental sustainability; the various news media; and the numerous politicians at the local, state, national, and international level.

We have to increase the numbers of professionally trained people working with the latest scientific knowledge and an understanding of successful programs utilizing Best Practices that have actually worked in the past and present. These techniques need to be used as examples for developing programs to help resolve current and future problems. The professionals then need to adapt this knowledge to meet the specific concerns of their community.

We need to share, adapt, and use the existing tools that we have by making available Best Practices and special resources for the multitude of existing and planned inter-related environmental programs.

We also need new tools to resolve these problems. We need to understand the major concerns of today and tomorrow and how to use available knowledge to fulfill the promise made in the Constitution of the United States of America that we all have the right to life, liberty, and the pursuit of happiness, which includes good health, the avoidance of disease and injury, and the protection and preservation of the environment for future generations.

MAJOR CONCERNS IN OUR SOCIETY TODAY

Eight major factors are interacting in our society today that require we make substantial changes in how we design practical programs, which prevent disease and injury, promote good health, protect the environment, and enhance environmental sustainability. These factors are: financial stress; regulatory reform; environmental justice; our changing planet; a multitude of environmental pollutants and their interactions; emerging and re-emerging infectious diseases, primarily of a zoonotic nature; where we live, work, play, and are educated; and the potential increase in acute and chronic disease in our mobile, growing, and aging population.

FINANCIAL STRESS

Financial stress at the personal level has led to increased homelessness and poverty. The largest numbers of new homeless are women and children. Hunger is another factor which contributes to disease, especially when there are negative environmental factors involved.

Financial stress at the state and local level of government has led to the reduction of many programs meant to protect the health and welfare of our citizenry in a time when more services are needed, not less. A research brief from the National Association of County and City Health Officials dated December 2011 stated that there has been a widespread reduction in essential services including environmental health services at the local and state level. In 2011, 55% of all local health departments cut at least one program protecting citizens in the communities. This affected 68% of the US population. Local health departments reduced environmental health services by 18% overall and by 21% excluding food safety programs. Epidemiology and surveillance was reduced by 9% and emergency preparedness was reduced by 20%. All of these losses are extreme.

There has also been a loss of trained professionals from local health departments. From 2008 through 2011, a total of 34,400 jobs were eliminated. Negative job impact has affected not only essential program areas but also the morale of working professionals. Emergency management is an example of this dire situation. Disasters and acts of terrorism can only be contained by highly trained individuals who need to be available to respond when the emergency occurs. The reality of disasters and the potential for terrorist acts has increased substantially. With the political situation constantly in flux, federal, state, and local officials must be prepared to respond immediately to the disaster to prevent loss of life and destruction of facilities and infrastructure, which leads to even further compression of the funds available to help the citizenry.

At the federal level, there have also been substantial changes in the budget for Environmental Health Programs. From 2010 to 2012, there was a 30% reduction in funding for the Centers for Disease Control and Prevention's (CDC's) National Center for Environmental Health and the Agency for Toxic Substances and Disease Registry, which provide funding and services to states and other groups. The new Healthy Home and Community Environments Program will be funded at a 51% decrease from the merged Healthy Homes and Lead Poisoning Prevention Program and the CDC's National Asthma Control Program. There will also be substantial reductions to the Basic Environmental Health and Radiation Preparedness Programs, Climate Change Program, Environmental Health Tracking Program, Healthy Community Design Program, and Safe Water, Food, and Waterborne Illnesses Programs. In the past, the federal government helped support state and local health departments through grants, loans, and tax subsidies, when there was a serious need, but now this help will be reduced. (See endnotes 1, 2, 3, 4, 5, 6.)

Preparation for bioterror and health emergencies has deteriorated across the country because of severe budget cuts. In 2011, the Trust for America's Health and the Robert Wood Johnson Foundation in their ninth annual report, *Protecting the Public from Diseases, Disasters, and Bioterrorism*, stated that 51 of the 72 cities in the Cities Readiness Initiative are at risk of being eliminated; 24 states are at risk of losing CDC epidemiologists; and all 50 states would lose the support of the CDC during a response to nuclear, radiological, chemical, and natural disasters. Although the public health and national security communities have repeatedly discussed the serious need for new funding for dealing with bioterrorism and other disasters and emergencies, funds continue to be cut in these areas. (See endnote 7.)

The President's budget for the remainder of the 2016–2017 budget year was issued in March 2017. The Department of Health and Human Services received a total cut of 18% of its existing budget, including the National Institutes of Health 20% cut of $5.8 billion; CDC loses funds to fight Ebola outbreaks research and operational funds. This is devastating for medical research.

REGULATORY REFORM IN ENVIRONMENTAL HEALTH AND PROTECTION PROGRAMS

America with its free-market system has always prospered because the nation has adopted common sense laws to help protect individuals and communities from environmental stresses and to resolve current and future problems. This has resulted in the implementation of many rules protecting younger people, workers, and citizens in general. Although a given rule may have a financial cost, over time the improvement in health and reduction in injuries to a given population has far exceeded the cost of resolving the problem. A good example is the removal of lead from gasoline and paint, and the reduction in lead poisoning in children. Further, the new rules and regulations have led to innovations and new products which have stimulated the economy.

The manufacturing, processing, use, and disposal of chemicals and materials by industry often affect environmental and occupational health, and result in safety hazards and damage to the environment, which can initiate or exacerbate various health problems from before birth to old age. All body systems are at risk. Potential toxins are not always tested before use and only come to the attention of public health authorities after the harm to people is proven and widespread. Further, air, water, and land become contaminated and a variety of ecosystems may be altered or destroyed. All of these issues need appropriate rules and regulations to protect the public from potential disastrous consequences which are an unwanted side effect of the manufacturing, processing, use, and disposal of chemicals and materials which help make our society better in some respect.

At times, a rule becomes obsolete and therefore should be removed. However, the elimination of rules and regulations for the sake of improving the economy may end up being very costly, as in shale oil extraction, because of the collateral damage done to the health and welfare of people and the destruction of the environment. Another poor approach is the defunding of programs and professionals who enforce the rules and regulations.

Cost-benefit analysis does not always apply in relation to regulations which improve the environment and/or the health and safety of individuals in the community. All costs are easy to express, but benefits may not be realized for many years. Benefits include improved quality of life and good health of the individual over long periods of time. Benefits also include reduction in medical expenses, greater productivity from individuals because of less lost time and higher performance, and the cost of long-term disability and the loss of a human life. The true costs of an industrial operation should include the damage to the environment by certain processes, the cost of clean-up of waste dumped into the air, water, and on the land, collateral damage to other parts of the community such as through acid rain, and societal costs to people, communities, and the physical environment. (See endnote 8.)

A proposed alternative to cost-benefit analysis is the Precautionary Principle, which states, "When the health of humans and the environment is at stake, it may not be necessary to wait for scientific certainty to take protective action." This principle was arrived at during the Wingspread Conference on January 26, 1998, at a special meeting of scientists, philosophers, lawyers, and environmental activists, at the Robert Wood Johnson Wingspread Conference Center in Racine, WI. The Precautionary Principle is necessary because: numerous toxic substances have been released into the environment, resources have been exploited, there are increasing rates of disease due to environmental actions, and environmental interactions of today will influence the environment of tomorrow. The Principle has been used as a basis for international agreements. The participants felt that cost-benefit analysis allowed new products and technologies which could cause disease and injury without considering the potential problems from the raw and finished materials. The Principle, which applies to human health and the environment, utilizes the ethical assumption that humans are responsible for protecting, preserving, and restoring all ecosystems in the world and preventing disease from occurring because of new technologies. (See endnote 9.)

Another technique for regulatory reform is to utilize the Environmental Public Health Performance Standards (v2.0) developed by the CDC in conjunction with the organizations which make up the nation's leadership in environmental health. The primary goal of these standards is to build capacity for essential services, build community accountability for environmental health services, and build consistency of services throughout the environmental health systems or programs. This goal, when met, will enhance communications, improve coordination of activities and resources, and reduce duplication of services. It will help identify the strengths and weaknesses of various programs. It will establish standards and the means of measuring them. It will provide information and data that can help provide for changes in policy or resources to improve community environmental health and the health and safety of the public and thereby the physical environment. (See endnote 10.)

ENVIRONMENTAL JUSTICE

Environmental Justice decrees by law that all people share an equally high standard of environmental protection and improvement of existing hazardous situations regardless of race, color, national origin, income, or education. This is necessary because many of the very worst environmental conditions are found in areas where the people live and work.

Executive Order 12898, "Federal Actions to Ensure Environmental Justice in Minority Populations and Low Income Populations," mandates that the implementation of environmental justice is part of all programs at the federal level. The goal of this Executive Order was to:

protect the environment and health in overburdened communities; empower communities to take action to improve their health and environment; and establish partnerships with local, state, tribal and federal governments and organizations to achieve healthy and sustainable communities.

This is leading to involving environmental justice in rulemaking, permitting, compliance, enforcement, and community-based action programs, and including environmental justice in all parts of the Environmental Protection Agency (EPA). Plan EJ 2014 is the EPA's roadmap for integrating environmental justice into all its programs and practices. It is based on good science, law, accurate information, and the use of appropriate resources. Examples of these programs are the Urban Waters Program, the Pesticide Worker Safety Program, the US-Mexico Border 2020 Program, the Community Engagement Initiative, and implementation of the Internal Technical Directive on Reviewing EPA Enforcement Cases for Potential Environmental Justice Concerns. (See endnote 11.)

Our Changing Planet

(See endnotes 12, 13, 14)

In 1989, the US Global Change Research Program was started as a Presidential Initiative and then mandated by Congress in the Global Change Research Act of 1990, which requires "a comprehensive and integrated United States research program which will assist the nation and the world to understand, assess, predict and respond to human-induced and natural processes of global change." Thirteen departments and agencies of the federal government work together to carry out the mandate of Congress, which is to build a knowledge base on human responses to climate and global climate change; to coordinate and regulate federal agencies on climate working in research, education, communications, and decision support; and to produce an annual report of their research findings on climate change, and have the President of the United States submit this report to Congress. In the last 20 years, the United States through the US Global Change Research Program has developed an understanding of the short-term and long-term changes in climate, the ozone layer, land cover, and ecosystems, and has made the world's largest scientific investments in the research of climate change.

Reports of global climate change impacts in the United States indicate at least ten key findings. They are:

1. Global warming is unequivocal and primarily human induced.
2. Climate changes are underway in the United States and are projected to grow.
3. Widespread climate-related impacts are occurring now and are expected to increase.
4. Climate change will stress water resources.
5. Crop and livestock production will be increasingly challenged.
6. Coastal areas are at increasing risk from sea-level rise and storm surges.
7. Risk to human health will increase.
8. Climate change will interact with many social and environmental stresses to substantially increase the level of these stresses.
9. Thresholds will be crossed, leading to large changes in climate and ecosystems.
10. Future climate change and its impacts depend on choices made today.

Despite year-to-year fluctuations in temperature and weather conditions, the overall trend is for warming, with all its attendant problems. Unfortunately, there is a huge political discourse surrounding the global warming topic, and in some areas there is a decision to ignore scientific data especially to make political points. This type of emotional response to global warming based on lack of scientific information can readily be found on the Internet today.

It is the increase in temperatures in the world that is helping to contribute to the potential for: the increase in emerging and re-emerging diseases, reduced crop yields, increased level of algae in water, increased water demands including for irrigation and use of the groundwater supply, severe droughts, and power outages. Increased temperatures also contribute to decreased ice packs in the Arctic and Antarctica and the increase in water levels around the coasts of the country leading to greater potential flooding of these areas. Also, increased ocean temperatures may lead to increased severity of storms.

It is projected that temperatures in the United States will rise on average in the next 100 years if nothing is done to reduce greenhouse gas emissions worldwide. (In fact, in 2012, the average temperature in the United States rose by 1° Fahrenheit.) This could cause extreme weather, both wet and dry. There were discussions at the 2012 annual meeting of the Geological Society of America about the potential problems created by climate change and how it affected the severity of hurricane Sandy when it reached the East Coast of the United States. Rising sea levels contributed to the storm surge record height. Sea surface temperatures contributed to the flooding. Increased temperatures over Greenland and increased temperatures in the Arctic may have had an effect on air masses and the position of the jet stream.

Global warming may result in a sharp increase in heat-related illnesses and deaths as well as respiratory and cardiovascular disease, especially in the most vulnerable populations of the elderly, the poor, children, and people suffering from other diseases. Increased temperatures would exacerbate problems of air pollution and health, ozone formation, amount of waterborne and foodborne diseases, amount of vector- and rodent-borne diseases, and amount of emerging infectious diseases.

Other changes affecting people and the environment include:

1. Deforestation which may result in loss of species of plants and animals, increase green-house gases, increase carbon when the trees are burnt, interfere with the water cycle, and cause soil erosion leading to increased silt entering bodies of water
2. Encroachment by agricultural areas which may result in loss of plant species, decreased water quality, loss of wildlife and wildlife habitat, loss of mineral sources, soil erosion, changes in groundwater hydrology, increased air pollution, and significant new noise pollution
3. Wetland modification which may cause a massive loss of fish and shellfish, increase vulnerability of the coastal areas to storm surge and flooding in commercial, industrial, and residential areas, and affect the navigation of ships in areas which are used for water transportation
4. Building of dams which may interrupt and destroy ecosystems, interfere with fish migration, alter water flows, retain sediments which may be necessary for ecosystems downstream, contribute to greenhouse gas emissions, etc.
5. Road construction may cause air pollution initially from the use of the various pieces of heavy equipment and the dust being created during the construction process; cause air pollution from the vehicles utilizing the new road; cause noise pollution initially from the construction work and later from the vehicles utilizing the road; cause water pollution from the rainwater and snow melt off the roads carrying gasoline, oil, heavy metals, and other pollutants; and create the potential for droughts, fires, and the spread of infectious diseases
6. Mining may cause erosion, acid mine drainage, contamination of soil and water with heavy metals, reduced surface and groundwater quality, deforestation, air pollution from dust and heavy equipment, and land pollution from tailings and slag heaps
7. Creation of new urban areas with all the attendant environmental and health problems mentioned throughout this book

ENVIRONMENTAL POLLUTANTS AND THEIR INTERACTIONS

(See the section on the chemical industry in Chapter 2, "Air Quality (Outdoor [Ambient] and Indoor)"

There is a complex interaction between human genetics, the environment, and environmental pollutants. In some situations, the same level of environmental pollutants will cause different levels of disease in different people because of their genetic disposition.

Humans are exposed to a variety of environmental pollutants, biological, chemical, physical, and radiological, either continuously, intermittently, repeatedly, or sporadically. These environmental stress factors can have an acute or chronic effect on the human body. They may affect the genetic, immune, and endocrine systems as well as all other systems of the body, causing fatal flaws, which can then in turn lead to cancer, as well as other potential diseases. Environmental chemicals modify gene expression and cause excessive stress and inflammation within the human body and the human brain. Endocrine disruptors also have inflammatory and metabolic effects.

Compounding the problem is the nature of the individual. The individual's age, sex, weight, genetic predisposition, previous illnesses, existing physical condition, level of nutrition, and exposure to various concentrations of pollutants contribute to the potential for acute and chronic illness caused by environmental pollutants. Further, the potential for disease and injury is complicated by ambient temperature and weather conditions.

Past exposure may have resulted in tissue damage or heightened the susceptibility of the individual to disease processes from the contaminant which has been retained in the human body or the damage done by previous exposures. The length of time of the exposures and the concentration of the exposures are very important. The exposures may be through inhaled air, ingestion of water and food, absorption through the skin, or a combination of these routes of entry into the body. Bioconcentration of chemicals may occur in the food chain when a higher form of life consumes a contaminated lower form of life, and then this food source may cause disease when ingested by people. Further, multiple contaminants may interact with each other to become more toxic to the individual. The human body may alter the contaminants and make them more toxic during the metabolic process. The liver and kidneys become special potential problem areas when the substances are processed and secreted.

Complicating the understanding of the severity of exposure to environmental stresses such as chemicals is the potential alteration of the chemicals in the environment when they are transported from the point of origin to the point of contact with people. Chemicals are deposited in the air, on the land, in the water, and taken up by plants and animal food sources. The original chemical may stay the same or may be transformed through the previous processes. In any case, the chemicals can be the direct or indirect cause of acute or chronic illness and potential physical injury. The chemicals may be precursors for cancer, create latent cancer cells, be direct carcinogens, alter, depress or enhance the action of enzymes, bind to cellular DNA, damage DNA, be direct toxins, or enhance the action of other chemicals in a negative manner for the human body.

Chemicals are rarely released in the pure form, but are rather a mixture of several different chemicals as well as the degradation stages of the original chemical and the mixture of chemicals. This makes it more difficult to remove the chemical from the source of pollution. Also, a chemical which is very useful for one type of pollution, such as methyl tert-butyl ether (MTBE) used as a gasoline additive to reduce air pollution, can cause another type of pollution, groundwater contamination. MTBE is often found together with other gasoline contaminants such as benzene, toluene, ethyl benzene, and xylenes.

Chemicals may contribute to increased levels of chronic disease. In a study in California, it was shown that children with type I diabetes were exposed to higher levels of ozone in the air than healthy children. These children were also exposed to higher levels of sulfate air pollution than the healthy children control group. In Montréal, Canada, a study of individuals with the autoimmune disease systemic lupus erythematosus had increased autoantibody levels when air particles in the 0.1–2.5 μm range increased beyond normal. Apparently these particles triggered an autoimmune response in these individuals. (See endnote 15.) Diesel exhaust particles have been shown to affect the development of the thymus during pregnancy and in early life, thereby affecting the development of the immune system.

ENVIRONMENTAL CANCER

(See endnote 33)

It is estimated that 41% of Americans will be diagnosed with cancer during their lifetime and that 21% will die from cancer. Environmental and occupational exposure contributes substantially to this problem. Sources of exposure may be found in: manufacturing; industrial products and processes, even those which have been banned and not used for many years; agricultural sources including insecticides, herbicides, fungicides, fertilizers, the active ingredients as well as the solvents, fillers, and inert ingredients; a variety of air pollutants from transportation sources, dry cleaning, particulates containing a variety of chemicals, burning of fossil fuels, etc.; medical sources including imaging, nuclear medicine, potentially new pharmaceuticals; military sources including weapons and improper disposal of chemical products; Superfund sites; cigarette smoking; contaminated drinking water; and natural sources such as radon gas and sunlight. There is a serious lack of funding and research to determine the specific levels and nature of environmental cancers.

EMERGING AND RE-EMERGING INFECTIOUS AND ZOONOTIC DISEASES

(See endnotes 16, 17, 18, 19, 20)

In the last 35 years, there has been a sharp increase in the number of new diseases that have been found globally and have contributed significantly to adult morbidity and mortality. In addition, there has been a sharp increase in the prevalence of infectious diseases that had been thought to be eliminated or reduced to such low levels that they would no longer be a problem for people. Many of these diseases have now re-emerged in forms that are resistant to antibiotics, and therefore even more difficult to cure with existing medications. The community response to the outbreaks have been basically on a case-by-case basis at the local level, which is very inefficient, rather than looking at the long-range problems and instituting preventive and control programs at the global level.

Diseases spread quickly from sparsely inhabited areas of the world to highly concentrated areas of population, by means of infected people, animals, insects, rodents, food, and water, which may be distributed by modern modes of transportation. People are displaced by natural disasters or wars and may spread disease from their area to new places. These evacuees are more susceptible to outbreaks of disease in the typically overcrowded conditions where they seek sustenance and shelter. The constant threat of the use of biological weapons by terrorists adds enormous complications to the control of emerging and re-emerging infectious and zoonotic diseases. The best way to protect a given population is to try to eliminate or control the disease at its source rather than trying to control it once it has moved to other areas of the world.

Globalization has meant that food production, processing, and distribution of meat, poultry, dairy products, fish, shellfish, fruits and vegetables, etc., instead of coming from local areas, which are supervised by health departments, now come from many different nations where the amount of regulatory control and the actual conditions of processing and handling of the food source may be very inadequate. There also may be frequent sources of contamination such as water, sewage, soil, insects, rodents, and pesticides. There may be a lack of proper sanitation measures which are used to control oral fecal route infections on farms and at other agricultural facilities.

Emerging infectious diseases from different parts of the world are being introduced into vulnerable populations elsewhere, where the agents can be spread rapidly from person to person and there is little or no resistance build-up among the people who may become affected. An example of a process of spreading zoonotic diseases to humans is the consumption of bush meat (the flesh of wild animals dead or alive) by natives in primarily tropical areas where there is a shortage of protein. This allows the microorganisms that are usually found in the wild animals to cross the animal–human barrier and bring disease to people. As these individuals move to new areas, they take the disease processes with them and spread them to new vulnerable populations. An example of this would be HIV which probably spread from chimpanzees to people. Other examples of recently recognized emerging diseases include acanthamebiasis, ehrlichiosis, hepatitis C, hepatitis E, Lyme's disease, and parvovirus B19.

Global warming is another concern for the spread of emerging infections or the re-emergence of infectious diseases. Although yellow fever and malaria have been virtually eliminated from the United States, increasing temperatures are making new areas of the country vulnerable to the growth of the mosquitoes which cause these diseases. In 2007, more than 100 residents of Ravenna, Italy, fell sick with the tropical disease chikungunya fever as the virus which typically was found in tropical regions around the Indian Ocean, had moved to northern areas because of the migration of the tiger mosquito.

The re-emergence of certain infectious diseases throughout the world is probably due to several factors including how humans live and utilize the environment; changes in technology that affect industry and the byproducts of industry; development and overuse of land; international travel and commerce; microbial adaptation and evolution; overuse of antibiotics; and a reduction in the use of public health controls. There has been a re-emergence of drug-resistant tuberculosis after more than 100 years of decline in the incidence of the disease. *Staphylococcus aureus* which was in decline because of the use of antibiotics has developed a highly resistant vancomycin-resistant strain of the organism. Other examples of re-emerging diseases include: anthrax, botulism, dengue fever, infection with *Escherichia coli*, hepatitis, shigellosis, etc.

Emerging and re-emerging infectious diseases may be delivered to unprepared areas by the use of biological weapons used by terrorists. Bioterrorism is the deliberate release of viruses and/or bacteria in order to cause sickness or death to people, animals, or plants. A variety of organisms may be utilized. See Chapter 5 on "Environmental Health Emergencies, Disasters, and Terrorism."

WHERE WE LIVE, WORK, PLAY, AND ARE EDUCATED

The amount of environmental stressors and pollutants that we are subjected to varies considerably with where we live, work, play, and are educated. There are some very special problems of the rural environment. (See endnote 21.) Programs are developed to control these environmental stressors and pollutants. (See Chapter 3 on "Built Environment-Healthy Homes and Healthy Communities.")

MOBILE, GROWING, AND AGING POPULATION

(See endnotes 22, 23, 24)

MOBILE AND GROWING POPULATION

The population of the world is projected to grow from 6.9 billion people in 2010 to 9.3 billion people in 2050. Population growth leads to significant long-term and environmental degradation. As consumption of everything including land, water, clean air, food, raw materials, and space increases, the environment supporting these resources deteriorates rapidly. Increased mobility because of people pressure and also because of modern transportation contributes to the rapid spread of people from one part of the world to another and therefore disease.

AGING POPULATION

The population of the United States of older adults is projected to increase to 71 million people by 2030. With age comes a deterioration of physical and mental capabilities. Many of these people have lifetime exposure to a complex set of environmental chemicals, infections, and other environmental stresses which create a body burden and leave the individual open to increasing levels of acute and chronic disease. Eighty percent of older adults have one chronic condition and 50% have two or more chronic conditions. The health of the individual further deteriorates because of infections and falls resulting in injuries.

Older people have more frequent falls and injuries because of problems related to vision, balance, medications they are taking, existing disease processes, level of nutrition and hydration, stress, environmental chemicals, and other environmental problems.

Environmental problems causing health threats to older people include indoor and outdoor air pollution, water pollution, and land pollution. Specifically, poor health has been attributed to power plant emissions, vehicle emissions, ozone, agriculture, industry, chemical contamination of land and water, and foodborne disease. The most frequent diseases identified as related to environmental concerns are pulmonary diseases and asthma. Also, cancer, heart disease, autoimmune disorders, and Parkinson's disease are related to environmental concerns. (Lead and PCBs may be related to dementia and the pesticide, rotenone as well as manganese may be related to Parkinson's disease.) Initial exposures may have occurred in the womb when the individual was a fetus and then accumulated during the ensuing years. Many environmental chemicals promote excessive oxidative stress and inflammation.

SUSTAINABILITY

Sustainability involves the implementation of a series of techniques and practices used for creating and maintaining conditions which allow humans to exist in productive harmony with nature, while reducing or eliminating destructive tendencies, actions, and substances which may adversely affect the natural environment and may lead to disease and injury in people. Sustainability allows for peaceful social interaction, reduction of poverty and hunger, economic growth and stability, appropriate use of resources, and other requirements needed by present and future generations to live a better life. Sustainability is an integration of all levels of economic, social, and environmental factors in the planning and implementation of programs to increase capacity and create new and better: infrastructure for transportation, education, communications, electrical usage, and the Internet; technologies for usage of natural resources including forests, minerals, fuel, air, land, and water; technologies for the removal and utilization or destruction of existing forms of contamination including toxic wastes; development of a green environment.

Sustainability is a combination of the prevention, minimization and control of pollutants, the preservation and proper use of resources, and the avoidance of damage to air, water, land, food, ecosystems, and people, to prevent disease and injury and to protect the natural environment. This is accomplished by the use of field-tested Best Practices in all inter-related environmental areas enhanced through the availability of excellent resources developed by specialists in their specific area of interest.

Sustainability is accomplished through the merging of the interests and efforts of strong governments at the local, state or provincial, national, and international level with highly trained technical experts, stakeholders who are citizens, organizations or business and industry, and the targeted populations. This group of individuals needs to establish an appropriate vision for a given situation, long-term and short-term goals and objectives, and actual programming based on the Best Practices of other areas, including necessary budgets, appropriate evaluation of the program, continued maintenance, and redirection were necessary.

HISTORY OF SUSTAINABILITY IN THE UNITED STATES

(See endnote 28)

INTRODUCTION

People and governments have been involved in preventing the contamination of water and land from the very beginnings of the country. In 1634, public sanitation regulations were passed in Massachusetts. The City of Boston ordered that people were not to leave any fish or garbage near the bridge or common landing between the two creeks to prevent polluting the water. There were about 3500 medical/sanitary practitioners in the colonies. Part of the work which they frequently did included:

- Enforcing cleanliness, and control of nuisances of filth and noxious trades
- Disposal of waste-garbage, excrement, offal, etc.
- Provision of pure water supplies and prevention of contamination of water by polluting substances
- Drainage of swamps, marshes, and stagnant pools of water

As the United States expanded westward in the 1800s, there was a free-for-all in the use of natural resources. Pollutants of all types were discharged into the air and water and on to the land. By the end of that century and the beginning of the 1900s, people began to realize that much of the natural resources had to be used more wisely and also preserved for future generations. This philosophy led to the Conservation Movement with Samuel P Hayes pushing for experts in the field to scientifically manage resources. President Teddy Roosevelt became one of the major people involved in the protection of water and the promotion of effective land development. He was influenced by the prospect that the country would run out of natural resources, especially trees, that the wilderness areas would disappear, and that pollution, especially in the big cities, would threaten our way of life. Roosevelt determined that the successful development of the West depended on conserving natural resources and using them wisely. Many other individuals, such as John Muir, and civic associations agreed with him including the Sierra Club.

In addition, anti-pollution programs assumed considerable significance in the period roughly from the end of the Civil War in 1865 to 1915, because of the severe environmental and living conditions that people were exposed to in cities. The governments built sewer systems and protected water sources, collected and removed solid waste, cleaned the streets, established parks, and started to regulate air pollutants, especially smoke. Studies were made of overcrowded and extremely poor housing conditions. The Rivers and Harbors Appropriation Act of 1899 prohibited the discharge of any kind of refuse into navigable waters. The Pure Food and Drug Act of 1905 protected people from contaminated or adulterated food and drugs. Subsequent amendments were the Federal Food, Drug, and Cosmetic Act of 1936, which also regulated pesticides, and the Food Quality Protection Act of 1996 that ensured that standards for pesticides in foods were established and enforced. In 1947, the Federal Insecticide, Fungicide, and Rodenticide Act was passed to control the sale, distribution and application of pesticides. This act was amended in 1972, 1988, and 1996. Because of the huge and profound dangers attached to nuclear energy and its use, the Atomic Energy Act of 1954 was passed establishing a regulatory structure for the construction and use of nuclear power plants and nuclear weapons facilities.

After World War II, people living in a booming society and now having more money, became less eager to accept the polluting effluents that were entering the air, water, and ground. With greater numbers and greater quantities of environmental pollutants being produced, used, stored, and put in disposal sites, there was a demand for stricter laws and programs and better enforcement. Especially visible where the vast clouds of black smoke coming from factories that caused people respiratory distress and blackened the environment.

The frustrations of constantly seeing people living with environmental pollutants were clearly established when some 20 million people throughout the country on April 22, 1970, created the first Earth Day. This was the people's way of saying enough is enough and we must now take control of our environment, our lives, and our health.

LAWS

Note that the various environmental laws will be discussed multiple times in the different chapters in order for the reader to understand the complexity of environmental issues and the interrelationship of environmental media, air, land, and water, and the highly significant environmental laws that apply to a multitude of different situations. The description of the laws may seem to vary, but actually for any given circumstance or situation, a portion of a very complex legal document may apply and this is what the reader needs to know. Listing and discussing these laws in various chapters also helps the reader understand that any given situation may be subject to a variety of different actions based on the laws that apply.

The United States established a national policy for environmental sustainability in 1969 when it passed the National Environmental Policy Act. This Act required the preparation of an Environmental Impact Statement when the environment was going to be affected by changes being sought, the development of a Council on Environmental Quality to Advise the President concerning Environmental Issues, the conducting of necessary research, etc. In 1970, the US EPA was created by bringing together programs from 15 different departments and agencies. Numerous additional laws were passed at the federal and state level to protect the environment and people. These laws included (see the appropriate chapters in this book for more detailed information about these laws):

1. The Clean Air Act of 1970, and its subsequent amendments, which established goals and standards for the quality of air in the United States

2. The Clean Water Act of 1972, and its subsequent amendments, which established goals and standards for water quality in the United States

3. The Coastal Zone Management Act of 1972 which allowed states and the federal government to protect the US coastal areas from environmentally destructive action and overdevelopment

4. The Marine Mammal Protection Act of 1972 which protected whales, dolphins, sea lions, manatees, seals, and other marine mammals that would be endangered

5. The Endangered Species Act of 1973, which protected endangered species of fish, plants, and wildlife in the United States and elsewhere by protecting habitats

6. The Safe Drinking Water Act of 1974, as amended in 1986 and 1996, established drinking water standards, and rules for groundwater protection and underground injection

7. The Federal Land Policy and Management Act of 1976, which provided for the protection of federal lands with scenic, scientific, historic, and ecological value

8. The Fisheries Conservation and Management Act of 1976 which helped maintain and restore healthy levels of fish stocks and prevented over-harvesting in the future

9. The Resource Conservation and Recovery Act of 1976 which helped prevent the creation of toxic waste dump sites by establishing standards for the management of the storage and disposal of hazardous waste

10. The Toxic Substances Control Act of 1976 gave the EPA the authority to regulate the manufacture, distribution, import, and processing of specific toxic chemicals

11. The Surface Mining Control and Reclamation Act of 1977 which was established to make sure that coal mining would not result in the destruction of the environment and that abandoned mine property would be used in the best interests of the public

12. The Comprehensive Environmental Response, Compensation, and Liability Act of 1980, and its subsequent amendments which required the clean-up of sites contaminated by toxic wastes, even those which had been placed there many years before

13. The Emergency Planning and Community Right-to-Know Act of 1986, which required companies to inform communities about toxic chemicals that they release into the air, water, and land

14. The Oil Pollution Act of 1990 which enacted a rapid and well-thought-out federal response to oil spills from oil storage facilities and vessels, while making the polluters responsible for the cost and damage to natural resources

15. The Federal Insecticide, Fungicide, and Rodenticide Act of 1947 amended in 1996

Since the inception of all the new environmental laws, a very substantial number of scientific studies, changes, actions, Best Practices, and programs have been initiated throughout the country to prevent disease and injury, protect the environment, and promote sustainability. To understand this material properly, read the remainder of this book under the appropriate chapter topics as need dictates. However, it is appropriate to discuss a very limited number of introductory subjects here.

PRESIDENT'S COUNCIL ON SUSTAINABLE DEVELOPMENT

The President's Council on Sustainable Development was established in 1994 by Executive Order. The national goals established if implemented would enhance the economy of the United States while preserving and protecting the environment and preventing disease and injury to people. Briefly, these goals state:

1. Every person should have access to clean air, clean water, and a healthy environment at home, at work, and in recreational areas.

2. Create and sustain a healthy economy providing good jobs and a high quality of life.

3. Ensure that all people can have affordable access to the law.

4. Protect, conserve, and restore natural resources as long-term commitments.
5. Take responsibility for all corporate, institutional, and individual decisions affecting the environment and others.
6. Develop sustainable communities that utilize resources efficiently.
7. Create an opportunity for everyone to work together for a better environment.
8. Become a leader in international responses to stabilization of the environment.
9. Create the opportunity for all Americans to have equal access to education and life-long learning. (See endnote 29.)

The Council made numerous excellent recommendations but unfortunately, few were adopted. The Congress was more interested in other issues.

RECENT EXECUTIVE ORDERS CONCERNING ENVIRONMENTAL SUSTAINABILITY

(See endnotes 30, 31)

According to Executive Order 13423 of January 24, 2007, the goals for the federal agencies were: to improve energy efficiency, to reduce greenhouse gas emissions, to require additional renewable energy, to reduce water consumption, to recycle and use recycled materials, to reduce toxic and hazardous chemicals, to improve disposal of solid waste, and to improve new and renovated buildings in keeping with reduced energy needs. The subsequent Executive Order 13514, October 5, 2009 (Federal Leadership in Environmental, Energy, and Economic Performance), added to the above specific deadlines as well as an inventory of absolute greenhouse gas emissions, means of improving water use efficiency, and promotion of pollution prevention and waste reduction techniques with appropriate deadlines.

HISTORY OF GLOBAL SUSTAINABILITY

(See endnote 26)

INTRODUCTION

The Greeks understood that human waste and solid waste could spread disease and therefore invented toilets, sewers, and dumps outside of the city. The Greeks outside of Athens developed the first dumps about 500 B.C.

In other ancient cities including Rome, air pollution from burning wood, solid waste, odors and runoff from garbage, sewage, and industries contaminated the air and the water, and of course the poor disposal of human waste contributed substantially to the problems.

In 1357, King Edward III of England passed a law prohibiting pollution of the river Thames. In 1366, in Paris, butchers were forced to dispose of animal waste outside the city. In 1388, in England, laws were passed prohibiting the throwing of filth and garbage into ditches, rivers, and waters.

There were many improvements in the generations and centuries to come in response to numerous outbreaks of disease and specific environmental problems. In modern times, there was a need to be concerned about environmental sustainability in the United States. Since the same need existed globally, the United Nations was used as a vehicle for promoting environmental sustainability. The material that follows is but a brief explanation of global interest and global actions.

UNITED NATIONS

The United Nations Conference on the Human Environment, better known as the Stockholm Conference, was held in Stockholm, Sweden in 1972. In preparation for this conference, a meeting was held in 1971 where the developed nations were concerned about the environmental consequences of global development and the less-developed nations were concerned about economic development. Sustainable development was a compromise between the environmental concerns and the economic concerns.

The United Nations Environmental Programme was formed with a mandate from the Stockholm conference to address issues of both sound environmental practices and sound developmental practices. In 1975, the International Environmental Education Programme was begun, followed in 1980 by the World Conservation Strategy. In 1983, the Prime Minister of Norway, Gro Harlem Brundtland, was asked by the Secretary-General of the United Nations to establish and become the chairperson of a special independent World Commission on Environment and Development. Its function was to re-examine critical problems of the environment and development throughout the world and propose realistic methods to deal with them. Second, it was the function of this commission to strengthen international cooperation on environmental and developmental issues. Third, it was to increase the level of understanding and commitment to sustainable development on the part of citizens and other stakeholders, various organizations, business and industry, as well as governments. An outcome of the report from the commission after numerous meetings was the establishment of the UN Conference on Environment and Development and then Agenda 21 produced at the 1992 Rio Earth Summit.

Agenda 21 Report, June 1992

Section I of the Agenda 21 report contained discussions on sustainable development in developing countries, poverty, consumption patterns, demographic dynamics, protecting and promoting human health, promoting sustainable human settlements, and integrating the environment into development decisions. The inter-relationship between the health and safety of people and the protection of the environment and how to meet each need while developing land and resources is apparent from the topics covered.

Section II reviewed ways of protecting the air, land, water, fragile ecosystems, and sustainable agriculture. It also provided discussions on environmentally sound management of toxic chemicals, trafficking in illegal international toxic and dangerous products, hazardous wastes, solid wastes, sewage related issues, and radioactive wastes.

Section III discussed the involvement of women, children, and youth in sustainable development, the role of indigenous peoples, the role of non-governmental organizations acting as partners for sustainable development, the role of workers and trade unions, the role of business and industry, and the role of the scientific and technological community.

Section IV discussed means of implementation. (See endnote 25.)

GLOBAL CLIMATE CHANGE

Global climate change has brought forth an immediate need by all countries to discuss the significant concerns affecting the world community. On September 16, 1987, the original Montréal Protocol on Substances that Deplete the Ozone Layer was signed and was effective as of January 1, 1989. One of the provisions of the protocol was to allow nations to make adjustments rapidly as new scientific information became available. Adjustments came into effect on March 7, 1991; September 23, 1993; August 5, 1996; June 4, 1998; July 28, 2000; and May 14, 2008. (See endnote 32.) In December 1997, over 150 countries adopted the Kyoto Protocol to protect the Earth's atmosphere and the climate. The nations agreed to establish legally binding limits on the admissions of heat-trapping greenhouse gases. The agreement also established substantial support for climate change research.

World Summit on Sustainable Development

The World Summit on Sustainable Development was held in Johannesburg, South Africa, and concluded on September 4, 2002. Various nations, United Nations agencies, financial institutions, non-governmental organizations, and citizen groups from all over the world discussed the priority issues of water, energy, health, agriculture, and biodiversity. All this activity was geared to increase the availability of safe water and clean energy, reduce hunger, improve agricultural areas, protect people from illness and injury, and protect the environment.

Report of the United Nations Conference on Sustainable Development, Rio de Janeiro, Brazil, June 20–22, 2012 (See endnote 27)

The nations of the world attending the conference reaffirmed their strong desire for a sustainable future, developmentally, economically, socially, and environmentally for present and future generations. The eradication of poverty, hunger, and disease while developing natural resources in an environmentally sound manner was of greatest significance. It was re-emphasized that people were at the center of sustainable development and that stakeholders, business, industry, and government were partners in the effort. It was re-emphasized that decisions should be science based. There was a renewal: of the political commitment to improvement and positive change in the actions of all; of the integration, implementation, and operational efficiency of various groups to achieve common goals and objectives; of engaging major groups and other stakeholders; and of working toward creating a green economy while eliminating poverty and improving sustainable development. There was an emphasis on creating and maintaining an institutional framework for sustainable development by governmental agencies sharing their knowledge and skills with each other and with others. There was once again recognition that there are inter linkages between various issues and various environmental media.

PROGRAMS WHICH ACHIEVE THE GOAL OF PREVENTING DISEASE AND INJURY AND PROTECTING THE ENVIRONMENT

THE PAST

How Programs Were Started

From pandemics to the dumping of pollutants into the air, water, and land, to the exposure of workers to metals and other environmental stresses, dangerous situations have existed in various forms for thousands of years. Thinking citizens, many times without scientific understanding, have used information based on experience and observations, when available Best Practices to try to determine the causes of the problems and find solutions to prevent disease and injury from occurring.

In the past 200 years, as people have developed various sciences and have become more aware of health and safety problems related to environmental circumstances, communities have tried to find better ways to try to resolve environmental and public health problems. The process of trying to find a solution usually involved a disease outbreak, most often cholera, or a serious problem, such as an air pollution episode. The community typically chose someone or a committee to study the problem and come up with potential solutions. These solutions included new laws resulting in the development of public health and environmental health programs, necessary personnel, and additional laws, rules, and regulations to resolve the problems. The personnel then conducted surveys and inspections, collected samples, and used special instrumentation to determine the nature and scope of the problem. The personnel prepared reports, used educational techniques to try to correct the problems, and then used enforcement procedures when necessary. The personnel also developed new techniques to try to prevent diseases and injuries from re-occurring and to lessen the impact of existing environmental conditions.

As environmental health problems became more complex, as the industrial revolution proceeded, and as new solutions to problems became available, professional organizations were formed to work with the individuals and their departments and industries to provide training and other resources to the personnel. This was necessary because initially individuals were hired to fill environmental health, safety, environmental protection, and sustainability positions for which they had little or no specific knowledge. They may have had considerable training in the various sciences. The health departments and other agencies had to provide in-service training for these individuals. The CDC developed many training tools and provided them to the health departments for use by their employees. Also, as an example, the National Environmental Health Association developed a series of training modules which were available to members of that association to increase their knowledge and skills in given areas. The organizations then developed performance standards and examinations for certification of the individuals who met the performance standards.

The professional organizations also were deeply involved in many aspects of promoting the profession and the professional, including supporting positive environmental health policies and positions established through governments and laws. The organizations worked at national, state, and local levels. The professional organizations also provided significant research in trying to upgrade the skills and knowledge of existing professionals.

Educational institutions recognized the need for new personnel and developed new programs in the specialty areas of concern to provide a high level of professionally trained people to enter the environmental health, public health, and environmental protection fields. The educational institutions also provided significant research to try to resolve existing and future problems.

THE PRESENT

It is incumbent upon us as professionals, as citizens of our community, as citizens of our country, and as citizens of the world, to utilize the substantial amount of knowledge that is currently available through the accumulation and analysis of successful environmental health and protection programs and their Best Practices, at the local, state, federal, and global level and in industry.

This evidence-based practice is currently being stressed by the CDC, educational institutions, various medically based professional organizations such as the American Psychological Association, the American Nurses Association, and the American Physical Therapy Association, as well as industry. The concept in fact is called Principles of Best Management Practices. What is being done in other fields needs to be adapted to the fields of environmental health science, safety, environmental protection, and sustainability.

COMPONENTS OF SUCCESSFUL ENVIRONMENTAL HEALTH, SAFETY, PROTECTION, AND SUSTAINABILITY PROGRAMS

A successful program typically has several components. They are:

- Scientific evaluation of the problems
- Discussion of the extent and nature of the problems
- Conducting necessary research for existing solutions to the problems
- Determining Best Management Practices to resolve the problems
- Determining who the various stakeholders are and meeting with them
- Establishing a preliminary plan
- Establishing the goals and objectives of the plan
- Determining the type and amount of people needed to carry out the plan
- Funding
- Carrying out the detailed program including actual work and type of communications
- Evaluation of the program
- Redirection where needed
- Publication of the content and results of the program in order that other professionals at the local, state, and federal level as well as business and industry can gain knowledge and skills in specific areas

EXAMPLE OF A SUCCESSFUL PROGRAM: COMMUNITY RODENT CONTROL, PHILADELPHIA, 1959–1963

In 1959, Dr. Walton P Purdom, Director of the Division of Environmental Health, Philadelphia Department of Public Health, and several of the section chiefs visited a new District Environmental Health Supervisor, Herman Koren, at Health District 5, located at 20th and Berks Streets. Their objective was to meet with the staff of the district health office including Dr. Rotan Lee, District Health Director to help in the orientation of the new supervisor to all of the facets of a complete environmental health and community health program. During the visit, Dr. Purdom suggested to Herman Koren and his staff that Health District 5 develop an innovative program in community rodent control in order to try to reduce the high levels of rats and rat bites within the district. This was the beginning of one of the most successful rodent control programs in the country and the basis for the federal War on Poverty rodent control effort.

Statement of the Problem

Shortly thereafter, two surveys were conducted by environmental health professionals. One was an actual survey of the physical conditions existing within the selected area and the second an informational study of the opinions and attitudes of the population. The survey area was 60 city blocks made up of three-story houses with multiple residences, owned by absentee landlords. These houses which had been homes for a single family of six to nine people, now housed six apartments with 50–60 people. The surveys were used: to determine the existence and level of rodent infestation within the communities; to determine in a more precise manner the number of rat bites that were occurring within a calendar year; and the interest of the citizens to work together to remove food sources and harborage and destroy a good portion of the rodent population. A huge amount of infestation was found along with substantial harborage and rodent entrances into houses. The number of rat bites reported to the Philadelphia Department of Public Health, were about 100 per year for the entire city. The number of rat bites found in the survey made by environmental health practitioners, numbered in the hundreds for the 60 city blocks. Obviously, there was an enormous amount of under-reporting of rat bites to doctors and to the Philadelphia Department of Public Health.

Solutions to the Problem

Although the Best Practices for resolving rodent control problems within a given structure are well known, that is the removal of food and harborage, poisoning, and rodent proofing, the rodent proofing portion was very difficult to carry out in an area so large, so old, so complex, and where most of the properties were owned by absentee landlords and the tenants were very poor. The structures were probably built in the early years of the 20th century and over time had deteriorated substantially. Pipes had burst. Basement walls had started to crumble. Floors had cracked. There were numerous openings to the outside.

It was decided that at best, food and harborage could be removed and a poisoning campaign instituted. It meant that continued surveys and continued maintenance of the individual city block programs would be necessary to keep a new rodent infestation from occurring.

Although some community program work had been done in the City of Baltimore prior to this time, information was not readily available on how to start such a program and therefore a new program had to be built from the very beginning.

Stakeholders

Various groups of people and organizations had to become involved in the community-oriented rodent control program if it was going to work. Several city agencies partnered with the Philadelphia Department of Public Health, including Licensing and Inspection, Police, Streets, and the Human Relations Commission.

A non-profit organization, the YWCA, had hired a community organizer to work in the area of the project to help people with various issues. This individual's work was then redirected to help with the rodent control project. A social scientist was also working within the overall area on a grant. The Spring Garden Street Civic Association, a group of influential citizens from the area who were trying to protect the area from deteriorating further, participated in many of the meetings but were not really involved in the activities.

Citizens groups within the individual blocks in the project area were formed by environmental health personnel, the public health educator, and the community organizer. The natural leaders of each city block were identified and then selected by the citizens to represent them. These block organizations were given information about the ensuing rodent control projects and the necessity for assistance by the citizens in the community. These individuals along with other volunteers participated in large numbers in helping clean up the community and in the rodent poisoning campaign.

The Actual Program Content

The program consisted of a series of community meetings, citizen participation, poisoning campaigns, and clean-up campaigns. The block leaders helped bring the citizens out on the appropriate days to work on the projects. Sound trucks went throughout the neighborhoods, and messages in Spanish and English were announced concerning what needed to be done on a given day.

Funding

Through an analysis of the work being done in the District Environmental Health office, it was determined that by making some changes, the environmental health practitioners could be more efficient and time could be released for the Community Rodent Control Program. In addition, field staff working on Saturdays or evenings received overtime pay, but in very limited amounts. All other expenditures for the program were absorbed by the individual city departments or were given on a voluntary basis.

Evaluation and Results

Surveys were conducted of the backyards, alleyways, basements, and homes in the communities to estimate the amount of garbage, trash, and junk that was in the area initially and that which was present after the clean-up. A large number of truckloads had been removed totaling over 3 million pounds. A considerable amount of Red Squill and then Warfarin were put out in a safe manner. Over 7500 dead rats were found and at a rate of three dead rats for each one found, it was estimated by the city entomologist that 30,000 rats had been killed.

Redirection

After the active part of the project was completed, it now became a matter of periodic re-assessment and continued maintenance in the program areas. Where additional facets of the program had to be carried out again, they were scheduled and completed.

Summary

What was learned from the community rodent control project was that by working with the community, other governmental agencies and various stakeholders, limited numbers of environmental health specialists can help interested communities successfully complete programs and reduce or resolve environmental problems. This project, which was carried out between 1959 and 1963, had all of the factors that comprise good programming in the present and for the future. In fact, the program was so successful that it became as previously stated a model for the War on Poverty Rodent Control Program approved by President Johnson, who was horrified by the fact that in our affluent nation children were being attacked, injured, and killed by rats. The City of Philadelphia received many millions of dollars not only to continue the efforts of this program but also to train others in community rodent control.

Also, the local block leaders became the initiating force that helped to successfully complete a program of immunization of children in the 60 city blocks that were covered by the rodent control effort. City buses were converted into mobile immunization units staffed by doctors and nurses who immunized 50,000 children over 2 days in the rodent control area.

CONCLUSION

We live in a society today where our priorities of reducing budgets at the local, state, and federal level and regulatory reform reduces our ability to control environmental factors which may lead to increased sickness and reduction in good health. Environmental pollutants and their interactions, emerging and re-emerging infectious diseases (primarily of a zoonotic nature), the potential increase of acute and chronic disease in our mobile, growing, and aging population cannot be wished away, but must be dealt with in an appropriate manner using scientific understanding of the problem and existing knowledge of Best Practices and programs that actually work. That is why this book tries to accumulate and present the recommended Best Practices developed and utilized by working professionals, in all environmental health and environmental protection areas. In addition, information on special resources and additional materials of excellence are provided in endnotes, many of which explain in far greater detail the subject being studied. This book helps eliminate the clutter of overwhelming amounts of information, accurate and inaccurate, and is a repository of knowledge and proven practices which can be used by professionals, students, and civic leaders to prevent disease and injury, and protect and sustain environmental quality.

ENDNOTES

1. Bell, Beverly. 2011. *An Impossible Choice: Reconciling State Budget Cuts and Disasters That Demand Adequate Management: The Book of the States.* The Council of State Governments, Lexington, KY.
2. Johnson, Nicholas, Oliff, Phil, Williams, Erica. 2011. *An Update On State Budget Cuts: At Least 46 States Have Imposed Cuts That Hurt Vulnerable Residents and Cause Job Loss.* Center on Budget and Policy Priorities, Washington, DC.
3. National Association of County and City Health Officials. 2011. *Local Health Department Job Losses and Program Cuts: Findings from July 2011 Survey: Survey Findings.* Washington, DC.
4. Meyer, Jack, Weisleberg, Lori, Health Management Associates. 2009. *County and City Health Departments: The Need for Sustainable Funding and the Potential Effect of Healthcare Reform.* Robert Wood Johnson Foundation and the National Association of County and City Officials, Naples, FL.
5. Robert Wood Johnson Foundation Trust for America's Health. 2015. *Investing in America's Health: A State-By-State Look at Public Health Funding Guarantee Health Facts.* Princeton, NJ.
6. Centers for Disease Control and Prevention. 2012. *2012 Budget Information: Overview.* Atlanta, GA.
7. Robert Wood Johnson Foundation. 2011. *Report Finds Preparedness for Bio-terror and Health Emergencies Eroding in States across the Country: Cuts to Keep Programs Could Hurt Ability to Detect and Respond to Crises.* Princeton, NJ.
8. Hinezerling, Lisa, Ackerman, Frank. 2002. *Pricing the Priceless: Cost-Benefit Analyses of Environmental Protection.* Georgetown Environmental Law and Policy Institute, Georgetown University Law Center, Washington, DC.
9. Science and Environmental Health Network. 1998. *Wingspread Conference on the Precautionary Principle.* Racine, WI.
10. Centers for Disease Control and Prevention. 2010. *Environmental Public Health Performance Standards (v 2.0).* Atlanta, GA.
11. US Environmental Protection Agency, Office of Environmental Justice. 2011. *EPA: Plan EJ 2014.* Washington, DC.
12. US Global Change Research Program. 2011. *Our Changing Planet: The US Global Change Research Program for Fiscal Year 2011.* Washington, DC.
13. US Environmental Protection Agency. 2012. *Climate Change Indicators in the United States, 2012,* Second Edition, EPA 430-R-12-004. Washington, DC.
14. US Environmental Protection Agency. 2013. *Climate Impacts on Human Health.* Washington, DC.
15. Bernatsky, Sash, Fournier, Michael, Pineau, Christian A. 2011. Association Between Ambient Fine Particulate Levels and Disease Activity in Patients with Systematic Lupus Erythematosus (SLE). *Environmental Health Perspectives.* 119, 45–49.

16. Brachman 2003. Infectious diseases past, present, and future. Editorial. *International Journal of Epidemiology*, Volume 32(5), 684–686.
17. Barrett, Ronald, Kuzawa, Christopher R, McDade, Thomas W. 1998. Emerging and re-emerging infectious diseases: The third epidemiological transition. *Annual Review of Anthropology*, Volume 27, 247–271.
18. National Institute of Allergy and Infectious Diseases. 2010. *Emerging and Reemerging Infectious Diseases, Research at NIAID*. Bethesda, MD.
19. Patz, Jonathan A, Daszak, Peter, Tabor, Gary M. 2004. Unhealthy landscapes: Policy recommendations on land-use change in infectious disease emergence. *Environmental Health Perspectives*, Volume 112(10), 1092–1098.
20. Centers for Disease Control and Prevention, National Center for Infectious Diseases. 2002. *Protecting the Nation's Health in an Era of Globalization: CDC's Global Infectious Disease Strategy*. Atlanta, GA.
21. Bailey, John M. 2009. *The Top 10 Rural Issues for Healthcare Reform*. Center for Rural Affairs, Lyons, NE.
22. Stein, Jill, Schettler, Ted, Rohrer, Ben. 2008. *Environmental Threats to Healthy Aging: With a Closer Look at Alzheimer's and Parkinson's Diseases*. Greater Boston Physicians for Social Responsibility in Science and Environmental Health Network, Cambridge, MA.
23. Wiener, Joshua M, Tilly, Jane. 2002. Population aging in the United States of America: Implications for public programmes. *International Journal of Epidemiology*, Volume 31(4), 776–778.
24. United Nations Population Fund. 2015. *Population and Sustainable Development in the Post-2015 Agenda: Report of the Global Thematic Consultation on Population Dynamics*. New York, NY.
25. United Nations: Sustainable Development. 1992. *United Nations Conference on Environment and Development, Rio de Janeiro, Brazil, June 1992, Agenda 21*. New York, NY.
26. US Environmental Protection Agency, Region 10: The Pacific Northwest. 2014. *History of Sustainability*. Seattle, WA.
27. United Nations. 2012. *Report of the United Nations Conference on Sustainable Development-Rio de Janeiro, Brazil 20–22, June 2012*. A/CONF. 216/16. New York, NY.
28. National Park Service. 2003. *Conservation, Preservation, and Environmental Activism: A Survey of the Historical Literature*. ParkNet, Washington, DC.
29. EPA. 2012. *Our Nation's Air: Status and Trends through 2010*. EPA. Washington DC.
30. Federal Advisory Committee Act. 1993. *President's Council on Sustainable Development*. Executive Order 12852. Washington, DC.
31. Office of the Federal Environmental Executive, Council on Environmental Quality, Executive Office of the President. 2012. *Instructions for Implementing Executive Order 13423, Strengthening Federal Environmental, Energy, and Transportation Management*. Washington, DC.
32. Office of the Federal Environmental Executive, Council on Environmental Quality, Executive Office of the President. 2012. *Executive Order 13514 Guidance and Reports*. Washington, DC.
33. United Nations Environment Programme, Ozone Secretariat. 2012. *Handbook for the Montréal Protocol on Substances that Deplete the Ozone Layer*, Ninth Edition. Nairobi, Kenya.
34. Reuben, Susanna H, for The President's Cancer Panel. 2010. *2008–2009 Annual Report: President's Cancer Panel: Reducing Environmental Cancer Risk*. US Department of Health and Human Services, National Institutes of Health, National Cancer Institute, Bethesda, MD.

BIBLIOGRAPHY

Centers for Disease Control and Prevention, National Prevention Council. 2011. *National Prevention Strategy*. Atlanta, GA.
Frist, Bill. 2002. Public health and national security: The critical role of increased federal support. *Health Affairs* 21:6.
Jackson, Richard J, Kochtitzky, Chris. 2003. *Creating a Healthy Environment: The Impact of the Built Environment on Public Health*. Sprawl Watch Clearinghouse, Washington, DC.
Koren, Herman. 2005. *Illustrated Dictionary and Resource Directory of Environmental and Occupational Health*, Second Edition. CRC Press, Boca Raton, FL.
Koren, Herman, Bisesi, Michael. 2003. *Handbook of Environmental Health: Biological, Chemical, and Physical Agents of Environmentally Related Disease*, Fourth Edition, Volume I. Lewis Publishers, Boca Raton, FL.
Koren, Herman, Bisesi, Michael. 2003. *Handbook of Environmental Health: Pollutant Interactions in Air, Water, and Soil*, Fourth Edition, Volume II. Lewis Publishers, Boca Raton, FL.
Reuben, Susana H. 2010. *National Institutes of Health, National Cancer Institute, 2008–2009 Annual Report for The President's Cancer Panel. Reducing Environmental Cancer Risk: What We Can Do Now*. National Institutes of Health, Bethesda, MD.

2 Air Quality (Outdoor [Ambient] and Indoor)

STATEMENT OF PROBLEM AND SPECIAL INFORMATION FOR OUTDOOR AIR QUALITY

Ambient air pollution (outdoor air pollution) consists of a complex mixture of gases, particulate matter, and liquids. The air pollutants include the six criteria pollutants (carbon monoxide, ozone, lead, nitrogen oxides, particulate matter, and sulfur dioxide), toxic (also known as hazardous) air pollutants, and greenhouse gas emissions, which affect human health, the environment including the various ecosystems, and the stratospheric ozone layer. The pollutants come from a variety of sources, general and specific, and are affected by weather and local topographical conditions. Indoor air quality is in a separate section in this chapter. (See endnotes 1, 17.)

Constituents of Air Pollution

Gases

The gases include sulfur dioxide, nitrogen oxides, ozone, carbon monoxide, volatile organic compounds (VOCs), hydrogen sulfide, various metals in gaseous form, etc. They come from a variety of different sources.

Particulate Matter

The particulate matter is of varying sizes and varying composition, and from varying sources, and may contain a multitude of chemicals, both as primary and secondary air pollutants. Fugitive dust which may be caused by wind erosion, disturbance of soil or movement of various products, vehicles or equipment, trucks, etc., may cause major visibility problems, detrimental effects to agricultural products by coating them and/or health effects depending on quantities and weather conditions.

Liquids

The liquids come from a variety of chemicals as well as water and may evaporate into the air in different sizes of droplets. These droplets can adhere to particulate matter and, depending on the size, be taken deeply into the lungs. They may also contribute to the production of acid rain and haze which reduces visibility.

Origins of Air Pollutants

Air pollution may be caused by anthropogenic (created by people) sources or natural sources. Air pollution may affect the health of people and the integrity of the environment. The various sources may interact and produce additional air pollutants. The pollutants are transported through the air from the source which may be a smokestack, etc., to the recipient which may be a person or the environment. The amount, nature, and direction of the wind have considerable bearing on the level of air pollution and the potential for damage. Also the presence of a temperature inversion (a weather-related event where the temperature of the atmosphere increases with altitude instead of normally decreasing, and cold air underlies the warmer air at higher altitudes thereby trapping contaminants coming into the air under what amounts to a lid of cold air) can enhance the effect of the pollutants.

This chapter will discuss the anthropogenic sources, including air toxics, and also the natural sources of air pollution which can be reduced to avoid disease, injury, and damage to the environment. It will also discuss primary air pollutants which are emitted from the anthropogenic and natural sources mentioned above. It will discuss secondary air pollutants that are formed by the reactions of the primary air pollutants and the atmosphere. Secondary air pollutants may include sulfuric acid, nitric acid, nitrogen dioxide, ozone, formaldehyde, peroxyacetyl nitrate, ammonium nitrate, and ammonium sulfate. It will discuss other pollutants including those that help cause climate change.

HEALTH EFFECTS

The health effects of the criteria air pollutants and the toxic air pollutants on humans have been determined by epidemiological and clinical evidence. These health problems occur either by causing new health conditions or exacerbating existing health conditions including: cardiovascular disease, cancer, impaired immune systems, birth defects, genetic mutations, asthma, chronic obstructive pulmonary disease (COPD), cystic fibrosis, diabetes, eye diseases, etc. All body systems including the lungs, heart, brain, nerves, liver, kidneys, skin, etc., can be challenged on a short-term or long-term basis by the various air pollutants. Confounding factors include age, sex, previous exposures and concentrations of the exposures, cigarette smoke, secondhand smoke, lack of exercise, existing health conditions, predisposition to health problems, prescription drugs taken, nutrition, and environmental exposures including the type, concentration, and amount of time of exposure, associated with the individual during all times of the day and night.

An example of deleterious health effects is the relationship of ambient air pollution to heart disease and stroke. Carbon monoxide, nitrogen oxides, sulfur dioxide, ozone, lead, tobacco smoke, and particulate matter, especially fine particles, are associated with increased hospitalization and deaths due to cardiovascular disease. This occurs both in short-term air pollution episodes and long-term exposure to air pollutants, especially from inhaling the byproducts of automobile combustion. Another example is COPD, which is especially aggravated by particulates and ozone.

ENVIRONMENTAL EFFECTS

Environmental effects include: acid precipitation damaging trees, soils, and bodies of water; eutrophication caused by the deposit of nutrients found in air pollutants, which enter water and cause growth of algae, fish kills, and destruction of plant and animal diversity; ozone depletion resulting in damaged crops such as soybeans and the reduction of crop yields; and potentially global warming.

Acid precipitation is produced, sometimes long distances from the source, when sulfur is released into the air as a pollutant, combines with oxygen to form sulfur dioxide and then in the presence of ozone or hydrogen peroxide becomes sulfur trioxide which dissolves in water and produces sulfuric acid. It also occurs when the nitrogen from the atmosphere is heated to temperatures found in steam boilers and internal combustion engines, combines with oxygen to form nitrogen oxide and nitrogen dioxide, and then dissolves in water to form nitric and nitrous acids.

Visibility, of great significance to people who are visiting national and state parks and wilderness areas, is affected by haze caused by air pollution from hundreds of miles away. Haze is created by light being absorbed or scattered by particles in the air, which then reduces clarity and may change the color of the surrounding air. The particles may also increase respiratory illness and decrease lung function, indicating that haze may not just be an aesthetic concern.

Air quality issues may also be associated with fracking activities used to find and utilize energy sources embedded in rock formations beneath the surface of the ground. There have been emissions into the air of methane, VOCs, hazardous air pollutants, and greenhouse gases. Diesel fuel used in fracking borings also contributes to air pollution.

Types of Air Pollutants

Criteria Air Pollutants

The criteria air pollutants are carbon monoxide, ozone, lead, nitrogen oxides, particulate matter, and sulfur dioxide. They come from a variety of sources and some of them are produced in the atmosphere. These pollutants were of greatest concern initially and still contribute substantially in many areas to the degradation of ambient air.

Toxic Air Pollutants

(See the special section on "Toxic Air Pollutants" later in this chapter)

Greenhouse Gases

(See the special section on "Greenhouse Gases" later in this chapter)

Stratospheric Ozone Layer Depletion

The stratospheric ozone layer is high above the earth and protects the health of humans and also the environment from the harmful ultraviolet radiation coming from the sun. It is a natural shield which is being affected by chemicals made by people. Chlorofluorocarbons used in various aerosols, such as coolants in refrigerators and air conditioners, rise into the stratosphere and reduce this necessary ozone protective layer. Other chemicals may also cause this problem.

Current Status

From 1990 to 2010 as reported in February 2012 by the US Environmental Protection Agency (EPA) in EPA. 2012. *Our Nation's Air: Status and Trends through 2010.* EPA. Washington DC, there has been a substantial improvement in reducing air pollution and increasing air quality. (See endnote 82.) This has been due to cleaner cars, cleaner industries, and the improved production of consumer products. The six common air pollutants are in decline as well as many of the toxic air pollutants. However, approximately 124 million people live in counties in the United States that exceed one or more of the National Ambient Air Quality Standards (NAAQS). (See endnote 58.)

There are still considerable problems with ground-level ozone and particulate pollutants, which are present in unhealthy amounts in many areas of the country. These pollutants can affect the cardiovascular system, cause premature death, and increase both emergency room and hospital admissions because of heart attacks and strokes.

Air toxics continue to be a problem especially in urban areas, around industrial facilities, and high levels of transportation. Small area sources continue to add to the overall volume of air toxics present.

Air pollution problems in one part of the country may be caused by industries in other parts of the country and therefore simply improving local situations may not be enough to achieve attainment areas. The taller smokestacks can move the pollutants into wind currents that can help them go hundreds or thousands of miles and contribute to smog, haze, and other air pollution episodes. The US EPA, although having made considerable progress in smokestack emissions control, is continuing to bring enforcement actions against large refineries, coal-fired power plants, cement manufacturing facilities, sulfuric acid and nitric acid manufacturing facilities, and glass manufacturing facilities for inadequate compliance with EPA regulations and the Clean Air Act.

International sources of air pollution, especially from China, are having an effect on the Western United States by increasing levels of ozone, mercury, sulfur, carbon monoxide, and other pollutants including very fine carbon particles carrying a variety of toxic substances. At present, these contaminants come from fumes and dust from industrial factories, energy sources created from coal, and the increased use of automobiles. Further, about a third of the pollution coming from Asia is dust which may in part be due to drought and deforestation. In part, it is due to the use of fossil fuels as economies improve and industrialization increases in scope. The air pollution in Beijing, China, was so intense before, during, and after the Summer Olympics of 2008 that there was major media coverage and an unusual level of concern by the International Olympic Committee.

Internationally, there is a concern for people in developing countries who live near resource extraction and processing industries, as well as in big cities, who are subjected to dust or hazardous fumes at the work site as well as in their homes. Many of the problems affecting the industrialized societies have now been transferred to the resource-rich Third World countries that are having difficulty coping with them. Cardiovascular and respiratory diseases have increased substantially with exposure to the additional particulate material which is now airborne. There is an increase in lung cancer and lung cancer deaths. There is increased mortality in children from air pollutants and the ozone exposure has triggered asthma attacks. Leaded gasoline is still a major factor in urban areas. The World Health Organization in 2002 estimated that in urban areas particulate matter caused as many as 5% of the global cases of lung cancer, 2% of the deaths from cardiovascular and respiratory conditions, and 1% of the respiratory infections. China and India are the two most affected countries. There have been signs of changes in learning ability, behavioral changes, and central nervous system damage from the increased amount of lead present in the air. (See endnote 9.)

CLIMATE CHANGE

Climate change is a reality and is linked to the accumulation of greenhouse gases. There have been five groups of observed changes including:

1. An increase in the amount of emissions of greenhouse gases in the United States, globally, and in the atmosphere
2. An increase in the average temperatures in the United States and globally, with an increase in the frequency of heat waves in the United States, an increase in drought in portions of the country, an increase in precipitation in the United States and worldwide, an increase in heavy precipitation in the United States, and an increase in the intensity of tropical cyclones in the Atlantic Ocean, Caribbean, and Gulf of Mexico
3. A decrease in the amount of Arctic sea ice, glaciers in the United States and globally, lake ice, snow cover in North America, and snow pack in the Western United States and Canada
4. Changes in society and ecosystems because of substantial numbers of heat-related deaths, the lengthening of the growing season, a shifting of the plant hardiness zones northward, a change in the dates of leaf growth and plants blooming, and a shift in the migration habits of birds. (See endnote 48)
5. An increase in the ocean heat, sea surface temperature, sea levels, and ocean acidity

In May 2010, the National Research Council of the National Academy of Sciences determined that climate change is occurring and it is caused largely by human activities and can affect people as well as various ecosystems. Carbon pollution which causes climate change may lead to more intense hurricanes, other storms, heavier and more frequent flooding, more drought, and more wildfires. It increases the acid level of the oceans, causes the sea levels to increase, increases storm surges, and is harmful to agriculture and forests. Climate change especially affects children, the poor, and the elderly. (See endnote 41.)

MEASURES OF AIR QUALITY

The EPA under the Clean Air Act has established air quality standards to protect the health of the public including sensitive populations such as individuals with asthma, children, and older people (these are primary standards). The Clean Air Act standards also decrease problems relating to public welfare by protecting ecosystems from harm including plants and animals, protecting against decreased visibility, and damage to buildings and other physical properties (these are secondary standards). The EPA has established NAAQS for the six major air pollutants known as criteria pollutants (the units of measure for the standards are in parts per million (ppm) by volume, parts per billion (ppb) by volume, and micrograms per cubic meter of air ($\mu g/m^3$)). Two types of air pollution trends are tracked by the EPA: actual air concentrations of pollutants in the outside air at monitoring sites throughout the United States, and the amount of emissions based on engineering estimates of the total tons of pollutants released into the air each year.

STANDARDS FOR CRITERIA POLLUTANTS

(See endnotes 2, 3)

- *Carbon monoxide*—The primary standard is 9 ppm for an 8-hour period or 35 ppm for 1 hour, not to be exceeded more than once per year. (The final rule was published on August 31, 2011.)
- *Lead*—On a rolling 3-month average, the primary and secondary standard of 0.15 $\mu g/m^3$ of air must not be exceeded. (The final rule was published on November 12, 2008.)
- *Nitrogen dioxide*—Averaged over a 3-year period, the primary standard must not exceed 100 ppb in a 1-hour period. The annual mean of the readings should not exceed 53 ppb for the primary and secondary standards. (The final rule was published on February 9, 2010.)
- *Ozone*—The 8-hour maximum concentration averaged over a 3-year period should not exceed 0.075 ppm for both the primary and secondary standards. (The final rule was published on March 27, 2008.)
- *Respirable particulate matter (PM$_{10}$)*—Measured over a 24-hour period, the primary and secondary standards must not exceed 150 $\mu g/m^3$ of air more than once per year on an average over a 3-year period. (The final rule was published on January 15, 2013.)
- *Fine particulate matter (PM$_{2.5}$)*—Measured as an annual mean and averaged over a 3-year period, the annual primary standard is 12 $\mu g/m^3$ of air, while the secondary standard using the same parameters is 15 $\mu g/m^3$ of air. The 24-hour standard is 35 $\mu g/m^3$ of air. (The final rule was published on January 15, 2013.)
- *Sulfur dioxide*—Measured as 1-hour daily maximum concentrations averaged over 3 years, the primary standard is 75 ppb, while the secondary standard using a 3-hour averaging time and not to be exceeded more than once per year is 0.5 ppm. (The final rule was published on June 22, 2010.)

National Air Toxics Assessments and Toxics Release Inventory

The US EPA through its National Air Toxics Assessment Program conducts a comprehensive evaluation of the major 33 air toxics found in the ambient air of the United States. This study helps determine which of the air toxics creates the greatest potential cancer risk or adverse non-cancer risk to health in the United States. These chemicals are measured by the best available scientific techniques, and represent the greatest health risk to people in state/local/tribal agencies. This information helped the US EPA establish priorities for improving data in various emissions inventories and work with communities in designing studies, and helps provide the priorities for expanding and improving the monitoring of air toxics. Knowledge of the toxic chemicals, concentration, and sources helps the agencies prioritize the establishment of effective programs.

The assessments can provide information about emissions, concentrations of chemicals in the ambient air, and exposures and risks across broad geographic areas at a given moment in time, and can help identify specific air toxics and sources such as stationary sources or mobile ones. However, they cannot identify exposures and risks for a specific individual or even a small geographic area such as a census block or hotspot. The most recent assessment of 2005 was made available to the public in early 2011.

The Toxics Release Inventory is a database of releases of toxic chemicals from manufacturing facilities which may pollute the air, water, and land. The individual can determine potential hazards by typing his/her ZIP code into the database information form. See endnote 33 for information on the most recent National Emission Standards for Hazardous Air Pollutants.

Ozone Action Day

An ozone action day or ozone alert can be declared by a local municipality, county or state at times during the summer months when weather conditions including heat, humidity, and air stagnation may cause health problems. This happens primarily in the Midwestern metropolitan areas of Chicago, Cleveland, Columbus, Detroit, and Indianapolis but could also happen in any of the major cities in the United States and certainly in other countries.

Air Quality Index

The EPA has developed a tool which is used by them and other agencies to provide the public with information about local air quality. This tool, which is the Air Quality Index for Health, is focused on health effects. When the Air Quality Index is 0 to 50, it is considered to be good and the color code is green. When the index is 51 to 100, it is considered to be moderate from a health concern point of view and the color code is yellow. When the index is 100 to 150, the air is unhealthy for sensitive groups and the color code is orange.

General Best Practices for All Sources of Air Pollution

There are several general Best Practices which can be applied to all sources of air pollution, whether they are criteria pollutants, major point sources of criteria pollutants and/or toxic air pollutants, area sources, mobile sources, or greenhouse gas sources. They are:

- Conduct an in-depth survey of the industrial facility to determine where the pollutants are being released into the air, the quantity, and at what special times.
- Establish and use an appropriate operating and maintenance practices schedule for all sampling equipment, operating equipment, storage vessels, control devices, and general housekeeping.
- Especially thoroughly clean all heating, ventilation, and air-conditioning systems on rooftops and replace necessary filters if they are there to prevent the growth and spread of *Legionella* organisms. This organism, which causes Legionnaires disease, although usually associated with indoor air and water mists, can be transmitted through dirty biomats used in the equipment.
- Check for leaks using proper leak detectors especially around valves, gaskets, seals, vents, ductwork, condensing coils, strainers, lint bags, exhaust dampers, pipe fittings, and sampling points.
- Replace filters and carbon absorbent material regularly.
- Train all employees to carry out their jobs in an appropriate manner and use continuing education to upgrade their skills on a periodic basis, then test their proficiency.
- Substitute less hazardous chemicals for more hazardous chemicals when appropriate.
- Recycle and reuse substances where possible.
- Clean up all spills thoroughly and as rapidly as possible.
- Keep written records of all maintenance of equipment and facilities, and record all unusual events and what was done to correct them.

POLLUTION PREVENTION AS PART OF BEST PRACTICES

Pollution prevention is a technique used in reducing air pollutants, hazardous wastes, solid wastes, and water pollutants by simply not creating the wastes or pollutants during industrial production. Pollution prevention helps reduce the amount and toxicity of pollutants by: substituting less hazardous materials for more hazardous materials; substituting reusable material for single use material where feasible; preventing spills and leakages; capturing, recycling, and/or treating leakages before they enter the air or water; efficiently using raw materials in a timely manner to reduce and avoid waste; cleaning and reusing solvents; inspecting all raw materials upon delivery and rejecting those that are not usable; establishing an appropriate inspection, maintenance and service program; and properly training, and retraining when necessary, all workers in their work assignments and in appropriate safety measures. Pollution prevention increases efficiency, is an excellent means of cost control increasing profits, and enhances environmental protection.

SUB-PROBLEMS INCLUDING LEADING TO IMPAIRMENT AND BEST PRACTICES FOR AMBIENT AIR QUALITY

CRITERIA POLLUTANTS

The discussion about the criteria pollutants will emphasize the pollutants more than the sources, because the pollutants come from a large number of sources and therefore the Best Practices are more general in nature. The discussion about the toxic air pollutants will be related to the source, because specific industries can use more specific Best Practices. The discussion on greenhouse gases will be more general in nature but will include Best Practices.

Particulate pollution and ground-level ozone, among the six criteria pollutants today in the United States, are the greatest contributors to impaired health. The level of these pollutants helps determine whether an area is in "attainment" or "non-attainment."

In a given geographic area where the air quality is better than the primary standards, it is called an attainment area. Where this is not so, it is called a non-attainment area.

THE SIX CRITERIA AIR POLLUTANTS

(See endnote 28)

1. *Carbon monoxide*

    Carbon monoxide is typically produced by the burning of hydrocarbon fuels. People are exposed to the pollutant: by inhaling indoor and outdoor air from a combination of residential, mobile (vehicular traffic, the most frequent source), and heavy industry (refineries, power plants using gas and coal, chemical plants, coke oven plants, etc.) sources; heavy equipment, farming equipment, and residential heating; and through the processing and storage of a variety of hydrocarbons, as well as in the occupational setting. Carbon monoxide may cause chest pain, headaches, aggravate heart disease, and result in emergency measures performed at a hospital. It can and does cause death.

    Best Practices for Carbon Monoxide
    - Utilize a program of motor vehicle inspection to determine if there is proper combustion of fuel.
    - Avoid cold and rapid starts of motor vehicles to reduce levels of carbon monoxide.
    - Utilize a program of motor vehicle maintenance to get maximum combustion of fuel in a proper manner.
    - Do not operate gasoline power tools and engines in confined spaces.
    - Establish programs of timely examinations of firefighters and other emergency response people and provide effective medical care when needed.

- Establish a program for the proper use and maintenance of fireplaces and wood burning heaters to make sure that there is appropriate combustion of the fuel.
- Enforce no smoking rules in public facilities to reduce the level of carbon monoxide inhaled by people.
- Substitute less hazardous paint strippers and other chemicals for more hazardous compounds.
- Publicize areas of high carbon monoxide potential to young families to try to encourage them to live in areas other than those related to industry and high traffic.
- Develop programs to more efficiently use fuel as a means of reducing carbon monoxide and other contaminants.
- Use continuous emission monitoring systems to record and report carbon monoxide from stationary sources.

2. *Ozone—Ground-Level*

Ozone (ground-level), a major component of smog, is typically formed when there is a chemical reaction in the atmosphere between a VOC and nitrogen oxides in the presence of sunlight. The VOCs may come from the burning of fuel, cars burning gasoline, chemical manufacturing plants, and petroleum refineries, as well as the use and storage of solvents, petroleum, and other hydrocarbons, as well as landfills. Inhalation leads to difficulty in breathing and lung tissue damage. Children are especially at risk because they play and exercise during the summer out-of-doors when the ozone levels are at their highest. Ozone may also damage rubber, some plastics, forests, and agricultural crops. The weather and topography is especially a problem if the person is in a depressed area such as a valley or in a city at street level between tall buildings. This type of configuration contributes to the concentration of ground-level ozone. The number of vehicles present adds substantially to the problems. (See endnote 8.)

Best Practices for Ozone—Ground-Level
- Reduce nitrogen oxide emissions from power plants and industrial combustion sources.
- Introduce low emission cars and trucks, use cleaner gasoline, and increase the efficiency of gasoline to get more mileage per gallon burned.
- Where feasible use solar, nuclear, hydroelectric, and wind power in place of fossil fuels.
- Reduce use of automobiles by carpooling, mass transit, and making less frequent trips for small things.
- Use low evaporation VOC solvents and paints in place of high evaporation VOC solvents and paints.
- Recover vapor during refueling of automobiles at service stations.
- Inspect all automobiles on an annual basis for emissions of hydrocarbons, nitrogen oxides, sulfur oxides, and carbon monoxide.
- Use special gas cans for refueling of lawn equipment to prevent spillage.

3. *Lead*

Lead poisoning may be the result of exposure in the indoor environment, outdoor environment, and the occupational environment. Children are especially vulnerable.

They may be exposed through ingestion and inhalation. Lead comes from resource recovery areas, deterioration of lead-based paint indoors and outdoors, waste incineration, battery manufacturing, and the use of leaded gasoline in piston engine aircraft. (The most common sources are mobile sources and the industrial processes.) Secondary lead smelters produce lead from scrap and are the primary means of recycling lead acid automotive batteries. The basic operations of breaking the batteries, smelting, and refining as well as fugitive sources release 1,3 butadiene, a known carcinogen, as well as lead compounds into the air. The particulate matter released can cause severe respiratory problems and the carbon monoxide can cause adverse health effects. In addition, lead damages the nervous system in children especially and the cardiovascular and renal systems in adults.

Best Practices for Lead (See Chapter 4, "Children's Environmental Health Issues," Best Practices for Lead section)

- Use fabric filters, wet scrubbers, or electrostatic filters to remove dust containing lead during the primary lead production process.
- Use baghouse filters to remove dust with lead from emissions in secondary lead production facilities.
- Control fugitive lead dust and smelting operations, and transportation to and from the smelter.

4. *Nitrogen Oxides*

The process of combustion should maximize fuel economy, minimize carbon monoxide and partially oxidize organic compounds, and minimize the volume of air in the furnace. This is typically accomplished by rapid and complete fuel air mixing with adequate time for oxidation. However, this also produces higher levels of nitrogen oxides. To minimize the nitrogen oxides, it is necessary to reduce the gas stream temperatures, minimize the amount of oxygen used, minimize the mixing rates of fuel and air, and have as large a furnace as possible for combustion. Nitrogen oxides are produced primarily from fuel combustion especially in electric utilities, industrial boilers, and all types of wood-burning fireplaces and appliances. Electrical utilities and industrial boilers create about 40% of the nitrogen oxides released to the air.

Further, although motor vehicles individually only contribute small amounts of nitrogen oxides, carbon monoxide, hydrocarbons, and particulate matter, as a group they are a very significant source of all of these air pollutants. Nitrogen oxides (especially nitrogen dioxide) cause irritation and damage to the lungs, lowering resistance to respiratory infections. Frequent exposure to high levels of nitrogen oxides increases the amount of acute respiratory illness in children.

Nitrogen oxides react in the atmosphere to form ozone and acid rain. Nitrogen oxides have caused eutrophication in the coastal waters of the United States, especially in the Chesapeake Bay. They are destructive to fish and other animal life.

Best Practices for Nitrogen Oxides (See endnotes 44, 62)

- Modify combustion systems to minimize peak temperatures and the amount of time necessary for combustion of the fuel at the peak temperatures at industrial sites.
- Use catalytic converters to reduce nitrogen oxides to nitrogen and oxygen. They also convert hydrocarbons into carbon dioxide and water, and carbon monoxide into carbon dioxide.
- Use low nitrogen oxides burners for industrial and utility boilers which utilize such techniques at low excess air, high-efficiency combustion, and recirculation of combustion gases in the flue gas.
- Use recirculation of exhaust gases in motor vehicles to reduce the nitrogen oxides.
- Use proper maintenance of motor vehicles to ensure appropriate air–fuel ratios, gas compression ratios, and spark timing.
- Avoid hard acceleration of motor vehicles to reduce nitrogen oxides, carbon monoxide, and hydrocarbons.
- Use energy conservation techniques in buildings to save energy and use less fuel.
- Use continuous emission monitoring systems to measure and record nitrogen oxides emissions from stationary sources.

5. *Particulate Matter*

Both respirable particulate matter (PM_{10}) (dust is the most common source) and fine particulate matter ($PM_{2.5}$) (dust is the most frequent source followed by fuel combustion in residential housing and electric power plants) increase respiratory problems and lung damage, may cause cancer and premature death, and are also responsible for surface soiling and reduced visibility.

All sizes of particulates may come from power plants, steel mills, chemical plants, grain elevators, and other industrial sources. They may also come from unpaved roads, parking lots, wood-burning stoves, fireplaces, and mobile sources. Typically, there is a continuum of sizes of the particles and therefore you will find both PM_{10} and $PM_{2.5}$ mixed together. There is a seasonal nature to $PM_{2.5}$ contamination. In the Eastern United States, there are more sulfates from sulfur dioxide contamination during July–September coming from electric power plants. In the Western United States, there is a greater concentration of $PM_{2.5}$ caused by fine particulate nitrates formed in cooler weather, and more carbon from wood stoves and fireplaces during cooler weather. Particulate matter can consist of fine dust, soot, smoke, and droplets from chemical reactions. The inhalable coarse particles range in size from 10 μm (PM_{10}) to 2.5 μm ($PM_{2.5}$), about the diameter of a human hair. (See endnote 14.)

Primary particles come directly from a source such as a construction site, unpaved road, field, smokestack, or fire. The secondary air pollutants are very small at the 2.5 μm or less level and typically are formed in the atmosphere from emissions from power plants, industries, and cars. The secondary air pollutants are primarily sulfates, nitrates, and carbon compounds. They may cause or contribute to a variety of illnesses, especially among the elderly, children, and those with lung diseases, heart diseases, or asthma. They also reduce visibility in national parks and wilderness areas. Fine particles can remain in the air and travel hundreds of miles or more.

Fugitive dust is particulate in nature and may originate indoors or outdoors from industrial or commercial operations and from unintended consequences of disturbing soil by wind or human activities. The dust particles may contain oxides of silicone, aluminum, calcium, and iron as well as sea salt, pollen, and spores. They may be in the PM_{10} range or $PM_{2.5}$ range. Some activities that create fugitive dusts in the industrial sectors of agriculture, mining, construction, manufacturing, transportation and utilities, wholesale trade, retail trade, and services include loading and unloading of material; disturbing the soil and site maintenance; equipment and truck use; storage and dispensing of products; and removal of waste materials. (See endnote 22.)

Best Practices for Particulate Matter
- Apply chemical dust suppressants or water to unpaved roads and bare soil in areas where there is traffic and construction.
- Choose cleaner fuels such as natural gas, which emits almost no particulate matter.
- Use low ash fossil fuels for combustion.
- Clean the coal before combustion to reduce ash.
- Use more efficient technologies for industrial processes.
- Use gasification products of coal instead of coal as a fuel.
- Use filters, dust collectors, and electrostatic precipitators, with scrubbers to remove dust particles from gas streams.
- Reduce dust in the cement industry by appropriate dust collection systems applied to the air as it leaves the processing and production areas.
- Reduce dust in coal processing by using a dust collection system using principles of ventilation, or use a wet suppression system through surface wetting or airborne wetting.
- Use appropriate pollution control equipment.
- Reduce fuel combustion by improving various technologies related to the product.
- Use continuous emission monitoring systems to record and report particulate emissions from stationary sources.

6. *Sulfur Dioxide*

Sulfur dioxide is produced during the combustion of fuel (most common source), especially high sulfur coal, by coal or oil burning electrical utilities, industries, refineries, and diesel engines. Ninety percent of the sulfur dioxide emissions come from power plants, petroleum refineries, fertilizer manufacturers, paper mills, copper smelters, and iron and steel mills. Sulfur dioxide aggravates asthma, increases respiratory problems, and reacts in the atmosphere to help form acid rain (see the section "Nitrogen Oxides" above).

Best Practices for Sulfur Dioxide (See endnote 63)
- Only use low-sulfur coal as a fuel.
- Prewash the coal before using as a fuel.
- Scrub the sulfur dioxide from the exhaust gases before releasing them into the atmosphere.
- Use natural gas instead of high sulfur coal.
- Use reactive lime to remove the elemental sulfur, or sulfates.
- Alter the industrial process where possible to remove the sulfur and sulfur compounds.
- Use continuous emission monitoring systems to record and report sulfur dioxide emissions from stationary sources.

In addition to the six criteria pollutants, there are other sources of pollutions as described below:

1. *General Pollutant Sources*

The six criteria pollutants, greenhouse gases and hazardous air pollutants come from stationary sources, mobile sources, and area sources, and may travel long distances rapidly, and may be deposited on land or water, dry or wet. They may interact with each other to form new even more complex pollutants, be highly bioaccumulative going from low levels of concentration in water contaminated by air pollution to high levels of concentration in animal tissue, be highly toxic at very low doses, and persist for long periods of time. They may interact with the environment, stick to surfaces to become available at later dates, or re-volatilize into the air and start the process all over again. Typically, pollutants are evaluated individually; however, they are usually present in various combinations and therefore the results of their presence and the potential damage they cause may be hard to predict. Pollutants listed under General Pollutant Sources may be criteria air pollutants, acid rain sources, greenhouse gases, and/or hazardous air pollutants. Specific high-polluting stationary sources and area sources will be discussed later in this chapter under Specific Sources.

2. *Stationary Sources* (See endnote 59)

The discussion of stationary sources, which are point sources, will be general in nature because the information applies to many of the specific areas listed, such as chemical plants, electric power plants, publicly owned wastewater treatment plants, etc. The individual utilizing this book should first read the stationary sources information and then go to the specific area of interest to enhance his/her knowledge.

Stationary sources typically are a variety of industries including major sources and smaller sources, which may start with the mining of the raw material, the storage of the raw material at the mining area, the storage of fuel to be used for process energy and heating (typically fossil fuels) at the industrial site, the storage at the industrial site of other raw materials to be used in the industrial processes at the plant, the storage of the finished products, and the storage and disposal of wastes. (The various stages of transportation are included below under Mobile Sources.)

Stationary sources emit a variety of inorganic and organic air pollutants as particulates and/or gases depending on the fuels being used in the industrial process of the plant. The industrial plant pollutants may also include sulfur oxides, nitrogen oxides, carbon monoxide, and carbon dioxide. The fuel being used may add the criteria pollutants and/or hazardous air pollutants. Other pollutants, especially hazardous air pollutants, may be created and released depending on the specific processes carried out at the individual facility.

These pollutants may come from the actual processes where incomplete chemical reactions and secondary reactions allow raw materials, impurities, and byproducts to be released into the air. Volatile organic solvents can escape as fugitive emissions from a number of places within the equipment used for the processing of the products. Bad odors may be released in the exhaust, and heating or drying processes. Flaring, which can be very useful for getting rid of excess gases, can also if operated improperly produce excessive amounts of sulfur dioxide and a group of hazardous air pollutants.

Four major locations in industrial production which may lead to discharge of air pollutants may be:

1. Process operations, such as emissions from the reactor vents, distillation systems, and spray drying
2. Fugitive sources, such as emissions from valves, pressure relief devices, and equipment cleaning and maintenance
3. Surface area sources, such as evaporation from holding ponds, cooling towers, wastewater treatment areas
4. Handling, storage, and loading areas, such as storage tanks and line venting

Best Practices for Stationary Sources (also see General Best Practices for All Sources of Air Pollution section) (See endnote 61)
- Eliminate the source of the problem in the industrial process.
- Change the industrial operation to reduce the level of air pollutants.
- Make the industrial operation more efficient, which also produces greater profits for the industry.
- Adjust various pieces of equipment such as burners, boiler doors, etc., to reduce nitrogen oxides emissions.
- Make industrial boilers using fossil fuels more efficient by proper maintenance, reducing excess air, eliminating leaks, insulating pipes, automatically controlling use of fuel, and switching from coal or oil to natural gas.
- Make electric motors for various industrial applications more efficient by using modern flat belts instead of V belts, adjust the components of the motor properly, or replace with energy efficient motors.
- Use cogeneration of heat and power to extract most of the energy that is available in the fuel.
- Use appropriate size smokestacks to dilute small amounts of pollutants but not carry them downwind to other areas.
- Pretreat raw materials before use to reduce the level of potential pollutants released during the actual industrial process.
- Substitute less polluting materials for more polluting materials.
- For particulate control, use wet and dry electrostatic precipitators, fabric filters, venturi scrubbers, cyclones, and settling chambers.
- To control gaseous emissions, use thermal oxidizers, catalytic reactors, flares, boilers and process heaters, carbon absorbers, absorbers, condensers, and biofilters.

3. *Mobile Sources*

Mobile sources are divided into on-road sources and off-road sources. Mobile sources regulated by the EPA include: commercial aircraft; heavy-duty vehicles, buses, recreational vehicles, and semi-trailers; light-duty vehicles, cars, minivans, and SUVs; diesel-powered engines on locomotives; motorcycles; a variety of boats and ships; construction and agricultural equipment; gasoline and propane industrial equipment; lawn and garden equipment; and snowmobiles, dirt bikes, etc. These mobile sources pollute the air through combustion of fuel and fuel evaporation. They produce carbon monoxide, hydrocarbons, nitrogen oxides, and particulate matter as well as air toxics and greenhouse gases. Mobile sources are the largest contributors to air toxics. Diesel exhaust from freight vehicles is a primary source of $PM_{2.5}$, air toxics, nitrogen oxides, and approximately one third of the greenhouse gas emissions in the United States. (See endnote 20.) In New Jersey, as of the US EPA's 1999 Air Toxics Inventory, on-road and non-road mobile sources accounted for 61% of the air toxics inventory. Area and other sources accounted for 27% of the air toxics. Only 12% were from major source contributors. (See endnote 16.)

An environmental justice issue is created by mobile sources in heavily industrialized areas of the city because the diesel semi-truck traffic around industrial plants use the same routes every day moving raw materials and finished products and therefore contribute heavily to air pollutants. This may lead to a substantial increase in disease for residents of the surrounding area.

Best Practices for Mobile Sources
- The US EPA established standards for specific pollutants being emitted by mobile sources including vehicles, engines, and equipment.
- The US EPA established sulfur dioxide emission standards for gasoline, on-road diesel fuel, and off-road diesel fuel which is being used in mobile sources.
- People selling a vehicle or engine within the United States have to show compliance with the Clean Air Act.
- Manufacturers, in order to meet the standards, design and implement efficient combustion systems, vapor recovery systems, computer technology to monitor and control the performance of the engines, catalytic converters, and particulate filters to remove the pollutants from the exhaust.
- Use reformulated gasoline to reduce emissions of benzene, toluene, and other toxic pollutants.
- Establish limits for tailpipe emissions of toxic pollutants and enforce them.

4. *Acid Rain*

Acid rain, acid snow, acid fog, acid mist, or dry forms of acid deposition are formed in the atmosphere and fall to the earth. The acid deposition changes the chemistry of the soil and bodies of water. It may destroy or alter ecosystems. The acid deposition causes health problems, hazy skies, and property damage, and can affect agricultural crops. The acid precipitation is formed when sulfur dioxide and nitrogen oxides are emitted to the air and react with water vapor and other chemicals to create acids.

Best Practices for Acid Rain
- Reduce the quantity of emissions of sulfur dioxide and nitrogen oxides from power plants and automobiles.
- Reduce the amount of electricity used by turning off lights, computers, and appliances that are not being used at a particular time.
- Purchase only energy-efficient appliances and phase out the older ones.
- Reduce the thermostat during the winter and raise it during the summer to decrease fuel use.
- Use appropriate maintenance programs for all motor vehicles and various motorized equipment and tools.

OTHER MAJOR SPECIFIC SOURCES OF AIR POLLUTION (INCLUDING TOXIC AIR POLLUTANTS)

1. *Agriculture* (See endnote 45) (See the section on "Greenhouse Gases" later in this chapter)

Agricultural sources contributing to air pollution problems include: concentrated animal feeding operations where a large number of animals are confined and fed or maintained for extended periods of time in small areas; production of considerable amounts of manure from the concentration of animals; improper application of fertilizers, herbicides, and pesticides; water pollutants that become airborne; gaseous emissions from the decomposition of manure; and particulate matter from soil erosion, unpaved roadways and other areas, agricultural equipment, fire and smoke, and bulk materials handling.

Worldwide, cattle, sheep, buffalo, and goats produce about 80 million tons of methane a year, which is about 22% of all the methane from human activities. Their digestive systems can use unusable plant materials for food and fiber, but also produce methane. The livestock production system can also produce nitrous oxide, ammonia, hydrogen sulfide, and particulate matter. Besides lowering air quality in the area of these factory farms and producing substantial amounts of methane (a greenhouse gas), nitrous oxide, which is also a powerful greenhouse gas, is also produced. Odors also reduce the quality of the air.

The animal feeding operations and especially concentrated animal feeding operations (factory farms) increase the level of asthma and bronchitis in neighboring communities, especially in children. The ammonia produced is rapidly absorbed in the body and can cause severe coughing, buildup of mucus, and scarring of the airways.

Best Practices for Agriculture (See endnote 46) (See the "Particulate Matter" section)
- When the land is prepared for growing crops (tillage), use techniques that reduce the intensity of the work and therefore retain the residue on the surface of the soil.
- Use as little vehicular traffic on fields as possible.
- Use mulching material when possible to retain the soil cover.
- Reduce agricultural field operations of tilling, planning, weeding, fertilizing, etc., to the minimum amount necessary to reduce dust.
- Reduce the amount of engine emissions through proper maintenance of motorized equipment.
- Modify the timing of operations where possible to achieve maximum productive effort with minimum soil removal as dust.
- Observe the level of air quality and take into account weather conditions when performing agricultural work.
- Use dust suppressants where possible on dirt roads.
- Use vegetative and artificial barriers to retard the flow of wind over unprotected surfaces.
- Retrofit existing equipment to prevent or reduce the emissions of particulate matter, VOCs, and nitrogen oxides.
- Use fire with great care and take meteorological conditions into account when controlling undesirable vegetation and plant disease, reducing fuel hazards, and improving plant productivity.
- Reduce the amount of methane produced by herds by improving grazing, supplementing diets with nutrients, and improving genetics and reproductive efficiency. These techniques allow for improved management of livestock with lower numbers needed to produce the same amount of meat and milk.
- Use an anaerobic manure digester (AgSTAR Program) to convert the manure into energy.
- Turn large quantities of manure in the form of methane found in biogas into renewable energy which can be used on the farm and/or be sold to electric utilities. (See endnote 47.)
- Evaluate the potential health effects of individuals who are close to large animal feedlots and provide necessary medical care for them.

2. *Aluminum Industry (Primary and Secondary)* (See endnote 68)
    The primary aluminum industry produces molten aluminum from ore and then the metal is used in a variety of products such as automobiles, trucks and other vehicles, packaging containers and foil, residential, industrial, commercial and farm structures, electrical appliances, etc. The ore called bauxite is refined into alumina, a feedstock. The alumina and electricity are then combined in a cell with a molten electrolyte called cryolite, which produces molten

aluminum metal and carbon dioxide. This process uses a substantial amount of electricity with the attendant air pollution problems based on the fuels being consumed. Other potential emissions to the air are: dust at the various production facilities from dryers, materials handling equipment, movement of trucks, and blasting; smoke; metal compounds; and VOCs including dioxins, carbon monoxide, nitrogen oxides, sulfur dioxide, chlorides, hydrogen chloride, and hydrogen fluoride. The process also produces perfluorocarbons, which are very long-lasting in the atmosphere and significant contributors to climate change.

The secondary aluminum industry recovers aluminum cans, foundry returns, and scrap. Toxics are released during the processing of the scrap through shredding, melting, and removing the coating on the material. The toxics include metals, organic compounds, hydrogen chloride, fluorine compounds, chlorine, chlorinated benzenes, dioxins, and furans.

Best Practices for the Aluminum Industry (Primary and Secondary)
- At the site of the bauxite mine wash, the ore to remove contaminants that can go into the air.
- Use dry scrubbers after the smelting systems with aluminum oxide as the adsorbent to capture gases containing fluorides and recycle the fluorides.
- Use low sulfur tars for helping to control sulfur dioxide emissions.
- Reduce the emissions of organic compounds from secondary aluminum production by removing coatings, paint, oils, and greases in a very efficient manner.
- Use electrostatic precipitators and baghouse dust collectors to remove the dust from the emissions from the dryers.
- Use hoods and enclosures at the site of conveyors and material transfer points to minimize dust which has been created by stockpiled materials.
- Use baghouses at the production plants and lime kilns to control bauxite and limestone dust.

3. *Cement Kilns*

The production of cement, called Portland cement, utilizes a substantial amount of electricity and produces about 22% of the carbon dioxide emitted by industries throughout the world. Cement is made commercially in over 120 countries and in every state in the United States and therefore contributes substantially to local air pollution. Cement is made by grinding and heating a mixture of materials such as limestone, clay, marl (an easily breakable earth deposit containing clay and calcium carbonate), iron ore, fly ash, slag from blast furnaces used by steel mills, etc., in a rotary kiln, which uses as its energy source coal, oil, gas, coke, or a variety of waste materials which may be the cheapest to buy and use but also produce the most pollutants. The product, which is called clinker, is cooled, mixed with a small amount of gypsum and then ground into a fine powder to produce cement. Concrete is then made by adding cement (about 10–15% of the concrete mix by volume) and water to sand, gravel or crushed stone and allowing the mixture to cure, thereby increasing its strength and durability. (See endnote 61.) Several harmful air pollutants are produced from making cement including air toxics especially mercury, hydrochloric acid, particulate matter, total hydrocarbons, carbon dioxide, nitrogen oxides, and sulfur dioxide. Many of these pollutants come from the firing of the substances and the materials used for heating, and are affected by the temperature of the flue gases. Mercury is found in small amounts especially in the raw materials and also in the fuels used in the cement industry, and the amount present varies widely from one place to another and from one country to another.

Best Practices for Cement Kilns
- Use high-efficiency fabric filters to control emissions of particulate matter from cement kilns, clinker coolers, material handling, product bagging, etc.
- Use continuous emission monitoring systems to record and report emissions from cement companies to meet US EPA regulations.

- Especially in older cement plants, check the accuracy of the bag leak detection system that is used to measure particulate matter loading in the exhaust of a fabric filter.
- Use a regenerative thermal oxidizer to destroy hazardous air pollutants by breaking the bonds of the hydrocarbon at high temperatures and then recombining the carbon and oxygen to form carbon dioxide and the hydrogen and oxygen to form water.
- Use powdered activated carbon injection technology for controlling mercury from cement plants.
- Reduce the amount of nitrogen oxides emissions through: control of temperature in the combustion zone, control of excess air through the process, use of low nitrogen oxide burners in the kiln, and use of efficient cooler systems.
- Use lime to help capture sulfur dioxide in the cement manufacturing process.
- Use a fuel that has particles that are small enough to burn quickly and easily.
- At the cement plant, put the cement kiln dust in enclosed, covered vehicles and conveyance systems.
- Compact the dust from the cement kilns and periodically wet it down at the disposal site in the landfills.
- Store cement kiln dust in enclosed tanks, containers, and buildings when preparing it for disposal or sale.
- Use a dry kiln instead of the wet process because it requires about a third less energy and therefore reduces greenhouse gas emissions.

4. *Chemical Industry* (See endnote 27)

The major industries engaged in chemical manufacturing transform inorganic and organic raw materials into finished products. They manufacture:

- Basic chemicals
- Resin, synthetic rubber, and artificial fibers
- Pesticide, fertilizer, and agricultural chemicals
- Pharmaceuticals
- Paint, coating, and adhesives
- Soap and other cleaning compounds
- Miscellaneous chemical products

There are five major steps in chemical manufacturing and each one may create contamination of the air, water, and land. They are:

1. Purification of the raw materials
2. Chemical reactions when the raw materials are converted into other products
3. Finishing operations such as purification of the products
4. Handling, storage, transport, and equipment cleaning
5. Disposal of unusable waste

Besides the inherent toxicity of the chemicals being produced by the manufacturing process, there are numerous highly significant heavy metals, VOCs, nitrate compounds, sulfur oxides, carbon monoxide, carbon dioxide, and particulate matter of all sizes produced and potentially released to the air. As an example, there are at least 131 organic air toxics found in the chemical manufacturing of synthetic organic chemicals. These toxics are emitted to the air from process vents, storage vessels, transfer racks and equipment leaks, and from wastewater treatment systems which have processed waste from the chemicals. Organic chemicals stay in the environment for long periods of time. They are deposited in water from the air and then bioaccumulate (the uptake, retention, and concentration of environmental substances by an organism) from low concentrations in water to high concentrations in animal tissue, and are highly toxic at low levels.

The chemical industry is responsible for releasing primary air pollutants into the air and creating secondary air pollutants which are formed from reactions between the air pollutants and atmospheric conditions. Particulate matter may either come from primary sources or be created by secondary sources. Organic chlorine compounds are of particular concern. They are found in large numbers of products including solvents, pesticides, plastics, disinfectants, etc., and may exist in the environment for long periods of time. Highly toxic byproducts are formed during the production of organic chlorine compounds. The burning of trash creates dioxins which may accumulate in the body fat and in the environment. Compounding the problem are temperature inversions which occur when the atmosphere is warmer as the altitude increases from the Earth. The temperature inversion creates a lid or blanket which forms over the area and the pollutants increase in concentration, thereby creating greater opportunities for illness among the residents as well as decreased visibility and deterioration of materials.

Best Practices for the Chemical Industry
- Automate processes to optimize the use of the raw materials and reduce waste.
- Improve catalysts to produce better products while reducing heavy metal use and reduce hazardous waste.
- Re-use byproducts instead of creating new products out of raw materials.
- Use a vapor recovery system to vent equipment to recover solvents.
- Substitute less toxic materials for more toxic materials.
- Keep amounts of raw materials to a minimum and use rapidly to avoid waste.
- Use recyclable containers.
- Use water-based biodegradable cleaners where possible.
- Avoid unneeded cleaning.
- Use dry clean-up methods where possible when spills occur.
- Use a precise inspection and preventive maintenance program on a regular basis.
- Cover chemical containers to reduce potential leaks, spills, and evaporation.
- Use a spill response team to accurately and quickly remove potential hazards.
- Train all employees in working skills and safety procedures.
- Develop appropriate timely inspection procedures for determining if equipment is leaking and make appropriate corrections.
- Keep appropriate records of the equipment, leaking times, and sites.
- Use preventive maintenance where appropriate to keep equipment from leaking.
- Determine if industrial flares are operating appropriately on a timely basis and take corrective action when necessary.
- Use the 12 principles of green chemistry: prevent waste; maximize consumption of all of the raw materials; use less hazardous chemicals; design safer chemicals and products; use less hazardous solvents; avoid unwanted reactions without using additional chemicals; make the process energy efficient; use renewable feed stocks; use catalytic reactions that are effective; design products to degrade to harmless substances over time; analyze the process to prevent pollution; and minimize potential for accidents. (See endnote 31.)

5. *Controlled Burns* (See endnotes 49, 50)

Controlled burns are used in forestry and agriculture to get rid of unwanted material in an efficient manner. Modern grasslands management includes prescribed burning which controls woody and herbaceous plants and improves the distribution of grazing areas, increases forage yield, and improves wildlife habitat. The time of the year of the burn is based on the types of grasses and undesirable plants and other materials that you want to get rid of quickly, and the type of grasses you are trying to stimulate. The time of the day of the burn as well as actual weather conditions can be contributing factors increasing air pollution problems. Controlled burns can cause forest fires when the conditions are such that they can no longer be controlled in an appropriate manner.

Burning creates particulate matter (smoke, ash, dust), nitrogen oxides, VOCs, and other air pollutants. Ozone becomes a key pollutant in downwind areas from the fires. Massive fires that may occur in oil refineries and other industries as well as train derailments involving tanker cars contribute immediate concentrated pollutants into the air. These may become an intense acute source of potential disease problems, and also become chronic in nature over time.

Best Practices for Controlled Burns
- Determine the time for the controlled burn of a given area based on what vegetative response is desired.
- Minimize smoke production by burning based on prevailing weather conditions, existing air quality, forecast air quality for the immediate area and areas downwind of the controlled fire, and the time of day.
- Do not start fires early in the morning or late in the evening.
- Take into account the amount of surface winds and transport wind (the rate at which the pollutants will be moved to other areas).
- Take into account the mixing height/dispersion potential for vertical mixing above the fire. Understand how relative humidity, fuel moisture, and air temperature can affect controlled burns and utilize this information to reduce potential for pollution.
- Understand ignition and burn techniques and the type of prescribed fire for safety purposes and pollution control.
- Contact appropriate fire authorities and advise them of the scheduled controlled burns.

6. *Erosion*

Erosion is the removal of soil and rock fragments by wind, water, snowmelt, flooding, rain, organisms, and gravity. It is also caused by fires, construction, agriculture, tree removal, drought, etc. (See the section on "Erosion" in Chapter 14)

7. *External Combustion Sources*

External combustion sources include: electric power plants using steam; industrial, commercial, and institutional boilers; process heaters; commercial and industrial solid waste incinerators; and combustion systems for commercial and domestic use. These systems use fossil fuel (coal, oil, both distillate oil and residual oil, and natural gas as their major source of fuel). Solid waste incinerators may use biomass, which can be wood and wood waste, municipal waste, corncobs, oats, etc., as their source of fuel.

The energy produced in external combustion units is used for generating electric power, process heating and space heating, and along with the production of energy are the unwanted pollutants including sulfur oxides, nitrogen oxides, particulates, carbon dioxide, mercury, hydrogen chloride, carbon monoxide, lead, cadmium, VOCs, arsenic, etc.

a.   *Electric Power Plants* (See endnotes 29, 51)

Electric power plants primarily use fossil fuels which are burned in a boiler to produce steam to turn the blades of the steam turbine that turns the shaft of the generator to produce electricity. In a nuclear power plant, the heat produced in the reactor is used to make steam. In a gas turbine, the combustion of the natural gas and distillate oil under high pressure produces hot gases, which spins the generator to produce electricity. In a combined cycle turbine, hot gases which have been used to spin one turbine generator go to a waste heat recovery system boiler where the water is heated again to produce steam and produces electricity. All of this uses one input of fuel to gain greater efficiency of production of electricity. In a hydroelectric generating system, the falling water or natural river current drives the turbine blades to cause the generator to produce electricity. Geothermal power uses the heat energy buried in the earth. Solar power uses the energy from the light and heat of the sun with photovoltaic conversion to generate electricity. The energy in wind power is converted to electricity.

Electric power plants, because of the use of fossil fuels as a source of energy, are a major contributor to criteria air pollutants, acid rain, greenhouse gases, and toxic air pollutants in North America, as well as throughout the world, especially in newly industrialized countries. In the United States, the Government Accountability Office (GAO), in 2010, conducted a study of 3443 electrical generating units, of which 1485 were older units (93% of the older units used coal) commissioned prior to 1978, and 1956 units were newer and more efficient. The older units using fossil fuel including coal produced 45% of the electricity used in the United States, but produced 75% of the sulfur dioxide emissions, 64% of the nitrogen oxides emissions, and 54% of the carbon dioxide emissions. For each unit of electricity produced by the older units, they emitted about 3.6 times as much sulfur dioxide, 2.1 times as much nitrogen oxides, and 1.3 times as much carbon dioxide as the newer units. This resulted in 90 times as much sulfur dioxide being released, twice as much carbon dioxide being released, and five times as much nitrogen oxides being released. (See endnote 11.)

Oil, used as a fuel, produces nitrogen oxides, sulfur dioxide, carbon dioxide, methane, and mercury compounds. The sulfur dioxide and mercury compounds vary widely with the sulfur and mercury content of the oil that is burned.

Natural gas, used as a fuel, produces nitrogen oxides, carbon dioxide, and methane (which is a natural component of the gas) in lesser amounts than coal or oil, but still in large enough quantities to pollute the air. Levels of sulfur dioxide and mercury are negligible.

Municipal solid waste has been used as a source of energy to produce electricity. This is still controversial because the burning of the waste produces nitrogen oxides, sulfur dioxide, and small amounts of toxic pollutants such as mercury and dioxins.

Landfill gas has been used as a source of energy to produce electricity. The burning of the landfill gas produces nitrogen oxides as well as small amounts of toxic materials depending on the waste that is burnt.

About 45% of the hazardous air pollutants identified by the EPA are found in coal, and approximately 40% of all hazardous air pollutants emitted from point sources come from coal usage. Coal-fired power plants are major sources of particulate matter, mercury, arsenic, etc. Noting the problems in the United States from coal-generated electric power, it is easy to understand the serious problems of newly industrialized societies, especially in China and India.

The Commission for Environmental Cooperation Council made up of cabinet level environmental health administrators from Canada, Mexico, and the United States issued a report in 2004 saying that electric power plants were the number one source of toxic air pollution in North America. They stated that 46 of the top 50 air polluters in North America were power plants. Hydrochloric acid and sulfuric acid were the most commonly released chemicals. (See endnote 12.)

b. *Industrial, Commercial, and Institutional Boilers*

Boilers are combustion units used to produce hot water or steam. The steam can run a variety of industrial processes and machinery, and produce heat or electricity. The boilers not only emit the pollutants from the fossil fuels, if they use them, but also from a variety of industrial processes. These may include nitrogen oxides, sulfur dioxide, particle pollution, carbon monoxide, formaldehyde, polycyclic aromatic hydrocarbons (PAHs), lead, hydrogen chloride, cadmium, mercury, dioxins, furans, greenhouse gases, etc.

There are approximately 1.5 million boilers in the United States. New emission standards issued in 2013 by the US EPA will apply to about 13% of all the boilers. These standards when enforced will lower emissions of hazardous air pollutants. These boilers in the United States are found in refineries, chemical plants, and other industrial facilities. These boilers are of greatest concern, contributing to the amount of hazardous air pollutants found in the air.

c.   *Process Heating Systems*

In a fuel-based process heating system, the heat is generated by the combustion of solid, liquid, or gaseous fuel and is transferred either directly or indirectly to the material being processed. The system needs to be energy efficient and product friendly. The systems may produce the same types of pollutants as in boilers as well as additional pollutants based on the type of industry involved. (See endnote 52.)

d.   *Municipal, Industrial, Commercial, Hospital Solid, and Hazardous Waste Incinerators* (See Chapter 12, "Solid Waste, Hazardous Materials, and Hazardous Waste Management")

Best Practices for External Combustion Sources (See endnotes 53, 54, 55, 56)

- Wash the coal to remove impurities including sulfur compounds or free sulfur, and hazardous air pollutants which are not bound to the coal carbon matrix, before using the coal as a fuel source for power plants.
- Treat flue gases to remove sulfur dioxide, nitrogen oxides, and mercury.
- Trap and then separate carbon dioxide from flue gases, compress the carbon dioxide usually in pipelines, and either inject deep underground for storage purposes or utilize a portion of the chemical in other processes or products.
- Use natural gas where feasible as the source of fuel for power plants since the byproducts of nitrogen oxides and carbon dioxide are lower than in other fossil fuels.
- Burn the natural gas fully and efficiently to prevent methane from entering the atmosphere.
- Evaluate the storage and use of fuel oil to prevent leaks, spillage, and other means of evaporation of the fuel into the atmosphere. Take necessary corrective action.
- Sample and evaluate the composition of flue gases to make sure that they meet the specifications and regulations of the US EPA.
- Use the desulfurization techniques including dry scrubbing and sorbent injection systems prior to the burning of coal.
- Use denitrification techniques including selective catalytic reduction and/or selective non-catalytic reduction.
- Use particulate collection techniques including electrostatic precipitators, fabric filters/ baghouses, and mechanical dust collectors.
- Contain ash and fly ash to keep it from contaminating the air, land, and water.
- Use condensing heat exchanger systems, where the latent heat in the flue gas is returned to the system, to reduce the emissions of hazardous air pollutants.
- Use activated carbon filters to collect hazardous air pollutants, especially mercury.
- Inject activated carbon powder onto fabric filter systems to reduce mercury.
- Use combined heat and power systems (cogeneration) where feasible in manufacturing plants, hospitals, colleges, and other large buildings. This is a much more efficient use of fuel, while producing electricity and utilizing the waste heat for either space heating or for process energy.
- Where coal is the fuel of choice, use a pulverized coal boiler where the burning is much more efficient than in the Stoker-type coal boiler and therefore less contaminants enter the air while more energy is being produced.
- Install emissions control systems on existing boilers where needed.

8. *Evaporative Sources*

These are volatile liquids that are not enclosed in a tank or other container and evaporate or release vapors to the air. They are typically liquids such as paints, solvents, pesticides, hair sprays, aerosol sprays, gasoline, etc. Evaporative sources are found in many of the major sources of air pollution and area sources of air pollution. (See individual sources of interest for Best Practices.)

9. *Fires*

Smoke from wildfires contains gases and fine particles from the burning of trees, grasses, and bushes. These fires also burn houses, businesses, and potentially industries. The contaminants include the criteria pollutants, hazardous air pollutants, and greenhouse gases. When houses and businesses burn, pollutants are created which are similar to those that are found in incinerators that do not have air pollution controls. These hazardous air pollutants include PAHs, dioxins; furans; hydrochloric acid; formaldehyde; heavy metals including lead, mercury, and arsenic; styrene from foam cups, meat trays, and egg containers; household chemicals including cleaning materials, pesticides, and fertilizers; etc.

The smoke can have a profound effect on an individual's health. It causes shortness of breath, chest pain, rapid heartbeat, fatigue, headaches, and the inability to breathe normally. People at greatest risk are the elderly, children, those with asthma, those with COPD, heart patients, etc. The smoke may also be a precursor to many diseases in the future.

Some people believe if they have a respirator or a mask that the smoke will not harm them. This may not be correct. The effectiveness of the respirator for the given problem depends on whether it will filter out very small particles, the type of chemical cartridge that is used, the fit of the mask on the individual, the amount of filtering capacity, etc.

Best Practices for Fires
- Follow the orders for evacuation issued by appropriate authorities.
- If you are elderly, very young, have asthma, or have other chronic diseases or allergies, leave the immediate area of the fires and go downwind from the fires as quickly as possible.
- Listen to local air quality reports and public health messages concerning your safety and act on them immediately.
- Stay indoors and make sure that you have clean air filters and the outside doors and windows are kept closed.
- Cease the use of wood stoves, smoke, candles, or fireplaces during fire emergencies.
- Delay the vacuuming of carpets during fire emergencies to prevent small particles from becoming airborne.
- A person with any health or physical problems should immediately call the doctor for specific directions on what to do because of the fire.

10. *Fugitive Dust* (See the section on "Particulate Matter" above)
11. *Hazardous Waste* (See Chapter 12, "Solid Waste, Hazardous Materials, and Hazardous Waste Management")
12. *Internal Combustion Engines* (See the section on "Mobile Sources" later in this chapter)
13. *Iron and Steel Manufacturers*

Iron and steel making consists of several processes which produce criteria air pollutants especially considerable particulate matter, sulfur dioxide, nitrogen oxides, hazardous air pollutants, and greenhouse gases. The greenhouse gases, primarily carbon dioxide, come from the production of coke; use of the blast furnace, boiler, process heater, reheating furnace, flame suppression system, annealing system, flare, ladle, re-heater, etc. (See endnote 7.) The processes, which produce numerous pollutants, include: sintering which recovers raw material from waste material; iron production from the original ore; iron preparation and removal of the sulfur; steel production; steel pickling which is the use of an acid solution to remove the oxides found in scale on the steel as it cools from its molten state; production of the semi-finished product; finished product preparation; and the handling and treatment of the raw materials, intermediate materials, and waste materials. (See endnotes 6, 70.)

Best Practices for the Iron and Steel Industry (See endnote 71)
- Remove waste gas from the sintering process and utilize modern electrostatic precipitators followed by fabric filters to trap dust, heavy metals, sulfur dioxide, hydrochloric acid, polyaromatic hydrocarbons, and organochlorine compounds.
- Use a wet scrubbing process with calcium oxide on waste gases to remove sulfur dioxide.
- Use encapsulated or semi-encapsulated housings for the steps of coke crushing, raw material handling, belt charging and discharging, and sintering to trap and remove dust and associated contaminants.
- Reduce the nitrogen oxides by rapidly cooling down flue gases.
- Recirculate waste gas if it can meet the quality standards of the sintering process.
- Reduce the heavy metals by using a wet scrubbing system.
- Remove dust from bag filters and recycle byproducts containing iron and carbon and then place remaining dust in a secure landfill.
- Use frequent and complete maintenance to keep equipment operating properly.

14. *Land Development and Building Construction*

Construction activities can produce substantial amounts of air pollution over short periods of time in a given area and therefore can have a substantial if limited local impact on people and the environment. Construction consists of site preparation, earthmoving including the hauling of the materials, the material used in paving of surfaces, tearing down of old buildings and structures, erecting new buildings and structures, waterproofing materials, paint for the structures, and tar for roofs.

The emissions coming from construction activities include particulate matter and fugitive particulate matter from concrete, cement, silica, disruption of the soil, various demolition activities, etc., and may have attached to them a variety of other contaminants; oxides of nitrogen and other contaminants from the combustion of diesel fuel and gasoline from large equipment, trucks used in construction activities, material delivery trucks, and the vehicles used by the workers; VOCs which come from paving activities and use of paint and tar, etc.; carbon monoxide from improper burning of fossil fuels; carbon dioxide from the burning of fossil fuels in diesel engines and construction equipment; asbestos previously used in a considerable amount of construction; mercury which can be found in demolition sites from switches, thermostats, electronic equipment, and batteries; lead which may be found in paints in older structures; and other toxic materials, such as arsenic, which was used previously to preserve wood.

The timing, magnitude, and duration of construction have to be taken into account when determining the problems related to release of criteria air pollutants, toxic air pollutants, and greenhouse gases. Weather conditions and existing levels of air pollution can complicate the problems of additional air pollutants released during destruction of facilities and preparation of the land and/or construction.

Best Practices for Land Development and Building Construction
- Develop a construction proposal which enumerates the duration, type, and size of the construction and its component parts. What are the types of pollutants which will be added to the air, how much, and for how long?
- Determine the countermeasures that will be utilized if equipment fails, such as dust collectors, etc., during the process of demolition or construction. Use controlled water spraying to mitigate dust problems.
- Use equipment that crushes material and produces as little dust as possible, and which uses crushing under pressure instead of pounding.
- Use equipment that contains dust traps which are cleaned and serviced regularly.
- Enclose the filling and emptying mechanism of structures that contain powdery materials and then filter the air within the structure where materials are stored.

- Use protective walls to prevent wind from blowing dust and other materials from road debris, etc.
- Restrict speed on construction sites and reduce the dust through compaction and water sprinkling.
- During demolishing of structures, break the demolition material into the largest pieces feasible to reduce dust.
- Keep Hot remix tar-based coating substances for surfaces and sealing purposes away from building sites.
- Use asphalt, pitch, or tar in an emulsion form instead of solution form, and use materials that have low emission rates of air pollutants.
- Reduce welding emissions by capturing and filtering the fumes.
- Use products which have less VOCs for painting, plastering, gluing, and sealing rather than more volatile ones.
- Use low emission explosives.
- Use low emission equipment, such as that with electric motors where possible.
- Inspect and maintain all gasoline or diesel engine equipment in a very careful manner to prevent pollutants from entering the air.
- Use dust abatement procedures such as wetting, trapping, suction, and filtering, for all dusty mechanical work.
- Cover trucks when hauling dirt or other materials.
- Stabilize the surface of dirt piles if they are not removed immediately.

15. *Natural Gas and Crude Oil Extraction Industry*

Natural gas and petroleum production from the well, processing, storage, transmission, and distribution, onshore or offshore, are the largest sources of methane emissions from industry in the United States. The oil goes from well extraction to the refinery for processing. The gas from the wells goes to gathering and boosting stations and then flows to processing plants, where impurities and water are removed and the gas can then be sent along pipelines to the ultimate user. All through this process, there may be a series of emissions to the air. If it is not cost-effective to use the gas associated with the oil, then it may not be handled properly and could become an air contaminant. All equipment used in drilling, other production operations, storage, and transportation become sources of air emissions.

Fracking

Hydraulic fracturing or fracking is a process in which a well is drilled and a steel pipe is inserted into the well bore with holes in the bottom and then a liquid under very high pressure, typically containing water, chemicals, and a proppant (usually sand), is inserted to overwhelm the natural pressure and fracture or crack rock underground to release gas or oil. During the stage of well completion known as "flowback," the fracturing fluids, water, and gas come to the surface at a high velocity and volume including VOCs, methane, and air toxics. The flowback lasts from 3 to 10 days and creates substantial amounts of air contaminants. Also when the borehole turns 90° to go into the shale formations, in some areas the concentrations of radon are considerably higher than others. This produces radioactive borehole cuttings that are brought to the surface. These are often sent to landfills or fill areas where they can become the source of radioactive contamination.

The fracking process, which is currently producing a large amount of natural gas, has several additional potential environmental concerns. They are air quality issues, including the release of VOCs, including methane; other hazardous air pollutants; and greenhouse gases released into the ambient air from the wastewater, from spills, from the gas or oil being recovered, and from the chemicals being used in the fracking process. There is also a problem of airborne dust coming from construction sites and traffic.

Glycol Dehydration Unit

The glycol dehydration unit which is used to remove water and hazardous air pollutants from the natural gas and oil is the primary source of emissions of air contaminants. Pneumatic devices used in the natural gas industry as liquid level controllers, pressure regulators, and valve controllers are also prime sources of methane escape from the systems. Part of the normal operation of pneumatic devices causes a release or bleeding of natural gas to the atmosphere. Storage tanks have potential for equipment leaks and flash emissions.

Natural Gas Flaring

Natural gas flaring is a controlled burn of natural gas used in oil/gas exploration, production and processing to: burn off gases during the testing process used to determine the pressure, flow, and composition of gas or oil; burn off gases during emergency situation when there is too much pressure in the pipes; burn off gases released during maintenance and equipment repairs; release small volumes of waste gases which cannot be captured efficiently; and burn off gases which accompany oil. Flaring may produce substantial heat and noise. (See endnote 67.)

Oil and Natural Gas Production Facilities

Oil and natural gas production facilities emit hazardous air pollutants and other VOCs, which may contribute to the health problems of the employees as well as the community. The chemicals emitted include benzene, toluene, ethyl benzene, xylenes, n-hexane, and VOCs. Benzene exposure in the short term may cause severe irritation of the skin, eyes, and upper respiratory tract. Over long periods of time, it may cause blood disorders, reproductive and developmental disorders, and cancer. Toluene long-term exposure may cause nervous system effects, irritation of the skin, eyes and respiratory tract, headaches, and birth defects. Ethylbenzene may cause short-term effects of throat and eye irritation, chest construction, and dizziness. It may cause long-term effects of blood disorders. Xylenes may cause short-term effects of nausea, vomiting, and neurological effects. Long-term effects may impair the nervous system. n-Hexane may cause short-term effects of dizziness, nausea, and headache. It may cause long-term effects of numbness in the extremities, muscular weakness, blurred vision, headaches, and fatigue. VOCs can help form ground-level ozone as part of smog, and cause breathing difficulties and severe respiratory problems in susceptible populations such as the young, the elderly, and those with asthma.

Best Practices for the Natural Gas and Crude Oil Extraction Industry (See endnote 65)
- Optimize the glycol dehydration system by modifying the process using diethylene glycol or ethylene glycol instead of triethylene glycol. Replace glycol dehydrators with desiccant dehydrators.
- Install VOC controls on all storage tanks used to temporarily hold liquids from the production and transmission of oil and natural gas, starting with those which are most prone to produce emissions of the VOCs and air toxics including benzene.
- Evaluate and eliminate all emissions to the air from valves, pumps, seals, pressure relief valves, and other equipment.
- Remove hydrogen sulfide, carbon dioxide, helium, nitrogen, and other substances from raw natural gas to produce pipeline quality dry natural gas.
- Use special equipment to separate and treat the gas and liquid hydrocarbons from the flowback and then prepare the products for sale instead of disposal.
- Use pit flaring, which is a horizontal flare burner that discharges into a pit, to burn VOCs and methane if they cannot be collected efficiently and it is not a safety hazard.
- Improve the compressors of natural gas by equipping them with dry seals.
- Use pneumatic controllers that are not gas driven.

- Strengthen the leak detection and repair requirements for natural gas processing plants.
- Improve the air toxics standards by doing a new assessment 8 years after the standard is issued and also a technology review every 8 years after the standard is issued.
- Replace the gas starters with air or nitrogen in compressors/engines.
- Conduct frequent inspection and maintenance at remote sites.
- Test and repair the pressure safety valves.
- Frequently inspect and maintain properly gate stations, surface facilities, compressor stations, processing plants, and booster stations.
- Inspect, service, and repair pipelines, pressure safety valves, and pipeline connections on a regular basis.
- Convert water tank blankets from natural gas to produced carbon dioxide gas and eliminate unnecessary equipment and/or systems.
- Install vapor recovery units on storage tanks and recover gases during condensate loading.
- Connect the casing of wells to the vapor recovery units.
- Install properly operating flares to destroy excess gases and toxic air pollutants.

16. *Natural Sources*

These sources vary by intensity and quantity of type of pollutants based on a variety of factors. They are:

- Bodies of water that produce methane through the digestive process of marine life and reaction within sediments
- Digestive gasses, such as methane, which are produced by the digestion of food in cattle and other animals
- Dust which comes from areas of little or no vegetation
- Forest fires caused by lightning or spontaneous combustion producing smoke, ash, dust, nitrogen oxides, carbon dioxide, and other air pollutants
- Geysers produce hydrogen sulfide, arsenic, other heavy metals, etc.
- Lightning converts atmospheric nitrogen into nitrogen oxides.
- Plants and trees, especially pine trees, which release VOCs
- Radioactive material which releases radon gas into the atmosphere from the Earth's crust
- Sea salt releasing sodium chloride and other particles into the air from the action of wind over the body of salt water
- Soil where microbial action in soils forms and releases nitrogen oxides
- Termites, which are the second-largest source of methane production because of their normal digestive process
- Volcanoes which produce smoke, ash, carbon dioxide, sulfur dioxide, and other air pollutants
- Wetlands in which microbial action produces and releases to the atmosphere a significant amount of methane.

Best Practices for Natural Sources

Basically there are no Best Practices for natural sources which would be practical other than carrying out clean-up activities where appropriate.

17. *Petroleum Refining*

Crude oil is processed to produce automobile gasoline, diesel fuel, lubricants, and other petroleum-based products. The toxic air pollutants including benzene come from storage tanks, equipment leaks, process vents, wastewater collection, and treatment systems at the facility. The emission of air toxics occurs during production, transporting, storage, separation, upgrading, leaks from pumps, compressors, valves, flanges, and other equipment used in oil and natural gas production, transmission, and storage. The air toxics include benzene and other VOCs as well as methane, an important greenhouse gas.

Best Practices for Petroleum Refineries
- Develop appropriate timely inspection procedures for determining if equipment is leaking and make appropriate corrections.
- Keep appropriate records of the equipment leaking times and sites.
- Use preventive maintenance where appropriate to keep equipment from leaking.
- Determine if industrial flares are operating appropriately on a timely basis and take corrective action when necessary.

18. *Power Plants* (See External Combustion Sources-Electric Power Plants above)
19. *Publicly Owned Treatment Works* (Public Sewage) (See Chapter 11, "Sewage Disposal Systems")

These plants treat wastewater from residential, commercial, and industrial sources. Depending on the types of commercial and industrial plants utilizing the publicly owned treatment works for disposal of their liquid waste and possibly from combined sewers on their properties, there will be considerable variation in the potential air toxics which will be emitted. Typically, there is a release of VOCs from the wastewater including xylenes, methylene chloride, toluene, ethylbenzene, chloroform, tetrachloroethylene, and naphthalene.

Best Practices for Publicly Owned Treatment Works (Public Sewage) (See Chapter 11, "Sewage Disposal Systems")

20. *Pulp and Paper Mills*

Wood and non-wood sources of fiber are turned into pulp by use of chemicals, mechanical grinding, or a combination of both. Pulp may also be produced from the re-pulping of recovered paper. The fibrous masses are washed, screened, and sometimes bleached. The individual fibers are then mixed with water and the slurry is sprayed onto a flat wire screen where the water is squeezed out and the fibers bond together to form paper of different qualities and thicknesses. The air emissions from the pulp and paper production contain process gases, which vary with the raw materials, and may include total reduced sulfur compounds, causing obnoxious odors, which are typically found in kraft and sulfite mills. The kraft chemical process, which uses sodium hydroxide and sodium sulfide to pulp the wood, is the most important process used in the world for virgin wood. In the kraft process, about half of the wood is dissolved and that solution plus the used sodium hydroxide and sodium sulfide becomes a substance called black liquor. The black liquor is then washed from the pulp and sent to a kraft recovery system where the two chemicals are separated for reuse while the dissolved organic material is prepared to be a fuel to make steam to generate electricity and heat to be used in the pulping process. The process consists of pulping chemicals, recovery boilers for dealing especially with particulate matter and sulfur dioxide, and lime kilns for reducing sulfur compounds.

Process gases also include particulate matter, nitrogen oxides, VOCs, fluorine, carbon dioxide, and methane. Other emissions may come from the materials used in the heating process and the energy producing process. Toxic air pollutants are produced including chloroform, fluorine, formaldehyde, methanol, acetaldehyde, methyl ethyl ketone, and metals. Other air pollutants include particulate matter and chlorine. (See endnote 5.)

The pulp and paper industry uses a very substantial amount of energy. This is over 15% of the manufacturing energy used in the United States. The use of energy sources significantly adds to the amount of criteria pollutants, greenhouse gases, and hazardous air pollutants. (See endnote 73.)

Best Practices Pulp and Paper Mills (See endnote 72)
- Collect and incinerate emissions to completely oxidize all reduced sulfur compounds coming from the process of working with black liquor, brown pulp, unbleached pulp, and condensate.
- Use a standby incineration system to destroy high concentrations of gases from condensates and digester events.

- Use a standby incinerator to incinerate low concentrations of gases when the industrial plant is close to a residential area.
- Use an oxygen-activated sludge process to capture odors from wastewater treatment plants.
- Ensure black liquor is oxidized prior to evaporation.
- Concentrate black liquor in the evaporator to reduce sulfur emissions.
- Control the combustion temperatures to properly incinerate sulfur emissions.
- Control excess air to reduce nitrogen oxide emissions.
- Collect sulfur dioxide emissions by absorption in an alkaline solution.
- Use low sulfur content fuel and control excess oxygen to reduce sulfur dioxide emissions.
- Use electrostatic precipitators in recovery boilers, auxiliary boilers, and lime kilns.
- Use an alkaline process with scrubbers to remove the acid gases.
- Incinerate VOC emissions from mechanical pulping of wood.

21. *Solid Waste Landfills* (See Chapter 12, "Solid Waste, Hazardous Materials, and Hazardous Waste Management")

Solid waste landfills produce landfill gas in the presence of microorganisms and the chemical reactions in the waste. This gas including methane, carbon dioxide, ammonia, and sulfides which enter the ambient air.

Best Practices for Solid Waste Landfills (See Chapter 12, "Solid Waste, Hazardous Materials, and Hazardous Waste Management")

22. *Solid Waste Incinerators* (See Chapter 12, "Solid Waste, Hazardous Materials, and Hazardous Waste Management")

Solid waste incinerators include waste incineration plants and waste-to-energy plants. Municipal, commercial, institutional, and industrial incineration is a means of reducing the quantity of solid waste to a point where the disposal of the remnants takes up a small amount of land. A number of pollutants may be released during this process, including cadmium, lead, mercury, dioxin, sulfur dioxide, hydrogen chloride, nitrogen dioxide, and particulate matter. These pollutants are a mixture of criteria pollutants and toxic air pollutants. Hospital/medical/infectious waste incineration is the destruction of solid waste materials used in all facets of diagnosis, treatment, or immunization of humans or animals in such a manner that the remnants can no longer potentially cause disease. This is accomplished in special incinerators.

Best Practices for Solid Waste Incinerators (See Chapter 12, "Solid Waste, Hazardous Materials, and Hazardous Waste Management")

23. *Spills and Run off from Stored Chemicals and Storage Tanks*

Ground or surface water is readily contaminated when petroleum products and chemicals are mishandled and spills occur, resulting in the chemicals seeping into the ground and to the water supply or becoming part of runoff and going into surface bodies of water. A single spill or leakage can spread out into the ground and become a source of contamination for many years. In numerous areas, there are a substantial number of underground storage tanks which have been used over many years for holding a variety of products including substances which are very hazardous. These tanks may be found in gas stations (the most common place), airports, dry cleaners, homes, agricultural areas, etc. An immediate problem occurs when there are leaks and spills because of poor housekeeping, overfilling of the tanks, sloppiness in loading and unloading the product, and poor maintenance and inspection of the facility. A long-term problem occurs when the tanks become corroded and start to leak product into the ground. A recent example of an overflow from the tanks into a drinking water supply occurred on January 9, 2014, when toxic chemicals were released from a storage tank spill into the Elk River which is the drinking water supply for Charleston, West Virginia.

This affected 300,000 people. The spill resulted in individuals going to the emergency room because of nausea, eye irritation, and vomiting. The chemicals released were also potentially carcinogenic.

Best Practices for Spills and Run off from Stored Chemicals and Storage Tanks

The Best Practices to prevent spills and leakage from aboveground storage tanks and underground storage tanks include:
- Inspect all pumps, hoses, and connections between pipes for leaks monthly.
- Check for loose fittings, worn gaskets, or damaged rubber nozzles monthly.
- Check underground storage tank equipment and dispensers for leaks and structural problems monthly.
- Make frequent inspections of all tanks and equipment during very cold weather and very hot weather.
- Inspect aboveground storage tanks weekly for leaks and monthly for deterioration.
- Check secondary containment areas for any sheen, which would indicate spillage.
- Keep an inspection log with the results of the inspections, dated and signed.

24. *Toxic Waste Storage, Transfer, Treatment, and Disposal Facilities*

These include facilities for industrial wastewater treatment, industrial waste, solvent recycling, and used oil recovery. Toxic air pollutants including chloroform, toluene, formaldehyde, and xylene are released from the tanks, from process events, through equipment leaks, from the containers, from surface impoundments, and through the piping system of the transfer, treatment, and disposal of the wastes.

Best Practices for Toxic Waste Transfer, Treatment, and Disposal Facilities (See Chapter 12, "Solid Waste, Hazardous Materials, and Hazardous Waste Management")

SUB-PROBLEMS INCLUDING LEADING TO IMPAIRMENT AND BEST PRACTICES FOR TOXIC AIR POLLUTION

Toxic or hazardous air pollutants are suspected of causing cancer, asthma, reproductive effects, birth defects, nervous system damage, immune system damage, skin, eye, nose and throat irritation, brain, lung, kidney and liver damage, respiratory and cardiovascular diseases, and/or adverse environmental effects. Some examples of the 187 substances and 33 major urban air toxins currently being monitored and controlled are acetaldehyde, aldehydes, arsenic compounds, benzene, butadiene, cadmium compounds, chromium compounds, coke oven emissions, ethylene oxide, formaldehyde, mercury compounds, pesticides, polychlorinated biphenyls, polycyclic organic hydrocarbons, solvents, and vinyl chloride.

Most toxic air pollutants come from anthropogenic sources including: mobile sources (the largest contributor of emissions) such as cars, trucks, and buses; stationary sources (typically industry, which may be called point sources) such as factories, chemical plants, power plants, and refineries; building materials and cleaning solvents; and accidental releases including leaks and spills. Natural pollutants come from volcanic eruptions, forest fires, etc. Major sources of release of toxic or hazardous air pollutants include chemical plants, steel mills, oil refineries, hazardous waste incinerators, etc. Area-wide sources, which are much smaller, are an accumulation of pollutants from dry cleaners, gas stations, etc. Also off-road equipment such as construction equipment, farm equipment, small boats and ships contribute to the problem, as well as secondary chemical reactions between compounds in the atmosphere.

After the toxics are released into the air, they are carried by the wind to places near and far depending on weather conditions, the topography of the land, the physical and chemical properties of the pollutant, and the changes to the chemical that may occur while airborne. They may be deposited on land or in the water, then evaporate and recycle again. Some of the air toxics such as heavy metals may remain airborne for indeterminate periods of time.

Depending on the concentration and length of exposure to the air toxics and other factors concerning the individuals who are exposed to these pollutants, a variety of health effects may occur. Of the 187 substances listed by the US EPA, 33 are considered very hazardous in urban areas.

To determine potential health effects of a given air toxic or combination of toxics, it is necessary to do an evaluation of the exposure using a five-step process as follows:

1. Identify the actual pollutants released.
2. Determine the patterns of release of each of the chemicals and how much is released in a 24-hour period.
3. Determine the amount of the pollutants that reach individuals and the concentration.
4. Estimate the number of people who have been exposed to the pollutants and at what concentrations for what period of time. (See endnote 10.)
5. Determine the number of slow-moving diesel trucks, their numbers of starts and stops and deliveries, in heavy industrial sections of the city where mostly poor people live. These underserved areas lead to greater exposure for the residents and it therefore becomes an environmental justice issue.

People are exposed to toxic air pollutants by:

- Inhaling quantities of contaminated air
- Eating quantities of contaminated food such as fish, meat, milk, eggs, fruits, and vegetables
- Drinking water contaminated by the deposition of toxic air pollutants
- Ingesting quantities of contaminated soil, primarily children
- Frequent skin contact with contaminated soil, dust, water, or air

Professional practitioners can then determine the potential health effects that individuals will have in reaction to the release of the contaminants. It is important to note that contaminants accumulate in the body and that mixtures of contaminants are much more frequent than single releases and this may well complicate the problems of diagnosis and prevention of short-term and long-term diseases.

To control toxic air pollutants, it is necessary to use pollution prevention measures such as product substitution, process modification, improved work practices, cleaning of coal, and cleaning of flue gas. Reduce emissions from cars, trucks, buses, and other vehicles. Use appropriate reformulated gasoline.

The major toxic air pollutants are (see endnotes 13, 18):

1. *Diesel Fuel and Emissions*

   Diesel fuel and emissions create about 80% of the total estimated cancer risk of all hazardous air pollutants. Diesel exhaust, which comes from on-road as well as off-road mobile sources, as well as stationary sources, is a complex mixture of gases, vapors, and fine particles containing arsenic, benzene, nickel, carbon monoxide, nitrogen oxides, sulfur dioxide, hydrocarbons, etc. It is known to increase respiratory and cardiovascular diseases, which may lead to death. There is a significant increase in lung cancer associated with people who constantly inhale the fumes.

   Best Practices for Diesel Fuel and Emissions
   - Utilize cleaner fuels such as ultra-low sulfur diesel, biodiesel, and/or liquid petroleum gas.
   - Retrofit the engine of older trucks and buses by installing emission-reduction equipment including catalysts.
   - Repair existing engines to make them the same as the original engine.
   - Use proper maintenance schedules to keep engines running properly.
   - Reduce idling time throughout the usage of the vehicle.
   - Increase the energy efficiency of the vehicle by using proper tires that are properly inflated.

2. *Acrolein*

   Acrolein is emitted by industrial plants where it is manufactured as an intermediate for other chemicals. It is also found in tobacco smoke, forest fire smoke, gasoline and diesel exhaust, paper mills, and other non-metallic mineral and wood products. It is a registered pesticide in California and is also used to control fungi and bacteria. It causes irritation of the eyes, nose, throat, and respiratory tract and potentially pulmonary edema.

3. *Benzene*

   Benzene is emitted mostly from gasoline motor vehicle exhaust, gasoline fugitive emissions, and to a much lesser extent, stationary industry sources. It is a widely used industrial chemical and is used in the manufacture of medicines, shoes, dies, detergents, explosives, etc. Benzene may also be found in tobacco smoke, heating and cooking systems, and evaporating from various products used within the home. It is a major contributor to overall cancer risk. It can cause central nervous system depression, nausea, tremors, drowsiness, intoxication, and unconsciousness.

4. *1,3 Butadiene*

   1,3 Butadiene is emitted from the incomplete combustion of gasoline and diesel fuels. It may also come from petroleum refineries, wearing of tires, residential wood burning, agricultural burning, and forest burning. It may be found in environmental tobacco smoke. It is used in the production of synthetic materials. It is irritating to the eyes and mucous membranes and may cause blurred vision, fatigue, headaches, and vertigo.

5. *Carbon Tetrachloride*

   Carbon tetrachloride in the past has been used for dry cleaning and as a green fumigant in the United States. Although it has been discontinued, it has an estimated lifetime in the atmosphere of 50 years and therefore can still be found upon testing. It affects the central nervous system and is a respiratory tract irritant as well as a toxin for various cell components.

6. *Chromium*

   Electroplating and anodizing operations are used to coat metal parts and tools with a small amount of chromium to protect them from corrosion and wear. Hexavalent chromium, which is released during the electroplating and anodizing processes, is known to cause cancer. It may also cause complications with pregnancy and childbirth.

7. *Coke Oven Emissions*

   Coke is used to extract metal from ores, especially iron, and is also used to make calcium carbide in the manufacture of graphite and electrodes. Coke oven batteries at steel plants convert coal into coke and in a blast furnace which converts iron ore to iron. The emissions contain benzene, which may cause cancer, conjunctivitis, severe dermatitis, and lesions of the respiratory and digestive systems.

8. *Formaldehyde*

   Formaldehyde is used in the production of dies, textiles, particle board, plywood, as an embalming fluid, etc. It is hazardous to the respiratory system and skin, and is highly toxic. It may cause cancer.

9. *Mercury*

   Mercury is a neurotoxin that bioaccumulates in the food chain. It is emitted to the atmosphere usually in its elemental form and remains viable for a long period of time, allowing it to be transported over great distances. During this time, it is oxidized and produces a reactive gaseous form which allows it to increase its rate of deposition in a variety of ecosystems.

10. *Polycyclic Organic Matter*

    Polycyclic organic matter is produced by the combustion of fossil fuels and vegetable materials. The compounds may be found in ambient air from cigarette smoke, asphalt roads, coal tar, hazardous waste sites, motor vehicles exhaust, smoke from wood burning in homes, fly ash from electric power plants using coal, petroleum refineries, and paper mills. They may cause health problems in the gastrointestinal tract, liver, skin, eyes, etc.

They may affect reproduction and also increase the risk of lung cancer. PAHs are part of this group and the most frequently monitored of these is benzo [a] pyrene, which has been found in urban air at twice the level found in rural air. (See endnote 26.)

SOURCES OF TOXIC AIR POLLUTANTS

Major Point Sources

These are stationary facilities or special processes, produce at least 10 tons of a single contaminant per year or at least 25 tons of a mixture of hazardous air pollutants per year. They include cement kilns, the chemical industry, the fertilizer industry, industrial, commercial, and institutional boilers, iron and steel manufacturing plants, petroleum refineries, power plants, pulp and paper mills, solid waste incinerators, and toxic waste storage, transfer, treatment, and disposal facilities. (See Other Major Sources of Air Pollution above for specific problems and Best Practices for each of these listed industrial operations.)

Area and Other Sources (Smaller Sources)

(See endnote 4)

Area and other sources are a series of small sources which by themselves would not be a significant problem, but when added together become a major concern. One of the greatest sources of emissions is the leaking of VOCs from equipment, tanks, and their components. This includes valves, connectors, pumps, sampling connections, compressors, pressure-relief devices, and open-ended lines. Between 90% and 95% of the leakage comes from the connectors and valves, with the majority of the problems due to seal or gasket failure because of normal wear or improper maintenance. These sources include but are not limited to those listed in this book. (To list all the potential sources would be beyond the scope and size of the book.) Some of the points mentioning sources below only discuss briefly some of the contaminants and sources of contaminants. To state all the Best Practices would be highly repetitive. (See endnote 21.) (See Stationary Sources and Best Practices for Stationary Sources.)

The Aerospace Manufacturing and Rework Industry

The industry produces or repairs aerospace vehicles and vehicle parts from airplanes, helicopters, space vehicles, and missiles. They produce toxic air pollutants such as methylene chloride and chromium which are released during paint stripping, cleaning, priming, grinding, and application.

Best Practices for the Aerospace Manufacturing and the Rework Industry
- Modify the facilities to recover copper sulfate from the etching and stripping process; recover metal from sludge production; substitute drip pans in place of rinse sinks.
- Improve management of material tracking, inventory, material usage, handling, and storage.
- Recover solvents during all processes.
- Use low VOC coatings and paints.
- Use water-based primers where possible.
- Use low residue mechanical methods for paint stripping.
- Where appropriate, use mechanical or dry means of cleaning instead of solvents.
- Substitute less hazardous chemicals for more hazardous chemicals.
- Use low-pressure spray guns to apply resins or paints.

Accidental Chemical Releases

This includes leaks and spills in various facilities. The EPA has established regulations under the Clean Air Act to utilize risk management programs to prevent this type of problem and if it occurs, special techniques to clean up the chemicals.

Best Practices for Accidental Chemical Releases (See Best Practices for the Chemical Industry)

Auto Body and Repair Shops
(See endnote 78)

Auto body shops and garages repair and/or replace various parts of vehicles, and activities may include welding, paint stripping, and painting. Degreasers are commonly used in the shops as well as in other small and large industrial processes. Degreasers remove old oil and other fluids from vehicle parts. The operations can be very dusty, producing a variety of particulates, and can also be the source of a series of hazardous air pollutants including: VOCs, lead, chromium, cadmium, etc. The shops also produce nitrogen oxides, sulfur dioxide, carbon monoxide, and carbon dioxide, which can lead to increased ground-level ozone.

Of special significance may be aboveground storage tanks, below ground storage tanks, use and disposal of absorbent material, disposal of waste antifreeze, used batteries, brakes, catalytic converters, repairs from accidents, floor drains, fuel, hazardous wastes, painting operations, cleaning of parts, use and disposal of shop rags and towels, tires, used oil and oil filters, untreated wastewater, and stormwater management.

Best Practices for Auto Body and Repair Shops
- Prepare and implement spill prevention, control, and countermeasure plans.
- Use aqueous cleaning processes where possible in auto shops in place of solvent cleaning.
- Use large spray cabinets and ultrasonic units to clean vehicles and parts.
- Place all solvents in leak-proof containers.
- Capture old oil and solvents in leak-proof containers for recycling.
- Absorbents used to clean up spills may be considered to be hazardous and must be disposed of appropriately.
- Waste antifreeze not only is toxic but also may contain lead, cadmium, and chromium in high levels and must therefore be disposed of as a hazardous waste.
- Batteries contain lead and other substances which are toxic, must be considered to be a hazardous waste and therefore should be either recycled or sent to special hazardous waste facilities for disposal.
- When working on brakes, if they contain asbestos fibers the technicians should wear special respirators and the used material should be disposed of as a hazardous waste.
- Catalytic converters, which reduce sharply emissions to the air, must be serviced by specially trained technicians on a regular basis.
- Repair of vehicles involved in collisions includes removal of fluids for appropriate disposal and removal of body parts which may be hazardous for appropriate disposal. It may also include sanding, grinding, paint stripping, and painting, which may contribute a variety of air pollutants and therefore must be carried out in special facilities with appropriate ventilation systems.
- Floor drains must be cleaned on a regular basis to remove engine oil, solvents, and other materials.
- If the garage provides fuel for vehicles, it must follow the recommendations under Gas Stations and Gasoline Distribution Systems.
- Solvents from parts cleaning, shop rags, and towels are also considered to be hazardous wastes and therefore must be disposed of in an appropriate manner.
- Stacked tires must be protected from the elements, either emptied of all water or treated for potential mosquito breeding.
- Used tires should be recycled instead of left on land where they can become mosquito breeding places or potentially catch fire and release toxic substances into the air.
- Used oil and oil filters must be treated as hazardous waste and removed to appropriate facilities for either recycling or disposal.
- Underground storage tank use and maintenance, etc., must be carried out according to the applicable rules and regulations (See Chapter 14, "Water Quality and Water Pollution").

- Stormwater should not be permitted to cross land on which washing, fueling, painting, changing oil, or any work done on vehicles has occurred to prevent contamination of the ground or surface water supply.
- Wastewater from any automobile repair activities must be separately collected and treated before being released to surface or groundwater areas.

Drycleaners

Drycleaners are the largest source of perchloroethylene emissions in the United States. Because of their locations and closeness to large numbers of people, they create a health hazard and may cause dizziness, nausea, headaches, and possibly cancer in people.

Best Practices for Dry Cleaners
- Use monitoring equipment to efficiently determine if there are any leaks or loss of solvent.
- Check the hoses, couplings, pumps, valves, and gaskets to detect leaks on a regular basis.
- Keep all containers covered to prevent evaporation and spillage.
- Keep lint screens clean.
- Extract the solvents from filters.
- Use water-based solvents when possible.
- Upgrade the equipment to use less hazardous solvents more efficiently.

Gas Stations and Gasoline Distribution Facilities

Leakage may occur from pipelines, bulk terminals, storage areas, transferring areas, tank trucks, rail cars, equipment, etc. Below the surface of gas stations, there are hundreds of thousands of underground gasoline storage tanks, many of them old, damaged, rusty, and poorly maintained, and all of them potentially can contaminate ground and surface water sources, and cause land and air pollution. Gas stations may have faulty, malfunctioning, or inadequately protected nozzles for dispensing gas. Other problems include gasoline spills, diesel spills, antifreeze and motor oil spills, etc. Ten toxic air pollutants including benzene and toluene are present in the gasoline vapor.

Best Practices for Gas Stations and Gasoline Distribution Facilities
- Use best available seals on storage tanks and pipeline facilities.
- Use proper fill pipes in the transfer of gasoline.
- Use vapor processors in all facilities where appropriate.
- Frequently test all parts of the distribution and storage system for leaks and make necessary corrections.
- Use appropriate housekeeping procedures to correct leaks and spills.
- Install new nozzles on gas dispensing equipment when needed.
- Use dry clean-up methods for spills of any kind and remove the material for hazardous waste disposition.
- Divert all wastewater of the gas station to a special place for decontamination.

Hospital Medical and Infectious Waste Incinerators

This solid waste is produced during diagnosis, treatment, immunization, or research on humans or animals and includes needles, gauze, boxes, packaging materials, dressings, and containers of various types. The material comes from hospitals, nursing homes, pharmaceutical research laboratories, veterinary clinics, and other medical and surgical facilities. The toxic air pollutants produced during the burning of this material include dioxins, mercury, lead, cadmium, etc. (See Chapter 12, "Solid Waste, Hazardous Materials, and Hazardous Waste Management.")

Lawnmowers and Other Garden Equipment

Lawnmowers and other garden equipment create about 5% of the air pollutants in the United States. This amount increases in urban and suburban areas. This equipment may emit carbon monoxide, carbon dioxide, particulate matter, sulfur dioxide, VOCs, and nitrogen oxides. Ozone may be formed in the air. Gardeners while filling machines spill a considerable amount of the fuel which then becomes an air pollutant hazard, a land pollutant hazard, and a water pollutant hazard.

Also debris is produced and has to be removed to landfills or to special plants for production of biogas or spread out on the grounds as mulch. Safety is a major concern since thousands of people go to hospitals for lawn equipment-related accidents and deaths also occur.

Best Practices for Use of Lawnmowers and Other Garden Equipment
- Use metal gas cans with spouts or funnels and pour slowly and smoothly while avoiding overfilling of the equipment or the gas can when filling it up.
- Tightly close the spout and vent hole after filling up the gas can.
- Transport and store gas cans and equipment away from direct sunlight and in cool places.
- Do not tip the equipment on the side for cleaning or maintenance purposes to avoid spills of gasoline or oil.
- Do not operate equipment during ozone alert days.
- Buy and use equipment approved by the EPA for meeting national emission standards.
- Keep all equipment in excellent working order through proper maintenance and repair.
- Use electrically powered equipment where possible instead of gasoline powered equipment. (See endnote 79.)

Off-Site Waste Operations
Off-site waste facilities are used for the treatment, storage, and disposal of hazardous wastes which come from various industries and other facilities. These facilities include industrial wastewater treatment facilities, solvent recycling facilities, used oil recovery facilities, etc. Numerous air toxics can be released from the tanks, process vents, equipment, containers, and surface impoundments.

Best Practices for Off-Site Hazardous Waste Operations (See Chapter 12, "Solid Waste, Hazardous Materials, and Hazardous Waste Management")

Printing and Publishing
This includes not only printers but also those facilities that produce paper products and packaging materials. VOC emissions are released from the cleaning solvents, inks, and wetting agents. The larger plants can be a source of nitrogen oxides and sulfur dioxide. The toxic air pollutants include toluene, xylene, methanol, and hexane.

Best Practices for Printing and Publishing
- Plan the job to only use the appropriate amount of paper and other supplies, correct inks, and best machinery and equipment.
- Inspect all materials received to determine if they meet the level of quality and consistency needed for projects and reject those that do not.
- Use less hazardous solvents.
- Inspect and maintain all equipment on a regular basis.
- Recycle paper and inks where feasible.
- Replace where feasible petroleum-based inks with vegetable oil-based inks or water-based inks.
- Use digital processes where possible.

Shipyards
Ship manufacturing and the maintenance and repair of commercial or military vessels is a major industry in certain parts of the country such as the Pacific Northwest and Norfolk, Virginia. Cleaning, stripping of old paint, and painting produce pollutants which contaminate air, water, and land. Coatings which are used on the hulls of ships contain heavy metals. Preparing the hull using

a blasting dry abrasive grit prior to painting increases the particulate matter in the air, as well as the pollutants found in paint and other substances. The painting process releases a variety of VOCs. The toxic air pollutants released include xylene, toluene, chlorinated compounds, solvents, etc. from the painting and cleaning operations. Also, products aboard the ships may release vapors while in storage or when being transferred to other storage or transportation facilities.

Best Practices in Shipyards
- Replace sandblasting with high-pressure water blasting and blasting with wetted grit chemically treated to bind heavy metals.
- Use Teflon or silicone on the hull to prevent the growth of marine organisms.
- Use advanced vapor control system technologies to capture vapors from various products stored on the ships or being transferred from the ships.

Small Chemical Plants
These produce a variety of solid, liquid, and airborne waste. Further chemicals that escape accidentally into the air increase the cost of the product as well as pose a threat to human health and the environment.

Best Practices for Small Chemical Plants
- Establish a waste reduction program.
- Conduct a waste survey identifying: the composition and source of the waste being generated and the potential for air pollution; the unintended escape of chemicals from valves and other places along the process, storage, and disposal lines.
- Keep all containers covered to prevent evaporation, spillage, or drying out of the material.
- Reuse all materials that can be reused.
- Separate all hazardous material from non-hazardous material.
- Improve scheduling to make sure that the final product is delivered rapidly, thereby avoiding potential air pollutants.

Surface Coating and Painting of Materials
Painting and coating is used to preserve and protect surfaces and products and for decoration of various materials including metal, wood, and plastics. Glue and adhesives are typically used in many of these operations to prepare the texture of the underlying surface. Frequently solvents are used to prepare the surfaces by cleaning them and removing unwanted previous material. Cleaning, painting, and coating can produce toxic air pollutants and VOCs.

Best Practices for Painting and Coating of Materials
- Use cleaners that have low toxicity and low VOC content for surface coating and painting of materials.
- Use coating techniques that do not require spraying.
- Train workers how to use minimal amounts of cleaning materials, paints, and varnishes.
- Where spraying is necessary, do so in a special booth with proper air systems and filters.

Wood Furniture Manufacturing
Wood furniture manufacturing facilities include cabinet shops and plants that make residential and industrial furniture. Toxic air pollutants including toluene, xylene, methanol, and formaldehyde are released during the finishing, gluing, and cleaning operations. These chemicals can cause eye, nose, throat, and skin irritation, and damage to the heart, liver, and kidneys.

Best Practices for Wood Furniture Manufacturing (See Guide Sheets for the Wood Furniture Manufacturing Industry, endnote 74.)

SUB-PROBLEMS INCLUDING LEADING TO IMPAIRMENT AND BEST PRACTICES FOR GREENHOUSE GASES

(See endnote 15)

Greenhouse gases are gases that trap heat in the atmosphere, thereby not letting the heat flow back into space and warming the Earth's surface. They include carbon dioxide (84% of the total), methane (9% of the total), nitrous oxide (5% of the total), and fluorinated gases (2% of the total).

Carbon dioxide is produced during the burning of coal, natural gas, oil, solid waste, trees, and wood products, and through certain chemical reactions such as during the production of cement. Carbon dioxide is sequestered or removed from the atmosphere when it is absorbed by plants and trees. (The destruction of forests greatly affects the amount of carbon dioxide in the air.)

Methane is found during the mining, processing, and transportation of coal, natural gas, and oil. It is also emitted by cattle and by organic waste decaying in municipal solid waste landfills. (The largest methane emitters are China, India, the United States, Brazil, Russia, Mexico, Ukraine, and Australia, which account for about half of all of the anthropogenic methane emissions in the world. The largest source of methane emissions in the United States is landfills.) The global warming potential of methane is very large, since it is 25 times greater than carbon dioxide. Methane contributes more than one third of today's global warming caused by people.

Nitrous oxide is produced by agricultural activities, industrial activities, and the burning of fossil fuels and solid waste.

Fluorinated gases, regulated by the Montréal Protocol and the Kyoto Protocol, come from industrial processes either as a final product or as a byproduct. Although they are released in small quantities, they have a very potent effect on global warming. Since chlorofluorocarbons deplete the ozone layer in the atmosphere, they had been replaced with hydrofluorocarbons, which are strong greenhouse gases. This is a prime example of trying to correct one environmental problem while causing another one. Sulfur hexafluoride is another potent greenhouse gas which is used primarily to insulate electrical equipment, since it is non-combustible and non-flammable. Leakage of the gas as well as emissions during equipment maintenance exacerbate warming trends in the area where this material is used.

Each of the greenhouse gases affect climate change depending on their concentration in the air, how long they stay in the atmosphere, and how effective they are in producing global warming. Some gases can remain in the atmosphere for many years. It should be recognized that even after there has been a substantial reduction in greenhouse gases, that the full effect of the emissions will be felt for many years into the future because of lag time. Water heats more slowly than the atmosphere and the greenhouse gases that have been trapped in the ocean will be released slowly over time back to the atmosphere.

Although ozone depletion is not a major cause of climate change, it still has some effect on the temperature balance of the earth. Atmospheric ozone absorbs solar ultraviolet radiation which heats the stratosphere. It also absorbs infrared radiation from the surface of the earth, which traps heat in the troposphere. Whereas ozone losses in the lower stratosphere from chlorine and bromine-containing gases cool the Earth's surface, surface pollution gases rising in to the troposphere can heat the Earth's surface thereby contributing to the greenhouse effect. (See endnote 80.)

General Best Practices for Land and Forests (See endnote 24)
- Increase or maintain carbon storage by transforming cropland into forest areas.
- Avoid destroying forest land.
- Reduce soil erosion in forests and farmlands to minimize soil losses which affects soil carbon storage.
- Plant appropriate vegetation that will grow rapidly on land where the soil has been disturbed and soil erosion has occurred.
- Encourage the growth of green areas in urban and suburban situations.

SOURCES OF GREENHOUSE GASES

(See endnotes 23, 81)

Electricity Production and Transmission

Energy-related activities such as the production, transmission, and distribution of electricity accounts for over 84% of the greenhouse gas emissions in the United States. Most of the greenhouse gases are carbon dioxide, but small amounts of methane and nitrous oxide are also produced, while less than 1% of the emissions is sulfur hexafluoride which is an insulating chemical used in transmission of electricity and distribution equipment. About 70% of the electricity is produced from the burning of coal and natural gas, both of which produce carbon dioxide in substantial quantities.

Best Practices for Electricity Production
- Increase the efficiency of existing power plants by using advanced technologies or substitute fuels that combust more efficiently. An example would be to convert a coal powered-turbine into a natural gas-powered turbine.
- Reduce the energy demand in residences, businesses, and industry. An example would be the use of ENERGY STAR products.
- Insulate all structures to gain maximum efficiency of power use.
- Capture carbon dioxide from the stacks of coal-fired power plants and then neutralize it.
- Use renewable energy sources instead of fossil fuel to generate electricity. An example would be to increase the total electricity generated by wind, solar, hydro, and geothermal sources.

Mobile

Motor vehicles account for about 28% of all greenhouse gas emissions from burning fossil fuels in on-road and off-road means of transportation.

Best Practices for Gas Emissions from Mobile Sources
- Fuel switching involves the use of fuels that emit less carbon dioxide such as biofuels, hydrogen, and electricity produced by renewable sources. An example of this would be to use buses that utilize compressed natural gas or electric or hybrid automobiles where the energy comes from renewable or low carbon fuels.
- Use advanced technologies, design, and materials to make more fuel-efficient vehicles. An example of this would be to use hybrid vehicles and to reduce the weight of the materials and reduce aerodynamic resistance. Both the use of alternative fuel and the changes in construction and design allow the vehicle to travel further on considerably less gasoline which produces the greenhouse gases.
- Minimize fuel use by reducing the taxiing time of aircraft, and engine idling in buses and trucks, and better maintenance. Also the purchasing of locally grown items of food creates significant fuel savings for transportation sources and enhances the local community while reducing greenhouse gases.

Industries

Industries account for about 14% of greenhouse gas emissions, primarily through the burning of fossil fuels. There are two types of emissions: direct, which are produced by burning fuel and chemical reactions, and indirect from leaks from equipment. (See endnote 81.)

Best Practices for Industry
- Identify the techniques which manufacturers can use to reduce energy for lighting and heating factories and for running equipment, and utilize this information to make necessary changes.

- Use natural gas rather than coal as a fuel to produce the energy needed to operate machinery more effectively.
- Install appropriate recovery equipment in industries to capture and recycle gas coming from tanks, pipelines, processing plants, equipment, etc.
- Monitor fugitive emissions through inspections of valves, connectors, open lines, and flares.
- Recycle metals to avoid using electricity to produce new metals.
- Recycle glass, paper, and plastics to reduce energy use.
- Provide adequate training to employees on how to prevent emission leaks from equipment and how best to handle and control various fluoride compounds.

Commercial and Residential

These sources account for about 11% of greenhouse gas emissions. The combustion of gas and petroleum products for heating and cooking produces carbon dioxide, methane, and nitrous oxide. Commercial and residential building sectors account for 39% of the carbon dioxide produced from fuel sources. The organic waste which goes to landfills from homes and facilities emits methane gas. The waste which goes to wastewater treatment plants emits methane and nitrous oxide. Fluorinated gases from air conditioning and refrigeration systems as well as some household products escape into the atmosphere.

Best Practices for Residential and Commercial Activities
- Reduce energy use through energy efficiency by retrofitting homes and commercial buildings by use of: better insulation, better windows and doors; more efficient heating, cooling, ventilation, and refrigeration systems; more efficient lighting and bulbs; more efficient electronic devices and appliances.
- Make the water and wastewater treatment systems more energy efficient.
- Reduce the amount of solid waste going to landfills by using recycling and solid waste compacting.
- Recycle organic material including food waste from homes and businesses and use composting as a means of reduction of quantity and providing material that can be reused in gardens and other green areas.
- Capture and use methane produced in landfills. (See Chapter 12, "Solid Waste, Hazardous Materials, and Hazardous Waste Management.")
- Use refrigerants that have less potential to create greenhouse gases.

Agriculture

Different agricultural producers account for approximately 8% of greenhouse gases, which come primarily from livestock, agricultural soils, and the production of rice. The management of agricultural soils is responsible for about 50% of the nitrous oxide emissions. Natural soil structure is disturbed by tillage, animals, weather, and mobile sources. The soil is broken into smaller pieces which can then cause soil erosion and the smaller pieces also may become airborne. Livestock always produce methane as part of their digestive process. This is about 33% of the methane produced by agriculture. Manure storage produces methane and nitrous oxide and accounts for 15% of the total greenhouse gas emissions from agriculture in the United States. Rice cultivation produces methane, and burning crop residue produces methane and nitrous oxide. Two weather-related events have the potential to increase the concentration of PM_{10}. They are high winds and stagnation of an air mass.

Best Practices for Agriculture (See endnote 25)

Note: Not all Best Practices will work equally well in all agricultural settings and for all agricultural categories such as tillage and harvest, non-cropland, and cropland. There are variations in wind, soils, moisture conditions, and management approaches.

- Discontinue all night tilling between 2 AM and 8 AM when stagnant air conditions are most prevalent.
- Apply the fertilizer, pesticides, or other agricultural chemicals through an irrigation system.
- Modify the equipment to prevent or reduce the amount of particulate matter.
- Harvest a forage crop (alfalfa, winter forage, silage corn) without allowing it to dry, which decreases the use of equipment that stirs up dust.
- Monitor crops for pests and incorporate biological processes to reduce spraying of chemicals.
- When the wind speed is 25 mph or more limit all tillage and harvesting.
- Use multiyear crops where possible.
- Plant specific crops that will accommodate to the amount of soil moisture present in the land being used.
- Reduce tillage to a minimum, and time the work properly.
- Use genetically modified crops that can reduce tillage and soil disturbance.
- Restrict or eliminate public access to non-croplands.
- Erect artificial wind barriers.
- Use the proper trees, shrubs, vines, grasses, or other vegetative cover on non-cropland areas.
- Reduce the vehicle speed of all farm vehicles to 20 mph or under.
- Use synthetic particulate suppressants such as calcium chloride, processing byproducts of soybean feedstock, lignin, polyvinyl acrylic polymer emulsion, etc.
- Plan strips of alternating crops or vegetative strips within the same field.
- Use crop or other plant residues on the soil surface to control erosion.

INDOOR AIR QUALITY

(See endnotes 75, 76)

STATEMENT OF PROBLEM AND SPECIAL INFORMATION FOR INDOOR AIR QUALITY

Indoor air frequently is more contaminated than the ambient air which entered the property originally from the outside. Additional pollutants are generated within the premises through many different sources. Indoor air has been contaminated as long as people have lived in enclosed areas, from the cave dwellers who breathed in smoke from cooking and heating fires, to the enclosed environments of the Industrial Revolution where products of combustion were present and a variety of chemicals were being used, to modern days where the most recent and most complex problems of contaminated indoor air were an unintended consequence of energy conservation, when structures were sealed to become airtight as a response to the energy crisis of the 1970s.

Indoor air quality in homes, offices, and other structures is affected by the accumulative effects of: air pollutants found in the ambient air and also those generated inside the structures including dampness, moisture, and flooding; biological agents, insects, and rodents; chemicals including solvents used for cleaning, air fresheners, pest control, cosmetics, hobbies, and those found in insulation, building materials, new carpeting, pressed wood products, etc.; waste materials created by renovation projects; cigarette smoke; particulate matter and other pollutants from malfunctioning and improperly vented fireplaces, stoves, furnaces, and space heaters; radon gas; improper and poorly maintained building ventilation, heating, cooling, and dehumidification systems; the accumulations of poor housekeeping which may occur in the inside of the structure; and the sealing of cracks and crevices in facilities. The amount of air flowing in and out of the structure may contribute to the level and concentration of pollutants in the structure.

HEALTH EFFECTS

People spend about 90% of their time indoors and therefore their exposure to indoor air pollutants is considerable. The effects of the indoor air pollutants may be immediate with eye, nose and throat irritation, headaches, dizziness, and fatigue, or it may be long term with increased symptoms of asthma, pneumonitis (lung inflammation), various types of sensitivities to biological and chemical substances, cancer or damage to various organs. The most vulnerable people are those with chronic illnesses and the elderly, especially those already suffering from respiratory or cardiovascular diseases. A separate discussion of indoor air pollution problems concerning children including mold, may be found in Chapter 4, "Children's Environmental Health Issues."

Indoor air pollutants add to the problems of short- and long-term health issues including allergic reactions, congestion, eye and skin irritation, fatigue, dizziness, and nausea. The pollutants can reduce the individual's ability to perform certain mental tasks. The pollutants will increase the speed of building deterioration and the decay of the contents.

People may become ill from diseases such as humidifier fever and sick building syndrome. Humidifier fever is caused by toxins from microorganisms that grow in ventilation systems in large buildings, home heating and cooling systems, and humidifiers. Sick building syndrome is a situation where occupants of the building experience acute health effects that seem to be linked to being within the building. The individuals complain of acute discomfort, headaches, eye, nose and throat irritation, dry coughs, dry skin, dizziness, nausea and fatigue, and being sensitive to odors. The causes of the symptoms are not known, but individuals seem to be suffering in increasing numbers as buildings have become airtight to be more energy efficient.

General Best Practices for Indoor Air Quality (also see Chapter 4 on "Children's Environmental Health Issues")
- Determine the source and nature of the indoor air pollutant and use source control procedures such as reduction of the emissions, and sealing or enclosing the area or equipment.
- Improve ventilation by opening windows and doors, using attic fans and local exhaust for contaminated areas, and vent these areas directly to the outside while frequently changing filters in the air systems.
- Use short-term immediate ventilation techniques when painting, welding, sanding, cleaning with solvents, etc.
- Use air-cleaning devices such as air cleaners with high-efficiency particulate air (HEPA) filters and clean them or change them frequently.
- Request help from local and state health authorities if symptoms of disease are present among the individuals within the enclosed environment.

SUB-PROBLEMS INCLUDING LEADING TO IMPAIRMENT AND BEST PRACTICES FOR INDOOR AIR QUALITY

The most prominent indoor air pollutants will be discussed.
1. Asbestos is a mineral fiber which when inhaled can increase the risk of disease. Asbestos may be found in the home and in some: roofing and siding shingles, insulation in houses built between 1930 and 1950, attic and wall insulation, textured paint and patching compounds used before 1977, artificial ashes and embers in gas-fired fireplaces, older stove-top pads, walls and floors around wood-burning stoves, vinyl floor tiles, around steam pipes or hot water systems in older houses, and older oil and coal furnaces and door gaskets. There is a long-term risk of chest and abdominal cancers and lung diseases. Smoking increases the risk of the diseases substantially. During removal of the asbestos products by cutting, sanding, etc., the fibers may become airborne and become a health hazard. Deterioration of the product may also create a problem. The very small fibers are inhaled most frequently and may accumulate in the lungs causing over time lung diseases and potentially chest and abdominal cancers.

Best Practices for Asbestos
- Do not cut, rip, or sand asbestos-containing materials.
- Utilize a professional company with trained contractors to remove any asbestos from the premises.
- Use appropriate personal protective equipment when working with asbestos.
- Isolate the work area for the removal of asbestos from the rest of the structure.
- Dispose of asbestos in an appropriate manner.

2. Biological Contaminants include bacteria, viruses, mold, mildew, animal dander, cat saliva, house dust mites, pollen, droppings from roaches, and droppings and urine from mice, rats, and domestic animals. Bacteria and viruses may directly cause disease from inhalation, ingestion, and skin contact. All the biological contaminants may help cause allergic reactions including hypersensitivity pneumonitis and allergic rhinitis, and may have a profound effect on asthma, which is the leading cause of absence from schools, lost workdays, emergency room visits, and hospitalizations. An especially high rate of asthma associated with allergens occurs in low-income, inner-city children. Allergens affecting humans include dust mites, dust, pollen, pet dander, insect, and rodent contaminant proteins found in urine, dander, saliva, and roach body parts and feces, and mold and mildew, which may release disease-causing toxins.

- Dust mites are microscopic eight-legged arthropods. They are non-parasitic and do not bite. They feed on shed human skin, pollen, fungi, bacteria, and animal dander. Their feces and body parts when inhaled cause sneezing, itchy, watery eyes, runny noses, and other respiratory symptoms as well as asthma.
- Dust is an accumulation of dust mite feces, cigarette ash, fibers from clothing and paper, animal and human hair, insect fragments, human skin, animal dander, soil, fungus spores, and debris from a variety of surfaces.
- Mold and mildew are fungi, which are found everywhere within the environment. The spores enter houses and grow at the temperatures most suitable for human life. Considerable moisture is necessary for growth, either 70% humidity or greater, where there has been recent water damage, or in bathrooms or under leaking sinks. Certain molds produce mycotoxins which cause adverse health effects including inflammation and injury to gastrointestinal and pulmonary tissues.
- Pollen is a fine powder-like material consisting of grains that are produced by seed plants and then carried by wind or insects prior to fertilization. Pollen enters the house through tiny cracks and through open doors and windows.
- Pet dander, an allergen, is composed of tiny flakes of dead skin, resembling dandruff, constantly sloughing off pets. It readily combines with oil and dirt to form a glue-like substance. Breathing in pet dander leads to allergic reactions in some people, causing inflammation, coughing, watery eyes, and runny nose. The human immune system function is operating as it should, in discovering foreign substances such as allergens, viruses, and bacteria and then removing them.

Best Practices for Biological Contaminants
- Use exhaust fans to vent kitchens, bathrooms, and clothing dryers to the out of doors to reduce levels of moisture and organic pollutants.
- Use exhaust fans to vent garages, hobby shops and paint and volatile chemical storage areas to the outside to reduce or eliminate chemical contaminants from the structure.
- Ventilate the attic and crawlspaces to help prevent or remove moisture.
- Empty water trays in air conditioners, dehumidifiers, and refrigerators on a frequent and regular basis.
- Thoroughly clean and provide fresh water daily for humidifiers according to the manufacturer's instructions.

- Thoroughly clean and dry water-damaged carpets or replace rapidly.
- Thoroughly clean the interior of homes and other structures on a regular basis.
- Clean and disinfect all drains in the house or structure, especially the basement, on a regular basis.

3. Carbon Monoxide is an odorless, colorless, and toxic gas which is frequently produced in homes where there are unvented kerosene and gas space heaters, leaking chimneys and furnaces, gas water heaters in need of service, wood stoves, fireplaces, gas stoves, gasoline generators, poorly vented attached garages and tobacco smoke. Low-level exposure may cause flu-like symptoms including headaches, dizziness, disorientation, nausea, and fatigue. High-level exposure may cause unconsciousness and death. All homes may be at risk. (Also see Carbon Monoxide above under Outdoor Air Pollutants.)

Best Practices for Carbon Monoxide
- Properly adjust all gas appliances.
- Use only vented space heaters.
- Use proper fuel for kerosene space heaters.
- Vent gas stoves to the outside.
- Install carbon monoxide detectors especially in the home.
- Have a trained professional inspect, clean, and tune up central heating systems yearly.
- Do not idle the automobile inside a garage.

4. Carpets: installation of new carpets may cause the release of chemicals into the air. The cleaning of older carpets may also release chemicals into the air. Either of these conditions can enhance existing symptoms of disease or create new symptoms. This is especially true for individuals with asthma, existing lung diseases, and allergic responses.

Best Practices for Carpets
- Unroll and air out carpet in a well-ventilated area before installing it in the facility.
- If adhesives are necessary, use those which emit the least amount of volatile material.
- Schedule installation of new carpeting when the individuals are off the premises and ventilate the area thoroughly before using the facility.
- Thoroughly ventilate the area after carpets are cleaned.

5. Environmental Tobacco Smoke, or secondhand smoke, a known carcinogen, is a combination of the smoke given off from the end of the burning cigarette, pipe, or cigar and the smoke which is exhaled by smokers. Secondhand smoke contains more than 4000 different substances, including at least 50 which can cause cancer. High doses of secondhand smoke exposure can increase the risk in children of asthma, sudden infant death syndrome, eye, nose, and throat irritation, lower respiratory tract infections, and middle ear infections. It is estimated that 11% of all children aged 6 years and under are exposed to environmental tobacco smoke on a regular basis of at least 4 or more days per week. Non-smoking adults can also be affected by the secondhand smoke which may result in an increase in disease, such as respiratory diseases, cardiovascular system diseases, cancer, and death.

Best Practices for Environmental Tobacco Smoke
- Stop all smoking in homes and facilities.
- Thoroughly clean all surfaces and ventilate as well as possible.

6. Formaldehyde is a colorless, pungent smelling gas that affects the eyes, nose, and throat and causes difficulty in breathing, asthma attacks, cancer, and death. It is found in new mobile homes, pressed wood products such as particle board and fiberboard, textiles including permanent press clothing, glue, furniture, insulation, household chemicals, preservatives, and combustion products including environmental tobacco smoke. In homes, the most common problem is pressed wood products used in subflooring, shelving, paneling, furniture, cabinets, plywood, and any other substances that contains resins with urea formaldehyde. The boards

produced for outdoor use usually contain phenol formaldehyde, which does not release as much of the formaldehyde into the air but is still a health problem. When the products are new and there are high indoor temperatures or humidity, there is a greater release of the formaldehyde than when the products get older. Formaldehyde is also a byproduct of combustion, especially from unvented fuel-burning appliances such as gas stoves or kerosene space heaters. Formaldehyde exposure can cause watery eyes, burning eyes and throat, nausea, and difficulty in breathing, and may cause asthma attacks. It may also potentially cause cancer.

Best Practices for Formaldehyde
- Where possible avoid the use of pressed wood products and other goods which may emit formaldehyde.
- Potentially coating pressed wood products with polyurethane may reduce formaldehyde emissions.
- Increase the ventilation of the area.
- Keep temperatures moderate and humidity levels low.
- Increase the rate of ventilation in the home.

7. Household Products usually contain chemicals that can be released to the air when stored, used, and discarded. These products may include paints, varnishes, candles, solvents, cleaning materials, disinfectants, aerosol sprays, air fresheners, stored fuels, dry cleaned clothing, hobby materials, etc. Elevated levels of the chemicals may persist in the air for extended periods of time and may be re-released from surfaces on which the chemicals have been deposited. Depending on the chemicals, there may be no effect clear up to dangerous effects. Typical acute symptoms include eye irritation, respiratory irritation, headaches, dizziness, visual problems, lightheadedness, coma, and even death. Long-term problems may be cancer and genetic effects. (Also see Volatile Organic Compounds.)

Best Practices for Household Products
- Follow the instructions on the labels for all household chemicals.
- Keep away from children and pets to prevent poisoning.
- Keep all household chemicals stored in a well-ventilated area.
- Discard all partially used containers of household chemicals and all old household chemicals in an appropriate manner on the special days when they are collected for proper hazardous waste disposal.
- Purchase limited quantities of paints, paint strippers, kerosene, or gasoline to be used for household chores.
- Reduce usage or eliminate usage of methylene chloride, since it may cause cancer in animals or be converted to carbon monoxide in the body.
- Reduce or eliminate exposure to benzene, which may be found in environmental tobacco smoke, stored fuels, paints, and auto emissions in garages.
- Air out all dry cleaned materials before bringing them into the home environment.

8. Lead may irreversibly affect brain and nervous system development and is found in approximately the 250,000 children in the United States aged 1–5 years who have blood lead levels greater than 10 µg/dL of blood. At high levels, the lead causes convulsions, coma, and death. At low levels, it affects the brain, central nervous system, blood cells, and kidneys.

The most common source of lead exposure for children is deteriorating lead-based paint from old houses typically constructed before 1978, and contaminated soil found in the immediate environment which has been exposed to past emissions of lead in gasoline, as well as discharges from industrial processes. Older structures may also contain leaded water pipes or lead used to solder various drinking water fixtures or pipes. It is estimated that 24 million homes may still potentially contain lead paint. Children may also ingest lead from toys and jewelry as well as eating utensils and cosmetics. (Also see Lead under Outdoor Air Pollutants above.)

Best Practices for Lead
- Clean all surfaces thoroughly where children can chew, touch, or lick the surface.
- Wash all toys and stuffed animals regularly.
- Make sure the children always wash their hands thoroughly and frequently.
- Where possible do not have children in structures which may have been painted with lead-based paints.
- Do not sand or burn off paint from older structures.
- Do not burn painted wood.
- If potentially lead-painted surfaces need to be scraped or sanded, have this done by professional contractors in an appropriate manner.
- Change all clothing before entering the home if the individual is involved in construction, demolition, painting, working with lead batteries, or any other situation where lead may have been used in the industrial process.
- Check the drinking water supply for lead.
- Make sure that the child gets adequate quantities of iron and calcium to reduce lead absorption.

9. Nitrogen Dioxide is a toxic gas, corrosive and a highly reactive oxidant. It is typically produced indoors from unvented combustion appliances such as gas stoves, kerosene heaters, and tobacco smoke. It is an irritant to the mucous membranes of the eyes, nose, throat, and respiratory tract. At high levels, it can cause pulmonary edema and lung injury.

Best Practices for Nitrogen Dioxide (See this topic in Outdoor Air Pollution above)

10. Pesticide and disinfectant residues are found indoors on surfaces and in the air. An example would be the use of Baygon in cracks and crevices for controlling bedbug infestations. (This pesticide is not approved by the US EPA for indoor use.) Although some of the pesticides are biodegradable, there is substantial persistence of the chemical on surfaces and toys for several days after application. Approximately 80% of human exposure to pesticides and disinfectants occurs indoors because of the chemicals that are used inside properties to control insects, rodents, fungi, bacteria, and viruses. Soil or dust contaminated with pesticides, agricultural chemicals, and chemicals deposited from air pollution also float into the property. These chemicals cause irritation to the eye, nose, and throat, damage to the central nervous system, liver and kidneys, and may increase the risk of cancer. Many thousands of children are poisoned by household pesticides, cleaning materials or disinfectants. The inert materials in pesticides may also be toxic to the child or older person. Cyclodiene pesticides, such as aldrin, endrin, and isodrin, in high concentrations cause headaches, dizziness, weakness, nausea, and potentially long-term damage to the liver and central nervous system.

Best Practices for Pesticides and Disinfectant Residues
- Never use a restricted pesticide in the home or outside.
- Follow instructions on the label when applying pesticides or using disinfectants inside structures or pesticides in outside areas.
- Use pesticides and disinfectants only in recommended quantities.
- Ventilate the area very well after pesticide use.
- Mix or dilute pesticides out of doors and in well-ventilated areas.
- Keep pesticides and disinfectants away from pets and children.
- Avoid using pesticides and use other approaches where appropriate.
- Choose a licensed pest control company when needed and check to see if they have the proper credentials and training.
- Dispose of unwanted pesticides in an appropriate manner as a household hazardous waste.
- Use moth repellents very sparingly and only when absolutely necessary.

- Be careful of the use of indoor air fresheners and ventilate the area immediately.
- Clean all surfaces after the use of disinfectants and other chemicals.

11. Radon is a tasteless, odorless, colorless, and invisible radioactive gas that comes from the natural decay of uranium found in the ground, in the soil, and rocks, and enters the home through cracks and crevices in the foundations, construction joints, walls, and floors. Radon may be found in the water supply. Concentrations of radon may build up within the structure.

The primary health effect is lung cancer. Cigarette smoking enhances the effects of the radon gas. It is believed that many thousands of deaths each year from lung cancer are caused by radon gas.

Best Practices for Radon Gas
- Use a state-certified test kit or a trained contractor to determine radon levels within your home and compare them to EPA guidelines.
- Read and use the EPA's *Consumer's Guide to Radon Reduction.*
- Make necessary corrections in the structure if the radon level exceeds 4 pCi/L.
- Use a certified contractor to make the necessary corrections, if it is beyond your capabilities to do so.
- Properly drain and seal foundations in new construction.
- Use mechanically assisted ventilation as the primary means of reducing levels of radon within a structure.
- Stop all smoking within the home or facility, clean thoroughly, and air out the entire area.
- Treat radon-contaminated well water.

12. Stoves, Heaters, Fireplaces, and Chimneys. Respirable particles and gases such as sulfur dioxide, nitrogen dioxide, and carbon monoxide are produced by sources of combustion including stoves, heaters, fireplaces, chimneys, and environmental tobacco smoke. They may cause eye, nose, and throat irritation, respiratory infections, bronchitis, dizziness, headaches, confusion, fatigue, irregular heartbeats, persistent coughs, and lung cancer. (See the section on "Outdoor Air" above)

Best Practices for Stoves, Heaters, Fireplaces, and Chimneys
- Install and use exhaust fans in hoods over gas burners and keep them properly adjusted.
- Ensure that there is a proper supply of air for combustion for all appliances.
- Never use a gas stove to heat the house.
- Make sure the flue is always open on a gas fireplace.
- Use the proper size wood-burning stove certified by the US EPA.
- Inspect at least yearly the central air handling system, furnaces, flues, and chimneys, and make necessary corrections and repairs immediately.
- Frequently check and replace air filters.

13. Volatile organic compounds are released from many substances used indoors including paints, varnishes, waxes, pesticides, cleaning supplies, clothing from the cleaner, air fresheners, glues and adhesives, printers, cosmetics, etc. They are also released from new carpets, new furniture, new walls, new floors, new shelving, etc. VOCs can enter the property from the outside air and also from the ground and water from contamination that surrounds the structure. These compounds are present in the structure at levels which are considerably higher than those found in the outside air. They cause eye, nose and throat irritation, headaches, loss of coordination, nausea, damage to the liver, kidney, and central nervous system, and at high enough levels over a period of time may cause unconsciousness and death. (Also see Household Products above)

Best Practices for Volatile Organic Compounds (See endnote 77)

LAWS, RULES, AND REGULATIONS

(See endnotes 34, 35, 36, 37)

What follows is an extensive discussion of the Clean Air Act and the various amendments, etc. There is also guidance for the readers wishing to expand their knowledge as needed in any special area of this law. The reason for this discussion is because the Clean Air Act in the United States is almost like the US Constitution. All the standards, rules and regulations enhanced by current science are based on the Clean Air Act and Its Amendments. The standards, rules and regulations are a living, evolving approach to the resolution of air pollution problems.

Since air pollution is a national and international problem as well as a state and local problem, the federal government passed laws and/or amendments in 1955, 1963, 1965, 1966, 1967, 1969, 1970, 1977, and 1990 to reduce air pollution and improve air quality to protect human health and the environment (welfare). The 1955 legislation, entitled the Air Pollution Control Act of 1955, emphasized that air pollution was a national problem and had to be dealt with on a national level. It stated that research and additional work was needed to improve the situation. The Clean Air Act of 1963 set emission standards for stationary sources such as power plants and steel mills. It did not recognize the problems of mobile sources. The amendments of 1965–1969 gave the Secretary of the Department of Health, Education, and Welfare the power to set standards for automobile emissions, expand local air pollution control programs, establish air quality control regions, and set air quality standards and compliance deadlines for stationary sources.

In 1970, Congress approved the enhanced Clean Air Act, expanding previous powers and programs and giving the authority and the program operation to the US EPA to administer and where necessary enforce scientifically established rules, regulations, standards, and pollution limits. It required the US EPA to establish NAAQS using the latest scientific information. It required the various states to develop and adopt enforceable plans to achieve the standards. It required states to clean up dirty air, protect the degradation of clean air, and prevent air pollution drift to other states and other areas. It required the EPA to review and approve the state plans. (It gave the states ample time to establish the state planning process program for the non-attainment areas.) New stationary sources had to be built with the latest technology to prevent and control air pollution. Appropriate permits had to be secured before construction of new industrial plants and also for the release of pollutants from older industrial plants.

The Clean Air Act of 1970 also provided authority for reducing:

- Hazardous air pollutants that created a health risk or endangered the environment
- Acid rain or acid deposition that damaged various ecosystems and could harm the health of people
- Chemicals that enter the air and deplete the stratospheric ozone layer
- Haze that impairs visibility
- Pollutants, such as greenhouse gases, that can cause climate change

The EPA determines scientifically if various areas of the country meet or miss the existing NAAQS. (The standards may change over time based on the latest scientific knowledge.) Those areas that meet the standards are called "attainment areas" and those that do not are called "non-attainment areas." An area can be an attainment area for one pollutant but not for one or more other pollutants.

The EPA's New Source Review program requires that a construction permit be obtained by a company that is going to either build a new plant or modify an existing plant so that air pollution emissions will not increase by a substantial amount. The EPA has established three different types of technology which can be utilized for these plants:

- *Reasonably Available Control Technology*—This is required if the plan is in a non-attainment area. It is a control technology that is reasonably available, technologically feasible, and economically feasible.

- *Best Available Control Technology*—This is required in major new or modified plants in attainment areas and limits emissions to the maximum degree of control that can be achieved. Control can be achieved through add-on equipment or process modification.
- *Lowest Achievable Emission Rate*—This is required in major new or modified plants in non-attainment areas and is the most stringent emission limits established in a State Implementation Plan or achieved in actual practice by a given source. (See endnote 64.)

In subsequent years, and with better scientific information and better technology available, the Clean Air Act was amended and expanded to continue to improve air quality in the United States. The EPA was required to set standards which could be met by the best available technology and that would reduce the risk of disease and deterioration of the environment. One of the programs added was the Prevention of Significant Deterioration Program where attainment areas would not be allowed to slip back to a non-attainment area. National standards or guidelines were set up for consumer and commercial products such as solvents, paints, and coatings which made up 80% of the VOCs that were involved in ozone forming at the Earth's surface. National standards were set for new vehicles (mobile sources), engines, and fuels. There was a usage drop in the amount of lead available in the atmosphere because of required reductions in the lead content of leaded gasoline and ultimately the sale of unleaded gasoline. The sulfur content of fuel was reduced substantially and lower volatility fuel had to be sold in the summer in areas where there were high levels of ozone pollution. Oxygenated gasoline had to be used in areas where there was substantial carbon monoxide pollution.

The EPA was also given the authority to regulate the emissions from any oil drilling and production and the associated vessels off the coast of the United States. Emission controls, use of permits, monitoring, testing, and issuing reports became required.

The Clean Air Act regulated hazardous air pollutants on a pollutant-by-pollutant basis based on risk. By 1990, Congress had listed close to 190 hazardous air pollutants but gave the EPA the authority to modify the list and issue "maximum achievable control technology" emission standards by categories, where the standards must reach the level of performance of the average of the top performing 12% of these sources. This allows industry the flexibility to decide the most cost-effective way to comply. The EPA was required to regulate the hazardous air pollutant emissions from electric utilities.

The US EPA has developed National Emission Standards for the sources being discussed in this chapter. Most of these standards have been updated very recently. Many of the Best Practices are a response to the standard.

The Clean Air Act required a national urban air toxics strategy for hazardous air pollutants including regulation of small area sources. The EPA was required to list and regulate enough of these sources to ensure that 90% of the emissions of the 30 pollutants causing the greatest threat to public health in the largest number of urban areas would be controlled. This would help reduce the health risks from mobile sources of air pollution.

The EPA had to issue new performance standards to control hazardous air pollutant emissions from solid waste incinerators and guidelines to satisfy the standards.

For mobile sources, toxic emissions had to be reduced based on the standards. These emissions included formaldehyde and benzene.

Provisions were established in law for the prevention of accidental releases of extremely hazardous air pollutants. The EPA is required to issue regulations for the prevention and detection of accidental releases from stationary sources. Risk management plans are required for these facilities and an independent investigative board was created to determine cause and effect.

A regional haze prevention program was established to protect visibility in park areas. The national goal was to prevent any future haze problems and correct existing visibility problems due to air pollution created by people.

The Clean Air Act insisted on reduction of the potential for producing acid deposition. Controls were established for the substantial reduction of sulfur dioxide emissions, mostly from power plants, on a market-based emissions trading approach. Companies had the flexibility to reduce their emissions to minimize their cost of compliance. Under the trading system, the EPA issued emissions allowances, with each allowance good for 1 ton of sulfur dioxide released to the air. At the end of each year, the company had to hold enough allowances to cover their annual emissions. If the company could not achieve this goal, it could purchase allowances from other companies which had gone past their goal. The means of reduction of the sulfur dioxide emissions was left to the company.

For control of nitrogen oxides from power plants, they were required to install low nitrogen oxide burners. Compliance was based on the average emissions of all their units.

The Clean Air Act protects the stratospheric ozone layer by phasing out the production and release of chemicals that can cause ozone layer problems. The provisions of the Clean Air Act implement the Montréal Protocol, which is an international agreement for protecting the ozone layer. Chlorofluorocarbons have already been phased out and now the EPA is phasing out hydro-chlorofluorocarbons, which were a temporary substitute for the chlorofluorocarbons. Allowances can be traded between companies. The EPA can issue regulations if any substance or activity can affect the ozone layer in the stratosphere.

Emissions of greenhouse gases that can cause or contribute to climate change are also regulated. There are now greenhouse gas regulations for vehicles and stationary sources. Carbon pollution standards are now being established for new power plants. A renewable fuel standards program has been established for gasoline and diesel fuel.

All major pollutant sources as well as others have to acquire operating permits to make sure they comply with the Clean Air Act requirements. The permits are usually issued by state and local agencies under EPA approval. The permit is issued for a period of time of up to 5 years and has to contain enforceable emission standards and limitations.

The Clean Air Act provides for enforcement through administrative or judicial methods of all rules and regulations established by the EPA.

The Clean Air Act allows for the various states to implement their own laws under the supervision of the US EPA. The states can adopt more stringent standards than the federal requirements except in the area of mobile sources. Tribal governments can also implement the Clean Air Act in their areas if they have the capability to do so. The EPA is authorized to pay 60% of the cost of planning, developing, and carrying out air pollution programs at the state level.

The EPA is given broad research authority in all facets of air quality control. This includes: conducting research; providing technical services to states; providing financial assistance; establishing technical advisory committees; and conducting and promoting training for air pollution specialists.

In 2014, the U.S. President Obama and People's Republic of China President Xi Jinping signed an agreement to reduce greenhouse gas emissions. This agreement has enormous consequences because for the first time China has recognized the huge significance of air pollution problems and their role in protecting the global society.

CLEAN AIR ACT AND THE ECONOMY

There is a requirement in the Clean Air Act Amendment of 1990 that the EPA periodically determines the effect of the law on public health, the economy, and the environment of the United States. The broad response is that the Clean Air Act has helped build the economy and create jobs while reducing pollution to protect the health of citizens and the environment.

Retrospective Study, 1970–1990

Following the requirements of the new law, a retrospective study entitled *The Benefits and Costs of the Clean Air Act, 1972 to 1990*, was sent from the EPA to Congress on October 15, 1997.

A brief summary of the results of the study indicated that from 1970 to 1990:

1. Direct costs for compliance with the Clean Air Act by businesses, consumers, and government agencies resulted in higher costs for goods and services because of the requirements to install, operate, and maintain necessary pollution abatement equipment.
2. Sulfur dioxide emissions were 40% lower, nitrogen oxides emissions were 30% lower, VOC emissions were 45% lower, and carbon monoxide emissions were 50% lower.
3. Levels of primary particulate matter which came from various stationary and mobile sources were 75% lower.
4. Emissions of lead were reduced by about 99%.
5. National average emission concentrations were reduced substantially and there was a sharp increase in good air quality.
6. Although ground-level ozone varies considerably from area to area depending on the relative amount of VOCs and nitrogen oxides as well as the weather conditions, there was about a 15% reduction in ozone levels.
7. There was a reduction in the amount of acid deposition depending on the area of the country.
8. There was a reduction in the amount of hazardous air pollutants.
9. There was a substantial improvement in the health of the population because of the reduction of air pollutants.
10. There was an improvement in various portions of the environment.

Despite an estimated cost over the 20-year period of $523 billion measured in the value of the dollar in 1990, the total benefits from a monetary standpoint were over $2 trillion. The Clean Air Act, from a monetary, health, and environmental point of view, to that point, was extremely successful. (See endnote 38.)

Prospective Study, 1990–2020
A brief summary of the proposed effects that will occur from the benefits and costs of the Clean Air Act is given in a report to Congress providing them and the public with information. The important points are as follows:

1. It is expected that the costs to the public and private areas to meet the requirements of the 1990 Clean Air Act Amendments will continue to rise and be about $65 billion annually by 2020.
2. The value to the economy due to the 1990 Clean Air Act for the year 2020 is expected to be about $2 trillion.
3. The value of the Clean Air Act Amendments far exceeds the cost.
4. Healthier living due to the reduction in particles and ozone will result in lower amounts of adult and infant mortality, bronchitis, heart disease, asthma, emergency room visits, lost school days, and lost work days. The Clean Air Act Amendments implementation will also save on medical expenses.
5. The EPA has projected that there will be an additional prevention of 230,000 early deaths in 2020 because of the Clean Air Act Amendments. (See endnotes 39, 40.)

RESOURCES

1. You can find the index of air quality for your area in the United States, on a daily basis, by accessing the California Air Pollution Control Officers Association page at https://www.airnow.gov/index.cfm?action=airnow.fcsummary&stateid=6.

2. Environmental Protection Agency. 2013. *Quality Assurance Handbook for Air Polluiton Measurement Systems*. EPA, Washington DC, can be used to help establish an ambient air quality monitoring network. (See endnote 69.)
3. The Centers for Disease Control has an Air Quality System Database which is very useful for the EPA, state, local and tribal agencies, and the public to have as a source of data on ambient air pollution. The United States has about 4000 monitoring stations, mostly in urban areas. The database can also be used for assessing which areas meet air quality standards, evaluating state air pollution plans, and determining trends of air pollution and health effects.
4. The US EPA has an Ambient Monitoring Technology Information Center as part of the Office of Air Quality Planning and Standards, at Research Triangle Park, North Carolina. Its websites include: Air Quality Analysis; Air Quality Systems; Air Quality Monitoring; Air Toxics; Clean Air Technology Center; Fate, Exposure, and Risk Analysis; National Ambient Air Quality Standards; Air Quality Models; etc. (See endnote 42.)
5. The US EPA has an excellent reference on air pollution measurements systems and ambient air quality monitoring programs. It discusses in very readable form ambient air quality monitoring networks starting with the individual air quality standard, all the necessary steps to follow, the state implementation plan and using air quality measurement on a continuous basis to determine if the standard is being met. It also discusses a series of monitoring systems which are most useful. (See endnote 43.)
6. US EPA—Natural Gas STAR Program. (See Oil and Natural Gas Construction Facilities and Best Practices above) (See endnote 66)

   The Natural Gas STAR Program is a voluntary working relationship between the various parts of the oil and gas industry and the United States government to encourage companies to implement technologies and practices that will reduce methane emissions and to then share the information that they accumulate concerning the process with other companies. This has led to a Global Methane Initiative and partnership which has been the basis of Natural Gas STAR International. The global program helps reduce methane emissions from oil and natural gas industries throughout the world. The program provides an overview of the various issues related to the production and processing of gas, its storage, transmission, and distribution. It also provides an overview of oil production. Because it is highly education oriented, it helps the individual look at recommended technologies and practices and those that have been implemented by the industry partners. The program provides comprehensive guides for implementing methane emission reduction technologies and practices. They are called Lessons Learned Studies. The program also provides Partner Reported Opportunities Fact Sheets, technical presentations such as Technology Transfer Workshops, and the Natural Gas STAR Partner Update newsletter. The newsletter lists the various industry and governmental agencies that are involved in trying to control methane and gives the reader an opportunity to quickly access the information.
7. The European Environment Agency has provided expert guidance in developing national emission inventories for air pollutants by publishing the European Environment Agency. 2013. *EMEP/EEA Air Pollutant Emission Inventory Guidebook 2013*. Copenhagen, Denmark. (See endnote 70.) It provides in Part A general guidance chapters on analysis and methods used, data collection, inventory management, spatial mapping of emissions, and projections. It provides in Part B specific chapters on combustion, fugitive emissions, fuels, industrial processes and product use, mineral products, the chemical industry, metal production, other solvents and product uses, agriculture waste, other sources, and natural sources. The document is Technical Report Number 12/2013 and was published by the European Environment Agency (Copenhagen, Denmark) on August 29, 2013.

8. Contact the US Environmental Protection Agency, National Radon Hotline on 1-800-767-7236 for 24-hour information.
9. Contact the US Environmental Protection Agency, National Lead Information Center on 1-800-424-5323 for 24-hour information.
10. Contact the US Environmental Protection Agency, National Pesticide Information Center on 1-800-858-7378 for questions on pesticides.

SUCCESSFUL PROGRAMS

(See endnote 19 for Greenhouse Gas Program)

GENERAL CONTROL STRATEGIES FOR DEVELOPING PROGRAMS FOR ALL SOURCES OF AIR POLLUTION

A control strategy is a group of specific techniques, when implemented, which will achieve a reduction in air pollutants. The control measure involves identifying abatement requirements, Best Practices and control technologies that will meet the goal of reducing the pollutants.

- Determine the nature of the problem including: the ambient air quality including the type, concentration, and nature of air toxics; the characteristics of the gas stream and control system and whether they could be sources of contamination; and the economic factors such as capital cost, operating cost, and equipment maintenance.
- Determine if substituting less toxic raw materials and using alternative manufacturing processes can reduce the potential air pollutants as well as the cost of the product.
- Determine which controls are necessary for major stationary sources, mobile sources, and area sources to make the program successful, and how to implement them.
- Use all of the control measures available in the program including those required by regulatory programs into the plan and make sure they are enforceable.
- Involve the various groups of stakeholders in the entire planning process to obtain their support for the project.
- Determine what resources are available at the local, state, and federal level to help with the control strategies being proposed.
- Ensure that the media are part of the entire process and will support the various changes needed in order to make the program successful.

Global Methane Initiative

The Global Methane Initiative was started in November, 2004 by 14 governments worldwide to create an international body to reduce the amount of methane going into the air from agriculture, coal mining, landfills, and oil and natural gas exploration and usage. Currently, it is a voluntary effort by the United States government (with the US EPA taking the lead) and 37 other governments as well as the private sector, academia, financial institutions, non-governmental organizations, etc. to share the technology and processes that work in methane reduction, recovery, and usage, and is supporting more than 300 projects worldwide using the resources and expertise of more than 1000 project network members. In the United States, specific programs are:

- *Agriculture: AgSTAR*—This is a voluntary effort sponsored by the US EPA, US Department of Agriculture, and the US Department of Energy. The program pushes the use of methane recovery technologies for concentrated animal feeding and manure management in liquid or semi-liquid form. Currently, there are over 125 such systems in the United States, which are working with the World Bank to help other countries develop such technologies.

- *Coal Mining: Coalbed Methane Outreach Program*—This is an EPA-sponsored voluntary program to reduce methane emissions from coal mining. The goal is to identify along with the coal mines and other industries methane emissions and implement methods of utilizing the gas as an energy source instead of making it a pollutant. The EPA is working with other coal mining countries especially China to utilize the latest technology to make methane an energy source.
- *Landfills: Landfill Methane Outreach Program*—This is a voluntary program to find, trap, and use landfill gas as a renewable, green energy source. The EPA is working worldwide to assist other countries with the technology needed to use the landfill gas as an energy source.
- *Oil and Gas Systems: Natural Gas STAR Program*—This is a voluntary partnership between the EPA and the oil and natural gas industry to identify and implement cost-effective technologies to reduce methane gas emissions. The EPA is working with a Mexican state-owned oil company to identify and implement a methane emission reduction project. (See endnote 30.)

Portland, Oregon, Air Toxics Solutions
(See endnote 32)

Even though air pollution in the Portland area has decreased substantially in the last 30 years, there is still a serious concern about air toxics. The Oregon Department of Environmental Quality working with the city and surrounding areas set up the Portland Air Toxics Solutions project. The advisory committee to the project composed of diverse stakeholders reviewed a technical study and recommended a framework for an air toxics reduction plan. The study was conducted in a scientific manner by technical environmental personnel and it estimated the air toxics concentrations for 19 different pollutants from hundreds of emission sources including industrial, mobile, and residential. Fourteen of the 19 were above the clean air health goals and eight of them caused the greatest risk to people. These were 1,3 butadiene, benzene, diesel particulates, PAHs, naphthalene, cadmium, acrolein, and formaldehyde.

The air toxics were found throughout the Portland region and were especially high in highly populated areas, near busy roads in areas where there was substantial business and industrial activity. A disproportionate amount of lower income minorities lived in these areas, and therefore it was also an environmental justice problem.

The Department of Environmental Quality and its advisory identified five specific emission categories for follow-up action as follows:

1. Residential wood burning
2. Cars and trucks
3. Heavy-duty vehicles such as buses and semitrailers
4. Construction equipment
5. Industrial metals facilities

Although only 2% of the homes in the area were heated by wood, many people either use wood stoves and fireplaces as an additional heat source or for aesthetic purposes. The old uncertified wood stoves in the typical fireplaces contributed most of the bulk of the toxic pollutants from residential wood burning. They also emitted fine particulate matter. The solution to the problem was to:

- Conduct a wood-burning survey to determine the numbers of units involved and the condition of the units.
- Provide an educational campaign to teach people how to use better burning techniques and improve their burning facilities.
- Provide financial assistance to low-income people to change their equipment to certified wood stoves.

- Find long-term funding to help citizens with the purchase of new wood stoves.
- Evaluate the effectiveness of ongoing programs used to improve wood-burning devices.

Light-duty cars and trucks typically use gasoline as the fuel source which produces air toxics and greenhouse gases. To reduce the air toxics from this source, the following recommendations were made:

- Reduce vehicle use by altering modes of transportation or car sharing.
- Improve traffic signals to keep traffic moving instead of stalled.
- Promote the use of electric vehicles by providing charging stations.
- Improve vehicle performance and reduce gasoline usage.

Heavy-duty vehicles are used throughout the Portland area, especially for deliveries and inter-state freight. Most of these vehicles use diesel fuel. The recommendations to decrease pollution from these vehicles were:

- Accelerate the use of cleaner diesel fuel and retrofit existing engines.
- Use the most effective means of education and outreach to the public and companies to get them to make changes in the diesel operation of heavy vehicles.
- Determine if public funding should be used for clean diesel fleets such as in school buses and in public transportation.

Construction equipment usually has diesel engines. Construction equipment is used throughout the area. Because of the low turnover of equipment, these engines may pose the most risk of emitting particulate matter and PAHs. In order to decrease pollution from this source:

- Conduct surveys to determine the amount of use of construction equipment as well as the age of the equipment.
- Accelerate the time used to retrofit the equipment with better and cleaner engines.
- Use alternative fuels where possible.
- Reduce idling time for construction equipment.

Industrial metals facilities contribute most of the cadmium, manganese, and nickel which are above the recommended benchmarks in the Portland area. These pollutants are usually found close to the industrial facilities.

To control these pollutants, make sure that all the industries comply with the permitting programs and all federal and state air toxics emissions limits.

For the present and future, the Department of Environmental Quality is working with local government and other partners to use air toxic considerations in the planning process for new transportation and for land use.

ENDNOTES

1. Koren, Herman, Bisesi, Michael. 2003. *Handbook of Environmental Health-Pollutant Interactions in Air, Water, and Soil*, Fourth Edition, Volume 2. Boca Raton, FL: Lewis Publishers, CRC Press.
2. US Environmental Protection Agency, Air and Radiation. 2012. *National Ambient Air Quality Standards*. Washington, DC.
3. US Environmental Protection Agency, Air and Radiation. 2013. *National Ambient Air Quality Standards for Particulate Matter*. Washington, DC: Federal Register.
4. US Environmental Protection Agency, Air and Radiation. 2000. *Taking Toxics Out of the Air: Summaries of EPA's Final Air Toxics MACT Rules*. Washington, DC.

5. US Environmental Protection Agency, Office of Water. 1997. *The Pulp and Paper Industry, the Pulping Process, and Releases to the Environment.* Fact Sheet EPA-821-F-97-011. Washington, DC.
6. US Environmental Protection Agency, Office of Air Quality Planning and Standards. 2001. *National Emission Standards for Hazardous Air Pollutants (NEDSHAP) for Integrated Iron and Steel Plants: Background Information for Proposed Standards: Final Report.* EPA-453/R-01-005. Research Triangle Park, NC.
7. US Environmental Protection Agency, Office of Air and Radiation, Office of Air Quality Planning and Standards. 2012. *Available and Emerging Technologies for Reducing Greenhouse Gas Emissions from the Iron and Steel Industry.* Research Triangle Park, NC.
8. US Environmental Protection Agency. 2012. *The Plain English Guide to the Clean Air Act: Cleaning Up Commonly Found Air Pollutants.* Washington, DC.
9. Kjellstrom, Tord, Madhumita, Lodh, McMichael, Tony. 2006. Air and Water Pollution: Burden and Strategies for Control. In Jamison, DT, Breman, JG, Meashman, AR, eds. *Disease Control Priorities in Developing Countries,* Second Edition, Ch. 43, 817–832. Washington, DC: World Bank.
10. US Environmental Protection Agency, Air and Radiation, Technology Transfer Network. 2012. *Evaluating Exposures to Toxic Air Pollutants: A Citizen's Guide.* Washington, DC.
11. US Government Accountability Office, Letter to Senator Sheldon Whitehouse. 2012. *Air Emissions and Electricity Generation at US Power Plants.* GAO-12-545 R. Washington, DC.
12. Commission for Environmental Cooperation. 2004. *Electric Power Plants Number One Source of Toxic Air Pollution in North America.* Montréal, QC.
13. Koren, Herman. 2005. *Illustrated Dictionary and Resource Directory of Environmental and Occupational Health,* Second Edition. Boca Raton, FL: CRC Press.
14. US Environmental Protection Agency, Office of Air Quality Planning and Standards. 2004. *The Particle Pollution Report: Current Understanding of Air Quality and Emissions through 2003.* EPA 454-R-04-002. Research Triangle Park, NC.
15. Choate, Anne, Freed, Randall, Gibbs, Michael, ICF Consulting. No Date. *California's Greenhouse Gas Emissions and Trends Over the Past Decade.* Washington, DC.
16. New Jersey Department of Environmental Protection. 2011. *Air Toxics in New Jersey.*
17. World Health Organization-Europe. 2005. *Air-Quality Guidelines: Global Update 2005.* Copenhagen, Denmark.
18. World Health Organization-Europe. 2000. *WHO Air-Quality Guidelines for Europe,* Second Edition. Geneva, Switzerland.
19. Lee, Barbara, Yu, John. 2009. *Model Policies for Greenhouse Gases in General Plans: A Resource for Local Government to Incorporate General Planning Policies to Reduce Greenhouse Gas Emissions.* California Air Pollution Officers Association, Sacramento, CA.
20. US Environmental Protection Agency. No Date. *Emissions Standards Reference Guide: Overview of Mobile Sources.* Washington, DC.
21. US Environmental Protection Agency. 1999. *Leak Detection and Repair: A Best Practices Guide.* Washington, DC.
22. Michigan Department of Environmental Quality. 2005. *Managing Fugitive Dust: A Guide for Compliance with the Air Regulatory Requirements for Particulate Matter Generation.* Lansing, MI.
23. US Environmental Protection Agency. 2013. *Climate Change: Overview of Greenhouse Gases.* Washington, DC.
24. US Environmental Protection Agency. 2013. *Inventory of US Greenhouse Gas Emissions and Sinks: 1990–2011.* Washington, DC.
25. Arizona Governors Agricultural Best Practices Committee. 2008. *Guide to Agricultural PM10 Best Practices: Agriculture Improving Air Quality,* Second Edition. Phoenix, AZ.
26. US Environmental Protection Agency, Technology Transfer Network. 2007. *Polycyclic Organic Matter.* Washington, DC.
27. City of Toronto, Canada. 2010. *CHEMTRAC: Resource for Greening Chemical Manufacturing Pollution Prevention Information,* Version 1.0. Toronto, ON.
28. Illinois Environmental Protection Agency. 2012. *Illinois Annual Air Quality Report 2011: Section 1: Air Pollutants: Sources, Health and Welfare Effects.* Springfield, IL.
29. Environmental Health and Engineering, Inc., for the American Lung Association. 2011. *Emissions of Hazardous Air Pollutants from Coal-Fired Power Plants.* EH and E Report 17505. Needham, MA.
30. US Environmental Protection Agency. 2012. *Global Methane Initiative.* Washington, DC.
31. Engler, Rich. 2013. *Basics of Green Chemistry.* US Environmental Protection Agency, Washington, DC.

32. Oregon Department of Environmental Quality, Air Quality Division. 2012. *Fact Sheet: Air Quality in Portland: Portland Air Toxics Solutions Report and Recommendations.* Portland, OR.

33. US Environmental Protection Agency, Technology Transfer Network. 2013. *National Emission Standards for Hazardous Air Pollutants.* Washington, DC.

34. US Environmental Protection Agency. 2013. *The Clean Air Act in a Nutshell: How It Works.* Washington, DC.

35. US Environmental Protection Agency. No Date. *The Clean Air Act: A Partnership among Governments.* Washington, DC.

36. US Environmental Protection Agency. 2013. *Overview: The Clean Air Act Amendments of 1990.* Washington, DC.

37. US Environmental Protection Agency. 2013. *The Clean Air Act: Solving Air Pollution Problems with Science and Technology.* Washington, DC.

38. US Environmental Protection Agency, Office of Air and Radiation. 1997. *The Benefits and Costs of the Clean Air Act: 1970 to 1990: EPA Report to Congress.* EPA-410-R-97-002. Washington, DC.

39. US Environmental Protection Agency, Office of Air and Radiation. 1999. *The Benefits and Costs of the Clean Air Act: 1990 to 2010: EPA Report to Congress.* EPA-410-R-99-001. Research Triangle Park, NC.

40. US Environmental Protection Agency, Office of Air and Radiation. 2011. *The Benefits and Costs of the Clean Air Act from 1990 to 2020: Summary Report.* Research Triangle Park, NC.

41. US Environmental Protection Agency. 2013. *Clean Air Act: Air Pollution: Current and Future Challenges.* Washington, DC.

42. US Environmental Protection Agency. 2012. *TTNWeb: Technology Transfer Network.* Research Triangle Park, NC.

43. US Environmental Protection Agency, Office of Air Quality Planning and Standards. 2013. *QA Handbook for Air Pollution Measurement Systems: Ambient Air Quality Monitoring Program*, Volume II. EPA-454/B-13-003. Research Triangle Park, NC.

44. Richards, John R., Schell, Robert, eds. 2000. *Control of Nitrogen Oxides Emissions: Student Manual: APTI Course 418.* US Environmental Protection Agency, Air Pollution Training Institute. Research, Triangle Park, NC.

45. Hribar, Carrie, Schultz, Mark, eds. 2010, *Understanding Concentrated Animal Feeding Operations and Their Impact on Communities.* Bowling Green, OH: National Association of Local Boards of Health.

46. US Department of Agriculture, Natural Resources Conservation Service, US Environmental Protection Agency. 2012. *Agricultural Air: Quality Conservation Measures: Reference Guide for Cropping Systems in General Land Management.* Washington, DC.

47. US Environmental Protection Agency, Ag STAR Program. 2010. *US Anaerobic Digester: Status Report.* Washington, DC.

48. US Environmental Protection Agency. 2012. *Climate Change Indicators in the United States, 2012.* Washington, DC.

49. K-State Research and Extension, Kansas Department of Health and Environment. 2011. *Kansas Flint Hills: Smoke Management: Fire Management Practices.* Manhattan, KS.

50. Kansas State University Agricultural Experimental Station and Cooperative Extension Service. 2012. *Prescribed Burning Notebook: One for 2012.* Manhattan, KS.

51. US Environmental Protection Agency, Office of Water. 2013. *Regulatory Impact Analysis for the Proposed Effluent Limitations Guidelines and Standards for the Steam Electric Power Generating Point Source Category.* EPA-821-R-13-005. Washington, DC.

52. US Department of Energy, Energy Efficiency and Renewable Energy, Industrial Technologies Program, and Industrial Heating Equipment Association. 2007. *Improving Process Heating System Performance: A Sourcebook for Industry*, Second Edition. Washington, DC.

53. Kansas State University Pollution Prevention Institute. 2003. *Colleges and Universities: Pollution Prevention for Power Plant Systems.* Manhattan, KS.

54. Kitto, JB. 1996. *Air Pollution Control for Industrial Boiler Systems.* BR-1624. ABMA Industrial Boiler Systems Conference, Alliance, OH.

55. US Environmental Protection Agency, Combined Heat and Power Partnership. 2013. *Fact Sheet: CHP as a Boiler Replacement Opportunity.* Washington, DC.

56. US Environmental Protection Agency, Office of Air Quality Planning and Standards. 2010. *Available and Emerging Technologies for Reducing Greenhouse Gas Emissions from Industrial, Commercial, and Institutional Boilers.* Research Triangle Park, NC.

57. US Environmental Protection Agency. 2011. *Improving Air Quality in Your Community: Outdoor Air: Industry, Business, and Home: Industrial, Commercial, and Institutional Boilers: Additional Information*. Washington, DC.

58. EPA. 2012. *Our Nation's Air: Status and Trends through 2010*. EPA. Washington DC.

59. US Environmental Protection Agency, Air Pollution Training Institute. No Date. *Principles and Practices of Air Pollution Control: Student Manual*. Washington, DC.

60. Portland Cement Association. 2013. *Cement and Concrete Basics: Frequently Asked Questions*. Skokie, IL.

61. Air and Water Management Association. 2007. *Fact Sheet: Air Pollution Emission Control Devices for Stationary Sources*. Pittsburgh, PA.

62. Neuffer, Bill, Laney, Mike. 2007. *Alternative Control Techniques Document Update: Nitrogen Oxides Emissions from New Cement Kilns*. EPA-453/R-07-006. US Environmental Protection Agency, Office of Air Quality Planning and Standards. Research Triangle Park, NC.

63. Miller, FM, Young, GL, von Seebach, M. 2001. *Research and Development Information: Formation and Techniques for Control of Sulfur Dioxide and Other Sulfur Compounds in Portland Cement Kiln Systems*. R&D Serial Number 2460. Portland Cement Association, Skokie, IL.

64. US Environmental Protection Agency, Office of Air and Radiation. 2011. *Oil and Natural Gas Sector: Standards of Performance for Crude Oil and Natural Gas Production, Transmission, and Distribution*. EPA-453/R-11-002. Research Triangle Park, NC.

65. US Environmental Protection Agency, Natural Gas STAR Program. 2012. *Recommended Technologies and Practices*. Washington, DC.

66. US Environmental Protection Agency, Natural Gas STAR Program. 2013. *Basic Information*. Washington, DC.

67. Ohio Environmental Protection Agency. 2012. *Understanding the Basics of Gas Flaring*. Columbus, OH.

68. Balco. Aluminum Production Process (PDF), Aluminum Production Technology. Available from balcoindia.com.

69. Environmental Protection Agency. 2013. *Quality Assurance Handbook for Air Polluiton Measurement Systems*. EPA, Washington DC.

70. European Environment Agency. 2013. *EMEP/EEA Air Pollutant Emission Inventory Guidebook 2013*. Copenhagen, Denmark.

71. European Commission. 2001. *Integrated Pollution Prevention and Control: Best Available Techniques Reference Document on the Production of Iron and Steel*. Brussels, Belgium.

72. World Bank Group, International Finance Corporation. 2007. *Environmental Health, and Safety Guidelines: Pulp and Paper Mills*. Washington, DC.

73. US Department of Energy, Energy Efficiency and Renewable Energy. 2005. *Industrial Technologies Program: Energy and Environmental Profile of the US Pulp and Paper Industry*. Washington, DC.

74. Missouri Department of Natural Resources, Environmental Assistance Office. 2005. *Preventing Pollution in Wood Furniture Manufacturing: A Guide to Environmental Compliance and Pollution Prevention for Wood Furniture Manufacturers in Missouri*. Jefferson City, MO.

75. US Environmental Protection Agency, Indoor Air Quality. 2012. *The Inside Story: A Guide to Indoor Air Quality*. Washington, DC.

76. US Environmental Protection Agency, Indoor Air Quality. 2012. *Indoor Air Pollution: An Introduction for Health Professionals*. Washington, DC.

77. Shaw, CY, Won, D, Reardon, J. 2005. *Managing Volatile Organic Compounds and Indoor Air Quality and Office Buildings: An Engineering Approach*. Ottawa, ON: National Research Council Canada, Institute for Research in Construction.

78. US Environmental Protection Agency, Region 2 Compliance. 2010. *Auto Repair Compliance Assistance Tools*. New York.

79. New York State Department of Environmental Conservation. 2014. *Reducing Air Pollution from Lawn and Garden Equipment*. Albany, NY.

80. Department of Commerce, National Oceanic and Atmospheric Administration, Earth Systems Research Laboratory Chemical Sciences Division. 1998. *Scientific Assessment of Ozone Depletion: 1998 Executive Summary*. Boulder, CO.

81. US Environmental Protection Agency. 2014. *Climate Change: National Greenhouse Gas Emissions Data*. Washington, DC.

82. EPA.2010. *Our Nation's Air—Status and Trends Through 2010*. EPA-454/R-12-001. Office of Air Quality Planning and Standards Research. North Carolina.

BIBLIOGRAPHY

British Columbia. 2008. *How Air Quality Affects Health*. Vancouver, BC.

Centers for Disease Control and Prevention. 2015. *Air Quality*. Atlanta, GA.

Consumer Product Safety Commission. No Date. *The Inside Story: A Guide to Indoor Air Quality*. Washington, DC.

Natural Resources Defense Council. No Date. *Air Pollution: Smog, Smoke and Pollen*. New York.

US Environmental Protection Agency, AirNow. 2015. *Air Quality Index (Basics)*. Washington, DC.

US Environmental Protection Agency. 2015. *Air and Radiation, Air Pollution Data Sources*. Washington, DC.

US Environmental Protection Agency. 2015. *Air Pollution Monitoring*. Washington, DC.

US Environmental Protection Agency. 2015. *Air Quality Planning and Standards*. Washington, DC.

US Environmental Protection Agency. 2015. *The Green Book Nonattainment Areas for Criteria Pollutants*. Washington, DC.

3 Built Environment— Healthy Homes and Healthy Communities

STATEMENT OF PROBLEM AND SPECIAL INFORMATION

The Built Environment is the sum of all the physical changes, both positive and negative, made to the natural environment by people. This includes roads and highways, homes and communities, schools and places of worship, agricultural areas, businesses, industrial areas, landfills, dams, etc.

The Built Environment causes the disruption of the natural environment including air, water, and land and the disruption of ecosystems. The preservation of natural lands and critical environmental areas protects water quality including appropriate quantities of high-quality drinking water as well as air quality. The natural environment helps regulate ambient air temperatures and the amount and frequency of precipitation, as well as helps prevent flooding.

Population growth has also contributed to the disruption of the natural environment. Planning agencies have difficulty in utilizing an area-wide approach to develop the land and water in the natural environment appropriately because there are a multitude of political jurisdictions with a huge number of potentially different laws, rules, and regulations. Zoning is also frequently a political issue instead of one which is in the best interest of the public.

It is generally understood that land use and decisions on types of transportation to be made available for people may affect environmental protection, public health, and the quality of life. On a regional basis, the location of types of transportation sources may influence housing patterns, other structures and services, and necessary infrastructure. On a local basis, the distance from places of employment and other services dictates the types of transportation used, whether it is walking, bicycling, using cars, or public sources such as buses and other vehicles. Short vehicle rides tend to release more pollutants into the air. Greater traffic increases the opportunity for noise. Street network connectivity is very important because it has been shown that where this is poorly designed and inadequate there are more accidents and more deaths from injuries. (See endnote 31.)

The Interstate Highway System has had a profound effect on where and how land is used for all types of purposes. It is now possible to have a core city with surrounding suburbs that utilize the resources and jobs in the city. Much of the land in the surrounding areas was never meant to be used for concentrated housing and therefore the land use has caused innumerable environmental problems in these communities. Also people may have to drive long distances to go to work and for other purposes.

The nature of the different types of housing available and how the communities are reconfigured or originally planned and constructed makes the difference between short distances to walking areas, bicycling paths, public transportation, or short drives to businesses, places of employment, schools, churches, and parks from homes, or long distances with resulting air pollution, excess energy use, additional cost to the individual, and waste of time by the public. The range of homes available to the elderly as well as the disabled increases their ability to remain within their own home and neighborhood instead of having to go to assisted living facilities or homes for the aged. Elderly people not only contribute financially to their community but also feel happier when remaining within familiar surroundings. Further, a mixed community of different ages of people is especially good for children. The children are constantly in contact with the elderly and this helps them establish appropriate values on learning to respect and help other people. They also have the opportunity to gain an enormous amount of knowledge from the elderly that cannot be taught in the schools or in textbooks. (See endnote 6.)

To be able to evaluate the Built Environment, the author has added to the potential concerns promoted by the European Green Group developed as indicator areas as follows: the heat island effect created by large structures and substantial paved areas crowded together in cities and how to reduce it; local means of preventing global climate change; condition and performance of local transport; the amount and distribution of green urban space; the amount of impervious surfaces built such as roads and parking lots as well as the areas surrounding buildings; the sustainability of current and future land use; the nature and diversity of the natural environment; the quality of the local ambient air as well as the indoor air of all structures; the quantity and level of noise pollution; the amount of solid and hazardous waste produced and how it is managed; the amount of water, especially drinking quality water consumed, and the quality and quantity of sources; the amount and concentration of biological and chemical pollutants as well as physical agents in wastewater, how it is treated and where it is ultimately disposed; the amount of electricity, gas and other forms of energy readily and economically available for use by people, business, industry and government; the level of environmental management carried out by government, industry and individuals; the type of communications strategy used and the distribution and quality of the communications network including the internet; the type, quantity and quality of emergency management used for all types of emergencies, disasters and terrorism; the positive and potentially negative effect of various facets of the Built Environment on the health, safety and general welfare of people. (See endnotes 4, 5.)

The nature of the Built Environment has a profound effect on all people, but even more so when it is badly substandard, helping create severe health, safety, and welfare problems for the citizens who live there because of release of pollutants from industries, motor vehicles, construction, and other sources as well as the poor neighborhood conditions and poor housing conditions present. Poor housing helps create social, physical, and mental health problems including anxiety, depression, substance abuse, crime and aggressive behavior, asthma, heart disease, obesity, etc. Many people continue to live in these types of conditions because other forms of housing are unaffordable. They do not have the money to both house their families and feed and clothe them.

An extremely significant part of the Built Environment not discussed above is the broad question of the types and conditions of structures that people use for homes. This area of housing and the many related problems will be better understood by reading the material below. A separate set of Best Practices will be described for the Built Environment and also for the Housing Environment including building location, design, maintenance, renovation or retrofitting, and sustainability.

SUB-PROBLEMS INCLUDING LEADING TO IMPAIRMENT AND BEST PRACTICES FOR THE BUILT ENVIRONMENT IN URBAN AREAS

HISTORY OF THE BUILT ENVIRONMENT AND HOUSING IN THE UNITED STATES

(See endnotes 1, 2)

The Built Environment is not a new concept. In fact, William Penn, in 1682, when he established the City of Philadelphia, laid plans that satisfied the needs of the then Built Environment. They provided for frequent open areas or green spaces called parks, appropriate size streets for major industry and commerce, commercial activities and residential activities, and space for the long-term growth of the community. A grid was used with the major highways and areas of major industry on the outer perimeter. These became sources of major transportation. Then wide roads were laid out followed by the smaller streets where lots were cut out that could be purchased for homes or small business activities. He specified that the city should be on the river so there would be an immediate water source as well as a major means of transportation. The rectangular grid pattern using letters for streets going East and West and numbered streets going North and South made it easier to find a given location.

From the inception of this country, there have been many problems with the housing of its citizens. In the 1600s, 1700s and 1800s, the greatest concern was fire. A prime example was the great Chicago fire of 1871 that burned from Sunday, October 8, to early Tuesday, October 10. About 300 people died, 100,000 were made homeless, and 3.3 square miles of Chicago, Illinois, were destroyed.

In the 1800s, especially after 1840, most immigrants came from Great Britain, Ireland, Germany, Scandinavia, and to a lesser extent from China, primarily because of economic reasons. With the beginning of the industrial expansion after the Civil War, there was a huge influx of people, primarily from Eastern Europe, into the United States and into the big cities. The immigrants came from Southern and Eastern Europe due to a combination of poor economic conditions, war, and religious persecution. These people were Greek, Italian, Jewish, Polish, Russian, Serbian, and Turkish. The biggest part of the immigration started in 1880 and continued in waves through 1920, with the largest number of these individuals moving into the poorest, most overcrowded, and structurally deteriorating areas of the big cities, such as New York. Increasing the pressure for housing in these cities in the last quarter of the 19th century was the movement of individuals from farms and small towns into areas where there were greater opportunities for employment.

In addition to the problems of potential fires, there were constant concerns about serious overcrowding, lack of cleanliness, poorly constructed and maintained structures, and inadequate or totally lacking groundwater protection because of the small housing lots used and contamination of wells from adjoining septic systems. Poor ventilation, inadequate or totally lacking proper lighting, lack of toilet and bathing facilities, bad odors, insects, rodents, trash and garbage, sewage, inadequate and contaminated water helped contribute to the spread of disease and very serious problems in housing units. The individuals were very poor, lacked adequate food, and suffered from many diseases, including substantial amounts of tuberculosis and other infectious diseases which could easily be spread to others in the compact living areas. This was the beginning of high-rise apartments on tiny pieces of land, with huge concentrations of people.

The Public Works Administration helped improve the infrastructure of the country during the Great Depression of the 1930s by building airports, electricity generating dams, railroads, and many other projects. The Wagner Steagall Housing Act encouraged governmental agencies to build high-rise buildings for public housing for poor people living in overcrowded urban slum areas. By the 1960s, many of these buildings in the major cities were already deteriorating. The structures were populated by large concentrations of individuals, especially large numbers of children, in small areas, which led to excessive vandalism and property destruction. Elevators in these buildings were a disaster because they frequently were broken and, where working, were a place where crime could be committed. It was difficult to move the children from their homes to the outside safely. Budget crises in the Housing Authorities contributed to very poor maintenance of the structures and equipment. Further, people were untrained in the use and maintenance of these high-rise buildings, leading to increased levels of vandalism, crime, disease, and injury, as well as very poor living conditions.

At the end of World War II, there was a growing need for housing for those returning from serving their country and their future families. There was literally an explosion of housing subdivisions in rural areas, outside of big cities, where the land had been utilized for farming. Whereas a single house sat on 40 acres and could easily utilize a well and on-site sewage disposal system, these 40 acres were now divided into quarter acre, half-acre or acre plots which were sold for home construction. Little or no attention was given to natural contours, creeks, ponds, etc. Now instead of one home with one well and one on-site sewage disposal system, the land was supposed to take care of 40–160 homes with separate wells and separate on-site sewage disposal systems. This led to many cases of overflowing sewage and contaminated wells.

Although the spread of infectious diseases in poor housing is less than it ever has been, an additional problem is the increase in chronic diseases, which may be attributed to inactive lifestyles, improper nutrition and exposure to many pollutants both within structures and outside of the structures. Injuries are a continuing problem. Land-use problems are of considerable concern, as well as the proximity of major polluting industries and high levels of traffic. Noise is a never-ending concern.

Sewage systems, water systems, and gas pipe systems may be extremely old and may cause serious health problems because of poor maintenance and simply because of age.

POVERTY AND HOMELESSNESS

(See endnote 3)

Homelessness has always been a problem among different groups of people in our society. New immigrants crowded into existing structures to have a place to sleep. During the Dust Bowl of the 1930s, vast numbers of people left the upper Midwest and moved to other areas to try to find jobs. They lived in tent cities. During this time of the Great Depression, men most frequently became hobos and traveled in an unsafe manner by railroad from community to community to find something to eat and a place to sleep.

In 1963, the Community Mental Health Centers Act was passed and along with it came the unintended consequences of certain individuals who had been de-institutionalized and could not adjust to society, and did not receive adequate supervision, care, and housing. This has led to homelessness, incarceration and violence among some of the approximately 1.8 million people who suffer from severe mental health problems. A portion of these individuals makes up the homeless of today.

The poor have always had higher rates of disease, limited access to health care, inability to pay for their basic needs, and a greater potential for helplessness and homelessness. They are mostly uninsured and therefore do not get preventive care, resulting in many serious chronic illnesses. The US Supreme Court ruled that each state can decide if it wants to extend Medicaid to additional individuals especially those without children. Some states such as Florida have not accepted an extension of Medicaid and therefore there are over 800,000 individuals without any type of insurance. These are basically the very poor. The poor live in neighborhoods which are overcrowded, deteriorating rapidly, and in close proximity to many environmental hazards. They are therefore more prone to become severely ill and die at earlier ages than would be anticipated.

Today, poverty is no longer just among the lower economic class but rather has moved into the middle economic class. Today's homeless are primarily women, children, and families and not the mentally ill or the street people who decide to live without a home. Some of the women and children have left abusive families and have no place to live. Others lost their jobs and then their homes. Past housing crises and foreclosure crises have contributed substantially to this problem. There are pregnant teenagers and also those who are drug addicted, who have been thrown out by angry parents, and senior citizens and veterans who no longer have a family to take care of them. Where a shelter is available, people typically leave them because of various fears. On any given night, 700,000–800,000 people are homeless and in any given year, 2.5–3.5 million people experience homelessness. The reduction or stagnation of income and the loss of jobs have created a new class of poverty, hunger and homelessness. In 2013, over 46.3 million Americans lived in poverty.

In addition, we have the service people who went to war to protect our country who for many reasons are now homeless and hungry.

In the United States today, over 5 million families with over 4 million children live in houses where there is overcrowding, very poor living conditions, and poor facilities. Despite government help in paying for shelter, there are millions of individuals who cannot find appropriate, affordable housing.

SUB-PROBLEMS INCLUDING LEADING TO IMPAIRMENT FOR THE BUILT ENVIRONMENT IN URBAN AREAS

When the larger cities started to become industrialized, the factories were typically located on rivers in the city and then additional factories joined the existing ones along these same rivers and other waterways. (These have always been major areas of insect and rodent infestation.) The waterway was utilized in manufacturing, as a means of transportation, and as a means of disposal of liquid, solid, and gaseous wastes. The land around the factories was typically very cheap and in many situations in poor condition. This was where housing was built for the workers, and both the structures and the surrounding environment were poorly constructed and poorly maintained. Generations of immigrants from abroad and migrants from farm areas were housed in these areas in extremely overcrowded, unsanitary conditions. In the better areas of the city, where the middle class and upper class lived, a single three-storey house typically was used by a family of six to eight people. As time went on and housing demands grew because of more poor people coming into the cities, these houses were sold to absentee landlords, who turned each house into six separate apartments and which now was the living facility for 50–60 people. The middle class and upper class moved to newer areas and into the suburbs. In these structures, the indoor air pollutants were greater and the level of asthma, especially among children, increased sharply. The children usually attended and still attend poorly constructed and/or older school facilities in close proximity to polluting facilities. In urban areas, poorer communities tend to be overcrowded and congested with higher levels of air pollutants from automobiles, incinerators, industries, and diesel buses. People tend to live in older houses with lead-based paints, substantial numbers of roaches, mice and rats, and poor housing and community conditions.

In general, urban areas are large population centers with old infrastructure and housing stock which utilizes substantial amounts of energy and water while contributing to a variety of air, water, and land pollution. The concentrations of people and facilities also lead to an increase in noise levels, respiratory illnesses, injuries, and violence. The results of these conditions are intensified by an aging population, many with chronic conditions, who are more susceptible to a variety of environmental problems.

Best Practices for Improving the Built Environment in Urban Areas (See endnotes 6, 7, 8, 9, 10, 11, 12)

- Establish a code enforcement regulatory agency which could be housed in either the environmental health division or in a special code enforcement division. There should also be a special prosecutor assigned to the code enforcement regulatory agency and a special housing court to quickly handle violations of the housing code when they affect the health and safety of the individuals or community. Fines should be utilized to enforce appropriate actions and, if necessary, the judge should issue an injunction which would mean that if the situation were not corrected within a given period of time, the owners of the property could go to jail.
- Establish appropriate laws, rules, and regulations concerning: zoning including all types of land use; housing codes; plumbing codes; fire codes; and use of various safety devices.
- Establish an overall master plan and strategy utilizing modern environmental technology for the entire area under consideration and for the subareas which will be worked on over time.
- Have the appropriate political entities to approve the necessary financing for a Planning Commission to survey specific areas of the urban setting and make determinations utilizing various professionals on necessary changes to make the area more livable, safer, and environmentally sound.
- Conduct an Environmental Impact Assessment to help predict the consequences to all aspects of the natural environment as well as social, economic, and cultural areas from specific development projects and take these into account when making land-use decisions.
- Establish a Planning Commission representing the various stakeholders in a given area to determine appropriate land-use planning of existing facilities and new use of land. This would include the various levels of government, business, industry, civic organizations, universities, political entities and others.
- Establish local committees of professionals on the use of land as well as building and maintenance of structures to advise the Planning Commission on all aspects of the urban plan.
- Determine if the plan is sustainable over a period of many years and the potential improvements in the community as well as the potential liabilities.
- Evaluate all of the current structures in a target area and determine how to make them more efficient and durable for at least the next 30 years.

- Provide a properly funded Environmental Management Office with appropriate staff and legal status to inspect all existing and future structures and prepare standards and Best Practices for compliance by the individuals in the structures to provide appropriate excellence in quality of life in these areas through local zoning and housing and other codes enforcement.
- Promote mixed development so that more people are closer to their jobs and services.
- Promote affordable housing for low-income families and newly established families and when individuals are forced out of their existing housing because of redevelopment. By law they should be given equivalent housing quarters in another acceptable area.
- Utilize existing land which is already serviced by water, sewage, gas, electricity, telephones, etc. before developing new land on the periphery of the urban area. This is far more efficient and cost-effective and avoids destroying natural habitats as well as creating new sources of pollution.
- All downspouts must discharge to the surface of the ground and not to the storm sewers.
- Do not put new industries that utilize or create hazardous materials within or close to people and their homes to avoid additional health and safety problems.
- Utilize space which is currently available in urban areas by retrofitting and converting old buildings and facilities for new uses, by filling in spaces between buildings where previous structures have been torn down, and by replacing parking lots with buildings containing parking garages and parks. The removal of parking lots and replacing them partially with parks will decrease air pollution and increase the groundwater supply.
- In newly developed land, create a cluster of structures which have smaller lot sizes and homes laid out efficiently with smaller square footage, thereby creating the idea of roominess. This allows for more people to live comfortably in a smaller area, conservation of energy and resources, and the provision of green areas which are necessary for reducing air pollutants, increasing groundwater supply, and providing a better and healthier environment for the people living there.
- Reduce impervious surfaces wherever possible to increase the groundwater supply and decrease the chance for wildfires especially in those areas where a substantial amount of land was used to create parking lots and roads.
- Avoid the destruction of existing wetlands.
- Clean up the contaminated areas of the urban environment called brownfields and use those properties for a variety of purposes including mixed residential, commercial and green areas.
- Establish specifications for energy-efficient buildings and energy conservation.
- Establish specifications for water conservation and wastewater reuse.
- Establish specifications and schedules needed to perform appropriate maintenance in all structures and associated equipment.
- Replace essential equipment within structures when the effective lifespan has been reached. Do this on a regularly scheduled basis.
- Reduce solid waste through waste minimization, use of composting and anaerobic digestion where feasible, recycling of materials, and turning waste into energy.
- Establish the roadways to be used to remove hazardous substances, materials, and waste away from residential areas.
- Reduce noise levels by enforcing the appropriate noise ordinances and providing specific assistance to areas of high noise to help them make appropriate reductions.
- Determine the current levels of: air pollution, outdoor and indoor; energy use and waste; groundwater contamination and depletion; destruction of historical structures and places; damage to the land and ecosystem; solid and hazardous waste from households and industries; water pollution and water depletion; and how the project will affect it. Make necessary adjustments if negative effects will occur.
- Determine the age and condition of the infrastructure under the various streets and roads involving water, sewage and gas, and establish a long-range plan to upgrade the systems to prevent unwanted breakdowns and potential hazards.

- Develop appropriate means of treating stormwater especially from areas of high contamination from roadways, other surfaces, and industries before the water is released into the watershed.
- Develop and enforce emissions standards for vehicles in the communities to reduce the potential for air pollutants going into the air and creating serious health effects.
- Determine the capacity and efficiency of the water treatment plants and sewage treatment plants and make necessary plans to upgrade them and expand them as needed.
- Monitor and evaluate the entire watershed to ensure that it will provide quality water in adequate quantities to sustain the urban area, and provide for intensive watershed planning and management.
- Establish a comprehensive solid and hazardous waste program using techniques of reduction of materials, reuse of bottles and other containers, recycling of paper, plastics, glass, and metal, and recovering energy from waste.
- Provide energy-efficient transportation which is convenient and economically viable.
- Utilize renewable energy sources whenever practical.
- Establish for the entire urban area a comprehensive energy reduction and energy-saving plan.
- Use compact land-use patterns where appropriate to open up green spaces for the community use as well as reduce carbon.
- Provide for a well-funded Public Health Department including a comprehensive Environmental Health Division that may help prevent disease and injury and promote good health and well-being in an expeditious and cost-effective manner.
- Provide for a well-funded Emergency Management Agency to be able to respond quickly and responsibly to hazardous events, emergencies, natural disasters, and acts of terrorism.
- Make available to all facets of the urban environment immediate access to the internet and electronic devices to utilize the various services and information available.
- Establish well-funded educational programs to help citizens help themselves through constructive actions in environmental efforts to reduce pollutants, reduce energy use, and increase the livability of the various structures.
- Encourage economic development through use of private funds.
- Ensure that stakeholders are constantly involved in all facets of the planning process and then the necessary developmental activities.

SUB-PROBLEMS INCLUDING LEADING TO IMPAIRMENT FOR THE BUILT ENVIRONMENT IN SUBURBAN AREAS

(See endnote 13)

Much of urban sprawl started to occur after World War II as the millions of service people came home from war and there was a huge baby boom. The Interstate Highway System was created and there was greater access to rapid means of traveling from outlying areas into the core cities. There was a huge economic expansion to satisfy the suppressed needs of a generation of young people as they started families and sought reasonably priced housing and new jobs. This resulted in a low average density of people in areas and therefore a substantial change in land-use patterns. Considerably lower land prices have fueled this expansion and allowed people to have large houses and large areas of land with relatively few inhabitants. Strip malls with large parking lots were created in place of green areas, affecting the natural environment. There was a new dependency on automobiles and this resulted in sharply increased air pollution. There was a substantial expansion of all types of utilities. The public investment in roads, public buildings, parks and green spaces, water, sewers, and other infrastructure did not meet the demands of the expansion and therefore has created additional problems. There is redundant cost for establishing multiple small municipalities which have a variety of political entities that may not interact easily or communicate well during emergencies and disasters. The various small groupings fragment open space and disrupt natural wildlife habitat while using up productive farmland and forest land. Throughout this process of development, there has been a lack of centralized or coordinated planning which has resulted in a large amount of unnecessary land consumption and disruption of the natural environment. This increases the cost of infrastructure which is underused, and therefore when problems occur, it makes it far more difficult to get adequate maintenance and repair.

Unfortunately, the unsettled problem of race relations caused a massive explosion and in 1967 the worst rioting in any American city ever occurred in Detroit, Michigan. It took the United States Army to quell the riots and resulted in a massive flight of the white population to surrounding counties and corporations with their numerous jobs followed. The city population shrank from almost 2 million to 680,000 people. As the city's population shrank, the physical structure deteriorated rapidly and the tax base diminished, leading eventually to the city defaulting on its debt. Only in the last couple years have there been the beginnings of a revival of the city and hopefully a re-emergence of this once important metropolis. Detroit has been an example of the problems of many of the inner city areas in the country.

Also, with all these individuals moving out to the suburbs, it reduces the tax rolls of the core cities and leaves those who are most disadvantaged, the poor, disabled, and elderly in situations where the cities may have to reduce services to them because of lack of proper funding.

Urban sprawl occurs when communities do not take into consideration appropriate land-use patterns and utilize formally rural areas for housing developments outside of the central cities. Land-use changes include deforestation; road construction; and encroachment on agricultural areas, areas of irrigation, coastal zones, wetlands, etc. The local ecosystems are affected or destroyed and numerous potential environmental protection and environmental health problems are created. Local farmers may sell their farms to developers in order to gain substantial profit for the land and find an easier lifestyle.

The level of emerging infectious diseases is exacerbated by the changes in the ecosystems. It has been shown that urban sprawl and loss of biodiversity are linked to an increase in Lyme disease in the northeastern United States. Nipah virus has emerged in Malaysia, and cryptosporidiosis in Europe and North America, and there has been a sharp increase in food-borne diseases. These diseases may be increasing because of the movement of wild animals into new and different areas. The importing of pets throughout the world is another factor in exposing our society to different microorganisms.

Air quality decreases and respiratory diseases increase primarily because of the increased use of automobiles which are necessary for the homeowners to travel to work, shopping, school, recreational areas, etc. Motor vehicle traffic is the main source of ground-level pollutants including carbon monoxide, nitrogen oxides, and 40% of the particulate matter (PM_{10}). Ozone and sulfur dioxide are serious respiratory irritants and asthma triggers.

Urban heat islands increase and with it there is an increase in heat-related illnesses and death. When the natural cooling energy of vegetation and trees is removed and the heat-absorbing surfaces such as streets, driveways and roofs are increased, there is an increase of 2–8°F in the area.

Uncontrolled growth reduces green space which drastically affects surface and groundwater quality. Rainwater, which would have percolated slowly through the soil, now washes over asphalt, concrete, rooftops, and other areas which are contaminated with residues from automobiles and from other sources. Large quantities of water go across the surfaces and move quickly to stormwater drains, thereby reducing the amount of water percolating down through the ground to renourish the groundwater supply. Newly created housing developments and golf courses typically have lawns on which fertilizer and pesticides are used. These chemicals wash away with the stormwater runoff, into surface receiving bodies of water and contaminate them.

Septic systems become the mode of choice for disposal of sewage because many of the housing developments are a substantial distance from public sewer systems. Improper operation of these systems and poor soil conditions lead to contamination of the land, groundwater and surface water. From 1955, when the author first started in the environmental health field and inspected on-site sewage systems and wells on pieces of land that ranged from 10,000 square feet to at most 0.5 acre, until the present, some builders have tried to utilize as little land as possible in order to maximize profits. This has resulted in a huge amount of pressure on the groundwater supply to handle the effluent from many inefficient and poorly maintained septic systems. Adding wells to these sites contributes to the problems and to the potential for groundwater contamination.

In order to avoid contamination of the groundwater supply by building houses on too small a lot, Oakland County, Michigan instituted a "Groundwater Protection through Density Control" policy which stipulated that a three-bedroom house had to be built on at least 1 acre of land and a four-bedroom house had to be built on at least 1.25 acres. The health department which was in favor of the rule prevailed, although the rule was challenged in the courts.

Risk of flooding increases because of the destruction of wetlands. Houses are built in areas where wetlands have been drained and also on flood plains. Other barriers to flooding such as trees and grassy areas have been removed for new developments.

Accidental injuries increase sharply because of the increased use of bicycles by the population. The bicycles are used because they are convenient and promote fitness. Also, additional injuries occur because many of the residential areas do not have sidewalks and people have to walk in the street in part because of a lack of proper planning. These areas are particularly difficult to utilize by older people and disabled people.

Best Practices for Improving the Built Environment in Suburban Areas (See endnotes 19, 20)
- Conduct a Housing Capacity Study to determine how much land is needed to provide additional projected housing for expansion of the given suburban community for many years into the future. The study should include all potential existing sites such as empty space above stores that can be used for apartments, underused parking garages where areas might be converted into inexpensive housing, development of land that would function best for additional housing, and provide incentives for carrying out these projects.
- Provide proper, reasonable cost and frequent public transportation to the higher density areas and natural commercial corridors, and effectively connect all of the suburban communities as well as the urban core.
- Develop a plan for better use of the commercial corridors which typically have large underused or low-value land that can be used for new purposes such as commercial, retail and entertainment areas.
- Reuse vacant buildings as quickly as possible to avoid deterioration for new purposes rather than demolishing them and building new properties.
- Reconfigure land usage on dying or obsolete shopping malls in suburban areas and turn the structure and huge parking lots into compact mixed-use communities including office buildings, college or university extensions, residential buildings, etc. where people will have all the amenities necessary including green spaces and also be close to jobs.
- Revamp and enhance suburban town centers including all the stores, restaurants, places of entertainment, and other services that people desire.
- Determine the demographics, preference and potential increase in growth of the population and where they would appear to want to live, and establish necessary plans for increased density mixed-use communities in those areas.
- Develop energy-efficient plans for existing structures and the use of underused space.
- Establish a council of local municipalities who will work together on infrastructure projects as the communities expand into one large group around the urban core city. This will make projects less costly and have the least chance of being repetitive in nature.
- Develop bike lanes, pedestrian trails and sidewalks along with green areas to help improve the health and safety of the population and reduce pollutants.

- Build a comprehensive approach to the infrastructure and access to reused land as well as new land for projects.
- Include in all planning for revitalizing areas adequate amounts of public space for trails, sidewalks, pedestrian walkways, and bicycle lanes where appropriate.
- Consider in all planning the necessity for appropriate oversight and management of the areas and adequate financing for programs of continued maintenance of all properties and infrastructure as well as replacement of facilities and equipment as needed.
- Develop a variety of financing tools and funding sources for construction of structures and necessary infrastructure. This topic is so vast that it cannot be covered in a book of this nature.
- Develop studies in communities to identify, analyze, and help make decisions on how best to improve pedestrian traffic based on pedestrian volume and other environmental aspects at different points along the roads and streets. (See endnote 18.)

Also see Best Practices for Improving the Built Environment in Urban Areas.

SUB-PROBLEMS INCLUDING LEADING TO IMPAIRMENT FOR THE BUILT ENVIRONMENT IN RURAL AREAS

Small towns and rural communities vary enormously across the country in natural resources, excellence of locations between larger areas, and natural environmental conditions. There are numerous challenges facing these communities ranging from job losses, to population losses or population gains due to newcomers seeking a quieter lifestyle, to poor transportation and poor roads, to development of natural resources, to access to jobs, services and transportation, to inadequate health care and healthcare resources, to developers pressuring to buy farms and turn them into residential communities with people living on large lots and with inadequate infrastructure, etc.

Means of transportation for people and agricultural products can be very challenging in rural areas because of: long distances between population centers; steep grades and mountain passes; very serious weather events in open areas; multiple governmental units taking care of certain roads and highways but not others; the cost involved in road clearing, road maintenance, and new road construction; and the high costs of delivery of services and materials to small communities. It is very difficult to get appropriate funding for necessary roads and maintenance services. There are over 450,000 rural bridges, many of them in very poor condition. About 50% of rural roads are not paved. (See endnote 28.)

Area planning and zoning may be limited or non-existent and therefore urban sprawl can easily occur. Typically, gateway communities which are next to recreational areas often struggle with seasonal sources of income and demand for services. Resource-dependent communities usually have a single industry and if the industry is challenged or leaves the area, there is a huge loss of jobs and serious problems for the small communities. Edge communities are on the fringe of metropolitan areas and connected by state or interstate highways and may be overwhelmed by a sudden influx of people who find the area an attractive place to live and easily accessible to jobs in the core city. They are not prepared for the increase in population and face numerous pressures including providing adequate housing and necessary infrastructure, plus schools, police, fire protection, emergency services, water, sewage, communications, and energy sources. The traditional main streets of the small communities are usually compact, historically significant, and easily accessible to transportation but still may struggle to get tenants for the stores when they compete with office parks, regional malls, and big-box stores. Second-home buyers in retirement communities have trouble keeping pace with the new growth while maintaining an excellent quality of life. Typically, there is a loss of forest land, especially in close proximity to metropolitan communities. There is also a loss of prime farmland. Where the rural communities are a distance from the urban centers and the transportation network is not well established, it is difficult for individuals to get to jobs, sources of education, and other services which may only be present in appropriate quality and quantity in the urban area. (See endnote 25.)

In the stable areas, there has been a dramatic demographic shift to an aging population. With the nation's population of 65 or older predicted to double in the next several decades, this will be the fastest-growing age group in the rural population. Low-income seniors may be below the poverty level and also lack affordable service options and housing options. Many more rural people have arthritis, asthma, heart disease, diabetes, hypertension, and mental disorders than populations in urban settings. This means that there are more disabled people who in fact are increasingly vulnerable to environmental pollutants. The numbers of healthcare professionals in rural areas are well below the amount that are needed to help prevent disease and injury, treat existing disease, and promote good health. (See endnote 22.)

The environment and ecosystems have also been affected by modern agricultural practices. There have been above-normal loads of phosphorus in bodies of water, which have increased the level of blue-green algae, for example, in Lake Erie. Another example is in North Carolina where the hog farming manure lagoons were dumped into the environment by a hurricane.

RURAL AREAS WITH POVERTY, LACK OF RESOURCES, AND JOBS

In the past, the conditions of rural America's housing and living situations in many areas of the country had presented a dismal picture of life. Many people lived in shacks or other structures without proper plumbing, sewage, safe water, heating, electricity, or kitchen facilities, and the upkeep was very poor. These families typically spent more than 50% of all their income for these housing facilities, which left little money for food, clothing, medical care, heating, etc. Many of the properties failed to meet the basic standards of the Department of Housing and Urban Development.

Agricultural productivity has increased substantially. Despite leading to increased sales of products, there has been a decrease in the total farm workforce, making jobs harder to find, resulting in lower incomes per family unit and increasing poverty. Educational levels have dropped accordingly. Health care is a constant challenge.

Today, even though there are many more structures available that are fine for residential use, there is a serious problem of affordability. Rural rental households typically have lower incomes than other parts of the country and have less experience with good housing. Over 9 million impoverished people live in homes and communities in rural areas that are moderately or severely substandard. Rural poverty varies considerably from one area of the country to another. Fewer jobs are available for poorly educated individuals in the traditional industries of mining, manufacturing, forestry and agriculture. Further, because of the physical characteristics of some rural areas such as being isolated, having poor infrastructure and limited economic development, there is little opportunity to improve these areas or the living conditions of the individuals. (See endnote 23.)

Rural homelessness is different than urban homelessness. Those affected are most likely to be white, female, married, working, and homeless for the first time. Families, single mothers, and children are part of this group. The homelessness may be caused by an extremely low-income base or domestic abuse. There is also a substantial amount of homelessness among Native Americans and migrant workers. Unfortunately, there are very few shelters that are available to help these individuals. (See endnote 24.)

In rural areas, the inhabitants of some places are highly transient. These areas typically have migrant labor living in poor housing with questionable water and sewage. Migrant labor camps which are provided frequently have highly congested and unsanitary living quarters for the workers and their children. There is a serious problem of a lack of communication skills in English for many of the rural individuals living in these areas. This has led to even more problems of disease and injury and an inability to get appropriate medical care of a preventive or curative nature. The migrant workers are subjected to high levels of agricultural chemicals including fertilizers and pesticides.

Best Practices for Improving the Built Environment in Rural Areas (See endnotes 21, 25, 26, 27)

- Create appropriate Planning Commissions with legal authority and taxing power to make and enforce decisions, while utilizing all significant stakeholders both public and private, for developing a comprehensive area plan which may include several political entities.
- Coordinate all planning activities as well as actual implementation with nearby towns and villages to be able to share resources, expenses, as well as experiences.
- Develop a comprehensive land-use map that indicates preferred development in areas and describe the type of uses to be carried out.
- Create appropriate annexation policies that conform to development standards and preserve the rural character of the area. This will help reduce the strain on necessary infrastructure and help prevent sprawl by adding communities or other structures that do not meet the overall comprehensive plan of the area.
- Conduct a comprehensive study of all roadways within the planning area and how best to extend them to major roads and highways where feasible. Determine which ones are unsafe and in need of increased widening and necessary improvement in structure. This is a function of the state highway department.
- Conduct a comprehensive study of sources of potable water supply currently in use and which will be needed for the future with an expansion of the community. Test all existing wells to determine if they are contaminated and make all necessary corrections.
- Determine if a public water supply in existence can be upgraded and extended or if a new water supply will be needed for the compact areas of the small town or city. In either case, necessary funding sources will have to be developed.
- Conduct a comprehensive study of all sources of sewage disposal and determine if they are working effectively and if they are not contaminating groundwater sources or surface water sources. Those which are not working properly will need to be corrected immediately. This is highly labor intensive work by environmental health professionals and will require special funding.
- Determine if a sewage treatment plant is in existence for the compact town or small city and if so, can it be extended or will a new sewage treatment plant have to be built. In either case, necessary funding sources will have to be developed.
- Install necessary stormwater systems, public sewers where feasible, and potable water systems, as part of any road construction or major repair.
- Conduct a comprehensive study of all means of solid waste disposal and hazardous waste disposal to determine if they are being conducted properly and if they may potentially contaminate the land, air, or water. All existing situations have to be corrected immediately.
- Develop additional means of removal and disposal of solid waste and/or hazardous waste to appropriate facilities to meet new demands of expansion of the community in such a manner that it does not create release of pollutants into the land, air, or water.
- Provide appropriate street lighting in all neighborhoods, business areas, and on roads.
- Create a street tree-planting program to help beautify the area while reducing potential air pollutants from motor vehicles.
- Develop appropriate funding of projects using a public–private mix of funds. In addition, public funding should be used for roads and highways, schools, emergency services, and necessary infrastructure.
- Make a comprehensive survey of all facilities and infrastructure to determine that which can be rehabilitated and that which can be used for new purposes and include in the master plan.
- Develop appropriate zoning laws, housing codes, fire codes, electrical codes, plumbing codes, etc. that are enforceable.
- Do not allow political influence or favoritism to create exceptions to zoning laws which will be in conflict with the overall plan for the area and may destroy valuable farmland or critical environmental areas such as wetlands.

- Develop a mixed land-use community incorporating the existing compact design, downtown area and associated new adjoining areas.
- Create desirable and safe walkways and bicycle paths.
- Utilize the existing charm of the community and coordinate new or reused structures with the existing look and feel of a rural downtown area.
- Preserve open space within the residential and commercial community for parks, green areas, and places for people to assemble as they desire.
- Preserve farmland and the natural beauty of the surroundings and protect critical sensitive environmental areas.
- Provide a variety of inexpensive but frequent means of transportation to all parts of the community and rapid transit to the core urban area or regional center where feasible.
- Provide a variety of communities with housing for various income groups including those who have low income and may be supported by various federal grants. The people in the communities need to have accessibility to jobs, various services, shopping, schools, and places of worship.
- Develop energy-efficient structures of all types including businesses and homes.
- Identify federal grants that may be used to assist in planned growth from federal departments including the US Department of Housing and Urban Development Community Development Block Grant Program, the US Department of Agriculture Community Facilities Direct Loan and Grant Program, and the US Environmental Protection Agency Brownfields Area-Wide Planning Pilot Program.
- Develop renewable energy sources by using wind farms, biomass, and other alternative energy sources to provide less expensive energy to the rural areas.
- Work with environmental groups and government to protect watersheds and enhance natural habitat to create opportunities for ecotourism and higher-paying jobs.
- Grow local fresh food as part of a regional food system which can be sold to suburban and even urban markets.
- Provide a broadband service to the rural communities to allow individuals engaging in electronic commerce to live in rural areas and thereby provide additional talents, skills and energy to help the community prosper.
- Plan for and encourage rural commercial development in small towns and cities by bringing in to the community in a special area, associated with the existing downtown corridor, new office buildings, stores, services, restaurants, medical facilities, entertainment facilities, educational institutions, etc.

SUB-PROBLEMS INCLUDING LEADING TO IMPAIRMENT FOR HOUSING

(See endnote 15)

SITE AND GROUNDS

The site of the structure, outside grounds and home have to be sound, safe, and in a good state of repair in order for it to be considered adequate shelter for individuals to have privacy and expect reasonable protection of their physical and mental health. There are numerous possible problems which can lead to potential disease and injury. They are related to the site which has been chosen, the building exterior, the building interior, the building systems, common areas, etc. Some problems of the outside of the structure include areas of erosion, holes in the ground, standing water, falling down fences, damaged retaining walls, cracks in pavement and fire ants; children's play area equipment, deteriorating paint, unprotected swimming pools, hanging or choking hazards, and damaged surfaces; loose railings or damaged outside steps; overflowing septic tank systems, stormwater systems that function improperly; and contaminated wells.

SUBSTANDARD HOUSING

Substandard housing conditions include poorly maintained housing and structural hazards such as: deteriorating walls, floors, ceilings with cracks and holes appearing; hazardous electrical and mechanical equipment; leaking pipes and leaking plumbing fixtures; mold and mildew; poor weather protection; infiltration of external air pollution; deteriorating and crumbling plaster, peeling paint and other coatings; and fire hazards from defective and/or poorly maintained equipment. There are frequent problems with: leaking roofs, walls, windows, doors; loose railings and loose steps; unsafe and improperly operating lights and wiring; inoperative or poorly maintained heaters and hot water heaters; and adequate supply of hot and cold running water. In some areas flooding or potential flooding is a great concern. Smoke detectors and carbon monoxide detectors are frequently either lacking or not operable.

BASIC SANITATION

Poor basic sanitation in areas of high population density and poor housing conditions is one of the contributors to the spread of disease and the increase of injury in the people living there. Basic sanitation refers to the quality and quantity of water supply, disposal of human waste and wastewater, solid and hazardous waste storage and disposal, control of insects and rodents, use of pesticides, personal hygiene, preparation and storage of food including refrigeration, and cleanliness of the structure.

Best Practices to Resolve Impairment in Housing

The Centers for Disease Control and Prevention and the US Department of Housing and Urban Development have worked together with many experts in various areas of housing and environmental health to develop a manual which incorporates all Best Practices in housing. The original manual was written in 1976, reprinted in 1988, and updated and revised in 2006. It is entitled *Healthy Housing Reference Manual* (Centers for Disease Control and Prevention, US Department of Housing, and Urban Development, US Department of Health and Human Services, Atlanta, Georgia, 2006). (See endnote 14.) It can be obtained from the Centers for Disease Control and Prevention by calling 1-800-CDC-INFO. It is current and easily usable. Obtain and utilize this document as needed.

The section below on "Specific Environmental Problems" will discuss the housing impairment problems and then reference will be made to indicate where in this publication the reader can find in-depth discussions as well as Best Practices.

TEMPERATURE EXTREMES

Temperature extremes especially affect the poor, children, elderly, and sick. The poor and sick live in structures that are not heated properly or air-conditioned, and in areas where there are high levels of crime and therefore windows and doors are kept closed and locked. This sharply decreases the amount of ventilation and allows the heat within structures to rise to dangerous levels. When it is very cold and there is insufficient heat, there is an increased risk of cardiovascular disease as well as pneumonia. Children are at very high risk from extreme weather conditions. When it is very cold, their resistance is lowered and they are more prone to upper respiratory diseases. When it is very hot, they breathe more rapidly for longer periods of time outdoor air including air pollutants. Older people have a significant problem since the individual's ability to regulate body temperature and adapt physiologically to the cold or heat decreases with age. In addition, older people are more prone to chronic disease and may be using medications which enhance cold or heat problems. All individuals are more vulnerable to infectious disease when their resistance is lowered by temperature extremes.

Best Practices for Protecting People during Temperature Extremes
- Provide air-conditioned facilities such as shopping malls, public libraries, or specially cooled shelters during periods of extreme heat especially for the elderly, those with chronic conditions, infants, and children.

- Drink cool, non-alcoholic beverages to keep well hydrated.
- Provide heated buildings for those subjected to extreme cold conditions, especially for the elderly, those with chronic conditions, infants, and children and those who have been exposed to substantial cold conditions.
- Drink warm non-caffeinated and non-alcoholic beverages.
- Avoid strenuous exercise in severe cold and dress appropriately in layers of clothing.
- Be aware of individuals who may not be able to take care of themselves in severe heat and extreme cold and provide necessary services to remove them to appropriate shelters.

SPECIFIC ENVIRONMENTAL PROBLEMS

AIR POLLUTION

People are exposed to many different types of air pollutants in the outside air depending on their location, time of year, weather conditions, and various environmental factors. They have an existing body burden of pollutants and when they enter their various residences they are exposed to a series of indoor air pollutants which increase their opportunity for disease. See Chapter 2, "Air Quality (Outdoor [Ambient] and Indoor)" for further details of air pollutant problems and potential health effects as well as Best Practices to follow to help reduce the problems.

ASBESTOS

(See endnote 36)

Asbestos is a group of naturally occurring silicate compounds found in the environment in bundles of fibers. Human exposure to the fibers when disturbed can result in asbestosis, non-malignant lung and pleural disorders, lung cancer, mesothelioma, and other cancers. Smokers are at greater risk of lung cancer when exposed to the disturbed fibers. Remodeling an older home containing asbestos products can result in disruption of the fibers and inhalation of them. Also crumbling drywall, insulation, certain roofing materials, textured paint and patching compounds, some vinyl floor tiles, hot water and steam pipe coverings, stovetop pads, and other materials containing asbestos can release the fibers to the air.

Best Practices in Preventing Asbestos Related Problems
- Do not disrupt intact non-damaged building materials containing asbestos fibers. Covering them appropriately will contain them.
- Do not saw, sand, scrape or drill holes in asbestos-containing materials and do not dust, sweep or vacuum any debris that contains asbestos.
- If asbestos problems do exist, use only licensed professionals to make the necessary inspections, removal if necessary and sealing of the surfaces where needed.

BROWNFIELD SITES

(See endnotes 32, 34)

A brownfield is a property which is contaminated with a variety of chemicals that are embedded either in the earth and/or water beneath it. A brownfield site is a property which is being considered for redevelopment or reuse. There are about 450,000 brownfields in the United States. A brownfield may be a very valuable piece of property which when properly evaluated for types and quantity of contamination and then safely cleaned up, could be used for reinvesting in the community, increasing the tax base, providing new sources of employment as well as new commercial and residential areas. Problems in converting the brownfield site include environmental liability for the past and the future of the property; financial barriers by private lenders because of the impaired land; the nature and cost of the clean-up operations; and the reuse planning which must be based on community goals, and sound economic and environmental information.

Best Practices in Reusing Brownfield Sites
- Search for the responsible parties that caused the contamination. Even if the companies are out of business, the successor companies are responsible for the clean-up costs of the brownfield sites.
- Conduct a comprehensive survey of the given community to determine the location, size, quantity and level of contamination of brownfield sites that could be redeveloped.
- Determine which of the brownfield sites should be given priority to fit into the overall community plan development and redevelopment and do environmental assessments to find out the specific problems that need to be rectified and the cost of this action. Also determine if a clean title to land can be obtained.
- Identify and involve critical participants including local public health officials in all phases of the brownfield planning, transaction and redevelopment and how best to effectively use the land, how best to get various types of financing, determining any tax credits, tax abatements or grants that can be used for the project. This would include property owners, public and private stakeholders, attorneys, local, state and federal government officials, and other interested people.
- Conduct a thorough cleanup of the site based on the environmental survey conducted. Redevelopment can go on at the same time as cleanup, to avoid downtime and additional cost.
- Establish a long-term management and maintenance program where needed, especially when water pumps and treatment systems are required to make sure that the site stays clean and uncontaminated.

CHROMATED COPPER ARSENIC

(See endnote 36)

Chromated copper arsenic is a chemical used in manufacturing wood products to resist decay and rot. The wood products are used for outdoor decking and children's play sets. The chemical is highly toxic and a human carcinogen. The products are no longer available for homes and other types of uses, however there may be a considerable amount of the products still in use in various parts of the country.

Best Practices for Eliminating Chromated Copper Arsenic
- Avoid being exposed to wood products which contain chromated copper arsenic. This is especially important for children.
- Where possible replace the chromated copper arsenic products with other types of materials.

COMBUSTION POLLUTANTS

(See endnote 36)

Combustion pollutants may be gases or particles, usually carbon monoxide, nitrogen dioxide, and particulates, and are created by burning organic materials in a variety of chambers including furnaces, fireplaces, ranges, and ovens. They are also produced by water heaters and clothing dryers.

Best Practices for Controlling Combustion Pollutants
- Use ENERGY STAR equipment, or the equivalent, which has sealed combustion units; use as well vented furnaces, boilers and hot water heaters.
- Use clean burning wood stoves and fireplaces certified by the US Environmental Protection Agency.
- Provide proper ventilation for all appliances and necessary exhaust fans.
- Clean all chimneys and other means of exhaust from the appliances on a regular basis to prevent buildup of deposits.

Drywall (from China)

(See endnote 35)

Owners of homes from 10 states and the District of Columbia, where drywall from China had been used, have complained about odors similar to rotten eggs, which in fact is a release of volatile sulfur compounds. This has caused copper water lines throughout the house to become corroded. This has also caused serious corrosion within electrical switches, wiring, and other places where copper is used. Further, people have been complaining about problems with asthma, respiratory irritations, breathing problems, eye irritations, and headaches. Various federal agencies including the Centers for Disease Control's Prevention Agency for Toxic Substances and Disease Registry conducted tests of Chinese drywall and found that sulfur was detected in the Chinese material but not in the American drywall. Also strontium was found at levels considerably higher than in the US drywall. However, one company in the United States has also had the same problems as the Chinese drywall companies.

Best Practices for Drywall from China
- Remove all Chinese drywall from homes and replace with that which is made in the United States.
- Replace all damaged copper including switches, wires and water lines.
- Thoroughly ventilate the entire premises.

Emergencies and Disasters

Many parts of the country frequently experience natural disasters such as serious heat conditions, winter storms, tornadoes, hurricanes, earthquakes, floods, wildfires, etc. Further, there is a serious potential for acts of terrorism which could kill and injure large numbers of people and cause substantial damage to property. (See Chapter 5, "Environmental Health Emergencies, Disasters and Terrorism for further information and Best Practices.")

Environmental Injuries

The home and its environment is typically a very unsafe place. Many unintentional injuries and deaths occur from falls, fires, drowning, poisoning, etc. (See Chapter 6, "Environmental and Occupational Injury Control for further information and Best Practices.")

Food Protection

Tens of millions of Americans get sick from eating contaminated or improperly stored or prepared food in the home. This is especially true of leftovers and because of improper refrigeration. Hand washing and proper cleaning and sanitization of surfaces are typically not performed properly, leading to outbreaks of disease. (See Chapter 7, "Food Security and Protection for further information and Best Practices.")

Formaldehyde

(See endnote 38)

Formaldehyde is a colorless flammable gas at room temperature with a pungent odor and causes burning to the eyes, nose, and lungs at high concentrations. It is produced by a variety of manufacturing processes and released to the outdoor air from power plants, manufacturing facilities, and automobile exhausts. In indoor air, it is released into the structure from building materials including: most types of particle board used for subflooring, shelving, and room paneling; finished products such as cabinets and furniture; and consumer products and tobacco smoke.

Best Practices for Eliminating Formaldehyde
- Where formaldehyde is an existing problem and the family cannot move to other facilities, use air conditioning and dehumidifiers to reduce the temperature and control the humidity. Heat and high humidity increase the amount of formaldehyde being released into the air.
- Continuously ventilate the property as much as possible.
- Where new construction is occurring within the structure, use only composite wood products which meet the standards of the American National Standards Institute 2009 or the California Air Resources Board Airborne Toxic Control Measures to Reduce Formaldehyde Emissions from Composite Wood Products.

HAZARDOUS HOUSEHOLD WASTE

(See Chapter 12, "Solid Waste, Hazardous Materials and Solid Waste Management")

There are numerous sources of household hazardous waste. They include anti-freeze; batteries of all types and sizes; compact fluorescent light bulbs; various electronic devices including old TVs; medical waste including bandages, discarded surgical gloves, discarded needles, infectious mucus, etc.; medicines of all types which should never be flushed down the drain because of contamination of the water; mercury-containing equipment including thermostats, thermometers, and other products; used oil; paints and varnishes; pesticides and fertilizers; solvents; etc.

Best Practices in Disposal of Hazardous Household Waste
- Use and store hazardous household materials in a safe manner away from food and in the original containers.
- Never mix remnants of hazardous household materials with other waste since it may corrode containers and cause fires or explosions and make the materials unable to be recycled.
- Take all hazardous household materials to permanent collection points or exchanges for appropriate disposal or recycling.
- Where permanent collection points do not exist, the community should establish special collection days to remove all household hazardous waste in a safe and appropriate manner and use proper disposal techniques.

INDOOR AIR POLLUTION

Indoor air pollutants such as allergens and other biological contaminants, asbestos, carbon monoxide, environmental tobacco smoke, formaldehyde, nitrogen dioxide, pesticide residues, radon, wood smoke, respirable particles from sources of combustion, and volatile organic compounds can cause headaches, nausea and fatigue, and can be either indirect or direct health hazards. A special problem is associated with the diisocyanates in polyurethane products. Polyurethane is used in floor finishes and also as a foam insulation material. Since contact with the vapors or particles if inhaled may be very dangerous, residents should leave the home or other area during the spraying operation and until all of the spraying material is removed from the air within the structure. (See endnotes 34, 39.) (See Chapter 2, "Air Quality (Outdoor [Ambient] and Indoor)" for an in-depth discussion of the problem and Best Practices.)

INSECTS AND RODENTS

(See Chapter 9, "Insect Control, Rodent Control and Pesticides")

LEAD

Housing that was built before 1950 or renovated before 1978 typically had lead-based paint on the walls and other surfaces both inside and outside. Over time the paint has deteriorated and there are both dust-containing lead-based materials and paint chips that are readily available for young children to ingest as they crawl along floors and put their hands into their mouths. Remodeling and repainting projects are of special concern because unless precautions are taken, the amount of dust containing lead increases substantially. Lead dust spreads readily through the structure or exterior soil and is difficult to adequately clean up.

There are also some children who chew on windowsills and other objects which may contain lead. Children playing outside also can ingest lead which has contaminated the soil around the structure. Over time the lead accumulates in the child's blood and soft tissue and creates toxic effects. The lead is very toxic to the brain, many organs and systems, and impairs neurological development.

Additional lead exposure may occur from old lead pipes or solder which is corroded by water and therefore creates additional hazards. Adults working in various industrial operations as well as in painting can bring residues of lead back to their homes on their shoes and on their clothing.

Best Practices in Preventing Lead Poisoning
- Select areas of communities where housing may likely contain lead-based paints and develop programs to protect children from lead poisoning.
- Determine from blood tests if children within the target area have elevated lead blood levels, which has now been set at 5 µg of lead/dL of blood, and provide necessary medical treatment plus appropriate removal of lead-based paints from the surfaces if the families cannot move to other quarters.
- Frequently remove dust including lead dust, paint chips, soil, and debris from the premises by using dust-free cleaning techniques.
- Educate the children in schools about the potential for lead poisoning and teach them to wash their hands thoroughly, wipe their feet after coming in from the outside and send home educational material to their parents in multiple languages if necessary concerning the dangers of lead.
- Teach parents when they attend public health clinics about the sources of lead poisoning, the symptoms that the children will be showing, and how to prevent the problem from occurring and where to get necessary help if the child is having problems.
- Use community-developed and community-based prevention/intervention strategies by community leaders to teach residents about the severe results of lead ingestion by children.
- Do not disturb paint surfaces that are intact.
- If a child shows that he/she is one that regularly puts unusual objects into the mouth, immediately seek medical assistance to determine if there is a lead-associated health problem.

MOLD

Mold is a highly significant public health problem within the indoor environment of all structures. It may cause discomfort or worse, including allergic reactions, asthma, respiratory problems, or forms of pneumonia. The mold whether dead or alive may become airborne and be inhaled by the individuals and workers removing it, and is especially a problem for children. Moisture of any sort when allowed to persist is the underlying support needed for mold growth. Dust within the premises may contain mold and readily become airborne.

Best Practices for Prevention and Removal of Mold
- Determine the extent of and rapidly correct the underlying moisture problems.
- Repair or replace all damaged building materials and contents.

- Establish a routine maintenance schedule to determine if there are any leaking pipes, leaking equipment, or potential leaks in the roof or around windows and doors.
- Remove all fungal contaminated material including the mold contaminants in dust in a safe and effective manner.
- Use appropriate personal protective equipment while removing mold contaminants.
- For porous materials such as carpet, upholstery, etc., bag or wrap in plastic and discard appropriately and then clean with HEPA vacuum cleaners to remove dust.
- For semi-porous materials such as solid wood furniture, resilient floor coverings, etc., use HEPA vacuuming, damp cleaning with soap, water, and disinfectant such as chlorine compounds, and then dry thoroughly.
- For non-porous materials such as metal, ceramic tile, porcelain, etc., use HEPA vacuuming, damp cleaning with a detergent solution and rapid drying. (See endnote 40.)

NOISE

(See endnote 41)

Environmental noise, which is unwanted sound, is found in many areas of the community. It is created on the highways, by railroads, by airplanes, and many other sources especially through construction. It may interfere with sleep, concentration, communications, and various activities of normal life. Hearing impairment is a frequent result of too much noise. The cardiovascular system may be disturbed and there may be an increase in blood pressure, heart rate, and constriction of blood vessels. Mental health may be affected and in severe cases the noise may exacerbate problems of hysteria and even psychosis. The ability to perform various tasks can be impaired by excess noise. Even at best, noise can be simply annoying and distracting. Those most affected may be individuals with a variety of existing acute and chronic diseases, hospital patients, fetuses, infants, young children, and the elderly.

The frequency and severity of the problem grows with an increase in population and urbanization. Home noises include lawnmowers, leaf blowers, garbage disposals, washers and dryers, air conditioners, swimming pool pumps, etc. Toys can be a considerable source of noise irritation. Blaring television sets or pounding radios can create enormous distractions and cause harm. Unfortunately, the accumulative effects of these noises from a variety of sources cannot be dealt with by regulations because it would affect the rights of individuals to operate the equipment any time they want to do so. It is only possible to control the level of noise coming from individual pieces of equipment and thereby regulate them.

Best Practices for Preventing and Mitigating Noise Problems in Residential Areas and Structures (See endnotes 42, 43)
- Develop and enforce noise control laws that utilize the most recent scientific data on the potential for affecting the health of the public.
- Establish noise control emission standards for road and off-road vehicles, equipment and industrial plants, and conduct noise monitoring to enforce the standards.
- Reduce speed limits in residential areas as well as around hospitals, and other healthcare facilities and those for the aging.
- Phase in appropriate new technologies for engines and road surfaces in order to reduce road noise.
- Establish specific roads for use by heavy trucks and other noisy vehicles outside of residential areas.
- Reduce railway noise at the source by appropriate maintenance of the rails and wheels of the trains. Mandate rubber wheels where possible.
- Reduce aircraft noise by utilizing the latest technology for aircraft and by altering takeoff and landing patterns, especially at night. The regulations would have to be set and enforced by federal authorities.

- Reduce noise from machines and equipment by replacing older equipment and utilizing proper preventive maintenance programs.
- Enforce community ordinances against individuals who have extremely loud music in automobiles. This should be a priority police function especially in quiet neighborhoods.
- For outdoor events instead of using a few very large speakers utilize numerous smaller speakers, and distribute them appropriately throughout the audience.
- Conduct a community-wide noise survey to determine hotspots where noise levels exceed those that are recognized as maximum for the health and safety of the public and act upon these areas initially.
- Reduce noise levels within communities and structures to the lowest possible level using cost-effective methods within specific situations and charge the polluter with the full cost of noise pollution including monitoring, management, lowering noise levels and supervision if the individual or company refuses to do so.
- Reduce noise levels at the source wherever possible by proper land-use planning and determining in advance through use of Environmental Impact Statements the sources, frequency, quantity and time of day of potential noise, and prevent or mitigate the problem.
- For indoor noise from various types of equipment, establish a routine maintenance program and replace older equipment with newer equipment.
- Sound proof ceilings, walls, doors, and windows within the structure to reduce noise levels.
- Establish a special noise control ordinance for indoor areas and develop a noise complaint program where individuals can contact an official agency in the event of excessive noise from neighbors or other sources.
- Develop educational programs to teach the public about the health impacts of noise and how to minimize sound levels by use of earplugs, earmuffs and proper insulation in structures, and use of sound reduction materials within buildings.

ON-SITE SEWAGE DISPOSAL

On-site sewage systems or septic tank systems are used in the United States by over a third of all homes. They are predominantly the main means of disposal of liquid waste in rural areas and in many suburban areas. The homes may be spaced at a sizable distance from each other or in housing developments use may be located away from normal public sewer pipes and outlets. As people move into more distant areas from central cities and towns the need for individual sewage systems continues to increase. Immediately after World War II with the return of the service people from abroad and discharge from the Armed Forces, the individuals married and started to create new families leading to a population explosion and a need for considerably more housing. With the advent of the Interstate Highway System, it was now possible to live outside of the core city and still get to work in a reasonable period of time. This led to the development of large tracts of farmland which used to support a single septic tank system and well for many acres of land and now had to support hundreds of septic tank systems and wells, resulting in massive on-site sewage system failures because the soil was simply not able to accept the high levels of liquid being produced. The soil became clogged as the natural fauna had been removed and replaced by houses; and the individuals who were city bred had no idea about the proper usage of water in this type of situation and overtaxed the sewage systems as well as the wells. Water softeners were frequently added as well as garbage disposal units and each of these systems had a serious effect on the on-site sewage disposal process. In colder climates, there were potential problems of freezing of these systems because of lack of snow cover, compacted soils, lack of plant cover, and pipes not draining properly. Of great significance was the type of soil being used for disposal of the sewage effluent, such as sand versus clay, and the necessity for proper permits from the local health departments to ensure that new on-site sewage disposal systems would be constructed properly. Recognize that the standard mentioned earlier utilized in Oakland County, Michigan of 1 acre of land per three-bedroom house and 1.25 acres of land per four-bedroom house helped to alleviate part of the problems of disposal of the on-site sewage if the soil allowed it, water usage was contained, garbage disposal units were not used, and maintenance of the system was performed on a regular basis. (See endnote 14.)

(See Chapter 11, "Sewage Disposal Systems" for further information and Best Practices in developing new systems as well as maintaining and rejuvenating old systems.)

On-Site Water Source

On-site water sources, especially wells, are used by about 15% of the population of the United States. These are especially prevalent where on-site sewage disposal systems are used. Many of the early wells after World War II were mass-produced in order to provide water for rural areas where housing developments were being built and no public water supply could reasonably be extended to the area. The wells were put on pieces of land that were parts of acres or even an acre or two and used an aquifer that had been the source of water for the single farm family for a few people and now had to be used for many people. This led to the drawdown of the water from the immediate area and extended the distances considerably, thereby potentially drawing in pollutants that had been either dumped on the land or buried. It also led to contamination by the various septic tank systems because of the small parcels of land that contain both sewage disposal and wells. Livestock yards, silos, petroleum tanks, manure storage, pesticides and other agricultural chemicals, and use, storage and handling areas also became potential sources of contamination of the water supply. (See endnote 14.)

(See Chapter 13, "Water Systems (Drinking Water Quality)" for further discussion on water supply and Best Practices in digging wells as well as maintenance and upgrading of existing wells and other water supplies.)

Pesticides

(See Chapter 9, "Insect Control, Rodent Control and Pesticides" for description of problems and Best Practices.)

Poisoning

Poisoning is of considerable concern especially for young children and the elderly and disabled. Children commonly ingest a multitude of substances including cosmetics, personal care products, cleaning products, pain relievers, prescription medicines, etc. Carbon monoxide poisoning occurs from poorly vented furnaces and fireplaces, appliances, gas generators, portable stoves, etc. Lead poisoning is an ever-present concern in older housing. Among the elderly, poisoning occurs because of poor vision, confusion and overuse of prescription medicines. (See Chapter 4, "Children's Environmental Health Issues," and Chapter 6, "Environmental and Occupational Injury Control" for further information and Best Practices.)

Solid Waste Disposal

(See Chapter 12, "Solid Waste, Hazardous Materials, and Hazardous Waste Management" for information and Best Practices for solid waste disposal in residential areas.)

Swimming Areas (Residential)

(See Chapter 10, "Recreational Environment and Swimming Areas" for information and Best Practices for residential swimming areas.)

LAWS, RULES, AND REGULATIONS

ZONING ORDINANCES

Any given state has police powers through their constitution to protect the health and safety of the community. This includes the issuing of zoning laws specifying which areas may be utilized to construct a variety of structures, and regulate building density and size. They also determine where land should be utilized for industrial parks, special types of manufacturing, and agricultural, commercial and residential areas. The various specifications for use of different parcels of land can be modified by the Zoning Board. Because of the large number of zoning ordinances based on local circumstances, this subject will not be discussed in a specific manner.

BUILDING AND HOUSING CODES

Building and housing codes are designed to make buildings safe, sanitary, and efficient in operation. Although there are innumerable building and housing codes throughout the country, there are some basic resources and documents that communities can utilize to bring their own codes to a satisfactory level. (See the section "Special Resources" below for details.)

SPECIAL RESOURCES

Centers for Disease Control and Prevention, US Department of Health and Human Services and the US Department of Housing and Urban Development, *Healthy Housing Reference Manual*. (See endnote 14.)

Centers for Disease Control and Prevention, US Department of Health and Human Services and the US Department of Housing and Urban Development, *Healthy Housing Inspection Manual*. (See endnote 15.)

Conference of Building Officials "Uniform Building Code," "International Residential Code," and "National Electrical Code," etc. (See endnote 16.)

Centers for Disease Control and Prevention Planning Tools for communities that integrate public health into all planning efforts. (See endnote 17.)

The US Department of Agriculture Rural Development Program provides housing and community assistance for individuals and families through loans, direct loans, and grants. It provides housing for the elderly and farm laborers, childcare centers, homeless shelters, domestic violence shelters, and other social service agencies in rural areas. (See endnote 30.)

The US Department of Housing and Urban Development provides numerous programs for various facets of housing.

The US Department of Housing and Urban Development Office of Native American Programs provides housing assistance to Native Americans. (See endnote 30.)

The US Department of Interior Bureau of Indian Affairs provides a Housing Improvement Program for housing repair assistance and for the homeless for Native Americans with incomes below 125% of the poverty level. The individuals need to either be homeless or living in substandard housing. (See endnote 30.)

The Department of Veterans Affairs Office of Rural Health provides access to quality health care to rural veterans and Native American veterans and tribal communities. (See endnote 30.)

SUCCESSFUL PROGRAMS

(See endnotes 20, 29, 33)

Dublin, Ohio, and the Bridge Street Corridor

Dublin, a suburb of Columbus, converted a 1000-acre area which was in the core of its community along Bridge Street from a low-density automobile-dependent one to a high density, mixed-use pedestrian friendly one with seven districts including new office buildings, retail centers, entertainment centers, and residential structures, which is easily accessible by major highways. Detailed transportation studies were made and appropriate new inexpensive transportation was added to the area. This new development provides places for younger workers to live and work and a place to live for those who want to retire to an active lifestyle in an urban area without having to move back into the core city. There is ample room for planned growth as more people decide to enjoy this lifestyle. As part of future planning, the city envisions constructing a pedestrian bridge across the Scioto River to connect to a park. Public and private financing has been used to establish this project.

Aurora Corridor, Shoreline, Washington

Shoreline, a suburb of Seattle, has changed the land use patterns of a 3-mile portion of State Route 99 from a highly congested complicated road which contributed to frequent accidents to a road with landscaped medians, sidewalks on each side of the road, a trail paralleling the road, new lighting, and a public plaza for recreational use and gathering of people. Public transportation has been upgraded. Business access lanes have been added to the road. The city upgraded all of the infrastructure including water, sewer, electricity, and communications. The individuals owning property on either side of the road have been encouraged to upgrade and enhance the properties and properly use unused or poorly used space. A 5-acre plot was set aside for senior housing, affordable housing and a new YMCA building. The transformation of the area has allowed the landowners to increase their rents and to make the locations even more appealing. All of this work has resulted in a sharp reduction in accidents and deaths along the strip of the road as well as an increase in revenues for the city.

Belmar, Lakewood, Colorado

A 103-acre site including a large mall which no longer operated well and had substantial parking areas, has been transformed into 22 urban-type blocks, based on a grid system, containing 880,000 square feet of retail space, 250,000 square feet of office space, 5000 public parking spaces, and 800 residential units both owner-occupied and rental, as well as two educational institutions and public plazas and parks as well as green spaces. Transportation to and from Denver was increased appropriately to meet the new and future demands. The diversity and attractiveness of the area as well as the ability to walk safely has already increased sales in the various stores and has created a perception of success.

City Centre, Houston, Texas

An obsolete mall property in the suburbs of Houston, Texas, was converted into a thriving community of offices, residential areas, and entertainment sources. The development has 700 residential units of all types. It is located at the intersection of major highways and is accessible to millions of people. It has 4000 spaces for parking but emphasizes pedestrian circulation through the entire area.

Indianapolis, Indiana

Two former gas stations, which were brownfield sites, had the underground storage tanks and also all other sources of contamination removed. The area has been redeveloped into residences, with storefront retail use on the lower level and condominiums on the upper level of the structures.

WORCHESTER, MASSACHUSETTS

A group of stakeholders with assistance from the federal government cleaned up and decontaminated a brownfield site of 30 acres, which had been a metal fabrication, auto dealer and repair shop, to provide better living conditions for a culturally mixed area. Vacant lots and substandard residential properties were converted into affordable housing where ENERGY STAR products were used and materials were environmentally friendly. The 7.4 acres of underutilized industrial buildings which had been contaminated were converted into a new Boys & Girls Club and outdoor athletic complex.

GREENVILLE, SOUTH CAROLINA

The City of Greenville in the foothills of the Blue Ridge Mountains from 1982–2004 added $160 million of new residential construction and has spent another $100 million in renovating older neighborhoods. This has resulted in a thriving downtown area and what is being called a "state-of-the-art" community in which to live. However, 1.2 miles from this downtown the West Greenville Center area of 230 acres consists of large numbers of people (mostly from minorities) living below the poverty level, high unemployment, and dilapidated and limited affordable housing with extremely poor infrastructure. The area has a large number of brownfields, mostly large, abandoned or barely used industrial operations. In addition, there are gas stations, dry-cleaning facilities, railroad properties, and other sites with unknown amounts of environmental contamination. The city, working with several Environmental Protection Agency programs and grants in partnership with Clemson University and others did an intensive study to determine the environmental problems existing in the various brownfield sites. Ground-penetrating radar was used as one of the techniques for determining underground storage tanks and contamination. Necessary cleanup for decontamination followed on several properties. One project involved collaboration between the Upstate Homeless Coalition of South Carolina, the South Carolina Department of Mental Health, the Greenville Mental Health Center, the Phoenix Center for substance abuse, community churches, the US Department of Housing and Urban Development, state housing programs, and the City of Greenville. A brand-new structure was built that provides a safe haven for the chronically homeless in Greenville and in South Carolina. The vast majority of the people seeking shelter are mentally ill as well as homeless. Another property was purchased by the Salvation Army of Greenville and it built a community center for disadvantaged residents and their families. Besides this group, a local Boys & Girls Club has access to and uses the facility for programs. A new elementary school is being built next to the community center. Critical housing and care will follow.

BALTIMORE, MARYLAND, HOUSING AUTHORITY

The Baltimore Maryland Housing Authority was established in 1937 when the federal government gave the states and localities the funds to fight urban slums. Baltimore was one of many cities that developed public housing programs and replaced slum tenements with clean, safe, and affordable housing for poor and working families. It ensures that all citizens in Baltimore have a decent place to live in a decent neighborhood. It has been extremely successful and is currently operating at a high level of performance. It serves over 20,000 people in some 10,000 housing units which are maintained properly for people who would otherwise be living in conditions that can be part of the underlying problems that lead to acute and chronic disease and injury. The Housing Authority also provides assistance to a variety of people regarding: valuing vacant lots and properties; code enforcement; community services; fair housing; green, healthy, and sustainable homes; land resources; resident services; housing voucher programs; permits; and plans and reports. (See endnote 44.)

ENDNOTES

1. Koren, Herman, Bisesi, Michael S. 2002. *Handbook of Environmental Health: Biological, Chemical and Physical Agents of Environmentally Related Diseases*, Volume 1, Chapter 8-Indoor Environment, Fourth Edition. Boca Raton, FL: CRC Press.
2. Perdue, Wendy Collins, Stone, Leslie A., Gostin, Lawrence O. 2003. The Built Environment and Its Relationship to the Public's Health: The Legal Framework. *American Journal of Public Health*, 93(9):1390–1394.
3. National Coalition for the Homeless. 2009. *Why Are People Homeless*. Washington, DC.
4. European Green Capital. No Date. *Environmental Best Practice and Benchmarking Report: European Green Capital Award 2012 and 2013*. Dublin, Ireland.
5. European Green Capital. 2014. *Expert Panel: Technical Assessments Synopsis Report: European Green Capital Award 2016*. Dublin, Ireland.
6. US Environmental Protection Agency. 2013. *Our Built and Natural Environments: A Technical Review of the Interactions among Land Use, Transportation, and Environmental Quality*, Second Edition, EPA 231K13001. Washington, DC.
7. Williams, Ronald A. 2000. Environmental Planning for Sustainable Urban Development. *Caribbean Water and Wastewater Association Ninth Annual Conference and Exhibition Chaguaramas*, Trinidad.
8. Union of British Columbia Municipalities. 2009. *Foundations for a Healthier Built Environment: Summary Paper*. Richmond, BC.
9. United Nations General Assembly. 1996. *The Habitat Agenda: Chapter IV: Human Settlements Development in an Urbanizing World*. New York.
10. Performance Based Building Thematic Network. 2001. *Domain 4-Performance-Based Built Environment-Urban Regeneration, Methods and Actions to Achieve the Urban Regeneration*. University of Reading, Berkshire, United Kingdom.
11. Performance-Based Building Thematic Network. No Date. *Domain 4-Performance-Based Built Environment-Recommendations, Urban White Paper and Urban Policy*. University of Reading, Brookshire, TX.
12. Dixon, Tim. 2011. *Sustainable Urban Development to 2050: Complex Transitions in the Built Environment of Cities*. WP2011/5. Oxford Institute for Sustainable Development, Oxford Brookes University, Readington.
13. Gray, Colin. 2005. *Built Environment-PeBBu Domain 4-Final Domain Report*. Performance-Based Building Thematic Network, University of Reading, Brookshire, TX.
14. Centers for Disease Control and Prevention, US Department of Health and Human Services, US Department of Housing and Urban Development. 2006. *Healthy Housing Reference Manual*. Atlanta, GA.
15. Centers for Disease Control and Prevention, US Department of Health and Human Services, US Department of Housing and Urban Development. 2008. *Healthy Housing Inspection Manual*. Atlanta, GA.
16. Centers for Disease Control and Prevention. 2009. *Standards and Organizations*. Atlanta, GA.
17. Centers for Disease Control and Prevention. 2014. *Health Planning Tools*. Atlanta, GA.
18. US Department of Transportation, Federal Highway Administration. No Date. *Pedestrian Safety Strategic Plan: Recommendations for Research and Product Development*. Washington, DC.
19. London Development Agency, the Urban and Economic Development Group. 2006. *Welcome to Tomorrow's Suburbs Best Practice Guide-Tools for Making London Suburbs More Sustainable*. London, United Kingdom.
20. MacCleery, Rachel, Peterson, Casey, Stern, Julie D. 2012. *Shifting Suburbs: Reinventing Infrastructure for Compact Development*. Urban Land Institute, Washington, DC.
21. Duncan, Cynthia M. No Date. *Community Development in Rural America: Collaborative, Regional and Comprehensive*. Federal Reserve Bank of San Francisco, Low Income Investment Fund, Investing in What Works for America's Communities, San Francisco, CA.
22. Housing Assistance Council. 2012. *Affordable Rural Housing for Seniors*. Rural Voices, 17:4. Washington, DC.
23. National Low Income Housing Coalition. 2011. *2011 Advocate's Guide to Housing and Community Development Policy*. Washington, DC.
24. National Coalition for the Homeless. 2012. *Rural Homelessness*. Washington, DC.
25. Mishkovsky, Nadejda. 2010. *Putting Smart Growth to Work in Rural Communities*. International City/County Management Association (ICMA), Washington, DC.
26. Nelson, Kevin. 2012. *Essential Smart Fixes for Rural Planning, Zoning, and Development Codes*. US Environmental Protection Agency, Washington, DC.

27. National Center for Appropriate Technology (NCAT). 2012. *National Conversation on the Future of Our Communities-Compendium*. (Supported by US EPA) Butte, MT.

28. US Department of Transportation, Federal Highway Administration. No Date. *Planning for Transportation in Rural Areas-Transportation System*. Washington, DC.

29. US Department of Agriculture, Rural Development. 2014. *Sustainable Rural Downtowns Case Studies*. Washington, DC.

30. United States Interagency Council on Homelessness. No Date. *Federal Resources for Rural Communities*. Washington, DC.

31. Blarch, Lawrence Frank, Kavage, Sarah, Devlin, Andrew. 2012. *Health in the Built Environment: A Review*. Canadian Medical Association, Ottawa, ON.

32. US Environmental Protection Agency. 2006. *Anatomy of Brownfields Redevelopment-Brownfields Solutions Series*. Washington, DC.

33. US Environmental Protection Agency, Office of Underground Storage Tanks. 2014. *Reusing Cleaned up Petroleum Sites*. Washington, DC.

34. Greenberg, Michael. 2002. Should Housing Be Built on Former Brownfield Sites. *American Journal of Public Health*, 92(5):703–705.

35. US Environmental Protection Agency, Environmental Response Team, Edison, New Jersey, personal communication to Lynn Wilder. 2009. *Drywall Sample Analysis*. Atlanta, GA.

36. US Environmental Protection Agency. 2012. *Green Building – Protecting Your Health*. Washington, DC.

37. Environment Canada. 2013. *Recommendations for the Design and Operation of Wood Preservation Facilities, 2013 Technical Recommendations Document*. Gatineau, QC.

38. Agency for Toxic Substances and Disease Registry. 2008. *Public Health Statement for Formaldehyde*. Atlanta, GA.

39. Jacobs, David E., Kelly, Tom, Sobolewski, John. 2007. Linking Public Health, Housing, and Indoor Environmental Policy: Successes and Challenges of Local and Federal Agencies in the United States. *Environmental Health Perspectives*, 115(6):976–982.

40. Minnesota Department of Health, Environmental Health Division, Indoor Air Unit. 2014. *Recommended Best Practices for Mold Remediation in Minnesota Schools*. St. Paul, MN.

41. Goines, Lisa, Hagler, Louis. 2007. Noise Pollution: A Modern Plague. *Southern Medical Journal*, 100, 287–294.

42. US Environmental Protection Agency. 2013. *Noise Guide for Local Government – Part 3 Noise Management Principles*. Washington, DC.

43. Berglund, Brigitta, Lindvall, Thomas, Shewela, Dietrich H., eds. 1999. *Guidelines for Community Noise*. World Health Organization, Geneva, Switzerland.

44. Baltimore Housing Authority. No Date. *Reflecting on Our past, Preparing or Future HABC Celebrates 75 Years of Public Housing in Baltimore*. Baltimore, MD.

BIBLIOGRAPHY

Brennan Ramirez, Laura K., Baker, Elizabeth A., Metzler, Marilyn. 2008. *Promoting Health Equity – A Resource to Help Communities Address Social Determinants of Health*. Centers for Disease Control and Prevention, Atlanta, GA.

Krieger, James, Higgins, Donna L. 2002. Housing and Health: Time Again for Public Health Action. *American Journal of Public Health*, 92(5):758–768.

National Alliance to End Homelessness. 2015. *The State of Homelessness in America*. Washington, DC.

National Center for Healthy Housing. 2009. *National Healthy Housing Policy Summit*. Washington, DC.

National Coalition for the Homeless. 2011. *Why Are People Homeless*. Washington, DC.

Raymond, Jamie, Wheeler, William, Brown, Mary Jean. 2011. *Inadequate Non-Healthy Housing, 2007 and 2009*. 60:1. Centers for Disease Control and Prevention Morbidity and Mortality Weekly Report. Atlanta, GA.

Rudolph, Linda, Caplan, Julia. 2013. *Health in All Policies – A Guide for State and Local Governments*. American Public Health Association. Washington, DC.

Saegert, Susan C., Klitzman, Susan, Freudenberg, Nicholas, Cooperman-Mroczek, Jana, Nassar, Salwa. 2003. Healthy Housing: A Structured Review of Published Evaluations of US Interventions to Improve Health by Modifying Housing in the United States, 1990–2001. *American Journal of Public Health*, 93(9):1471–1477.

US Department of Housing and Urban Development. 2014. *Vacant and Abandoned Properties: Turning Liabilities into Assets, Evidence Matters.* Washington, DC.
US Department of Housing and Urban Development. 2014. *Strategic Plan 2014–2018.* Washington, DC.
von Hoffman, Alexander. 2013. *The Past, Present and Future of Community Development – The Changing Face of Achieving Equity in Health, Education, and Housing in the United States.* The Voice of Community Development. National Housing Institute, Montclair, NJ.

4 Children's Environmental Health Issues

STATEMENT OF PROBLEM AND SPECIAL INFORMATION

INTRODUCTION

This chapter on children's environmental health issues of necessity will have to have a lengthy introduction, some historical facts, and a more complete problem and special information statement than the rest of the chapters. It will be necessary to learn about the distinctive qualities and issues of the child versus the adult. Also, instead of having a separate chapter on the school and preschool environment, these topics are included in this chapter.

This chapter differs from the others because we are speaking about a specific group of people and how the various environmental pollutants and other social issues affect them, instead of learning about an overriding environmental concern and its subissues and the effects on humans and ecosystems.

If we do not protect and promote the health and welfare of our children, we lose our future. There is a huge economic cost to our society from the unintended consequences of building our global economy, through the production, use, and improper disposal of chemicals and chemical products, which have a profound effect on our children and their lives. (See endnote 23.)

Throughout recorded history, there has been the push and pull of using our children for working in industry as if they were small adults and could help sustain the family. This is still true in some emerging nations, as well as in industrialized nations, where children are exposed on a daily basis to chemicals known or suspected of causing cancer, developmental damage, reproductive damage, and neurological damage. Many of these chemicals persist in the environment for long periods of time and are carried to many distant areas from the point of origin, by water, air, and food.

PAST

(See endnote 42)

Throughout the industrial revolution and even afterwards in industrialized nations, children wearing highly flammable clothing were forced to work under intolerable conditions, such as long hours in filthy sweatshops, under extremes of temperature and poor lighting, and around open fires, and this led to injuries, disease, and early death. Children working in mines and as chimney sweeps were special problems, with enormous numbers of injuries, illnesses, and deaths. The ratio of women and children, because they were cheap labor, to men working in factories was four to one. The development of the steam engine allowed factories to be moved from areas around streams and rivers into the cities.

Women and children, especially, although men were also involved, had inadequate food, poor clothing, sparse medical care if any, and therefore a very short life span. They were afflicted with typhus fever, typhoid fever, diphtheria, rickets, tuberculosis, scarlet fever, smallpox, and of course cholera. The living conditions found by Chadwick, Shattuck, Griscom, and others were simply intolerable. (See endnotes 26, 27.) This was true of the 19th century, 20th century, and in some cases still occurring in the 21st century. Further, environmental disasters which occurred and affected large numbers of people had even more profound effects on children. Coal smoke and other industrial pollutants caused substantial deaths from respiratory disease and asthma. In 1900, in Australia, there was an epidemic of lead poisoning among children who ingested lead-based paint. In the 1950s, in Minimata Bay, Japan, methyl mercury in fish, ingested by children, caused limb defects and mental retardation. In the 1960s and 1970s, pregnant women who ingested polychlorinated biphenyl (PCB)-contaminated rice bran oil had fetuses that became small and sick children. Tobacco use has resulted in premature babies and compromised respiratory systems.

There are highly significant racial and ethnic differences, as there always have been (read the book *The Jungle* by Upton Sinclair, about the lives of ethnic minorities in Chicago at the beginning of the 20th century) in the rates of disease and disabilities from environmental toxicants, pollutants, and microorganisms. Apparently, this is due to these groups living in close association with areas of work, areas of significant pollution, in overcrowded housing in inner cities, or in agricultural areas where the individuals and children are in close contact with agricultural chemicals. Children were also potentially exposed to serious disease from microorganisms in childcare settings, at home, and at play.

The conditions of the 19th century and 20th century led to a variety of commissions that reported that the rate of sickness and death among children needed to be sharply reduced.

New laws were passed and working conditions for children improved. (See endnotes 28, 32.)

Child labor laws (part of the Fair Labor Standards Act of 1938) and mandatory school attendance helped reduce illness, injury, and death among children. The conditions of urban housing and neighborhoods were examined in great depth and numerous changes were made and continue to be made in the housing environment, school environment, work environment, and recreational environment. Further, many professional associations, civic organizations, the local and state government, the federal government, businesses and industries, and civic-minded individuals worked individually and at times together to improve the life of the child. (See endnote 12.)

Present

In the developing world, there has been a failure to establish the foundations of good public health practice. There are highly persistent problems of water and airborne infections, diseases spread through mosquitoes and flies, poor environmental sanitation, overwhelming amounts of fecal material and solid waste, poor nutrition, and extremely poor housing. Children are most affected by these circumstances. In the industrialized world, even though some of these situations occur, children are most affected by where they live and environmental pollutants.

Children have always been exposed to diseases from microorganisms. However, with the onset of the use of antibiotics and successful vaccination programs, many of these diseases appeared to be a thing of the past. This is no longer true in many instances. Complacency, lack of good sanitary practices, contaminated water and food, overuse of antibiotics, refusal to vaccinate children based on highly erroneous so-called scientific data, and the emergence of new microorganisms have created the potential for very serious outbreaks of disease. Once again there is serious concern about the spread of cholera, cryptosporidiosis, hepatitis A and B, meningitis, pneumonia, rotavirus, shigellosis, tuberculosis, *Escherichia coli* 0157:H7, vancomycin resistant staphylococcal infections, hantavirus, *Streptococcus* infections, etc.

In addition to social media being used to spread erroneous information concerning the health effects of vaccinations and the determination of some people to try to convince others not to vaccinate their children, other types of erroneous information have been spread. For instance, there is supposed to be a "5-second rule, 10-second rule, or 15-second rule" depending on the part of the country that you live in which states that "if an object or food falls on the floor, if you snatch it up immediately within the appropriate time, the individual can give it back to the infant or small child to consume or play with." This is entirely and completely wrong and based on information that was made up by someone. Anything that's on the floor for any amount of time is contaminated and every floor is contaminated no matter how well it is cleaned. Children are particularly susceptible to disease and therefore even greater care must be taken about anything that they will ingest or put into their mouths, as children do.

In the past, there have been many special commissions to improve the health of children. In April 1997, President Clinton issued Executive Order 13045, Protection of Children from Environmental Health Risks and Safety Risks, which directed federal agencies to make a concerted effort to deal with the health issues of children because of their increased susceptibility to toxic chemicals and air pollutants. The Executive Order established an interagency task force, chaired by the administrators of the US Environmental Protection Agency (EPA) and the US Department of Health and Human Services. It also created in the EPA an Office of Children's Health Protection, and the Children's Health Protection Advisory Committee.

The World Health Organization Task Force for the Protection of Children's Environmental Health organized the Fourth Ministerial Conference on Environment and Health in Budapest, Hungary in 2004 to be totally devoted to children's health issues. (See endnote 30.)

Globally, there has been a substantial reduction in the number of children dying because greater attention has been given to the prevention of disease and injury. However, a great many children's deaths still occur in sub-Saharan Africa and Southeast Asia. Diarrheal-type diseases, typically caused by contaminated water or food, continues to be the second leading cause of death in children under 5 years of age globally. Many of these children are malnourished, have impaired immunity, and may be living with people who have HIV and therefore are even more susceptible to the effects of diarrheal-type diseases, especially dehydration which can lead rapidly to death.

The child mortality rate in industrialized countries is much lower than the previously mentioned regions of the world. Many of the deaths are related to poor maternal care, poor nutrition, second-hand smoke from tobacco, diarrhea, pneumonia, malaria, and measles.

Infant mortality rates in American cities especially among African-American children are much higher than among non-African-American children. Although poor prenatal care and unsafe sleeping practices may be part of the problem, environmental factors have to be considered. Also there is a vast environmental health concern about lead-based paint in the older housing stock found in the inner cities where especially the underserved minorities live, which creates serious risk for children for elevated blood lead levels.

The Commission for Environmental Cooperation working with public health organizations and the governments of Canada, Mexico, and the United States released the first ever report on children's health and environmental indicators in North America on January 26, 2006. The report presented 13 indicators of environmental problems that affected children. They included outdoor air pollution, indoor air pollution, asthma, blood lead levels, lead in the home, industrial releases of lead, industrial releases of selected chemicals (153), pesticide residues on foods, drinking water contamination, general sanitation including lack of sanitary sewers, and waterborne diseases. Older homes in all three countries contribute to the lead poisoning problem. Mexico has the worst challenge regarding water and basic sanitation, which leads to disease and injury. Approximately 123 million children are at risk in the three countries. (See endnote 15.)

As an example of the current problems of exposure to toxic chemicals or environmental pollutants, which may harm the health of 74 million children in the United States, the Government Accountability Office states that 66% of the children live in counties which exceed the allowable levels for at least one of the six principal air pollutants that cause or aggravate asthma. The medical costs alone are estimated to be $3.2 billion per year.

There is a disparity between Latino children and other children in the United States. Although, Latino children only make up 16% of the total population of children, they suffer disproportionately from exposure to air pollutants, pesticides, toxic industrial chemicals, and lead and mercury from candy, traditional folk remedies, religious practices, etc. In addition, a substantial number of these children tend to live either in agricultural settings, with all the problems of agricultural chemicals and migrant labor habitats, or in crowded urban settings and poor housing. An advisory group has lowered the reduced action blood lead levels to 5 μg/dL, however no blood lead level is acceptable.

Latino children appear to be more susceptible than the general population to environmental toxicants. Multiple exposures to environmental toxicants are added to preexisting disease, poor nutrition, poor healthcare, lack of food, and other problems, which increases the potential for serious acute disease, chronic disease, currently and later in life, and injuries. There is also a huge barrier to protecting the children because the parents and children lack understanding of the significance of environmental exposures. English as a second language may also be a serious barrier.

In the United States for children over 1 year of age and adolescents, injuries are the leading cause of death and sustained disability. Globally, the injury problem is substantial, with many injuries due to road traffic crashes. The response by various governments to this severe problem of preventable injuries is underwhelming regarding the provision of funds, research activities, and other resources.

There are many organizations and industries working to improve the health of children through correction or amelioration of the problems of the environment. These organizations (professional, civic, foundations, etc.) and how they interact with governmental entities will be discussed in various parts of this presentation.

UNIQUENESS AND STAGES OF DEVELOPMENT OF THE CHILD AFFECTED BY ENVIRONMENTAL STRESSES

(See endnotes 2, 11)

This presentation will examine, without going into great specifics, the unique physiological and behavioral traits of the fetus and child and how they respond to environmental pollutants; the several different environments the child is exposed to and subsequent potential increase in disease and injury; and various problems and sub-problems of high-level and low-level exposure to environmental contaminants.

The child is recognized as a distinct entity with varying responses to the environment based on time of exposure, physiologic maturity, differential reaction to a given exposure, and age of the child. The exposed fetus or infant may have many more years than the adult to develop a particular disease due to an exposure to an environmental contaminant. For the infant, toddler and child, differences in his/her physiology, metabolic rate, mobility, increased surface area of body to in relation to body mass, as well as body weight, consumption of water and food and intake of air, especially close to the ground, potentially create additional burdens from levels of environmental contaminants that might even be safe for adults. The differences in behavioral patterns of children, such as considerable hand-to-mouth contact, eating of dirt, constantly exploring the environment and tasting inedible things, may lead to the ingestion of environmental contaminants. Chronic low-level exposures to chemicals and metals further complicate the potential for disease in children. An infant's respiratory rate is twice that of an adult. In the first 6 months of life, children consume seven times as much water per kilogram of body weight as an adult does. From 1 to 5 years of age, children consume three to four times more food per kilogram of weight than an adult. Children have fewer dietary choices. Some toxicants can penetrate a child's skin much more readily than that of an adult.

Exposure of the child to toxic substances which may lead to disease is determined by the toxicity of the substance, the route of exposure, and the various host factors. In order for a disease to occur as a result of a given environmental contaminant, it must:

- Be dispersed into the environment
- Be in a medium to which the person is exposed
- Be in a medium which enables biological uptake of the substance
- Be absorbed into the body after exposure
- Cause a biological change in the person
- Affect a target organ adversely
- Create clinical symptoms of disease in a person

Environmental contaminants are found in air, water, soil, and food sources. The route of exposure is through breathing, skin, eating, and drinking, through the placenta or intravenously. The contaminants may be found in: agriculture, such as pesticides and fertilizers; occupational settings, such as solvents and other chemicals; incinerators, such as a variety of hazardous air pollutants and particulate matter; mobile sources, such as particulate matter and gases such as nitrogen oxides, sulfur oxides, and carbon monoxide; in folk medicines and special cosmetics, such as lead and other heavy metals; in the home, school, and recreational environment; and brought home from the occupational environment. A new potential threat is the terrorist attack, which can create a multitude of very serious air pollutants which can be inhaled by people within a reasonable distance of the attack, can be consumed in food or water, or contaminants which can be absorbed through contact with various parts of the body.

It is estimated that in the first 5 years of life, a child will consume an estimated 50% of the pesticides that the individual will ingest over a lifetime. About 80% of the individual's lifetime exposure to damaging ultraviolet light occurs before 18 years of age. Despite the reduction in lead poisoning from leaded gasoline, over 1 million children in the United States have elevated blood lead levels, and another approximately 2 million children are at risk of lead poisoning, which can be congenital in nature. Asthma, the most significant of the chronic diseases in children under 18 in the United States, affects approximately 5 million children.

The differences in the child's daily environment compared to that of the adult, such as in the home, play areas, daycare environment, schools, and school buses, may contribute to the spread of disease and injury, based on exposure to other children and the peculiarities and dangers of those environments. (See Chapter 6, "Environmental and Occupational Injury Control" for additional information on children's injuries.)

PRECONCEPTION

The egg and/or sperm can be compromised by environmental conditions which cause genetic damage that can lead to a viable but defective fetus. Preconception, mutagenic effects on either the paternal or maternal side can be caused by environmental contaminants causing chromosomal anomalies and new mutations. Environmental tobacco smoke has been associated with spontaneous abortion, premature delivery, and low birth weight and is considered to be a human growth retardant. Cocaine and other drug use can lead to an infant with addictions, congenital abnormalities, low birth weight, and slowed neurological and behavioral development. Chronic alcoholism can lead to spontaneous abortion, mental retardation, congenital heart disease, and slow growth. Poor maternal nutrition potentially can compound the effects of environmental hazards which can cause problems at conception and also in the fetus.

PLACENTA

The placenta is a semi-permeable membrane whose circulation is established around day 17 after fertilization. It becomes a very important route of exposure for the fetus to a variety of contaminants. During the 6-week period that occurs after the beginning of the placental circulation, exposure to a variety of environmental chemicals can cause profound systemic damage, well beyond that which would be expected from the usual response to a given dose of the chemical. This is a particularly vulnerable time for exposure to environmental pollutants, typically when the woman may not know that she is pregnant and therefore would avoid specific environmental hazards.

Low-molecular-weight compounds such as carbon monoxide, fat-soluble compounds such as polycyclic aromatic hydrocarbons and ethanol, and other compounds such as lead, easily cross the placental barrier. These maternal contaminants can have come from the past and been stored in the body or they can be part of the current exposure. The placenta has a limited ability to detoxify chemicals. Some water-soluble and high-molecular-weight compounds can also be a problem, because they may cross the placental barrier. An example would be *bisphenol A* (BPA), a chemical compound which has been used in producing plastics, as a fungicide, as an antioxidant, as a flame retardant, in rubber, and as a polyvinyl chloride stabilizer. This chemical migrates into the environment, especially in water. The journal *Reproductive Toxicology* published a warning in August 2007 that BPA can likely cause various human reproductive disorders. A National Institutes of Health (NIH) panel in August 2007 stated that BPA could cause some risk to the neurological development of infants and children. (See endnotes 67, 68.)

FETUS

Between the time of conception and birth, there is a huge vulnerability to the effects of environmental pollutants, which may be either short term or long term. The fetus undergoes rapid growth and organ development. The actual effect of a given contaminant depends on the exact sensitive time periods of exposure and the concentration of the environmental agent. The majority of these sensitive time periods are found in the first trimester of the pregnancy. Environmental exposures can reduce birth weight and cause premature birth or lead to certain birth defects. These babies are more likely to die in infancy. If they do survive, they are at high risk of brain, respiratory, and digestive problems in their early lives, as well as learning disabilities.

The fetus may be more prone to genetic damage from much lower concentrations of chemicals than the adult. The fetus clears the toxicants much less efficiently than the adult does. The developing nervous system is extremely sensitive and the specialized organs and tissues are highly vulnerable to deficits in oxygen and nutrients, as well as to toxic chemicals. A number of polycyclic aromatic hydrocarbons are reproductive and developmental toxicants, as well as mutagens and carcinogens. They can bind to the DNA and damage it. They can disrupt the endocrine system by altering the metabolic pathways of the natural hormones or interfere with their activities. Polycyclic aromatic hydrocarbons are found in ambient air indoors and outdoors, attached to particulate matter, at the workplace, in food, in water, and associated with a variety of activities.

The fetal brain is unusually vulnerable because it lacks a blood–brain barrier and so does not have the ability to detoxify chemicals. Learning and behavioral disabilities may be the result of a complex interaction of genetic, social, and environmental factors, including low-level exposure to toxic chemicals.

Exposure to lead within the uterus during this time period causes more damage to the nervous system than at any other time of development. The lead exposure may cause hyperactivity, compulsive behavior, reduced IQ, and aggression. Exposure to ethanol can lead to brain malformation and fetal alcohol syndrome. Exposure to methyl mercury has caused cerebral palsy and severe mental retardation. Even small amounts of mercury can impair IQ, language development, visual-spatial skills and memory. Elevated levels of manganese can cause hyperactivity and are associated with Parkinson's disease. PCBs can decrease reflexes and IQ, and delay mental and motor skill development. Tobacco smoke and nicotine can reduce IQ and lead to learning disorders.

Other chemicals are endocrine disruptors, altering and interfering with normal hormonal function by binding to receptors, blocking them or interfering with proteins which regulate the production, transport, metabolism, and activity of hormones. Endocrine disruptors can work at very low dose levels. The effects of the chemicals may not be seen for years. The fetus can experience acute toxicity to carbon monoxide at levels that are harmless to healthy children. Other chemicals, which have teratogenic properties, can cause birth defects, congenital anomalies and congenital malformations, and abnormalities of structure, function, or metabolism.

Fetal exposure to environmental pressures such as heat, noise, and ionizing radiation may occur from outside of the placental barrier. Also, exposure of the parents prior to, or at the time of conception, to anesthetic gases and some solvents can produce adverse reproductive effects. Exposure to lead within the uterus during this time period causes more damage to the nervous system than in any other time of development.

EARLY CHILDHOOD (0–2 YEARS)

Air toxics and other environmental chemicals can impair the development of the brain, lung, and neurological and immune systems of the infant and small child, as well as have a direct effect on these biological systems. Infants are highly susceptible to chemical exposure because of the premature development of the body's chemical detoxifying mechanisms, such as the liver and metabolizing enzymes. Air toxics can exacerbate many existing health problems, especially asthma. Infants who are born health compromised, premature, and with low birth weight are particularly vulnerable. Infant mortality is exacerbated by high concentrations of nitrogen dioxide, which apparently is a contributor to sudden unexpected infant death syndrome. High levels of carbon monoxide and PM_{10} also contribute to elevated infant mortality. Environmental tobacco smoke, excessive exposure to sunlight, pesticides and other chemicals, environmental contaminants brought in from the occupational environment, etc., profoundly affect the infant and small child.

The growth rate of tissues and organs, up to about the first 9 months, is faster than during the rest of life, making these tissues more vulnerable to carcinogens. Cancer is the fourth leading cause of death of this age group. A variety of factors, including genetic abnormalities, ionizing radiation, viral infections, certain medications, industrial and agricultural chemicals, and exposure to alcohol and tobacco products, can be involved in the development of childhood cancer. Childhood cancer increased by 13% from 1973 to 1997. The other concern is that the early exposure may increase the risk of cancer over a person's lifetime.

Since there is a small amount of body fat in children, this leads to a concentration of lipid-soluble chemicals in smaller areas of the body. Plasma protein binding is reduced because of a lower plasma albumin concentration, causing higher chemical levels and potential toxicity. It is the unbound fraction of the chemical which has the pharmacological effect. The thick keratin layer of the skin, which protects the adult when he/she comes in contact with a toxic substance, is incomplete in the small child. Infants and toddlers are frequently placed on the floor, carpet, or grass, where they have much greater exposure to chemicals that have been used on these surfaces, such as multiple organic chemicals from carpet, pesticide residues, cleaning compounds, and fertilizers. Also, the height of the child is significant if the chemical is heavier than air and may be found in the breathing zone. Toddlers (1–2 years of age) are also vulnerable because of their level of absorption, detoxification, and organ development. The younger the child, the higher the child's respiratory rate, and therefore the greater the effect of the inhalation of air contaminants such as air pollutants, dust mites, cockroach antigens, viruses, etc. There is a greater potential for lung problems and asthma.

The small intestine of a newborn absorbs nutrients at a high level. If the child is exposed to lead, the lead would compete with calcium for rapid transport. Although breast-feeding is considered to be the optimal form of infant nutrition, a baby is vulnerable to the current and historic maternal exposure to fat-soluble chemicals such as dioxins, other chlorinated pesticides, PCBs, and lead. Formula feeding involves large quantities of water which may contain heavy metals and nitrates that are not eliminated by boiling water. Toddlers eat a diet rich in fruit, grains, and vegetables with a greater risk of food-borne pesticide residues. The ingestion of soil by toddlers can result in lead, pesticides, mercury, lawn chemicals, or floor cleaning products also being ingested.

The infant uses milk as his/her primary source of nutrition. Although the addition of drinking water to powdered milk products can be hazardous in some instances based on the level of contamination in the water, breast milk cannot be considered to be totally safe if the mother is exposed to substantial quantities of chemicals. Because of bioaccumulation of a variety of chemicals which persist in the environment, the infant may be exposed to them in the breast milk. These chemicals may include organochlorine pesticides, PCBs, polychlorinated dibenzodioxins, polybrominated diphenyl ethers, polycyclic aromatic hydrocarbons, mercury, lead, nicotine, and some solvents.

Young Child (2–6 Years)

The young child is involved in substantial mobile activities, which leads to an expansion of the area that the child normally resides in and the risk of exposure to environmental contaminants and injury causing accidents increases. Lead poisoning is a serious problem in children who eat foreign objects. This is an eating disorder called pica where kids will put paint chips from peeling areas inside the house into their mouth, chew on painted surfaces within the house, or eat dirt outside which may contain lead. The child spends considerably more time outdoors.

Younger School-Aged Child (6–12 Years)

The child spends more time outdoors, in school and after-school environments, and in sports activities. Outdoor air pollution becomes a serious concern. Normal outdoor activities could include the ingestion or inhalation of arsenic, mercury, and other environmental toxicants. There is a greater propensity for unregulated sports injuries. Football is particularly a potential contributor to numerous injuries in the present and future.

(The school environment will be discussed separately for all age groups.)

Adolescent (12–18 Years)

The preteen and teenager is involved in increasing potential exposure to all types of contaminants. Because of risk-taking behaviors, these individuals, especially boys, get involved in going into environments where industrial waste, abandoned buildings, and hazardous situations may exist. They may also start experimenting with psychoactive substances, such as hard drugs, prescription medicines, and sniffing of solvents. Children may start smoking and use alcohol. Children become sexually active and are exposed to a variety of diseases. Gangs, especially in poor areas, contribute to a considerable amount of violence not only among children but also among all age groups in society.

The child may be involved in hazardous jobs or activities where he/she is exposed to workplace hazards found in the adult occupational environment. The child's hobbies or actual job may lead to serious problems of safety or environmental hazards, which may cause disease or injury. The motor vehicle accident and injury rate for adolescents aged 16–19 is considerably higher than for drivers aged 20 and older.

A lack of maturity and a wish to explore new things, without using proper judgment, can result in short-term or long-term injury or illness. It can also lead to a reduction in the proper operation of the immune system and make individuals more susceptible to disease.

Special Circumstances and Risk Factors

(See endnotes 4, 31 and Introduction)

The problems of environmental pollution and its effects on children will increase in the years to come because of environmental pollution. This is especially true of the older cities where industries built their factories near the rivers for a source of water and disposal of waste of all types. Cheap housing was built next to the factories to provide accommodation for the necessary workers to build the products that were being made as a result of the industrial revolution. This housing stock was poorly built and proper sanitary services were rarely provided. In addition, because they were so close to the factories' air pollutants as well as the pollutants buried in the ground close to the factory, location became a serious source of potential disease for this very poor and usually immigrant population. The current urban poor typically live in this type of housing or facilities close to or on contaminated land. This then becomes a problem of environmental justice because the enforcement of environmental laws tends to be considerably more lax than in more prosperous areas. The concept of environmental justice states that regardless of race, color, national origin, or income, all citizens should be given the same opportunity to live in a clean environment free of pollutants and other problems which may affect the residents.

The release of pollutants is a byproduct of the improvements in agriculture and industrial processes which enhance our economy and make our lives better. The major question will be how to accomplish these improvements without adversely affecting certain select communities.

It is estimated that in the first 5 years of life, a child will consume an estimated 50% of the pesticides that the individual will ingest over a lifetime. Despite the reduction in lead poisoning from leaded gasoline, over 1 million children in the United States have elevated blood lead levels, and another approximately 2 million children are at risk of lead poisoning. Asthma, the most chronic of children's diseases in the United States, affects approximately 7 million children below the age of 18 (see Introduction). About 80% of the individual's lifetime exposure to damaging ultraviolet light occurs before 18 years of age.

SOCIETAL RISK

(See endnote 24)

THE ECONOMY

States and local communities counting on federal stimulus money will always be in danger of making sharp cuts in programs, including those for children, who are at greatest risk during times when there is a poor economy, falling revenue, and people have increased needs due to unemployment and recession. At a time when there is a better understanding of the harmful effects of environmental pollutants on children, and the need for diagnosis and treatment of these children, there will be a substantial reduction for those on Medicaid. Further, denial of additional unemployment benefits will have a disastrous effect on children, good nutrition, and housing. Provisions must be made for putting special funds into a rainy day account during the prosperous years to meet the needs in times of recession which occur during various business cycles.

HOMELESSNESS

An estimated 2.3–3.5 million Americans are homeless at least once a year. People of all ages, geographic areas, occupations, and ethnic groups are involved. Unfortunately, homelessness has a disproportionate effect on women and children. It is estimated that families with children comprise one third of the homeless. Many of these families are led by a woman who has limited education and limited potential for earning, if a job could even be found. Almost 1 million of the children enrolled in public schools were homeless during the 2009–2010 school year. The economic downturn, loss of jobs, foreclosures, and lack of housing for those who are at the poverty level contribute to this substantial problem. In addition, there are an estimated 575,000 to 1,600,000 runaway youth, who then become homeless.

Homelessness, especially for children, leads to hunger, poor physical and mental health, missed schooling, and disruption of schooling because of frequent moving. Many of these children already have problems which are now exacerbated by the homelessness. They frequently have asthma and have been exposed to lead. Where living accommodation can be found, typically it is in an area at high risk of environmentally adverse situations. Even though these children have the greatest need, frequently they receive fewer services than other children.

HUNGER

In 2010, over 48 million Americans including many millions of children had inadequate food or at times no food on a regular basis. Hunger increases the potential serious effects of environmental contaminants. In eight states and the District of Columbia, over 20% of the child population consistently go hungry. About 20% of Americans use at least one of the US Department of Agriculture nutrition assistance programs every year. About 50% of infants born in the United States get support from the Women, Infants, and Children Program. Some 55% of schoolchildren participate in the National School Lunch Program and School Breakfast Program, with 50% of the lunches and 71% of the breakfasts given free to the children. The hunger creates a huge cost to society as well as all the potential problems for our citizens who are food insecure. Although hunger is detrimental to all people, it is especially so to the developing child. (See endnote 33.)

Maternal undernutrition during pregnancy increases the risk of poor birth outcomes, including premature birth, low birth weight, smaller head size, and lower brain weight. Premature babies are at greater risk for a series of health problems and learning problems when they reach school age. During the first 3 years of life the brain develops very rapidly. If there is too little energy provided, or a lack of protein and nutrients, there are deficits in cognitive, social, and emotional development. This affects 5–10% of American children under the age of 3. Hungry children in school show 7–12 times as many symptoms of conduct disorders such as fighting, having problems with teachers, not following the rules, stealing, etc. as children who are properly nourished.

POVERTY

The current economic situation severely affects the lives of millions of Americans, including a large number of children. Those most affected are minority groups including African-Americans, Hispanics, and American Indians. Poverty has a dramatic effect on the academic outcomes of children, especially during the early years. Schools which the children attend typically are underfunded and out of date. The neighborhoods tend to be unsafe and areas where crime is more prevalent. Children suffer from both physical and emotional health problems. Accidents and related injuries are common.

POOR MEDICAL CARE

The high cost of health care has a very negative effect on a population of young children. The parents are unable to provide medical care or prescription medicines for themselves or their children. Many parents have a choice of providing food and housing for their children, or proper medical care, and food and housing has to take priority. Approximately 90% of these children are in households where there is some insurance, but still the amount is inadequate. Where a child under 3 does not receive adequate health care, the child is more likely to be hospitalized. The child is then potentially at risk for developmental delays, fair or poor health, and hunger. The mother is more likely to be depressed and to be in fair or poor health herself.

LANGUAGE BARRIERS

There is a serious concern about the effect of language barriers on the accumulation of information about environmental health hazards in minority communities. Cultural diversity may be a serious barrier to health. For example, immigrants from the Middle East and the Horn of Africa distrust government and do not allow females to be addressed by other people. This may also apply to other groups. Also, when specific instructions or information needs to be given to a group of citizens in order to avoid environmental health hazards, language barriers may cause the citizens to misunderstand or not utilize the information to protect themselves. Community leaders may not have the resources or the information to teach the non-English-speaking members of their society how to avoid these hazards. Unfortunately, many minority groups including Latinos are living in areas with poor environmental conditions. Latino children disproportionately have asthma which has been caused or exacerbated by air pollution, both indoor and outdoor. Pregnant Latino women are subjected to more of these pollutants.

While pollution in the United States is a health risk for everyone regardless of race, ethnicity, language, or country of origin, a high percentage of Latinos are exposed in urban and agricultural areas to higher levels of air pollution, unsafe drinking water, pesticides, and lead and mercury contamination. Approximately 1.5 million US Latinos live in unincorporated communities with substandard housing along the US–Mexico border. There is a lack of potable water and sewage treatment and therefore a serious risk of waterborne diseases such as giardiasis, hepatitis, and cholera. More than one third of US Latinos live in the Western states, where arsenic, industrial chemicals, and fertilizers can readily contaminate local drinking water supplies. About 88% of farm workers are Latinos who are exposed to pesticides and other agricultural chemicals, which can lead to cancer and other health effects. Hispanic children have blood lead levels twice as high as other children. (See endnote 34.) Some other cultures may promote the use of eye makeup for children. Unfortunately, lead has been found to be one of the ingredients in the makeup.

Mercury is a special problem because of consumption of large quantities of fish not only commercially caught but also individually caught. Because of language barriers, many of the individual fisherman are catching fish in contaminated waters, since these fishermen do not either hear about fish advisories or do not understand them. Certain religious and cultural practices create another route of exposure for mercury, which may be sprinkled indoors for religious reasons. Folk remedies, especially for indigestion or gastroenteritis, and cosmetics to make skin lighter may contain mercury compounds.

CLIMATE CHANGE

(See endnote 14)

Climate change may affect the health of children, and cause serious illness and injury and even death. Climate change can result in poorer air quality, especially in cities. Ozone levels can rise. Climate change may affect the growth, survival, transmission and distribution of disease-causing organisms. This is a particular problem regarding the ingestion of food and liquid, which may be contaminated. Infectious diarrhea, the second leading cause of death in young children under age 5, is associated with contaminated water.

Climate change can increase sharply the numbers and diversity of insects and arachnids. This can lead to various mosquito-borne diseases and also tick- and louse-borne diseases. Tropical diseases such as dengue fever and chikungunya may now be transmitted in the continental United States. Flies are more prevalent and also contribute substantially to disease outbreaks.

Children have a smaller body mass to surface area and therefore are at greater risk of heat-related illnesses, such as heatstroke and heat exhaustion, which can lead to serious illness or death. Extreme temperatures and extreme weather compound the health and safety effects of large areas that are highly contaminated and are substantially cluttered with all types of solid and hazardous waste including building materials. There is also an increase in infectious diseases, mental health problems, and behavioral problems. Children are more vulnerable in severe weather than adults, not only because of their size and inexperience but also because they rely upon the adults to take care of them and provide for them. The children's lives are totally disrupted by severe weather and they cannot understand what has happened, why it happened, and what to do to make it right.

ECONOMIC EFFECTS OF POLLUTANTS

(See endnote 35)

The US EPA's National Center for Environmental Economics is the key agency that provides technical expertise to all federal agencies, Congress, universities, and other organizations in analyzing the economic and health impacts of environmental regulations and policies. It has produced a book entitled *Children's Health Valuation Handbook* as a reference tool to help conduct economic analyses of EPA policies that affect risks to children's health. It is used to help estimate the value of the health benefits to children from an environmental improvement and incorporates this value into the benefit–cost analysis of the proposed rule. It utilizes the following information: hazard identification, dose–response evaluation, risk characterization, quantification specifying the ways in which changes in children's health affect their welfare, and a monetary valuation of the welfare effects. An example of how this information is used would be when the Food and Drug Administration conducted an economic analysis while developing final regulations concerning the safe and sanitary processing of fruit and vegetable juices. There was a determination made of the long-term toxic effects in children due to lead and also illnesses because of *E. coli* and the potential long-term cost involved to rectify these situations. Another example would be the estimated cost of the developmental delays due to prenatal secondhand smoke exposure in New York City. The estimate of this was $50 million dollars a year.

The economic cost of disease and disabilities caused by environmental contaminants is usually measured using an "environmentally attributable fraction" (EAF) model. This model estimates the proportion of the cost attributed to environmental exposures to contaminants. The cost of illness estimates is based on direct healthcare costs, including hospital and nursing home care, prescription drugs, home care, physician care, and other related services, and indirect costs from lost productivity due to disease and premature death. Costs not included, which may be very significant, are those related to the psychological and emotional loss to patients, families, friends, and communities.

SPECIAL PROBLEMS OF WHERE WE RESIDE AND RISK FACTORS

All children and their families need assistance from schools, pediatricians, and other medical and social work personnel. However, inner-city children and agricultural children also need specific nursing and environmental health assistance to help determine existing, problems related to the home and home environment, and how to resolve them.

There are at least three potential home and neighborhood environments that need to be examined. These are the urban environment, suburban environment, and agricultural environment.

URBAN ENVIRONMENT

The urban environment for the purposes of this discussion is the overcrowded, older, congested neighborhood with older homes, especially those that were originally single-family and now are broken up into a multitude of apartments. A house that once held from two to nine people now holds 60 or 70.

The schools tend to be considerably older and have the types of problems that come with older school structures. The components of the structures that cause major health, safety, and learning problems are: inappropriate temperatures, indoor air quality, and outdoor air quality; poor lighting and acoustics; inadequate and outmoded science laboratories, workshops, and equipment; and overcrowded school buildings. In addition, the condition and operation of the equipment in the food service areas, the drinking water supplies and equipment, the number and condition of lavatories, the sewage disposal systems, the solid and hazardous waste materials disposal systems, and the hazardous materials storage areas may be questionable and cause additional hazards. Older facilities that are improperly maintained become an underlying factor in contributing to a variety of injuries. The oldest school buses with the greatest number of problems contributing to air pollution tend to be found in these areas. Fire safety is a serious concern. There is a lack of good communications systems for emergencies and security systems to protect the students, faculty, and other individuals within the structure. Many studies have shown that children in this type of environment perform at a lower level academically than children in better facilities. (See endnote 36.)

Socioeconomic factors cause more children to live in poverty than any other age group in the United States. Their families are more likely to live in public housing or older rundown housing in parts of the city which are in close proximity to industry and overcrowded conditions. The amount of environmental contamination is increased substantially because of locally heavier air pollution and increased levels of contaminants from automobiles and trucks. Benzene and particulate matter are especially of concern in these settings. Further, children living in poverty tend to underutilize healthcare services, and if they have asthma or other respiratory-type diseases, they have further complications. One of the triggers for asthma is exposure to roaches and mice and their droppings.

Over the last several years, studies were conducted by the Columbia University Center for Children's Environmental Health in Northern Manhattan/South Bronx, New York involving African-American and Latino pregnant women and their children. Their health and conditions within their environment were measured on a regular basis and a determination was made concerning the environmental agents that affected the fetus, newborn, and small child. These research study participants were exposed on a regular basis to multiple common environmental pollutants that can affect fetal and child development and respiratory health, or increase cancer risk. They were exposed to polycyclic aromatic hydrocarbons in indoor air and outdoor air, pesticides, especially chlorpyrifos and diazinon, used in homes, phthalates, secondhand smoke, etc. The findings indicated that polycyclic aromatic hydrocarbon exposure resulted in reduced birth weight, reduced head circumference, especially in African-American babies, and lowered IQ. Pesticide exposure resulted in a delay in psychomotor and cognitive development, attention deficit hyperactivity disorder, and personality disorders. Secondhand smoke resulted in reduced cognitive development and potentially asthma. Overall, a risk of asthma and increased risk of cancer from exposure to environmental pollutant could be found in this group. The individuals within this study represent people living in these types of inner city areas everywhere. (See endnote 19.)

SUBURBAN ENVIRONMENT

The second major environment is the suburban one, where homes are typically separated from each other by a reasonable amount of land, and they tend to be newer and the schools are newer. These are usually middle-class, upper middle-class and upper-class neighborhoods where there is a considerable amount of education and financial support. The children who live in more spacious or suburban environments are subjected to air pollution and special problems based on the location. (See Chapter 3, "Sub-problems, Factors Leading to Impairment for the Built Environment in Suburban Areas," for a much more detailed discussion of this topic.)

AGRICULTURAL ENVIRONMENT

(See endnotes 6, 13)

The third major type of environment is the agricultural environment, where there is a mixture of middle-class and lower-class areas based on the occupations of the parents. This is where you find a substantial amount of migrant labor with all the problems associated with this group including inability or low ability to work within the English-speaking world. This latter group lives in considerable poverty and has a very insecure base for children's health and growth. The children who live within the agricultural communities are most vulnerable to agricultural chemicals and also poor living conditions if they belong to migrant worker families.

Latino children are particularly at risk because they represent one in six children in the United States but are mostly concentrated in areas of substantial exposure to air pollutants, pesticides, and other toxic chemicals. Latino children have a higher rate of injuries, asthma, and lead and mercury poisoning and are subjected to higher levels of cancer, which may also be caused by extensive ultraviolet radiation exposure from the sun. Behavioral and developmental problems are greater in this group of children who typically have preexisting conditions and suffer from poor nutrition, poor housing, inadequate medical care, and an inability to use existing resources because of language barriers. Latino children along the US–Mexico border have additional problems because of environmental contamination due to poor water, inadequate sewage disposal, hazardous waste disposal, and poor solid waste disposal.

Farming and other agricultural work is associated with high rates of occupational injury, disability, and illness in all people, especially children, who have a disproportionate amount of serious injuries and deaths.

Weather conditions are an added factor causing injury, illness, and death in the agricultural setting for children. Typically, they work 10–12 hours a day in many cases in 100°F under a very hot sun. This leads to many heat-related conditions, dehydration, and reduction in the normal body's resistance to disease. The sun can also be a cause of skin cancer at various times in the person's life.

Drowning rates for all age groups of children are three times higher in rural areas than in urban areas. Rural areas tend to have more bodies of water which are unsupervised, including farm ponds, irrigation canals, and small streams. Also, liquid manure areas can easily become a place of drowning for the small, inexperienced child. Abandoned wells which are not properly secured create another potential hazard. Among older children, the consumption of alcohol before or during water activities leads to drowning.

The machinery and tools that the children utilize are by nature very dangerous. Although they may not be old enough to drive a car, they may be driving farm equipment which can cause severe injuries and death. Many of the accidents are due to inexperience, lowered ability to recognize potentially hazardous situations, and inadequate training. The resulting muscular and bone injuries can affect the child for the rest of his/her life.

Modern farming in the United States has become in many ways a very large, resource-intensive business. These farms tend to have higher production per acre, a reduction in biodiversity, and increased dependence on fertilizers and pesticides that are not part of the normal environment of the area and therefore are not easily assimilated into the local ecosystem. There is little understanding of the true costs of production, the severity of the impact caused by environmental toxicants, and the potential increase in disease levels, both short term and long term. Understanding of ergonomic factors related to farming procedures, prevention of injuries on the farm, and promotion of agricultural worker health is of great significance in developing sustainable agricultural practices.

Children are exposed to a variety of chemicals, since 75% of all pesticides used in the United States are used in the agricultural area. Especially significant is the high level of occupational lung disease from exposure to organic dusts in many agricultural environments. Also present in the dusts are endotoxins, bacteria, viruses, and fungi. In one community in the state of Washington, which was close to pesticide-treated orchards, the levels of pesticide metabolite concentrations in the urine of agricultural children were nine times higher than in other children. The children are exposed by a multitude of different routes. Pesticides are found in the air, food, water, dust, and on clothing, tools, vehicles, parents' skin, shoes, and other surfaces.

The simple act of protecting oneself against disease by hand washing is typically not available to the children or the adults and therefore they have far greater opportunity for spread of disease and ingestion of chemicals.

The use of antibiotics in the agricultural setting has created in humans, especially children, resistance to various microbes. This has also led to multidrug-resistant bacteria. In the past 10–15 years there has been a rapid acceleration of the emergence of multidrug-resistant pathogens, especially *Campylobacter* and *Salmonella*. Almost 20% of *Campylobacter* infections from food and 33% of non-typhoidal *Salmonella* infections are found in children younger than 10 years of age. The rate of infection of *Campylobacter* in children less than 1 year of age is twice as high as in the general population. The rate of infection of non-typhoidal *Salmonella* in infants is 10 times as high as in the general population. Children, particularly very young children, are at high risk of developing infections with drug-resistant organisms linked to the agricultural use of antibiotics.

FACTORS LEADING TO IMPAIRMENT AND BEST PRACTICES TO REDUCE HAZARDS FOR SPECIAL ENVIRONMENTS

HOME ENVIRONMENT

(See Chapter 3, "Built Environment—Healthy Homes and Healthy Communities")

The home environment is discussed in various sections of this presentation. It is listed separately here to give the reader a better understanding of the total picture of exposures to a variety of hazards that children face in their daily lives. One of the major concerns of the home environment is childhood asthma which will be discussed below.

FACTORS LEADING TO IMPAIRMENT AND BEST PRACTICES FOR CHILDHOOD ASTHMA

(See the endnotes 3, 5, 8, 40, 41)

Asthma is a chronic disease of the airways carrying air to the lungs. The airways become constricted and the linings become swollen, irritated, and inflamed. The developing lung is a special target of environmental contaminants. The disease is caused by a combination of exposure to allergic substances, environmental substances, infections and a genetic predisposition, and is found more frequently in areas of lower socioeconomic conditions. The triggers for the disease may be allergens and/or environmental conditions, including tobacco smoke, ozone, sulfur dioxide, inorganic acids, particulate matter, dust, molds, pollen, roaches, etc. Children, because of their unique physiology and behavioral traits, react to much smaller doses of various asthma triggers than adults. The condition, which has reoccurring episodes, can range from mild to life threatening and is more prevalent in African-American children than white children. Asthma, which affects over 9 million children and is very costly, is the most common chronic disorder of childhood. It is on the increase, with outdoor air pollution and indoor air pollution being major factors. Immediate medical care and environmental management by removal of triggers are absolutely essential.

Best Practices for Preventing Childhood Asthma
- Install exhaust fans close to sources of contamination such as fuel-burning appliances, wood stoves, kerosene or gas space heaters, fireplaces, etc., and then vent to the outside.
- Do not use gas cooking appliances as a heating source.
- Change clothing when coming home from work.
- Avoid using products that produce smoke, perfumes, various sprays, solvents, etc.
- Use high-efficiency particulate air filters in vents in rooms and in vacuum cleaners.
- Keep children away from trucks and buses that are idling.
- Monitor outdoor air quality while keeping children indoors during an unhealthy range of air pollutants.
- Keep pets out of a child's bedroom.
- Remove carpets from rooms and use other floor surfaces.
- Use a portable air cleaner with high-efficiency particulate air (HEPA) filters in a child's bedroom.
- Do not use ozone generators in living quarters.
- Use traps and baits to control mice and rats in living quarters.
- Encase all pillows and mattresses that a child sleeps on with non-permeable material and wash all bedding in hot water weekly.
- Remove stuffed toys on the pillow of the child.
- Use a damp mop or rag to remove dust.
- Keep home and car totally smoke-free.

- Reduce potential for mold/mildew by controlling leaks, venting bathrooms, clothes dryers, and other sources of water.
- Keep the child away from air conditioning in the home and car when it is first started.
- Do not allow the child to rake leaves, mow lawns, or work with mulch.
- If mold or mildew appears, remove the child from the area and wash the area with chlorine bleach solution.

FACTORS LEADING TO IMPAIRMENT AND BEST PRACTICES FOR SCHOOLS

PRESCHOOL ENVIRONMENT

About one half of the children in the United States are cared for by people other than their immediate family. Two thirds of two-parent homes have both parents working and therefore child care is needed. Single-parent homes have 20% of the children in the United States who need child care. The preschool or childcare setting may be unsafe and contribute to health problems. The most prominent hazards for children are: toys and equipment; chemical hazards, such as cleaning materials and disinfectants; biological hazards such as airborne and blood-borne infections; handling and moving of equipment and children; unattended children; security of entry points and exits; availability of drugs and other medication; and visual or hearing impairment. There is a need for good personal hygiene and good facility sanitation practices, infectious disease control, injury prevention techniques, illness management, and emergency preparedness.

Bacteria, viruses, parasites, and fungi are readily spread by young children through close personal contact, hand-to-mouth behavioral traits, and inadequate personal hygiene. Infections can be spread readily from person to person by means of hands and bodies. Infections may be spread by the oral–fecal route, respiratory route, and blood, urine, and saliva. Many surfaces, such as diaper changing tables, food tables, bathroom surfaces, fabrics, toys, and books, are contaminated and can easily be the reason for the transmission of disease. (See endnote 23.)

Children under age 4 are especially at risk for injuries. Children are prone to drowning, motor vehicle injuries, burn injuries, poisonings, choking, and suffocation.

There are a range of environmental contaminants found in early learning and childcare environments. The contaminants come from: outdoor air pollution; indoor and outdoor pesticide use; inadequate ventilation, and the inhalation of carbon monoxide and especially other contaminants from painting, cleaning, and art supplies; dust carrying chemicals from furniture, carpets, televisions, computers, plastic toys, and cleaners; dust tracked in from outside with all the usual contaminants; mold from water-damaged drywall and carpets; lead from old paint, toys, and objects illegally painted with lead-based paints; mercury from broken old glass thermometers and potentially from fish; radon from surrounding soil; tobacco smoke from previous smoking areas; fragrances found in air fresheners, cleaning agents, and perfumes; and fumes from disinfecting or cleaning products. (See endnote 1.)

Best Practices for the Preschool Environment (See endnote 1)
- Conduct a comprehensive study of the preschool environment to determine the types of pollutants and microorganisms which may potentially cause disease and injury to children and the appropriate measures needed to prevent this.
- Check the Air Quality Index and weather conditions each day before allowing children to be out of doors. (This is especially important in areas where there is frequent smog, fires, and excessive heat or excessive cold.)
- Locate children's activities out of doors away from motor vehicles and parking areas.
- Do not allow motor vehicles to idle near the facility.
- Reduce the potential for motor vehicle exhaust to enter the building.
- Children should not be allowed to play in areas where pesticides have been used recently, near stagnant or standing water, or during peak mosquito biting times.

- Play equipment must not be treated with chromated copper arsenate.
- The children under supervision must wash their hands after using the toilet facilities, playing with toys, touching equipment, and before eating.
- The adults must frequently wash their hands to reduce the chance of spreading disease.
- Children must be protected against the effects of the sun especially between 10 AM and 2 PM.
- Provide adequate and appropriate ventilation for all areas of the facility.
- Determine if mold or mildew is present and remove it immediately and also remove the source.
- Remove dust frequently with damp mops or treated cloth.
- Vacuum all carpets and rugs utilizing equipment with special filters.
- Periodically test the facility for levels of radon.
- Do not use air fresheners or fragrances in the air.
- Store all cleaning products, disinfectants, and sanitizers away from children.
- Utilize cleaning products, disinfectants, and sanitizers only after the children have left the facility.
- Make sure that all toys, arts, and crafts materials, etc. are safe, age appropriate, lead free, sanitized, and used under the supervision of an experienced adult.
- Maintain kitchen, food preparation, and food serving areas at the highest standard. (See Chapter 7, "Food Security and Protection.")
- Continuously evaluate storage temperatures, cooking temperatures, serving temperatures of food, and the temperatures of equipment used in the food preparation process.
- Be careful of the temperature of the food that the child is consuming so that the child will not be burned.

SCHOOL ENVIRONMENT

(See endnote 44)

In many ways the problems of the school environment are similar to the preschool environment but at a much higher level because of: the many millions of children in schools who may be potentially sick; the time they spend each day, 5 days a week in the facility; the numbers of school buildings that are old and dilapidated, especially in the inner cities; size of the facility; the numbers and types of programs, and sheer volume of people, equipment, chemicals, storage, and the disposal areas; and the continuous complex interaction between these factors and people. Chemical management is of great concern because of the vast numbers of different chemicals and the quantities being used within the facility. Chemicals being used are stored in science laboratories, preparation areas, closets, disposal areas, etc. Chemicals being used include art supplies, paints, stains, inks, glazes, cleaning products, pesticides, fertilizers, deicers, paint removers, degreasers, lubricants, cleaning agents, adhesives, laboratory grade chemicals for teaching and learning, drinking water and swimming pool chemicals, old discarded used and unused chemicals, etc., and chemicals collected by untrained students, teachers and administrators from the community for later disposal. Closets used to store chemicals for laboratories, shops, art facilities, cleaning, and other purposes may typically contain old and abandoned substances which can be hazardous. Saving energy by making schools more airtight may have contributed to an increase in indoor air pollution caused by chemicals and other substances.

Older schools may be more likely to have the potential to be contaminated with lead-based paints and asbestos. Some older buildings from the 1950s through the 1970s may contain material contaminated with PCBs.

Key potential violations of federal EPA regulations in K-12 schools are:

- Asbestos violations
- Chlorofluorocarbon violations

- Combustion source violations
- Hazardous waste violations
- Oil storage tank violations
- Underground injection control violations
- Pesticide violations
- PCBs violations (See endnote 46)

Accident hazards also become a serious problem within the school, on the school grounds, and during transportation back and forth from school to home. In many schools there has been a severe deterioration of the facilities and equipment which may lead to injuries. (See below for section on "Injuries to School Children.")

In the school environment, disease can be spread from the bathrooms and the water fountains as well as through contaminated hands which were not washed after use of the facilities. Respiratory infections spread rapidly because of the closeness of the students in various classrooms and other activities and because of the nature of the spread of airborne infections through ventilation systems. The potential for the spread of blood-borne disease exists because the children may already be infected and then come into a very close environment where the nature of the activities including sports can lead to transfer of the microorganisms. Food-borne disease is always a potential problem because of the mass feeding operation and because students bring packaged lunches from home and keep them in their lockers instead of refrigerated areas. Children frequently trade items from lunches back and forth.

Asthma, which is a non-contagious disease, can be triggered in the school facilities through a variety of indoor air pollutants. The children are already very susceptible.

The school typically has eight major service areas, each with its own environmental problems and Best Practices:

1. Administrative/business which has very few environmental concerns
2. Academic/vocational including personnel and equipment in all science areas with associated laboratories and classroom experiments, all art classes with associated work areas, technology areas with associated shops, swimming pools, all chemicals and other materials and storage, use and disposal
3. Facilities operation and maintenance areas including personnel and equipment for building maintenance, heating and cooling, construction/renovation, solid and hazardous waste management, and the cafeteria
4. Grounds maintenance including personnel and equipment for mowing and landscaping, irrigation, snow removal, maintenance of outside areas, collection of trash and yard waste, pesticide use, fertilizer use, and deicer/salt use
5. Custodial areas including personnel and equipment for cleaning, storing, handling, using and disposing of commercial cleaning products, and collection, storage, and removal of solid waste from the school building
6. Transportation including personnel and equipment for the operation, inspection, maintenance, repair, and storage of school buses and other vehicles
7. Nursing stations/infection control including personnel and equipment for health service, physical education, and athletic programs
8. Printing facilities including personnel and equipment for all types of printing

Stormwater management is a serious concern because of the contamination of water from school facilities due to the presence of septic tank effluents, vehicle wash wastewater, improper oil disposal, improper radiator flushing, spills from accidents, improper disposal of toxics, accumulated automotive wastes on parking lots, and wash from use of pesticides and fertilizers.

Best Practices for School Environment (See endnote 16)
- Trained professionals should conduct an annual risk assessment analysis of all areas and make the necessary adaptations as needed.
- Each school or preschool should have an emergency plan and practice it in case of an emergency or disaster. The school emergency plan should include evacuation and returning to the structures; lockdown procedures and when to use them; the roles and responsibilities of the school staff; means of communicating with the staff, students, and family; emergency equipment and supplies and how to use them; usable maps of the facility distributed to the appropriate authorities in advance of a problem; student release procedures; and how to document all actions. Training sessions should be held for all of the staff and drills should be conducted periodically at random. There should be continuous assessment of the school environment and emergency equipment. Each of the drills should be evaluated by the appropriate authorities to make sure that the children are safe. There should be drills including all proper emergency response people to determine if appropriate action is taken as rapidly as possible to secure the facility, personnel, and students.
- To provide Best Practices for each of the individual classrooms and laboratories and other areas of the school would be overwhelming and often repetitive. Instead major concerns are being chosen and Best Practices proposed for these concerns identified in some of the eight major service areas of a school facility listed above. (See specific concerns listed below.) Additional information is also available in various other parts of this book by topic area.

Best Practices for Indoor Air Quality
- Provide emergency gas shut-off valves for all laboratories.
- Immediately contact housekeeping staff when a spill occurs, especially a chemical spill.
- Clean and dry all areas immediately to prevent growth of mold.
- Clean ventilation system affected if fumes from spill enter the system.
- Dust and vacuum all rooms with wet or dust-free techniques.
- Remove trash from all areas on a daily basis.
- Inspect all pipes and plumbing fixtures for leaks on a regular basis.
- Inspect all fume hoods for airflow to make sure they work effectively.
- Train students and have them closely supervised in the use of fume hoods.
- Clean heating and cooling coils, handling units, and ducts on a periodic basis.
- Replace all air filters at appropriate times.
- Have the fans operate continuously when people are present in the building.
- Check the mechanical operation of all heating and cooling systems.
- Make sure that all air supply vents and return vents are open, clean, and not blocked.
- If moisture condensation occurs in areas, raise indoor air temperature, improve air circulation, remove any sources of water, and use dehumidifiers where appropriate.
- Ventilate all areas where chemicals are being used and can become air contaminants.
- When stripping off old paint, prevent airborne dust and lead dust.
- Ventilate all areas where painting is being done.
- All areas of construction and renovation must be physically separated from students and properly ventilated.
- Test for the presence of radon gas and take necessary precautions to protect against its movement into the structure.

Best Practices for Air Quality Management (See Chapter 2, "Air Quality (Outdoor [Ambient] and Indoor)")
Best Practices for Stormwater Management (See Chapter 14, "Water Quality and Water Pollution" for Best Practices)

Best Practices for Chemical Management (See Chapter 2, "Air Quality (Outdoor [Ambient] and Indoor)" for Best Practices) (See endnote 45)

Best Practices for Septic Tank Management (See Chapter 11, "Sewage Disposal Systems")

Best Practices for Solid Waste Management and Recycling (See Chapter 12, "Solid Wastes, Hazardous Materials, and Hazardous Waste Management")

Pollution Prevention/Waste Reduction (See Chapter 12, "Solid Waste Management, Hazardous Materials, and Hazardous Waste Management")

SPECIAL EQUIPMENT AND FACILITIES IN THE SCHOOL

Conduct preventive maintenance and monitor and/or control equipment and facilities in their specific locations based on the manufacturer's recommended standards and the standards of the appropriate licensing authorities.

Cooling and Ventilation Systems

Conduct preventive maintenance; monitor and control airflow; replace air filters; and clean, repair, and replace air handlers, exhaust fans, and hoods on a regular schedule and then as needed.

Drinking Water

Conduct periodic water sampling for bacteria and chemical contamination, and clean water fountains several times a day.

Plumbing System

Evaluate system for cross connections and submerged inlets especially in laboratories, and carry out system maintenance on a regular basis.

Fume Hoods/Spray Booths

Evaluate airflow and filtering systems, clean frequently to prevent buildup, and utilize all appropriate occupational health and safety procedures to prevent contamination of the facility and contamination of the worker.

Electrical Shop

Electrical facilities maintenance includes electrical systems, fire alarms, security systems, and a variety of electrical equipment. Utilize all current requirements for the disposal of hazardous wastes since hazardous waste may be generated from the use and disposal of electrical equipment components such as: thermostats, switches, level gauges, and relays that contain mercury; telephone panels, fire panels, motion detectors, and emergency lights which may use sealed lead-acid or lead-cadmium batteries; other batteries that may contain mercury, cadmium, silver, or lithium; fluorescent lights containing neon, mercury vapor, high-pressure sodium, and metal halide light bulbs which may contain mercury; and old electrical equipment manufactured before July 1979 which may contain PCBs or PCB-contaminated items. (Also see Chapter 12, "Solid Waste Management, Hazardous Materials, and Hazardous Waste Management")

KEY POTENTIAL VIOLATION AREAS IN SCHOOLS

- *Asbestos*—Develop, maintain, and publicize an up-to-date asbestos management plan, train custodial and maintenance staff, take and analyze bulk samples, post suspicious areas, conduct routine and special inspections, and advise the EPA of renovation or demolition of potential asbestos-containing material.
- *Chlorofluorocarbons*—Use certified personnel to service air conditioning and refrigeration equipment containing these chemicals and recover them from the equipment in an appropriate manner.

- *Combustion Sources*—Obtain appropriate permits for removal of combustion products from boilers and report installation of new boilers.
- *Hazardous Waste*—Determine the amount and type of hazardous waste, properly enclose the waste, label it, and dispose of the waste in an appropriate manner.
- *Oil Storage Tanks*—Monitor underground storage tanks, upgrade, replace, or close them when appropriate; provide secondary containment; and develop and provide a plan for spills, prevention, control, and countermeasures.
- *Underground Injection*—Use appropriate septic tank systems for effluent from the bathrooms, kitchens, etc., divert effluent from shops in laboratories to special systems, and divert outdoor stormwater away from the systems.
- *Pesticides*—Use appropriately trained and certified personnel to apply pesticides according to prevailing weather conditions and air quality conditions.
- *Polychlorinated Biphenyls*—Manage appropriately and dispose of properly PCB oil or PCB-contaminated wastes from pre-1980 electrical equipment.

FACTORS LEADING TO ILLNESS AND INJURIES AND BEST PRACTICES FOR THE RECREATIONAL ENVIRONMENT FOR CHILDREN

The recreational environment ranges from playground and playground equipment to sports programs to swimming areas to artificial sun tanning to body piercing and tattooing which may lead to scarring and infections. Small children may become infected when they play with animals. Skin eruptions of all types occur from being in the outdoor environment and exposed to a variety of plants which the sensitive individual normally does not encounter.

Obviously, the problems which have been identified in other environments, especially air pollution and water, soil, and land contamination, also apply to the recreational environment.

Faulty equipment, improper surfaces, and careless behavior contribute to more than 200,000 children being treated in the emergency room of hospitals on a yearly basis. About 45% of playground-related injuries are severe including fractures, internal injuries, concussions, dislocations, and amputations. About 75% of playground injuries are caused by equipment in public playgrounds, mostly schools and daycare centers. There is a serious concern about blood-borne infections including hepatitis, HIV, and methicillin-resistant *Staphylococcus aureus*.

About 3.5 million sports injuries occur each year in the 5–24-year-old age group. These injuries are caused by improper sports gear, inadequate physical preparation including warm-up, temperature, and the risky behavior of young people.

Recreational water-related illnesses and injuries are prevalent among children. Microorganisms are spread by swallowing contaminated water, breathing in aerosols or physically coming in contact with contaminated water in swimming pools, hot tubs, water parks, etc. Recreational water illnesses include gastrointestinal, skin, ear, respiratory, neurological, and wound infections. Most frequently the children will have diarrhea, which may be caused by several different microorganisms. Drowning can occur when swimming and boating. Non-fatal drowning can cause brain damage and associated problems. The most frequent injury category among 1 to 4-year-old children is drowning. (See endnote 51.)

Artificial sun tanning, even as few as 10 times per year, can increase the chance of melanoma during a lifetime by sevenfold. Approximately 60,000 people are diagnosed with melanoma each year and at least 8000 of them die. Squamous cell carcinoma and basal cell carcinoma are also associated with tanning beds. Each year, 250,000 people are diagnosed with squamous cell carcinoma, and at least 800,000 people are diagnosed with basal cell carcinoma yearly. The problem of tanning beds is especially significant in teenage girls.

Dogs bite more than 4.7 million people a year, 800,000 of whom are treated in an emergency room. About half of these people are children, typically aged between 5 and 9 years.

There are health hazards associated with tattooing and body piercing. It is estimated that 15–20% of the young adult and teenage populations in the United States have tattoos or body piercings. There is a serious concern about blood-borne infections, HIV transmission, cardiac problems, skin disruption, and skin infections. Serious scarring and disfigurement can be an additional concern in some people.

Best Practices for Children in the Recreational Environment (See Chapter 10, "Recreational Environment and Swimming Areas")

FACTORS LEADING TO IMPAIRMENT AND BEST PRACTICES FOR THE OCCUPATIONAL ENVIRONMENT

Exposure to a variety of occupational risks may cause serious damage before pregnancy, during pregnancy, during infancy, and during the remainder of the individual's life. Before conception, the future father may be exposed to estrogens, heat stress, lead, ionizing radiation, carbon disulfide, pesticides, vinyl chloride, etc., and this may lead to changes in genetic material causing birth defects. Before conception, future mothers may be exposed to ionizing radiation, carbon monoxide, ethylene glycol ethers, solvents, lead, mercury, pesticides, antimony, arsenic, cadmium, carbon disulfide, chlorinated hydrocarbons, nitrous oxides, vinyl chloride, etc., and this may lead to changes in genetic material causing birth defects.

As an example, pesticide exposure before or during pregnancy may cause an increased risk of death of the fetus, spontaneous abortion, and/or early childhood cancer. Pesticides may also cause premature birth, retardation of the growth of the fetus, low birth weight, and congenital malformations.

Four major occupational hazards during pregnancy include:

1. Chemical risks discussed above
2. Physical agents such as x-rays, gamma rays, noise, heat, pressure, and extreme cold
3. Biological agents such as HIV, rubella, hepatitis viruses, etc.
4. Strenuous physical labor such as prolonged standing, heavy lifting, twisting of the body, staying in one position for an extended period of time, etc.

Occupational exposure can occur during the growth and development of the infant. The exposure may occur through the use of breast milk when the mother is exposed to PCBs, polybrominated biphenyls, pesticides, organic solvents, mercury, lead, methylene chloride, etc., and transfers the chemicals in the fluid to the infant. There is a potential for exposure from the parents' clothing which may transfer industrial chemicals, pesticides, fibers, metal dust, and microbiological agents brought home from the workplace. For the work-at-home parent or hobbyist, there is the potential for the transfer of a vast number of contaminants to the infant.

Child labor can create an immediate and far-reaching problem. Hazardous work includes agriculture, factories, difficult service areas such as heavy domestic work, and construction, scavenging, or recycling. Children are more susceptible to hazards than adults. They are subjected to: mechanical hazards such as falling, being caught between objects or equipment, and being cut or burned; biological hazards such as insects, exposure to bacteria, viruses and parasites, and dangerous animals; chemical hazards such as dangerous gases, liquids, dusts, solvents, explosives, flammable, or corrosive materials, etc.

Their lack of physical and mental maturity may cause them to take unnecessary risks, misjudge situations, ignore appropriate safety considerations, etc. They also are more susceptible to a variety of chemical, physical, and biological agents. They may not be able to use appropriately personal protective equipment that was designed for adults. Poorly designed working environments, lifting and carrying heavy and bulky loads, driving or using machinery, or repetitive motion can lead to short-term and chronic injuries. They are less able to cope with extremes in temperature, noise, vibration, and radiation. Stress and continuous monotonous work may cause them to become less observant and more susceptible to injuries.

Children frequently receive injuries from cuts, fractures, burns, scalds, and insect and animal bites, and trauma to the soft tissues or skeletal system. These injuries may be temporary in nature or may cause long-lasting chronic effects in later life. Drowning is also a concern in the occupational environment for children.

Poisoning is a frequent problem of the children's occupational environment. They are exposed to cleaning solutions, volatile solvents, and pesticides/fertilizers. They may be poisoned through inhalation, ingestion, or contact with the skin or eyes.

Chronic exposures which may affect children as adults include:

- Hearing loss at a young age due to excessive noise
- Contact dermatitis from exposure to corrosive/irritant substances, excessive cold or heat and humidity
- Infectious and parasitic diseases which may either reoccur or have a chronic aftermath
- Respiratory diseases from exposure to dust, nitrous fumes, phosphorus dust, silica, coal, asbestos, etc.
- Neurological effects from lead, mercury, carbon monoxide, etc.
- Anemia from lead, benzene and malnutrition from lack of food or poor eating habits
- Cancer from a vast variety of exposures to carcinogens either separately or in combination (See endnote 48)

The National Institute for Occupational Safety and Health (NIOSH) reported that from 1998 to 2007, a total of 5719 younger workers died from occupational injuries and an estimated 7.9 million non-fatal injuries to younger workers were serious enough for the individuals to have to go to the emergency room of a hospital. (See endnote 47.)

Best Practices for the Occupational Environment (See endnote 49)
- Train medical personnel how to determine the effects of occupational and environmental conditions on the individual and the possible cause or causes of the problem.
- Learn to identify specific hazardous situations, estimate patient's exposure and obtain professional help from occupational health experts.
- Plan pregnancies if possible so that the future father and future mother will not be exposed to occupational problems that can affect the fetus or child.
- Evaluate past histories of individuals who had occupational exposures and other exposures to harmful substances and advise them concerning current working conditions.
- Determine the amount and type of potential exposures or actual exposures to substances that may harm children in the workplace.
- Do appropriate blood analysis to determine lead levels, pesticide exposure, etc.
- Determine if children work with hazardous equipment in agriculture.
- Determine the amount and type of cleaning agents used in occupational areas, schools, homes, etc.
- Determine the amount and type of pesticides that are applied to lawns, farms, buildings, etc.
- Avoid the use of benzene by children.
- Determine if children are exposed to biological hazards.
- If there is a severe problem or potential problem from occupational exposure, reduce the time spent in the area or change jobs temporarily or permanently.
- Reduce potential exposure to chemicals in breast milk by removing the individual from areas where there may be contamination from PCBs, polybrominated biphenyls, organic solvents, pesticides, methylene chloride, phthalates, perchloroethylene, mercury, and lead.
- Change clothing at the occupational setting in a clean area before leaving the premises to keep from contaminating vehicles and the home setting if there are children present.

- Use special training materials to make sure the young people understand the best approach to various situations and the occupational environment. (See endnote 50.)
- Avoid sprains and strains by using appropriate techniques and equipment to lift heavy or bulky items.
- Avoid cuts, bruises, and fractures by keeping all work place spaces neat and by removing all slick materials from floors.
- Stabilize ladders before utilizing them.
- Follow instructions carefully when utilizing and working around cleaning and disinfecting material, pesticides, fertilizers, and all other chemicals.
- Provide direct supervision for all children working in various occupational areas.

FACTORS LEADING TO IMPAIRMENT AND BEST PRACTICES TO REDUCE HAZARDS FOR SPECIFIC COMMUNITY AND HOME ENVIRONMENTAL RISKS

Note: Material in this section may repeat that in earlier sections and chapters. (Where data seem to differ it is because different sources are used and the data change over time and with the source.) However, this information is necessary at this place in the book chapter in order to relate to the reader the full story of environmental problems and their inter-relationships as they affect children.

TOXIC CHEMICALS

(This is listed first because it impacts many of the areas described below.) (See endnote 7)

It is estimated that there are approximately 82,000 chemicals which are used in our society today. Fewer than 200 chemicals are required to be tested under current law. Reports from the National Pollutant Release Inventory and the Toxics Release Inventory indicate that 0.5 million tons of cancer-causing chemicals, 0.5 million tons of developmental and reproductive damage–inducing chemicals and 2 million tons of suspected reproductive and/or neurological toxicants were released into the environment and transferred to other locations in the United States and Canada over a 1-year period. The most recent report on North American pollution for 2005, issued by the Commission for Environmental Cooperation in 2009, reaffirms that there are massive amounts of potential hazardous chemical releases. The recorded list of toxic substances released is less than the actual amount because only a small group of chemicals are actually monitored.

The risk of a toxic chemical, during human exposure, causing an unwanted reaction, may vary with the individual based on route of entry, genetic differences, age, gender, pregnancy status, dietary or nutritional status, existing disease or health status. It also varies with the action of the toxicants within the body and may be increased or decreased by the route of entry, type of absorption and excretion, the time it takes for the toxic action to start, the level of bioaccumulation, the biotransformation of the substance within the body, the metabolites produced, and the duration of the effect of the chemical and metabolites.

Gastrointestinal, respiratory, cardiac, renal, liver, and thyroid disorders can increase the toxicity of environmental chemicals, since the body does not efficiently remove the toxicants from tissues and organs.

Typically, the fetus or child is exposed to more than one chemical at a time. There is poor understanding of the synergistic effect of a mixture of chemicals on the human body.

The fetus as well as the child are especially vulnerable to the various synthetic chemicals that enter the body through ingestion, inhalation, or the skin. Many of the chemicals are easily absorbed. There are a substantial number of chemicals available, as well as their waste and byproducts, to contaminate the air, water, or soil. Just one type of chemical which is released, pesticides, has a profound effect on the brain and nervous system. Over 1 million pounds of pesticides are used in the United States each year.

The chemicals may add another complication if they have immunosuppressive properties. In addition, the concentration of the chemical, the length of exposure, the temperature during the exposure, the amount of moisture, the existence of other chemical contaminants as well as other environmental pressures, contribute to the ultimate effect on the organs and tissues. Bioaccumulation of the chemical may occur if it is in the food chain, so that small amounts of the chemical which may be non-toxic or less toxic increase to substantial amounts which may now be more toxic or hazardous. In the 1990s, it was found that exceedingly low levels of environmental toxicants could be associated with lower intelligence, diminish school performance, increase rates of behavioral problems and asthma, etc. Some of these chemicals have low-level toxicity for adults, but apparently in combination with ambient air pollutants such as nitrogen dioxide, acid vapor, and fine particulate matter, can produce profound effects in children.

There is an ever-evolving concern about children being exposed to very low levels of pesticides. These pesticides, which are commonly used in or around homes or on pets, or may be found as residues on fruits and vegetables consumed by the children, may interfere with immune, thyroid, or neurological and respiratory processes in children. The pesticides include organophosphates, organochlorines, carbamates, and pyrethroids.

An example of a pesticide that works extremely well on bedbugs is the organophosphate Proxor. This is a nerve agent and has now been banned for indoor use because little kids crawl along floors and touch cracks and crevices where this chemical has been applied for residual control of bedbugs and now these areas have become a source of contamination for the child.

Best Practices for Toxic Chemicals (See Chapter 2, "Air Quality (Outdoor [Ambient] and Indoor)")

CHEMICAL AND BIOLOGICAL RELEASES INCLUDING TERRORISM

(See endnote 9)

Children are particularly vulnerable to a group of chemical and biological agents that may be released into the environment by the military during warfare, by terrorists and through criminal acts by single individuals or groups, and by accidental releases. On August 21, 2013, the Syrian military fired rockets containing sarin nerve gas into civilian urban areas in the suburbs of Damascus, killing 1429 people including 426 children and injuring at least 2200 others. From April to July 2012 in several provinces in Afghanistan, terrorists released pesticides in girls' schools at least 16 times, injuring 1383 people including 1355 children. From April to August 2010 in Afghanistan, terrorists carried out 20 gas attacks on girls' schools, injuring 672 people including 636 children. In Bhopal, India on December 2–3, 1984, an accidental release of methyl isocyanate gas killed 3787 people and injured 558,125 people including 200,000 children. (See endnote 10.) The chemical groups of greatest concern are: nerve agents that can be absorbed through intact skin, vesicants released as aerosols causing severe skin burning and wounds; choking agents causing pulmonary injury; cyanogens causing respiratory distress, coma and death; and incapacitating agents, which are easily obtainable and designed to frighten and hurt people. Biological agents include smallpox, ricin, etc.

Exposure to chemical or biological agents may be through airborne releases from crop-dusting airplanes and ventilation systems in closed areas. Exposure may be through contaminated water supplies. Exposure may be through contaminated food.

Children are particularly vulnerable because of their developmental and physiological differences from adults. (See the section "Uniqueness and Stages of Development of the Child Increased by Environmental Stresses" above) For the same amount of contaminants and the same length of exposure, children will suffer disproportionately more than adults. Substances in the air which are heavier than air will be more concentrated at the lower regions both on the inside and outside. Consequently, children will breathe in greater levels of the contaminants. Once children are exposed to the contaminants (they are more vulnerable and less able to understand danger and leave the area, and their immune systems are immature), they dehydrate much more rapidly, and there is a greater risk of posttraumatic stress disorder.

Best Practices for Biological and Chemical Releases Including Terrorism (See endnote 9)
• Create an all-hazards response system to chemical and biological releases.
• Establish a specially trained HAZMAT team to respond to anticipated terrorist attacks.
• Train medical personnel to recognize the symptoms of biological and chemical warfare agents and report them immediately to the all-hazards response system.

FACTORS LEADING TO IMPAIRMENT AND BEST PRACTICES FOR CHILDHOOD LEAD POISONING

Child lead poisoning is a completely preventable condition caused by a variety of environmental factors, most usually lead-based paint, contaminated household dust found on floors and windowsills, old pipes with lead-based solder, contaminated water, contaminated soil, contaminated industrial areas, air pollution, vinyl mini blinds, imported candy, lead-glazed pottery, fishing tackle, homemade medicines and toys especially antique toy soldiers and other war toys, cosmetics, or workers bringing home contamination on their clothing or in scraps of material from their jobs or hobbies. A child can absorb twice as much lead from the gastrointestinal tract as an adult. The problems of lead are multiplied when children live in older housing stock and industrial areas.

Much lead contamination is ingested by children, but may also be inhaled. Schools that use wells are particularly susceptible to lead problems. The total annual cost in 2011 for lead poisoning in children in the United States was $50.9 billion, and $977 billion in low- and middle-income countries in the world. (See endnote 56.) Lead poisoning can harm the child's brain, kidneys, bone marrow, and other body systems. Low levels of lead in the child can reduce intelligence and cause disruptive behavior. At high levels, lead poisoning can cause coma, convulsions, and death.

The EPA reported that in 2005 about 250,000 American children aged 1–5 years old had elevated blood lead levels of 10 μg/dL, whereas the current standard is 5 μg/dL, although in fact no lead level in the blood is safe. However, recent studies have shown that blood lead levels well below this federal standard can lower IQ and have effects on behavior in children. An additional 6.4% of children aged 1–5 in the United States have a blood lead level higher than 5 μg/dL. Five times as many foreign-born children have elevated blood lead levels compared to US-born children.

Best Practices for Childhood Lead Poisoning (See endnotes 21, 52)
• If you live in an area built before 1978, or are in a situation where a child may have consumed lead, ask a physician or a public health clinic to do a blood test to determine if the child has any level of lead in the body.
• Utilize federal grants for low-income housing in assisting private and non-profit organizations in removing lead paint hazards in houses and other facilities.
• Check to see if there is any recall of toys that may contain lead.
• Do not drink, cook, or make baby formula using water from the hot water faucet.
• Run the cold water for 15–30 seconds first thing in the morning to flush out any lead in the system.
• Wash the child's hands frequently, especially before eating.
• Remove peeling paint from surfaces in all areas where children can have access.
• If the soil is contaminated, plant grass so that the child will not have soil available to eat or inhale.

- Do not store food in open cans, especially those which are imported.
- Do not use pottery or ceramic ware that is improperly fired or highly decorative for food service or storage.
- Test all children who have newly come to the United States, between the ages of 6 months and 16 years, for potential lead problems.
- Only use state-certified companies to remove lead from surfaces, especially painted surfaces, and to perform dust handling after renovation.
- Only use state-certified people for conducting any form of lead-based paint abatement activities.
- Conduct research to improve and make less costly the removal of lead-based paints from structures.

CHILDHOOD POISONING

Over 86,000 child poisonings were treated in the United States in hospital emergency rooms in 2004. About 70% of these poisonings were in children 1–2 years of age. About 60% of the poisonings involved oral prescription drugs, non-prescription drugs, or supplements. Cleaning products, cosmetics, plants, toys, pesticides, art supplies, and alcohol contributed to most of the other poisonings. Another source of poisons for all individuals, but especially children, is illegal controlled substances. Children are being exposed to heroin, cocaine, and a variety of painkillers including oxycodone. In states where marijuana is legal, it may be incorporated in food as well as smoked, and this could be a serious problem for young children who acquire it, especially within the home.

About 55% of the poisonings came from products that were in child-resistant packages under the Poison Prevention Packaging Act. The American Association of Poison Control Centers reported 1.2 million cases of childhood poisoning in 2001 in children less than 6 years of age. In 2011, the 57 poison control centers in the United States handled 4 million calls including 2.4 million human poison calls.

Best Practices for Childhood Poisoning (See endnote 53)
- Post the poison control number by the main telephone in the home, or on the refrigerator.
- Store all medicines, household products, and personal care products in cabinets with locks.
- Make sure that none of the plants in the yard are poisonous if ingested or by contact or inhalation.
- Determine if visitors bring medicine into the house.
- Use carbon monoxide monitors near the bedrooms.
- Check all heating appliances to make sure they function properly.
- Determine if there is any lead-based paint in the house.
- Read directions before using any chemicals and use them judiciously.
- Take medicines appropriately in proper dosages.
- Do not mix household cleaners or other chemicals.
- Do not burn fuel, charcoal, or use gasoline-powered engines in confined spaces.

If poisoning occurs immediately contact the Poison Control Center and 911.

ENVIRONMENTAL (SECONDHAND) TOBACCO SMOKE

Environmental tobacco smoke contains more than 7000 different chemicals including hundreds that are toxic and 70 which cause cancer in people and animals. Environmental tobacco smoke causes about 150,000–300,000 lower respiratory tract infections each year in children under 18 months of age. It may cause ear infections, serious respiratory symptoms, and respiratory infections including bronchitis and pneumonia, and increase the frequency of severe asthma attacks. It has an immediate effect on the cardiovascular system and increases the risk for heart attacks. It may cause cancer and sudden infant death syndrome. It may result in low birth weight and make the babies and children more frequently sick. Smoke from a single cigarette in a room can be present for hours.

Best Practices for Environmental (Secondhand) Tobacco Smoke
- Do not permit smoking in cars.
- Do not permit smoking in homes.
- Do not permit smoking in day care settings.
- Do not allow anyone to smoke around children.
- Do not go to restaurants or other facilities that allow smoking.
- Teach children the dangers of smoking and to avoid areas where people are smoking.

FOOD

By 2009, all 50 state public health agencies and federal agencies had issued fish consumption advisories because of mercury contamination. The individuals most affected by this problem are low-income, low-education and limited English proficiency immigrants, since they are often unaware of the nature of advisories. In fact, these individuals have been taught to consume more fish because omega-3 fatty acids and other nutrients in fish are associated with better birth and developmental outcomes. Many of the children bring packed lunches and store them in lockers without proper refrigeration. The overall topic is large and complex. (See complete discussion in Chapter 7, "Food Security and Protection.")

INDOOR AIR POLLUTION

(See Chapter 2, "Air Quality (Outdoor [Ambient] and Indoor)")

Indoor air pollutants affecting children, especially those with existing respiratory and allergic problems, include asbestos, carbon monoxide, formaldehyde from particle board, household cleaners, consumer home products, fire retardants, lead (especially in pre-1950 built homes) from lead-based paint, molds, particulate matter, pesticides (especially in crowded lower income housing typically because of severe roach infestation), radon gas, secondhand smoke (especially a very serious problem in crowded lower income housing), phthalates, burning of candles and incense, recent indoor painting, plastics emitting plasticizers, and other volatile organic compounds. Also of great concern is the inhalation of a mixture of these contaminants.

Environmental tobacco smoke is a problem for 43% of children from 2 months of age to 11 years of age. Environmental tobacco smoke is a mixture of some 4000 chemicals, some of which are extremely hazardous to the health of a small child. Children who are exposed to these chemicals tend to have greater problems with bronchitis, pneumonia, respiratory infections, middle ear infections, and asthma. Typically, the frequency of these health problems is related to the amount of smoke that is present in the child's environment.

Allergens are found in the indoor air including those that come from dust mites, cockroaches, pet dander, pollen, molds, spores, bacteria, and viruses. These allergens combined with the environmental tobacco smoke and other chemicals found in the environment have a profound effect upon the child and serve as triggers for asthma, which is especially prevalent in minority children living in the inner city. The economic impact of this is considerable, resulting in lost school attendance, lost workdays for the parents, and a substantial number of emergency room visits which are very costly.

Volatile organic compounds including cleaning products, adhesives, paints, and dry-cleaning fluids are continuing problems in the child's environment.

Nitrogen oxides from fuel combustion may be found in the indoor air as well as the outdoor air.

Illegal drug laboratories, especially used to produce methamphetamines, use a variety of toxic chemicals during the cooking process. A methamphetamine residue coats all the surfaces in the property.

Building materials can be hazardous to children if they are contaminated and if they release chemicals such as formaldehyde. Recently, drywall from China has been suspected of containing corrosive materials and also elevated levels of mercury. Another potential problem is the presence of asbestos.

Inadequate ventilation in the structure, especially close to the ground, may contribute to the inhalation of contaminants from cleaning and disinfecting products, air fresheners, and perfumes from adults.

Contaminated dust is present in the small child's environment and is inhaled by the child. The chemicals come from furniture, carpets, plastic products, flame-retardants and pesticides, and are carried in from the outside or stirred up inside by people walking on surfaces. Mold spores from water-damaged walls or carpets add to this problem. Lead may be part of the dust from peeling paint in structures built before 1976 or from lead-based painted toys or lead toys. (See Chapter 2, "Air Quality (Outdoor [Ambient] and Indoor).")

INJURIES

(Also see the section on "The Recreational Environment" above)

Unintentional injuries are the leading cause of death to children from age 1 to age 19 in the United States. On a yearly basis, 9000 children in the United States from birth to age 19 die as a result of their injuries, 30,000 are permanently disabled, and 8.7 million need emergency medical care for their injuries, with 225,000 needing hospitalization costing $87 billion. Males are at greater risk than females. Poverty, hunger, overcrowding, the age of the mother, location of housing, etc., contribute to an increase in injuries. The most common causes of injuries are: drowning, falls, fires or burns, poisoning, suffocation, sports or recreation, and injuries related to transportation. (See endnote 38.)

Motor vehicle injuries increase in number with the age of the child. In 2010, about 1200 children under 14 years of age died in motor vehicle crashes and 171,000 were injured. A study indicated that 618,000 children in a single year were riding without either a child safety seat, a booster seat or a seatbelt. About two thirds of the children killed were in car crashes with drunken drivers. (See endnote 37.)

Young worker safety and health is another concern. Young workers (aged 15–24 years) make up 14% of the United States labor force and are at high risk for injury on the job. They are injured while driving motor vehicles as part of their job, working in agricultural areas, working in landscaping, greenhouses and nurseries, working in restaurants, etc. (See endnotes 19, 20.)

The 10 most hazardous household items leading to childhood injuries include:

1. Cords leading to asphyxiation
2. Bathtubs leading to drowning
3. Small toys leading to choking deaths, and riding toys (58,000 injuries reported in emergency rooms yearly)
4. Climbing on dressers or other furniture, leading to crushing injuries
5. Falling out of open windows, especially those with screens in them
6. Cribs and crib bedding leading to strangulation and suffocation
7. Adult exercise equipment (25,000 emergency room visits yearly)
8. Hot stoves leading to numerous burns and over 600 children a year dying
9. Ingestion of medicines, vitamins, and other pills especially found in grandparents' homes
10. Ingestion of cleaning supplies, cosmetics, and other household chemicals

Best Practices to Prevent Injuries
• Develop a data and surveillance system to determine the types and frequencies of various injuries in a given area.

- Conduct appropriate research on the prevention of these childhood injuries, such as the use of bike helmets, pool fencing and booster seats, and use of smoke alarms.
- Establish concussion guidelines for all sports.
- Determine best teen driving policies and teach them to the drivers.
- Establish a variety of communication processes concerning injury prevention in children and utilize them with targeted populations.
- Utilize educational and training techniques to teach people, including professionals, the best way to prevent various types of injuries.
- Develop written policies and provide leadership training to prevent injuries.
- Create health systems and healthcare practices where the professionals are well trained in a variety of injuries, and the appropriate treatments to be used are identified, and then determine how best to prevent the injuries from reoccurring.
- Use child safety seats, booster seats, and seatbelts to prevent injuries.
- Establish rules for teenage driving.

METHAMPHETAMINE-CONTAMINATED HOUSES

(See endnote 25)

Houses that have contained methamphetamine laboratories can be found in all parts of the country and at all income levels. All of these houses put current and future occupants, especially children under the age of 2, at extremely serious risk of a multitude of health hazards or death. The chemicals, including acetone, phosphine, iodine, hydrogen chloride, anhydrous ammonia, etc., used in the production of methamphetamine, their precursor chemicals, and their byproducts are extremely toxic at the time of production and for many years to come. The cooking process aerosolizes the chemicals and allows them to penetrate many surfaces within the structure. There is a constant hazard of potential explosions due to the use of the chemicals. Police and firemen are also exposed when they come to the house. Sinus problems, respiratory problems, migraine headaches, skin irritation, and burns are common. Prenatal exposure can result in placental problems, fetal distress, and postpartum hemorrhage.

In Missouri alone, 12,354 methamphetamine laboratories were found by authorities and closed between 1998 and 2010. Home-cooking methamphetamine spreads toxins to all parts of the house including carpets, walls, furniture, drapes, air ducts, anything found in the house, and even the air.

Before an individual purchases a property, he/she should inquire about the history of the property and if necessary have a certified inspector evaluate the potential for methamphetamine having been produced in the structure. Where methamphetamine is found, an extremely thorough process must be used for removal of all chemicals to make the structure safe. This is an expensive process.

Best Practices for Cleanup of Methamphetamine Contaminated Houses
- Ask law enforcement to accompany the individuals preparing to clean up methamphetamine-contaminated houses.
- Determine the types, concentrations and quantities of chemicals used in the structure and the areas which have been contaminated.
- Clean until there are no remnants of the chemicals in any areas of the structure.
- Repair or replace appliances and surfaces which have become contaminated. Remove contaminated items to secure landfills.
- Thoroughly air out the building and clean all of the vents and heating and air-conditioning ducts.
- Dispose of all carpets, drapes, clothing, furniture and other substances which may have been contaminated.
- After thoroughly washing the inside of all the walls, floors, and windows, air out the structure and repaint all the surfaces.

- Check all plumbing, septic tank systems and other potential underground disposal areas for chemicals and thoroughly clean the areas.
- Retest the inside of the structure after the cleanup and repair.
- Avoid having young children or pregnant women move into the structure wherever possible.

OUTDOOR AIR POLLUTION

(See Chapter 2, "Air Quality (Outdoor [Ambient] and Indoor)")

Urban air is contaminated by dense street traffic in congested city areas. The air acts as a stream carrying the pollutants from automobiles, residential boilers, bus depots, sewage treatment plants, industry, power plants, and waste disposal sites. These pollutants have certainly contaminated the adjacent land which is now usually very cheap and has become the area of residence of the poor. This creates an environmental justice problem because most municipalities are concerned about building or renovating the downtown areas and spend huge sums of money doing this while neglecting the areas of the city where the poor live and which are obviously substantially contaminated by pollutants.

Further, outdoor air pollutants, as well as fumes from heating of the properties, cooking and smoking in other areas of the building, can easily move through apartments and other areas creating potential health effects for the residents, especially the children. Outdoor air pollutants include arsenic, benzene, diesel exhaust, dioxins, endocrine disruptors, mercury, nitrogen oxides, ozone, particulate matter, polybrominated diphenyl ethers, PCBs, sulfur dioxide, ultraviolet radiation, and volatile organic compounds.

Along with the previously mentioned air pollutants, increased particulate matter from fires, volcano eruptions or explosions set off by terrorists can result in acute and chronic illnesses of the respiratory system, especially in children. There is an increase in respiratory symptoms, decrease in lung function, exacerbation of asthma and development of chronic bronchitis. The polycyclic aromatic hydrocarbons can reduce fetal growth, increase the risk of developmental and behavioral problems, reduce IQs and add to the precursors of asthma. High particulate levels have been associated with increased numbers of preterm births, low birth weight and increased infant mortality. Climate change can increase the risk of mosquito-borne disease, the production of ground-level ozone, and the risk of hazardous air pollutants.

Best Practices for Air Pollution (See Chapter 2, "Air Quality (Outdoor [Ambient] and Indoor)")

PESTICIDES

In the United States, 78 million homes use home and garden pesticides including many that are toxic. When children are exposed to these pesticides, the risk of childhood leukemia is increased by about seven times. These chemicals can also increase the numbers of miscarriages, suppress nervous, endocrine and immune systems, and increase the risk of children developing asthma.

Chlorpyrifos, which is banned for household use in the United States but still used in agriculture, is associated with early childhood developmental delay. Exposure to this chemical results in mental and physical impairments and increases the risk for attention deficit/hyperactivity disorder in children, especially in low-income areas, where the chemical was previously used to control insect infestation. Diazinon, which also had been used widely in housing, had adverse effects on children. Young children have greater exposure to the chemicals because they play on the floor or in the grass, and they frequently put things in their mouth. (See endnote 39.)

Pesticides found in food and water, especially in rural areas, even in low concentrations, may affect the health of children. It is suspected that pesticides can contribute to three major developmental disabilities: autism, cerebral palsy, and severe mental retardation. The Food Quality Protection Act of 1996 requires that a 10-fold factor be in place in risk assessments for pesticide residue for food consumed by infants and children.

Accumulations of toxins from a variety of sources are considered to be a very significant problem for children.

Endocrine disruptors are chemicals that interfere with the natural function of hormones that control natural development in the body. They may be present in very small quantities but still cause harm. They are especially problematic in pregnancy and in the fetal, infant, childhood, and adolescent stages. The endocrine disruptors can come from municipal waste disposal and incineration, from agriculture (with the weed-killer atrazine being a primary example) or any other exposure to chemicals which may be found in cosmetics, plastics, perfumes, etc. Phthalates and BPA are two of the chemicals most incriminated in causing health effects. The Endocrine Disruptor Screening and Testing Program was required by the Food Quality Act and the 1996 Safe Drinking Water Act.

Chromated copper arsenate, a compound used as a chemical wood preservative containing chromium, copper, and arsenic, has been used out of doors to protect wood, including play equipment, from insects and microbes. Railroad ties are treated with this preservative. When they are removed from areas where there are no longer being used they can leave behind high levels of arsenic in the soil. Also if they are recycled and used in gardens or as retention walls, they can easily contaminate the new areas. This substance is highly dangerous for children. (See Chapter 9, "Insect Control, Rodent Control, and Pesticides.")

Consumption of Unpasteurized Milk

Recently there has been a move by certain parents to give their children unpasteurized milk as a better way of providing a healthy food. Unfortunately this fad, which has been spread by social media, as well as the fad of not immunizing children, not only endangers the child but also the entire community. There are far too many years of experience with outbreaks of disease and what constitutes good practices to allow these types of fads to endanger children. In Minnesota as an example, there have been numerous outbreaks of foodborne illness from the consumption of unpasteurized milk. People who buy a share of a dairy cow and then milk it and do not pasteurize it are definitely endangering their family. (See Chapter 7, "Food Security and Protection" for further details.)

Soil

(See endnote 54)

Soil or dirt, the upper layer of earth in which plants may grow, may be contaminated by a variety of chemicals and microorganisms. The contaminants may leach into water, volatilize into the air, or bind to soil particles. The bioavailability of the substance to humans is determined by the various soil conditions. The contaminants rarely are distributed uniformly in the soil because the amount being distributed at any given time varies and the movement of air and water. The contamination may be from the residue of air pollution, the storage and disposal of solid and hazardous waste, the results of manufacturing and usage of chemicals especially pesticides, and on/in industrial/commercial properties, high-traffic areas, treated lumber, automobile repair shops, junk vehicle storage areas, and furniture refinishing facilities. They may also come from landfills, fires, use of fertilizers in agricultural areas, construction, mining, oil/chemical storage, transportation or spills, flooding, natural sources, etc. An example of natural arsenic contamination is found in the soil of Hawaii. Arsenic is a naturally occurring element in the earth's crust. However, elevated levels of arsenic have been found in the soil of former sugarcane fields, pesticide storage or mixing areas, sugar plantation camps, wood treatment plants, etc. In the past, the practice was to put the waste material on the land of the industry, bury it beneath the surface of the land, or dump it into a body of water. Also, lead may be present because of the use of lead paint in commercial, industrial, and residential properties.

All sources of contamination become a concern when small children are present because the contaminant may enter the child's body primarily through inhalation of the dirt, ingestion by eating the dirt, and skin absorption through contact with the dirt. Food and water may become contaminated and then consumed by the child. The potential health effects depend on all the previously mentioned contributors to disease in children and: the types of chemicals, the dose, the duration of exposure, the frequency of exposure and the previous experience with this chemical and others.

Microbial contamination including bacteria, viruses, protozoa, fungi including molds, and helminths, which are parasitic worms, cause a series of illnesses ranging from mild to severe including potentially death, and may be found at various times in soil especially in the aftermath of flooding. The microorganisms and helminths may survive in the soil for an extended period of time.

Best Practices for Limiting Soil Exposure (See endnote 55)
- Do not allow contaminated land to be used for the sites of schools, other facilities, and homes where children may be exposed to the contaminants.
- All contaminated land, including by radiation, must be remediated to remove or encapsulate all forms of contamination to prevent exposures in surrounding areas.
- Industries must not contaminate the land site or off-site with their products, byproducts, raw materials, and waste materials.
- All remediation activities must be licensed and inspected by the appropriate authorities.
- A planning authority must be established to determine the appropriate use of land and the removal and prevention of potential contamination.
- Follow all Best Practices as established in: Chapter 14, "Water Quality and Water Pollution"; Chapter 2, "Air Quality (Outdoor [Ambient] and Indoor)"; and Chapter 12, "Solid Waste, Hazardous Materials, and Hazardous Waste Management."
- Prevent disease from microbial contamination and helminths by use of the following procedures.
  - Practice the highest levels of proper hand-washing techniques.
  - Use extensive educational efforts to teach people to avoid standing water, ground saturated with flood water, and areas of debris.
  - Use signs to indicate areas of potential contamination.
  - Scrub and decontaminate all playground equipment and areas where children will be involved in activities.
  - Decontaminate all areas near septic tanks.
  - Put new soil top on the affected soil and compact it.
  - Plant new grass.
  - Use dust suppression techniques.

SOLID AND HAZARDOUS WASTE DISPOSAL

(See Chapter 12, "Solid Waste, Hazardous Materials, and Hazardous Waste Management")

It is estimated that 3–4 million children in the United States live within 1 mile of at least one hazardous waste site, which may or may not be a Superfund site. As of 2008, approximately 800,000 children lived within 1 mile of a Superfund site that had not been cleaned up or controlled since 1990. Although this is an improvement over past years, still too many children are at risk. In the United States, approximately 60% of African-Americans live in communities with uncontrolled toxic waste sites. Also, typically low-income minority communities are located close to chemical storage and disposal sites and in rural areas where toxic waste is generated. Large numbers of African-Americans live in communities close to toxic waste areas called brownfields. These wastes include chemical waste, waste from fossil fuel power plants and municipal incinerators, and solid waste landfills. Examples of these facilities can be found in North Carolina, Jersey City, New Jersey, and elsewhere. Schools have been found immediately adjacent to chemical plants, such as the one in Framingham, Massachusetts. Three of the five largest hazardous wastes landfills are located in predominantly African-American or Latino communities.

Love Canal was the ultimate example of how people, especially children, can be endangered for a lifetime because of exposure to hazardous chemicals. Approximately 100 homes and a school were built on the land of a former chemical company that buried its waste there. This industrial dump contained 82 different compounds, 11 of which were suspected of being carcinogens. Children playing at the school who were in the immediate area had burns on their hands and faces. There was a substantial number of birth defects and a high rate of miscarriages in the individuals living in the area. Eventually, the school was abandoned and 221 families were moved to new housing areas.

Another dangerous practice is the collection of hazardous and toxic wastes on school property by teachers and children. They are certainly not trained to deal with this type of activity and even though it is considered to be a public service and good for the community, the location of the storage of the hazardous waste at the local school is very poor practice. Spillage of the hazardous waste would not only constitute an immediate hazard but also a long-term hazard.

UNSAFE DRINKING WATER

(See Chapter 13, "Water Systems [Drinking Water Quality]") (See endnote 43)

Between 15% and 20% of US households including millions of children, especially in rural areas, are at substantial risk from drinking from water supplies drawn from wells which may be contaminated by overflowing, improperly operated septic systems, and chemicals and microorganisms from a variety of other sources. Children have potentially multiple opportunities to drink contaminated water since thousands of schools across the country have been found to use water with unsafe levels of lead, arsenic, pesticides, agricultural chemicals, and many other toxins. There is a serious concern about the mishandling of chemicals within school laboratories.

Wells are implicated in this situation. Supervision of water supplies is spread among numerous local, state, and federal agencies, which leads to confusion and the under-reporting of potential hazards. Older buildings with older plumbing, homes, or schools, show increased levels of lead in the water.

Best Practices for Protecting Drinking Water Supplies
- Test a well before purchasing a new home, then on an annual basis or if the well is suspected of becoming contaminated.
- Require that all records (well logs) are made available prior to the purchase of the home and that they show that there is a pitless adapter, which has been certified, and the well was properly grouted.
- Test a well before purchasing, renting, or leasing an existing home with a well, then on an annual basis, or if the well has been suspected of becoming contaminated.
- Test water in vacation homes, camps, etc. for bacteria and chemical contamination before using the water.
- Test wells for childcare facilities and schools before use, periodically and if the wells are suspected of becoming contaminated.
- Thoroughly decontaminate wells that have been contaminated and then retest them.
- If ever in doubt about the safety of a water supply, use bottled water from approved sources for drinking, cooking, and washing of dishes.

INSECTS, RODENTS, AND PESTICIDES

(See Chapter 9, "Insect Control, Rodent Control, and Pesticides")

VECTORS OF DISEASE

Climate change has special disease-related concerns including the creation of potential epidemics of vector-borne disease, which is especially a problem in children. Diseases such as malaria, dengue fever, and hantavirus pulmonary syndrome, which are not commonly found in the United States, can increase sharply with increased warming conditions. A good example of this would be the sharp increase in West Nile virus in the United States.

FACTORS LEADING TO IMPAIRMENT AND BEST PRACTICES FOR EMERGENCY ENVIRONMENTAL RISKS

(See Chapter 5, "Environmental Health Emergencies, Disasters, and Terrorism")

FLOODS, TORNADOES, AND HURRICANES

Children have additional problems in the aftermath of floods, tornadoes, and hurricanes. Children are typically more vulnerable to chemicals and/or microorganisms because of children's behavioral traits and physiology. Potential hazards to children after floods, tornadoes, and hurricanes increase substantially. Some of these hazards include contaminated food; a sharp increase in mold in carpets, drywall, and other porous areas; carbon monoxide from poorly vented emergency energy sources or damaged energy sources; contaminated drinking water; disrupted sewage systems; solid waste, hazardous waste, and debris; houses and household items contaminated by flood water and torn apart by tornadoes; and schools, school grounds, or playgrounds contaminated by flood waters or torn apart by tornadoes. It is highly recommended that children be the last individuals back into areas contaminated by floodwaters, or the after effects of tornadoes and hurricanes.

EXTREME HEAT

Heat-related illnesses have a more profound effect on children than other people. Children have a smaller body mass to surface area ratio than adults, can lose fluid more rapidly, and tend to play outdoors in the extreme heat. They also do not have the capacity to make judgments about protective measures in extreme heat. Children may show the following symptoms: behavioral changes, dizziness or fainting, rapid breathing and rapid heartbeat, extreme thirst, hot dry skin, headaches, reduced volume of urine which may be dark yellow, etc.

Best Practices for Extreme Heat
* Re-hydrate the child over a period of time.
* Keep the child in a cool and shaded area.
* Have the child wear light clothing.
* If the child is lethargic, not responsive, has a fever or continues to be dehydrated, seek immediate medical attention.

AFTERMATH OF WILDFIRES

(See endnote 17)

Wildfires create numerous environmental hazards, especially for children because of their unique physiology and size. These hazards include fire, extreme heat, and smoke; combustion of wood, plastics, and other materials in homes and other structures, which release a large number of hazardous chemicals as well as various sizes of particulate matter; landslides; and acute stress from the emergency and severe emotional reactions.

The smoke consists of very small inorganic and organic particles, liquid droplets, carbon monoxide, carbon dioxide, and volatile organic compounds. Eye irritation and respiratory tract irritation are common. Lung function is reduced substantially and there is a worsening of preexisting lung or cardiovascular disease, especially asthma.

Ash becomes a serious potential hazard because it is very irritating to the skin, nose, and throat and may cause severe coughing. Ash and dust from burnt structures may contain a series of toxic and cancer-causing chemicals including asbestos, arsenic, lead, formaldehyde, and a multitude of other substances depending on what was in the buildings.

Debris can cause physical and physiological problems. Broken glass, exposed electrical wires, nails, wood, metal, plastics, and other objects can cause wounds, especially puncture wounds, electrical injuries, and burns.

During the recovery phase, the children may be exposed to a whole new environment to which they are unaccustomed which can lead to additional injuries and accidents. The soil may be contaminated with hazardous materials such as lead. There may be persistent hotspots which can burst into flame without warning.

Best Practices for Aftermath of Wildfires
- Drinking water systems, electrical systems, and sewage disposal systems must be back in proper operating order before children can re-enter the area.
- The structures must be inspected and found to be sound and habitable before children can re-enter the area.
- Ash and debris must be removed or at least isolated from areas where children will be living, going to school, or playing.
- All roads and approaches to various facilities and homes must be cleared and cleaned.
- Never use gasoline or kerosene appliances including generators in enclosed spaces as a source of electricity.
- Be careful to avoid ash pits or burned stumps of trees, which may be very hot.
- Wear protective facemasks and goggles as well as protective clothing when cleaning up ashes and debris.
- Wet down ashes before removing.
- Wash all homegrown fruits, vegetables, and children's toys before using.
- Remove all water-damaged building materials which could result in the growth of mold and check frequently for mold growth.
- Discard all foods which have not been stored in proper temperatures during the shutoff of electricity and refrigeration.
- Discard all foods which have been exposed to heat, fire, smoke, or water.
- Check the child's behavioral patterns frequently to determine if professional assistance is needed to deal with the aftermath of the trauma.

LACK OF CONSISTENT APPROACH TO CHILDREN'S ENVIRONMENTAL HEALTH ISSUES

The problem is that there is no consistent approach to the resolution of hazards affecting children from state to state and locality to locality. Unfortunately, children's programs typically do not produce revenue and therefore, especially in times of serious budget shortfalls, children's issues are not given the resources that are necessary to avoid injury and illness and promote better health. In addition, where local and state health departments receive federal funding for specific programs, the programs cease to exist typically when the funds run out. Also, many local and state funding agencies use the federal funds as an excuse for reducing the regular budget for other environmental health programs. For any given locality or state, it is necessary to contact the appropriate health department, over the internet or by phone, to determine what programs or educational materials are available to help resolve children's environmental health issues. Federal internet sites are also very useful for obtaining educational and training materials.

INADEQUATE TRAINING OF PROFESSIONALS IN CHILDREN'S ENVIRONMENTAL HEALTH ISSUES

Numerous organizations have stated that because of the presence of environmental health hazards in communities, health professionals must be prepared to diagnose, treat, and prevent health conditions caused by environmental exposures. Unfortunately, a survey of environmental medicine content in United States medical schools showed that environmental medicine content over a 4-year period in 75% of the schools amounted to a total of 7 hours. A Migrant Clinician Network study found that about half of the doctors had no training in environmental or occupational health. A survey of chief residents of the United States pediatric residency programs showed that less than half of the pediatric programs included pediatric environmental health issues in their curriculum except for lead poisoning. The lack of environmental education in the training of physicians, nurse practitioners, physician assistants, nurses, nurse midwives, and community health workers is a serious shortcoming in the protection of children. Numerous associations of health professionals have indicated strong support for better training of health workers in children's environmental health issues and solutions.

Pediatricians

The role of the pediatrician since the September 11, 2001 terrorist attacks has changed in some respects. Prior to that the pediatrician was concerned about keeping a healthy child healthy, and diagnosing the cause of an illness and making the child healthy again. Most of the illness problems appeared to revolve around bacterial and viral infections, although some of the concerns were about chemicals and injuries. Now there has been a dramatic increase in concern from parents, the community, and society at large about the potential for acute and long-term problems related to environmental conditions in homes, schools, recreational areas, occupational areas, the outside environment, the indoor environment, and terrorism. In a special study, 86% of a group of pediatricians reported that their medical training had not prepared them sufficiently to deal with problems of the environment or terrorism.

Clinicians, through the use of self-administered environmental investigation forms and nurse evaluation forms, should consider the potential for environmental conditions causing the observed symptoms. For instance, seizures can be the result of lead poisoning or carbon monoxide intoxication, learning disabilities can be caused by intrauterine alcohol exposure or lead intoxication, eczema can be aggravated by solvents, etc. The clinician should determine if symptoms subside or worsen in a particular location, during weekdays or weekends, at a special time of the day or during certain activities, and are other children associating with the child having the same symptoms. The clinician should determine if there are clusters of cases of the symptoms and work with local public health authorities and the Centers for Disease Control and Prevention (CDC) to share knowledge and gain more information about the specific problem. Parental occupational exposure to environmental factors, which may cause the symptoms experienced by the children, should always be investigated and evaluated.

The American Academy of Pediatrics and local chapters are emphasizing the extreme importance of the community in pediatric practice. This perspective enlarges the pediatrician's focus from one child to all children in the community and includes the many problems related to the environment. There is an understanding that many factors affect the health of children including the family and social, political, cultural, economic, and environmental forces. In order to truly help the child, the pediatrician needs to combine good clinical practice with public health principles and work with the family, school, and community. There are many resources that need to be accessed and utilized appropriately.

To better prepare the pediatrician for his/her new role, it has been recommended by the American Academy of Pediatrics and local chapters that these doctors receive training and utilize information in the following areas:

1. Utilize epidemiological, demographic, and economic data to better understand the effects of poor health and social risks on child outcomes.
2. Work with public health departments and colleagues in other professions to identify and decrease barriers to the health and well-being of children.
3. Interact with childcare facilities, schools, and youth programs to help them understand how they affect child health.
4. Help provide health care for all children in the community (unfortunately politics in certain state and local areas have interfered with the care for all children).
5. Utilize community resources to promote good health.
6. Educate residents and medical students in community settings.
7. Learn about environmental concerns and children's health and implement a program of environmental history taking within the office.
8. Educate parents about environmental problems and how they affect the health of their children.

Nurses

Nurses are exceedingly important in the evaluation of environmental stresses in children. Nurses see the children in the doctors' offices, in the schools and in the community as public health nurses or visiting nurses. Unfortunately, the level of environmental education given to nurses in their training is very limited. The understanding of environmental conditions and their effect on children must always be considered in evaluating the child's health because of existing symptoms or as a preventive measure in special situations.

Nurses are uniquely qualified to teach parents and children about the potential for environmental problems causing disease and injury. They may be working on a one-to-one basis and therefore can have a more positive effect on the family's thinking and resolve to correct serious environmental hazards.

Registered Environmental Health Specialists/Registered Sanitarians

Registered Environmental Health Specialists/Registered Sanitarians are experienced, highly trained, well-educated, applied scientists and leaders in recognizing, evaluating, and resolving environmental health and community health issues. They utilize both educational and enforcement tools to reduce and remove environmental health hazards and disease causing problems from a variety of situations. They work closely with other health professionals, civic and professional associations, governmental agencies, schools, preschools, and community leaders to effect positive change in the environment. They help promote better health and injury- and disease-free situations for all citizens, children and adults alike. These individuals work in the governmental sector, private sector, and academia, but the overwhelming number are in the public sector.

Environmental Health Aides

Environmental Health Aides/Community Health Aides are individuals, preferably from the local community, who have a minimum of a high school diploma or equivalency with good knowledge of high school level math, some science, and excellent communication skills in English. Typically, the individual should also have good communication skills in the language primarily used in the given community area. The individual is then given specialized training in the collection of samples of air, water, soil, sewage, paint, and other materials that could be causing environmental problems. The individual investigates complaints, writes simple reports, and teaches the residents how to reduce exposure to a variety of insects and rodents, environmental toxicants, potential injury-causing circumstances, and infection control problems. The aide is under the direct supervision of Registered Environmental Health Specialists.

CRITIQUES OF CHILDREN'S ENVIRONMENTAL HEALTH EFFORTS

(See endnotes 61, 62)

FEDERAL LEVEL

Government Accountability Office

The Government Accountability Office in September 2008 reported that the EPA was failing on children's environmental health issues. The EPA had rolled back or not acted on standards for dangerous chemicals, such as perchlorate, mercury, and lead. The advisory committee, which was made up of public health officials from government, non-profit organizations, academia, industry, and healthcare organizations, had met more than 30 times in 10 years, but the EPA rarely sought out the committee's advice and recommendations in developing regulations, guidance, and policies concerning children's health, despite the fact that the advisory committee had sent over 600 recommendations for EPA action. Further, the last time cabinet-level officials met to discuss children's environmental health issues was in October 2001. The task force expired in 2005 and there was no further mandate to coordinate the federal response to children's health issues. The task force no longer reported its results to the President. The Government Accountability Office in a report dated January 2010 to several congressmen recommended improvements to help the EPA protect children and the EPA agreed to implement them. The Government Accountability Office also suggested that Congress consider reinstating the Children's Health Protection Advisory Committee.

Environment and Public Health Committee of the United States Senate

On March 17, 2010, Dr. Cynthia Bearer, Children's Environmental Health Network Board Chairperson, testified before the Environment and Public Health Committee of the United States Senate. She urged the committee to include the basic concepts of pediatric environmental health into all of its policies and programs. She said that today's standards, regulations, and guidelines are based on the adult healthy male, rather than considering the unique problems of children. The US EPA regulates thousands of industrial chemicals through the Toxic Substances Control Act. She recommended that the Toxic Substances Control Act be replaced with a new statute that provided for:

- Health protection of children, pregnant women, the fetus, nursing women, and women of childbearing age, as a basis for regulating chemicals
- Strong safety standards, which protect children's health and other vulnerable populations
- Making industry responsible for demonstrating the safety of a chemical
- Protecting children from chemicals that interfere with their hormone systems
- Protecting children from new technologies such as nanotechnology

Dr. Bearer urged the committee to focus on the home environment including the pollutants and products to which children are exposed. Further, millions of preschoolers go to childcare centers for at least 40 hours a week and the centers need a substantial amount of monitoring. During the course of a week, 54 million children and about 7 million adults spend a substantial time in the school environment and therefore this environment needs to be closely monitored. Particular attention must be given to the many sources of indoor air pollution which can affect the child.

Dr. Bearer commended the new EPA administrator for improving the EPA's response to children's environmental health issues, including protections for farm workers and farm worker children from pesticides and improving pesticide labeling. Also, there has been an increased effort in improving the environmental health of schools. Dr. Bearer emphasized that there had been a tremendous loss in the abandonment of Presidential Executive Order 13045. She urged the committee to pass a law reestablishing the interagency task force, composed of agencies including the EPA, Department of Health and Human Services, Department of Housing and Urban Development, Department of Education, and others, that worked effectively on children's environmental health and safety issues like asthma, lead poisoning, childhood cancer, injury prevention, etc.

Also, on March 17, 2010, John B. Stephenson, Director, Natural Resources and Development, Government Accountability Office, reaffirmed Dr. Bearer's testimony and added that the Government Accountability Office was issuing a series of recommendations to the EPA in a special report that would help the EPA re-issue a child-focused strategy with specific goals, objectives, and targets. It asked the EPA Administrator to maximize the agency's concerns in all areas of environmental health hazards related to children. The report also urged Congress to establish, in law, President Clinton's Executive Order on children's environmental health.

LAWS, RULES, AND REGULATIONS

These laws include (see the appropriate chapters in this book for more detailed information about these laws):

1. The Clean Air Act of 1970, and its subsequent amendments, which established goals and standards for the quality of air in the United States
2. The Clean Water Act of 1972, and its subsequent amendments, which established goals and standards for water quality in the United States
3. The Safe Drinking Water Act of 1974, amended in 1986 and 1996, which established drinking water standards and rules for groundwater protection and underground injection
4. The Resource Conservation and Recovery Act of 1976, which helped prevent the creation of toxic waste dump sites by establishing standards for the management of the storage and disposal of hazardous waste
5. The Toxic Substances Control Act, which gave the EPA the authority to regulate the manufacture, distribution, import, and processing of specific toxic chemicals
6. The Comprehensive Environmental Response, Compensation, and Liability Act of 1980, and its subsequent amendments, which requires the cleanup of sites contaminated by toxic wastes, even those which had been placed there many years before
7. The Hazardous and Solid Waste Amendments of 1984 (HSWA-Underground Storage Tanks)
8. The Emergency Planning and Community Right-to-Know Act of 1986, which requires companies to inform communities about toxic chemicals that they release into the air, water, and land
9. The Asbestos Hazard Emergency Response Act of 1986
10. The Oil Pollution Act of 1990, which enacted a rapid and well-thought-out federal response to oil spills from oil storage facilities and vessels, while making the polluters responsible for the cost and damage to natural resources
11. The Federal Insecticide, Fungicide, and Rodenticide Act of 1947 amended in 1996

SPECIAL RESOURCES

CHILDREN'S ENVIRONMENTAL HEALTH CENTERS

(See endnotes 58, 59)

Scientific research into the various facets of children's environmental health problems is being conducted at the Children's Environmental Health Centers created by the US EPA through a special program. The goals of the Children's Center Program are to: provide multidisciplinary research on environmental contributions to children's health and disease through interactions between basic, clinical, and behavioral scientists; establish research/prevention centers to pursue high quality research with clinical applications; utilize exposure assessment and resulting health effects in risk management and disease prevention; establish a national network of centers to rapidly share research findings and innovative approaches to controls; and understand the impact of chemical and other environmental exposures on the fetus and child. These centers include:

1. *Columbia University School of Public Health*
   The Columbia Center for Children's Environmental Health conducts scientific studies on the links between common pollutants in the environment and health risks. The research results are used to help educate people on how to reduce children's exposure to harmful pollutants. Special studies have been done on the exposure of the fetus to high levels of pollution from fuel burning, pesticides, and secondhand smoke. The center's largest study involves 725 African-American and Latino pregnant women and their children, living in low-income neighborhoods, with the children being monitored from birth through age 11. The exposures studied include polycyclic aromatic hydrocarbons in ambient air pollution, secondhand smoke, pesticides, endocrine-disrupting chemicals such as phthalates and BPA, and indoor pest allergens.

2. *Duke University*
   The Duke University Southern Center on Environmentally-Driven Disparities in Birth Outcomes attempts to understand how environmental, social, host, and genetic factors interact to affect the fetus and newborn. These stressors include children living in situations where there is deteriorating housing, poor access to health care, poor schools, high unemployment, high crime, high poverty, homelessness, poor nutrition, and increased environmental exposures to toxic chemicals. They study normal birth outcomes, racial disparities in birth outcomes, and the effect of prenatal environmental exposure and neonatal respiratory health.

3. *Johns Hopkins University Schools of Medicine and Public Health*
   The Johns Hopkins Center for Childhood Asthma in the Urban Environment studies how exposures to air pollutants and allergens can induce airway inflammation in increased asthma in children with the goal of developing effective strategies to reduce the disease.

4. *Mount Sinai School of Medicine*
   The Mount Sinai Center for Children's Environmental Health and Disease Prevention Research works on the neurodevelopment impacts of pesticides, lead, and PCBs.

5. *University of California, Berkeley*
   The University of California, Berkeley Center for Children's Environmental Health Research primarily conducts its research in the Salinas Valley of California, which is an agricultural region where about 38,000 farm workers, mostly Hispanic, work and live. More than 0.5 million pounds of organophosphate pesticides are used in this area yearly. The research attempts to determine pesticide and other environmental exposures in pregnant women and young children and the effects of these exposures on childhood growth, neurodevelopment, and respiratory disease. Community-based programs are developed to try to reduce take-home pesticide exposure in the children of farm workers.

6. *University of California Davis*
   The University of California Davis Center for the Study of Environmental Factors in the Etiology of Autism studies chemicals known to be toxic to the nervous and immune systems and attempts to determine if this leads to abnormal development of social behavior in children. This knowledge would lead to strategies for prevention and intervention.

7. *University of Southern California and University of California, Los Angeles*
   The Children's Environmental Health Center at the University of Southern California and the University of California, Los Angeles Children's Environmental Health Center focus on the study of asthma and allergic airway disease as it relates to ambient air pollutants, environmental tobacco smoke, traffic density, and proximity to homes.

8. *University of Cincinnati*
   The Children's Environmental Health Center at Cincinnati Children's Hospital Medical Center conducts research and training to reduce disease and disability in children caused by environmental hazards. The center studies the behavioral effects of prevalent toxicants.

9. *Harvard School of Public Health*
   The Center for Children's Environmental Health and Disease Prevention Research at Harvard School of Public Health was established to understand the real problem of the Tar Creek Superfund Site in northeastern Oklahoma and the effect of the metal contamination in the mining waste on the Native American residents. The goal of the program is to determine the potential hazards of the exposure to the mixed metals on child health and development.

10. *University of Illinois at Urbana-Champaign*
    The University of Illinois Fox River Environment and Diet Study of the Children's Environmental Health Center investigates the interactive effects of PCBs and methyl mercury on cognitive, sensory, and motor development in children. The study is being conducted on Hmong and Laotian refugees who came to northeastern Wisconsin after the Vietnam War and regularly consumed fish contaminated by PCBs and methyl mercury from the Fox River.

11. *University of Iowa*
    The University of Iowa Children's Environmental Airway Disease Center conducts research on the etiology and pathogenesis of airway disease in children from rural communities. The research indicates that the incidence and severity of asthma in rural children is similar to that in urban children. Apparently viral infections increased sensitivity to environmental exposures. The products of bacteria and fungi are frequently found in the rural environment and apparently may act as triggers for asthma.

12. *University of Medicine and Dentistry of New Jersey*
    The University of Medicine and Dentistry of New Jersey Center for Childhood Neurotoxicology and Assessment's main focus is to examine the effects of exposure to environmental chemicals on neurological health and development potentially leading to learning disabilities and autism.

13. *University of Michigan*
    The University of Michigan Center for the Environment and Children's Health investigates the characteristics of the household, school, and neighborhood environment of children with asthma; the environmental factors and cofactors contributing to the exacerbation and progression of childhood asthma; and the cellular and molecular targets which modulate immune system responses to environmental contaminants. They use a community-based approach to intervene at a household and neighborhood level to reduce exposure to environmental hazards and improve the child's health status.

14. *University of Washington*
    The University of Washington Center for Child Environmental Health Risks Research studies the biochemical, molecular, and exposure mechanisms of children's susceptibility to pesticides and how the pesticides affect normal growth, development, and learning.

15. *National Institute of Environmental Health Sciences*
    The National Institute of Environmental Health Sciences, Centers for Children's Environmental Health and Disease Prevention Research studies the effects of environmental exposures on children's health. It uses basic, applied, and community-based

participatory research. The center's long-range goals include stimulating new research and expanding existing research on the role of the environment and the causation of disease and dysfunction in children; developing unique intervention and prevention strategies; and using the knowledge for intervention and prevention of disease. The number of research grants given out throughout the country are simply too numerous and complex to discuss in a book of this nature. Some of the research areas include:

1. Asthma symptoms related to maternal use of acetaminophen during pregnancy
2. Phthalate exposure and children's behavior and cognitive abilities
3. Use of beauty chemicals by small girls in early puberty
4. External exposure to urban air pollutants and children's cognitive abilities
5. Effects of environmental exposures on child health
6. Bisphenol A and children's health

OTHER RESOURCES

GOVERNMENTAL

World

Fourth Ministerial Conference on Environment and Health
(See endnote 30)

The World Health Organization held the Fourth Ministerial Conference on Environment and Health, entitled Children's Environment and Health Action Plan for Europe, on June 23–25th, 2004. The ministers recognized that improvement in the health of children as it related to the environment was very uneven across Europe because of the consequences of poor environmental conditions, poverty, disruption of social protection and health systems, armed conflict and violence. The ministers recognized that all children suffered from the consequences of polluted and unsafe environments. The ministers stated that: injuries were the first cause of death from birth to age 18 years; contaminated water, air, food, and soil caused gastrointestinal disease, respiratory disease, birth defects, and neurotoxic and developmental disorders; there was a serious need for safe and balanced nutrition; and that many chemicals had the potential to cause long-term toxicity and carcinogenic, neurotoxic, immunotoxic, genotoxic, endocrine-disrupting, and allergenic effects. They recommended the following priority goals:

1. Prevent and reduce morbidity and mortality from gastrointestinal disorders and other health effects related to water and sanitation for children
2. Prevent and substantially reduce health consequences from accidents and injuries
3. Prevent and reduce respiratory disease due to outdoor and indoor air pollution and reduce asthma attacks
4. Reduce the risk of disease and disability from exposure to hazardous chemicals

Fifth Ministerial Conference on Environment and Health
(See endnote 60)

The World Health Organization on March 10–12, 2010, held the Fifth Ministerial Conference on Environment and Health with its focus on protecting children's health in a time of changing environmental problems and issues. One of its major concerns was the effect of climate change on the health of children since climate change contributes to the increase in the amount of natural disasters and this has a more profound effect on children than adults. The conference stated that a considerable number of Europeans were suffering from health problems that were linked to environmental conditions and among the vulnerable groups were children, pregnant women, and the socially disadvantaged.

The conference recommended that environmental and health issues should be considered one of the priorities in establishing policies in transportation, energy, industry, and agriculture. It stated that the solution to these problems could boost innovation and competitiveness and create a better economy. It also stated that many environmental effects could be controlled through well-tested health system interventions, specifically the use of primary health care. The various nations attending pledged to work together to help resolve environmental and health issues.

World Health Organization Environmental Health Activities

The World Health Organization environmental health activities are examples of how a world organization works with national environmental health programs to help improve the environment and prevent disease and injury. In 2005, the World Health Organization stated that more than 3 million children under 5 years of age died each year globally from environment-related causes and conditions. This was 30% of the total of 10 million children dying each year. It also was an important factor in the well-being of the mother. The four major causes of death were: diarrhea from unsafe water and poor sanitation; indoor air pollution producing severe respiratory effects, especially from the burning of biomass fuels; malaria from mosquitoes breeding in areas with poor water management and storage, as well as inadequate protection and housing; and unintentional physical injuries in the home and other environments. In addition, lead in air, paint or dust, mercury in food, and other chemicals could pose severe potential risks for the fetus, infant, and older child.

In June 2009, at the Third World Health Organization International Conference on Children's Health and the Environment in Korea, the World Health Organization was asked to help develop a global plan of action to improve children's environmental health, monitor the plan, and report on the plan's progress. The plan was to provide a comprehensive approach for the World Health Organization, governments, intergovernmental and non-governmental organizations, and all people seeking to improve the health of children.

To meet these priority goals, the ministers established a series of programs including technical support for countries, groups, and individuals to help reduce children's environmental health hazards. They also established collaborative research efforts and appropriate exchanges of information. They set forth further goals to develop and start implementing national children's environment and health action plans and cooperation with the World Health Organization.

Federal

A large number of school environmental health activities have been established by the following departments of the federal government: the US Department of Agriculture, US Department of Education, US Department of Energy, US Department of Health and Human Services (a substantial number), US Department of Housing and Urban Development, US Department of Labor, US Consumer Product Safety Commission, US Department of the Interior and US EPA (a substantial number).

Some of the specific programs are discussed below.

Agency for Toxic Substances and Disease Registry

The Agency for Toxic Substances and Disease Registry (ATSDR) is a federal agency which determines the effects of hazardous substances in the environment on the health of people. It assesses waste sites, carries out health consultations concerning specific hazardous substances, carries out health surveillance, and responds to emergency releases of hazardous substances. ATSDR provides a series of educational products to communities, health educators, healthcare providers, environmental health professionals, and others on community environmental health topics such as exposures to hazardous substances, roots and pathways of exposure, health effects, treatment options, and techniques to minimize exposure to hazardous substances.

America's Children and the Environment

America's Children and the Environment is a program of the US EPA which provides access to information to show trends in the levels of environmental pollutants in air, water, food, and soil, the concentration of contaminants in mothers and children, and the level of childhood diseases which may be related to environmental contaminants. By tracking the environmental contaminants, the EPA through the America's Children and the Environment group tries to determine how varying concentrations have changed over time and the percentage of children who may still be at risk from the contaminants. They also get involved in the emerging issues, such as mercury in fish caught by individual fisherman and the potential for attention deficit/hyperactivity disorder.

Association of Maternal and Child Health Programs

The Association of Maternal and Child Health Programs is a non-profit group that partners with the CDC to work closely with state maternal and child health programs in order to improve the environment and reduce the potential for disease and injury

Centers for Children's Environmental Health and Disease Prevention Research

Centers for Children's Environmental Health and Disease Prevention Research is a federal program of the National Institute of Environmental Health Sciences which examines the effect of environmental exposures on children's health using a multidisciplinary research technique involving basic, applied and community-based participatory research.

Centers for Disease Control and Prevention—Environmental Health Services Branch

The Centers for Disease Control and Prevention—Environmental Health Services Branch is a federal program used to revitalize, enhance, and strengthen local, state, tribal, and national environmental health programs and the professionals working within the programs to reverse environmental exposures and control the consequences for human health.

Consumer Product Safety Commission

The Consumer Product Safety Commission is an independent federal regulatory agency which protects the public from products that cause electrical, chemical, or mechanical hazards including toys, household items, furniture, bicycles, and other products. They develop standards, recall products, conduct research, and educate the public. In recent years, the Commission has recalled very small rare-earth magnets; toys, jewelry, lunch boxes, and clothing containing lead; craft kits which may contain very small objects; cookware; electrical devices; cribs and strollers; etc.

EPA's Children's Health Protection Website

The EPA's Protecting Children's Health website is dedicated to a discussion of the most recent information concerning children's environmental health issues including: lead, drug laboratories, pesticides, mold, radon, drinking water, asthma, secondhand tobacco smoke, indoor air pollution, poisons, various materials and consumer products, toxins in schools, pesticides, asthma, diesel fuel, eco-friendly child care and schools, hazardous waste sites, fish contaminants, blue-green algae, and beach water quality information.

EPA's Healthy School Environment Resources

The EPA's Healthy Kids, Healthy Schools is a one-stop location for information and links to school environmental health issues divided into 14 major categories including:

1. Chemical use and management
2. Design, construction, and renovation of physical facilities
3. Energy efficiency

4. Environmental education
5. Facility operations and maintenance
6. Indoor environmental quality
7. Legislation and regulation
8. Outdoor air pollution
9. Portable classrooms
10. Safety/preparedness
11. School facility assessment tools
12. Waste
13. Waste reduction
14. Water

A total of 77 topics are covered. Each one can be accessed simply by the click of a mouse. In addition, there is immediate access through a click of the mouse to 11 EPA programs for schools, 10 related EPA programs, and 10 resources outside of the EPA.

National Library of Medicine Environmental Health and Toxicology
The National Library of Medicine Environmental Health and Technology resources include comprehensive environmental health and toxicology website access to databases, bibliographies, tutorials, and other scientific and consumer-oriented resources. TOXNET is a toxicology data network. Tox Town is a consumer website which provides an introduction to toxic chemicals and environmental health risks in the home and community.

National Center for Environmental Health
The National Center for Environmental Health is a federal agency that is part of the CDC and provides information on air pollution, asthma, lead poisoning prevention, public health tracking, and other environmental health issues.

National Environmental Education Foundation
The National Environmental Education Foundation is a foundation chartered by Congress in 1990 to advance environmental knowledge and action. It is a complementary organization to the US EPA, extending the EPA's ability to improve environmental literacy through EPA's own resources and also private funds which may have been made available for this purpose. Its goal is to:

1. Expand environmental education in the schools
2. Establish basic environmental education for the adult public
3. Develop an environmental education program for health professionals
4. Develop environmental education for business managers
5. Work to overcome issues of environmental health in children

An example of a children's program is the Pediatric Asthma Initiative, a long-term project used to integrate environmental management of asthma into pediatric medical and nursing education and practice.

National Institute for Occupational Safety and Health
NIOSH is a federal agency that is part of the CDC and conducts research to reduce work-related illnesses and injuries and to promote safe and healthy workplaces. NIOSH has prepared a series of resources for young worker safety and health. These training programs, reports, and publications can be accessed through the NIOSH website.

Other Organizations

There are numerous other national, state, and local associations who simply cannot be named here. When working with a local health department or the regional EPA, it is possible to get a list of local partners in resolving children's environmental health issues.

State and Local

The following organizations are examples of state and local groups that work to prevent disease and injury and promote health, especially for children. The overall list of these organizations would be far too long for this presentation so therefore a small sample is being utilized. When working in a given state or local area, it is wise to contact the local or state health department, professional health associations, or regional EPA, to get a better understanding of the many organizations that work together to try to improve the environment of the child.

Improving Kids Environment (IKE)

Improving Kids Environment is a not-for-profit corporation based in Indianapolis, Indiana, that works to reduce environmental threats to children's health and provides the family with information to protect children. It promotes practical means to remove or reduce recognized serious threats. It has been involved in helping provide funds to remove lead from properties where children live. It publishes information on various environmental health issues for children through a newsletter entitled *Children's Environmental Health Newsletter.*

Michigan Network for Children's Environmental Health

The Michigan Network for Children's Environmental Health is a coalition of health professionals, health-affected groups, environmental organizations, and others who are interested in reducing toxicity for children and improving their environment. The organizational members include state chapters of national organizations such as the American Academy of Pediatrics, National Association of Pediatric Nurse Practitioners and American Nurses Association; local and state health departments; local and state civic associations; children's health associations; and other interested citizens.

PROFESSIONAL ASSOCIATIONS AND OTHER ORGANIZATIONS

American Academy of Pediatrics Council on Environmental Health

The American Academy of Pediatrics Council on Environmental Health is an organization made up of pediatricians who are concerned with issues related to environmental health and exposure to toxins by children and how to protect the children. The Council works on legislative issues and educational initiatives, and has developed the Academy's well-received book entitled *Pediatric Environmental Health Manual*, second edition. This guide promotes the identification, prevention, and treatment of pediatric environmental health hazards and helps parents understand what is happening and what should be done.

American Public Health Association—Public Health Nursing Section

The American Public Health Association—Public Health Nursing Section is a professional organization of nursing practitioners and educators who maintain the highest standards of public health nursing through nursing practice, education, and research. In response to the 1995 report of the Institute of Medicine Committee on Enhancing Environmental Health Content in Nursing Practice, the Public Health Nursing Section established an Environmental Health Task Force made up of nurses and environmental health practitioners. This group developed and published a series of environmental health principles for public health nurses and environmental health recommendations for public health nursing in practice, education, research, and advocacy. Their ongoing work is a huge resource for nurses, doctors, environmental health practitioners, and the general public.

Canadian Association of Physicians for the Environment

The Canadian Association of Physicians for the Environment is an organization of over 4700 physicians and concerned citizens from Canada focusing on educating health professionals on environmental issues. As one of its accomplishments, the organization produced a 322-page handbook on environmental threats and how citizens can effectively counteract them. The book is entitled *The Canadian Guide to Health and the Environment.* (See endnote 63.) The group is deeply involved in children's environmental health issues.

Canadian Partnership for Children's Health and Environment

The Canadian Partnership for Children's Health and Environment is an affiliation of groups with overlapping missions to improve children's environmental health in Canada. The Canadian Partnership for Children's Health and Environment is a consortium of 16 organizations made up of physicians, childcare experts, environmental law experts, learning disability experts, medical schools, public health officials, and pollution specialists in collaboration with the Canadian Institute for Public Health Inspectors Ontario Branch and the Ontario Association of Supervisors of Public Health Inspectors.

They produce documents of great value to practitioners. *Advancing Environmental Health in Childcare Settings—A Checklist for Child Care Practitioners and Public Health Inspectors* is an outstanding example of their work. (See endnote 64.)

Children's Environmental Health Network

The Children's Environmental Health Network is a federal and state tax-exempt non-profit national organization with a multidisciplinary mission involving protecting the fetus and children from environmental health hazards and promoting a healthy environment. It promotes sound child-focused national policy; raises public awareness of environmental hazards to children; educates health professionals and policy makers; and stimulates prevention-oriented research. It provides the *Training Manual on Pediatric Environmental Health* and also a resource guide of related programs, projects and organizations. The Children's Environmental Health Network has a resource guide on children's environmental health organizations, which includes approximately 140 entries.

Healthy Schools Network

The Healthy Schools Network is a national not-for-profit environmental health information, education and research organization dedicated to making the environment safe and healthy for children in the school setting. It partners with a variety of other organizations to carry out its mission.

National Association of City and County Health Officials

The National Association of City and County Health Officials is an organization serving the needs of approximately 3000 local health departments nationwide. Its function is to provide and promote the local perspective on national health programs and policies. It has conducted nationwide education programs about the problems and solutions to children's environmental health issues.

National Association of School Nurses

The National Association of School Nurses is a professional organization of school nurses who provide continuing education programs at the national, regional, state, and local community level to teach nurses about a variety of health-related topics including children's environmental health issues and solutions.

National Children's Center for Rural and Agricultural Health and Safety

The National Children's Center for Rural and Agricultural Health and Safety is part of the Marshfield Clinic Research Foundation, Marshfield, Wisconsin. The center attempts to improve the health and safety of all children exposed to hazards in agricultural work and rural environments. In 2008, it received \$4.6 million in a competitive grant from NIOSH to expand its programs related to injury prevention on farms. The projects funded include research, education, intervention, prevention, translation, and outreach to enhance the health and safety of children exposed to hazards.

National Environmental Health Association—Children's Environmental Health Technical Section

The National Environmental Health Association—Children's Environmental Health Technical Section is part of a professional association which is an advocate for the improvement of the child's environment by use of highly technical environmental health practitioners, educators, and researchers. The association and the technical section provide a considerable amount of educational experiences for environmental health professionals in order to utilize their skills in helping resolve and prevent environmental problems related to children. The National Environmental Health Association (NEHA) issued a comprehensive document stating the great significance of the movement to try to improve children's health for a better environment. This was issued on July 2, 1997, and was entitled *NEHA Position on Children's Environmental Health*. (See endnote 65.) The Children's Environmental Health Technical Section was also formed to provide necessary assistance for the present and future, and the Association continues to work toward improvement of the health of children.

National Safety Council—Environmental Health Center

The National Safety Council—Environmental Health Center is a unit of the National Safety Council that attempts to teach the public about significant environmental health risks and problems facing families and society. It uses a variety of activities and techniques including public education and outreach, emergency planning and management, and environmental journalism to protect against lead poisoning, radon gas, indoor air pollutants, and hazardous chemicals in the community and other environments where children live.

Natural Resources Defense Council, Children's Environmental Health Initiative

The National Resources Defense Council, Children's Environmental Health Initiative is a division of the Natural Resources Defense Council, which works to reduce the potential for children to be subjected to various toxins. This group became focused on children's environmental health problems because of the publication of a report entitled *Intolerable Risk, Pesticides in Our Children's Food*. (See endnote 66.) The group works toward the elimination or reduction of hazardous substances in food, air, water, and other media.

The Coalition to End Childhood Lead Poisoning

The Coalition to End Childhood Lead Poisoning is a national non-profit organization that creates, implements, and promotes programs and policies to eliminate childhood lead poisoning and make homes healthier and safer. It was founded in 1986 as a voluntary citizens' effort to eliminate lead poisoning. The Coalition works with families, community organizations, educators, government agencies, insurers, property owners, and healthcare providers to make homes free of lead and safe.

The Collaborative on Health and the Environment

The Collaborative on Health and the Environment is an international partnership of over 3500 individuals and organizations in 45 countries and 48 states and includes scientists, health professionals, health-affected groups, non-governmental organizations, and other concerned citizens who seek to improve human and ecological health. The Initiative on Children's Environmental Health, formerly the Institute for Children's Environmental Health, is now part of the Collaborative group.

On October 1, 2010, a conference entitled Promoting Ecological Health for the Whole Child was held at the University of San Francisco. The topics discussed included the complex interacting factors of nutrition, education, socioeconomic status, exposures to toxic chemicals, and access to preventive health care. Childhood cancer had become the leading cause of childhood death. Adult onset diabetes, once very rare in children, had increased by 50% over previous existing rates. Speakers at the conference, while recognizing the importance of prenatal care, emphasized the need for examining environmental situations the young child is exposed to, as well as looking at and addressing all aspects of children's health. (See endnote 57.)

Pediatric Environmental Health Specialty Units

Pediatric Environmental Health Specialty Units are a source of medical information and advice on environmental conditions that can influence children's health. They are concerned with the prevention, diagnosis, management, and treatment of environmentally related health effects in children. They are involved in community education and outreach, the training of health professionals, and consultation and referral. These groups are typically based at university medical centers and are located in the United States, Canada, and Mexico. They are part of the Association of Occupational and Environmental Clinics, founded in 1987. This group received significant financial support through a multi-year agreement with ATSDR and NIOSH.

PROGRAMS

STATE AND LOCAL LEVEL

State Programs

There is a huge variation in state programs addressing children's environmental health issues. Most states have programs to correct lead-based problems in houses to reduce levels of lead poisoning in children. Most states have educational programs which vary from providing internet sites for securing information on asthma and lead poisoning to providing comprehensive sites for a variety of potential hazards to children. This includes information on: air quality, asthma, carbon monoxide, hazardous pollutants, nitrogen dioxide, ozone, particulate matter, volatile organic compounds, asbestos, lead-based paints, indoor air pollution, allergens such as dust, mold at home and in schools, pet dander, chemicals from fragrances and cleaning products, insects and rodents, pesticides, healthy homes, diesel exhaust, arsenic found in soil, bacteria found in fresh recreational water, algae, pet wastes, BPA, mercury in fish, lead in children's jewelry and toys, aftermath of flood waters, noise and hearing loss, PCBs in schools, pressure-treated wood in playground equipment, fly ash, hazardous waste sites, secondhand smoke, drinking water contaminants, contaminated food, injury control in the home, school and playground, agricultural chemicals, lead in home remedies and cosmetics, and mosquito and tick control. Some states have more active programs since they provide limited funding for such things as removing lead-based paint from homes to reworking diesel engines on school buses. Several states even have a formal Children's Environmental Health Program.

The problem is that there is no consistent approach to the resolution of hazards affecting children from state to state and locality to locality. Unfortunately, children's programs typically do not produce revenue and therefore, especially, in times of serious budget shortfalls, children's issues are not given the resources that are necessary to avoid injury and illness and promote better health. In addition, where local and state health departments receive federal funding for specific programs, typically when the funds run out, the programs cease to exist. Also, many local and state funding agencies use the federal funds as an excuse for reducing the regular budget for other environmental health programs. For any given locality or state, it is necessary to contact the appropriate health department, over the internet or by phone, to determine what programs or educational materials are available to help resolve children's environmental health issues. Federal internet sites are also very useful for obtaining educational and training materials.

Local Programs

King County, Washington

Despite the fact that King County, Washington, has worked diligently to try to improve the health of children through correction of environmental hazards, it is still an example of the problems found in children's environmental health programs at the local level. The many issues that this local program has include the following:

- Isolated projects and programs with different funding are used instead of a coordinated delivery stream.
- There is no uniform way, across various governmental entities, to identify the needs of children to deal with environmental health issues.
- Funding for effective programs related to children has been severely diminished because of ongoing economic situations.
- Children's environmental health programs typically do not provide billable fees.
- There is a lack of community knowledge and participation in children's environmental health issues.
- There is a problem relating to the school districts because of unfunded mandates.
- There is a lack of local funding for children's environmental health issues.
- There is a lack of knowledge by parents concerning the environmental health needs of the child.
- There is no comprehensive local needs assessment related to the environmental health of the child.

OTHER PROGRAMS

AGRICULTURAL ENVIRONMENT

Migrant Clinicians Network

The Migrant Clinicians Network is an international non-profit organization of over 5000 health professionals who work to eliminate health problems among migrant and seasonal farm workers and other mobile underserved populations. A significant part of their work has been in occupational and environmental health efforts focused on children. As an example of their work, in Virginia, the Migrant Clinicians Network has provided educational materials and sessions to healthcare providers, head start workers, outreach workers, migrant farm workers, and the community about children's environmental health issues and how to minimize potential hazardous exposures. Over a 2-year period, the group has worked with 600 migrant farm workers and their families. This has improved the knowledge of these individuals concerning environmental hazards and has helped minimize exposure.

URBAN ENVIRONMENT

Community-based Intervention Research Project—University of Michigan

This research project was used to try to reduce the amount of asthma in a given community and also to reduce the severity of the disease by carrying out asthma screening (by use of questionnaires, pulmonary function tests, and skin tests to determine specific allergies) and environmental testing for dust samples that were analyzed for cockroaches, which has high potential as an asthma trigger, dust mite, cat, dog, and mouse or rat allergens and provided for intervention by trained Community Environmental Specialists (CES). These individuals were all residents of Detroit, had a minimum of a high school education, and completed a 4-week training course before working with the community members. The project was composed of a group of children from 44 schools in areas in Detroit.

The study involved a group of caregivers of children, the local health department, community organizations, and a group of research scientists.

The household intervention consisted of the CES making a minimum of nine visits to each of the homes of the children involved in the project. The CES took annual dust samples, conducted annual surveys for the caregiver and the child, and on the other occasions provided educational materials and services to reduce the exposure to asthma triggers. The CES also made referrals for a group of other issues including medical care and tenants' rights. Undermedicated children were found and this was corrected. Caregiver depression decreased because of the personal assistance given to them. Household-level intervention did work because the individuals reduced indoor exposures to environmental hazards by vacuuming and other cleaning techniques. Asthma-related health status improved. The neighborhood-level intervention project was not carried out because of a 10% reduction in the original grant funds.

The program involving CES in households to try to reduce asthma triggers is an excellent idea. By working with community groups, additional help is provided in education and support services for the child caregivers. The problem with the approach is that funds need to be provided by local entities with help from federal and state authorities to provide the necessary workers to carry out an actual program of this nature.

Healthy Homes Project Pilot, Baltimore

The Baltimore pilot project sought to increase the capacity of the Healthy Homes Project, through a grant from the CDC to transition from a lead-based paint program to a more inclusive healthy homes program. The project, which attempted to improve the living conditions of all individuals in inner-city housing situations, actually was most helpful in improving the urban environment for children by studying and then working with the residents to eliminate the following concerns:

- Lead exposure from deteriorating lead-based paint, cultural use of lead, renovation of structures, and occupational exposure of parents to lead
- Carbon monoxide exposure
- Fire hazards
- Inadequate or lack of smoke alarms
- Moisture, mold, and other urgent triggers
- Rodents and roaches
- Hazardous or harmful household products
- Smoking and secondhand smoke
- Inadequate ventilation, heating, and cooling
- Visible physical hazards

Public health professionals including environmental health practitioners, public health nurses, and public health investigators were involved in helping the families. The families were provided with proper educational tools, supplies, referrals for a variety of health and housing problems, and in some cases assistance to make home repairs and other modifications to the property.

Healthy Children, Healthy Homes—North Miami, Florida

An educational program for nurses was developed in both English and Spanish in the general elementary school community in North Miami, Florida. A study had indicated that black and Latino children had worse problems with asthma and less use of preventive asthma medications than Caucasian children. (See endnote 69.) The study demonstrated the importance of identifying and developing effective preventive techniques for asthma by reducing exposure to indoor environmental asthma triggers such as dust mites, roaches, animal dander, and indoor tobacco smoke. The demonstration consisted of the use of a community-based asthma education program which expanded the knowledge of individuals to household asthma triggers and means of counteracting these triggers. The program was in both English and Spanish. Registered nurses provided two 90-minute educational sessions which were shared with 10 potential community contacts, both individuals and groups. The community leaders followed up on the presentations to the citizens. Elementary school children participated in much shorter educational sessions on asthma, asthma triggers, and asthma prevention. This effort appeared to have some success.

EPA Rules on Contractors Making Renovations

On April 22, 2010, new rules went into effect for contractors who make renovations on properties where there are children less than 6 years of age or a pregnant woman. Although lead was banned as an ingredient in paint in 1978, in older properties there still is a potential for a substantial amount of lead on surfaces. The contractors will have to use plastic sheeting to seal off the rooms that are being worked and work areas will have to be cleaned with HEPA-equipped vacuum cleaners. The cleanup will require wiping down walls, surfaces, and floors using multiple buckets of cleaning fluid and electrostatic dusters. Warning signs will have to be posted outside the work area. The contractor will have to take a 1-day certification class.

Integrated Pest Management

Integrated pest management helps reduce pest infestations and the use of insecticides in homes. It includes: use of sticky traps, bait stations and gels; repairing leaks and holes; and proper cleaning of the premises. This is an effective way to reduce pest infestation including roach infestation, without the use of very toxic chemical pesticides. The level of chemicals found in indoor air samples and maternal cord blood will decrease significantly.

Lead Poisoning

In April 2000, in Philadelphia, a Community Childhood Lead Poisoning Prevention Project consisting of a consortium of 44 different groups of children's citizens groups, business and industry (several hardware stores and paint stores participated), housing, health, and other governmental organizations worked together to provide appropriate health education material, free blood lead testing in clinics and other locations, campaigns in schools, and advertisement through the local media. The program also consisted of arranging inspections, risk assessment, and lead-hazard abatement of privately owned housing for children under 6 years old who had been identified as lead poisoned or at risk for lead poisoning. At least 2175 people in Philadelphia were served by the program.

By October 2000, the District of Columbia Coalition to End Childhood Lead Poisoning working in high risk areas of childhood lead poisoning established a day care education pilot program and a health educator's program, and provided education materials to area schools and housing organizations.

Also, by October 2000, the Pittsburgh Lead-Safe Coalition, a group of 50 organizations dedicated to preventing lead poisoning, mailed the new Pittsburgh Childhood Lead Screening Guidelines to over 2000 physicians. In 1999, the group taught 300 people how to prevent lead poisoning.

PRESCHOOL ENVIRONMENT

Children's Environmental Health Network

In 2006, the Children's Environmental Health Network started a program, using Alameda and Contra Costa counties in California as the pilot, to train and educate childcare providers and administrators on creating a healthier and safer environment. The program entitled "Healthy Environments for Child Care Facilities and Preschools" teaches childcare providers and administrators to:

1. Understand how vulnerable the child is to environmental health exposures
2. Identify a variety of environmental health hazards in the facility, surrounding grounds and community

3. Learn how to remove the environmental health hazards
4. Utilize communications strategies which will help parents to eliminate environmental hazards at home
5. Obtain accurate scientific data, which can be applied readily by facilities to remove hazardous substances and situations

The California Childcare Education Project did the above and also provided a Childcare Inspection Checklist for easy use by providers and administrators. This list helped identify gaps in knowledge.

The Children's Environmental Health Network provided a series of fact sheets, on significant topics, for use by the attendees of the California program including:

1. Chemicals in art supplies
2. Safe cleaning alternatives
3. Pesticides and integrated pest management
4. Lead poisoning
5. Air pollution from nearby traffic
6. Air quality
7. Indoor mold

In addition to discussion on these topics, the instructors added material on sun exposure, mercury, arsenic in playgrounds, and other current topics.

Because of the success of the California program, the state of Georgia initiated a similar program for childcare resource and referral agencies, head start programs, and preschools. The education modules were prereviewed and the training curriculum was approved by the Georgia Department of Early Care and Learning. Training began in 2008. The topics taught were expanded from those in California. The topics included: air quality, diesel motor vehicles, radon gas, the built environment, indoor mold, asbestos, mercury reduction, arsenic, physical education and nutrition, noise pollution, chemicals in art supplies, safe cleaning alternatives, and pesticides and integrated pesticide management.

In the Fall of 2007, the Children's Environmental Health Network expanded this program to 20 childcare facilities in Washington, DC. In 2008, an additional 20 childcare facilities were taught the substance of the program. Subsequent to the program, 95% of the trainees were able to recognize environmental hazards and do what was necessary to remove them. Training programs have been coordinated and funded by the CDC.

Environmental Protection Agency

The US EPA through its Outreach and Partnerships programs is: providing information and tools to the public such as the Children's Health Podcast Series; supporting community actions to protect children through giving awards to individuals and communities who have successful children's environmental health programs; helping increase the ability of healthcare providers to identify, prevent and reduce environmental health threats to children through grants and support; and working with states to develop programs to address children's environmental health issues.

Canadian Partnership for Children's Health and Environment

(See endnote 1)

The document *Advancing Environmental Health in Childcare Settings—A Checklist for Child Care Practitioners and Public Health Inspectors* (see endnote 1) is an excellent tool for environmental health practitioners and childcare practitioners in evaluating the environment of the childcare setting. The tool is used to understand the problems and help correct them in the areas of: outdoor air quality, outdoor areas, sun safety, indoor air quality and dust, cleaning and disinfection, activity, learning and play areas, kitchen and food preparation areas, areas of renovations, surrounding sources of chemical emissions, and general concerns about energy efficiency, conservation of water and promoting good environmental use. Special attention is paid to the problems of: outdoor air pollution, indoor and outdoor pesticide use, inadequate ventilation, dust especially contaminated with flame retardants, phthalates, mercury, lead, mold, radon, fragrances, and disinfecting/cleaning products.

SCHOOL ENVIRONMENT

The US EPA has developed a unique software tool to help school districts evaluate and manage the key environmental, safety, and health issues in their school facilities. This program, Healthy Seat, helps track over 400 items which can be measured within the school programs and facilities. It can be customized to any specific school district. Some of the problem areas and programs discussed include integrated pest management in schools, asbestos, school chemical clean-out campaigns, indoor air quality tools for schools, problems of exposure to sunlight, school bus retrofit and anti-idling programs, lead in drinking water, lead-based paint programs, asthma programs, mercury, pesticides, and other toxic chemicals.

Areas of evaluation in the school environment include chemical management, asbestos management, carbon monoxide, lead-based paint, lead in water, mercury, PCBs, radioactive materials, hazardous materials, non-hazardous materials, pesticide use, drinking water, wastewater, general injuries, traffic problems, occupational injuries and illness, fire safety, custodial areas, laboratories and science classrooms, hazard communications, ventilation systems, food protection, copy rooms, building, roof and floor structure, playgrounds, construction and renovation areas, buses, and bus safety.

PROPOSED CHILDREN'S ENVIRONMENTAL HEALTH PROGRAM

General Issues

All programs designed to help eliminate potential environmental hazards for children must include a number of preventative programs. They are:

1. Appropriate health education experiences for female and male high school students concerning the potential disastrous effects of environmental hazards and the special needs of women prior to pregnancy, during pregnancy, and after pregnancy.
2. Proper prenatal care from physicians, nurses, and public health clinics.
3. Use of women's, infant's and children's programs, where appropriate to obtain appropriate nutrition and other services as needed.
4. Reduction of environmental pollutants in the air, water, and soil, to bring levels of hazards down to the quantities that children can be exposed to, without present and future damage or disease.
5. Appropriate training of pediatricians, nurses, and other medical professionals to recognize the symptoms of diseases and conditions related to environmental contaminants and microorganisms.
6. Provision of health education materials to school children and parents concerning specific environmental health issues and how to correct them.
7. Provision of health education materials to school officials, nursery schools, preschools, and other institutions where children spend part of their day. Specific environmental hazards should be identified and specific preventive/corrective actions should be taken.
8. Self-inspection programs and self-inspection forms should be introduced into the preschool and school environments. Specialized training should be given to administrative, maintenance, and housekeeping personnel.

9. Special training for recreational personnel to help them reduce the levels of injuries and possible health hazards from dangerous situations, poor equipment, and potential environmental situations.
10. Special training for individuals in agricultural settings to help them understand the potential health hazards from agricultural chemicals and injuries due to equipment and the surrounding environment.
11. Elimination of environmental tobacco smoke (secondhand smoke) in all homes and other buildings.
12. Rapid phasing out of hazardous waste sites, especially those close to housing and schools.
13. Reduction of air pollutants especially from sources close to schools and housing and affecting them.
14. Establishing National Ambient Air Quality Standards which are applicable to children.

Content of Educational Programs

The content of educational materials must address problems and suggested solutions to environmental health concerns that affect children. These concerns are as follows:

1. Secondhand smoke—Smoke away from children in the out of doors.
2. Contaminants brought inside from out of doors, such as lawn chemicals, lead dust, and other chemicals found in the air, water, or naturally occurring in the soil—Use a doormat and leave the shoes at the door.
3. Drift from chemical applications especially in agricultural areas—Close windows at least 30 minutes before chemical treatment and then reopen 30 minutes after completion of spraying.
4. Gasoline and kerosene fumes—Store equipment in outside sheds or garages and keep gasoline and kerosene in proper storage containers.
5. Chemicals and smoke from backyard burning—All wastes need to be removed to the proper disposal sites and backyard burning is never permissible.
6. Breathing in chemicals from paints, wood finishing products, cleaning products, new carpets, building materials, glues and solvents, from the home, garage, or workshop—Air out all areas where the previous substances have been used prior to the time that children come into that environment.
7. Contaminants in water or on food either bacterial or chemical—Wash all foods before processing and then consuming, and only use water which has been treated.
8. Toys and surfaces contaminated with chemical dusts and residues—Clean all toys and surfaces using a wet cleaning technique.
9. Contamination of hands of small children with chemicals or microorganisms—Wash hands carefully with soap and water and then rinse thoroughly.
10. Potential for poisoning by a variety of household substances—Lock all household cleaners, pesticides, and other chemicals in cabinets which are not accessible to children.
11. Discussion and demonstration of asthma triggers and techniques utilized to eliminate them.
12. Discussion of the effects of ultraviolet rays on children and the potential for cancer in the future.

Homes and Communities

Since there is a higher proportion of environmental health problems in the highly concentrated urban area, most of the resources, financial, and personnel, will be typically utilized in a series of programs within these areas. These programs will include:

- Lead poisoning prevention from paint program
- Healthy homes inspection program

- Rodent and insect control programs inside the home and the immediate environment
- Pesticide control programs
- Injury control programs
- Poison control programs
- Indoor air quality control programs including secondhand smoke
- Mold control programs
- Appropriate housekeeping programs

In the agricultural setting special programs, especially where migrant laborers are used, will be directed at the control and proper disposal of agricultural chemicals, safe drinking water, safe and healthy housing, and injury control especially as it relates to agriculture. Since migrant labor housing may be in older buildings, lead-based paint could be a serious concern.

SCHOOLS AND PRESCHOOLS

Indoor Air Quality Program for Schools

Establish an indoor air quality program for the school or preschool and utilize the EPA guide entitled *Indoor Air Quality Tools for Schools Action Kit.* (See endnote 70.)

Day care centers should utilize the 2-hour training program on indoor air quality and asthma. Schools should provide, for children with asthma in the third, fourth, and fifth grades, the interactive asthma curriculum entitled "Open Air-Ways for Schools."

Urban Schools

Urban schools are usually older and present numerous problems which contribute to indoor air quality concerns. Specifically, structural improvements should include removal and replacement of moldy walls and ceilings, repair and upgrade of ventilation systems, proper removal of lead-based paints, proper removal of asbestos, repair of steam system and plumbing leaks, and comprehensive cleaning and removal of dust and dirt throughout the school. Insects and rodents are a serious health problem and should be eliminated. The droppings can become asthma triggers and can be the source of bacterial diseases. The chemicals used for treatment must not add to the indoor air quality problem.

Site Assessment for Schools and Preschools

Existing schools and preschools as well as planned schools and preschools must have a site assessment to determine if there is a potential for exposure to hazardous substances in the air, soil, and water. Specific existing problems and potential hazards should be eliminated as quickly as possible. For planned schools, if the site analysis identifies serious problems, the site should be changed.

Chemicals

Hazardous chemicals including laboratory and shop chemicals, pesticides, and cleaning compounds need to be removed from schools in an appropriate manner. Trained personnel should be involved in this task and not school children.

Injury Control Programs

Comprehensive injury control programs must be instituted for the school and preschool. Proper supervision of the children, evaluation of equipment used by the children, and evaluation of the overall environment, are essential.

School Environmental Health and Safety Inspection Program

The state of Ohio produced a school environmental health and safety inspection manual which can be used as a model for all states and localities for conducting comprehensive school inspections to reduce or remove children's environmental health hazards. In March 2006, the state legislature passed a law that requires a more comprehensive health and safety inspection by local health department sanitarians. It also requires school districts to respond to the reports prepared by the local health departments and provide a written plan of abatement for all problems found.

ENDNOTES

1. Canadian Partnership for Children's Health and Environment. 2010. *Advancing Environmental Health in Childcare Settings: A Checklist for Child Care Practitioners and Public Health Inspectors.* Ontario, Canada.
2. Agency for Toxic Substances and Disease Registry. 2012. *ATSDR Case Studies in Environmental Medicine: Principles of Pediatric Environmental Health-Course: WB 2089.* Atlanta, GA.
3. Akinbami, Lara J., Moorman, Jeanne E., Garbe, Paul L., Sondik, Edward J. 2009. Status of Childhood Asthma in the United States, 1980–2007. *Pediatrics,* 123(Suppl 3):S131– S145.
4. US Environmental Protection Agency, Children's Environmental Health Network. 2010. *An Introduction to Children's Environmental Health.* Washington, DC.
5. Arizona Department of Health Services, Bureau of Epidemiology and Disease Control, Office of Environmental Health. 2003. *Arizona's Children and the Environment: A Summary of the Primary Environmental Health Factors Affecting Arizona's Children.* Phoenix, AZ.
6. Hess, Benjamin. 2007. *Children in the Field: An American Problem.* Association of Farmworker Opportunity Programs. Washington, DC.
7. Natural Resources Defense Council. 1997. *Her Children at Risk the Five Worst Environmental Threats to Their Health.* New York.
8. Indiana State Department of Health, Indiana Department of Environmental Management, Indiana Joint Asthma Coalition. 2004. *A Strategic Plan for Addressing Asthma in Indiana.* Indianapolis, IN.
9. American Academy of Pediatrics, Committee on Environmental Health and Committee on Infectious Diseases. 2006. Chemical-Biological Terrorism and Its Impact on Children. Policy Statement. *Pediatrics,* 118(3):1267–1278.
10. Johnston, William R. compiler. *Summary of Historical Attacks Using Chemical or Biological Weapons.* http://www.Johnstonsarchive.net. Accessed on: 30 November 2016.
11. Canadian Association of Physicians for the Environment. 2000. *Children's Environmental Health Project.* Toronto, ON.
12. Michael, Jerrold M. 2011. *The National Board of Health.* Public Health Reports.
13. Carter-Pokras, Olivia, Zambrana, Ruth E., Poppel, Carolyn F., Logie, Laura A., Guerrero-Preston, R. 2007. The Environmental Health of Latino Children. *Journal of Pediatric Healthcare,* 21:307–314.
14. Sheffield, Perry E., Landrigan, Philip J. 2011. Global Climate Change in Children's Health: Threats and Strategies for Prevention. *Environmental Health Perspectives,* 119(3):291–298.
15. World Health Organization, Commission for Environmental Cooperation. 2006. *North American Report on Children's Health and Environment Indicators: A Global First.* Montréal, QC.
16. Ohio Department of Health, Indoor Environment Section. 2007. *School Environmental Health and Safety Inspection Manual, Ohio School Inspection Program.* Columbus, OH.
17. Seltzer, James M., Miller, Mark, Seltzer, Diane L. 2007. *Environmental Hazards for Children in the Aftermath of Wildfires.* American Academy of Pediatrics, Pediatric Environmental Health Specialty Units. Elk Grove Village, IL.
18. Mailman School of Public Health, A Conference Report. 2009. *Translating Science to Policy: Protecting Children's Environmental Health, Columbia Center for Children's Environmental Health.* New York.
19. Centers for Disease Control and Prevention. 2013. *Young Worker Safety and Health: Reports and Publications.* Atlanta, GA.
20. Centers for Disease Control and Prevention. 2010. *Occupational Injuries and Deaths among Younger Workers: United States, 1998–2007.* Morbidity and Mortality Weekly Report, 59:15 Atlanta, GA.
21. North Carolina Department of Health and Human Services, Childhood Lead Poisoning Prevention Program. 2012. *N. C. Childhood Lead Testing and Follow-Up Manual.* Raleigh, NC.
22. Landrigan, Philip J., Schechter, Clyde B., Lipton, Jeffrey M. 2002. Environmental Pollutants and Disease in American Children: Estimates of Morbidity, Mortality, and Costs for Lead Poisoning, Asthma, Cancer, and Developmental Disabilities. *Environmental Health Perspectives,* 110(7):721–728.

23. Delaware Division of Public Health, Department of Health and Social Services. 2013. *Infectious Diseases in Childcare Settings: Informational Guidelines for Directors, Caregivers, and Parents*, Third Edition. Dover, DE.
24. American Psychological Association. 2010. *Effects of Poverty, Hunger and Homelessness on Children and Youth*. Washington, DC.
25. Wisconsin Department of Health Services. 2012. *Cleaning up Hazardous Chemicals at Former Meth Labs*. Madison, WI.
26. Shattuck, Lemuel. 1850. *Report of a General Plan for the Promotion of Public and Personal Health*. Appointed Under a Resolve of the Legislature of Massachusetts Relating to a Sanitary Survey of the State. Boston, MA.
27. Catholic Publication House. 1869. *The Sanitary and Moral Condition of New York City*. Catholic World, 9:53. New York, pp. 553–566.
28. de Moura, Tereza Soares, Berkelhamer, Jay E. 2012. Overview of the Global Health Issues Facing Children. *Pediatrics* (1):1–3.
29. Calman, Kenneth. 1998. The 1848 Public Health Act and its Relevance to Improving Public Health in England. *BMJ* 317(7158):596–598.
30. World Health Organization. 2005. *Fourth Ministerial Conference on Environment and Health: Final Conference Report*. EUR/04/504-6267. Budapest, Hungary.
31. Minnesota Department of Health. 2011. *Children's Environmental Health: Background*. St. Paul, MN.
32. Kramer, Howard D. 1948. Effect of the Civil War on the Public Health Movement. *Mississippi Valley Historical Review* 35:449–462.
33. Shephard, Donald S., Setren, Elisabeth, Cooper, Dominic. 2011. *Hunger in America: Suffering We All Pay For*. Center for American Progress. Washington, DC.
34. Quintero-Somaini, Adrianna, Quirindongo, Mayra. 2004. *Hidden Danger: Environmental Health Threats in the Latino Community*. Natural Resources Defense Council, Washington, DC.
35. Dockins, Chris, Jenkins, Robin, Owens, Nicole. 2003. *Handbook on Valuing Children's Health*. US Environmental Protection Agency, The National Center for Environmental Economics, Washington, DC.
36. Earthman, Glenn I. 2004. *Prioritization of 31 Criteria for School Building Adequacy*. American Civil Liberties Union Foundation of Maryland. Baltimore, MD.
37. Centers for Disease Control and Prevention. 2013. *Child Passenger Safety: Fact Sheet*. Atlanta, GA.
38. Centers for Disease Control and Prevention, National Center for Injury Prevention and Control. 2012. *National Action Plan for Child Injury Prevention: An Agenda to Prevent Injuries and Promote the Safety of Children and Adolescents in the United States*. Atlanta, GA.
39. Columbia University, Mailman School of Public Health. 2011. *Prenatal Exposure to Common Insecticide Linked to Decrease in Cognitive Functioning at Age Seven*. New York.
40. US Department of Health and Human Services, National Institutes of Health, National Heart, Lung and Blood Institute. 2007. *Guidelines for the Diagnosis and Management of Asthma, National Asthma Education and Prevention Program Expert Panel Report 3*. NIH number 08-5846. Bethesda, MD.
41. The National Environmental Education Foundation. 2005. *Environmental Management of Pediatric Asthma: Guidelines for Healthcare Providers*. Washington, DC.
42. Jackson, R.H. 1995. The History of Childhood Accident and Injury Prevention in England: Background to the Foundation of the Child Accident Prevention Trust, A Guest Editorial. *Injury Prevention*, 1:4–6.
43. Committee on Environmental Health and Committee on Infectious Diseases. 2009. Drinking Water from Private Wells and Risks to Children. *Pediatrics*, 123:e1123–e1137.
44. US Environmental Protection Agency Region 2. 2006. *Environmental Compliance and Best Practices: Guidance Manual for K-12 Schools*. Long Island University, Brooklyn, NY.
45. US Environmental Protection Agency. 2008. *Building Successful Programs to Address Chemical Risks in Schools: A Workbook with Templates, Tips and Techniques*. EPA 530-K-08-003. Washington, DC.
46. US Environmental Protection Agency. 2009. *Key Potential Violations of Federal EPA Regulations at K-12 Schools*. New York.
47. CDC. 2010. Occupational Injuries and Deaths among Younger Workers: United States, 1998–2007. *Morbidity and Mortality Weekly Report*, 59(15):449–455.
48. World Health Organization, Global Occupational Health Programme. 2009. *Occupational Risk and Children's Health: Training for the Health Sector*. Geneva, Switzerland.
49. US Department of Labor, Occupational Safety and Health Administration. 2010. *Reproductive Hazards*. Washington, DC.
50. Washington State Department of Labor and Industries. No Date. *Youth Job Safety Resources: Job Safety Resources for Teens, Parents and Employers*. Olympia, WA.

51. Borse, Nagesh N., Gilchrist, Julie A., Dellenger, M. 2008. *CDC Childhood Injury Report: Patterns of Unintentional Injuries among 0–19 Years Olds in the United States, 2000–2006.* Centers for Disease Control and Prevention, National Center for Injury Prevention and Control, Atlanta, GA.
52. Centers for Disease Control and Prevention. 1991. *Preventing Lead Poisoning in Young Children: Appendix II.* Atlanta, GA.
53. Minnesota Department of Health. 2004. *Best Practices to Prevent Poisonings.* St. Paul, MN.
54. Shayler, Hannah, McBride, Murray, Harrison, Ellen. 2009. *Sources and Impacts of Contaminants in Soils.* Cornell Waste Management Institute, College of Agriculture and Life Sciences, Cornell University, Ithaca, NY.
55. Centers for Disease Control and Prevention, National Center for Environmental Health. 2011. *Guidance on Microbial Contamination in Previously Flooded Outdoor Areas.* Atlanta, GA.
56. Attina, Teresa M., Trasande, Leonardo. 2013. Economic Costs of Childhood Lead Exposure in Low-and Middle-Income Countries. *Environmental Health Perspectives,* 121(9):1097–1102.
57. Shabecoff, Alice. 2010. *Children's First: Promoting Ecological Health of the Whole Child.* Osher Center of Integrative Medicine of the University of California San Francisco, the Whole Child Center, and the Collaborative on Health and the Environment. San Francisco, CA.
58. National Institute of Environmental Health Sciences. No Date. *Centers for Children's Environmental Health and Disease Prevention Research.* Research Triangle Park, NC.
59. National Institute of Environmental Health Sciences. No Date. *Grantees-Centers for Children's Environmental Health and Disease Prevention Research.* Research Triangle Park, NC.
60. World Health Organization. 2010. *Fifth Ministerial Conference on Environment and Health: European Governments Adopt Comprehensive Plan to Reduce Environmental Risk to Health by 2020.* Copenhagen, Denmark.
61. Government Accountability Office. 2010. Environmental Health: High Level Strategy and Leadership Needed to Continue Progress toward Protecting Children from Environmental Threats. GAO-10-205. United States Report to Congressional Requesters.
62. Stephenson, John B. 2010. *Environmental Health: Opportunities for Greater Focus, Direction, and Top-Level Commitment to Children's Health at EPA: Testimony before the Committee on Environment and Public Works, United States Senate.* Washington, DC.
63. The Canadian Guide to Health and the Environment. 1999. Guidotti, Tee Lamont, Gosselin, Pierre. The University of Alberta Press.
64. Canadian Partnership for Children's Health and Environment. 2010. *Advancing Environmental Health in Child Care Settings: A Checklist for Child Care Practitioners and Public Health Inspectors.* Toronto, ON: CPCHE.
65. Thacker, Laura, Gist, Ginger L., NEHA Position on Children's Environmental Health. *Journal of Environmental Health*; Denver 60.3 (Oct 1997):20, 21+.
66. Sewell, Bradford H., Whyatt, Robin M. 2011. *Intolerable Risk, Pesticides in Our Children's Food.* NRDC. New York.
67. Vandenberg, Laura N., Hauser, Russ, Marcus, Michele, Olea, Nicolas, Welshons, Wade V. 2007. Human Exposure to bisphenolA (BPA). *Reproductive Toxicology.* 24:139–177.
68. Perera, Frederica., Vishnevetsky, Julia, Herbstman, Julie B., Calafat, Antonia M., Xiong, Wei, Rauh, Virginia, Wang Shuang. 2007. Prenatal Bisphenol A Exposure and Child Behavior in an Inner-City Cohort. *Environ Health Perspect*; DOI:10.1289/ehp.1104492.
69. Brooten D., Youngblut, JM, Royal, S, Cohn, S, Lobar, SL, Hernandez L. 2008. Outcomes of an asthma program: Healthy Children, Healthy Homes. *Pediatr Nurs.* Nov-Dec;34(6):448–55.
70. EPA. 2009. *Indoor Air Quality Tools for Schools Reference Guide.* EPA

BIBLIOGRAPHY

Agency for Toxic Substances and Disease Registry, Environmental Health and Medical Education. 2012. *Principles of Pediatric Environmental Health.* Course WB 2089. Atlanta, GA.
Arkansas Department of Environmental Quality. 2009. *A Healthy Future for Arkansas Children: How to Safely Handle Lead-Based Paint.* North Little Rock, AR.
Cabral, Sergio A., Now, Anna, Calman, Kenneth. 1998. Overview of the Global Health Issues Facing Children. *British Medical Journal,* 317.

Chew, G., Carlton, E., Kass, D. 2006. Determinants of Cockroach and Mouse Exposure and Associations with Asthma Among Families and the Elderly Living in New York City Public Housing. *Annals of Allergy, Asthma and Immunology.*

Children's Health Watch. 2009. *Policy Action Brief: Affordable Health Care Keeps Children and Families Healthy.* Boston, MA.

Florida Department of Environmental Protection. 2010. *Yellow Buses in Ten Panhandle Counties Roll Out Green Retrofits During Environmental Education Week.* Tallahassee, FL.

Government Accountability Office. 2010. *Environmental Health: High Level Strategy and Leadership Needed to Continue Progress toward Protecting Children from Environmental Threats.* GAO-10-205. United States Report to Congressional Requesters.

Koren, Herman, Bisesi, Michael. 2003. *Handbook of Environmental Health: Biological, Chemical and Physical Agents of Environmentally Related Disease,* Volume 1, Fourth Edition. Lewis Publishers, Boca Raton, FL.

Koren, Herman, Bisesi, Michael. 2005. *Illustrated Dictionary and Resource Directory of Environmental and Occupational Health,* Second Edition. Lewis Publishers, Boca Raton, FL.

Landrigan, Philip J., Schechter, Clyde B. 2002. Environmental Pollutants and Disease in American Children: Estimates of Morbidity, Mortality, and Costs for Lead Poisoning, Asthma, Cancer, and Developmental Disabilities. *Environmental Health Perspectives,* 110(7):721–728.

Massachusetts Department of Public Health. 2009. *An Information Booklet Addressing PCB-Containing Materials in the Indoor Environment of Schools and Other Public Buildings.* Boston, MA.

Rivara, Frederick P. 2009. The Global Problem of Injuries to Children and Adolescents. *Pediatrics,* 123.

Stephenson, John B. 2010. *Environmental Health: Opportunities for Greater Focus, Direction, and Top-Level Commitment to Children's Health at EPA. Testimony before the United States Senate's Committee on Environment and Public Works.* Washington, DC.

Victorian Work Care Authority. 2010. *Keeping Children Safe in the Workplace: A Handbook for Workplaces,* Second Edition. Victoria, Australia.

Wattigney, Wendy A., Kaye, Wendy E., Orr, Maureen F. 2007. Acute Hazardous Substance Releases Resulting in Adverse Health Consequences in Children: Hazardous Substances Emergency Events Surveillance System, 1996–2003. *Journal of Environmental Health.*

5 Environmental Health Emergencies, Disasters, and Terrorism

STATEMENT OF PROBLEM AND SPECIAL INFORMATION

Environmental health emergencies, disasters, weather-related incidents, and acts of terrorism create situations where normal problems are intensified, existing environmental health problems are magnified, and new problems are created. The level of concern varies with the scope and nature of the situation which has occurred. For instance, a fire in a food warehouse would be an emergency, whereas an uncontrolled wildfire would be a disaster. (See Sub-Problems Including Leading to Impairment, and Best Practices for Environmental Health and Protection Emergencies, Disasters, and Special Issues section below for specific environmental concerns regarding food, water, housing, sewage disposal, solid waste, etc. These specific problems and Best Practices relate to most of the industrial accidents leading to disasters, natural disasters, and acts of terrorism discussed in this chapter and therefore to avoid repetition are being put in one special section.)

Responsibility for reaction to emergencies, disasters, and acts of terrorism may be part of the operational procedures of many different governmental and voluntary agencies. Numerous times, especially around large urban areas, there is a multitude of governmental entities, all with responsibility for the same type of problem. The environmental affects and health effects on people caused by an emergency, disaster, or act of terrorism are not confined to the boundaries set by specific governmental entities. Many times the types of communications systems utilized within communities and between communities are not compatible and therefore one entity cannot communicate effectively with another entity. Emergencies, disasters, and acts of terrorism demand a very rapid response but this may be affected by a lack of communication and coordination of response efforts.

After the immediate problems of the disaster or terrorist act are over, there is typically an overwhelming need for assistance for all individuals, especially the chronically ill and the elderly. People are displaced by the disaster and cannot typically live within their own environment, leading to increased injuries and potential emotional problems. There may be a lack of safe food, water, housing facilities, and bathroom facilities. The individuals may not have their medications available and therefore may deteriorate or die from existing diseases. Older people may tend to have reduced capabilities in their smell, touch, vision, and hearing and in their mobility. Many people, especially the elderly, may have limited financial resources, a reluctance to seek assistance, inability to complete necessary paperwork, and a lack of good transportation. Nutrition may become a serious issue and such things as meals-on-wheels may not be available. Fraud and abuse may be part of the aftermath of the disaster since the individual may be defenseless.

Evacuation centers may be overcrowded and become a means of transmission of acute diseases. The level of noise and overcrowding may profoundly affect the senior adult who is used to be quiet and being involved with very few people. Individuals, especially older ones, may not evacuate from hazardous situations for fear of leaving their pet behind or losing their pet.

In all cases, there is a sudden occurrence which demands an immediate, but thoughtful, coordinated response. An advance plan on how to work together and how to use the resources which are available for the protection of the community is essential. Appropriately staffed, well-prepared, and well-coordinated command-and-control groups working with a multitude of government agencies and citizens' groups and utilizing good sources of communications, are essential to prevent loss of life and property.

Special attention needs to be given to the environmental conditions within evacuation centers and to the needs of certain people, especially the young, the elderly who typically represent the largest number of people who become ill, injured or die, and the handicapped, pregnant women, and those with existing infectious and/or chronic diseases. Special pet shelters need to be provided. Special care, equipment, and training must be given to all volunteers and first responders.

Industrial accidents leading to disasters may occur in: the defense industry such as explosions at various weapons storage areas, and severe contamination of land, water (especially drinking water), and air from improper storage and disposal of hazardous wastes; the energy industry such as the Chernobyl disaster where the reactor went out of control and there was a nuclear meltdown, the oil spill into the Gulf of Mexico after an explosion on an oil rig, the spilling of coal ash into a river contaminating the water supply, etc.; the food industry with explosions in grain mills and severe outbreaks of foodborne disease; the manufacturing industry with numerous fires killing many people because of improper building construction and maintenance as well as extremely poor working conditions; chemical industry with accidental releases of hazardous chemicals into the air, water, and land during production, transportation, storage, or handling of the hazardous chemicals, and fertilizer plant explosions; the mining industry with improper enforcement of safety standards and improper storage of mine waste which leaks into the water and contaminates the land as well as the air; and the transportation industry with derailments of trains especially those carrying large amounts of crude oil which will catch fire. (See endnote 93 for further information on crude oil transportation by railcar.) This list is but a brief survey of an unusually large potential for disease and injury and severe contamination of the environment. The subject of industrial accidents is extremely complex and therefore needs to be the topic of a separate book. However, many of the Best Practices utilized in other types of disasters will also be utilized to resolve the negative results of industrial accidents. Of great concern is the type of contamination that is released to the air, water, and land, and special engineering controls and other techniques used to control the hazards. Best Practices developed from new knowledge will also be utilized.

Disasters, weather-related incidents, and acts of terrorism may cause contamination of the air, water, land, food, housing, and other facilities. Some of the issues which will be discussed include biological hazards; carbon monoxide; chemical hazards; cleaning and sanitizing of contaminated structures after floods and other events; non-safe housing; electrical hazards; foodborne disease; hand hygiene; heat exposure; heating, ventilation, and air-conditioning (HVAC) systems; mold; respiratory protection; etc. Also, during disasters and weather-related incidents, substantial amounts of solid waste including hazardous waste are produced leading to an increased potential of disease and injury and a sharp increase in insects and rodents. This increases the potential for disease. Collateral damage from disasters includes loss of human and animal habitat, food and water sources, sewage systems, electricity, and communications.

Many of the previously named public health, environmental health, and personal health problems need to be handled efficiently and expeditiously by various health departments at the local, state, and federal level as well as environmental management, environmental protection, and environmental sustainability departments working in a coordinated manner with a large group of medically oriented and public health-oriented partners. These health-oriented people and facilities are tasked to: carry out tests, provide laboratory analysis, and control safety for food, water, air, hazardous chemicals, biological agents, etc.; plan and carry out emergency responses to all types of health problems; study and detect outbreaks of disease; provide emergency medical treatment; etc.

Unfortunately, in recent years there has been a decrease in the funding which is necessary to promote key programs to detect and respond to various emergencies, disasters, and acts of terrorism. Some of the finest career epidemiologists, environmentalists, and others working in preventive medicine are seeking other opportunities because of severe budget cuts at the federal, state, and local level. Also, there is decreased funding for all of the laboratory analysis which needs to be done to make determinations and provide quick responses to serious problems such as the spread of microorganisms.

SUB-PROBLEMS INCLUDING LEADING TO IMPAIRMENT AND BEST PRACTICES FOR ENVIRONMENTAL HEALTH AND PROTECTION EMERGENCIES, DISASTERS, AND SPECIAL ISSUES

PREVENTION, CONTROL, AND MITIGATION OF INJURIES AND ILLNESS WHILE PROTECTING THE ENVIRONMENT

The primary objective of the environmental health and protection programs is to protect the health of the community while also protecting the environment. This is accomplished by carrying out the traditional role of the environmental health staff of investigation and control of a variety of licensed facilities, the prevention or minimizing of the spread of disease, and the prevention and minimizing of injuries. In addition, the professionals help promote public awareness by providing instructions for the public on how to avoid disease and injury, providing technical assistance to reduce hazards or threats of hazards, and promoting good environmental health practices in emergency facilities. In addition, environmental protection specialists also conduct studies of contamination of air, water, and land and work together with the environmental health professionals in resolving these issues, which also helps mitigate injuries to and illnesses of people.

Specific environmental health emergency programs and/or specific environmental protection programs include evaluations of air, land, and water to determine levels of contamination and how best to mitigate these circumstances; foodborne disease outbreak investigation; waterborne disease outbreak investigation; ensuring safe, potable water; ensuring a safe food supply in the absence of refrigeration and by condemning and having food destroyed that has been subjected to microbial and fungal growth, spoilage, smoke, chemicals, and contaminated water; ensuring that portable toilets are put in areas where they will not contaminate water but still will be available to the population where sewerage systems have been disrupted; supervision in mass care facilities of food, water, housing, waste disposal, etc.; control and reduction of insects and rodents to prevent disease and control nuisances; supervision of removal of solid waste to solid waste facilities and emergency dump sites; determining if hazardous waste generators and facilities are affected by the disaster and, if so, supervising safe removal of the substances to secure sites; supervising the removal of household hazardous waste to secure sites; evaluating and determining if healthcare facilities are damaged and are a danger to patients and staff; evaluating the removal of infectious wastes from healthcare facilities; evaluating the disposal of medical/infectious waste; evaluating chemical releases that may endanger the health and safety of the community; determining if a radiological release is hazardous to the community; evaluating residences to determine if they are safe for human habitation; evaluating worker safety and personal protective equipment to determine if the employees involved in studies being conducted as well as in removal of hazardous materials are protected properly from environmental situations; evaluation and control of outbreaks of emerging diseases; etc.

TOPIC AREAS

Air Quality (Outdoor and Indoor)
(See Chapter 2, "Air Quality (Outdoor [Ambient] and Indoor)")

In addition to the problems found on any day in the air quality of a given area, there is a substantial amount of additional hazardous pollutants added to the air which may affect not only the emergency responders and other workers, but also the citizens within the community who have been exposed to the substances. The hazardous air pollutants may include a variety of chemicals, biological agents, physical agents such as dust from pulverized buildings or land, pollutants as byproducts of fires, and material in the air as a result of wind, etc., which are directly related to the disaster. Indoors the air may be contaminated with a vast amount of pollutants from the existing contaminants and new contaminants in the outside air, and the existing contaminants and new contaminants from inside the structure which have become airborne. The heating, cooling, and ventilation systems may be severely contaminated. Further, in the event of terrorist acts, specific highly dangerous substances may have been added to the air.

Best Practices for Air Quality (See endnotes 48, 49)
- Utilize all best practices related to personal protective equipment listed in the Occupational Health and Safety Practices for First Responders, Construction Workers and Volunteers Section in Chapter 8.
- Keep all unauthorized people and other citizens from the disaster area to prevent command-and-control problems as well as reducing the potential for these individuals to inhale hazardous substances.
- Keep vulnerable people, such as the very young, the very old, those with chronic diseases, those with respiratory problems and heart problems, etc., indoors in safe facilities away from the disaster scene.
- Determine the air flow patterns and dynamics of the HVAC systems for various types of existing buildings and how contaminants can be introduced into the structures.
- Determine the mechanical condition of: the HVAC systems; the filtration systems and their efficiencies; the functioning of the dampers related to outdoor air, return air, fire, and smoke; and the seals around all portions of the systems.
- Do not permanently seal outdoor area intakes, alter or interfere with fire protection and safety, or modify the HVAC systems without the express permission of the appropriate authorities and the concurrence of knowledgeable engineering staff.
- Prevent access to areas where mechanical equipment is present or air intakes are stationed. Only appropriately identified maintenance and service people should have this access.
- Determine how to contain these potential hazards and provide appropriate assistance to building owners to upgrade their facilities.
- Design new buildings to be more secure and reduce the opportunity for the introduction of potential hazards into the structure.
- When the HVAC system has been damaged or contaminated turn off the system, discard all filters, and utilize professional services to inspect the system and clean all vents and air ducts before restarting it.

Asbestos
(See specific indoor air pollutants in Chapter 2, "Air Quality (Outdoor [Ambient] and Indoor)")

Additional Best Practices for Asbestos
- Do not remove asbestos sheeting unless it is broken or damaged.
- Do not use power tools or abrasive materials on asbestos surfaces.
- Clean the surface of asbestos material carefully and to not use high water pressure equipment.
- Wet asbestos materials before removing them.

Biological Hazards
(See specific indoor air pollutants in Chapter 2, "Air Quality (Outdoor [Ambient] and Indoor)")

Biologic and Infectious Wastes

(See Chapter 12, "Solid Waste, Hazardous Materials, and Hazardous Waste Management")

Carbon Monoxide

(See specific indoor air pollutants in Chapter 2, "Air Quality (Outdoor [Ambient] and Indoor)")

Chemical Hazards

(See Household Products in the Sub-Problems, Factors Leading to Impairment, and Best Practices for Indoor Air Quality section in Chapter 2, "Air Quality (Outdoor [Ambient] and Indoor)")

Debris

(See endnote 52)

Note: The removal and disposal of debris varies enormously from one disaster situation to another and is also extremely complex depending on the composition of the material and the ability of communities to handle large volumes within reasonable traveling distances, especially when there are already existing problems of solid and hazardous waste disposal. The discussion below will be limited for this reason.

Typically, the removal of the debris is a substantial item in the budget of all communities involved in disasters. Large quantities of trees, shrubs, sediment, soil, rubble from structures, metal, concrete, and asphalt need to be hauled away to disposal sites or for recycling purposes.

There is an increased potential for hazardous waste generation and special techniques are needed for removal and disposal. Some of the potentially hazardous materials include asbestos found in fire proofing, thermal and acoustical insulation, flooring tiles, and roofing material; pool chemicals, household chemicals, fertilizers and pesticides, and contaminated prescriptions; tires, batteries, and automobiles; explosives including ammunition, fireworks, and explosive chemicals; metal or plastic fuel containers, pressurize gas cylinders and underground storage tanks; electrical transformers utilizing PCBs; air-conditioning equipment utilizing Freon; containers of chemicals and various petroleum compounds; medical waste; radioactive wastes; etc.

Best Practices for Debris
- Establish a master plan for potential disposal of debris from a disaster and determine how to implement it.
- Establish a master plan for potential disposal of a variety of hazardous wastes from a disaster and determine how to implement it.
- Develop a recycling program prior to a disaster to establish the techniques, facilities, and equipment needed to recycle materials from the event. Shredded trees and shrubs can be recycled into compost or mulch. Concrete and asphalt can be crushed and used as a sub-base for roads. All types of metals can be recycled and produce funds for other purposes. Dirt can be used as a landfill cover.
- Do not allow open burning of any debris since it causes air pollution problems and is potentially hazardous to the health of many people.
- Obtain, read, and then utilize the *Debris Management Guide* mentioned in endnote 52 for detailed explanations of how to deal with debris in a disaster setting.

Drinking Water (On-Site Water Supply and Public Water Supply)

(See Chapter 13, "Water Systems (Drinking Water Quality)")

Additional Best Practices for Drinking Water (On-Site Water Supply and Public Water Supply)
- Appropriate authorities should issue drinking water advisories when water is thought to be contaminated.
- Determine if local water can be made safe for public usage with proper decontamination and chlorination techniques and if not provide a safe supply of water to the community.

- Notify all appropriate state and federal authorities concerning the conditions of all water-producing facilities.
- Establish secondary sources of communications with appropriate authorities if the primary sources are disabled.
- Determine in advance where to obtain necessary chemicals, equipment, and heavy equipment to make repairs to water and sewage systems.
- Determine in advance how to obtain substantial quantities of safe water as rapidly as possible in the event of an emergency.
- Schedule all specialized emergency and clean-up crews to be brought in rapidly as floods subside.
- Establish means of transportation for workers to be moved rapidly into the water and wastewater treatment plants.
- Train all personnel on how to rapidly shut down water and wastewater treatment plants to avoid damage to the equipment.
- All pump stations should be in well-drained areas and above flood levels.
- Secure all fuel tanks and chemical tanks including those holding chlorine.
- Check all electrical equipment to determine if they are functioning and if they are hazardous.
- Sandbag all critical areas, shut down exposed pipes, and park all vehicles and essential equipment on high ground.
- Wash all hands and forearms thoroughly after coming in contact with any form of contamination.
- Inspect all wells and pumps to determine if there has been damage to the equipment, electrical, and other, or well construction including the casing, and make necessary corrections before disinfecting and potentially using the water supply.
- Disinfect all wells, cisterns, and other water sources in areas which may have been flooded, run the water for at least 48 hours and then submit the water for bacteriological testing before using the water source. **Note:** Chemically contaminated water cannot be made into drinking water quality for public consumption or other usage in this manner.

Electrical Systems and Appliances

Power outages are not only inconvenient but also create many potential environmental problems. Food may spoil and become inedible or potentially cause disease. Water may become contaminated because of loss of function of water purification systems. Extreme heat or cold may cause discomfort and severe conditions leading to illness and death. Portable generators if used within structures may cause carbon monoxide poisoning. Dangerous situations may occur when working in the vicinity of downed wires and lead to electrocutions, which are a common cause of occupational deaths.

Best Practices for Electrical Systems and Appliances
- Extreme caution must be used when working in the vicinity of downed power lines which may or may not be de-energized.
- Specialists working on power lines must assume that all lines are energized and de-energize them by making a visible open point between the load and supply sides before conducting any type of repair.
- All electrical lines must be properly grounded.
- All workers must wear appropriate personal protective equipment.
- If a worker is shocked, he/she must immediately receive emergency medical help.
- If a power line falls on a car, do not touch the car or leave the car until the local utility company has shut off the power.

- Avoid carbon monoxide poisoning by using gasoline generators out of doors and away from confined areas.
- Do not siphon gasoline from motor vehicles for use in gasoline generators.
- When power is restored, do not touch electrical equipment or start heavy equipment in the occupational setting until experts determine that the equipment is safe.
- Before plugging in electrical equipment within the house, make sure that the equipment has not been wet or damaged and that it is being plugged into a grounded receptacle.

Food

(See Chapter 7, "Food Security and Protection")

Additional Best Practices for Food
- Discard any food that is not in water-proof containers including all fresh vegetables and packaged goods that have come in contact with flood waters or fires.
- Discard all cans that are dented, creased, swollen, rusty, or without labels.
- Where water is added to baby formula, utilize bottled water sources only that have not come in contact with the floodwaters.
- If the electricity had been shut off for less than 4 hours, the refrigerator has not been opened and is now back in service, check all food in the refrigerators and freezers and make sure the temperatures are below 40°F. If the temperature is higher than 40°F, discard the food in a safe manner. If the temperature is below 40°F, you may use the food. However, because meat, poultry, fish, both raw and cooked eggs, and leftovers are highly perishable and frequently are the source of the growth of microorganisms that cause foodborne disease, it is best to get rid of them in a safe manner instead of using them.
- Prior to the anticipated electrical outage, freeze meat and poultry to help keep their temperature below 40°F if there is a brief service interruption.
- Freeze containers of water into blocks of ice and have them ready to insert into the refrigerator as soon as there is an interruption to the electrical service to help maintain the temperature of the refrigerator below 40°F.
- Thoroughly wash all surfaces, equipment, pans, dishes, and utensils with detergent and hot water, thoroughly rinse off the detergent and then either use chlorine or a quaternary ammonium chloride compound to decontaminate the surfaces and allow them to air dry.
- Determine if individuals are having symptoms of foodborne or waterborne diseases from the consumption of contaminated food or water by checking if two or more of the people have consumed food or water at the same establishment and now have symptoms of diarrhea, bloody diarrhea, and/or vomiting, and have approximately the same period of incubation. If so, immediately conduct comprehensive epidemiological studies and take necessary actions to stop the immediate outbreak and prevent future ones.
- Conduct thorough inspections including flowcharts of preparation procedures of hazardous foods of all facilities which have been damaged by flood waters and other sources of disasters and take necessary action to correct potential problems.

Household Items and Clothing Potentially Contaminated with Flood Waters

(See endnote 57)

Since household items and clothing may be contaminated with sewage, chemicals, and biological agents including mold spores, they may potentially cause disease or an allergic reaction.

Best Practices for Household Items and Clothing Potentially Contaminated with Flood Waters
- Keep children and pets out of the affected areas until all cleaning, disposal of contaminated items, and repair has been completed.

- Wear appropriate personal protective equipment while completing the above tasks in the cleanup.
- Discard all personal service items, furniture, stuffed animals, paper goods, carpeting and padding, etc. in a safe and appropriate manner.
- Wash all non-porous items that can be salvaged in hot soapy water by hand, rinse thoroughly, and then place in a disinfectant and air dry.
- Wash all clothing and other washable materials in hot water and detergent and repeat a second time before drying.
- Thoroughly clean and wash hands and arms followed by rinsing with cooled boiled water and then take a bath before putting on clean clothing after completing the household cleaning and disposal of materials.

Insects and Rodents

(See Chapter 9, "Insect Control, Rodent Control, and Pesticides")

Disasters typically lead to situations where there is a sharp increase in insects, especially mosquitoes and flies, and also rodents. This increases the potential for disease in people and animals. Frequently, there is a disruption of on-site sewage systems and there are substantial quantities of standing water from rain or flooding. Rodent harborage may be disrupted and the rodents are scattered widely. There are also now numerous sources of new food and harborage for flies and rodents. The loss of solid waste collection including garbage, sewage treatment and animal control, and dead animals and humans contribute substantially to the large pest control problem.

Best Practices for Insects and Rodents
- Advise the public that there may be a sharp increase in insect and rodent problems, especially mosquitoes and flies, and that they should report significant increases to their local public health department for immediate treatment.
- Advise the public about using insect repellents, especially on children when they are out of doors.
- Immediately schedule special mosquito control spraying in areas where potential problems may exist.
- Work with all waste collectors to get back into operation as quickly as possible and also to advise the health department if they see problems of mosquitoes, flies, rodents, etc. and the exact locations of the situation.

Lead-Based Paint

(See specific indoor air pollutants in Chapter 2, "Air Quality (Outdoor [Ambient] and Indoor)")

Molds and Other Fungi

(See endnote 57)

Molds which are part of the natural environment out of doors may produce spores which are transported indoors and may grow in wet or damp areas. Molds may produce allergens which can become irritants or potentially toxic when inhaled or touched, especially if the individual is sensitive to them. Molds can irritate the eyes, skin, nose, throat, and lungs. Molds have an excellent opportunity to grow especially after any type of flooding or where high humidity is present.

Best Practices for Molds and Other Fungi
- Remove all visible water and dry out moist areas within structures as rapidly as possible.
- Fix all leaking pipes, equipment, and appliances immediately.
- Vent all dryers directly to the outside air.
- Discard all absorbent or porous material including carpets that have been exposed to water.

- Scrub mold off of hard surfaces with detergent and water and thoroughly dry the surface.
- Use the special N-95 Particulate Filtering Facepiece Respirator available at hardware stores, etc., to avoid breathing in mold or mold spores. (See endnote 55.)
- Use long gloves when removing mold to avoid contact with the skin and then wash your hands and forearms very carefully and rinse them thoroughly.
- Wear goggles without ventilation holes to avoid contact between mold and your eyes.
- Reduce humidity within the structure.
- Avoid all fungi by reducing repeated or prolonged contact of the skin with flood waters and the continuous sweating which may occur in special personal protective equipment.
- Seek immediate medical attention if skin eruptions or flu-like symptoms occur.
- Use professional mold removal services if the area which needs to be treated is extensive or if it is suspected that the mold is in hidden places such as drywall that is covered with vinyl wallpaper or wood paneling, which therefore are vapor barriers and contain the moisture for mold growth. (See endnote 56.)
- Wear appropriate personal protective equipment when investigating or removing mold from the area.
- When you use biocides or disinfectants to control and remove mold, ventilate the area. (See endnote 64.)

Septic Tank Systems and Public Wastewater Systems
(See Chapter 11, "Sewage Disposal Systems")

Additional Best Practices for Septic Tank Systems and Public Wastewater Systems (See Drinking Water (On-Site Water Supply and Public Water Supply) section above)
- Do not drink any well, cistern, spring, or surface water on the property or within the vicinity until the water has been tested for microorganisms and chemicals and the water has been found to be consistent with drinking water quality regulations.
- Do not use the on-site sewage system until the water level in the ground is beneath the drainage area of the system and the ground has dried out.
- Have professionals in on-site sewage systems determine if the system is operable or if it has been damaged and needs to have necessary repairs made.
- If sewage has come to the surface of the ground or backed up into the structure, use appropriate cleaning and decontaminating procedures while wearing necessary personal protective equipment.
- Flush out any silt or debris which may have entered the system prior to using it.
- Examine all electrical equipment to determine if it is functional.
- Secure the septic tank manhole cover and all inspection ports.
- Have the septic tank and other ancillary equipment pumped out if needed by trained professionals.
- Where sewerage systems or the network of pipes carrying the wastes to the sewage treatment plant have been damaged, it may be necessary to temporarily install portable chemical toilets in a given area. These units should be cleaned thoroughly frequently and kept under constant supervision. The individuals using them should have access to immediate hand-washing facilities in order to prevent the spread of oral–fecal route diseases.

Shelters
Emergency shelters and other necessary support are provided by governmental agencies and nongovernmental agencies to masses of individuals in need of temporary safe housing, food, clean and adequate bathroom facilities, and medical care during times of disasters. Unfortunately, in some instances such as in Hurricane Katrina, poor planning and poor decision-making created huge

numbers of problems. The shelters were inadequate to house the large numbers of people who had to be evacuated from highly dangerous areas. The shelters would not accept pets and since there was no provision for pet evacuation, many older people refused to leave their homes resulting in unneeded deaths. There was inadequate transportation to get people out of dangerous areas and into safe facilities. Shelters of last resort such as the Superdome and New Orleans Convention Centre did not have sufficient capacity, equipment, or supplies. There was a total lack of organization within the Superdome, and it was very difficult for the police and military to control the masses of people. There was no compiled list available showing all shelters. There were extreme problems of lack of adequate clean and equipped sanitary facilities. The opportunities for crime were rampant. There was poor record keeping and it was difficult to determine who was in what shelter and how they could be reached. The removal of individuals to more permanent housing was complicated by numerous factors including a lack of safe facilities that could be moved to the vicinity of the disaster area. The enormity of the discussion of these problems is beyond the scope of this book. (See endnote 58.)

Of major concern is the potential for infections spreading in an overcrowded special area where large numbers of people who have been displaced are seeking shelter. The people arriving are in various stages of health or sickness, acute and/or chronic. They are of all ages. They have all levels of susceptibility to disease and probably all levels of immunizations including no protection necessary to prevent certain diseases. This complex group of people, who are under extreme stress thereby lowering their ability to resist disease, may spread microorganisms through the enclosed air of the facility, in water, food, fecal matter, bedding, and other inanimate objects including toys, and by personal contact especially by contaminated hands. In the past, some of the worst epidemics were caused by this type of situation and therefore great precautions need to be taken to prevent the addition of these burdens to all of the problems caused by the disaster.

Best Practices for Shelters
- Determine the needs for various types of shelters for different levels of disasters for a given area.
- Determine the amount and type of transportation needed to remove people from areas where disasters have occurred to shelters.
- Determine how and when to remove homeless people from disaster areas to shelters.
- Establish special-needs shelters and means of transportation to access them for the very old, incapacitated, very young, chronically ill, etc.
- Professionals should inspect the proposed facility to ensure safety and control of infectious and communicable diseases.
- Establish communications and alternate communications with the disaster control team of the community and with the media.
- Develop a comprehensive shelter plan with the proper management and special services needed to perform at a high level to sustain life in an appropriate manner. This plan should include a shelter manager, assistant manager, shift supervisors, and all necessary personnel to run the facility efficiently; a registration coordinator who supervises the registration of all individuals utilizing the structure and present at any given time including a sign-in and sign-out log; appropriate food facilities; appropriate sleeping and living facilities; first aid services; mental health services; recreational services; childcare services; pet care services in an enclosed area away from the facility; and appropriate security and safety. (See endnote 59.)
- The staff should have all of the appropriate vaccinations and updates as needed.
- Use appropriate surveillance procedures when individuals enter the shelter to determine if they are sick and if they may have an infectious disease. If so, they need to be seen by a trained medical professional and isolated from the rest of the population. If the problem

is significant, there needs to be a means of transfer to a healthcare facility for appropriate follow-up. The staff should use proper hand-washing techniques and use personal protective equipment at all times. (See endnote 60.)

- Trained environmental health personnel should inspect and supervise all areas of: environmental decontamination; drinking water and ice; food safety and mass feeding facilities; housekeeping; personal hygiene and bathing; sleeping and living facilities, where no food is allowed; removal of soiled linen and clothing as well as laundry services; indoor air quality; vector- and pest-related problem areas and practices as well as the use of insect repellents especially on children; water management; removal of soiled diapers to appropriate containers for disposal; solid and hazardous waste management; disposal of sewage; and the handling and cleaning of toys and other objects shared by people.

Solid and Hazardous Waste Removal

(See Debris section above) (See Chapter 12, "Solid Waste, Hazardous Materials, and Hazardous Waste Management")

During and after the advent of a natural disaster or act of terrorism, there is disruption of the infrastructure of the community including the routine work of storage, collection, and disposal of solid waste as well as the storage, collection, and disposal of hazardous waste. Areas of the community become inaccessible and areas of storage and disposal may become inaccessible and also disrupted by the events occurring. In addition, there is a huge increase in waste from the debris caused by the events which have occurred, which is potentially able to cause disease and injury and an assortment of environmental problems. (See Debris section above.)

Medical and infectious wastes produced by hospitals, other healthcare facilities, veterinary facilities, mortuaries, and research facilities may contain substances that can transport bacteria and/or viruses, sharps, and packaging. There will be an increase in medical/infectious waste because of illnesses and injuries to people in the community. Healthcare, other facilities, and equipment may be damaged and therefore may increase the level of medical/infectious waste.

The typical home contains a large number of household hazardous chemicals which may need to be removed. During the course of a hazardous event, many of these chemicals may become unusable or potentially modified by flood or fire. These substances may include solvents, gasoline, pesticides, drain cleaners, bleach, oil-based paints, lighter fluid, and a large assortment of other necessary household chemicals and cleaners.

Best Practices for Solid and Hazardous Waste Removal

- Prior to the potential occurrence of different types of disasters, establish a plan of what actions to take and what equipment and people need to be provided if an event occurred and a portion of the community was unavailable for normal solid and hazardous waste removal and disposal. Test the plan and make sure that it is operational.
- After the disaster event occurs, determine the type and extent of disruption to the normal solid and hazardous waste removal systems and how best to accommodate them utilizing the previously established plan of operation.
- Determine the potential public health concerns and work with the appropriate groups to mitigate them.
- Provide assistance to homeowners, businesses, and industry on how best to remove the existing material and where to take it for proper and safe disposal.
- Re-establish the normal operating waste disposal removal system as soon as possible.
- Provide technical assistance while observing all appropriate regulations for the collection, storage, separation, and disposal of mixed wastes including hazardous materials.

- Determine if hazardous waste producers or storage and disposal facilities have been affected by the event and if so take necessary steps to correct the situation.
- Utilize source reduction techniques for the hazardous waste producers.
- Establish a reuse, recycling, and recovery system for all solvents and specialized chemicals.
- Determine if the hazardous waste treatment process has been affected by the event and make necessary corrections to prevent air pollution, land pollution, and water pollution.
- Determine if facilities dealing with medical/infectious waste are operational and can accept this material from other facilities.
- Determine if medical waste transporters are operational and can ensure proper storage, management, and disposal of medical wastes.
- Determine if the sewerage system is operational and can receive sterilized liquid medical waste.
- Ensure that all biohazardous waste is decontaminated prior to the time that it enters the solid waste stream.
- Ensure that special attention is given to the decontamination of all laboratory wastes, blood, other special body fluids, sharps, and research animal waste including the carcasses to prevent outbreaks of disease.
- Provide information to the community about the handling, use, storage, disposal, and nature of the various household substances which can be considered to be hazardous waste. Discuss the interruption of service for these hazardous substances and when and where the facilities will be available and open to receive household hazardous materials.
- Do not allow the dumping of hazardous materials into the public wastewater system.

Structural Safety

See Best Practices for the "Health and Safety of Personnel Involved in Emergency/Disaster Cleanups and Repairs" section below and endnote 53.

Underground Storage Tanks

Underground storage tanks may be displaced, damaged, or destroyed by floods, earthquakes, mud or landslides, wildfires, etc. Particularly in flooding, the system may release its contents into the ground, air, or water. The tank and its associated pipes: may float free because of the pressure of the water in the soil; may become inundated by flood waters; may be loosened from its supports by erosion of the soil by flood waters; may have extensive electrical system damage; and may lose its product through openings in pipes, vents, gaskets, loose fittings, covers, sump pumps, and damaged areas of the container.

Best Practices for Underground Storage Tanks
- In flood-prone areas, anchor the underground storage tank to prevent it from floating when the soil become saturated and increase the depth of the tank in the ground while also increasing the amount of concrete above it.
- Equip all fuel lines with automatic shut-off valves.
- Extend the vent pipe above known flood levels.
- When flood warnings are issued: turn off the electricity to the underground storage tank; fill the tank with product and determine how much is there; secure all openings on the top of the tank; make sure all seals and valves are operational; temporarily cap off the vent pipe; put sandbags or other heavy objects on the concrete platform.
- After a flood: make sure that the electrical power is off until the system is inspected and determined to be safe; make sure that all systems are operational; clean and empty any areas where flood waters have been; perform an underground storage tank system tightness test before placing the system into operation.

- If a tank floats free: turn off the electricity if it is on; rope off the area and keep people away; notify the appropriate governmental agencies and the fire department; remove all contents of the tank under the supervision of trained personnel; and use professional help to dismantle the system, clean it, repair it, or replace it.
- Use daily inventory control to determine if there are leaks in the system. (See endnote 63.)

SUB-PROBLEMS INCLUDING LEADING TO IMPAIRMENT AND BEST PRACTICES FOR NATURAL AND WEATHER-RELATED DISASTERS

(See endnotes 6, 7, and Chapter 4, "Children's Environmental Health Issues")

INTRODUCTION

In 2011, in the United States, there were 1091 weather-related deaths, 8830 reported weather-related injuries or illnesses, and $23.9 billion in damages to property and crops. Natural disasters included fires, tornadoes, thunderstorms, floods, drought, excessive heat, hurricanes, and earthquakes. (See endnote 1.) In Europe there was excessive cold leading to hundreds of deaths. In Japan there was a major earthquake and tsunami, with water 33 feet high, leading to massive destruction including the Fukushima nuclear power facility. The potential for a nuclear disaster had been created. There were approximately 23,000 dead and missing people from this incident.

INCIDENTS OF NATURAL DISASTERS AND WEATHER-RELATED IMPAIRMENT

Drought

Drought is caused by an unusual severe lack of rain in a given area over one or more years compounded by the use of excessive quantities of water. It is a deficiency in precipitation compared to a normal level that a given area receives. Factors involved in various stages of drought include the rate of all types of precipitation; the amount of snow pack available; the rate of stream flow; the lake and reservoir storage levels; the groundwater levels; the amount of evapotranspiration from the soil and plants; the demand for water by the plants growing there; and the amount of water usage by people, business, and industry. The degree of drought varies from area to area and from time to time, but is always affected by the rate of usage of the water resources.

Drought crosses political boundaries, therefore all individuals within a given water basin should be involved in drought management and contingency planning. Drought also is not a stand-alone problem but rather may be related to several different types of hazardous situations, such as fires, mudslides, destruction of crops, destruction of the built environment, etc. It should be part of multi-hazard planning which should be part of all municipal planning processes.

Best Practices Related to Drought
- Map the entire water supply system including surface water, groundwater, reservoirs, pipes and pumping systems, tunnels, aqueducts, wells, etc., determine where there are leaks and potential problems, make necessary repairs, and then use continued maintenance programs and determine if each of these segments of the system meet the appropriate criteria established at the state and federal level.
- Develop a comprehensive drought management and contingency plan which provides for the public health and safety of the population being served, protecting and minimizing impacts on economic activity while utilizing responses to at least four major stages of drought including: drought watch, the least severe, where conservation of public water supplies is started on a voluntary basis; drought warning, where water conservation is voluntarily intensified and local plans are implemented to decrease water usage; drought

emergency, where the governor of a state orders mandatory restrictions on the use of water sources to protect public health and allow for essential uses; drought disaster, where the water usage is further restricted and the governor may ask the federal government to declare a state of emergency and provide federal funds.

- Use a water supply management strategy which involves monitoring current conditions; utilizing the best forecasting processes from the National Weather Service and others; communicating and coordinating with other agencies within the community and state to best utilize water resources; making necessary operational adjustments; curtailing water usage to various groups; and using alternate water supplies or emergency water supplies when necessary. (See endnote 3.)
- Build a sustainable network of drought communities with open lines of communication, where there are common interests and knowledge of common problems and solutions consisting of Best Practices.
- Develop guidance documents which can be used to resolve specific issues in a given water basin.
- Develop regulations for zoning, subdivisions, landscaping, grading, building, and water conservation which includes drought considerations. (See endnote 4.)
- Ensure that water and wastewater pumps and systems are working efficiently.
- Utilize recycled water for lawns and other plantings to reduce drinking water needs in a community.
- Evaluate all water use systems to determine if there are leaks and correct them.
- Upgrade and retrofit plumbing systems with new technologies to reduce water consumption.
- In new construction only use the most water efficient systems available.
- Assess, maintain, and repair existing irrigation systems and make sure they function as required.
- Use native plants which are accustomed to the local water conditions instead of plants which require substantial amounts of water.
- Educate the community about water usage and how to conserve. (See endnote 5.)
- Capture and utilize rainwater where appropriate.
- Use water-saving pool filters and cover pools and spas when not in use.

Earthquakes
(See endnotes 11, 12 for more details)

An earthquake occurs when two sides of a fault slip suddenly against each other and produce vibrations. A fault is a fracture in the rock or the crust of the Earth where the two sides are displaced against each other in a minor or major way. The slippage causes a shaking or vibration because of a sudden release of stored elastic energy from within the Earth and seismic waves (waves of energy) may be produced. Seismic shaking is not uniform. As the waves approach the ground surface and potentially travel through areas of loose soil, the amplitude of the waves will increase producing larger waves that may be far more damaging to structures than the original waves. Other earthquake hazards include ground failure and/or rupturing of the fault resulting in the creation of cracks, settlement of the soil, vertical shifting of the ground after large earthquakes, soil liquefaction, where the soil, in an area of sand and silt as well landfill areas, takes on the characteristics of a liquid during the shaking of an earthquake and flows like a liquid; this can cause enormous damage to any structures that are within this area. Aftershocks, which may occur for periods of time after the initial major shock, can be extremely damaging to structures and can cause additional injuries and deaths. (See endnotes 6, 8, 10.) Thousands of faults are currently present in California, the United States, and the rest of the world. There are thousands of earthquakes each year throughout the world.

The amount and severity of structural hazards caused by earthquakes are typically determined by: the amount of energy released and the intensity of the energy at a given place; the distance from the epicenter to the structure; the design of the structure; the condition and type of surface material at the site of the damage; the nature and condition of the various structures; and the quality and quantity of any earthquake resistant provisions. The epicenter is on the surface of the Earth directly above the hypocenter of an earthquake, the location where the eruptions within the Earth start.

Damage can typically occur to: buildings of all types from single homes to large factories, schools, and hospitals; ports and harbors; highways; railways; bridges; dams; electric transmission systems; power plants; water treatment systems and supply lines; wastewater treatment systems and supply lines; natural gas and oil systems and supply lines; communication systems; etc. Landslides and fires may occur. Structural damage of any type, collapsing walls and falling interior ceilings, pipes and light fixtures can cause serious injury and loss of life, as well as loss of critical facilities. Roadways and other means of transportation may be severely damaged and impede rescue efforts.

The seismic waves, which are the vibrations coming from the earthquake and cause the damage, are recorded by seismographs, which help determine the time, locations, and magnitude of the event. Different scales may be used in determining the nature and size of the earthquake.

The Richter scale, a common measurement scale for earthquakes, has certain limitations for very large earthquakes. It is a mathematical technique used to compare the size of earthquakes. The magnitude of the earthquake is reported in whole numbers and decimal fractions. It is based on a logarithmic scale and therefore an increase of 1 is actually 10 times greater rather than the initial number (e.g., a reading of 6 would be 10 times greater than a reading of 5). The Moment Magnitude Scale is more precise for large earthquakes because it is related to the total energy released during the earthquake. Intensity measures the amount of shaking at a given location, while moment measures the size of the earthquake at its source. (See endnote 9.)

Best Practices Related to Earthquakes
- Gather and analyze seismic data and data concerning the amount and type of destruction which has occurred previously in the area of concern for which information and planning is needed to mitigate the effects of future earthquakes.
- Gather and analyze data concerning the type of soil that various structures rest on in the area of concern.
- As in California, use the State Geologists' Seismic Hazards Mapping Program which provides maps that identify areas of amplified shaking, liquefaction, and landside hazards to establish appropriate building, zoning, and other codes and plans to reduce future problems.
- Use custom mapping and analysis tools including custom hazard mapping and interactive hazard maps provided by the US Geological Service to refine any data related to earthquakes within the study area.
- Use the Hazard Risk Assessment Program (geographic information system (GIS)-based software) from the National Institute of Building Sciences to estimate potential building and infrastructure losses from earthquakes, floods, storm surges, and hurricane winds. (This program was developed for the Federal Emergency Management Agency (FEMA).) (See endnote 13 and the "Resources" section.)
- Model building codes for new commercial or institutional structures after the International Building Code from the International Code Council, a non-profit organization whose objective is to protect the health, safety, and welfare of people by creating safe buildings and communities. All 50 states and numerous federal agencies use the applicable building codes for structures under their approval or supervision.
- Model local and/or state building codes for new residential areas after the International Residential Code.

- Model local and/or state building codes which apply to the alteration, repair, addition, or change in occupancy of existing structures to the International Existing Building Code. (See endnote 14.)
- Provide seismic code provisions for all building codes, with consensus-approved modifications by experts, written in user-friendly up-to-date designs and information from recommended sources such as the Building Seismic Safety Council. This helps ensure that the structures and the attachments can adequately resist the forces related to earthquakes to: avoid serious injury and loss of life; avoid loss of function of critical facilities; and minimize structural and non-structural repair costs. (See endnote 15.)
- Provide adequate budget and stringent enforcement policies to ensure that the appropriate code for the structure is utilized effectively to make sure that the buildings and occupants are protected from the effects of the earthquakes.
- Prioritize first retrofitting critical facilities for first responders, and then critical infrastructure providing electricity, water, communications, and heat to the community.
- Make older buildings safer by determining how best to prevent the effects of earthquakes and then retrofitting the buildings to protect the structure and the non-structural components such as suspended ceilings, non-load-bearing walls, utilities, etc.
- Strengthen the core infrastructure such as roads, bridges, railroad tracks, etc.
- Do not use liquefaction-susceptible soils for new construction of buildings or other facilities.
- Where liquefaction-susceptible soils are present but the areas around them are very stable, excavate the problem soils and replace with compact material.
- When grading for new construction, stabilize all slopes.
- Lower the groundwater level by installing a wellpoint system with suction pumps.
- Prepare the facilities and the occupational setting to become more earthquake damage and disruption resistant by utilizing appropriate experts to evaluate these structures and then use appropriate cost-effective retrofitting procedures.
- Anchor, brace, reinforce, and secure all non-structural items such as light fixtures, suspended ceilings, windows, furnishings, equipment, and supplies in all occupational settings.
- Develop a mandatory earthquake plan and use unanticipated drills on a periodic basis to test the knowledge and skills of the employees before, during, and after earthquakes or other disasters.
- Train employees in cardiopulmonary resuscitation (CPR) techniques, first-aid techniques, and use of fire extinguishers.
- Establish secure telecommunications systems and supply chains, store data away from the site, and set up a special command center to direct employees after a damaging earthquake.
- Establish emergency response and recovery plans for the workplace in conjunction with community plans.
- Evaluate school facilities and non-structural building components and prepare them for earthquakes including necessary retrofitting.
- Develop an earthquake safety program for the school and use practice drills frequently.
- Prepare the schools to become shelters for local citizens in the event an earthquake destroys their homes and other facilities.
- Assess the structure and contents of homes and reinforce or secure all areas and contents.
- Prepare an earthquake plan for the home and neighborhood by: determining which room in the house is most secure and then practice dropping to the floor, covering your body, and holding on during an earthquake; preparing a household emergency kit including a first aid kit, all necessary medicines, and 3 days' worth of food and water; prepare a list of people to stay in contact with and test the plan periodically; being trained in first aid and CPR; preparing secure quarters for all pets; practicing emergency family drills; and sharing the plan with immediate neighbors.

Extreme Temperatures (Heat and Cold)

Heat Stress

Extreme temperatures can cause illness, injuries, and deaths as well as damage to infrastructure such as roads, bridges, dams, etc. Agriculture and fisheries are specifically dependent on certain climate conditions. Excess heat events can impact crops, livestock, fisheries, and shellfish, and thereby reduce vital food supplies and increase prices. Electrical transmission systems may be affected by sagging power lines which may short out because of high temperatures. The unusual demand for excess energy may lead to blackouts and blackouts may lead to increased crime, particularly in large cities. Increased air pollution may be associated with the high temperatures and the heat island effect of urban areas. (See Chapter 2, "Air Quality (Outdoor [Ambient] and Indoor).") Transportation may be impacted because of the frequency and intensity of extreme weather events. Higher temperatures can cause roadways to soften and expand, especially in high traffic areas, and place extra stress on bridge joints. Railroad tracks may expand and buckle. Airlines may face cargo restrictions and flight delays. The demand for water increases sharply and may result in shortages in various areas.

Extreme heat causes more deaths each year than the combined mortality due to hurricanes, lightning, tornadoes, floods, and earthquakes. In 1995, there were 465 heat-related deaths in Chicago. From 1999 to 2010, there were 7415 heat-related deaths in the United States. (See endnote 16.) The number of heat-related deaths is rising because the number of heat-related events is increasing.

Some people are more vulnerable to heat than other people. Vulnerability can increase because of the degree of exposure, length of exposure, and the individual sensitivity to heat. Vulnerability can increase because of the individual's socioeconomic status, degree of overcrowding in the structure and community, accessibility to air-conditioned facilities, and level of literacy in the English language. Isolation especially among the elderly can lead to serious problems related to excess heat especially because of restricted physical capabilities, a lack of communications with others, and a lack of an understanding of what is occurring and how to rectify the problem.

Heat stress is of greatest concern for the very young, the elderly, the overweight, the chronically ill, individuals on certain types of medications, and individuals who are working in the extremely hot environment. The young child or baby is very sensitive to high temperatures and can quickly get stressed although not showing immediate symptoms. They dehydrate very rapidly. The elderly are at increased risk because the body may be unable to cope with heat as a younger person could and cannot cool itself normally. Additionally, the individual may have: reduced mobility; confusion or mental problems; chronic medical problems that require medication that can affect the ability of the body to maintain a proper fluid level such as diuretics; pharmaceuticals that limit or reduce the amount of perspiration and therefore deprive the body of a natural cooling mechanism; pharmaceuticals that can increase the risk of sunburn when exposed to the sun for limited periods of time; and kidney conditions which can be affected by the amount of fluid intake. In addition, there may be environmental problems because of the lack of air circulation and/or air-conditioning during times of high heat levels. Severely overweight individuals not only have a greater potential for a variety of diseases including heart disease, high blood pressure, cancer, renal disease, diabetes, etc. but also may have profuse sweating and fatigue leading to heat emergencies. The chronically ill have already been weakened by their conditions and are typically not very mobile or even immobile and therefore are at much greater risk of serious problems caused by sustained heat events. They may also be seriously dehydrated. People in an urban setting without green areas may be more susceptible to heat than others. (See endnote 17.)

Individuals working in very hot environments are exposed to all forms of heat illnesses. This is especially true of young people starting on new jobs in the hot summer outside or in very hot areas of businesses or factories. Normally, they have not had enough time to be acclimated to the higher temperatures in their surroundings.

Cold Stress

Winter weather can bring severe cold stress to infrastructure as well as humans. Winter storms, high winds, snow, sleet, ice, and dangerously low temperatures can damage infrastructure as well as humans and animals. Pipes may burst and roadways may be damaged. Cars, buses, and other means of transportation may fail more often than normal. Frequently, there is a loss of heat and power, and communications and roads may be blocked causing people to be isolated from essential services. Structures may be damaged by fallen trees and downed wires. There is often a sharp increase in traffic accidents and falls leading to serious and even fatal injuries.

People in homes where the electricity and/or heat have been shut off may suffer from an exacerbation of their diseases because of the disturbance in the normal environment. This is especially true of the elderly, the very young, and chronically and acutely ill individuals who have exceptional trouble with the cold. The dangers of carbon monoxide poisoning increases substantially because in trying to keep warm, people may use poorly functioning or unvented furnaces, generators, and gasoline, propane, or charcoal-burning units and wood-burning stoves. This leads to over 500 deaths a year from carbon monoxide poisoning and thousands of people going to the emergency room for treatment. Serious house fires may occur as a result of the improper use of these heating and cooking appliances.

Large areas of communities may be affected by severe winter storms which cause problems exceeding the capacity of first responders and other personnel to respond to the emergency needs of the individuals, thereby compounding the potentially hazardous situations.

Humans and animals are affected when there is a substantial drop in temperature below normal and the wind speed increases or there is immersion in cold water. The body can lose heat faster than it can be replaced and will over time use up all the stored energy of the individual. This may cause the body temperature to be lowered in an unnatural way and may cause tissue destruction and death.

Best Practices for Extreme Temperatures
- Determine the names, addresses, and phone numbers of those individuals who are most vulnerable to extreme temperatures, especially the elderly who typically are most affected, and the isolated.
- Establish a plan to have volunteers or governmental employees visit and assess the needs of the most vulnerable individuals and provide supportive services during times of extreme temperatures.
- Coordinate the efforts of all governmental and voluntary service providers to take care of those in need while working efficiently.
- Provide a heat alert and response system to inform the public of the potential danger of present and near future heat events.
- Provide bottled water and cool-down places for individuals who are vulnerable to excess heat.
- Have public health personnel visit all places where the poor and chronically ill congregate or live and take whatever actions are necessary to cool these individuals down.
- Have the homeless seek cool areas for shelter, especially during the hot part of the day.
- Provide interpreters for areas where English is not the primary language in order to make people aware of the hazards of excess heat and what to do to resolve the problem.
- Instruct individuals to wear lightweight, light-colored, loose-fitting clothing.
- Teach individuals to slow their pace of activity and seek shady and cool areas as frequently as possible.
- Regulate the environment of infants and young children to allow minimal exposure to high temperatures.
- Teach the sick, the chronically ill, the elderly, the very young, and the overweight to reduce their activity level in hot environments.

- Avoid hot and heavy foods and/or drinks.
- Do not leave infants, children, or pets in cars or other vehicles.
- Watch all individuals within the hot environment carefully to determine if they have symptoms of dizziness, nausea, confusion, unconsciousness, rapid heartbeat, red skin, or high blood pressure and then take effective medical-based actions as needed.
- Acclimate all young workers especially in hot environments for short periods of time while observing their physical condition and make appropriate changes in time spent within the environment and encourage appropriate cooling techniques.
- Provide appropriate cold weather and storm alerts.
- Prepare for winter storms by stockpiling salt, sand, snow-removal equipment, snow shovels, heating oil for isolated facilities or homes, quantities of mandatory medicines, adequate warm clothing and blankets, necessary food and water, and appropriate shelters for pets.
- Provide carbon monoxide detectors in the basement and on every floor in the structure.
- Evaluate all heating devices to make sure that they are properly vented and cleaned on a regular basis, at least annually.
- Never use gasoline or propane grills or charcoal inside of the house or structure for warmth.
- Do not warm up the automobile in an enclosed garage.
- Do not use generators in the home or garage.
- Wear appropriate insulated clothing when working outdoors in severe weather.
- Protect ears, face, head, hands, and feet in extremely cold weather.
- If someone has been submerged in cold water, immediately remove wet clothing, warm the body tissue, and take the individual to first aid facilities or an emergency room.
- Have immediately available clean dry loose clothing, blankets, chemical hot packs, and thermoses full of hot liquids to be used to protect exposed individuals.
- Move individual into a warm locations as quickly as possible.
- Work in extremely cold temperatures for limited time periods with periodic substantial breaks in warm areas.
- Do not drink any alcohol in either extremely hot or extremely cold environments.

Floods

Floods are the most frequent and costly natural hazard in the United States, leading to many deaths and, as a result of the severity, result in the most frequent presidential disaster declarations for natural or weather-related events. Millions of people live in flood-prone areas in the United States, where repetitive flooding causes additional damage in already damaged areas. There are very large population concentrations in many of these flood-prone areas and therefore the opportunity for huge costs and substantial injuries and potential deaths continues to increase. Destruction of wetlands in order to erect buildings in prime areas has contributed substantially to this problem. Large numbers of people live in areas that may either flood occasionally or not flood at all, but because of new conditions in the area such as the aftermath of wildfires or because of changing weather patterns and conditions, flooding may now occur. Floods may be caused by: tidal surge, severe thunderstorms, oversaturation of soils, and spring thaw of frozen land, snow, and ice; heavy rains from tropical storms, hurricanes, tornadoes, and other severe weather events; and the topping or destruction of levees and dams. A flash flood, which may be caused by a slow-moving storm, is a rapid filling of a low-lying area in less than 6 hours, however it can occur in a matter of minutes. The force of the water can cause landslides, move boulders, rip out trees, and damage or destroy various structures, roadways, and bridges. It can also inundate water and sewage systems, electrical systems, communications systems, and all means of rescue. Areas of wildfires denude the land and make it vulnerable to mudslides and flooding when heavy rains occur.

There are numerous potential health and safety consequences of floods. They include drowning or near drowning; substantial injuries from structures, debris, etc.; electrical hazards from downed power lines and wet appliances; potential explosions and fires from disrupted gas lines;

hypothermia from working in a wet environment; infectious diseases; chemical contamination from household chemicals, industrial chemicals, fertilizers, and pesticides; carbon monoxide poisoning from unvented gas-powered equipment; respiratory problems from mold and exposure to wet conditions; bites and transmission of disease by rodents and sick animals; exacerbation of chronic diseases such as asthma, allergies, and ear, nose, and throat infections; mental health problems especially in children; antisocial and violent behavior and rioting; and poor nutrition due to lack of safe food and water supplies.

Best Practices for Floods
- Develop a comprehensive watershed management program for each individual watershed including: point sources of pollution and mitigation techniques; non-point sources of pollution and mitigation techniques; provision for sustained torrential rains and mitigation techniques; determination of the floodplain and mitigation techniques; appropriate storage and delivery of water; coordination with all local agencies and governmental entities; and supervision and management of the program by highly professional competent individuals.
- Promote and enforce stringent zoning regulations, codes and standards, and regulations for permits for developments and housing in floodplains and other frequently flooded areas.
- Promote the increased elevation of existing properties in floodplains that have been repeatedly damaged.
- Governmental agencies should provide easily accessible information on flood-prone areas for individuals who are seeking to purchase property or upgrade structures.
- Governmental agencies should restore watersheds, wetlands, sand dunes, beaches, and wildlife habitat to mitigate the effects of floods.
- Inspect and repair all levees and dams prior to the local onset of severe weather conditions.
- Avoid building homes and other structures in a floodplain.
- Elevate existing structures above the 100-year flood level.
- Elevate the furnace, air conditioner, water heater, and electrical panel in the home above high flood risk level.
- Install check valves to prevent flood water from backing up in the drains.
- Communicate to the public information about flood watches, flood warnings, and flash flood warnings.
- Move to higher ground when information is communicated about floods which may occur rapidly.
- If ordered to evacuate, secure the home and outdoor furniture, turn off utilities at the main switches or valves, and disconnect electrical appliances.
- Restore basic utilities and public services as quickly as possible including a safe water supply, electricity and gas, a communication system, reliable food supplies, and health services.
- Have experts inspect and declare safe all structures which will be used by individuals.
- Have properly trained individuals remove all materials including wallboard, insulation, flooring, furniture, etc. which have been flood damaged and dispose of them in an appropriate manner. Clear and clean all outside areas which are needed for recreational purposes by individuals, especially children.
- Never walk through moving water unless it is a life-and-death situation.
- Do not drive through unknown depths of water after a disaster or heavy rains.
- Avoid flood waters which may be contaminated with chemicals, oil, and gasoline or raw sewage.
- Avoid downed power lines. (See endnote 18.)

Landslides and Mudslides

Landslides are created by the downward movement of unstable materials on slopes caused by weakening of the soil structure over time, construction of houses and roads thereby destroying natural slopes, weathering of the ground materials, earthquakes, droughts, fires, and then substantial rainfall, rapid snow melts, and the effects of gravity. Landslides may be caused by earthquakes, storms, and volcanic activity. Rapidly moving water and debris can cause injury, death, and destruction of property, utilities, roads, railroads, etc.

Best Practices for Landslides and Mudslides
- Create and utilize proper land planning, zoning, and professional inspections in mountains, canyons, and coastal regions to avoid new construction in areas subject to wildfires, steep inclines, and potentially unstable land.
- Avoid building in areas near drainage ways or natural erosion sites.
- Determine the history of an area concerning landslides, debris flow, and wildfires before considering purchasing the land and building structures.
- Plan groundcover on slopes and build appropriate retaining walls.
- Contact local authorities to find out about emergency and evacuation plans in the event of a landslide or mudslide.
- Listen for rumbling sounds that may alert individuals to impending landslides or mudslides.
- Evacuate all sites immediately when given orders to do so by the appropriate authorities.
- Do not re-enter damaged sites until given permission to do so by the appropriate authorities.

Tsunamis

Tsunamis are a series of extremely large waves created by sudden disturbances in the ocean or on the ocean floor caused by earthquakes, volcanic eruptions, or landslides in the ocean. The effects can be felt locally and very rapidly or hundreds or thousands of miles away. Extremely large waves moving through the water suddenly appear and then breach coastal areas at a speed of over 600 mph. Tsunamis create vast numbers of deaths and injuries and huge amounts of destruction to property. It may take weeks or months to clean up the debris and re-establish a place to live, if it is at all possible.

Best Practices for Tsunamis
- Determine if the area which is being inhabited is tsunami prone.
- Prevent construction in a zone at or near the coast which may become inundated from a tsunami.
- Establish an early warning network and system to advise governmental agencies and people of an oncoming tsunami and evacuate the individuals to much higher ground. The saving of lives is of greatest importance.
- Alert the appropriate governmental agencies and people to earthquakes, volcanoes, or landslides which appear in the ocean.

Volcanoes

A volcano is an opening or vent in the Earth's upper mantle which allows magma and gases produced by the melting of rock below the surface to be discharged. The volcanic eruption is caused by the buoyancy of the magma, the pressure of the gases, and new magma being injected into a chamber which is already full of existing magma. (See endnote 25.) Volcanoes may produce ash, toxic gases, flash floods of hot water and debris, lava flows, and fast-moving flows of hot gases and debris resulting in death by suffocation, infectious disease, respiratory illness, burns, injury from falling, etc. Volcanic ash is gritty, abrasive, corrosive, slippery, and may cause injury or exacerbate chronic conditions in individuals who have existing lung and heart diseases as well as harm infants, the elderly, the infirm, and those with impaired immune systems. The gases, primarily carbon dioxide and hydrogen sulfide, are found in low-lying areas, and in large concentrations can injure and kill people. At low levels they are very irritating to the eyes, nose, and throat and can cause rapid breathing, headaches, dizziness, and spasms of the throat. (See endnote 26.)

Best Practices for Volcanoes
- Determine if you are in an area containing a volcano and if so determine how and when to evacuate your family and pets to a safer area if there is a volcanic eruption.
- Always have the car filled with gasoline when living in this type of area in preparation for a rapid evacuation.
- Obtain and utilize an air purifying respirator (N-95) if volcanic ash is present.
- Close all openings to the outside in homes and other structures and turn off all fans, heating, and air-conditioning systems.
- Seek shelter indoors away from bodies of water which may rise rapidly as a result of the volcanic eruption.
- Stay away from all volcanic ash unless properly protected with respirators and special clothing. (See endnote 27.)

Wildfires

Wildfires are triggered by lightning, people being careless in the use of fire, and arson. Most of the wildfires are caused by people, while some are due to lightning and occasionally lava flows. The wildfires spread quickly because of dry conditions and substantial fuel provided by brush, trees, and homes, which may be located in woodland settings, near forests, and in rural areas, mountainous areas, and gullies. Windy conditions exacerbate the problem. High temperatures and low humidity are contributing factors. The time of year is significant because the composition of the fuel and moisture level help determine the speed of the fire and how hot the fire gets. The chemical composition of the fuel and the amount of oils or resins present and the density of the material also help determine how fast the fire will spread.

There is a constant conflict between the desire to build new homes in high-risk areas and the danger of increasing the potential for wildfires. Houses and urban life have expanded rapidly into wildlands and have created a substantial potential for destruction by wildfires, not only because of the increase in potential fuel from the structures, but also because of a poor level of knowledge by individuals of their new environment. Wildfires may ignite one house or hundreds of houses depending on a variety of existing conditions including the weather. This concentration of homes also creates the necessity to use scarce budgetary sources for protection of the homes when the funds should be used for the prevention, mitigation, and control of the wildfires.

There has been and still is an ongoing discussion about the value of allowing small wildfires to burn out instead of trying to suppress them. Suppression not only requires additional funds but also may create the potential for larger and more dangerous wildfires in the future because small fires use up fuel and burn out naturally but if suppressed may leave much larger sources of fuel for wildfires in the future. There is also an effect on the ecosystem when small fires are suppressed. Although fires may be devastating, they can also be beneficial. They can destroy insect pests, exotic or non-native plants, remove undergrowth to encourage native plant growth, and add necessary nutrients from the ashes for native plants and trees. (See endnote 29.)

The ultimate wildfire or series of wildfires is the *firestorm*. A firestorm acts like a hurricane or tornado. The intense fire can create its own weather conditions with powerful winds and many tornadoes of spinning flames. It can have as much energy as a thunderstorm and as the hot air rises it sucks more oxygen and debris into the flames and intensifies the fire. The firestorm, which is an explosion of fire, can rapidly destroy large areas of structures and fields as well as kill many individuals through suffocation and burning.

The National Fire Danger Rating System is used by various resources to indicate the potential level of fire danger as follows:

- *Low (Green)*—These fires are relatively easy to control.
- *Moderate (Blue)*—These are wildfires with moderate problems.

- *High (Yellow)*—These are wildfires which may be difficult to control especially if it is windy.
- *Very High (Orange)*—These wildfires are very easy to start and spread rapidly.
- *Extreme (Red)*—These wildfires start and spread very rapidly and may become extremely dangerous. (See endnote 28.)

Best Practices for Wildfires
- Evaluate and record short-term and long-term wildfire risk for communities and individual structures in various subareas within the given area of study using a variety of scenarios related to population growth, global warming, seasonal variations, type and quantity of fuel available, previous history of wildfires and extent of damage, degree of high-growth urban sprawl, and the amount of fireproof or fire-resistant material utilized in building structures. (See endnote 30.)
- Develop and implement a Wild Fire Prevention and Mitigation Management Program coordinating the activities and resources of federal, state, and local authorities to minimize the impact of wildfires and the potential loss of life, injury, and destruction of property by: conducting the previously mentioned studies and drawing usable conclusions for immediate and long-term actions; developing and utilizing appropriate zoning codes and building codes for new developments and existing housing; communicating to individuals and communities the cause and nature of wildfires and how to prevent them and protect properties and lives; reducing and removing hazardous fuel loads which can be used to start the fire or enhance it; rapid suppression of the fire depending on the size, nature, and location using the most modern available technology; encouraging the use of fire-resistant or fireproofed material in structures; enhancing the firefighting capabilities and communication systems of fire departments and other resources used to keep the fires from spreading; developing appropriate legal tools to deal with recalcitrant individuals for the good of the community; rehabilitating the site of the wildfire where feasible and if of value to the community using federal and/or state funding, etc. (See endnotes 31, 32, 33, 34.)
- Use fuel management techniques, usually controlled burning and mulching and chipping of materials prior to burning, to reposition the potential fuel for wildfires closer to the surface of the ground.
- Use herbicides where there is potential live fuel in a given area.
- Communities should not allow houses to be built in high-risk wildfire areas.
- Communities should supervise the removal of excess fuel and the thinning out of trees and bushes to reduce the potential for serious wildfire problems.
- Develop and present to the public uniform clear messages in all media and teaching formats concerning: closures of areas and restrictions; campfires; smoking; storage, use, and disposal of fireworks; trash burning and debris burning; restrictions on vehicles being parked on dry grass or other dry material; and the severe penalties involved, including being indicted for murder, if someone dies from a fire set purposely and considered to be arson.
- Develop an appropriate fire evacuation plan for the community and encourage families to develop their own evacuation plan.
- Homeowners should take responsibility for making their land and home fire resistant and defensible by designing into the structure effective fire control materials and measures.
- Use materials and plants to landscape around the home that will not readily support fires.
- Use fire-resistant or non-combustible materials on the roof and in all portions of the structure.
- Within a 150-foot radius of the house, prune all trees and bushes, remove all dead branches and fallen trees, and remove all debris from the site.
- Do not stack wood for fireplaces next to the house.

- Clean and inspect all chimneys, the roof, and gutters on a regular basis.
- Install smoke and carbon monoxide sensors on all levels of the house near bedrooms and test monthly.
- Clear all potentially flammable material from the house.
- Never utilize open burning techniques for trash or debris in areas vulnerable to wildfires.

Wind-Related Storms (Tropical Storms, Hurricanes, Tornadoes, Severe Winter Storms, etc.)

These storms create varying degrees of wind and wind damage and varying degrees of rain and flooding. The problems of flooding and Best Practices have already been discussed in the "Floods" section. This discussion will be about wind, wind damage, and means of mitigation.

Severe winds may cause high levels of water to accumulate and inundate areas. Wind depending on the speed, type, direction, and length of time of the event, may cause serious injuries and deaths as well as devastating destruction to communities, road systems, utilities, schools, and other public buildings, homes, and businesses.

There are various types of meteorological events which produce high and sustained winds including:

1. *Mid-Latitude Cyclones*—These are extremely large weather systems with low-pressure areas, most frequently occurring during the winter, around which the winds flow counter-clockwise in the northern hemisphere and clockwise in the southern hemisphere. North Easters and blizzards may be caused by these weather systems and may create extremely hazardous conditions for extended periods of time. (See endnote 21.)
2. *Tropical Cyclones*—These are various low-pressure systems with maximum winds of 73 mph.
3. *Tropical Depression*—This is a low-pressure tropical system that forms over warm water with maximum winds of 38 mph and may produce extremely large amounts of rain leading to severe flooding.
4. *Tropical Storm*—This is a tropical system with sustained winds of 39–73 mph that may produce extremely large amounts of rain leading to severe flooding while causing some beach erosion and damage. (See endnote 20.)
5. *Hurricane*—This is a tropical cyclone or system which has defined wind circulation that develops over the tropics drawing its energy from the warm water, producing winds of 74 mph or greater and causing heavy sustained rain, destructive winds depending on the speed, and storm surges which may inundate and destroy property. Differential pressures from the storm may destroy windows and doors especially garage doors, roofs, and walls. Severe injury and death may be the result of the force of the winds, the debris found in the winds, falling trees, downed electrical wires, drowning from the flooding, etc. The gale force winds can extend hundreds of miles from the center which is called the circular eye. There are five categories of hurricanes listed on the Saffir-Simpson hurricane wind scale as follows:
   a. *Category 1*—These are dangerous sustained winds of 74–95 mph which may produce some damage to roofs, siding, trees, and power lines.
   b. *Category 2*—These are extremely dangerous winds of 96–110 mph which may cause serious damage to roofs, siding, and trees, and near total power loss.
   c. *Category 3*—These are extremely dangerous winds of 111–129 mph which may cause devastating damage to roof decking and gabled ends of housing, trees, electricity, and water.
   d. *Category 4*—These are extremely dangerous winds of 130–156 mph which cause catastrophic damage to houses, trees, and power poles, and create power outages that could last for weeks.

    e. *Category 5*—These are extremely dangerous winds of 157 mph or higher causing catastrophic damage including destroying homes, trees, and power poles with resultant power outages lasting weeks to months. The affected area they become uninhabitable for long periods of time. (See endnote 22.)
6. *Thunderstorms*—These may be extremely dangerous weather systems that affect small areas, usually 15 miles in diameter, lasting about 30 minutes, and composed of moisture and unstable air that is lifted skyward by cold or warm fronts, sea breezes, air warmed by the sun's heat, or mountains. Although there are about 100,000 thunderstorms each year in the United States, all of which are considered to be dangerous, about 10,000 are classified as severe and are very dangerous. Along with the wind and rain, thunderstorms create lightning which is extremely hazardous because it causes human fatalities and injuries, wildfires, and destruction of property at a cost greater than $1 billion in insured losses each year. Straight-line winds, which are not associated with the rotation of a tornado but associated with the thunderstorm, can exceed 125 mph and cause massive damage.
7. *Tornados*—These are violently rotating columns of air which extend from a cumuliform cloud, such as a thunderstorm, to the ground. The funnel becomes visible when dust and/or debris are picked up by the storm. The amount of destruction of the tornado is based on numerous factors of time on the ground, size of the tornado, condition of properties, etc. and the 3-second wind gust as measured by the Enhanced Fujita Scale (EF-Scale) used by the National Weather Service as follows:
    a. EF-Scale Rating 0: 65–85 mph
    b. EF-Scale Rating 1: 86–110 mph
    c. EF-Scale Rating 2: 111–135 mph
    d. EF-Scale Rating 3: 136–165 mph
    e. EF-Scale Rating 4: 166–200 mph
    f. EF-Scale Rating 5: over 200 mph (See endnote 23)

A waterspout is a weak tornado that forms over water usually over the Gulf Coast. Usually it dissipates over land but may cause damage and injuries.

Best Practices for Wind-Related Storms (See "Floods" section above)
- Develop a comprehensive plan for removal of people and pets to appropriate secure and safe areas in occupational settings, school settings, and home settings, and use practice drills to determine the efficacy of what is being done. Each setting should have emergency phone numbers, cell phones, smoke alarms, carbon monoxide alarms, fire extinguishers, water, emergency first aid kits, food, flashlights, and cleaning supplies for people and facilities.
- Develop and frequently test a disaster plan for external community disasters where people have to go to hospitals and other healthcare settings and a disaster plan for serious damage to the hospital and other healthcare settings where people have to be moved to other facilities.
- Develop a plan for each school which ensures that everyone takes cover in a safe area as determined by registered engineers or safety specialists within 60 seconds in the event of a tornado or other severe weather system. Delay lunches or assemblies in large rooms because the roofs are very vulnerable. Keep children at the school even beyond normal school hours until the emergency has passed.
- Communities should develop a comprehensive search, rescue, cleanup, and people assistance plan with alternative approaches in the event that parts of the plan are inoperable.
- Develop an up-to-date list of individuals who are very young, disabled, or elderly and need assistance in the event of an emergency and have close-by friends, neighbors, or family prepared to help the needy individual.

- Communities should test and utilize a severe weather notification system by means of all types of communication systems including sirens, special weather radios, announcements from the National Weather Service, the internet, and police and firemen going door-to-door depending on the type of weather emergency, etc.
- Where feasible in areas where there are regular periodic severe weather systems, utilize if possible an underground shelter, safe room or area of the house where there are no windows and the structure is built in a substantial manner.
- Immediately leave mobile homes when instructed to in the event of a tornado, hurricane, or flood warning and go to the nearest substantial shelter.
- If caught outside during severe weather, make sure that your seatbelt is buckled and pull-over into a safe area until the storm passes.
- Evacuate all areas when ordered to do so by the appropriate authorities prior to the onset of a hurricane.
- Do not go outside unless it is a life-and-death situation prior to, during, and in the immediate aftermath of a hurricane.
- Provide food, medicine, etc. for all pets and secure them.
- Evaluate before storms the potential for windstorm damage to the home including windows, doors, garage doors, roofs, etc., utilizing the Homeowner Windstorm Damage Mitigation Checklist. (See endnote 24.)
- In severe winter weather, dress appropriately with layered clothing and protection for the head and extremities and remain in the low temperatures for very short periods of time.

SUB-PROBLEMS INCLUDING LEADING TO IMPAIRMENT AND BEST PRACTICES FOR TERRORISM

(See endnote 71)

The main objective of terrorism is to influence the behavior of the public and frighten them by inflicting serious injury and death in a spectacular manner. Therefore, the terrorists will utilize whenever they can, weapons of mass destruction, whether they be biological, chemical, radiological, or large quantities of explosives.

In many areas of the world, typical warfare with an army against an army has been replaced by ideological extremists who use acts of terrorism which may affect millions of people in order to achieve their political goals. They typically utilize unconventional weapons to kill or injure civilians in particular to gain attention, status, funds, and new recruits to their cause, and instill fear in the population. Terrorism may be state-supported and therefore may include a vast variety of weapons and funding sources. Within the terrorist organization there are: leaders who provide a direction, policy goals, and objectives; active members or zealots who carry out the actual work of the group including homicide (suicide) bombings; active supporters who are involved in fundraising and the political and information part of the program; and individuals who are sympathetic to the goals and intentions of the terrorists but don't necessarily take action. (This group might be like a Trojan horse ready to act when the individual decides to become active instead of passive). The acts of terrorism are typically well-planned and by individuals who may be well-educated and have been raised in a middle-class or upper-class family before they assumed their current ideology. The United States and other democracies are seen as natural enemies of terrorism and therefore are potentially excellent targets for these groups.

There are many potential forms of terrorism including the use of: biological agents; chemical agents; explosives and nuclear bombs; radiological agents; arson; agricultural terrorism; intentional release of hazardous materials; and destruction of infrastructure such as the electrical grid, the components of the internet, etc. For the purposes of this book, only the Best Practices dealing with the resolution of problems created by bioterrorism, chemical terrorism, radiological materials, massive use of explosives, food terrorism, and water terrorism will be discussed. The entire range of terrorism activities and responses including prevention, mitigation, and control by all levels of government and others is well beyond the limitations of this book.

BIOTERRORISM

Bioterrorism is the intentional release of biological agents to injure or kill civilians, in particular to achieve political goals and objectives. Biological warfare has been practiced in one form or another for at least 2000 years. The Romans, the Tartars, and others used the blankets of the dead and even the dead bodies to help spread a given disease among their enemies. Native Americans, when the first Europeans came to the New World, as they had no experience of, or resistance to, smallpox, measles, plague, typhoid fever, and influenza, died in large numbers. The first work on the creation of weapons using biological agents was done by the Japanese army during World War II, when prisoners of war were used in biological agent research.

Today there are various groups of people as well as nations who may want to create weapons that can spread various diseases rapidly to an unsuspecting population. Some have attempted to alter biological agents in order to increase the level of disease that will occur and/or find a rapid means of dispersal of the microorganisms. Also, there are numerous nations that either have or have had biological weapons, which may be spread through air, water, or food. Terrorists are constantly looking for a way to possess and utilize these agents.

There are three primary categories of priority pathogens that can cause disease and death in people, either through traditional spreading of disease or by creating weapons with dangerous organisms. They are:

- Category A Priority Pathogens, including those that cause botulism, plague, smallpox, tularemia, viral hemorrhagic fevers, etc., create the highest risk to national security and public health by being easily disseminated and transmitted from person to person, resulting in panic and high levels of death.
- Category B Priority Pathogens, including those that cause Q fever, brucellosis, typhus fever, food and waterborne diseases, encephalitis, etc., create the second highest risk to national security and public health by being moderately easy to disseminate, resulting in moderate levels of disease and low levels of death.
- Category C Priority Pathogens, including those that cause tick-borne hemorrhagic fevers, yellow fever, rabies, etc., create the third highest priority, but are of great concern because they are readily available, easily produced, and can become emerging pathogens by re-engineering the organism for mass dissemination. (See endnote 73.)

The Centers for Disease Control and Prevention, Emergency Preparedness, and Response has developed a series of fact sheets for a variety of microorganisms that may possibly be used as bioweapons. The fact sheets discuss the microorganism, people at risk, the organism used in bioterrorism, and various resources. (See endnote 74.)

Best Practices for Bioterrorism
- Determine the level of preparedness and the competencies of personnel to resolve problems related to bioterrorism.
- Determine the level of competencies of laboratories to recognize the most common probable microorganisms which would be used by bioterrorist and the origin of the infection.
- Have the Centers for Disease Control and Prevention working with other professional groups and associations assist local areas to enhance their ability to rapidly detect, diagnose, and manage outbreaks of disease.

- Strengthen local public health response capacities to control and contain outbreaks of disease.
- Develop and utilize an information technology infrastructure to rapidly transfer information to all significant sources at the federal, state, and local level to immediately respond to an outbreak of an unusual disease.
- Provide appropriate vaccines at the national level which can be immediately delivered to a local area for distribution and usage.
- Develop an appropriate response system at the national level to send to local areas within a 12-hour period all necessary medicines and supplies to counteract a bioterrorist event.
- Train a large number of people under the direction of an epidemiologist to respond rapidly to gather essential data concerning a bioterrorist event.
- Prepare hospitals and clinics to assume the responsibility for using appropriate infection control measures to keep a highly contagious disease from spreading. (See endnote 82.)

CHEMICAL TERRORISM

Chemical warfare is as old as at least the 5th century BC when the Spartans hurled burning bombs of sulfur and pitch at their enemies. In World War I, chlorine and mustard gas were used to decimate armies. In 1995, a Japanese cult used sarin nerve gas to harm civilians. Chemical warfare agents have been made by many countries and are a constant threat for use by terrorists. In 2015, it was reported that various terror groups in Syria as well as the Syrian government had used chemical warfare against their enemies as well as the civilian population.

Chemical terrorism is a deliberate release of chemicals which may poison people, animals, plants, or the environment, and cause fear and panic in the general population. Children may be affected more readily and rapidly than adults. The chemical agents may be delivered in sprays or bombs and can be in the form of vapors, aerosols, liquids, and solids.

Unexplained animal, fish, bird, and insect deaths in a given area may be an indicator of either a chemical attack or an accidental chemical release. Also, be aware of symptoms of serious health problems such as blisters, rashes, burning skin, sudden running nose and drooling, sudden headache, irritation of the eye, nose, and throat, sudden blindness, nausea, disorientation, breathing difficulties, and convulsions. Determine if there are unusual odors, liquid droplets, or a film of oily droplets in a given area where trees and vegetation may also have an unusual appearance.

The Centers for Disease Control and Prevention has identified several categories of chemical agents that if delivered in lethal concentrations can either disable or kill someone within a few seconds, hours, or couple of days. These categories include:

- Anticoagulants which may cause uncontrollable bleeding
- Biotoxins which come from plants or animals
- Blister agents which causes blisters on the eyes, skin or throat, and lungs
- Blood agents which are absorbed into the blood
- Caustics (acids) which burn on contact
- Choking agents which affect the pulmonary system
- Incapacitating agents which alter consciousness and thinking
- Metallic poisons
- Nerve agents which interfere with the workings of the nervous system
- Organic solvents which damage living tissues by dissolving fats and oils
- Tear gas and riot control agents
- Toxic alcohols
- Vomiting agents (See endnote 75)

Each of these categories list specific chemicals and each of the chemicals have fact sheets available. As an example, sulfur mustard or mustard gas is a blister agent which was used as a chemical warfare agent in World War I. The fact sheets provide information about the compound, how it is used, what happens to exposed people, how the compound affects individuals, signs and symptoms of those exposed, long-term health effects, means of protection against exposure, and how the individual is treated medically. (See endnote 76.)

Best Practices for Chemical Terrorism (See Best Practices for Bioterrorism)
- Immediately remove individuals as rapidly as possible from areas where there may have been a chemical discharge.
- Go indoors and close all windows and doors and other openings to the building.
- Shut out any air that may be entering the building through ventilation systems, air conditioning systems, heating systems, etc. Use only air which has been re-circulated within the structure.
- Have individuals wash their eyes and faces with quantities of uncontaminated water without rubbing.
- Remove all contaminated clothing as quickly as possible, decontaminate the individuals, and dispose of the clothing in an appropriate manner.
- Determine what chemicals have been released into the air and the quantity that may be affecting the individuals.
- Use appropriate countermeasures including antidotes for chemical exposure.
- Have all first responders use appropriate personal protective equipment including respirators which can be utilized for the given chemicals in the air.
- Seek trained medical care as rapidly as possible while avoiding becoming further contaminated by the chemicals.
- All medical personnel need to be using appropriate personal protective equipment when dealing with affected individuals.
- Do not touch contaminated surfaces without appropriate personal protective equipment.

RADIOLOGICAL TERRORISM

Radiological material including that which is used for nuclear weapons is available in several places in the world. There is an ever present concern about the material being used by terrorists. Terrorists might consider attacking a nuclear power plant or the spent fuel pools which contain a large amount of radioactive material. They may try to purchase existing nuclear weapons or create an improvised nuclear device. A dirty bomb (radiological dispersal device) is a combination of conventional explosives with radioactive material. Whereas the explosion of a nuclear bomb or nuclear weapon is far more powerful than that of a dirty bomb and could spread over large areas of land and effect innumerable people, the dirty bomb depending on the amount of explosives, type of radioactive material, how it is dispersed and local weather conditions could contaminate individuals and equipment but would be most effective in terrifying the community. Radioactive material can be found in hospitals, research facilities, industrial facilities, and construction sites. There is a risk of cancer to individuals who are contaminated by the radioactive sources. The level of potential disease for individuals depends on the amount and type of radiation involved in the exposure. (See endnote 77.)

Best Practices for Radiological Terrorism
- Provide special training to emergency medical services staff and other medical staff in the recognition of symptoms of radiological contamination, methods of decontamination, and methods of treatment.
- Provide appropriate notification to the public as quickly as possible about any radiological emergency and how to respond to it based on the distance to the potential problem.

- Provide information on the safety of food, water, shelter, indoor and outdoor air, etc.
- Perform environmental monitoring and personal monitoring within the exposed area and determine radiation exposure based on time of exposure, distance from the source, shielding from the source, and the results of environmental and personal testing.
- Stay within the existing facility if so ordered by governmental agencies and close all openings to the outside.
- Remove your outer layer of clothing which may have been contaminated with radioactive material and place in a plastic bag out of the way.
- Decontaminate your body by washing all areas of the exposed skin with soap and lukewarm water to remove contamination. Do not scrub or shave which may cause additional injury.
- Seek medical attention as quickly as possible if exposed. (See endnote 84.)

EXPLOSIVES

The most frequently used weapons of mass destruction are some form of explosive, whether it be an airplane full of fuel, a truck or car wired with explosives, booby-trapped old shells, and explosives buried alongside a road, or a person wearing an explosives vest. The improvised explosive device (IED) has caused large numbers of deaths in Iraq and Afghanistan of military personnel and also civilians. IEDs can be made from existing explosive devices, artillery shells, mortar bombs, or various commercially available materials such as nitrate-based agricultural fertilizers or hydrogen peroxide. Much suitable material along with detonators, detonating cords, and plastic explosives can be found on the black market where it is circulating freely having been stolen from unsecured stockpiles. The threat of the use of IEDs is extremely high and will continue well into the future. (See endnote 78.)

Blast terrorism and explosions cause in people: primarily tissue damage especially in the lungs and interfaces between tissue and gas, from the blast wave; secondly, penetrating or blunt injury from shrapnel and debris; and thirdly, injury from the acceleration of the blast wind and then rapid deceleration.

Best Practices for Explosives Terrorism

Note: Best Practices are basically related to avoidance of these explosive devices and then necessary treatment of those who have been injured. Treat all threats as potential acts of terrorism, vacate areas immediately, and contact the proper authorities. First responders must be aware of secondary explosives that have been placed in areas which can harm them as they respond to the terrorist act.

FOOD TERRORISM

(See endnote 79)

The World Health Organization recognizes a very serious concern about the possibility of biological, chemical or physical agents, or radioactive materials being deliberately used to contaminate food to injure or kill civilian populations. This act of terrorism would put an overwhelming burden on existing public health systems, and cause disruption to the social, economic, and political stability of the areas affected. It would cause food prices to rise, reduce availability to necessary foodstuffs, and potentially threaten food security leading to unrest and hunger in the population.

Whereas previously much of the food consumed by a community was grown in the immediate vicinity or at least within the country for specialized types of food, today countries import vast amounts of food from other areas along with potential contamination from those areas as well as potential contaminants added by terrorists. The easy accessibility to various foods and the previous experience of groups who have wanted to contaminate the food supply have been recorded. As an example, in 1984, members of a religious cult contaminated salad bars in the United States causing 751 cases of salmonellosis. This was a trial run for future contamination of the food supply. Other outbreaks of foodborne disease have occurred and although not caused by terrorists, could easily have been. In 1985, 170,000 people were contaminated with *Salmonella* from a US pasteurized milk plant. There was an outbreak of hepatitis associated with clams in China where almost 300,000 people were affected. In 1996, about 8000 children in Japan were infected by radish sprouts contaminated with *Escherichia coli*. In Germany, *E. coli* caused sickness in over 4300 people and killed 50 in 2011. Besides potential health effects causing illness or death, there are many other consequences of food being contaminated by terrorists. They may be economic consequences where the disruption will cause huge financial problems as happened in Israel when citrus fruit exported to European countries was purposely contaminated with mercury leading to trade disruption, and in Chile where grapes were contaminated purposely with cyanide, etc.; an impact on public health services such when nerve gas released in the Tokyo subway system caused the death of 12 people and harmed 5000 others who needed to be treated, thus completely inundating emergency services; and the imposition of fear and anxiety on the population, leading to a sense of paralysis within the community.

Although microorganisms have and will continue to cause serious outbreaks of foodborne disease, chemicals may be easier to obtain and can cause equal or even greater problems. Contamination with radio nucleotides is also a serious consideration. In any case, time of exposure and concentration of the contaminant at the point of insertion of the substance into the food supply is a great concern and, in the case of microorganisms, also the temperature of the food.

Best Practices for Food Terrorism

(See Chapter 7, "Food Security and Protection" for an in-depth discussion on the principles and practices of food safety throughout the food chain and the Food Safety Modernization Act. The first step in prevention is to enhance and properly utilize existing food safety programs at the governmental and industry levels, thereby enhancing all food safety efforts through improved and recognized programs in food monitoring, surveillance, inspection, foodborne disease detection, education, and training for professionals and others. This will then lead to not only improved food safety for the general population but also help in the event of a terrorist attack.)

WATER TERRORISM

(See endnote 80 and Chapter 14, "Water Systems [Drinking Water Quality]")

Water is essential to the operation of society and the health and welfare of the population. Water may be used as a political or military target to cause physical or psychological damage and therefore help the terrorists achieve their goals of the disruption of society to impose their will upon others. The water infrastructure can be damaged or destroyed and therefore produce too much water at one place such as when a dam is blown up, or too little or no water in another place when the facilities processing the water are disrupted. Water may be contaminated with large quantities of microorganisms, chemicals, radioactive material, or with silt or other physical material. Reservoirs, pipes, and treatment plants are highly vulnerable to attack. In fact, in Philadelphia during World War II, the perimeters of the reservoirs were off-limits to all individuals and were patrolled by soldiers from the United States Army to ward off any such terrorist attacks. Although a given contaminant might not cause widespread illness or poisoning, it would probably cause massive problems by instilling considerable hysteria in the population. As an example, in 1993, an outbreak of cryptosporidium in Milwaukee killed over 100 people and affected the health of over 400,000 others. The degree of harm that can be caused by contaminated water depends on a variety of factors such as quantity, concentration, and treatment for contaminants through the normal water treatment process, which might work on microorganisms but not on chemicals. There is inadequate information available to determine the level of threat for water resources, but the potential exists. However, it is well known from outbreaks of waterborne disease, that if pressure differentials can be overcome, then wastewater can be made to flow into drinking water instead of the other way around. This has happened in the past quite frequently and continues to do so today. Once the wastewater has contaminated the drinking water supply, it can be distributed rapidly to large populations that utilize the water source and potentially cause considerable illness and possibly death. A few gallons of highly toxic chemicals if distributed properly within the water system could cause large amounts of illness.

The existing water quality infrastructure in many cases dates back 75 or 100 years or more and is in very poor condition. Not only is there a problem of the collapse of the system but also contamination from sewage and waste-bearing areas entering drinking water quality pipelines.

Best Practices for Water Terrorism
- Teach the public the ramifications of the effects of water terrorism.
- Encourage the public to immediately contact the proper authorities if they observe anyone tampering with the water supply by: dumping or discharging material into the water; climbing or cutting utility fences; parking unidentified vehicles near water sources; tampering with manhole covers or equipment within buildings; and unidentified vehicles hooked up to water hydrants, etc.
- Report any unusual tastes, odors, or color in the drinking water supply.
- Determine if any industrial plants are contaminating either the raw water supply or the drinking water supply from wastes or other problems within the industry.
- Protect all water supply sources from outside intrusion by people or animals by using proper fences, lighting and surveillance systems, and motion detectors, and securing all chemicals on-site.
- Evaluate the infrastructure of the water quality system including the raw water supply, the water treatment plants, the piping system, finished water storage areas and all dams, and make necessary corrections where necessary.
- Advise the public immediately of any terrorist acts and what potentially has been used to contaminate the water. Emphasize that boiling the water cannot remove any chemicals that are present.
- Several times a day test and evaluate what is present within the water and take immediate action if it is hazardous.

DISASTER SITE MANAGEMENT, COMMAND-AND-CONTROL PROBLEMS FOR EMERGENCY, AND OTHER PERSONNEL, SUPPLIES, EQUIPMENT, AND ENFORCEMENT

(See endnotes 36, 40, 41)

When a disaster occurs, there is an almost instantaneous response from the media and therefore information, accurate as well as inaccurate, is transmitted extremely rapidly throughout the community, the country, and the world. Large numbers of people, substantial quantities of equipment, and huge amounts of supplies start to flow initially and over the early days into the area. A vast number of governmental, voluntary, and private organizations and other people may respond in order to get help to the seriously injured and protect property. As an example, during the aftermath of Hurricane Andrew there were so many semitrucks with supplies and clothing going to South Florida that the roads were clogged and there was no place to put it all. Also there was difficulty in distributing that which had arrived. Further, how do you house, feed, and provide medical care for the governmental workers from the local, state, and federal programs as well as the private sector and other volunteers.

Each group arriving to help in the aftermath of the disaster may be bringing its own personal protective equipment, which may not be effective in the particular disaster situation. The standards for equipment as well as the performance of tasks may vary substantially or may be lacking entirely. Alarms and devices used by the guest first responders and construction crews needed to track or identify the location of individuals may malfunction or be so bulky that they interfere with the actions of the first responders and therefore are turned off or discarded in the hurry to save lives

This chaotic response may interfere with a well-thought-out and regulated team effort controlled by a unified command under a single controlling authority which can make decisions and issue orders to the experts in the unified command and then have them transmit the needs and orders to actual working groups trying to resolve problems caused by the disaster. However, typically even this command structure is lacking and in fact this was one of the most serious problems encountered by first responders when the World Trade Center was destroyed. Part of the problem was the loss of key leaders during the building collapse and the lack of a plan to respond in the event certain leaders were disabled or killed during the disaster. In addition, there was a lack of awareness of the extent, nature, and dangers of the incident, a serious breakdown in communications between various entities, a lack of clarity regarding the most significant risk factors and tasks to be performed, and a lack of training for a disaster of this size and scope.

On-scene incident commanders complain that orders and responses may be unclear, repeated numerous times and conflicting leading to uncooperative workers who are at times totally dysfunctional. People in fact ask who is in charge and why was this order issued. Since there are various agencies responding to the disaster at the local, state, and federal level, time after time in emergency and disaster situations each agency follows its own plan, causing confusion and duplication of effort and conflicts between the agencies, which often diminishes the effect of work being done by the field professionals.

The destruction of physical facilities including communication systems, without contingency back-up provisions, is a major challenge during large disasters. Emergency responders and citizens have difficulty coordinating because of the lack of a reliable communication system and everyone tends to go his/her own way.

Evacuation planning typically does not account for large numbers of people moving out of a given area into a safer area. Roads get clogged and in fact become parking lots, while the impending severe conditions continued to increase in scope. Provisions may have not been made for the poor who may not have access to transportation, the elderly, the handicapped, and the very young. Where will you put all these people and who will provide the necessary emergency housing, food, water, bathroom facilities, and safety of person and personal effects? Since major disasters involve multiple jurisdictions, typically these governmental bodies have not considered the broad picture of large area evacuation of people and what to do with them.

There seems to be a problem with poor public relations, since the governmental agencies depend primarily on the news media to get the messages out and the information released may be incomplete because of lack of time allowed by the media. Individuals may have difficulty in understanding directions and not observe the instructions from governmental units. There may be a substantial problem with understanding the language used, especially in areas where there are large immigrant populations.

For a large-scale, long-duration situation, there is an inordinate demand on resources including personnel, equipment, supplies, housing, etc. Who is going to authorize and pay for all of these resources on an emergency basis thereby avoiding the usual bidding process? How quickly will everything arrive because the need is for immediate delivery? How will the resources be managed and the materiel be stored and accounted for? Who will determine if the materiel is usable for the particular incident? Who will control the use of donated resources? How will unusable materiel be transported away from the scene and properly disposed of? Who will supervise various groups of volunteers and how will they be coordinated with the professionals working at the site? Who will verify the credentials of the volunteer specialists arriving at the scene?

When it is not apparent that a disaster has occurred, who determines that a routine fire is actually a far greater problem and a different level of response and command-and-control is necessary in the ongoing situation because of the potential level of unknown risk involved. Risk evaluation procedures, the training of risk evaluation specialists, and risk reduction behaviors may be inconsistent, lacking, or totally absent because of the variety of first responders coming from different parts of the country. Poor communications between various groups causes innumerable misunderstandings and interferes with the progress of the work needed at the disaster site. Further, the actual communication systems of the various groups working on the disaster may be incompatible and therefore, as an example, the police may not be able to speak to the fire officials or rescue officials, and so valuable time may be lost in resolving problems and protecting life. Further, who has the authority to decide if the site of the disaster now includes biohazards, chemical hazards, or severe physical hazards and determines the types of personal protective equipment including appropriate respiratory devices needed by the first responders to protect themselves while appropriately completing their work.

At what point do you transition from a rescue effort to a recovery effort, and therefore utilize different techniques and different work assignments and allowable time periods for a given task. A delayed, improper, or untimely transition may create unnecessary risks taken by the workers and may cause immediate or long-term health hazards to them.

How and when do you establish external perimeter or scene control of the area and thereby remove unauthorized people while isolating the area of the disaster? Who identifies and enforces the distribution of people within the various levels of risk within the disaster site?

Ultimately, there is a necessity for learning from actual situations that have occurred during various disasters that tend to repeat themselves. Unfortunately, there is no universally accepted means of developing reports concerning what happened and therefore it is difficult to compare one agency to another agency and one approach to another approach when they are dealing with the same problem. Various groups tend to write their own reports without consulting the others. Candor may be lacking because of the stakes or because inaccuracies have occurred but no one wants to own up to them. The level of detail may be so insufficient as to make the information unusable. Typically reports focus on what has gone wrong rather than what was right. Information on both is important, but something which goes right may become a Best Practice and be utilized by others in the same type of situation. When determinations are made as to a Best Practice, it is then necessary to continuously test the situation and upgrade the knowledge and practices as needed.

Best Practices to Resolve Problems for Disaster Site Management Command-And-Control for Emergency and Other Personnel, Supplies, Equipment, and Enforcement

Note: This is a very complex topic well beyond the scope of this book. The Best Practices listed will be but a small number of the actual recommendations to correct the problems found in management and command-and-control. (See endnote 41 for in-depth information.)

- Determine what are the most significant types of disasters, through the use of risk analysis techniques, that typically affect a given area.
- Develop a community-based disaster plan for different levels of potential disasters by involving all organizations, governmental bodies, business and industry, and other interested parties in the planning process.
- Integrate all plans from all levels of government, non-governmental organizations, business and industry, and other sources while recognizing that the initial emergency response is usually from the local group and then depending on the size of the incident may spread outward thereby requiring a new command-and-control group under a very competent individual with total operational authority to determine and direct the use of appropriate resources.
- Determine the level of resources needed and develop appropriate plans for the use of people, equipment, and supplies, and frequently carry out drills to determine if the procedures and plans are being carried out appropriately with successful measurable results of disaster prevention, mitigation, and control.

- Use the lessons of past disasters gleaned from briefings and reports to improve the response of the various organizations and coordination of their efforts to protect life and property.
- Develop a uniform method to train people in the Best Practices to use in various aspects of the prevention, mitigation, and control of areas involved in disaster situations.
- Enforce the use of recognized personal protective equipment standards and the use of safe techniques in task performance, while practicing appropriate decontamination procedures.

OCCUPATIONAL HEALTH AND SAFETY PRACTICES FOR PERSONNEL INVOLVED IN EMERGENCY/DISASTER CLEANUPS AND REPAIRS

Note: This is a very complex topic well beyond the scope of this book. The Best Practices listed will be about dealing with debris and damaged buildings. There will not be any recommendations to correct the problems related to such cleanups as massive oil spills, hazardous chemicals, severe biological hazards, radiological hazards, etc. The reader should contact the state Occupational Safety and Health Administration (OSHA) for guidance on how to correct these situations. There is a considerable body of information available from the National Institute of Occupational Safety and Health and OSHA.

There are significant potential health, environmental health, and safety challenges to personnel as well as the general public in the cleanup and repair of homes and businesses. There may be immediate hazards such as fires, leaking gas lines, downed electrical wires, carbon monoxide, other respiratory hazards, extreme heat or cold and high humidity, sharp objects, stored hazardous chemicals, microorganisms from sewage and contaminated objects, unstable structures, a vast variety of potential air pollutants, etc. The structures may be infested with insects, rodents, or snakes. Workers may suffer from being in confined spaces which can have a buildup of dangerous gases or be oxygen deficient, extreme fatigue, musculoskeletal injuries from lifting heavy materials, and potential chronic diseases from asbestos, lead, and other hazardous substances which may cause cancer. (See endnotes 43, 44.)

Best Practices for the Health and Safety of Personnel Involved in Emergency/Disaster Cleanups and Repairs
- Utilize experts in structural safety to determine the structural damage which has been done to various buildings and whether it is safe to enter to make further determinations concerning the structure and contents.
- After damaged buildings have been cleared for entry, enter them with great care, and determine potentially serious conditions visually and by use of testing equipment to identify all hazards including the levels of risk. Determine appropriate controls and the types of workers needed to carry out the assignments. Always use the buddy system.
- Establish a comprehensive plan to carry out in a safe manner the necessary removal of damaged materials or portions of the structure, cleanup, and necessary repairs. Take into account any potential additional damage and new risk created by this work.
- Determine if all workers working on the site are qualified to carry out their duties, using appropriate personal protective equipment, and being properly supervised by very knowledgeable and experienced people.
- When working at any heights, use appropriate handrails and security belts.
- Determine if there are possible combustible or explosive gases present and leave the building immediately if they are suspected.
- Open all windows and doors in the building to air it out and do not smoke, light matches, operate electrical switches, cell phones or other phones, or do anything that could create a spark.

- Do not use any fuel-burning devices within the property to avoid the potential of carbon monoxide poisoning.
- Use considerable caution in the vicinity of downed electrical power lines and only allow highly trained personnel to deal with the problem.
- Have knowledge of all overhead or underground electrical power lines to avoid striking them or disrupting them during cleanup.
- Avoid injuries by using power lifting of heavy objects instead of hand lifting.
- Allow for frequent breaks and ample fluids for workers to avoid dehydration, reduce potential for fatigue, and other physical symptoms of stress.
- Make note of the problem areas and use defensive tactics to avoid injury to the personnel.
- Keep all children and pets away from damaged buildings.
- Determine if there are leaking or spilled chemicals and use appropriate clean-up techniques and personal protective equipment.
- Dispose of chemicals in an appropriate manner at a hazardous waste site.
- Do not burn any household chemicals or dump them into the drains, storm sewers, or toilets.
- Use appropriate personal protective equipment when dealing with insulation, fireproofing material, floor tiles, roofing material, and other substances which may contain asbestos as well as other types of fibers.
- Wear appropriate personal protective equipment while removing standing water as quickly as possible to reduce the potential for microbiological problems.
- Wear appropriate personal protective equipment while removing all building contents composed of paper, cloth, wood, and other absorbent or porous material that can contribute to the growth of mold.
- Wear appropriate personal protective equipment when cleaning and sanitizing the interior or exterior of contaminated equipment and structures.
- Wash all surfaces and equipment inside or outside the structure with detergents and warm water, rinse thoroughly, and then disinfect with household bleach using appropriate quantities to destroy potential for growth of mold and disease-producing organisms. Allow the surfaces to air dry.
- Utilize appropriate respiratory equipment when exposed to water vapors formed from contaminated water within the structure.
- Never mix household cleaners or disinfectants because of the possibility of producing toxic fumes.
- Determine if all electric fans used to help dry out a structure are properly grounded and connected to a grounded electrical source.
- Use appropriate respiratory equipment where there is a potential for airborne asbestos dust and lead dust.
- Use only specially trained and licensed individuals who are properly equipped with personal protective equipment to remove asbestos or lead-based paint and who will dispose of them in an appropriate and safe manner.
- Frequently wash hands to avoid contaminating areas which are clean and functioning.
- Temporarily move all the people to safe sanitary facilities until renovation can be accomplished.
- Do not permit any open burning of materials from the structures.
- When demolishing the structure, follow the appropriate federal regulations and/or Best Practices identified for asbestos demolition, lead-based paints, removal and disposal of electrical equipment containing PCBs, other hazardous chemicals, and storage tanks. (See topic areas above.)
- Always use wet demolition procedures to reduce airborne dust.
- Workers should always wash their hands and arms frequently especially if they are using toilet facilities and before eating or drinking fluids.

- Keep, evaluate, and use when necessary appropriate fire extinguishers on-site.
- Preferably use bottled water from authorized sources for drinking water purposes.

OCCUPATIONAL HEALTH AND SAFETY PRACTICES FOR FIRST RESPONDERS, CONSTRUCTION WORKERS, AND VOLUNTEERS

(See endnotes 36, 37, 38, 39)

September 11, 2001, was the beginning of a new era in the dangerous lives of first responders. A fire or a building collapse may also be a terrorist event and there may be large numbers of unknown hazardous factors that will put the first responders' lives at great risk. Of the approximately 3000 people who died that day when the World Trade Center was destroyed, 450 were first responders. Construction workers and other tradespeople were put at inordinate risk by the many physical and chemical agents present in the air and in the debris.

Multistory building structural collapse was not understood well. It created a large number of physical, chemical, and biological hazards. The physical hazards included potential electrical shocks from electrical equipment; extreme noise; danger of injury from vehicles and heavy equipment; and potential for injury from sharp objects, falling objects, uneven or unstable working surfaces, floors collapsing, etc. The chemical hazards included: a variety of chemicals stored within the building; byproducts of the fires; and the disintegration of building materials especially concrete, and a large number of building contents. In addition, there were numerous potential chemical hazards from the fuel aboard the jet aircraft that crashed into the buildings. The biological hazards included blood-borne pathogens such as HIV, hepatitis B virus, and hepatitis C virus if individuals came in contact with body fluids; a variety of other infectious diseases which might occur through contact or if airborne; and waterborne organisms from sewage and other contaminated water sources.

Professional first responders included firefighters, emergency medical technicians (EMTs), paramedics, police officers, nurses, and doctors. When a disaster occurs, whether it be weather related, a natural event such as an earthquake, or caused purposely by people such as acts of terrorism, large numbers of people, structures, utilities, etc. may be at risk and this situation would be beyond the capacity of the local professional first responder.

Also, especially in very large disasters, a vast variety of people of all kinds of professions become volunteers and may rapidly enter the disaster area to help save lives, property, and protect people from the consequences of the event. The professional first responders receive training in dealing with the day-to-day emergencies of the various communities that they serve, whereas the volunteers may have limited or no training. The professionals have equipment and communication systems which may work extremely well for emergencies, but in the event of disasters they may be inefficient or lacking. The volunteers' equipment and communication systems may be totally inadequate for the response to the disaster.

Search and rescue operations at the destruction of the World Trade Center were hampered by unstable structures, intense fires, unknown hazards, and the slow implementation of command-and-control operations. Because the urgent part lasted for many days and the recovery part lasted for many months, there was a huge amount of stress placed on the rescuers and their equipment. The personal protective equipment was not designed for continuous use and after a while hampered the work of the first responders. The garments became wet and the footwear caused blisters on the feet. The first responders not only faced the usual flames, heat, and byproducts of combustion and smoke, but also air which was heavily contaminated with fine particles of concrete, metals, a variety of chemicals, disintegrated human remains, etc. The firefighters had to perform unaccustomed tasks including the breaking up of concrete, which was very demanding physically. Mentally they were put at extraordinary risk because of the scope of the disaster and the finding and removal of body parts, decayed corpses, the loss of their friends and colleagues, and the sheer enormity of what had happened. Their training had not prepared them for such an event. For many there was a potential for posttraumatic stress disorder. Many of the first responders were without previous experience in this type of disaster and had to improvise as they went along because their supervisors also lacked this type of experience. This increased potential danger to the individuals as they tried to help others.

Search and rescue dogs were put in unusually dangerous situations because of the overwhelming size of the disaster, numerous sharp objects and edges, and unstable environmental situations. Protective booties or other equipment could not be used because they hinder the dogs and can create more problems than they solve. Dogs also can suffer from posttraumatic stress disorder.

Construction workers were immediately put into hazardous situations without appropriate training and in many cases necessary personal protective equipment. They were subjected to the same conditions as the first responders, but without years of experience in emergency situations. They were exposed to untold hazards and severe emotional situations.

The personal protective equipment varied enormously in composition and effective use. Head protection and high visibility vests appeared to work quite well, whereas protective clothing and respirators had considerable problems. The bottles of oxygen ran out quickly and had to be replenished. The respirators reduced the field of vision and the faceplates frequently fogged up. There was a substantial problem with the types of filters that were available for the air-purifying respirators. The protective clothing did not typically protect the individual against biological agents and infectious diseases, the extreme heat of the fires, and the demanding work needed in the unstable rubble. They were neither light nor flexible. The eye protection while usually able to handle impact injuries, did not protect against the dust which was present everywhere. Some people abandoned their personal protective equipment because it hindered them in search and rescue activities. Because of the vast amount of different types of personal protective equipment, it was extremely confusing to determine what replacements were needed and how to obtain them. There was a severe lack of training of personnel in actual on-site health and safety activities to protect themselves.

A report issued by the Office of the Mayor New York City indicated that there were numerous injuries to rescue and recovery workers beyond those who died in the destruction of the World Trade Center. The Fire Department of New York first responders showed poor health effects shortly after the terrorist attack: 99% of the exposed firefighters experienced at least one respiratory symptom within 1 week. The pulmonary function of the exposed firefighters in the first year deteriorated 12 times as rapidly as would be expected for individuals their age. After 8 years, respiratory conditions still continued to be high among these individuals. Other elevated health conditions were noted. Symptoms of posttraumatic stress disorder increased notably in the fire department force, but less in the police responders and other rescue recovery workers. Construction workers involved in the cleanup had increased respiratory problems. Injuries at the site included broken bones or burns and increased risk of chronic disease years later. (See endnote 42.)

Best Practices for Occupational Health and Safety for First Responders, Construction Workers, and Volunteers
- Develop a command-and-control center to immediately go into operation in the event of a major disaster and test the system frequently to determine if there are shortcomings that need to be corrected.
- Develop for each community and region, a community-based disaster plan based on recent experience for different potential disasters and different degrees of devastation by involving all organizations, governmental bodies, business and industry, and other interested parties in the planning process.
- Utilize professionals who can quickly recognize health and safety hazards and report them immediately to the command-and-control center.
- Recognize the symptoms of chemical, biological, and/or radiological substances that can be used by terrorists and immediately report them to the command-and-control center.
- Constantly assess the content and concentration of hazardous substances in the air that can impact first responders and the type of air filters that they use, and advise them immediately about the findings.

- Ensure that there is a constant stream of equipment and materiel that is needed by personnel at the scene of the disaster.
- Determine if the personal protective equipment to be used at the scene of the disaster is appropriate for the situation.
- Emergency medical service personnel must be made aware of the potential for the spread of a variety of diseases from the individuals who are being rescued and therefore use appropriate infection control techniques and protective clothing in these situations.
- Develop new personal protective equipment including respirators that can be used in a disaster where there are multiple hazardous situations.
- Utilize where possible supplied-air respirators since they require less effort by the individual to use them.
- Position substantial quantities of personal protective equipment in a variety of sizes in areas which are subject to potential disasters and make the equipment available to various groups of professionals and volunteers as needed.
- Develop specialized training programs for all first responders and the types of volunteers who come to the scene of the disaster for the use of personal protective equipment.
- Provide hazard information on a routine basis to first responders and teach them how to deal with the hazards.
- Develop specialized emergency response training programs and require that all first responders take them and refresher programs on a periodic basis.
- Utilize barcode identification cards for all individuals working on the site of the disaster to be able to keep track of them.
- Recognize that all terrorists' acts are considered to be crime scenes and utilize the techniques that are required by the appropriate police authorities when gathering evidence.
- Develop back-up communication systems if the original systems in use are disabled or destroyed.
- Provide alternate sources of electricity when the normal sources are disabled or destroyed.
- Be aware of the potential for secondary explosive devices to disable or kill first responders.
- Map out the community and determine where there are hazardous or explosive chemicals or devices and alert first responders to the problems before they enter the disaster area.
- Use extreme caution in the event of downed power lines.
- Use extreme caution in working around piles of rubble and debris because of unstable surfaces, sharp objects such as rebars and glass, and collapsing surfaces.
- Develop a uniform set of tested scientifically valid guidelines for lightweight comfortable personal protective equipment for individuals involved in specific types of disaster responses that last for a considerable period of time.
- Develop a uniform set of tested scientifically valid guidelines for lightweight easily used personal protective equipment for biological hazards, chemical hazards, radiological hazards, and physical hazards (such as disintegration of concrete).
- Equip first responders with National Fire Protection Association approved personal protective equipment when dealing with blood-borne pathogens.
- Utilize the equipment above and take extra precautions such as wearing gloves that are resistant to viral pathogens under more durable gloves, goggles, or face shields when treating victims or working with human remains.
- Use water-resistant clothing and boots when working in areas that have contaminated water or dust contaminated with waterborne pathogens. Decontaminate the clothing and boots as soon as possible.
- Monitor frequently the air in the environment that the firefighters will be working in to determine the appropriate type of respirators which should be worn by the individuals to avoid serious health consequences.

- Do not allow anyone to enter collapsed building areas without proper respiratory, head, eye, and skin protective clothing and equipment.
- Use lightweight helmets and protective clothing which is not cumbersome to prevent exhaustion and injuries, where fires are not occurring and individuals are part of search and rescue teams.
- Use supplied-air breathing equipment in oxygen-deficient environments.
- Use self-contained breathing equipment when working around active fires and where there is a potential for carbon monoxide, organic chemicals, and hazardous byproducts of combustion.
- Use close supervision for all individuals involved in active fires where there is a potential for carbon monoxide, organic chemicals, and hazardous byproducts of combustion.
- Develop lightweight, comfortable air-purifying equipment for various major hazards and make sure they are not used beyond their capacity without changing the equipment and/ or air filters.
- Do not allow anyone to enter the disaster area without appropriate air-purifying equipment that can be used as needed.
- Do not use an air-purifying respirator in an oxygen-depleted environment.
- Recognize that the heat intensity may be so great that it will destroy the personal protective equipment.
- Have safety supervisors do ongoing risk analysis at each of the rescue scenes to determine if the workers need to be moved to a safer location and communicate this information immediately.
- Have first responders take appropriate breaks from various tasks to avoid severe heat-related conditions.
- Have supervisors recognize extreme stress in the workers and remove them from the site in order to avoid bad judgment and injuries.
- Be aware of the potential for building collapse and provide immediate information to all emergency personnel.
- Determine and carry out effective means of issuing more uniform personal protective equipment to all individuals working at the scene of a disaster and provide back-up supplies when additional equipment is needed.
- Develop a means of rapidly communicating information and eliciting responses from the individuals concerning hazards to people working at the site of a major disaster.
- Use hearing protection devices when heavy equipment is needed to remove parts of the structure and debris.
- Ensure that all first responders are trained in the use of their personal protective equipment before allowing them to work within the primary disaster area.
- Train first responders how to avoid excessive risk-taking behavior.
- Conduct drills of first responders to determine their ability to respond to various types of disasters and correct shortcomings.
- Establish medical screening programs to evaluate the personal fitness of individuals to perform extremely hazardous and stressful work prior to the time of their use in disasters. Evaluate physical fitness and mental stability and establish a baseline of various health parameters and health concerns including preexisting conditions and familial problems. Determine pregnancy status if it applies. Make sure that individuals take with them an appropriate amount of pharmaceuticals that they typically use. Collect personal identifying information and appropriate contacts in the event of an emergency. Determine the previous amount, time, and level of exposure to a variety of situations such as chemicals, biological agents, radiological agents, heat and cold stress, repetitive motion, air contaminants, etc., to decide if the individual can be stressed further by these various agents. (See endnote 46.)

- Require all individuals to have received tetanus vaccinations within the past 10 years and a hepatitis B vaccine series.
- Establish medical postexposure screening programs to evaluate workers' health after working at a disaster site which was physically demanding, very dirty, unstable, and where there was exposure to numerous contaminants. Determine if the various stresses caused injury or illness in the short term, either mental or physical, related to the work exposure. Follow-up at a later date to determine if chronic conditions have been established and long-term problems created. (See endnote 47.)

LAWS, RULES, AND REGULATIONS

(This section will be limited to major federal laws, rules, and regulations. To include local and state laws, rules, and regulations is well beyond the scope of this book. The laws, rules, and regulations found in the other chapters on specific areas such as water, air, food, land, etc. are also applicable to each of the situations during emergencies, disasters, and acts of terrorism. Therefore, go to the appropriate chapter in the area of interest.)

The Pandemic and All-Hazards Preparedness Reauthorization Act of March 2013 utilizes the existing work of the US Department of Health and Human Services to advance national health security. It authorizes funding for public health and medical preparedness programs by amending the Public Health Services Act to give state health departments flexibility in how to determine and utilize staff and other resources to meet community needs in the event of a disaster and authorizes special funding through 2018 to buy special equipment under the Project BioShield Act. It also enhances the authority of the US Food and Drug Administration (FDA) to support rapid responses in the event of public health emergencies. (See endnote 92.)

The Robert T. Stafford Disaster Relief and Emergency Assistance Act, as amended April 2013 is the most current law for use in major disasters and when specialized emergency assistance is needed. It is based on the congressional findings that declare that disasters often cause loss of life, human suffering and loss of property, disrupt normal governmental and community efforts and responsibilities, and impede the effort of providing aid, assistance, and emergency services as well as the reconstruction or rehabilitation of the devastated areas. The Act provides for disaster preparedness and mitigation assistance, disaster planning for preparedness, disaster warnings, and financial assistance. (See endnote 87.)

The Emergency Planning and Community Right-to-Know Act was passed by Congress in 1986 in response to the massive number of people who died and were injured in Bhopal, India, from the release of a hazardous chemical into the air. This disaster was followed by one similar to it in West Virginia about 6 months later. The law established requirements for federal, state, and local governments, Native American tribes, and industry to carry out emergency planning and provide information to the community about the potential hazards to people and the environment if there was an accidental release of a chemical. There are four major provisions of the law covering: emergency planning; notification of emergency releases; reporting requirements for hazardous chemical storage; and a toxic chemical release inventory. The emergency response plans have to: identify facilities with hazardous substances; describe transportation routes; describe emergency response procedures; appoint a community as well as facility coordinator to work on the plan; provide emergency notification procedures; determine the estimated affected area and population; describe the local emergency equipment and facilities needed; establish an evacuation plan; provide specialized training for emergency responders; and provide how emergency response plans will work and what schedules they will be following. It also provides information on the various chemicals and their use, the level of health hazard (immediate or delayed), their fire hazard, the effect of a sudden release of pressure, and their ability to react with other substances as well as the air and water. (See endnote 88.)

The Pets Evacuation and Transportation Standards Act of 2006, an amendment to the Robert T. Stafford Disaster Relief and Emergency Assistance Act of 2006, was a response to the loss of life especially of older people who refused to leave their pets behind during Hurricane Katrina. FEMA is responsible for developing emergency preparedness plans and making sure that they are integrated into the state and local emergency plans to provide for pets and other animals during disasters. (See endnote 89.)

PROPOSED RULE—MEDICARE AND MEDICAID PROGRAMS; EMERGENCY PREPAREDNESS REQUIREMENTS FOR MEDICARE- AND MEDICAID-PARTICIPATING PROVIDERS AND SUPPLIERS

The Centers for Medicare and Medicaid Services proposed on December 27, 2013, a new rule for Medicare and Medicaid Programs providers and suppliers for emergency preparedness including appropriate planning in a vast area of the medical, public health, and healthcare fields. This was done with the cooperation and input of its public, non-profit, collegiate, professional, and private partners. It meets the standards of all agencies. The new rule emphasizes risk assessment and planning; appropriate policies and procedures for emergencies; developing a back-up communications plan; and appropriate training and testing of people and facilities during the worst possible situation. It discusses: the role of presidential directives; the role of the Office of the Assistant Secretary for Preparedness and Response; Centers for Disease Control and Prevention state and local preparedness; hospital preparedness; the Office of the Inspector General; statutory and regulatory background, etc. (See endnote 91.)

RESOURCES

1. The National Drought Mitigation Center which is located at the University of Nebraska-Lincoln was established to help communities reduce their vulnerability to drought. The center houses the US Drought Monitor which provides a weekly map that shows drought conditions throughout the country. The US Drought Monitor is a partnership with the US Department of Agriculture, the National Oceanic and Atmospheric Administration, and about 350 experts around the United States. Drought relief for agriculture in part is based on information provided by the US Drought Monitor. The center works with state and tribal governments and national governments around the world to develop risk management strategies related to monitoring, early warning, and planning. (See endnote 2.)

2. The California Seismic Safety Commission in Sacramento, California, through their publication entitled *A Safer, More Resilient California: The State Plan for Earthquake Research* (CSSC Publication 2004-03, June 2004) (See endnote 95) discusses how best to mitigate the effects of earthquakes by encouraging earthquake research as part of its 5-year hazard reduction plan. It recommends and makes available information on: coordination of research activities in California; research priorities such as improving hazard assessments and seismic monitoring; cost-effective mitigation strategies; postearthquake investigations; social and economic vulnerabilities; new product development; turning research into practice by encouraging the adoption of relevant findings; and methods of cost-effective research.

3. Hazus-MH is a GIS-based regional loss estimation tool. It was developed by FEMA and the National Institute of Building Sciences. It provides loss estimates for earthquakes, hurricanes, and floods. Its disadvantage is that does not provide details of infrastructure at the local level. However, it identifies critical response facilities such as fire stations, police stations, hospitals, city emergency operations centers, and evacuation centers. It identifies and discusses utility infrastructure such as potable water systems, electric power systems, wastewater systems, oil refineries, natural gas systems, and communication systems. It identifies freeways, streets, bridges, railroads, airports, and harbor facilities.

4. The National Earthquake Hazards Reduction Program was established by Congress to coordinate the activities of the National Science Foundation, National Institute of

Standards and Technology, US Geological Survey, and FEMA. The Agency supports basic earthquake research related to earth sciences, social sciences, and engineering as well as empirical research carried out after earthquakes. It uses the information to develop better earthquake practices, performance-based tools, guidelines, and standards to improve risk assessment and provide better techniques of mitigation. FEMA is the lead group in providing and disseminating a variety of publications to the public and others.

5. The Centers for Disease Control and Prevention have issued a series of announcements and podcasts concerning floods and what to do in the aftermath. The topics covered under "Floods: Public Service Announcements and Podcasts" (Atlanta, Georgia, November 1, 2012) include mold, chainsaw injuries, hand washing, electrical safety, re-entry to damaged buildings and facilities, preventing tetanus, electrical safety, safe food, pharmaceuticals exposed to flood waters, potential drowning especially for children, identifying and treating hypothermia, dog safety, rodent control, driving through water, disaster distress hotline, and coping with depression. (See endnote 19.)

6. The Centers for Disease Control and Prevention have provided links to resources for emergency responses concerning: Hurricane Recovery; Generator Safety; Medical Recommendations for Relief Workers and Emergency Responders; Interim Assessment Tools for Occupational Safety and Health in Hospitals, Health Departments, and Shelters Involved in Hurricane Response; Air Quality; Carbon Monoxide; Confined Spaces; Cold Stress; Disaster Site Management; Electrical Hazards; Falls; Fire; Hazardous Materials; Healthcare Workers; Heat Stress; Identifying and Handling Human Remains; Motor Vehicles and Machines Safety; Musculoskeletal Hazards; Protective Equipment and Clothing; Stress and Fatigue; Other Stress and Fatigue Resources; Tree Removal/ Chainsaws; West Nile Virus; Interim Guidance on Health and Safety Hazards When Working with Displaced Domestic Animals; Pet Care; and Radiation. (See endnote 45.)

7. The Centers for Disease Control and Prevention, National Center for Environmental Health, Division of Emergency and Environmental Health Services, in conjunction with federal, state, and local health agencies and the National Environmental Health Association present a series of courses, thereby training existing professionals in special courses including Environmental Health Training in Emergency Response and the specific areas of: food safety; potable water; wastewater; shelters; vector control; responder safety; disaster management; solid waste and hazardous materials; building assessments; and radiation. (See endnotes 50, 51.)

8. The Centers for Disease Control and Prevention provides a series of resources for coping with disasters for individuals, parents and families, teachers and schools, health professionals, and local and state health departments, in a document entitled *Coping with a Disaster or Traumatic Event.* (See endnote 54.)

9. The International Federation of Red Cross and Red Crescent Societies provides a series of modules in disaster preparedness that can be utilized by communities in addressing the problems. These modules include information on: disasters; disaster needs assessment; response based on assessments; reporting; how to conduct assessments visually, through interviews, using sampling techniques, and relying on other sources; avoiding assessment bias; different assessment tools; finding recurrent patterns in various natural and weather connected disasters, as well as chemical and industrial accidents.

10. The American Red Cross responds to about 70,000 disasters in the United States yearly. They provide mobile emergency response vehicles, food, water, shelters, trained volunteer workers, health and mental health contacts, and other supplies both personal and for cleanup and temporary housing, etc.

11. The Corporation for National and Community Service provides a document entitled *Disaster Relief Agencies* (see endnote 96), which lists and gives a short statement on the services provided by a large number of volunteer and church groups which will help in a variety of ways during and after a disaster.

12. The "Environmental Health Emergency Response Guide" with project leadership by Brian R. Golob was developed by the Twin Cities Metro Advanced Practice Center in cooperation with the Centers for Disease Control and Prevention and the National Association of County and City Health Officials. It is an excellent source of information for environmental health practitioners to add as a resource to their knowledge and existing disaster plans within their communities. (See endnote 61.)

13. OSHA requires that employers comply with hazard-specific safety and health standards to ensure that workers receive necessary protection to avoid injuries and illness, short-term and long-term. Although some of the states have their own OSHA-approved State Plan, all states have to meet the requirements for the safety and health of workers involved in disaster response, cleanup, recovery, or any other situations that occur as a result of the disaster. OSHA is involved in: preparing for catastrophic events; the crisis management phase of the event; the recovery phase of the event; and the re-occupancy of the structures. They have highly trained response teams that can go into operation as needed and their laboratories and engineering services are always available and on call.

14. OSHA provides specialized training courses for workers and for trainers of workers in all phases of disasters. They offer a series of training courses in disaster work including the 15-Hour Disaster Site Worker Course Number 7600 which provides specialized information to those involved in utilities, demolition, debris removal, or use of heavy equipment. This course emphasizes knowledge, precautions, and personal protection by discussing hazard recognition, traumatic stress from the disaster, personal protective equipment to be used, and decontamination. There is an emphasis on the use and maintenance of proper respirators for the situation involved. Another course is HAZWOPER for individuals who are involved in chemical and oil spills and radiological disasters as well as the remediation of hazardous waste sites. (See endnote 62.) OSHA also provides a series of health and safety guides on a variety of topics. OSHA carries out on-site inspections and is involved in many other ways in worker safety and health issues.

15. The United Nations Office for the Coordination of Humanitarian Affairs has prepared a set of guidelines for disaster waste management. They discuss the framework for disaster waste management including the four major phases: emergency phase, early recovery phase, recovery phase, and contingency planning. They present a series of tools that can be used in the work. (See endnote 65.)

16. FEMA is involved in all phases of disasters and disaster relief. When a major disaster occurs, the President of the United States makes a determination of what has occurred and then may declare a major disaster with the funding coming from the President's Disaster Relief Fund which is managed by FEMA and disaster aid programs of other federal agencies. The local government is helped by numerous volunteers and if overwhelmed asks the state for assistance. The state helps with the National Guard and various state agencies. Damage assessment is made by local, state, federal, and volunteer groups which determine the types and amount of losses and recovery needs. The governor of the state requests a Major Disaster Declaration based on the facts that have been discovered and the resources available. FEMA evaluates the request and the recommendations and the community and state's ability to recover and advises the White House. The President then approves the request or denies it, and FEMA contacts the governor. This process can take hours or weeks depending on the situation. The disaster aid programs include individual assistance; disaster housing; disaster grants; low-interest disaster loans; and other disaster aid programs which involve crisis counseling, unemployment assistance because of the disaster, legal aid and help with income taxes, and social security and veterans' benefits. Applications need to be filled out for all types of aid and then they are evaluated and granted or rejected based on the situation. FEMA then helps the community with hazard mitigation studies and Best Practices for all types of situations. (See endnote 94.)

PROGRAMS

CITY OF LOS ANGELES HAZARD MITIGATION PLAN

(See endnote 66)

The City of Los Angeles Hazard Mitigation Plan supplements the City of Los Angeles Emergency Operations Plan, which has been developed to preserve life and property since that is the responsibility of various levels of government, and therefore knowledgeable personnel are trained about the various facets of protecting the public's health and welfare. The purpose of the plan is to give specific guidance to the public and utilize all available resources from all levels of government in an effective and immediate response to emergencies and disasters. The plan identifies all forms of resources, governmental, non-profit and private, industries, community-based organizations, and educational institutions involved in disaster relief and the mitigation process. The plan establishes: three types of risk (high risk, moderate risk, and low risk); the types of hazards which may occur and how best to assign a degree of risk to each of the hazards; how to measure and evaluate risks; and measures needed to determine mitigation programs and projects to lessen various categories of risk. The planning process includes appropriate goals, objectives, and recommendations for various techniques of mitigation. Public funding is made available through the Disaster Mitigation Act of 2000. There is a substantial degree of effort to involve all public agencies, residents, non-profit organizations, businesses, and industry in an approach which is united in scope and effective in operation. Specific hazards, vulnerability to the hazards, and risk assessment addressed in the document cover earthquakes, terrorism, wildfires, floods, public health hazards, hazardous materials, drought, severe weather, landslides, etc.

STATE OF CALIFORNIA MULTI-HAZARD MITIGATION PLAN

(See endnote 7)

The state of California has developed a Multi-Hazard Mitigation Plan which recognizes that mitigation is a means of reducing risk to people and property and then integrates the planning efforts for all types of hazards and mitigation efforts in such a way as to avoid duplication and allow for speedy resolution of concerns. The public and a variety of other partners have had an opportunity to review the details and to modify the plan as needed to make it more effective. Each specific hazard is identified and then the actions necessary to carry out a valid mitigation program are divided into component parts to make it easier to evaluate them and then act on the findings. For example, for floods, the following subcategories exist: identifying the flood hazard and how frequently it has reoccurred; profiling the different phases of the flood hazard; determining flood vulnerability for each area; estimating flood losses for each area; determining the strategies and priorities for mitigation measures for each of these flood areas; and determining the mitigation efforts that have already been implemented and how effectively they are working.

The plan establishes Standardized Emergency Management Systems which makes it much easier and more efficient for multiagency responses to a given disaster and coordinates the efforts of all individuals and groups. It also provides for far better communications between various agencies and other groups. This program has minimized deaths and injuries, reduced losses to structures and infrastructure, minimized the impact to the environment, and reduced the work of the emergency responders.

The plan requires that all entities within the state regulate in a consistent manner land use, zoning, housing, open space, and infrastructure as well as building codes, retrofitting techniques, and fire codes to protect the health and safety of the population and their property and mitigate potential disasters. As a result of this, a solid foundation has been constructed for mitigating the impact of floods, fires, earthquakes, and other disasters in new housing and business developments. The plan also provides for using common terminology, enhancing data systems, GIS modeling, progress and communication of Best Practices from one community to another, and a tracking system to check for levels of completion of mitigation plans. Successful programs in different parts of the state of California are identified as examples for others.

Hazus-MH (Multi-Hazard), furnished by FEMA, is a federal government database and software which helps identify various elements of essential infrastructure and provides a GIS-based regional loss estimation tool for earthquakes, hurricanes, and floods. This helps provide necessary data to various parts of California for plan implementation. Since this does not cover very specific areas, the City of Los Angeles has improved upon the risk assessment by identifying critical response facilities, critical infrastructure facilities, and transportation infrastructure. Response facilities include fire stations, police stations, hospitals, city emergency operations centers, and evacuation centers. Utility infrastructure includes potable water systems and their components, electrical power systems and their components, wastewater treatment systems and their components, oil refineries, natural gas systems, and communication systems. Transportation infrastructure includes freeways, streets, bridges, railroads, airports, and harbor facilities. A determination is made of the quantity of water which is available and the quality of drinking water that can be utilized. A determination is made of the economic losses due to the disaster and the number of people killed, injured, and displaced.

STATE OF NEW YORK COMPREHENSIVE EMERGENCY MANAGEMENT AND CONTINUITY OF OPERATIONS PLAN

(See endnote 67)

The state of New York has developed and put into use a comprehensive emergency management plan which includes a 10-step planning process with the main categories of hazard analysis, risk reduction, assessment of capabilities, recovery techniques, and community involvement. There are separate categories for: emergency response, risk management, dam failure, radiological emergencies, school safety, in-house hospital disaster plan, nursing homes, adult care facilities, electrical utility failures, solid waste management, hazardous waste and radioactive waste facility management, water supply, pet and service animals, etc. The plan goes into effect when the governor declares that a disaster is either an imminent threat or has already caused widespread or severe damage, injury, or death. Besides natural emergencies, the plan provides for: collapse of structures; power failure; transportation accidents which involve large numbers of people or substantial damage to property; explosions which cause substantial damage; leaking of hazardous material; large industrial accidents; nuclear reactor accidents; and terrorism and civil unrest. A national emergency may be declared by the president when federal assistance is needed to supplement state and local efforts and capabilities to save lives and protect property.

FEDERAL EMERGENCY MANAGEMENT AGENCY COMPREHENSIVE PREPAREDNESS GUIDE ON DEVELOPING AND MAINTAINING EMERGENCY OPERATIONS PLANS

(See endnote 68)

FEMA has prepared and updates frequently a Comprehensive Preparedness Guide which provides assistance in the development of the fundamentals of planning for emergency operations plans to be used in all areas of the country. It discusses the fundamentals of planning principles and analytical problem-solving which should address the complexity and uncertainty found in traditional disasters and catastrophic events. It discusses strategic plans, operational plans, and tactical plans. Strategic plans are based on policy established by senior management and the political entity in order to meet emergency situations and homeland security situations. Operational plans describe the roles, responsibilities, tasks, and how to integrate the actions of various departments and agencies during emergencies to obtain the quickest action in the most efficient manner possible. The tactical plans describe how to manage personnel, equipment, and resources that are directly involved in preventing, mitigating, and resolving the disastrous event or catastrophe. The plans need constant re-evaluation and adjustment to meet the realities of the problems within the community.

The guide utilizes practical information from what has been learned from actual disasters, major incidents, assessments, and grant programs. It provides techniques that might be used by the appropriate planners in: working with the whole community and all organizations in establishing an emergency operation plan; determining the level of specific risk in each situation and the amount and type of resources which will be needed to mitigate the risk and resolve the problem; utilizing and adapting the plan to a given disaster event; and coordinating the efforts of the various groups working to resolve the disaster event. It explains various planning approaches and common planning pitfalls. It discusses the relationship between federal plans and state emergency operations plans and how best to resolve problems and conflicts between the plans to get immediate assistance to the disaster area; it helps explain how to coordinate and integrate all plans and the actual operational capabilities of all levels of government, non-governmental organizations, universities, religious organizations, business and industry, individuals, and families. It provides information on regulatory requirements for given disasters.

FEDERAL EMERGENCY MANAGEMENT AGENCY

The federal government has been involved in disaster relief at the federal level since the Congressional Act of 1803 was passed to provide assistance to a New Hampshire town that had been devastated by a fire. In 1979, President Carter of the United States established by Executive Order the Federal Emergency Management Agency which was brought together from a group of separate disaster relief programs throughout the federal government into a single agency. In November 2007, FEMA became part of the Department of Homeland Security. Among its many functions, it is responsible for: service to disaster victims; integrating the preparedness of federal, state and local governments, volunteer agencies, private partners and the American public in response to terror attacks, major disasters, and other emergencies; operational planning and preparedness, providing 24-hour-a-day support in order to move swiftly in all hazardous situations; disaster logistics and procurement systems; hazard mitigation; emergency communications; public disaster communications; supporting the Integrated Public Alert and Warning System; protecting communities through the National Flood Insurance Program; and providing funding for a variety of hazard mitigation programs. Specifically, its Multi-Hazard Mitigation Planning Program helps state and local government by: identifying cost-effective means of risk reduction; focusing resources on the most prominent risks; creating the environment for various groups to work together; improving necessary education and awareness of hazards; and helping communities establish and communicate priorities. FEMA's mission is to develop and apply knowledge to actual hazardous situations based on research, engineering, analysis of previous events, and the establishment of Best Practices, (See endnote 69.)

THE OFFICE OF PUBLIC HEALTH PREPAREDNESS AND RESPONSE

The Office of Public Health Preparedness and Response is part of the Centers for Disease Control and Prevention. Its mission is to safeguard health and save lives through appropriate public health preparedness and emergency response. It has four major divisions. The Division of Emergency Operations is responsible for coordinating all of the efforts of the Centers for Disease Control and Prevention in preparedness, assessment, response, recovery, and evaluation before and during an event of public health emergency status. It is responsible for the Centers for Disease Control and Prevention Emergency Operations Center. The Division of Strategic National Stockpile manages and maintains critical medical assets that are needed at the site of a national emergency and provides technical assistance to state and local areas. The Division of State and Local Readiness has provided billions of dollars to public health departments throughout the country to upgrade their ability to respond to hazards including infectious diseases, natural disasters, and biological, chemical nuclear and radiological incidents. The Division of Select Agents and Toxins regulates all units that possess, use or transfer biological agents or toxins to make sure that everything is done safely and in compliance with regulations. (See endnote 70.)

TEXAS COLORADO RIVER FLOODPLAIN COALITION MULTI-JURISDICTIONAL HAZARD MITIGATION PLAN UPDATE 2011–2016

The Texas Colorado River Floodplain Coalition Multi-Jurisdictional Hazard Mitigation Plan Update 2011–2016 is an excellent plan and presentation describing how large groups of entities can work together to protect people from injury, illness, and death and protect property from damage and destruction. It includes: the planning process; a regional profile covering population, demographics, land use, the major lakes and dams, the economy and industry; an overview of hazards and the level of risk to a large number of areas and communities within the overall project based on past experience and new projections; a discussion of a large number of potential hazards, the hazard profile, and how to use Best Practices to manage it; a discussion of the potential hazards due to floods, hurricanes, tropical storms, thunderstorms, tornadoes, hail, winter storms, drought, extreme heat, wild fires, dam failure, hazardous materials, pipeline failure, and terrorism; discussion of the means and capability of dealing with the disaster; a mitigation strategy for each of the regions including a variety of counties within different states; and a discussion of repetitive loss properties and the National Flood Insurance program. This Multi-Jurisdictional Hazardous Mitigation Plan Update demonstrates the validity of working with a large group of partners in trying to help each other mitigate the effects of a variety of disasters and then take needed action when the problems occur. (See endnote 72.)

PUBLIC HEALTH EMERGENCY RESPONSE GUIDE

The Centers for Disease Control and Prevention have produced a Public Health Emergency Response Guide for state, local, and tribal public health directors. In its 2.0 version it provides considerable information to state, local, and tribal health departments and the many things that they need to do and who to contact in all emergency preparedness and response situations for all types of hazards. This guide helps the health department personnel initiate an immediate response within the first 24 hours and then additional responses to the disaster. The guide supplements emergency plans by providing key reminders of what needs to be done. (See endnote 81.)

CENTERS FOR DISEASE CONTROL AND PREVENTION—BIOTERRORISM RESPONSE DOCUMENTS

The Centers for Disease Control and Prevention have produced a document which can be utilized to quickly access a variety of topics which have been prepared in depth including Anthrax Vaccine, Guidance for Protecting Building Environments, Bioterrorism Readiness Plan, Public Health Response to Biological and Chemical Terrorism, etc. (See endnote 83.)

RADIOLOGICAL EMERGENCY RESPONSE-RELATED LINKS

The US Environmental Protection Agency has developed a document entitled "Radiological Emergency Response Related Links" which provides information on how to get assistance for radiation protection in emergencies from a variety of federal and state organizations. (See endnote 85.)

CENTERS FOR DISEASE CONTROL AND PREVENTION HEALTH ALERT NETWORK (HAN)

The Centers for Disease Control and Prevention provides to the public an information-sharing system concerning all forms of emergency public health concerns and information. This system, which is called HAN, provides constant updates, health alerts on time-sensitive information, and health advisories on specific incidents or problems. It tracks diseases in all parts of the world and potential exposure for other people. It provides information to the healthcare providers on the severity of the disease, the symptoms to look for, how best if necessary to establish isolation procedures, and how to treat the individuals. It emphasizes that a health alert is the highest level of importance and should be given immediate attention.

INTERIM PLANNING GUIDE FOR STATE AND LOCAL GOVERNMENTS FOR MANAGING TERRORIST INCIDENTS

FEMA developed a planning guide for state and local governments on how to manage the emergency consequences of terrorist incidents because of the terrorist attacks on the World Trade Center on September 11, 2001, anthrax attacks, and the previous bombing of the Alfred P. Murrah Federal Building as well as the mass shootings at Columbine High School in Colorado. This document based on the experience of many people including first responders, presents a framework for emergency operational plans involving the resources of state, local, and federal terrorism management activities. The guide helps readers understand the problems related to weapons of mass destruction and other acts of terrorism. It emphasizes that terrorists are flexible and use the element of surprise to achieve their goals and that planners must consider this as part of how to respond to a sudden action. Planners must use a regional approach to planning; coordinate local state and federal plans of response; develop a strong uniform communications network with all involved and back-up systems if the network collapses; provide for contingencies for personal protection, alternate emergency facilities, and loss of command-and-control facilities and personnel; provide emergency public information to the media as rapidly as possible; and develop a large support service including volunteers and a variety of other specialized groups. The plans must include potential targets, initial warning systems, techniques of initial detection, and how to investigate and contain hazards. It must also discuss any protective actions that can be taken, mass care, use of health and medical resources, management of supplies, techniques of recovery, urban search and rescue, etc. (See endnote 86.)

FEDERAL COMMUNICATIONS COMMISSION'S EMERGENCY COMMUNICATIONS GUIDE

The Federal Communications Commission has prepared a guide for the country's communication systems during times of emergencies. Some of the very worst problems during the course of disasters, especially in the terrorist act involving the World Trade Center in New York and Hurricane Katrina in Louisiana and elsewhere, have been connected directly to an inability to communicate between various agencies and others. This totally breaks down the command-and-control needed in these serious events. The guide discusses the various components of the emergency communications components, including 911 calls by cell phone and landline; the Emergency Alert System; radio, TV, and computer updates of weather and other types of alerts; accessibility of emergency information for visual- and hearing-disabled people; network and power outages; and emergency preparedness. (See endnote 90.)

ENDNOTES

1. National Weather Service Office of Climate, Water and Weather Services. 2012. *Summary of Natural Hazard Statistics for 2011 in the United States.* Silver Spring, MD.
2. University of Nebraska-Lincoln. 2012. *National Drought Mitigation Center.* Lincoln, NE.

3. City of New York Department of Environmental Protection. 2012. *Drought Management and Contingency Plan.* New York.
4. National Drought Mitigation Center Workshop Report. 2011. *Building a Sustainable Network of Drought Communities.* Chicago, IL.
5. National League of Cities Sustainable Cities Institute. 2013. *Water and Wastewater Systems Best Practices.* Washington, DC.
6. The Governor's Office of Emergency Services. 2004. *State of California Multi-Hazard Mitigation Plan.* Rancho Cordova, CA.
7. California Governor's Office of Emergency Services. 2013. *2013 State of California Multi-Hazard Mitigation Plan.* Mather, CA.
8. Incorporated Research Institutions for Seismology. 2014. *Animations: Earthquake Faults.* Washington, DC.
9. US Geological Survey. 2014. *USGS FAQs.* Reston, VA.
10. US Geological Survey. No Date. *USGS Earthquake Hazards Program: Liquefaction Susceptibility.* Reston, VA.
11. US Geological Survey. No Date. *USGS Earthquake Hazards Program-Earthquake Hazards 101-the Basics.* Reston, VA.
12. US Geological Survey. No Date. *USGS-Earthquake Hazards Program-Earthquake Hazards 201-Technical Q&A.* Reston, VA.
13. National Institute of Building Sciences. 2014. *Security and Disaster Preparedness: Hazard Risk Assessment Program/HAZUS.* Washington, DC.
14. International Code Council. 2014. *About ICC.* Washington, DC.
15. Building Seismic Safety Council (A Council of the National Institute of Building Sciences). 2009. *NEHRP (National Earthquake Hazards Reduction Program) Recommended Seismic Provisions for New Buildings and Other Structures,* 2009 Edition, FEMA P-750. Washington, DC.
16. Centers for Disease Control and Prevention. 2013. *Extreme Heat.* Atlanta, GA.
17. Gower, Stephanie, Mee, Carol, Campbell, Monica. 2011. *Protecting Vulnerable People from Health Impacts of Extreme Heat.* Toronto Public Health, Toronto, ON.
18. State of California, CA.Gov. 2011. *Floods.* Sacramento, CA.
19. Center for Disease Control and Prevention. 2012. *Floods: Public Service Announcements and Podcasts.* Atlanta, GA.
20. National Oceanic and Atmospheric Administration-National Weather Service Forecast Office. 2005. *Tropical Definitions.* Boston, MA.
21. Strahler, Alan, Wiley, John. 2013. *Introducing Physical Geography,* Sixth Edition. Hoboken, NJ.
22. National Oceanic and Atmospheric Administration-National Weather Service-National Hurricane Center. 2013. *Saffir-Simpson Hurricane Winds Scale.* Miami, FL.
23. US Department of Commerce, National Oceanic and Atmospheric Administration, National Weather Service. 2013. *Thunderstorms, Tornadoes, Lightning: A Preparedness Guide.* NOAA/PA 201051. Washington, DC.
24. University of Florida, UFAS Extension Service. 2007. *Homeowner Windstorm Damage Mitigation Checklist.* Gainesville, FL.
25. Kiline, Attila. 1999. What Causes a Volcano to Erupt How do Scientists Predict Eruptions. *Scientific American* VII:2.
26. Centers for Disease Control and Prevention. 2012. *Key Facts about Volcanic Eruptions.* Atlanta, GA.
27. Centers for Disease Control and Prevention. 2012. *Volcanoes.* Atlanta, GA.
28. US Department of the Interior, National Park Service, Fire and Aviation Management. No Date. *Fire in-Depth.* Washington, DC.
29. US Department of the Interior, Office of Policy Analysis. 2012. *Wildland Fire Management Program Benefit-Cost Analysis, A Review of Relevant Literature.* Washington, DC.
30. Bryant, BP, Westerling, AL. 2012. *Scenarios to Evaluate Long-Term Wildfire Risk in California: New Methods for Considering Links between Changing Demography, Land Use, and Climate.* CEC-500-2012-030. California Energy Commission, Sacramento, CA.
31. Hydrosphere Resource Consultants, Incorporated and Anchor Point Group, LLC, BLM. 2003. *Final-Upper San Pedro Watershed Wildfire Hazard Assessment and Mitigation Plan Summary Report, Cochise County, Arizona: A Wildland Urban Interface Communities-at-Risk Program.* Order Number: AAD 020144. US Department of the Interior, Bureau of Land Management, Tucson, AZ.
32. Bracmort, Kelsi. 2014. *Wildfire Management: Federal Funding and Related Statistics.* R-43077, Congressional Research Service. Washington, DC.

33. National Interagency Fire Center. No Date. *Communicator's Guide: Wildland Fire, Chapter 8-Fire Prevention*. Boise, ID.
34. National Interagency Fire Center. No Date. *Communicator's Guide: Wildland Fire, Chapter 9-Fire Mitigation*. Boise, ID.
35. Marshall, Douglas J., Wimberley, Michael, Bettinger, Pete. No Date. *Synthesis of Knowledge of Hazardous Fuels Management in Loblolly Pine Forest*. General Technical Report SRS-110, United States Department of Agriculture, U.S. Forest Service, Southern Research Station, Asheville, NC.
36. Jackson, Brian A., Peterson, DJ, Bartis, James T. 2002. *Protecting Emergency Responders: Lessons Learned from Terrorist Attacks*. RAND Science and Technology Policy Institute, Santa Monica, CA.
37. LaTourrette, Tom, Patterson, DJ, Bartis, James T. 2003. *Protecting Emergency Responders: Community Views of Safety and Health Risks and Personal Protection Needs*, Volume 2. RAND Science and Technology Policy Institute, Santa Monica, CA.
38. Jackson, Brian A., Baker, John C., Ridgely, Susan. 2004. *Protecting Emergency Responders: Safety Management and Disaster and Terrorism Response*, Volume 3, MG 179. RAND Science and Technology Policy Institute, Santa Monica, CA.
39. Willis, Henry H., Castle, Nicholas G., Sloss, Elizabeth M. 2006. *Protecting Emergency Responders: Personal Protective Equipment Guidelines for Structural Collapse Events*, Volume 4. RAND Science and Technology Policy Institute, Santa Monica, CA.
40. Donahue, Amy, Tuohy, Robert. 2006. Lessons We Don't Learn: A Study of the Lessons of Disasters, Why We Repeat Them, and How We Can Learn Them. *Homeland Security Affairs*, VII(2), Washington, DC.
41. Department of Homeland Security, Federal Emergency Management Agency. 2010. *Developing and Maintaining Emergency Operations Plans: Comprehensive Preparedness Guide* (CPG) 101. Version 2.0. Washington, DC.
42. City of New York, Office of the Mayor. No Date. *Rescue Recovery Workers: What We Know from the Research*. New York.
43. US Environmental Protection Agency. 2013. *Natural Disasters and Weather Emergencies: Dealing with Debris and Damaged Buildings*. Washington, DC.
44. Centers for Disease Control and Prevention. 2012. *Worker Safety during Fire Cleanup: Fact Sheet*. Atlanta, GA.
45. Centers for Disease Control and Prevention. 2013. *Emergency Response Resources: Storm/Flood and Hurricanes/Typhoon Response*. Atlanta, GA.
46. Centers for Disease Control and Prevention. 2013. *Storm, Flood, and Hurricane Response: Guidance for Pre-Exposure Medical Screening of Workers Deployed for Hurricane Disaster Work*. Atlanta, GA.
47. Centers for Disease Control and Prevention. 2013. *Storm, Flood, and Hurricane Response: Guidance for Post-Exposure Medical Screening of Workers Leaving Hurricane Disaster Recovery Areas*. Atlanta, GA.
48. Department of Health and Human Services, Centers for Disease Control and Prevention, National Institute for Occupational Safety and Health. 2002. *Guidance for Protecting Building Environments from Airborne Chemical, Biological or Radiological Attacks*, DHHS (NIOSH) Publication Number 2002-139. Washington, DC.
49. Department of Health and Human Services, Centers for Disease Control and Prevention, National Institute for Occupational Safety and Health. 2003. *Guidance for Filtration and Air-Cleaning Systems to Protect Building Environments from Airborne Chemical, Biological, or Radiological Attacks*. DHHS (NIOSH) Publication Number 2003-16. Washington, DC.
50. Centers for Disease Control and Prevention, National Center for Environmental Health, Division of Emergency and Environmental Health Services. 2011. *Environmental Health Training in Emergency Response (EHTER)*. Atlanta, GA.
51. National Environmental Health Association. 2012. *NEHA CERT Online Education: Courses-Emergency Response: CDC 1202: Environmental Health Training in Emergency Response (EHTER) Awareness Level*. Denver, CO.
52. Federal Emergency Management Agency. 2007. *Debris Management Guide*. Washington, DC.
53. Centers for Disease Control and Prevention. 2012. *Building Environment*. Atlanta, GA.
54. Centers for Disease Control and Prevention. 2012. *Coping with a Disaster or Traumatic Event: Mental Health Resources for Traumas and Disasters*. Atlanta, GA.
55. Centers for Disease Control and Prevention, The National Personal Protective Technology Laboratory. 2014. *NIOSH-Approved N95 Particulate Filtering Facepiece Respirators*. Atlanta, GA.
56. US Environmental Protection Agency. 2013. *Mold and Moisture: A Brief Guide to Mold, Moisture, and Your Home*. Washington, DC.

57. Brandt, Mary, Brown, Clive, Burkhart, Joe. 2006. *Mold Prevention Strategies and Possible Health Effects in the Aftermath of Hurricanes and Major Floods: Recommendations and Reports.* Morbidity and Mortality Weekly Report, 55 (RR08).

58. Patel, Neil S. 2005. *Shelter and Housing: Long-Standing Weaknesses and the Magnitude of the Disaster Overwhelmed FEMA's Ability to Provide Emergency Shelter and Temporary Housing.* Email to Charles P. Durkin Destined for J Lewis Libby Junior, Chief of Staff to VP Cheney.

59. Alameda County, California. 2003. *A Guide for Local Jurisdictions and Care and Shelter Planning.* Alameda County Operational Area Emergency Management Organization, Alameda County, CA.

60. Rebmann, Terri, Wilson, Rita, Alexander, Sharon, 2007/2008. *Infection Prevention and Control for Shelters during Disasters.* APIC Emergency Preparedness Committee, Arlington, VA, pp. 1–52.

61. Golob, Brian R. 2007. *Environmental Health Emergency Response Guide.* Twin Cities Metro Advanced Practice Center, Hopkins, MN.

62. Occupational Safety and Health Administration. 2011. *Disaster Site Worker Procedures.* Arlington Heights, IL.

63. US Environmental Protection Agency. 2010. *Underground Storage Tank Flood Guide.* EPA-510-R-10-002. Washington, DC.

64. US Environmental Protection Agency. 2008. *Mold Remediation in Schools and Commercial Buildings.* EPA 402-K-01-001. Washington, DC.

65. United Nations Office for the Coordination of Humanitarian Affairs, Emergency Preparedness Section. 2011. *Disaster Waste Management Guidelines.* Geneva, Switzerland.

66. City of Los Angeles Emergency Management. 2011. *Hazard Mitigation Plan.* Los Angeles, CA.

67. New York State Office of Emergency Management. 2011. *Comprehensive Emergency Management and Continuity of Operations Plan: A Tutorial Sample Plan.* Albany, NY.

68. Federal Emergency Management Agency. 2010. *Developing and Maintaining Emergency Operations Plans: Comprehensive Preparedness Guide,* (CPG 101-Version 2.0). Washington, DC.

69. Federal Emergency Management Agency. 2008. *FEMA-Prepared, Responsive, Committed,* FEMA B-653. Washington, DC.

70. Centers for Disease Control and Prevention. 2014. *Office of Public Health Preparedness and Response.* Atlanta, GA.

71. Joint Chiefs of Staff. 2009. *Counterterrorism.* Washington, DC.

72. Texas Colorado River Floodplain Coalition–H2O Partners. No Date. *Multijurisdictional Hazard Mitigation Plan Update 2011–2016.* Austin, TX.

73. Centers for Disease Control and Prevention-Emergency Risk Communications Branch. 2007. *Bioterrorism Overview.* Atlanta, GA.

74. Centers for Disease Control and Prevention. No Date. *Bioterrorism Agents/Diseases.* Atlanta, GA.

75. Centers for Disease Control and Prevention. 2013. *Chemical Categories.* Atlanta, GA.

76. Centers for Disease Control and Prevention. 2013. *Facts about Sulfur Mustard.* Atlanta, GA.

77. US Nuclear Regulatory Commission. 2014. *Fact Sheet on Dirty Bombs.* Rockville, MD.

78. Bevan, James, ed. 2008. *Conventional Ammunition in Surplus: A Reference Guide.* Small Arms Survey, Graduate Institute of International Studies, Geneva, Switzerland.

79. World Health Organization, Department of Food Safety, Zoonoses and Foodborne Diseases. 2008. *Food Safety Issues: Terrorist Threats to Food-Guidance for Establishing Strengthening Prevention and Response Systems.* Geneva, Switzerland.

80. Gleick, Peter H. 2006. *Water Policy 8: Water and Terrorism.* Pacific Institute. Oakland, CA.

81. Centers for Disease Control and Prevention. 2011. *Public Health Emergency Response Guide for State, Local, and Tribal Public Health Directors,* Version 2.0. Atlanta, GA.

82. Jeffrey Koplan. 2001. CDC's Strategic Plan for Bioterrorism Preparedness and Response. *Public Health Reports,* Supplement 2, Volume 116, 9–16. Bethesda, MD.

83. Centers for Disease Control and Prevention. 2013. *Preparation and Planning for Bioterrorism Emergencies.* Atlanta, GA.

84. New York State Department of Health. 2013. *Radiological Terrorism and Rapid Response Card.* Albany, NY.

85. US Environmental Protection Agency. 2013. *Radiological Emergency Response Related Links.* Washington, DC.

86. Federal Emergency Management Agency. 2002. *Managing the Emergency Consequence of Terrorist Incidents: Interim Planning Guide for State and Local Governments.* Washington, DC.

87. Federal Emergency Management Agency. 2013. *The Stafford Act: Robert T Stafford Disaster Relief and Emergency Assistance Act, as Amended.* Washington, DC.

88. US Environmental Protection Agency, Office of Solid Waste and Emergency Response. 2012. *The Emergency Planning and Community Right-to-Know Ac.* EPA 550-F-12-002. Washington, DC.
89. United States Congress. 2006. *Pets Evacuation and Transportation Standards Act of 2006.* Public Law 109-308, 120 Stat. 1725. Washington, DC.
90. Federal Communications Commission. No Date. *Guide: Emergency Communications.* Washington, DC.
91. The Federal Register, Daily Journal of the United States Government. 2013. *Proposed Rule: Medicare and Medicaid Program; Emergency Preparedness Requirements for Medicare and Medicaid Participating Providers and Suppliers.* Washington, DC.
92. United States Congress. 2013. *Pandemic and All-Hazards Preparedness Reauthorization Act of 2013.* Public Law Number 113–5. Washington, DC.
93. US Department of Transportation. 2014. *Emergency Order: Petroleum Crude Oil Railroad Carriers.* Docket Number, DOT-OST-2014-0067. Washington, DC.
94. Federal Emergency Management Agency. 2015. *The Disaster Process and Disaster Aid Programs.* Washington, DC.
95. CSSC. 2004. *A Safer, More Resilient California: The State Plan for Earthquake Research.* CSSC Publication. Sacramento.
96. DeGraff, Kelly, Murphy, Jen. 2014. National Service Assets in Times of Disaster. CNCS Disaster Services Unit. Washington DC.

BIBLIOGRAPHY

Centers for Disease Control and Prevention. 2000. *Biological and Chemical Terrorism: Strategic Plan for Preparedness and Response, Recommendations of the CDC Strategic Planning Work Group.* Atlanta, GA.

Centers for Disease Control and Prevention, National Center for Infectious Diseases. 2002. *Protecting the Nation's Health in an Era of Globalization: CDC's Global Infectious Disease Strategy.* Atlanta, GA.

Committee on Environmental Health and Committee on Infectious Diseases. 2006. Chemical-Biological Terrorism at Its Impact on Children. *Pediatrics,* 118(3):1267–1278.

Minnesota Department of Health, Twin Cities Metro Advanced Practice Center. 2007. *Environmental Health Emergency Response Guide.* Hopkins, MN.

National Institute of Allergy and Infectious Diseases. 2010. *Emerging and Reemerging Infectious Diseases.* Bethesda, MD.

National Institute of Allergy and Infectious Diseases. 2011. *Bio-Defense and Emerging Infectious Diseases.* Bethesda, MD.

Shantz, Peter M., Tsang, Victor C.W. 2003. Protecting the Nation's Health in an Era of Globalization: CDC's Global Infectious Disease Strategy. *Acta Tropica,* 87.

Southern Illinois University School of Medicine, Department of Internal Medicine, Division of Infectious Diseases. No Date. *Overview of Potential Agents of Biological Terrorism.* Springfield, IL.

US Environmental Protection Agency. 2010. *Children's Health in the Aftermath of Floods, Children's Health Protection.*

6 Environmental and Occupational Injury Control

STATEMENT OF PROBLEM AND SPECIAL INFORMATION

Injuries may be divided into two major categories: intentional and unintentional. Intentional injuries occur when the individual or individuals want to injure either someone else or themselves. Intentional injuries consist of all forms of violence including those related to assault, suicide, homicide, sexual situations, intimate partner situations, child mistreatment, youth violence, elder mistreatment, and bullying including electronic and/or social media aggression, especially in schools and school-related activities, etc. There will be a limited discussion in this area since an entire book will be written about the topic of violence and intentional injuries which are a major public health concern, with tens of thousands of deaths and millions of injuries. The injuries discussed in this chapter are mostly of the unintentional type. (See endnote 17, 29.)

Note: Wherever data are used within this chapter, they may vary depending on the source and date of the document, especially the data that are issued over the Internet. In fact, since the implementation of WISQARS (Web-based Injuries Statistics Query and Reporting System), an interactive database system, there has been a change in the database because there has been a change in the population used for compilation of the data. This change in the reference population started in September 2012. (See endnote 28.) Other data used will come from the report from the Trust for America's Health and the Robert Wood Johnson Foundation of May 2012 entitled "The Facts Hurt: A State-by-State Injury Prevention Policy Report." (See endnote 29.)

Injuries are not accidents. They can be prevented. An injury is the stress upon an organism that disrupts the structure or function and results in a pathological process. The stress may be: physiological due to age or chronic/acute disease, fatigue, mental, or physical infirmities such as deafness, blindness, poor balance, etc.; physical due to weather conditions, heat, cold, inadequate light, noise, and vibration; chemical due to pollutants, household substances, and pharmaceuticals; spatial due to misplaced objects or toys; environmental such as slippery surfaces and disrepair; and alcohol, excessive risk-taking, inexperience, and inattentiveness. Injury prevention is the recognition, evaluation, and control of hazardous situations or substances. (See endnote 1.)

Over 180,000 people die each year in the United States from unintentional injuries and violence. Over 26,000 people die from unintentional falls. Over 33,500 people die from traffic-related incidents. About 33,000 people die from unintentional poisoning. Over 31,500 people die from firearm discharge. Injuries are the leading cause of death for people between the ages of 1 and 44. It is estimated that injuries cost more than $400 billion a year in lost productivity and medical care. Because of injuries, violence, and the adverse effects of medical treatment, 41 million people annually utilize very expensive health care by being treated in an emergency department. There are a substantial number of medically untreated injuries that occur in all areas. These can lead to further difficulties either initially or long-term. Chronic illness or lasting disability from injuries may affect millions of people for a good portion of their lifetime and may reduce their normal lifespan. (See endnote 2.)

Each year in the United States over 60% of injuries to people involve the musculoskeletal system, including sprains, strains, fractures, open wounds, cuts and punctures, contusions, and bruises. While 10% of injuries occur during sports activities and 10% are automobile or pedestrian injuries, over 50% are believed to occur in homes and there is belief among experts that this number is under-reported. Falls represent the leading cause (29%) of the nonfatal musculoskeletal injuries. Among people aged 65 or older, 63% of the injuries are musculoskeletal. Over 57 million of these injuries yearly are seen in healthcare settings. (See endnote 5.)

A traumatic brain injury is an injury that disrupts how the brain works. It may be due to various reasons but the most frequent causes are falls, motor vehicle crashes, various contact sports, firearms, explosions, or intentional battering of the head. Males, young children, and older people are at greatest risk. About 1.7 million traumatic brain injuries occur in the United States each year with about 52,000 deaths, 275,000 hospitalizations, and 1,365,000 emergency department visits. The number of people who have traumatic brain injuries who do not seek medical care is unknown but is thought to be substantial. Weeks, months, or even years later traumatic brain injuries can lead to a variety of serious chronic conditions and early death. (See endnote 18.)

Injuries are the leading cause of death in children, with over 12,000 children each year in the United States, from birth to 19 years of age, dying from unintentional injuries. The ratio of male to female deaths is 2:1. Injuries from motor vehicles or bicycles lead to the highest number of deaths. Each year, 9.2 million children go to the emergency room for an unintentional injury, with 2.8 million of these children having experienced injuries from a fall. The cost for injuries is over $2 billion a year. (See endnote 3.)

Although behavior, age, physical fitness, and certain medications are important causes of injuries, there are many environmental factors that can be altered, such as storage and use of poisons, faulty heating equipment, electrical wiring, lack of smoke alarms and carbon monoxide detectors, slippery surfaces, improperly constructed physical facilities, etc.

UNINTENTIONAL INJURIES

The major categories of unintentional injuries which will be discussed are related to: the built environment (home and community); transportation; school facilities and people; sports and recreation facilities and people; healthcare facilities and people; and occupational facilities. (See endnote 16.)

SUB-PROBLEMS INCLUDING LEADING TO IMPAIRMENT AND BEST PRACTICES FOR THE BUILT ENVIRONMENT (HOME AND COMMUNITY)

(See Chapter 3, "Built Environment—Healthy Homes and Healthy Communities")

Injuries in homes cause over 18,000 deaths, over 7 million people being disabled for at least 1 full day, and a financial loss of $100 billion dollars per year. The top five causes of home injuries are falls, poisoning, fire and burns, airway obstruction, drowning, and submersion in water.

General Best Practices for the Built Environment (See endnote 4)
- Utilize a home safety checklist such as that developed by the New York State Health Department in every area of the residence especially in the kitchen, bathroom, nursery, bedrooms, basement, garage, on the stairs, and in the out of doors, to determine potential hazards. (See endnote 4 for a comprehensive list for all areas.)
- Rapidly correct all hazards found in the survey listed above.
- Remove all potential impediments from walkways inside and outside of the structure.
- Building inspectors should inspect the bracing and condition of all balconies protruding from buildings to determine if they are safe, properly maintained and secure, and a limited number of people should be allowed on the balcony at any given time. There definitely must be weight restrictions.

- Before remodeling properties built prior to 1978, determine if there is lead paint present in the remodeling areas.
- Use damp mops or damp or treated cloths to remove dust from surfaces to prevent individuals from inhaling potentially contaminated material.
- Avoid use of solvents and other hazardous materials in enclosed areas.
- Provide adequate light in all areas.
- Have a licensed electrician inspect all wiring, receptacles, and appliances on a periodic basis and make sure that all appliances are grounded.
- Properly install, operate, clean, maintain, and ventilate all wood-burning equipment and chimneys.
- Eliminate all smoking from the indoor and immediate outdoor areas of the structure.
- Establish a household emergency and escape plan and practice it for the potential rapid removal of people from the premises.
- Provide, inspect, and maintain smoke alarms and carbon monoxide evaluation units.
- Wash hands thoroughly before preparing food, after using bathroom facilities, after changing diapers, after handling solid and hazardous materials and wastes, and after helping the sick.
- Properly ventilate all areas of the structure and change air filters on a regular basis.
- Only purchase cleaning materials, pesticides, and other hazardous materials that are in childproof containers.
- Store all cleaning materials, pesticides, and other hazardous materials in well-ventilated areas away from children.
- Use safety gates and safety latches in all areas where children or older people can harm themselves by means of falls, especially around swimming pools.
- Use safety rails and grab bars where appropriate to assist people in maintaining their balance.
- In areas where radon gas is a problem in structures, periodically test for this substance and then take necessary action if the results are positive.
- Properly maintain and utilize bicycles and other moving conveyances.
- Properly train individuals utilizing bicycles and other moving conveyances on how to prevent injuries.
- Fence in all swimming areas and lock all doors leading to the pool.
- Set the hot water heater at a maximum of 120°F.
- Put signs out when washing floors and only wash one half at a time, allowing a dry surface for individuals to walk on in the structure.
- Place the Poison Control Center phone number by each telephone in the house and enter it on each cell phone for immediate access: 1-800-222-1222.

FALLS

There are three principal risk factors involved in falls among older people:

1. Biological risk factors which include problems of mobility caused by muscle weakness or improper balance; chronic health conditions such as arthritis, strokes, post-polio syndrome, etc.; vision changes and loss; loss of feeling in the feet; and inner ear balance problems
2. Behavioral risk factors which include inactivity; use of alcohol; taking of medications and possible interactions; lack of knowledge of causation of falls and means of prevention; and reckless behavior
3. Environmental risk factors which include home and outdoor environmental hazards such as clutter, poor lighting, throw rugs, holes and step downs out of doors, and slick and

slippery surfaces; steep inclines; poorly maintained railings and staircases, etc.; incorrect size, type or use of walkers, canes, crutches, wheelchairs, and other mobile devices; and poorly designed interior and exterior areas

In children, the principal risk factors are related to the child's age, state of development, type of activity, and level and intensity of activity. Also, children fall more frequently in poor neighborhoods with minority groups and older rental housing.

One third of adults aged 65 and older fall each year; 20–30% of these individuals suffer moderate to severe injuries which may affect their ability to live independently and may be a cause of early death. Falls in older adults result in hospitalizations five times more than other types of causation of injuries. Falls are the leading cause of fatal and non-fatal injuries in this age group. People who fall will frequently fall again within a short period of time. In 2010, there were 2.3 million non-fatal fall injuries among older people which were treated in emergency departments and more than 662,000 of these people were hospitalized. The direct medical costs were $30 billion. About 21,700 older adults died from unintentional fall injuries, with men having a considerably higher rate. People age 75 or older fall four to five times more often than those aged 65–74 and are therefore more apt to be placed in a long-term care facility. Falls in older people account for over 95% of hip fractures as well as numerous traumatic brain injuries. An unwanted complication of the falls for the individuals is the fear of falling again, which may limit the persons' lifestyle and creates a greater dependency on family as well as other people. (See endnotes 6, 7, 8.)

Best Practices for Preventing Falls
- Conduct a home survey which includes all potential tripping hazards, level of lighting in various rooms in the house, and potential places to place grab bars or other means of support.
- Modify the home to meet the needs of the individuals using the structure.
- Older adults need to make sure they are getting adequate calcium and vitamin D and be screened for osteoporosis.
- Safety rules should be published and enforced for all playground and other areas for children. Children's playground equipment should meet safety standards and maintenance standards.
- Join an exercise program such as Tai Chi to increase leg strength and improve your balance.
- Use trained professionals to teach individuals how to compensate for balance problems and determine the need for an assistive device.
- Have the doctor or pharmacist review all medications, including herbal supplements, used by the person, and the interactions between them to determine potential side effects such as dizziness or drowsiness and advise how to prevent or control these conditions.
- Have a professional evaluate the footwear of the individual to determine if it is appropriate for the individual's physical condition and activities.
- At least once a year, have the eye doctor evaluate vision, explain to the individual his/her limitations, and prescribe necessary glasses that will correct vision problems.
- Develop a Community Fall Prevention Program utilizing the resources of the agencies in the public health system to: assess community needs, potential funding, and physical and staff resources both volunteer and professional; determine the major risk factors which will be addressed; develop a significant education program which will reach the target population; utilize existing exercise programs; provide professional assistance to individuals on review of medications; provide vision screening at least once a year; provide hearing screening at least once a year; conduct home safety evaluations; and teach people how to properly use various assistive devices.

POISONINGS

"A poison is any substance, including medications, that is harmful to your body when ingested, inhaled, injected or absorbed through the skin." EPA (see endnote 9)

This Centers for Disease Control and Prevention definition does not include an adverse reaction to a medication when the medication is taken appropriately. Virtually any substance taken in large enough quantities could be poisonous to an individual. An "unintentional poisoning" includes an overdose of drugs or chemicals used for recreational purposes. It also occurs when a chemical is taken accidentally by an individual, such as when a toddler swallows drain cleaner. An "intentional poisoning" occurs when the individual is attempting to do harm to him- or herself. This would include the categories of suicide and assault. (See endnote 9.)

Most poisonings are caused by drug overdoses, where deaths have been rising over the last 20 years. Among people aged 25–64, drug overdoses are now the leading cause of injury death, even more than automobile crashes, in the United States. It is estimated that the cost of prescription opiate abuse exceeds $55 billion. Men are almost twice as likely to die from injuries than women are, with the highest death rate among Native Americans/Alaska Natives. The highest death rate age bracket is 45–49 years of age. Children are most involved in injury from chemicals in the home including household cleaners, pesticides, and parents' and grandparents' medicine. In 2010, there were over 38,000 drug overdose deaths in the United States. (See endnote 10.)

Narcan (naloxone) is a pure opioid antagonist. It should be kept in all emergency drug kits where people respond to drug overdoses.

Since the above data seem to vary from year to year and from reporting agency to reporting agency, it is best to use it as general information instead of for study purposes. For any given community, a study should be made prior to the inauguration of a program of prevention and control. A subsequent study using the same parameters should then be conducted after a specific time, and a statistical comparison should be made to determine the effectiveness of the program. All outside factors should be accounted for in this study.

Best Practices for Preventing Built Environment Poisonings
- Develop an educational program that emphasizes the proper use of prescription medications including proper lighting to determine dosage used, avoidance of concurrent herbal medicine, avoidance of alcohol, keeping medication in original containers, monitoring the use of medication by children and teenagers, and proper storage and disposal of all medicines.
- Store all household products used for cleaning, lawn care, and insect and rodent control away from small children.
- Have the local pharmacy monitor the amount and types of medications prescribed to patients and how frequently they use them to prevent overdose problems.
- Pass state laws to reduce or eliminate doctor shopping for special prescriptions of painkillers.
- Use protective clothing when utilizing any chemicals inside or outside of the house.
- Do not mix chemicals together, and store and use all chemicals in a well-ventilated area.
- Place the Poison Control Center phone number (1-800-222-1222) in or near every telephone within the house including cell phones for general information.
- Call 911 if the individual is exhibiting serious symptoms or if you have any doubt about the nature of the severity of the poisoning. Provide information such as the person's age and weight, what is on the label of the container, approximate quantity of material taken into the body, the time of the poison exposure, and your address and phone number.
- Local and state government should identify a single agency to coordinate response to drug overdoses and other poisonings.

FIRES

Deaths caused by fires and burns are the third leading cause of fatal home injuries. In 2010, 2640 people (not including firefighters) died and 13,350 were injured by fires. Those people died from inhalation of smoke or toxic gases. Smoking is the top cause of fire-related deaths followed by residential cooking. Fire and burn injuries cost $7.5 billion a year. The groups most at risk are children age 4 and under, older adults age 65 and over, poor people, African-Americans and Native Americans. Over one third of the deaths due to home fires are in facilities without smoke alarms. Alcohol is a contributing factor in 40% of people dying in residential fires. (See endnote 11.)

Best Practices to Prevent Fires
- Especially in multifamily dwellings, have the fire department or another governmental agency conduct fire and carbon monoxide poisoning surveys on a periodic basis.
- Utilize an educational home fire prevention program to teach students in the schools and the overall community the dangers of: leaving food unattended on a hot stove; wearing loose fitting clothing while cooking; placing flammable objects near cooking areas; smoking in bed; placing portable space heaters near flammable materials such as drapes; and storing matches and lighters within reach of children.
- Install an appropriate number of smoke alarms and carbon monoxide alarms and test them frequently.
- Install sprinkler systems where appropriate.
- Prepare a family fire escape plan from the residence and test it periodically.
- Avoid the use of alcohol when working with flammable substances or cooking.

DOG BITES

About 4.5 million people are bitten by dogs each year in the United States. About 20% of them need medical attention and about half of those are children. In 2012, over 27,000 people had to undergo reconstructive surgery because of dog bites. Children, especially those aged 5–9, are most frequently bitten, followed by adult males. (See endnote 12.)

Best Practices to Prevent Dog Bites
- Have a dog evaluated and properly trained and socialized before bringing it into the home, especially if young children are present. Meet the dog's parents if possible.
- Do not buy or adopt breeds that are known to be aggressive.
- Evaluate if the dog is stressed or anxious and may be getting ready to bite by looking at the various body parts such as: ears pulled back and tense or forward; darting eye movements; tight lips, closed mouth, teeth bared; low drooping position of the head; paws held close to the body or toes curled tightly; tail low or between the legs; and a curled up or crouched-down body posture.
- Spay or neuter dogs to reduce aggressive tendencies.
- Never approach a dog you do not know, and assume that it will bite.
- Do not disturb strange dogs that are sleeping, eating, or raising puppies.
- If a strange dog approaches an individual, remain still, don't make any sudden sounds or movements, and back away slowly.
- If attacked, try to place objects between the dog and the person while protecting the face and neck.
- Never leave infants or young children alone with a dog.

DROWNING

(See Chapter 10, "Recreational Environment and Swimming Areas")

SUB-PROBLEMS INCLUDING LEADING TO IMPAIRMENT AND BEST PRACTICES FOR HOSPITALS AND OTHER HEALTHCARE ENVIRONMENTS

(See Chapter 8, "Healthcare Environment and Infection Control" for potential injuries of all types including occupational and Best Practices in preventing, mitigating, and controlling them)

SUB-PROBLEMS INCLUDING LEADING TO IMPAIRMENT AND BEST PRACTICES FOR TRANSPORTATION

The US transportation system is a hodgepodge of multiple policy inputs at the local, state, and national level. Many times decisions are made because of politics rather than essential needs. The infrastructure is in serious disrepair throughout the country. Roads, railroads, bridges, dams, etc. and the buried pipes used for transportation of water, sewage, natural gas, oil, etc. are frequently out of date and need replacement. The transportation system is essential to society because it is used to move people and goods in an efficient manner and therefore may have a profound effect on the health, safety, and quality of life of all people.

The use of red light cameras for traffic violations is very controversial. There are questions about the technology and whether or not this is a system to reduce accidents or bring in new revenue for the political entity. Because of confusion on the part of drivers, it may lead to additional rear-end collisions.

Motor vehicles account for over 33,000 deaths and 2.6 million drivers and passengers being treated in emergency rooms annually. Most of the deaths in those aged 1–34 occur as the result of motor vehicle crashes, with 16- to 19-year-olds involved in the most crashes. Alcohol is a factor in almost one third of motor vehicle deaths and in 47% of pedestrian deaths from motor vehicles. (See endnote 13.)

Best Practices for Preventing Injuries in Transportation
- Gather and analyze data on mechanical problems, roadway problems, weather-related problems, environmental problems, and behavioral problems, for a given area to determine the primary causes of injuries from transportation sources.
- Use current research to determine the best approach to improve the mechanical parts of the automobile.
- Develop programs unique to certain age groups such as teenagers and first-time drivers, middle-age, and older people.
- Develop and enforce restraint laws including seatbelt laws for all people within a vehicle and also for pets.
- Develop and enforce laws for infant carriers, child car seats, and booster seats, and insist that the children be seated in the back.
- Develop special programs to evaluate and reduce motorcycle crashes and injuries.
- Develop and enforce a helmet law for motorcycles.
- Develop and enforce a helmet law for bicycles, scooters, and skateboards.
- Enforce all motor vehicle laws that apply to bicycles. Individuals who run stop signs or red lights should receive the same ticket as those using an automobile.
- Develop and enforce mandatory ignition interlocks for all people convicted of drunk driving.
- Develop and enforce a graduated driver license law for teenagers based on age and experience.
- Discourage teenagers from driving other teenagers, which frequently leads to unintended injuries in car accidents.
- Require vision tests for renewal of driver's licenses for older drivers.

- Require all drivers who receive speeding or reckless driving tickets to take special driver education courses.
- Enforce all laws related to drunk driving and minimum age legal driving.
- Develop special programs for individuals who have received warnings or tickets for aggressive, inattentive, or sleepy driving.
- Develop and enforce distracted driving laws which cover the use of cell phones and texting while driving. All drivers must be banned from texting while driving.
- Develop and enforce alcohol impaired driving laws which include the use of car breathalyzers for ignition for individuals convicted of driving while intoxicated.
- Enforce and strengthen driver education requirements, periodic testing, and licensing systems.
- Analyze community design in high accident areas to determine how to make engineering changes to promote vehicle movement while promoting pedestrian and bicycle safety measures.
- Provide sidewalks where feasible.
- Provide safe roadway crossings for pedestrians with flashing yellow lights or other lights.
- Reduce vehicle speed in residential areas.
- Promote public transportation to reduce the number of vehicles on the roadways.
- Provide safe routes to schools and recreational sources for pedestrians and bicycles.
- In new construction in the community, provide for various pedestrian and bicycle routes as part of the planning measure.
- Improve air quality for pedestrians by: retrofitting diesel engines; mitigating traffic congestion; and utilizing stricter emission regulations for railroads, ships, buses, semitrailers, and other sources of transportation.
- Develop and provide educational programs in injury control for all school levels. (See endnotes 14, 15.)

SUB-PROBLEMS INCLUDING LEADING TO IMPAIRMENT AND BEST PRACTICES FOR SCHOOL FACILITIES AND PEOPLE

The top five sources of injuries in schools are trips and falls (40%), playgrounds (25%), gym class (12%), industrial arts (8%), and science laboratories (6%). Some 200,000 children aged 14 and younger are treated at emergency hospitals for playground-related injuries each year. Approximately 45% of the injuries are severe. Many of non-fatal injuries are related to the playground equipment in public playgrounds, while most occur at schools and day care centers.

There are intentional injuries, many of them due to bullying and other forms of violence, and also unintentional injuries. Bullying causes a significant number of students to seek emergency room help for physical injuries while much is unknown about the mental problems caused. (See School Environment in Chapter 4, "Children's Environmental Health Issues" for an extensive discussion of the nature of the school injury problem for children and the potential problems for adults working within this occupational environment. There is also an extensive discussion of Best Practices in that chapter.) (Also see endnote 19.)

SUB-PROBLEMS INCLUDING LEADING TO IMPAIRMENT AND BEST PRACTICES FOR SPORTS AND RECREATION FACILITIES AND PEOPLE

(See endnote 14) (Most of the discussion on injuries in the recreational environment will be found in Chapter 10, "Recreational Environment and Swimming Areas.")

Exercise is essential to prevent a variety of chronic diseases and for healthy living. The consumption of excess quantities of calories and the lack of physical activity have become growing problems in our society today, leading to an increase in diabetes, heart disease, strokes, lung disease, obesity, etc. To counter this serious health concern about lifestyle, people have been urged to participate in a variety of physical activities. Yet because of poor understanding of the exercise process, overstressing systems of the body which are not prepared for the exertion involved, the inherent dangers resulting from various types of sports activities, and environmental problems, short-term and long-term injuries may occur. For example, 45% of playground-related injuries include fractures, internal injuries, concussions, dislocations, and amputations. Many of the nonfatal injuries are related to playground equipment, which is frequently in disrepair, used at schools, day care centers and public playgrounds. About 20% of all emergency department visits are due to sports and recreation injuries. (See endnote 24.)

Each year millions of teenagers are involved in high school sports which may result in injuries. Although the teenage injuries occur at about the same rate as professional injuries, among young children the effects may be far greater because children are still growing. Children are more susceptible to muscle, tendon, and growth plate injuries because of the uneven growth occurring. Acute injuries may be due to a sudden trauma caused by players colliding or suddenly hitting the ground. Injuries may be due to overuse of various body parts. Concussions or mild traumatic brain injuries occur quite frequently, and repetitive ones may lead to long-term disability. High school football accounts for 47% of all sports concussions, followed by ice hockey and soccer. Growth plate injuries in children occur in the developing cartilage tissue near the ends of the long bones.

Best Practices for Sports and Recreational Facilities
- All activities should be age-appropriate.
- All children involved in high school or intra-mural sports activities should receive a physical examination by appropriate medical personnel prior to the activities.
- All individuals should receive proper conditioning, training, and equipment before participating in any type of sports activity.
- Upon receiving an injury, the individual should be immediately examined by a doctor and then appropriate diagnostic studies should be carried out to determine the extent and nature of the injury.
- Treatment for injuries should be thorough and completed before resumption of activities.
- After an injury, physician authorization should be required to grant the individual the opportunity to once again resume the activity.
- Do not overstress young people by having them participate in too many physical activities. (See endnote 25.)
- Use proper equipment to help prevent traumatic brain injuries.
- If an individual is suspected of having a concussion, he or she should be removed from the activity and immediately referred to professional medical personnel for diagnosis and management of the injury. Before returning to the activity, the individual has to be cleared by the medical team.
- See Chapter 10, "Recreational Environment and Swimming Areas" for drowning and other water-related problems.

SUB-PROBLEMS INCLUDING LEADING TO IMPAIRMENT AND BEST PRACTICES FOR OCCUPATIONAL FACILITIES AND PEOPLE

(This will be a very limited discussion since there will be an entire book written about Best Practices in Occupational Health and Safety in this series.)

Although there has been a sharp reduction in the number of people who die and are injured in an occupational setting since the implementation of the Occupational Safety and Health Act over 40 years ago, still about 4500 people a year die and over 4.1 million workers have a serious job-related injury or illness each year. In 2009, it was estimated that workers' compensation costs were $74 billion a year. (See endnote 20.) Major workplace hazards may include violence, tripping and falling hazards, toxic substances, harmful physical agents, harmful chemical agents, harmful biological agents, hazardous wastes, electrical hazards, unprotected equipment, ergonomic situations, etc.

(See Chapter 4, "Children's Environmental Health Issues"; Chapter 5, "Environmental Health Emergencies, Disasters, and Terrorism"; Chapter 7, "Food Security and Protection"; Chapter 8, "Healthcare Environment and Infection Control"; Chapter 10, "Recreational Environment and Swimming Areas"; and Chapter 12, "Solid Waste, Hazardous Materials, and Hazardous Waste Management" for a limited discussion and Best Practices for occupational injury prevention in those select areas.)

Best Practices for Injury Prevention in Occupational Settings. (This is a very limited presentation.)
- Appoint a safety manager for each specific area.
- Examine every case of violence and determine the best way to prevent future problems.
- Develop an injury and illness prevention program focused on finding all hazards in the specific workplace and then develop a plan for preventing and controlling them. (See endnote 21.)
- Conduct inspections and surveys on a routine, periodic, and as-needed basis when a hazard is discovered or an injury occurs.
- Prioritize hazards to eliminate or control those first which are of greatest significance.
- Determine if all control measures are being implemented effectively and in a timely manner.
- Develop a list of knowledge, attitudes, and practices needed by the workers for specific types of work situations.
- Provide initial specialized training programs and continuing education opportunities for the workers to implement the knowledge, attitudes, and practices required for a given work situation.
- Train all immediate supervisors in the intricacies of the operation of all phases of manufacturing and transportation under their control
- Train all immediate supervisors and first-line managers in the basic skills of supervision and management of people.
- Have the supervisors evaluate the competencies and skills of the workers and use continuing education techniques and experiential teaching techniques to upgrade them to appropriate levels.
- Post explicit rules and procedures to avoid injuries to the worker or others.
- Ban the use of alcohol, tobacco, and drugs from the workplace.
- Immediately investigate any situations that can or already have caused injuries, and find and implement methods of correcting them.
- Gather new data on standard forms and evaluate the program to determine effectiveness.

SUB-PROBLEMS INCLUDING LEADING TO IMPAIRMENT AND BEST PRACTICES FOR VIOLENCE-RELATED INJURIES (INTENTIONAL INJURIES)

A discussion of intentional injuries will briefly cover: intimate partner violence; teen dating violence; homicide, assault, and suicide; teen violence; child abuse; and abuse of the aging. Almost

17,000 people in the United States were murdered in 2009 and more than 37,000 committed suicide. Assaults account for more than 1 million injuries each year. Homicide is the second major cause of death and suicide the third among teens and young adults. About 750,000 emergency room visits by children and teenagers each year are due to violence. More than 33% of women and 25% of men in the United States have either been raped, exposed to physical violence, or stalking by an intimate partner during their lifetime. The cost of violent death is $47 billion, while the cost of suicides is $26 billion and the cost of homicides is $20 billion. Teen violence includes that which happens at school, gang-related violence, and bullying. Child abuse and abuse of the elderly contribute to this severe situation that needs to be corrected in our society.

Best Practices for Violence-Related Injuries Prevention
- Gather and analyze accurate data on all forms of violence in order to determine causes and then determine means of prevention to reduce the intentional injuries.
- Develop effective services with federal, state, and local governments working with voluntary agencies for victims of violence including providing shelters, legal aid, counseling, and financial assistance where needed.
- Develop a plan to obtain rapidly protection orders for the abused and enforce them rigorously.
- Provide screening and counseling for individuals involved in intimate partner violence, with special attention to teenagers, with no out-of-pocket expense according to the Affordable Care Act.
- Allow minors to get civil protection orders and special health services as needed.
- Require that schools teach Best Practices in the prevention of teen dating violence and other teen violence including all types of physical, verbal, and electronic bullying.
- Establish appropriate rules and regulations concerning all types of violence to be followed by the young people and then use strict enforcement when the rules and regulations are violated.
- Educate the public about the safe storage and use of guns.
- Provide suicide hotlines and teach young people that if one of their friends appears to be severely depressed and is contemplating suicide, they should discuss this with the proper medical authorities to preserve life and prevent injury.
- Teach in the schools and strictly enforce the message that violence, harassment, bullying, and intolerance will not be accepted and that there will be severe penalties for the individuals exhibiting this type of behavior.
- Develop a team of outreach workers respected by the community who are highly trained and who can detect and interrupt violent actions before they occur and when they start.
- Provide afterschool programs for children to keep them off the streets and doing something productive instead of getting into trouble.
- Provide adequate amounts of quality food for every child so that no one goes hungry during the course of the day.
- Equip young people with necessary work skills and appropriate language skills to obtain afterschool jobs and provide work for them to make these young people productive members of society. They will not only contribute to the growth of the economy and help extend the life of the Social Security program through their taxes and, most importantly, will have a feeling of achievement which will help keep them out of trouble.
- Provide Best Practices programs to prevent child abuse and neglect by parents, caregivers, and others. Teach appropriate parenting skills and provide information on what resources are available when the situation becomes one that is difficult to cope with.
- Implement home visitation programs by public health workers where there are at-risk family situations especially in families having financial and health difficulties and there are pregnant women or young children present.

- Establish a program to prevent shaken baby syndrome.
- Where children are removed from the home and placed in the care of the state, there must be strict follow-up and supervision by highly qualified individuals to ensure the health and safety of the child.
- Public health workers should be trained to see the signs of elder abuse. Where this occurs, the situation must be immediately reported to the appropriate authorities and action taken to protect the health and welfare of the older person. Special training should be given to all caregivers of older people including those in nursing homes and other facilities. If the situation is too difficult for the caregiver, other resources must be made available to protect the health and safety of the individual.

LAWS, RULES, AND REGULATIONS

The number of local, state, and federal laws rules and regulations for the six major areas of injury control discussed are so extensive that they are well beyond the scope of this book. However, if an individual is interested in a specific topic within one of the given areas such as electrical safety, he or she should contact the local government and find out the necessary rules for the installation or upgrading of electrical services. If the individual is interested in a specific portion of transportation, then he or she should contact the local, state, and federal transportation authorities and request the information.

RESOURCES

1. The *CDC Injury Fact Book* from the National Center for Injury Prevention and Control, Centers for Disease Control and Prevention, Atlanta, Georgia, November 2006 is a comprehensive view of the injury problem in the United States and how the Centers for Disease Control and Prevention (CDC) is trying to develop research and prevention programs for all types of injuries. (See endnote 30.)
2. The National Center for Injury Prevention and Control, CDC, Atlanta, Georgia, has a mission to prevent and control injuries and violence and to reduce their consequences to reduce the public health burden to society. It leads injury research through interdisciplinary means by improved data collection and analysis, information sharing, and developing relationships with the various states to analyze their problems, and try to come up with appropriate programs. The Center is the focal point for the public health approach to utilizing the latest science by helping determine logical and reasonable approaches for programmatic activities, better known as Best Practices. They also provide funding and technical assistance to the states.
3. The US Consumer Product Safety Commission can provide information on a variety of products which may be used to prevent potential injuries.
4. The Occupational Safety and Health Administration (OSHA) can provide information and assistance on occupational injuries.
5. The American Association of Poison Control Centers is available on a 24-hour basis to assist individuals with necessary information in the event of a poisoning.
6. OSHA provides various programs, both mandatory and voluntary, to help industries improve their workers' health and safety while increasing their own profits. These results are achieved through the "Injury and Illness Prevention Programs" established by OSHA. (See endnote 20.)
7. The National Institute for Occupational Safety and Health (NIOSH) is the federal agency that conducts scientific research and makes recommendations to determine causes of worker injury and illness and techniques for prevention and control. NIOSH provides practical solutions to identified problems. It provides large amounts of highly specific

information for a series of potential hazards through its A–Z index. The Institute works with people inside and outside of government and supports the training of occupational health and safety professionals. By moving research into practice, it not only provides for a safe and healthy environment for workers but also adds substantially to the economy of the United States.

8. The National Occupational Research Agenda (NORA) is a partnership program to stimulate innovative research and improved workplace practices. It has national sector agendas in major areas of the occupational environment including: agriculture, forestry, and fishing; construction; healthcare and social assistance; manufacturing; oil and gas extraction; public safety; services; transportation; warehousing and utilities; and the wholesale and retail trade. (See endnote 27.)

PROGRAMS

Injury control has several components: collection of data to determine the nature and cause of injuries occurring in a specific community; specific studies to determine environmental causes that can be changed or eliminated; monitoring and correction of environmental problems; education used as a tool to reduce injuries; helping coordinate a group effort to resolve problems; necessary enforcement procedures; rapid and thorough treatment procedures to reduce the intensity of the injury; and research to determine how best to eliminate environmental causes and how to best treat individuals with injuries. Environmental health practitioners are involved in a portion of these efforts while participating in helping develop operational plans for the entire effort. The practitioners specifically help collect data, determine specific environmental causes, assist in correction of environmental problems, are involved in educational efforts, and may be involved in enforcement efforts.

Environmental health practitioners, trying to reduce unintentional injuries, typically carry out studies in homes, food facilities, schools, nursing homes and hospitals, recreational settings and occupational health settings, as part of their normal program activities, and in communities and buildings after disasters. Their major areas of interest are: falls, especially among the elderly and children, residential fire-related injuries, poisoning, dog bites, and water-related injuries.

NATIONAL ACTION PLAN FOR CHILD INJURY PREVENTION—CENTERS FOR DISEASE CONTROL AND PREVENTION, NATIONAL CENTER FOR INJURY PREVENTION AND CONTROL

The CDC working with 60 different stakeholders has developed a National Action Plan to reduce childhood injuries and deaths including those which are sports and recreation related. There are six major parts of the plan:

1. *Data and Surveillance*—This includes the systematic collection, analysis, and interpretation of child health data to plan, implement, and evaluate injury prevention techniques.
2. *Research*—This includes evaluating the gaps in knowledge and creating priorities in risk identification, techniques of intervention, program evaluation, and publication of Best Practices.
3. *Communication*—This includes targeting special audiences through relevant channels to reach those who are most vulnerable to specific types of injuries.
4. *Education and Training*—This includes physician training while in medical school and continuing education training for physicians, teachers, coaches, officials, and administrators in all phases of injuries, injury prevention, injury treatment, and injury control.
5. *Health Systems and Health Care*—This includes the health infrastructure needed to deliver appropriate care to the injured person, preventive services, parent counseling on how to prevent injuries in sports and recreational activities, as well as what is appropriate for a given child at a given age and physical condition.

6. *Policy*—This includes laws, regulations, administrative actions, and voluntary practices needed to support evidence-based practices to prevent and control injuries during sports and recreation. (See endnote 26.)

Occupational Health Surveillance Program—New York Department of Health

New York State provides an Occupational Health Surveillance Program for workplace injuries, illnesses, and deaths in various industries by providing epidemiologists, industrial hygiene professionals, special evaluators, and outreach specialists. They track patterns of work-related injury and illness at the fatalities. They investigate and intervene in situations to find appropriate means of correcting significant hazards. They monitor immediate and long-term health problems related to various occupational exposures. They provide education to the medical community and others in techniques for diagnosing of various illnesses and how best to treat them. They provide occupational health registries in a variety of areas including heavy metals, pesticide poisoning, occupational lung disease, etc. (See endnote 22.)

Workplace Injury and Illness Prevention Program—California OSHA

By law in California every employer has a legal obligation to the employees to provide and maintain a safe and healthy workplace. In keeping with this obligation, the California OSHA developed a manual that describes the employer's responsibilities in establishing, implementing, and maintaining an employee injury and illness prevention program. The program consists of: fact-finding through specialized surveys; the establishment of the program including hazard assessment and control; the scope and nature of the management team; the investigation of all injuries; planning for rules and work procedures to prevent and control injuries; the communication of all necessary information to the workers; and practical training and generalized continuing education. (See endnote 23.)

ENDNOTES

1. Koren, Herman. 2005. *Illustrated Dictionary and Resource Directory of Environmental and Occupational Health*, Second Edition. CRC Press, Boca Raton, FL.
2. Centers for Disease Control and Prevention. 2014. All Injuries. *FastStats*. Atlanta, GA.
3. Borse, Nagesh N., Gilchrist, Julie, Dellinger, Ann M. 2008. *CDC Childhood Injury Report: Patterns of Unintentional Injuries among 0–19-Year-Olds in the United States, 2000–2006*. Centers for Disease Control and Prevention, Atlanta, GA.
4. New York State Health Department. No Date. *Home Safe Home—A Home Safety Checklist*. Albany, NY.
5. American Academy of Orthopedic Surgeons. 2009. Half of All Musculoskeletal Injuries Occur in Homes. *AAOS Now*. Rosemont, IL.
6. Centers for Disease Control and Prevention. 2013. *Falls Among Older Adults: An Overview*. Atlanta, GA.
7. Centers for Disease Control and Prevention. 2014. *Cost of Falls among Older Adults*. Atlanta, GA.
8. National Center for Injury Prevention and Control. 2008. *Preventing Falls: How to Develop Community-Based Fall Prevention Programs for Older Adult*. Atlanta, GA.
9. Centers for Disease Control and Prevention. 2008. *Poisoning in the United States: Fact Sheet*. Atlanta, GA.
10. Centers for Disease Control and Prevention. 2014. *Drug Overdose in the United States: Fact Sheet*. Atlanta, GA.
11. Centers for Disease Control and Prevention. 2011. *Fire Deaths and Injuries: Fact Sheet*. Atlanta, GA.
12. Centers for Disease Control and Prevention. 2000. *Dog Bites*. Atlanta, GA.
13. Centers for Disease Control and Prevention. 2012. *Impaired Driving: Get the Facts*. Atlanta, GA.
14. Centers for Disease Control and Prevention. 2009. *CDC Injury Research Agenda—2009–2018*. Atlanta, GA.

15. Centers for Disease Control and Prevention. 2011. *CDC Transportation Recommendations.* Atlanta, GA.
16. Department of Health and Human Services, Centers for Disease Control and Prevention, National Center for Injury Prevention and Control. 2002. *CDC Injury Research Agenda.* Atlanta, GA.
17. Gallagher, Susan S., Shepphard, Monique A. 2005. *Bridging the Gap: Bringing Together Intentional and Unintentional Injury Prevention Efforts.* National Association of Children's Hospitals Related Institutions—Spring Conference, Alexandria, VA.
18. Faul, M., Xu, L., Wald, M.M. 2010. *Traumatic Brain Injury in the United States: Emergency Department Visits, Hospitalizations and Deaths 2002–2006.* Centers for Disease Control and Prevention, Atlanta, GA.
19. California Department of Industrial Relations, The Commission on Health and Safety and Workers' Compensation. 2010. *SASH-School Action for Safety and Health-Promoting Injury and Illness Prevention Programs for California's School Employees.* Oakland, CA.
20. United States Department of Labor, Occupational Safety and Health Administration White Paper. 2012. *40 OSHA-Healthier Workers, Safer Workplaces, A Stronger America—Injury and Illness Prevention Programs.* Washington, DC.
21. United States Department of Labor–Occupational Safety and Health Administration. 2012. *Job Hazard Analysis.* Washington, DC.
22. New York State Department of Health. 2008. *Occupational Health Surveillance Program.* Albany, NY.
23. California/OSHA. 2014. *Guide to Developing Your Workplace Injury and Illness Prevention Program with Checklist for Self Inspection.* Oakland, CA.
24. Centers for Disease Control and Prevention. 2012. *Playground Injuries: Fact Sheet.* Atlanta, GA.
25. American Academy of Orthopedic Surgeons. 2012. High School Sports Injuries. *Orthoinfo.* Rosemont, IL.
26. Centers for Disease Control and Prevention. 2012. *National Action Plan for Child Injury Prevention—An Agenda to Prevent Injuries and Promote the Safety of Children and Adolescents in the United States.* Atlanta, GA.
27. Centers for Disease Control and Prevention. 2013. *The National Occupational Research Agenda.* Atlanta, GA.
28. Centers for Disease Control and Prevention. 2014. *Welcome to WISQARS™.* Atlanta, GA.
29. Trust for America's Health, Robert Wood Johnson Foundation. 2012. *The Facts Hurt—A State-by-State Injury Prevention Policy Report.* Washington, DC.
30. National Center for Injury Prevention and Control. 2006. CDC Injury Fact Book. CDCP. Georgia.

BIBLIOGRAPHY

Centers for Disease Control and Prevention. 2007. *Injuries among Older Adults.* Atlanta, GA.
Centers for Disease Control and Prevention. 2011. *Home and Recreational Safety.* Atlanta, GA.
Centers for Disease Control and Prevention. 2014. *Injury: The Leading Cause of Death among Persons 1–44.* Atlanta, GA.
Centers for Disease Control and Prevention—National Center for Injury Prevention and Control. 2006. *Preventing Injuries at Home and in the Community.* Atlanta, GA.
National Food Service Management Institute, University of Mississippi. 2002. *Injuries in the Childcare Setting.* Oxford, MS.

7 Food Security and Protection

STATEMENT OF PROBLEM AND SPECIAL INFORMATION

Food security and protection is: the prevention of contamination of the food flow process by biological, chemical, physical agents, or radionuclear materials, intentionally or unintentionally; the prevention and control of the growth of microorganisms causing foodborne disease once the food stream has become contaminated; and the prevention and control of the spread of foodborne disease when it has occurred. This chapter will bring together considerable information about how to prevent and mitigate these problems by using a systems approach to the flow of the products and a determination of the hazardous points of contact and how to create appropriate interventions. It will not give you all of the technical details which may be found in *Handbook of Environmental Health—Biological, Chemical, and Physical Agents of Environmentally Related Diseases*, Volume 1. (See endnote 1.)

Food protection has been a concern for at least the last 12,000 years when people came together to plant crops and grow sufficient food to keep the clan from starving. Communities grew and society developed along with the increased quantity of food available. The unintended consequence of this action was the increase in disease potential and the spread of microorganisms from animals to people, since the animals shared habitats with the people. Disease then spread from people to people. The movement of people and resources, throughout recorded time and from place to place, has always resulted in increased levels of disease within the population, sometimes of a pandemic nature. In modern times, even more than in previous times, the flow of food from production to consumption to disposal has led to a substantial increase in the numbers of individuals exposed to foodborne disease, other infectious diseases, and severe environmental concerns related to the air, water, land, and a variety of chemicals.

It is estimated that yearly 48 million Americans became ill with foodborne disease and it was reported that 28,000 of these people were hospitalized and 3000 died. (See endnote 2.) Norovirus is the most frequent organism involved in foodborne disease followed by non-typhoidal *Salmonella*. (See endnote 2.) It is only because of chance that the numbers are not far higher than estimated yearly. Further, in the last 20–30 years, there has been an increased failure to control the classic zoonoses, such as *Brucellosis* caused by a bacterium which frequently comes from animals, increased foodborne infections, and livestock pollutants in rapidly growing urban and suburban areas. A potential emergence of new zoonoses, such as bovine spongiform encephalopathy, increasing globalization and trade in livestock and livestock products, greater livestock concentrations in production areas, more varied and intensive feed sources, and a growing interest in raising market animals in suburban and urban lots has developed and poses potentially new hazards to people.

Then add recent major veterinary public health problems as they relate to animals and people including: Nipah virus in Malaysia resulting in human deaths caused by direct contact with infected bats, infected pigs, or people infected with the Nipah virus through close contact; RVF virus causing Rift Valley fever, a mosquito-borne viral zoonotic disease in Egypt causing 200,000 human cases and 600 deaths in 1977; Crimean–Congo hemorrhagic fever, caused by the bites from infected ticks of the *Hyalomma* species, affecting workers in abattoirs (although this is not a food-transmitted disease, it is an occupational hazard to people working in the food industry); anthrax among domestic and wild animals and causing human disease and death; avian influenza with epidemics affecting poultry flocks and humans with disease and death; and *Salmonella* Enteritidis, verotoxigenic *Escherichia coli* and listeriosis, however, most of the human infections come from direct or indirect contact with the blood or organs of infected animals (See endnote 4). Although many of these diseases are appearing more frequently in poor countries of the Third World, it is only a matter of time before they arrive in the United States. (See endnote 3.) In fact, the current H1N1 influenza virus has both avian and pig elements in its genetic code.

In the last 20–25 years, the food supply chain has become increasingly global in nature. In many foreign countries, there is: a lack of awareness of food safety concerns; a lack of integration of food safety into the primary care system; inadequate information and ability to diagnose potential disease and contamination problems; inadequate food laws; and inexperienced and poorly trained personnel to evaluate food safety activities. All of these have led to an increased potential for food-borne and waterborne disease.

Since food, its ingredients, and the water used to process it may reach a mass audience, it is a perfect vehicle for terrorists to disseminate biological, chemical, or physical agents, or radionuclear materials. (Water in bottles is included as a food.) It seems logical that purposely contaminated food could lead to massive illness and death. An example of a bioterrorist attack in the United States occurred in 1984 when followers of the Indian guru Bhagwan Shree Rajneesh were in conflict with local people politically and caused a food poisoning outbreak affecting 751 people in Dallas, Oregon, by introducing *Salmonella* contamination into salad bars at 10 local restaurants, to keep people away from the polls.

Unintentional contamination of food has led to massive outbreaks of disease, such as *Salmonella typhimurium* contamination of milk causing over 150,000 illnesses in 1985 in the United States (See endnote 5) and *E. coli* O157:H7 contamination of radish sprouts in Japan causing 8000 illnesses with several children dying. Toxic chemicals including pesticides, mycotoxins, heavy metals, and other toxic substances can cause severe reactions and death. Physical agents such as glass, needles, metal fragments, etc. can cause disease or injury. Radionuclear agents are typically radioactive chemicals capable of causing injury or death when present at hazardous levels or lower levels over a period of time.

Such contamination could also cause enormous financial costs and panic in a given population, thereby spreading fear locally, regionally, nationally, and globally. Public health systems would then be inundated and have great difficulty in functioning in an effective manner.

In addition, it is estimated that 20% of the population is over the age of 60. These people are more susceptible to certain types of infections because of chronic illness and a reduced immune response due to the use of certain prescription drugs, radiation, therapy and existing infections.

The public is consuming far more prepared foods outside the home in restaurants and other facilities. New microorganisms will be found in the global food supply. Further, there is a greater emphasis on the use of fruits and vegetables than in prior years and they are apparently becoming more contaminated.

There is a constant turnover of personnel in many food facilities because these are minimum-wage jobs that require low-level skills and education. Workers are not given paid sick leave and therefore come to work when they could infect the food supply. Individuals with little training or understanding of the intricacies of food production and preparation can easily contaminate various parts of the food process including preparation and serving, since some of the workforce may not be trained effectively to prevent disease and injury.

The problems identified by the Food and Drug Administration (FDA) in a recent report on food manufacturing establishments and similar to the ones found in retail food establishments are basically the same as those that the author encountered when he first started in the environmental health field as a practitioner in the state of Pennsylvania and then the City of Philadelphia in 1955 and 1956. The major differences are concerns about allergens affecting people and a sharp increase in the globalization of raw foodstuffs and finished products entering the country and the food supply. There has also been increased concern about the use and abuse of a variety of chemicals including pesticides, fertilizers, etc. Other problems include condensate on pipes and other equipment; contamination of reused products; contamination of raw materials; contamination of the food process line and equipment; use of contaminated water for processing and cooling; hard-to-clean equipment; improper cooling, heating, and time control; improper cleaning and storage of containers; improper or totally lacking regular maintenance and cleaning and sanitization schedules; improper or inadequate cleaning and sanitization materials and techniques; improper design and use of aged equipment; improper pest control; poor facility design; poor ventilation system design and maintenance; improper plumbing, and submerged inlets and back flow; contamination of the finished products; poor personnel hygiene; lack of necessary training and retraining of personnel; lack of adequate supervision and management training for first-line supervisors and managers; inadequate numbers of and training for individuals who need to respond to outbreaks of foodborne disease to contain them; etc. The two leading causes of foodborne illness are the abuse of time and temperature activities and improper hand washing. (See endnotes 6, 7.)

A potentially extremely serious problem revolves around protecting the food supply during temporary events which have massive audiences. This includes football games, baseball games, other sporting events, state fairs, etc. There are a large number of eating facilities in areas where there may be very poor control of refrigeration, water supply, preparation, and serving by temporary staff, insect, and rodent problems, and sewage problems from a substantial amount of human waste produced in a short period of time or, in the event of a fair, the use of portable toilets, and solid waste disposal.

BEST PRACTICES IN CONTROLLING TIME/TEMPERATURE EVENTS AND OTHER FOOD PROBLEMS THROUGH THE USE OF HAZARD ANALYSIS CRITICAL CONTROL POINTS (HACCP) PRINCIPLES

The FDA's definition of the Hazard Analysis and Critical Control Points (HACCP) system is as follows: "HACCP is a management system in which food safety is addressed through the analysis and control of biological, chemical, and physical hazards from raw material production, procurement and handling, to manufacturing, distribution and consumption of the finished product." (See endnote 60.) This definition came out of the work between government and industry to prepare food for space exploration individuals that would not cause foodborne disease.

HACCP is a system used to identify and prevent disease and injury from microbial contamination and other hazards in all types of food production as well as in the preparation of foods and a variety of retail outlets. There are seven major principles involved in this approach as follows:

- Perform a hazard identification and analysis review of the raw materials areas and the food facility whether production or retail to determine the severity and nature of all potential food safety hazards and appropriate methods of preventing, mitigating, and controlling them. Prepare a written plan that can be evaluated.
- Identify the critical control points (a point in the process or a procedure which may lead to disease) where food safety hazards may be occurring and prevent, mitigate, or eliminate the situation to produce food which can be safely consumed.
- Establish in a scientific manner a critical limit for each critical control point beyond which the food may be unsafe for human consumption and apply appropriate means to alleviate the problem. An example would be a time and temperature range.
- Establish specific monitoring requirements for the limits of the critical control points beyond which potential health problems may occur, and enforce them strictly.
- Establish and strictly enforce appropriate corrective actions to be taken when problems occur or reoccur before the end of production or preparation of the food, and then dispose of that which is contaminated and record the results in writing.
- Establish verification procedures for testing, sampling, and other means of monitoring, and document in writing the HACCP plan, as well as the results of monitoring and correction of problems.

- Establish procedures that will validate in writing all of the problems which have occurred, the means of handling the problems, as well as the adequacy of the plan and if it is working properly. (See endnotes 11, 12.)

General Best Practices for Food Security and Protection
- Use the seven HACCP principles as modified for all types of food production and food facilities.
- Produce safe quality food for the consuming public by providing highly trained supervisors and managers to effectively oversee a variety of procedures which helps quickly recognize and control potential sources of contamination. Training must consist of at least in-depth knowledge of means of contamination of food and Best Practices in avoiding them including all means of food protection; employee health and personal hygiene, especially hand washing procedures; in-depth knowledge of all facets of each of the production and preparation processes and Best Practices; in-depth knowledge and practical application of supervision, training, and management techniques to ensure that the workforce is knowledgeable and competent in carrying out food safety requirements.
- Conduct background checks of all prospective employees.
- For temporary food venues such as fairs and festivals, immediately check all semitrucks and other storage areas to ensure that food is kept under proper refrigeration and that the storage areas are clean and not subject to any type of contamination including flooding and breach of security in locked units.
- For temporary food venues in fixed facilities such as stadiums, all locations must be licensed and this should be done prior to the athletic or other event. During the event, all available environmental health personnel should be inspecting each of the food serving units as a priority and then the drink serving units secondarily if time permits. The local and/or state government should grant the authority to the environmental health personnel to immediately shut down any unit that has serious environmental health and safety problems in the area of food protection.
- For temporary food venues and non-fixed facilities such as fairs and festivals, prior to the opening of the event, the sponsor of the event must inform in writing all food and drink vendors of the rules and regulations concerning food protection at the event and that a serious violation of them will result in immediate closure and denial of future right to operate at any event within the jurisdiction of the community. All available environmental health personnel should make complete inspections including of refrigeration, source of water supply, sewage and solid waste disposal practices, insect and rodent control techniques, etc. prior to the opening of the event. During the event, all available environmental health personnel should make frequent multiple spot checks of the food operations which are most likely to result in a potential for foodborne disease and also to check the water supply. The environmental health personnel should have the same authority to shut down units that are potentially hazardous.
- For temporary food venues which are mobile food delivery systems typically by trucks, all these units must be given a full inspection at the commissary before they are permitted to go on the road. All environmental health personnel despite their assignments, unless it is an absolute emergency, should stop and inspect all mobile food venues. They should have the same authority to shut down the units if they are potentially hazardous and escort them back to the commissary or if necessary call the police.
- Conduct self-inspections or third-party audits of all raw materials and raw material producers, facilities, processes, and personnel on a regular basis and make immediate corrections were problems occur.
- Ensure that all interior surfaces within buildings are impermeable, easily cleaned, and that there is no peeling paint.

- Upgrade facilities and production methods as needed to continue producing safe quality food.
- Use a safe water supply with adequate quantities of water for growing of crops; production, processing, and preparation of foodstuffs; and cleaning and sanitizing of all equipment, surfaces, and utensils.
- Frequently clean ice production and storage facilities and do routine bacteriological and chemical testing.
- Store all chemicals and other potentially hazardous materials away from food processing, food preparation, and food storage areas.
- Transport and then store all foods at proper temperatures in clean, pest-proof vehicles, containers, and facilities.
- Prevent contamination by sewage of all food growing, processing, storage, preparation, and serving areas.
- Discard all chipped, cracked, and marred items which have surfaces that may come in contact with food and replace these items or equipment immediately.
- Only purchase equipment that has been approved by an accrediting authority such as the National Sanitation Foundation.
- Teach the public how to prevent disease by proper use of time and temperature controls for all foods, as well as proper hand washing techniques when handling food.
- Do not permit food from unsafe food sources to enter the food production or food preparation system.
- Determine and enforce strictly all appropriate hot and cold temperatures and time periods for storage, processing, cooking, holding, preparation, and serving of food, and make necessary immediate corrections where needed, and destroy the food that has been handled improperly.
- Consider all equipment prior to usage and after usage to be contaminated and therefore clean and sanitize in an appropriate and timely manner.
- Enforce strictly all personal hygiene rules, especially the appropriate washing of hands and the removal of sick employees from the food processing and food preparation areas.
- Provide appropriate clean clothing for all employees to help prevent contamination of food.
- Keep written records of all major potential breaks in technique and exposures to microbiological, chemical, or physical contaminants, review them frequently and establish protocols to prevent them from reoccurring.
- Provide frequent training and retraining for all personnel.
- Use microbiological testing procedures as required and self-inspections to: determine areas of potential contamination and the type and amount of contamination; validate decontamination techniques and success rate; validate the effectiveness of training, and use of appropriate procedures especially concerning high-risk foods.
- Calibrate all instruments frequently to ensure their accuracy for microbiological and chemical testing.
- Use first in, first out procedures for all raw materials and finished products.
- Certify all vendors prior to use.
- Develop and strictly adhere to an allergen management program including: specific training for processing and supervisory personnel; segregation of food allergens during processing and storage; proper cleaning procedures to remove potential allergens; prevention of cross-contamination between allergens and allergen-free foods; review of allergen-free products labels which must state the potential contamination of any products; and determine if the ingredients supplied by the provider are certified as allergen free.
- Establish a regular cleaning and maintenance schedule for all buildings and facilities to prevent deterioration and contamination which may affect the health and safety of employees as well as the food for human consumption.
- Establish a regular maintenance, cleaning, and sanitizing schedule for all equipment with food contact surfaces.

- Establish an integrated pest management program including pest traps and screening of all openings to control insects and rodents by eliminating food, harborage, and breeding areas and then applying targeted specialized chemicals when necessary to destroy the pests.
- Regularly review all production and process controls to ensure that the food will not become contaminated. (See endnotes 6, 15, 16.)

DETERMINING THE SCOPE OF THE PROBLEM AND SPECIFIC BEST PRACTICES THROUGH USE OF THE SYSTEMS (PROCESS) APPROACH

Use of the systems approach in problem solving and resolution is a technique for determining, in a chronological manner, the complexity, parameters and various components of the entire food system, the linkages and interactions between the components, the critical points of potential failure, and the successful methods used to avert disease and injury and promote good health. The system consists of two basic components: *elements* and *processes*. The *elements* include the actual flow of food from origin to consumption and the potential hazards at each step along the way. The *processes* include the various people, situations, activities, restrictions, and interactions that can affect the flow in a negative or positive manner.

ELEMENTS

To understand food protection at the local level, it is necessary to understand the entire flow of food from production to consumption and the special problems related to each part of the flow process. The food flow includes production, processing, packaging, transportation, storage, preparation, consumption, and disposal of the food and its byproducts.

SUB-PROBLEMS INCLUDING LEADING TO IMPAIRMENT AND BEST PRACTICES FOR PRODUCTION OF FOOD

In general, the production of food is a very complex process during which disease organisms can be readily transmitted and allowed to grow in various environments because of deficiencies in: protection of raw products; protection of finished products; use of poor water sources; improper time and temperature controls; poor, outdated, and hard-to-clean equipment; production process control; buildings and facilities; personal hygiene of the employees; lack of appropriate training of employees, supervisors, and managers; improper use of partially used product; improper handling of high-risk foods; and lack of appropriate review to evaluate programs established to prevent disease and injury and protect the public. This situation is made even more complex when there is a demand on wholesalers of food to meet deadlines that are sometimes almost impossible to accomplish. A rushed situation many times leads to accidents and mishandling of foodstuffs.

There have been many recent examples of contaminated food entering the food stream in the United States. This food has come from numerous countries. A challenge of the global market relates to food that has come from foreign countries and which has been contaminated with sewage and has caused outbreaks of foodborne disease in the United States.

Specifically, in the production of food (eggs, fish and shellfish, fruits and nuts, grains, meat and meat products, milk and milk products, produce), the land and water utilized can be contaminated by people, animals, naturally occurring substances, and air. The contaminants of all raw food can be microbiological, chemical, and/or physical. Microorganisms, which cause disease in people, can be added to the raw food product by people and animals, either directly or through contaminated water. Chemical contaminants may include pesticides, fertilizers, and heavy metals from land and water pollution, natural sources, mining operations, aerial sources from smelting operations, etc. Chemicals, especially heavy metals that are bioaccumulative, can be concentrated in the food chain, therefore producing a greater effect in people than the original quantity of chemicals in the environment. Antibiotics are frequently added to feed to make chickens grow bigger and to help preserve fish and meat. Extraneous materials such as insect parts and rodent hairs are frequently found in raw agricultural material. Radioactive fallout can contaminate food products. Foods coming from foreign sources contain the contaminants from the areas of growth or production. In some cases, chemicals have been added intentionally to increase the value of the food. However, these chemicals can be extremely hazardous.

Industrial farm animal production systems create a substantial amount of air, water, and land pollution. Land is contaminated with large quantities of concentrated manure and urine, large quantities of solid waste such as feed, bedding, etc., agricultural chemicals, and dead carcasses, especially from chickens. Water sources, both surface and groundwater, are contaminated with runoff from the manure and urine, nutrients, industrial and agricultural chemicals, microorganisms, and heavy metals. The air is contaminated with methane gas, ammonia, carbon dioxide, etc. (See endnotes 18, 19.) In addition, poor indoor air quality is among other things a direct hazard to workers. The environmental problems may be highly detrimental to fish, other wildlife, and people. (See endnote 19.)

Best Practices in Production of Food

Refer to each of the individual categories in the Processing of Food section below for specific Best Practices. Also for specifics on environmental issues and Best Practices see Chapter 2, "Air Quality (Outdoor [Ambient] and Indoor)"; Chapter 11, "Sewage Disposal Systems"; Chapter 12, "Solid Waste, Hazardous Materials, and Hazardous Waste Management"; and Chapter 14, "Water Quality and Water Pollution."

PROCESSING OF FOOD

Bottled Water

Bottled water is simply water taken from a variety of different sources and put into a bottle for sale to the public. Bottled water in the United States is the fastest growing drink and the public spends billions of dollars each year to purchase it. Bottled water may come from artesian wells, artesian springs, and other water sources and may be purified, filtered, sterilized, or distilled. The waters differs in taste depending on the minerals that are present and the quantity. The type of water and source is very significant because as in all other water used for drinking purposes, the source may be contaminated with a multitude of organisms which could lead to disease, and/or a multitude of chemicals which could be harmful to the individual. Always read the label to determine the contaminants in the water and the quantities. The FDA set standards for bottled water based on the Environmental Protection Agency (EPA) standards for drinking water from the tap. (See endnote 59.)

Eggs

Eggs are an excellent medium for the growth of microorganisms. Contamination may be introduced through *S*. Enteritidis-contaminated chicks, improper handling, unhealthy workers, unclean equipment, dirty facilities, soiled uniforms, insects, rodents, and the presence of general debris. Unsafe water supplies and the waste from the chickens, including broken eggs, are also sources of contamination. Rats, stray cats, other animals, flies, and other insects are attracted to the poultry and egg producing and processing areas and readily spread many microorganisms. These organisms may penetrate the egg shell and grow rapidly. The major problem is *S*. Enteritidis, although other microorganisms can grow as well. Further, chemicals, especially fertilizers and pesticides, can contaminate the eggs and penetrate the shells, but this appears to be a limited problem. (See endnotes 13, 19.)

Best Practices in Processing of Eggs
- Utilize the voluntary "5-Star" Egg Safety Program of the Egg Safety Center, the seven principles of HACCP, and the Safe Quality Foods Program for all eggs as well as egg products supplier companies. (See Programs below and see endnote 10.)

- Wash and sanitize eggs prior to packaging.
- Purchase chicks from *S*. Enteritidis-free breeder flocks only.
- Limit the number of visitors and individuals that can go into different poultry houses.
- Carefully monitor how and when equipment is moved between different poultry houses and make sure the equipment has been cleaned properly.
- Prevent birds, cats, rats, other animals, and flies from entering poultry houses.
- Do not use chemical pesticides in poultry houses but use appropriate trapping and other devices for removal of insects and rodents.
- If a poultry house has to be depopulated because of *S*. Enteritidis-positive testing, remove all visible contaminants, remove all dust, feathers and old feed, clean completely, disinfect the premises, and air out very thoroughly.
- Complete all environmental sampling, testing, egg sampling, and further testing as required by The Egg Safety Final Rule issued by the FDA.
- Eggs must be kept at 45°F or less.
- Prepare and maintain all required written records including: a prevention plan; documentation of purchase of pullets; biosecurity measures; pest control measures; cleaning and disinfection procedures; and all sampling and testing records as well as temperature records. (See endnote 9.)

Fish and Shellfish

Fish and fishing may cause considerable problems to the environment through the production of substantial quantities of solid waste and byproducts, much of which may be organic in nature, waste water, water consumption, air pollution, and the use of substantial quantities of energy. Solid waste produced includes: inedible fish parts and endoskeletons; shell parts; inedible fish and other sea life; and other wastes including internal organs, etc. Fish and other seafood processing requires large quantities of uncontaminated potable water. The waste water has a high organic content because of the presence of blood, tissue, and dissolved proteins. It also contains high levels of nitrogen and phosphorus. The detergents and disinfectants used to clean and sanitize the facilities contribute to the high level of contamination in the waste water stream. Odors may quickly become a substantial problem because of the amount of organic matter present in the facility. Other potential air pollutants are exhaust gases of carbon dioxide, nitrogen oxides, and carbon monoxide from the various heating and power generation systems. Particulates may be generated during the fish smoking process. High levels of energy are used for producing hot water, steam, electricity, air conditioning, cooling, freezing, and production of ice. (See endnote 17.)

Fish and shellfish are exposed to microorganisms in polluted water, harvesting equipment, processing equipment, or from infected people. Insects, rodents, and birds contribute considerable quantities of contamination to fish and shellfish areas. Fish and shellfish provide an excellent medium for the rapid growth of these microorganisms. Shellfish obtain their food by filtering it out of water and at the same time take in microorganisms, chemical contaminants and natural toxins and also concentrate these substances within their bodies, thereby increasing the risk for disease. The consumption of raw products, such as oysters, adds considerably to potential health problems. Disease outbreaks have included hepatitis A, salmonellosis, typhoid fever, and cholera. Botulism type E, coagulation-positive *Staphylococci*, *Shigella*, *Vibrio vulnificus*, *Vibrio parahaemolyticus*, and other enteric organisms have also caused disease. Certain fish species have a tendency to bioaccumulate heavy metals and PCBs to dangerous levels. Some fish species deteriorate rapidly and produce histamines which causes histamine fish poisoning. Using contaminated ice and water for refrigeration and processing may add substantially to the problem. Microorganisms causing spoilage may contribute to the health problems especially in the histamine-producing fish such as tuna and mahi-mahi. Chemicals may become problems especially those which are cleaners and sanitizers. Extraneous items found along with fish may cause serious physical problems to individuals consuming this fish product. (See endnotes 14, 15.)

The incidence of disease is expected to increase because a substantial amount of seafood consumed in the United States is imported from a variety of countries where inadequate surveillance techniques may be utilized. The fish and/or shellfish may be grown in small ponds (aquaculture) or in areas where untreated animal or human waste is found. Further, wild animals can contaminate these areas. Rapid refrigeration and freezing at proper temperatures is essential but may be compromised and therefore create the necessary conditions for disease organisms to grow.

Best Practices in Processing of Fish and Shellfish (See endnote 15)
- Establish a food safety plan and management program using the HACCP system in a systematic, comprehensive, and thorough manner for each product.
- All critical control points must be identified and procedures established and followed in a very strict manner.
- Determine the accuracy of thermometers for refrigerators, freezers, and heat processors and frequently check the time and temperatures in the various parts of the process.
- Consider potential hazards of environmental contaminants such as agricultural chemicals that may have run off into the water as well as runoff from manure piles or other sewage systems.
- Monitor for residues of banned antibiotics.
- Utilize complete hand washing and sanitizing procedures especially after use of toilet facilities and before touching any part of the food or equipment used in the food process, as well as use foot baths where deemed necessary.
- Have the HACCP plan evaluated frequently to make sure that it is working properly and make necessary changes immediately.
- The facility must have proper control over every part of the food process that is outsourced.
- Provide a facility layout which enhances the cleaning of all parts of the food processing plant, provides immediate access to hand washing facilities, etc., while minimizing traffic through the processing plant.
- Use cold storage areas that are clean and dry.
- Use only well-serviced equipment which has been certified by a rating agency such as the National Sanitation Foundation.
- Clean all ventilation equipment and filters on a regular basis.
- Evaluate the incoming air for potential pollutants and take defensive measures against contamination of the food process.
- Establish a comprehensive pest control program that includes the following: removal of all litter and old equipment; installing seals or screens on all windows, doors, walls, floor drains, and other openings to the facility; establishing pest traps in areas away from the food processing areas and frequent inspection of them; and checks to determine if there are any rodent signs in the facility.
- Maintain the outdoors in such a manner that there are no puddles or muddy areas.
- Separate restrooms from food production and processing areas and provide hot and cold running water, hand detergent, and forced air dryers.
- Separate the finished product from the raw product areas to avoid cross-contamination.
- Protect all overhead lights with shields.
- Inspect all roofs to make sure that there are no leaks into the food processing or storage areas.
- All walls, ceilings, and floors must be made of easily cleanable, non-permeable materials and need to be cleaned frequently with the floors done daily and more frequently if necessary.
- All food contact surfaces must be cleaned very thoroughly and sanitized after each process or more frequently if necessary.

- Establish a routine maintenance and cleaning plan which is used at the end of each shift and more frequently when necessary.
- Store all packing materials off the floor and in clean and dust-free areas.
- Protect all food and raw ingredients during cleaning and sanitizing processes.
- Establish a mandatory personal hygiene program for all employees, provide necessary training and continuing education on a regular basis, and enforce the program strictly.
- Train all supervisors how to observe potential signs of disease in employees and immediately remove them from the food processing areas.
- Monitor all employees for signs of contagious diseases including skin sores and remove them from work immediately.
- Do not permit food, drinks, chewing gum, or tobacco in processing, packaging, or storage areas.
- Ensure that all workers wear appropriate protective clothing including clean aprons, facemasks, boots, and hair coverings.
- Provide a proper source of high-quality water in adequate quantities to perform all necessary functions for the growing, processing, and preparation of food, as well as all cleaning functions.
- Routinely test water and ice for potential contamination from microorganisms and chemicals, and evaluate for adequate chlorine residual in the water.
- Prevent the potential for backflow into the potable water supply by use of check valves, proper hose storage, and elimination of potential submerged inlets.
- Train all employees who use chemicals on proper dosages, use and storage away from food processing and preparation areas, and how to dispose of them properly.
- Provide an appropriate ventilation system and clean and service it regularly to avoid condensation on packaging materials or food contact surfaces.
- Use appropriate quality control procedures to ensure the safety of the finished product. (See endnote 15.)

Fruits, Vegetables, Juice, and Nuts

The substantial increase in the consumption of raw vegetables, fruits, and unpasteurized juices in the last several years has caused a significant increase in foodborne disease. Produce is now available in all seasons of the year because of increasing imports from many foreign countries. *Listeria monocytogenes*, *Clostridium botulinum*, and *Bacillus cereus* are naturally found in the soil and can readily contaminate produce. *Salmonella*, *E. coli* 0157:H7, *Campylobacter jejuni*, *Vibrio cholerae*, *Cryptosporidium*, parasites, and viruses are frequently found on fresh produce because of manure, contaminated irrigation water, contaminated processing water, mammals, reptiles, birds, insects, rodents, and contaminated human hands or bodies. A combination of sick people, healthy carriers of disease, lower sanitation standards locally, nationally, or internationally, and high mobility among seasonal workers helps spread a large variety of foodborne diseases.

Fruits and vegetables are usually sold fresh, receive minimal care, and may become contaminated in the fields during the growing season, during harvesting, during processing, and during preparation or use. Besides surface contamination, microorganisms may enter through the stem scar and breaks in the skin of the fruits or vegetables. The cutting of the products releases plant cellular fluids which may act as an excellent medium for the growth of microorganisms. The contaminants may be caused by microorganisms or agricultural and other chemicals. Sources of contamination may be in the air, soil, water, from sewage, from crops being flooded, from insects, rodents, birds, and other animals, and through the harvesting and transportation equipment as well as the people working in the fields. The contamination may occur in all parts of the produce system. This is compounded by potential contamination from water used in processing and during preparation from slicing, chopping, or shredding, which may readily spread the microorganisms from the equipment or simply through the mechanism of producing smaller pieces of the product and presenting larger surface areas for the growth of microorganisms. The fruits and vegetables may become contaminated with *E. coli* O157:H7, yeasts and molds, and other organisms. The fruits or vegetables may also become contaminated by chemicals in the air, soil, or water. The chemical contaminants may be spread throughout the products through the processing procedures. (See endnote 20.)

Seeds either fresh or cooked may contain low levels of pathogens such as *Salmonella* which may multiply rapidly during the sprouting process. Sprouts are seeds that have just started to grow. They are usually stored longer than the seeds used to grow fruits and vegetables. They are grown in warm, watery, and dark conditions, which is perfect for the growth of microorganisms. They may become exposed in storage as well as during the growth process to dust, contaminated water, animals, and/or people. The microorganisms can exist in or on the sprouts for about a year. The potential for disease caused by *Salmonella*, etc. is substantial as has been shown by previous outbreaks of disease. Sprouts may be especially a problem if the seeds and the seedlings are consumed raw. (See endnote 27.)

Juice may contain biological, chemical, and physical hazards. It may also be involved in causing allergies through cross-contamination with peanuts, soybeans, milk, eggs, fish, tree nuts, and wheat. Some acidic juices with a pH of 4.6 or less have been shown to contain *E. coli* O157:H7, various *Salmonella* species, and the protozoan parasite *Cryptosporidium parvum*. These organisms may be found in the intestinal tract of animals, their manure, or feces. Enteric organisms causing most of the foodborne illness outbreaks associated with juice may be found in the low acidic juices with pH greater than 4.6. Juices may be contaminated with viruses from sick farm workers or food handlers and cause outbreaks of foodborne disease.

Chemical hazards may come from the air, soil, and water and may include agricultural chemicals, especially pesticide residues, and other hazardous substances. Of considerable concern is patulin, a mycotoxin produced by fungi commonly found on improperly stored or damaged apples. Lead contamination of produce, such as lead arsenate from past use in agricultural settings, may come from contaminated soil, equipment, and vehicles that used leaded gasoline. Tin, which is frequently used as a coating in metal cans which are not lacquered, may leach into juice.

Physical hazards may include glass fragments from broken containers and metal fragments from cutting equipment. Other contaminants may be dirt, etc. (See endnote 24.)

Foodborne illness may be caused by *Salmonella* in low moisture products such as nuts or peanut butter made from the nuts. Outbreaks of disease have also been traced to chocolate, infant cereals, milk powder, powdered infant formulas, and other types of snacks and cereals containing nuts or nut products. Chemical or physical hazards may be present caused by contamination and/or adulteration of the products. Of considerable concern is allergen management since so many people have allergic responses to different types of food, especially peanuts. Contamination in all parts of the production system including the farm, through transportation and the shellers, hullers, processors, and manufacturers can lead to foodborne disease or allergic responses. Chemical hazards on nuts may include mycotoxins, antibiotics, pesticides, and sulfites. Physical hazards include all extraneous materials.

Floodwaters which contact the edible portions of crops present a significant biological and chemical hazard and therefore a serious health risk. All flooded foodstuffs must be considered to be contaminated and therefore must not enter the food processing chain. Disposal must be handled in such a manner that it will not contaminate other food crops and will not create land, water, or air pollution problems and not increase the levels of insects and rodents present in the area. (See endnote 26.)

Best Practices in Processing Fruits, Vegetables, Juice, and Nuts (See endnotes 22, 23, 24, 25)
- Follow the Best Practices in Processing of Fish and Shellfish shown above for processing of fruits, vegetables, juice, and nuts.
- Establish a food safety plan and management program using the HACCP system in a systematic, comprehensive, and thorough manner for each product.

- All critical control points must be identified and procedures established and followed in a very strict manner.
- Determine the accuracy of thermometers for refrigerators, freezers, and heat processors and frequently check the temperatures of these very significant pieces of equipment.
- Consider the potential hazards of environmental contaminants such as agricultural chemicals that may run off into the water as well as runoff from manure piles or other sewage systems.
- Inspect all fruits and vegetables throughout the processing stream to determine if there is gross contamination present and make immediate necessary corrections.
- Remove from the processing line all damaged or decomposed fruits and vegetables.
- Remove and destroy diseased or damaged seeds and sprouts that may be used for human consumption.
- Keep seeds that are used for production of sprouts in dry storage areas and in rodent-proof containers.
- Treat seeds and sprouts with approved methods using chemicals that will reduce the level of pathogens present. (See endnote 28.)
- Test seeds and sprouts for the presence of microorganisms that may cause disease in humans.
- Carefully treat and test all process water which will be reused for additional processing to avoid cross-contamination.
- Treat all contaminated water before releasing into a body of water.
- Use appropriate temperatures of wash water on the fruits and vegetables to produce an appropriately clean product.
- Establish a potential allergen control program for all foods but especially for nuts.
- Prevent cross-contamination from potential allergens to foods that are not typically causing allergic responses.
- The fruit used for citrus juices must be picked from trees, cleaned, and culled (the removal of any damaged fruit) prior to any type of treatment process.
- Processed juices must be tested for levels of generic *E. coli* in the finished product, and if one sample is positive the entire process must be reviewed for breaks in techniques.
- Shelf-stable juices and concentrates must be heat treated at 194°F or 90°C for 2 seconds and then the containers must be filled at 185°F or 85°C and then held for 1 minute at this temperature.
- Treat the surface of the fruit either with ultraviolet radiation, pulsed light, or a chemical treatment approved by the FDA.
- Pasteurize juices at 160°F or 71°C for 3 seconds.
- Control patulin in apple juice by not using any fallen fruit or fruit with visible damage, and proper storage of apples to avoid improper temperatures, core rot, and contamination from insects and fungi.
- Rigorously control the suppliers of juice concentrate when making juice from concentrates.
- Primarily focus on contamination by *Salmonella* in raw nuts because of how nuts are cultivated and harvested.
- Use heat treatment such as oil roasting, dry roasting, or steam pasteurization to control *Salmonella*, which is not eliminated during refrigeration, freezing, or drying.
- Carefully desegregate all crops affected by flood waters from crops that have been flood free.
- Do not use farm equipment that has either been contaminated by floodwaters or has been utilized to remove crops which have been damaged by flood waters when harvesting non-floodwater crops. Decontaminate all of this equipment as quickly as possible.
- Establish a 30-foot buffer zone between flooded areas and those areas containing crops that will be used for human consumption.

- Conduct a survey of the wells providing water to the process and determine if they have become contaminated by flood waters and if so follow all necessary procedures for decontamination before utilizing the water for crops or processing of food for human consumption.

Grains, Soybeans, and Hay

Grains include corn, millet, barley, buckwheat, wheat, oats, rice, sorghum, and malt. Grains can become contaminated with molds during the storage process. Dust from the grains can become explosive during storage. Grains may be contaminated by pathogens present in the soil, in agricultural water or processing water, in manure used as fertilizer, and from animals in or near the fields or harvesting or storage areas. *Fusarium* mycotoxins may be produced on various cereals and other crops and cause illness in people and animals. (See endnote 29.) Worker health and proper sanitation is also a potentially serious problem.

Other major crops are soybeans and hay. Soil preparation for the crops may result in major soil erosion, water erosion, and wind erosion, leading to problems in surface bodies of water. The use of genetically modified seeds which may make crop production more efficient may also increase resistance to pesticides as well as the development of weeds. The management of nutrients from chemical fertilizers, manure, and sewage sludge if done improperly can lead to substantial contamination of surface bodies of water and also groundwater supplies. Sprayers used to disperse chemicals, although efficient, can cause toxic chemicals to enter the ambient air.

Irrigation systems may operate inefficiently and produce waste water while also creating drainage problems. Improper drainage may contaminate the land, water, and wetlands. Improper harvesting may create substantial dust clouds and lead to contamination of other crops, land, water, and air. Storage of fertilizer and chemicals including pesticides can lead to contamination of air, water, and land.

Best Practices in Processing of Grains, Soybeans, and Hay

- Hold the managers and owners of food processing, production, preparation, and serving companies criminally responsible for purposeful acts of negligence resulting in the injury or death of people, as was done on September 21, 2015, when a federal judge for the first time gave the chief executive officer and other top level administrators of a US peanut butter company a substantial prison sentence for outbreaks of foodborne disease due to massive and known acts of negligence based on fraud, conspiracy, and the introduction of adulterated food into interstate commerce.
- Reduce the potential for increased mycotoxins through: rotation of crops; reduction of prior crop debris; appropriate weed and insect control; application of the correct amounts of fertilizer; appropriate use of fungicides, and very importantly, adequate drying. (See endnote 29.)
- Remove damaged grains and chaff from the grain stream to remove concentrated contaminants which would increase mycotoxins.
- Harvest grain as soon as possible when ripened and cool and dry as soon as possible.
- Reduce erosion through the use of: conservation tillage, contour farming and strip cropping, cover crops, grassed waterways, terraces, windbreaks, grass barriers, and snow fences. (See endnote 30.)
- Avoid insect resistance to the Bt gene (This is a natural toxin produced by the bacterium *Bacillus thuringiensis.* This product is frequently used by organic farmers.). Plant strips of non-genetically modified seeds between areas of modified seeds.
- Conduct regular soil tests on each area to determine levels of nutrients present and amount of nutrients which needs to be used.
- Determine through testing manure the amount and types of nutrients present and how much to use in different areas.

- Use conservation tillage and other erosion control practices to minimize the loss of phosphorus from the land and into the water.
- Determine the proper timing of the application of fertilizer to maximize its use.
- Use acceptable integrated pest management practices to obtain maximum results while applying minimal quantities of chemicals.
- Use the pesticide to match the problem and minimize contamination of the environment.
- Evaluate weather conditions before applying pesticides and avoid times of substantial rainfall or windy conditions to prevent contamination of bodies of water and areas not meant to be sprayed.
- Carefully determine the amount of chemicals needed and adjust the calibration of the sprayer to perform at maximum capacity.
- Do not spray chemicals in areas where there are sinkholes, depressions, wells, surface water, schools, businesses, etc.
- Minimize water usage in all types of irrigation systems while providing good drainage.
- Store fertilizers away from pesticides (both should be stored in dry, secured well-managed areas and containers) and prevent accidental spills while quickly containing those that have actually occurred.
- Train personnel in all aspects of the preparation, use, storage, and disposal of fertilizers and pesticides.
- Secure petroleum products in aboveground and underground storage systems while preventing overfilling, pipe leaks, and tank corrosion. (See endnote 30.)

Meat

(See endnotes 6, 21, 31, 32)

Most meat today comes from animals that are raised on feedlots, with all the attendant environmental problems of the spread of disease from animal to animal and disposal of large quantities of waste. Meat is exposed to a large variety and quantity of microorganisms from people, surfaces, and equipment, throughout processing. The health of the animals can be compromised. The stockyards, slaughterhouses, meat-processing plants, retail cutting operations, storage, and display areas are subject to substantial levels of contamination. There are large quantities of fecal and urine waste to be removed, which are serious fly and rodent attractants. Offal, blood, and other wastes are present in large quantities. Odors are intense. During the slaughtering process, microorganisms can be added frequently or continuously by equipment, appurtenances, the physical structure, people, and improper cleaning techniques. Cutting instruments and food contact surfaces are of great significance and frequently are not cleaned and sanitized in an appropriate manner or in a timely fashion. The presence of abscesses in the carcasses is always a problem and many meat packing plants have too few federal meat inspectors to ensure a production line is shut down when an abscess is encountered.

Bench trim (the fat and meat trimmed from roasts and steaks), which is mixed throughout the ground beef, appears to be a special problem helping cause the spread of microorganisms. There are the potentially severe problems of moving the slaughtered meat rapidly into appropriate refrigeration and freezing units that maintain safe temperatures determined by finely calibrated instrumentation. Recall programs fail when there is slow determination of the presence of contaminated products and they are not taken off of the shelves of the retailers rapidly enough. Meats contain *Brucella, Salmonella, Streptococcus, Mycobacterium, Pseudomonas, Staphylococcus, E. coli*, etc. Pork may contain *Trichinella spiralis*. Chemical contamination can occur from cleaning and sanitizing materials, the presence of pesticides, indiscriminate spraying of the premises, introduction of allergens from added materials, etc.

Language barriers may lead to misinterpretations of information in all of the food trades and therefore may complicate the necessary training that should be given to all employees in the prevention of the spread of disease and contamination by chemicals during cleaning, storage, and disposal processes.

Ready-to-eat products which are consumed without further processing by heating are used in very large quantities. These food products include deli-style meats, snack foods, buffet foods, and various platters and spreads that are prepared for mass feeding programs. These are very frequently improperly refrigerated, handled improperly, and easily contaminated.

Best Practices in Processing of Meat and Meat Products

Follow the Best Practices in Processing of Fish and Shellfish for all meats and meat products.
- Establish a food safety plan and management program using the HACCP system in a systematic, comprehensive, and thorough manner for each product.
- All critical control points must be identified and procedures established and followed in a very strict manner.
- Determine the accuracy of thermometers for refrigerators, freezers, and heat processors and frequently check the time and temperatures in the various parts of the process.
- Consider potential hazards of environmental contaminants such as agricultural chemicals that may have run off into the water as well as runoff from manure piles or other sewage systems.
- Determine the quality of the air which will be going through the slaughter plant and if necessary, use appropriate filters and air treatment procedures before use within the plant.
- Control air movement throughout the slaughter plant with the air flowing from clean areas always to dirty areas and not vice versa.
- Separate physically all dirty areas from clean areas and always have the flow of people and materials as well as products go from clean to dirty.
- Ensure the integrity of the structure and make sure there are no leaks from roofs or other areas into the clean areas and processing areas especially.
- Use proper drains in all processing areas, and clean and maintain them on a regular basis.
- The mid-shift should perform complete cleanup of all areas surfaces and equipment as well as coolers while preventing splashing and aerosols from getting into the air and contaminating exposed products.
- All reused water including re-circulated warm water and that used in thermal pasteurization units must meet the standards of potable water.
- If the machinery breaks down, all products must still be maintained at appropriate temperatures and for appropriate time periods.
- Holding pens and troughs must be kept as clean as possible.
- Reduce potential pathogens on hides before the de-hiding process by washing the animals down thoroughly while reducing airborne dust, animal waste, and dirt.
- Remove the hides in as clean a manner as possible to reduce potential problems later in the process.
- Use sanitized knives with color-coded handles, which are constantly rotated, for the slaughter of the animal and do not reuse on another animal without proper cleaning and sanitizing in water at a temperature of at least 180°F.
- Eviscerate the animal with sanitized equipment in as clean a manner as possible to keep from contaminating the carcass.
- Wash the carcass thoroughly with potable water and chill rapidly within 1 hour of the bleed out.
- Frequently during the day determine, record, and alter if necessary the time and temperature controls for all critical control points from the slaughtering process through the finished products being sent out for retail sales.
- Evaluate carefully all raw materials coming into the meat processing plants including checking the physical condition of the meats, potential breakdown in refrigeration temperatures, and visible contamination, and reject them before they enter the food processing chain.

- On a daily or more frequent basis check all sources of potable water, cleanliness, and use of properly sanitized tools and equipment, and cleanliness of all facilities from the unloading bay to the storage rooms to the preparation areas to the final product production including all forms of heat treatment and when necessary cold storage.
- At a change of product labeling and more frequently if necessary, clean and sanitize the labeling machine since it may become a major source of cross-contamination or introduction of substances that can cause allergic responses.

Milk and Dairy

Environmental problems on dairy farms include poor ventilation, poor lighting, poor physical facilities, waste handling, flies, rodents, improper handling of chemicals and pesticides, improper cleaning up and sanitizing of pipelines, bulk tanks and milking equipment, inadequate refrigeration and improper storage areas. Of particular concern are: the safety of processing water; the condition and cleanliness of food contact surfaces; and the potential for cross-contamination. The protection of the milk and milk products, including the packaging material, the control of employee health and personal hygiene, and the exclusion of insects, especially flies, and rodents from animal areas and food processing areas are also of paramount concern. The amount of foodborne disease involving milk has dropped dramatically from 25% in 1938 to less than 1% today. Although milk and milk products may be pasteurized, they still can become contaminated as the product is used or actually misused. Unpasteurized milk and milk products continue to lead to outbreaks of foodborne disease.

In some states, individuals are allowed to purchase raw milk because some people think that it is healthier for children. People are allowed to also purchase a part of the cow which is a way of getting around the requirement for pasteurization. This is an extremely dangerous procedure. There are frequent outbreaks of disease from unpasteurized milk. Unpasteurized milk does not have any health benefits beyond those conferred by pasteurized milk and even if it had, it would not be worth the risk of giving children foodborne diseases.

Although *L. monocytogenes* is not supposed to grow at freezing temperatures, an outbreak of this organism resulting in sickness and death occurred in 2015. This resulted in a massive recall of ice cream by the company producing it and rigorous cleaning of the entire facility, equipment, and piping systems and an intensive sanitizing program. Testing protocols, policies, and procedures and intensive employee training programs should be revised. Studies were conducted concerning the cause of the outbreak and potential or actual sources of the microorganism. An independent microbiological expert had been contracted to establish and review all controls to prevent further similar outbreaks of disease. The state agencies were notified immediately of any problems that had been found at the plant and how the plant responded. If a positive test for the microorganism or others that may cause disease was found for any batch, the batch was immediately retained and disposed of properly. (See endnote 57.)

Best Practices in Processing Milk and Dairy

The Best Practices in all aspects of the processing of milk and dairy have been evolving and documented in great depth by the United States Public Health Service since 1924. The document is entitled "Grade 'A' Pasteurized Milk Ordinance" (2011 Revision) and was revised in 2013. It provides for states and localities as well as others the most up-to-date in-depth document on milk sanitation and production from the cow to the consumer. It would be a duplication of effort to list other Best Practices. (See endnote 33.)

Poultry

Poultry processing areas can be highly unsanitary because of the presence of substantial quantities of blood, dirt, and feathers. In addition, fecal waste and other solid wastes are a constant and substantial problem. Rats, flies, roaches, and other insects may be found in substantial quantities. The slaughtered birds must be chilled immediately. Antiquated processing methods that include the use of a common rinsing submersion after evisceration have been implicated in the spread of multistate outbreaks of multidrug-resistant *Salmonella* organisms. Major concerns relate to the feed and water as well as the presence of sick birds in the slaughtering process.

Best Practices and Poultry Processing
- Follow all Best Practices in Processing of Meat and Meat Products (including endnotes).
- Keep birds well hydrated and reduce stress.
- Eliminate all sick or injured birds from the slaughtering process.
- Accurately monitor water temperatures in the slaughtering process as well as in the pre-chilling tank.
- Add clean ice made of potable water to keep the birds chilled to an internal temperature of 40°F for storage, transportation, and handling.
- Sell as rapidly as possible, but not later than 2 days after slaughter if fresh and not frozen.
- Protect all sources of surface and ground water supplies from poultry manure litter and waste water by using proper disposal and/or treatment techniques.
- Keep all poultry confinement areas clean and with dry litter.
- Thoroughly clean all areas especially where de-feathering occurs and where blood is present.

SUB-PROBLEMS INCLUDING LEADING TO IMPAIRMENT AND BEST PRACTICES FOR PACKING OF FOOD

Packaging is used to contain, preserve, and protect the food product against contamination. It also provides necessary information to trace the food products and remove them rapidly from the food distribution chain when contaminated. The packaging may make the product easier to handle, be tamper resistant, and improve the shelf-life of the product.

Glass is usually chemically inert, odorless, and rigid, and is used for certain types of food and as a container for food processing. It can be made in different colors and therefore protects the contained foods from light. It can be recycled and therefore not contribute to the solid waste stream. However, some glass may contain lead which may leach into the food. Substances used to seal the bottles and jars may also leach into the food.

Metal including aluminum and tinplate is very versatile and provides excellent physical protection from contaminants as well as the ability to have the food processed within the containers. The metal containers may be recycled and therefore reduce substantially the solid waste stream.

Plastics are typically chemically resistant, inexpensive and lightweight, and easy to use. However, the use of recycled plastics in food packaging may be problematic. Plastic bottles can absorb the substances that they contain and therefore may later contaminate food. Paraffin from packaging materials may be found in the enclosed food. In Europe, studies have shown that migration of chemicals from food packaging is unacceptably high. In Asia, phthalates, a PVC plasticizer which is banned in several countries, migrate into food from plastic food jars.

Paper and paperboard frequently used for containing foods have poor barrier properties and are not heat sealable. Paper food packaging may contain printing ink from recycled paper products and become a source of phthalates. The coated paper lining in the inside of cereal boxes may become a source of contamination as it leaches into the food.

Packaging materials frequently become serious environmental problems because of disposal quantities and the nature of the materials, especially plastics, which may not be biodegradable or very slowly biodegradable. While appropriate food packaging can reduce bacterial contamination and extend the shelf-life of the product, their production utilizes raw materials and considerable energy, and may cause land pollution, water pollution, problems with sea life, and air pollution. (See endnote 34.)

Best Practices in Packaging of Food

- Implement appropriate source reduction practices such as using lighter weight and reusable materials in the manufacturing of various packaging containers for food in order to reduce later the amount of the solid waste stream.
- Develop legislation that encourages or makes mandatory certain types of recycling of food and beverage containers.
- Develop an effective system to prevent the use of contaminated, damaged, or defective containers for food.
- Determine that all incoming materials used in packaging are from an approved source and are not contaminated.
- Utilize food containers only for their intended purpose and not other types of storage.
- Evaluate labeling on food containers to make sure that it is accurate and informative. (See endnote 35.)

SUB-PROBLEMS INCLUDING LEADING TO IMPAIRMENT AND BEST PRACTICES FOR TRANSPORTATION OF FOOD

Yearly a huge amount of food is transported around the world. This food while waiting for shipment and en route may be unprotected from purposeful contamination, improper holding temperatures and time periods, cross-contamination from other products, and storage in areas which are not meant for food transportation. Another complicating factor is that various foods have various temperature and time control requirements which the system does not always provide for. Pest control is always a serious concern because of the presence of organic material.

Food may be transported in bulk or in packages by means of trucks, trains, ships, and planes. The cleanliness of the transporting vehicles and food contact surfaces is essential to prevent contamination of the food supply. Chemicals must not be transported in the same vehicles as food. The transportation vehicles must be free of insects and rodents. Inadequate refrigeration or poor refrigeration contributes to the growth of microorganisms. The time elapsed during transportation can affect the quality of the products as well as the potential for a foodborne disease. Also, one of the weakest links in the transportation process is the final step, which is the receiving and moving into appropriate storage food products at the retail delivery site. These include refrigerated, frozen, or dry storage areas. Care must be taken that the product is received intact at the proper temperature and properly stored at once.

Best Practices in Transportation of Food

- Transport all finished products at appropriate refrigeration or freezer temperatures, keep thermometers (recording and indicating) accurate and immediately available, and use the vehicle only for the type of product indicated after thorough cleaning and sanitizing between different batches of the product.
- Circulate air uniformly through transport vehicles that involve refrigeration or freezing.
- Ensure that loading and unloading of the transport vehicles is done efficiently and quickly and that the temperature of the product does not change because of the movement in and out of facilities.
- Do not use transport vehicles carrying food of any type for other purposes such as the transport of animals or materials placed in the empty vehicles as they return to the original sender.
- Keep accurate records of all substances placed in the trucks and have them ready for inspection by the appropriate authorities. (See endnote 36.)
- Use preventive control requirements proposed under the Food Safety Modernization Act—Proposed Rule on Sanitary Transportation of Human and Animal Food, which has specific requirements for preventive controls and standards for human food, animal food, produce

safety and foreign supplier verification, and a program for accreditation of third-party auditors. (See endnotes 37, 38.)

SUB-PROBLEMS INCLUDING LEADING TO IMPAIRMENT AND BEST PRACTICES FOR STORAGE OF FOOD

Food may be stored either in dry form, refrigerated form, or frozen form. In dry form, the hazards may include insects, rodents, birds, dirt, chemicals, poor ventilation, flooding, improper rotation of food, etc. In refrigerated form, the hazards may include inadequate cooling, poor circulation of air, timeliness in cooling, oversized containers, contaminated food contact surfaces, improper storage of food, and improper rotation of food. In frozen form, the hazards may include inadequate and slow cooling, poor air circulation, contaminated food contact surfaces, defrosting and then re-freezing of food, and improper rotation of food. An additional major concern is the outbreak of a fire that can contaminate the food in all three types of storage with water, smoke, and heat.

Best Practices in Storage of Food
- Develop and utilize a program of food rotation so that the first in to storage is the first out to be used.
- Keep storerooms used for dry storage well ventilated, well lit, at temperatures between 50°F and 70°F in low humidity away from direct sunlight, and thoroughly cleaned on a regular basis.
- Store all foods on pallets off floors and away from walls.
- Use continuous inspection procedures to determine the presence of insects and/or rodents and eliminate them as soon as they are discovered.
- Store all chemicals separate from foodstuffs in a different area.
- Observe and note the temperatures for all refrigerated and frozen items and observe proper time sequences as determined by the latest rules and regulations of the US FDA as well as those published by local and state authorities.
- Use by designated use date as listed in the endnote 39 or by tables published by the proper authorities.
- Cool down and heat up products based on the time sequences allowed by the proper authorities. (See endnotes 39, 40.)

SUB-PROBLEMS INCLUDING LEADING TO IMPAIRMENT AND BEST PRACTICES FOR PREPARATION OF FOOD

Preparation and serving of food is highly labor intensive and therefore can easily become a major source of contamination potentially causing outbreaks of foodborne disease. Food preparation may follow a complex pattern from point of delivery of raw ingredients or finished products that are put into storage or taken immediately to the preparation area. The ingredients or products may be prepared in numerous ways including mixing, grinding, chopping, combining, and cooking at a variety of temperatures and holding times, thereby spreading existing and new contaminants throughout the food. A substantial amount of different types of equipment may be used and not properly cleaned and sanitized. Serving may be from steam tables or cold tables, thereby exacerbating the potential for disease spread because of the presence of consumers along the food line. Other types of serving by wait staff at tables may be poor and cause potential contamination. Menus become a source of cross-contamination since they are handled by numerous people. Cleanup and removal of dirty dishes and utensils as well as partially eaten food may be handled poorly. A contaminated rag may be used to wipe off the tables and also the seats of the chairs or booths. Dishwashers tend to operate in many cases at improper temperatures. There are potentially serious insect and rodent infestations, poorly cleaned facilities, improper use of sanitization techniques, etc. (Further, there are many other items which could be listed as potential sources of contamination during the preparation and serving of food. It would take a complete inspection of the establishment conducted by a well-qualified environmental health person to determine all of the problems.)

Because of the size and nature of the food preparation and serving industry, there is a constant turnover of large numbers of untrained or poorly trained people and low-level managers who actually carry out the various functions of the establishment. Many of these people, although well-meaning and eager, are very inexperienced and very young and without proper guidance. A lack of off- and on-site training as well as a lack of continuous close supervision unfortunately has become part of the problem. When you find serious problems in a food service establishment, it is necessary to immediately recognize they may be caused by inadequate supervision and managerial control. It is the responsibility of supervisors and managers to make sure that all necessary actions are taken to prevent the contamination of food and the spread of foodborne disease. Scheduling is also a serious concern because a tired worker may become ineffective and therefore more readily spread disease.

Best Practices in Preparation of Food
- See Food Problems through Use of Hazard Analysis Critical Control Points (HACCP) Principles above.
- See General Best Practices for Food Security and Protection above.
- When making an inspection or special study, follow the flow of the food through delivery, storage, use of equipment, preparation, serving, cleaning and storage, and waste disposal.
- Determine critical points of potential infection or contamination and evaluate these using appropriate time and temperature controls and determination of physical exposure to the food. (See endnotes 8, 41.)

TYPES OF PROCESSES AND METHODS OF INTERVENTION

PROCESSES

Interventions

An intervention is a means of either preventing an environmental problem from occurring or resolving the problem once it has occurred by the use of Best Practices. Regulatory agencies at the federal state and local level and laws, rules, and regulations including the Food Safety Modernization Act which was signed into law on January 4, 2011, are used to protect food. The law provides for: inspections, surveys, sampling, laboratories, and epidemiological studies; databases, and technical and general information from the Centers for Disease Control and Prevention (CDC), US Department of Agriculture (USDA)/Food Safety and Inspection Service (FSIS), FDA, and EPA; industries, industry associations, and professional associations at the local, state, and national level; educational procedures; enforcement procedures; Voluntary National Retail Food Regulatory Program Standards; and the Presidents' Food Safety Councils.

USE OF SYSTEMS APPROACH WITH CRITICAL CONTROL POINTS TO PREVENT AND MITIGATE POTENTIAL FOODBORNE DISEASE

See Food Problems through Use of Hazard Analysis Critical Control Points (HACCP) Principles above.

Best Practices in Using Systems Approach with Critical Control Points

See Food Problems through Use of Hazard Analysis Critical Control Points (HACCP) Principles and General Best Practices for Food Security and Protection above.

USE OF THE REGULATORY PROCESS TO PREVENT AND MITIGATE POTENTIAL FOODBORNE DISEASE

The goal of the regulatory agencies is fourfold:

1. Preventing harm to consumers
2. Providing effective food protection inspections and enforcement which depends upon good data and analysis
3. Identifying outbreaks of foodborne illness quickly and stopping them
4. Creating a transparent system and advising the public on the results of food establishment inspections and surveys

However, the goal is only partially being met, because the regulatory system in the United States varies from community to community and is haphazard, repetitive, lacking in appropriate current laws, or enforcement of them, and lacking sufficient trained personnel, financial support, and other resources for inspections, investigations, controls, and appropriate legal actions, including mandatory recalls. Several major federal agencies and numerous state and local agencies are involved in overlapping and underfunded food safety control. A multitude of inspection forms is utilized and therefore data collection and analysis is problematic. As with the interface between any number of governmental bureaucratic institutions, turf battles exist and need to be addressed.

Currently, the federal government has greater responsibility for imported foods, food production, food processing, and interstate food transportation, while state governments are responsible for food processing, intrastate transportation, and some retail food. Some state health departments and/or agricultural departments are charged with the oversight of local food protection programs while others are not, thus leading to inconsistencies in food protection from one location to another.

Local government, although it may be involved in production and processing programs, essentially works with the retail food industry. All agencies have some enforcement power with the weakest being at the federal government level and the strongest at the local government level. All agencies are involved in foodborne disease outbreaks, with the federal government having the greatest expertise and resources (but even these resources are far too few to carry out the type of programs that are necessary to keep the food protection system from being, in fact, a food contamination system), followed by the state government and finally the local government with the least physical resources, including epidemiological support, and funding.

The federal government is responsible for setting up the recommended standards for food protection through its model, the current edition of the FDA Food Code, which is based on sound scientific, technical, and legal determinations to regulate restaurants, grocery stores, institutions, etc. Most of the states and territories, and therefore local entities, utilize some form of the Food Code. However, this still means that there are at least 50 food safety systems in the United States instead of one. In total, the lack of good communications despite the efforts of the CDC and other groups, the varying abilities of the individuals conducting the food protection programs, and the lack of funding from federal sources especially to local programs, has created a severe national health and safety problem.

Best Practices in Regulatory Process (See endnotes 53, 54, 55)
- Establish uniform standards for all localities at all levels by adopting immediately the latest version of the FDA Food Code, put them immediately into operational form, and make revisions in the uniform standards as new versions of the code are produced and put into operation.
- Develop immediately a system-wide regulatory food safety reform program in the areas of surveillance, foodborne outbreak response, inspection, regulations, and program enhancing.

- Enhance legal authority for all entities by modeling this authority on the Voluntary National Retail Food Regulatory Program Standards which requires that critical violations which may cause foodborne illness be corrected in a timely manner.
- Improve communication and coordination between all entities to promote a rapid response to foodborne illness outbreaks and food recalls.
- Improve the sharing of information between all entities in order to prevent, mitigate, or control an outbreak of foodborne disease rapidly.
- Use a standardized national model for all entities in identifying, reporting, and tracing back foodborne illness cases to the source food.
- Reduce the number of pathogens in raw and processed foods.
- Provide appropriate budgetary support at all levels, especially the local and state level, where a very large proportion of the actual work is carried out.
- Determine the competencies needed by all levels of governmental program employees and prepare effective continuing education programs as well as certification programs to bring individuals up to appropriate levels of competency.
- Provide adequate numbers of appropriately trained individuals to carry out all food protection activities in an efficient and timely manner.
- Provide food protection professionals with proper and up-to-date equipment including accurate thermocouples, test kits, pH meters, light meters, cameras, etc.
- Provide computers and appropriate software to quickly record necessary findings and allow them to be printed out immediately for the use of managers in various food service organizations.
- Model local food safety programs after those which have been highly successful, for example, recent recipients of the Samuel J. Crumbine Award, such as the City of Columbus, Ohio; Sacramento County, California; Multnomah County, Oregon; Lincoln-Hamilton County, Nebraska; Hamilton County, Ohio; City of Toronto, Ontario, Canada; and Salt Lake Valley, Utah. Modifications can be made to meet local and regional practices and sensibilities. (For the requirements for the award, see endnote 58.)
- Develop an integrated national food protection system which will act as a full partnership between federal, state, local agencies, and industry, with financial support from the federal government along with the current financial support from local government. A properly operating, financially secure, food safety system is an integral part of a healthy economy and the homeland security of the United States of America.
- Provide well-equipped rapid response epidemiological support to help resolve outbreaks of foodborne and waterborne disease.
- Provide well-equipped rapid response laboratory support to help resolve outbreaks of foodborne and waterborne disease.

USE OF VERIFICATION OF SAFETY STANDARDS FOR IMPORTED FOOD TO PREVENT AND MITIGATE POTENTIAL FOODBORNE DISEASE

(See endnote 42)

The United States imports food from more than 150 different countries which arrives at more than 300 ports of entry into the country. About half of the fresh fruit eaten comes from abroad. The sheer numbers are beyond the capacity of the United States to regulate properly at this time. The World Health Organization has identified foodborne disease outbreaks and incidents as a major global public health threat for the 21st century.

Beyond the very serious problems of food protection in many of the producing countries, there is the increasing threat of harmful chemicals intentionally added to the food supply to increase profits and/or for terrorism. In China, companies added melamine (a very cheap but highly dangerous organic compound used to bulk up food including milk products) to infant formula which has caused infant deaths and potential lifelong health problems.

Best Practices for Verification of Safety Standards for Imported Foods (See endnote 42)

- Frequently update all practices and procedures related to imported foods by utilizing the recommendations and specific short- and long-term action steps identified in the current "Good Importer Practices" document of the Office of Policy and Planning of the FDA. These recommendations come from a working group of experts from eight different federal departments meeting together as a committee.
- Establish a products safety management program for each company importing food into the United States.
- Understand the product being imported and the appropriate requirements of the United States for the production, transportation, storage, and delivery of the product and its usage.
- Verify and document in writing that the product meets and the company follows the standards throughout the product lifecycle and throughout the supply chain as established above.
- Immediately take corrective and preventive action when the product is not in compliance.
- Establish a clear management structure for product safety including all necessary training in evaluating risks associated with the product, means of disease prevention, mitigation and control, and disposal when necessary.
- Especially in relation to drugs, beware that the product may be counterfeited and therefore a threat to health and safety.
- Determine if the product has been exposed to pesticides or other hazardous chemicals during any of the facets of growth, production, transportation, and storage and if so condemn and destroy the product.
- Determine when a safety recall is necessary, immediately inform the public and remove the product from distribution or sales as quickly as possible.
- Periodically inspect the foreign companies that are exporting products to the United States and make sure that they are complying with all appropriate laws, rules, and regulations.
- Compile and constantly update a master list of companies that have been found in noncompliance on multiple occasions.

FUNCTION OF INDUSTRY IN THE PREVENTION AND MITIGATION OF POTENTIAL FOODBORNE DISEASE

In the 1950s and early 1960s, industry did a remarkable job in correcting many of the problems that led to foodborne disease and therefore the greatest concern was typically at the local level. However, in subsequent decades, with globalization and the increase in vulnerable populations in the United States and the rest of the world, there was a significant increase in foodborne disease problems related to original sources of food and their processing. Some corporations elected to be excellent corporate citizens, while others were strictly concerned with profits and ignored the potential for outbreaks of foodborne disease.

In 2009, a very serious outbreak of foodborne disease occurred, with its origin in a peanut butter processing plant. Nine people were reported to have died and 22,500 people were reported ill. The plant had many serious environmental health problems ranging from rodents to various types of other contamination. Although the facility had been inspected by a company paid by the facility and problems had been noted by the individual carrying out the inspection, corrections were not made and the product was sent out to other manufacturers as an ingredient in the other manufacturers' products. Industry-type good citizenship was totally lacking. In addition, the reliance of regulatory agencies on certain industries to evaluate and control ongoing problems without proper supervision or with limited supervision, in some areas, because of lack of funding and/or incompetence or neglect, is a serious concern.

This is but an example of the potential for outbreaks of foodborne disease from a large variety of industries today. To determine some of the problems and their extent, it is necessary to examine data gathered, analyzed, and put out for consumption on a regular basis by the CDC.

Regulatory agencies alone cannot accomplish the goal of excellent food protection without the full cooperation and intense level of preemptive programs carried out by highly qualified professionals in environmental health working in industry. This programming is not in lieu of regulatory work, but is rather a highly significant portion of the total effort to protect the consumer at the national, state, and local level. After all, the regulators may see a given food operation once a year, maybe twice a year, or even four times a year in some instances, but the industry supervises the food protection operation on a daily basis.

A serious current problem in a multistate investigation of foodborne illness is a lack of uniform protocol and forms issued by regulatory authorities. This leads to confused and inaccurate data collection and therefore hinders resolution of the disease outbreak.

Best Practices in Industry
- Work cooperatively with other parts of the industry, including competitors and suppliers, to determine potential sources of contamination and techniques for removing them from the food processing chain.
- Utilize highly professional environmental health staff with excellent credentials and substantial field experience within the industry to control potential disease outbreaks and maintain product quality.
- Use outside experts to resolve specific problems which cannot typically be handled by in-house staff.
- Work closely with local, state, and federal food protection programs to resolve problems before they become substantial in size and costly.
- Recognize that the CDC has the ultimate scientific knowledge and experience and is the authority regarding foodborne disease outbreaks, nationally and internationally. They provide a series of highly effective programs to prevent disease in various parts of the system including the grower, supplier, distributor, and food service outlet.
- Develop a produce safety initiative including: field inspections of the suppliers; testing for pathogens in water and on produce; enhanced pathogen elimination in the processing facility; and verifying and tracing where the product comes from in order to immediately discontinue its use if necessary.
- Conduct field audits to look for animal intrusion, flooding, use of manure, and other environmental hazards before planting fields and 2–7 days prior to harvest with primary focus on *E. coli* 0157:H7 and *Salmonella*.
- Establish supplier programs including mandatory HACCP programs including taking immediate corrective actions, and annual, or more frequently if necessary, food safety assessments for all suppliers.
- Establish distributor programs including: semi-annual audits, mock recalls after hours on a regular basis, product traceability within 2 hours, and monitoring the temperature of sensitive products at the distribution center and during transportation to restaurants.
- Establish restaurant food protection programs including daily restaurant checklist evaluation in writing performed two to three times daily, an effective pest management program, use of a food safety pocket guide, a semi-annual food safety audit conducted by a third party, and semi-annual corporate-level audits.
- Provide a crisis management team which can go immediately into operation on a 24/7 basis when there are concerns about product quality, communicable disease, foodborne illness, product contamination or foreign material in food, and when an employee shows

symptoms of foodborne illness. The team needs the authority and capability to advise all stores within the corporation about the problems and how to immediately prevent them or correct them.

- Develop a quality assurance outline with an 800 or 866 prefix available 24 hours a day, Provide a hotline with an 800 or 866 prefix so that employees and stores can immediately reach the crisis management team.
- Provide a nutrition and allergens program including information to the public on a variety of allergens (milk, eggs, fish, shellfish, tree nuts, wheat, peanuts, and soybeans) and sensitivities to gluten and monosodium glutamate (MSG).
- Mandate that all managers have food service management certification, which is recertified at least every 3 years, with at least one certified manager present in each restaurant at all times.
- Provide all employees with intense basic food safety training during orientation, including hands-on work.

FUNCTION OF PROFESSIONAL ASSOCIATIONS IN THE PREVENTION AND MITIGATION OF POTENTIAL FOODBORNE DISEASE

The National Environmental Health Association located in Denver, Colorado, provides a prime example of the mission and functions of professional associations as they work to help create a better life for all people. The Association "facilitates the prevention of disease and injury and the promotion of good health while protecting and sustaining the natural environment for future generations through the control of physical, biological, chemical and radiological factors and by working with programmatic areas as well as personnel while conducting significant research to improve basic practices." (The Association functions are listed under Best Practices below.)

Best Practices for Professional Associations
- Serve as the public voice and as an advocate for the profession in all topic areas at all levels of local, state, and national decision-making to provide accurate information and attempt to influence public policy in a positive manner concerning all areas and aspects of the environment and the health and safety of people.
- Improve the quality of environmental health services through developing the knowledge and skills of environmental health professionals.
- Protect and promote the profession by establishing appropriate standards for professional practice.
- Promote the sharing of Best Practices through work with technical and professional groups in developing those which are based on practical evidence-based techniques.
- Build coalitions with other professional associations, business, industry, and governmental agencies to improve access to quality, cost-effective environmental health services.
- Promote standardized educational efforts that respond to ongoing and future environmental health issues.
- Improve strategic planning and decision-making through the use of a variety of conferences and special meetings of experts.
- Build the professional environmental community through the Annual Educational Conference, *Journal of Environmental Health*, various educational programs, and consultations with members or others working in or interested in various aspects of environmental improvement.
- Establish and maintain a highly recognized credentialing and licensing program for environmental personnel who seek to be at the top of their professional field and which is necessary for maintaining employment and/or promotion. The premier credential issued and recognized internationally is the Registered Environmental Health Specialist/Registered Sanitarian certification.

- Provide continuing education opportunities in various environmental areas to enhance the knowledge and skills of environmental personnel as well as provide required knowledge for maintaining credentialing or licensing.
- Become involved in program evaluation and accreditation of collegiate programs to ensure that these programs meet the most recent standards of performance needed by students when they enter the professional field.
- Engage the global environmental health community in the exchange of pertinent information, new research, and Best Practices to resolve problems in an effective and inexpensive manner.
- Partner with other agencies to develop effective environmental health training in emergency response and preparedness.
- Provide extensive training and credentialing in food safety from the National Environmental Health Association utilizing highly qualified nationally registered environmental health personnel.
- Utilize research and development tools in a vast variety of environmental health areas to help produce actionable information that can be immediately utilized.
- Respond to immediate emergencies such as the recent outbreak of bedbugs by providing accurate information and resources to individuals, industries, and communities.
- Provide integrated pest management initiatives to help reduce potential vector-borne diseases while preserving and protecting the environment.
- Inform the public and train personnel in the large number of issues related to children's environmental health and how best to prevent conditions that may affect these young people throughout their lives.
- Develop programs on the air, water, land use, energy, institutional environment, built environment, infection control, sewage, swimming areas, etc.
- Provide members' services which provide information rapidly and efficiently as well as establish an ongoing training schedule to ensure employees are kept up to date concerning the latest Best Practices.
- Develop an institute food management training program provided by various professional associations and other sources.

The National Restaurant Association's Serve Safe® program along with other food training programs are of great value to the industry and to the public. They are: Learn 2 Serve® from 360 Training; Food Protection Management Certification Program® from the National Registry of Food Safety Professionals; and Food Protection Manager Certification Program® from Prometric.

Best Practices for Special Food Management Training Programs
- All of the above programs should require that the individuals upon completion pass the comprehensive examination of the Conference for Food Protection/American National Standards Institute (CFP/ANSI).

Note: The FDA Food Safety Cooperative Agreement to Build Capacity for State and Local Regulatory Agencies has been awarded to the National Environmental Health Association.

This agreement, which is for a 5-year, $5 million cooperative effort, was announced on August 26, 2015. Its function is to develop and implement training which is badly needed by state, local, territorial, and tribal food and feed safety agencies. The National Environmental Health Association will also research the training needs for the Integrated Food Safety System inspectors and regulators and train instructors to meet the FDA Best Practices program. This is a major movement toward standardizing as well as improving the efforts of all regulatory agencies in food protection to prevent foodborne disease outbreaks and promote good health.

THE PROBLEMS RELATED TO ADEQUATE PERSONNEL AND BUDGETS TO PREVENT AND MITIGATE FOODBORNE DISEASE

Federal funding is essential for the maintenance of many state and local public health and environmental health programs. The budget for the CDC has decreased from $7.07 billion in fiscal year 2005 to $5.98 billion in fiscal year 2013 and may be lowered further for 2016–2017. This has not only resulted in severe cuts at the CDC in programs and preparedness as well as the loss of highly skilled people with huge amounts of practical experience, but also severe cuts in federal/state and local programs resulting in once again losses of highly skilled individuals with excellent knowledge and long-term experience. Forty-eight states, three territories, and Washington, DC, have reported budget cuts and serious job losses. Instead of practicing prevention as the most effective and common sense approach to reduce potential disease and injury, the various departments have to resort to putting out the fire after the problem has occurred. Some 75% of the CDC budget is distributed to states, local agencies, and other public and private partners and is used to support services and programs. (See endnote 50.)

In addition, the state and local public health workforce is declining at a substantial rate each year. According to the Association of State and Territorial Health Officials, approximately 20% of the various health agencies workforces are eligible to retire within the next 3 years and there do not appear to be sufficient specially trained new people to fill their positions. Much of this is due to the relatively low salaries paid to people in the environmental health and public health fields although the qualifications for the positions and the necessary credentials or licensure are very strict. In all cases, appropriate continuing education is required to maintain the credentials or licensure.

Prevention appears to be a nasty word until we have an outbreak of an exotic disease such as Ebola and then everyone demands to know why public health officials cannot solve the problem immediately and prevent the spread of disease. It is the easiest budget to cut because it is rather difficult to prove in many instances that prevention has saved lives and money.

Because of restricted budgets, local units of government have furloughed huge numbers of public health staffers, including environmental health professionals. Others have been forced to accept unpaid furlough days, pay reductions, and across-the-board program eliminations and reductions.

Environmental health professionals especially have a serious public relations problem. When the newspapers refer to them as inspectors, the work that these individuals do is demeaned. It reinforces the idea that in a budget crisis, the first place to cut is in public health, especially environmental health.

At budget cutting time, the last to go are the police and firemen, because of the potential for disasters involving the public. The same should also be true of the environmental health practitioner, who helps prevent disasters involving food, water, air, hazardous chemicals, hazardous wastes, housing, terrorist acts, serious outbreaks of disease, etc.

Best Practices for Personnel and Budgets (Also see Best Practices in Avoiding Problems in Politics below)
- Develop and utilize a comprehensive fee-for-service program to obtain appropriate funding for a comprehensive environmental health services program including adequate and properly educated and trained credentialed environmental health practitioners receiving appropriate salaries and using state-of-the-art equipment. Fees should be charged for virtually all services including those related to: camps and campsites; water supplies of all types; all types of food service establishments; mobile home parks; subdivision plan reviews; individual sewage disposal systems; sanitary surveys; all forms of swimming pools, spas, and bathing beaches; tanning facilities; tattoo and body piercing facilities; hotels, motels, and boarding houses; all forms of solid and hazardous waste disposal; hospitals, rehabilitation centers, nursing homes, retirement homes, etc.; all forms of pet businesses where there is necessary disposal of wastes; plumbing and gas piping fees; food establishment plan reviews; school plan reviews; temporary foods and special events; food

manager training programs; mobile homes or RV parks; child care facilities of all types; etc. It is then essential that adequate funds are provided for response to different types of environmental health and public nuisance complaints, fires, and other emergencies or disasters, as well as outbreaks of disease. It is necessary to convince the political entities that all of these funds which go into a general fund be allocated solely for the operation of the various environmental health programs in the community.

- Determine the actual costs of each of the services provided above and add a percentage for administrative fees to cover all of the other services which should be provided free to the public in the event of special problems, emergencies, or disasters.
- Develop a concerted united effort using all forms of communication including all types of internet platforms, by all agencies and all professional associations such as the National Environmental Health Association to teach the public the serious potential for disease, injury, and environmental disasters that are occurring currently and may accelerate in the future, and the extreme value of the highly trained professional environmental health specialist to help prevent, control, and mitigate the effects of these occurrences.
- Develop and distribute professionally prepared materials to various media including the various internet platforms and also to all portions of the public and political groups, brief descriptions of the qualifications of the applied scientists called "environmental health specialists" and also describe the nature of his/her work including prevention, mitigation and control of factors leading to disease and injury and the protection and sustainability of the natural environment.
- Using Ebola as an example, develop a concerted effort to discuss with all national, state, and local legislators the great significance of expanding budgetary and other resources for environmental and public health to prevent the potential effects of other outbreaks of exotic diseases as well as those which are occurring regularly now to prevent future disasters from occurring. Prevention is far cheaper than treatment or control at a later date.
- The United Nations has to empower and fund the World Health Organization at such levels that when there is just the hint of an outbreak of a highly infectious disease, adequate numbers of highly trained individuals utilizing the latest equipment and resources will move into the area and help the particular country stop the outbreak before it becomes an epidemic or a pandemic, thereby affecting other countries.
- Discuss with your colleagues of similar size communities what techniques you used to help persuade legislative bodies to provide funding for necessary programming.
- Establish a special group of senior environmental health administrators and retirees who can give guidance to individual communities on how best to get appropriate budgetary support and best utilize the funds to achieve appropriate goals.

THE PROBLEMS RELATED TO POLITICS IN THE PREVENTION AND MITIGATION OF FOODBORNE DISEASE

Even though environmental health people do not carry out their work as members of a political party, they still are constantly involved with the public, politicians, and the media and have to take this into account when seeking funds for programs. An example of what can go wrong is as follows:

- In County "A" in Illinois, the new Director of Public Health eagerly started a variety of environmental health and public health programs. He generated a huge amount of media publicity. His work was so good that the County Board of Health gave him an increase in salary. However, the County Board of Commissioners fired him. The problem was that he did not understand the concerns of the Commissioners, which were more important to them than environmental health or public health problems. You must understand your local political environment.

- In County "B" in Illinois, the new Director of Public Health wanted to enhance the programs of the department. He started with the Maternal and Child Health Program. He formed an advisory committee of citizens including several of the wives of the County Commissioners. The advisory committee moved the Maternal and Child Health Program forward because of the needs of the community and because it was non-controversial. This was then followed by additional public health and environmental health programs.

These two incidents actually occurred in the past and as a result of the miscalculation of the needs of the political entity in County "A," there was a sharp reduction in budget for essential programming instead of an increase. This is a prime example of the need to understand political factors when instituting new programming and requests for budget.

Best Practices in Avoiding Problems in Politics
- Make one authoritative individual responsible for all contacts with the media.
- Provide to the media time and authoritative personnel needed to explain various environmental health issues and programs especially during times of great interest such as outbreaks of disease.
- Make the media your friend wherever possible by providing good and interesting stories that will not only teach proper techniques but also can be used to highlight specific efforts in prevention of disease and injury or resolution of ongoing and/or new problems.
- Learn the current concerns of the public as well as various political factors and try to adapt the needs of specific programs to meet these concerns.
- Arrange for members of the various budgetary committees in government to get frequent one- or two-page briefings on the problems, potential solutions, and budgetary needs for personnel and equipment to resolve current problems and potential serious problems in the future. Substantiate the briefings with well-thought-out, concise, easy-to-read-and-understand documents which can be presented to these individuals if they so desire.
- Never practice partisan politics.
- Let the media or political entities take credit for whatever successes occur during the operation of various environmental health programs.

THE NEED FOR TRAINING FOR ENVIRONMENTAL HEALTH PROFESSIONALS

(See endnotes 51, 52)

There is substantial deficiency in the size, training, and qualifications of the available environmental health workforce at all levels of government to carry out all existing programs of prevention, mitigation, and control as well as emergency programs to prevent disease and injury and protect and sustain the natural environment. Further, because of the many specific science-based competencies needed to carry out this work in an effective, cost-effective, and efficient manner, programs become inefficient and potentially hazardous situations are not appropriately solved. This will lead inevitably to very serious consequences of disease and injury and destruction of environmental resources. These harmful situations may be very costly and could have been prevented.

The environmental health workforce has grown substantially older and as a result of this there will be a large number of retirements of highly qualified professionals from all levels of government in the next few years. Because of many factors including limited salaries and opportunities, as well as severe budget cuts in working departments, many young people are choosing other areas of endeavor for their career paths. Retention of qualified individuals has become a challenge.

Many of the new individuals being hired have general knowledge but not the specific knowledge, attitudes, practices, and experience to carry out their role in determining the causes of many of the potential hazards and the necessary solutions and to teach others how to make appropriate changes, or the necessary communication skills to prevent disease and injury and promote health through successful implementation of the various programs of the departments. Even the brightest individuals and those most motivated to achieve appropriate goals can only do as much as they are equipped to do if they do not have the appropriate college education in the sciences and environmental health sciences and experiential work through internships.

Another concern is that enrollment at the nation's accredited environmental health programs has been in decline over the past decade.

Best Practices in Training for Environmental Health Professionals

General

- Provide closely supervised paid internships to colleges and universities for environmental health students as a means of securing new trained personnel and covering essential services especially during the summer months when the demands increase sharply.
- Determine the nature and scope of various environmental problems within communities and utilize Best Practices to resolve them.
- Determine the various competencies needed by environmental health personnel to effectively carry out Best Practices to resolve ongoing and emergency problems.
- Pretest all new personnel to determine the effective level of various competencies prior to training programs.
- Develop a comprehensive training program to prepare personnel to effectively operate in a variety of different situations.
- Train all supervisory personnel in all aspects of each of the environmental areas under their supervision and routinely certify them and recertify them in order to ensure that they in turn evaluate and retrain the personnel under their supervision.
- Closely supervise all new personnel in field situations and help them make appropriate improvements where needed.
- Closely supervise all personnel in a variety of communication techniques to make sure that their findings are transmitted simply to the individuals they are dealing with and that they are able to implement necessary corrections.
- Posttest all new personnel after various courses and 6–12 months of closely supervised field experience to determine current level of various competencies needed to carry out the various types of programs within the department.
- Provide an appropriate orientation course for all personnel describing: the nature of specific groups within the community as well as their ability to communicate effectively in English; natural leaders identified in these communities that can be of assistance to environmental and public health personnel in achieving appropriate objectives; the various programs within the department; the appropriate laws which are utilized by personnel; departmental benefits and expectations; opportunities for licensing, credentialing, and also advancement; etc.
- Develop problem solving skills as well as human relations skills for all individuals since books and education may give you the framework for resolving situations but only the individual on-site can actually apply this knowledge in a meaningful way and help individuals prevent disease and injury, promote good health, and protect and sustain the environment.
- Use supervisory evaluations which include a day in the field with the individual at least twice a year for all existing staff and more frequently for probationary staff with the supervisor making separate evaluations or inspections and then comparing the results with the field person. This will be a wonderful learning experience for the field person especially if the supervisor explains carefully what the differences are in his/her findings from those of the field person and why. Also the supervisor should conduct an audit at least twice a year of a previous day's work of an individual and then explain to the person the differences in findings and why. The supervisor must remember that things do change from one day to another and be careful about how the evaluation is carried out.

Specific

- Provide closely supervised field experience for the graduate of an accredited environmental health science program for a period of time and then have the person work on his/her own program areas with periodic review by the supervisor.
- Provide specific training in environmental health issues and then closely supervised field experience for the graduate of an environmental science program and then have the person work on his/her own program areas with periodic review by the supervisor.
- Provide specific training in all environmental areas including environmental health issues for the graduate of a general science program and then closely supervise him/her in the field for an extended period of time.
- Provide specific training in the general sciences needed for environmental health issues and also the specific areas of the environment including environmental health issues for the graduate of a non-related college program. This individual will need to be under close supervision for long periods of time and should initially be assigned simple tasks such as gathering samples and routine inspections.
- Insist that all employees upon being hired have taken courses in communication skills at the college level and have successfully completed them.
- Present short but frequent programs on the use of a variety of communication skills including role playing, report writing, oral presentations of written material in a brief but understandable manner to a non-scientific audience, and proper use of emails and letters. Give individuals assignments and then evaluate them immediately.
- Make the content of all short courses immediately usable by the participants and have them test the information under supervision and actual field situations.
- Train personnel how to effectively evaluate all situations, and make appropriate recommendations for correction and transmit the information in an effective manner to the individuals responsible for the change.
- Emphasize prevention, mitigation, and control over the enforcement of laws, rules, and regulations. Consultation and assistance will have much better outcomes than the threat of enforcement actions.
- Use the reward of registration and credentialing to acknowledge the proficiency of the individual and also a substantial increase in salary.
- Utilize the National Environmental Health Association (see Best Practices for Professional Associations above) nationwide network of certified food safety trainers for information concerning the training of trainers, managers, and employees in appropriate food service techniques.
- Use the National Environmental Health Association Epi-Ready Team Training program developed in conjunction with the CDC.
- Use the National Environmental Health Association Food-Safe Schools Program developed in conjunction with the CDC and USDA.
- Use the program presented by the National Environmental Health Association entitled Industry—Foodborne Illness Investigation Training.
- Use the CDC Health Practices on Cruise Ships: Training for Employees, for food protection presentations.

A GLOBAL RESPONSE IN THE PREVENTION AND MITIGATION OF FOODBORNE DISEASE

The World Health Organization, Department of Food Safety, Zoonoses, and Foodborne Diseases, Cluster on Health Security and Environment has worked with member states to provide advice and assistance on the prevention and control of outbreaks of foodborne disease and food terrorism.

It has established a preventive approach to food safety with more frequent surveillance of food sources and potential causes of foodborne disease and a more rapid response to actual outbreaks and incidents of contamination. Rapid diagnosis and appropriate rapid available treatment of those who are ill is encouraged. Then it is necessary to have swift communication of information to all parts of the world.

LAWS, RULES, AND REGULATIONS

This section will be limited to the federal government only, since the use of state and local government laws, rules, and regulations would make this document overwhelming. A model food code is also discussed.

FOOD CODES

The US Public Health Service starting in 1924 issued its first recommended milk ordinance and then in 1926 issued a recommended milk ordinance and code which could be adopted by various communities. Then, in 1934 they issued a set of model restaurant sanitation regulations and later upgraded both types of model codes periodically and also included all types of food delivery systems in their model regulations. The current Grade A Pasteurized Milk Ordinance and Code came out in 2013. (See endnote 48.) The current Food Code came out in 2013. (See endnote 49.)

The FDA Food Code

This is a model code that can be adopted by states, counties, Native American tribes, and other governmental agencies to help: reduce the risk of foodborne disease; provide uniform standards for retail food establishments; remove redundant materials and inspection and survey sheets; and establish a standardized approach to inspections and special surveys of all food-type establishments. It covers many topics: improper holding temperatures, inadequate cooking, refrigeration, and storage temperatures; contaminated equipment and how to decontaminate it; food from unsafe sources; poor personal hygiene; facilities that can contribute to outbreaks of disease; etc.

It is essential that all providers responsible for the control of potential foodborne disease in the more than 3000 state, local, and tribal agencies supervising over 1 million establishments have the competencies to carry out their assigned roles and understand the Best Practices involved in protecting the health and safety of the consumer. Understanding and use of the model FDA Food Code is a starting point for all agencies with primary responsibility to regulate retail food and food service establishments.

FDA Food Safety Modernization Act

The FDA Food Safety Modernization Act passed by Congress on December 21, 2010, and signed into law on January 4, 2011, is a response to the increasing burden of foodborne disease. It provides for the FDA a legal mandate to require comprehensive, prevention-based controls throughout the entire food supply chain. Now the FDA has the power to enforce preventive measures in all aspects of the food chain, instead of having to respond to outbreaks of disease and then still have limited power to do that which is necessary. The various food operations must evaluate all potential hazards, come up with a working plan to prevent or ameliorate them, implement the plan, and monitor the results. The FDA now has to establish science-based standards for the safe production and harvesting of fruits and vegetables to minimize the risk of disease. The seafood industry is being held accountable for preventing contamination and protecting the public by modernizing their systems. Inspections are conducted on a risk-based approach. This is essential to the proper operation of the various parts of the food system. Imported food safety is a significant part of the law which authorizes the FDA to refuse any imported foods when necessary. Proper recordkeeping is highly important so that appropriate agencies can determine what is going on when they are not present. All testing of products must be done by accredited laboratories. To clarify the various provisions of this Act, the FDA issued a series of proposed rules on Produce Safety, Preventive Controls for Human Food, Preventive Controls for Animal Food, and Foreign Supplier Verification Programs. These rules are open to comment by the public and then final rules will be issued in order to obtain appropriate compliance with the Act and prevent disease and injury to people and animals. (See endnotes 43, 47.)

Egg Products Inspection Act

The most recent Egg Products Inspection Act became effective on October 17, 1979. Congress determined that eggs and egg products are an important source of food for the entire country and that it was in the public interest that they not be adulterated, become a source of foodborne disease or be labeled inappropriately to protect the health and safety of the American public. (See endnote 44.)

Prevention of *Salmonella* Enteritidis in Shell Eggs during Production, Storage, and Transportation This rule came into effect September 8, 2009. (See endnote 8.)

Poultry Products Inspection Act

The most recent Poultry Products Inspection Act became effective on August 18, 1968. Congress determined that poultry and poultry products were an important source of food for the entire country and that it was in the public interest that all poultry and poultry products be wholesome, not adulterated, properly labeled, and packaged, and not a source of foodborne disease to protect the health and safety of the American public. (See endnote 45.)

Federal Meat Inspection Act

The most recent Federal Meat Inspection Act became effective on December 15, 1967. Congress determined that meat and meat products were an important source of food for the entire country and that the health and welfare of consumers be protected by making sure that all meat was wholesome, not adulterated, properly marked, labeled, and packaged and that meat and meat products would not be a source of foodborne disease. (See endnote 46.)

OTHER SIGNIFICANT REGULATIONS AND LAWS

Seafood Hazard Analysis Critical Control Point Regulations were produced to promote the safe and sanitary processing of fish and fish products including seafood which has been imported. This regulation came out in 1995.

The *Food Quality Protection Act* upgraded the Federal Insecticide, Fungicide, and Rodenticide Act and the Federal Food, Drug, and Cosmetic Act to resolve inconsistencies in the two major pesticide statutes. Now the EPA had to establish consistent regulations on pesticides usage and residues left in or on foods based on sound scientific data. It also required a periodic re-evaluation of pesticide intolerances based on sound scientific data. This became law on August 3, 1996.

Juice Hazard Analysis Critical Control Point Regulations were produced to promote the safe and sanitary processing of juice and the placing of warning labels on unpasteurized juice. This regulation came out in 1998.

The *Food Allergen Labeling and Consumer Protection Act* required that all ingredients from eight allergenic foods be described on the label. This went into effect in 2006.

The *Egg Safety (final) Rule* established requirements for control of *S.* Enteritidis in eggs from production through distribution. This was issued in 2009 and implementation for large producers was July 2010 and for small producers 2012.

RESOURCES

GOVERNMENTAL

Federal Level

US Department of Agriculture

The USDA has several food programs including: the FSIS which is responsible for meat, poultry, and egg products; the USDA Food Emergency Rapid Response and Evaluation Team which responds to food emergencies; and the USDA Food and Nutrition Service which reduces food insecurity by helping feed the needy and implementing the Farm Bill.

US Department of Health and Human Services

The Department of Health and Human Services has several food programs under the FDA and is responsible for the safety of all foods including shell eggs moved in interstate commerce, except for meat, poultry, and egg products regulated by the FSIS. The FDA is responsible for identification of the source of a product and the extent of its distribution in an outbreak of foodborne disease. It also coordinates voluntary recalls of food and a variety of field investigations and laboratory support functions. The CDC works closely with state and local public health epidemiologists, laboratories, environmental health personnel, and medical and nursing personnel to identify illnesses and clusters of illness related to food, and rapidly promotes controls and adds significantly to future prevention of the disease. The Health and Human Services Department is also involved in emergencies and in the case of overwhelming foodborne disease outbreaks.

US Department of Defense

The Department of Defense has several units working on food-related diseases and prevention within the department. They are the United States Army Veterinary Crops; the United States Army Center for Health Promotion and Preventive Medicine; the United States Air Force Biomedical Science Corps; and other military agencies who are involved in public health training, food safety, epidemiology, inspections, testing of food, and reviewing processing and delivery systems of food. Preventive Medicine/ Environmental Health Officers conduct investigations of foodborne disease throughout the world.

US Department of Homeland Security

The Department of Homeland Security under Presidential Directive 9: Defense of the United States Agriculture and Food, 2004, established a national policy to defend the agriculture and food system against terrorist attacks, major disasters, and other emergencies. It was noted that the United States agriculture and food systems were vulnerable to disease, pests, and poisonous agents occurring naturally, unintentionally introduced into the food supply, or intentionally delivered as an act of terrorism. The fact that the entire food system is extensive, open, diverse, and very complex, making it an easy target for terrorists, could have serious consequences.

US Environmental Protection Agency

The EPA responds to waterborne disease outbreaks, from drinking water, at the state or local level, when the CDC makes the determination that it may be a multiple state outbreak. The EPA also becomes involved in problems of pesticides and other toxic substances.

Environmental Health Specialist Network

The Environmental Health Specialist Network (EHS-Net) is a collection of environmental health specialists who improve environmental health by working with epidemiologists and laboratories to identify and prevent the environmental factors that contribute to foodborne and waterborne disease. The environmental assessment of a foodborne disease investigation allows people to better understand the contributing factors such as workers' health, poor hand washing practices, improper temperatures, presence of insects or rodents, presence of hazardous chemicals, complexity of food handling procedures, numbers of meals served, numbers of certified, well-trained food service supervisors, cleaning policies, sick leave policies, hand sink with hot and cold running water availability and usage, etc.

CDC's OutbreakNet

The CDC's OutbreakNet team works with the national network of epidemiologists and other public health practitioners to investigate outbreaks of foodborne, waterborne, and other illnesses in the United States. It helps ensure rapid, coordinated detection, and response to multistate outbreaks of enteric diseases. It develops a short list of suspect foods or other exposures, identifies food exposure associated with illness, and determines how the food became contaminated. This information can then be utilized in not only stopping the outbreak of disease but also in training professionals how to prevent the outbreak.

Foodborne Diseases Active Surveillance Network

The Foodborne Diseases Active Surveillance Network (FoodNet) is the primary section for foodborne disease of the CDC's Emerging Infections Program. It is a special project using the information and skills of the CDC, 10 state health departments, the FSIS of the USDA, and the FDA. Its function is to provide a place for information to be gathered and then disseminated to interested individuals concerning foodborne diseases and their sources. It can rapidly pick up outbreaks which may be of a national nature and therefore advise the entire public health community about specific foodborne disease potential in their areas.

PulseNet

PulseNet is a national network of public health and food regulatory agency laboratories coordinated by the CDC. PulseNet agencies perform molecular subtyping of bacteria which cause foodborne disease. This helps to determine which strains of a given organism are involved and then helps to determine the sources of infection.

Governmental Manuals

The US FDA has published two manuals which are of great help to the overall food industry and to the regulators involved in managing food safety using HACCP principles. They are *Managing Food Safety: A Manual for the Voluntary Use of HACCP Principles for Operators of Food Service and Retail Establishments* (College Park, Maryland, April 2006) (see endnote 61) and *Managing Food Safety: A Regulators Manual for Applying HACCP Principles to Risk-based Retail and Food Service Inspections and Evaluating Voluntary Food Safety Management Systems* (College Park, Maryland, April 2006) (see endnote 62.)

Other Resources

ETP Food Protection—EHS-CDC

This internet page, last updated on October 16, 2015, is a compilation of various sources outside the CDC that provide a variety of information on various food safety issues in a variety of training techniques. It also briefly mentions some of the CDC food safety program areas.

STATE LEVEL

At the state level, the Departments of Health, Agriculture, and Environment may all be involved in outbreak surveillance, investigation, and response. This makes for confusion, duplication of effort, and unfortunately a waste of precious resources.

LOCAL LEVEL

At the end of the 1800s and early 1900s there were overwhelming problems with milk and various types of meat and other foods. By the 1960s, it was a rare occurrence to find an outbreak of disease caused by problems on the farm or with the food processor. Today, there is increased concern about the production end of food because of outbreaks of a variety of diseases caused by contamination on farms, in imported foods, and in food processing facilities. These outbreaks usually are documented in the national press and receive a considerable amount of publicity and therefore community interest.

The analysis of programs, in this presentation, will be at the local level, since thousands of health departments are responsible for foodborne disease outbreaks, at least initially. The greatest proportion of these outbreaks are related to food preparation facilities, since 57% of current foodborne disease outbreaks occur in these facilities, whereas 29% occur in homes, 6% in schools, and 8% in other settings. In 2006, Norovirus (Norwalk virus) accounted for 54% of the outbreaks and some 12,000 cases, whereas *Salmonella* accounted for 18% of the outbreaks. Poultry was responsible for 21% of the outbreaks, various vegetables for 17% of the outbreaks, and fruits and nuts for 16% of the outbreaks. Norovirus is a serious problem probably attributed to the personnel who are handling the food at the point of service. (As noted previously in this book dates change frequently depending on the source, therefore this information is only a means of giving the reader some indication of the size of the problem.)

At the local level, the health department typically is the lead or only agency involved in the response to a foodborne disease outbreak. However, in some local governmental units, such as Sacramento County and San Diego County in California and Maricopa County in Arizona, the environmental health agency is not part of a local health department, but rather a free-standing and equal, separate agency. In those situations, the environmental health unit must work closely with epidemiologists assigned to the health department to investigate foodborne illness outbreaks. The local environmental health unit is typically the first responder. It is responsible for food inspections in restaurants, grocery stores, daycare facilities, hospitals, schools, prisons, and food manufacturing plants, and for answering complaints concerning food or facilities. It provides food protection education to food handlers and managers, helps educate the community about foodborne disease and prevention, and works with establishments on critical control points along the food processing and preparation chain.

INDUSTRIES, INDUSTRY ASSOCIATIONS, PROFESSIONAL ASSOCIATIONS, AND PROFESSIONAL JOURNALS

Council to Improve Foodborne Outbreak Response
The Council to Improve Foodborne Outbreak Response (CIFOR) was formed by the Council of State and Territorial Epidemiologists and the National Association of County and City Health Officials to improve foodborne outbreak response, to develop guidelines for studies and solutions, to provide a list of resources and tools, and to provide means of measuring levels of performance in solving enteric illnesses. The National Environmental Health Association has made recommendations to improve the effectiveness of state and local food protection programs. These recommendations are of considerable significance because many of the members of the National Environmental Health Association are practitioners at a local or state level and have a fine understanding of the practical issues involved in food protection in a vast variety of local and state locations. The specific recommendations are in the categories of surveillance, inspection, regulation, food safety and food defense, emergency response, and leadership.

PROFESSIONAL ASSOCIATIONS—NON-GOVERNMENTAL

(For further information see Best Practices for Professional Associations outlining good practices which can then be modified for professional and nongovernmental associations. Several of the associations affiliated with environmental health people are listed below:)

- American Public Health Association
- Association of Food and Drug Officials

- Association of Nutrition and Foodservice Professionals
- Association of Public Health Laboratories
- Council of State and Territorial Epidemiologists
- International Association for Food Protection
- International Food Protection Training Institute
- National Environmental Health Association

PROFESSIONAL JOURNALS

Peer-Reviewed Journals

There are numerous peer-reviewed journals and other journals and magazines which contain many articles on a variety of environmental issues. Some of them are excellent, while some of them carry material that is controversial and not substantiated by scientific data and deliberations. The author at one time was the Associate Editor of the *Journal of Environmental Health* and because of an understanding of the publication and its professional means of operation and because the articles published are of value to field practitioners, has selected it as a prime example of an excellent journal for environmental personnel. This does not exclude other journals including the *Journal of Public Health* from having the same level of excellence as the *Journal of Environmental Health*.

Journal of Environmental Health

The *Journal of Public Health* (*JEH*) is a peer-reviewed technical journal that has been published for 78 years. The *JEH* provides an avenue for environmental practitioners and academics to publish articles on studies they have conducted that relate to the science of environmental health, safety, protection, and sustainability. The published articles provide a scientific resource base for the environmental health field, as well as for other related environmental fields including epidemiology, toxicology, industrial hygiene, and public health. The *JEH* provides authors with a needed medium to publish their research, thus offering professional development through publication. The *JEH* also provides its readership with much needed scientific findings and information on current practices and lessons learned, thus offering professional development through readership as well as providing a basis for developing Best Practices and determining competencies needed by personnel to carry out effectively these Best Practices. The *JEH* also publishes columns, guest commentaries, and news articles/information from the National Environmental Health Association that provide insight, knowledge, and resources on current issues in environmental areas.

EDUCATIONAL PROCEDURES

(See the section "Columbus Public Health Food Protection Program" later in this chapter)

ENFORCEMENT PROCEDURES

(See the section devoted to "Philadelphia Department of Public Health and Columbus Public Health Food Protection Programs" later in this chapter)

PRESIDENTS AND FOOD SAFETY

In 1998, President Clinton formed the President's Council on Food Safety by Executive Order. He was concerned about the safety of the food supply and wanted science-based regulations, well-coordinated inspections, proper enforcement, excellent research, and education programs. In 2000, President Bush formed the President's Food Safety Initiative. It recommended an intergovernmental

Foodborne Outbreak Response Coordinating Group (FORCG). Its goal was to improve the approach to interstate outbreaks of foodborne disease by all federal, state, and local agencies. The group developed standard operating procedures for the rapid exchange of data and information related to a given outbreak, and provided a blueprint for local entities to follow to expedite the investigation of the outbreak and get help at various levels. In 2009, President Obama created the President's Food Safety Working Group. The Working Group has recommended a new public health focused approach to food safety:

1. The regulatory agencies should become proactive and utilize methods of prevention of disease especially prioritizing those areas that are most critical and protecting individuals who are most vulnerable, instead of waiting for an outbreak of disease to occur. Rigorous standards need to be established and enforced in a timely manner.
2. High quality data analysis should become the basis for determining which foods are at risk and which situations contribute to the level of risk. This information should be provided to the regulatory agencies in order for them to make excellent risk-based inspections, provide appropriate solutions to problems, assist industry in correcting the situations in a timely manner, and evaluate the results.
3. Outbreaks of foodborne illness should be rapidly identified and stopped. An effective food tracing system helps shorten the outbreak, detection, solutions, and preventive measures to avoid future outbreaks.
4. The federal government should provide necessary resources to help local health departments, state health departments, and industry. A sharp increase in funding would provide the effective level of highly competent personnel that would be needed to carry out the programs and protect the public from foodborne disease outbreaks. In addition, the Congress needs to provide the necessary funds for sharp increases in personnel and programming at the local and state levels.

PROGRAMS

Voluntary National Retail Food Regulatory Program Standards
(See endnote 56)

- *Standard 1*—Regulatory Foundation
  - This standard covers: public health interventions, control measures for risk factors for foodborne disease, good retail practices, and compliance and enforcement found in the Food Code.
- *Standard 2*—Trained Regulatory Staff
  - This standard covers: appropriate educational requirements for the individuals to have the necessary knowledge and skills to perform their duties, completion of 25 joint field training inspections followed by 25 independent inspections, completion of an FDA-type standardization process, and completion of 20 contact hours of continuing education every 3 years.
- *Standard 3*—Inspection Program Based on HACCP Principles
  - This standard covers: implementing an inspection form that identifies and documents risk factors, a specific grade for each risk factor, interventions, and compliance, helps assign a risk category to each establishment, encourages on-site corrective actions, and establishes written policies for all facets of the inspection process.
- *Standard 4*—Uniform Inspection Program
  - This standard covers: an on-going quality assurance program, appropriate documentation of all risk categories, proper interpretation of laws, regulations and policies, and effective long-term control of risk factors, especially those which have been repeated.

- *Standard 5*—Foodborne Illness and Food Defense Preparedness and Response
  - This standard covers: an established process to detect problems, collect data, and investigate and respond to complaints and emergencies about foodborne illness and injury from intentional and unintentional food contamination, a written operating procedure on epidemiological investigations and databases, procedures for notifying state and federal agencies, and subsequent appropriate measures to contain the outbreak and/or prevent another one from occurring.
- *Standard 6*—Compliance and Enforcement
  - This standard covers: all voluntary and regulatory actions needed to achieve compliance with the rules and regulations, a written detailed procedure describing compliance and enforcement, inspection forms and other documentation, and compliance and enforcement actions and results.
- *Standard 7*—Industry and Community Relations
  - This standard covers: appropriate interactions with industry and consumer groups, educational outreach by all forms of media and personal contact, and provision of regulatory personnel to assist industry, schools, and communities as needed
- *Standard 8*—Program Support and Resources
  - This standard covers: one full-time equivalent person working in the food area for every 280–320 inspections including all facets from routine to training to disease investigation outbreaks, grouping of establishments by risk, inspection equipment, computers and software, administrative staff support, regulatory staff trained in HACCP principles, foodborne disease, compliance and enforcement, program assessment, and an accredited laboratory.
- *Standard 9*—Program Assessment
  - This standard covers: the program manager conducting an initial self-assessment within 12 months of the date of enrollment in the National Registry, thereafter a verification self-assessment every 36 months, and a survey of the occurrence of risk factors at least every 5 years.

UEP "5-Star" Egg Safety Program

This program, which is voluntary in nature, was developed by the United Egg Producers (UEP) for egg producers throughout the country as a way of monitoring and controlling *S. Enteritidis* on the farm. It goes beyond the FDA Egg Safety Final Rule. It covers all areas from the procurement of the chicks to cleaning and disinfection of poultry houses, refrigeration, environmental and egg testing, laboratory standards, processing plant sanitation, etc. (See endnote 10.)

Columbus Public Health (Ohio) Food Protection Program 2009

The Columbus Public Health (Ohio) Food Protection Program 2009 is compared below to the Philadelphia Department of Public Health Food Protection Program 1961 (both Samuel J. Crumbine Consumer Protection Award Recipients).

Good food handling practices, especially proper hand washing procedures, are essential and that is why the ensuing discussion will be about two Samuel J. Crumbine Consumer Protection Award Programs which have been shown to be effective. Also, both programs are mentioned in "A National Strategy to Revitalize Environmental Public Health Services" as innovative programs today. However, in the Philadelphia program of 2008, where the self-inspection process is used in restaurants within the city, there is still a need to provide proper regulatory inspection and supervision of this industry program of self-assessment. Self-inspection should be added to the programs of the regulatory agency and be not in lieu of having adequate personnel to carry out necessary services to prevent foodborne disease. The Columbus Public Health Food Protection Program was the recipient of the 2009 Samuel J. Crumbine Consumer Protection Award, while the Philadelphia Department of Public Health was the 1961 recipient of the Samuel J. Crumbine

Consumer Protection Award. The Columbus program meets the Voluntary National Retail Food Regulatory Program Standards of the FDA. The Philadelphia program met the standards of the Ordinance and Code Regulating Eating and Drinking Establishments, Recommended by the United States Public Health Service, 1943, FSA, Public Health Bulletin Number 280 (republished in 1955, DHEW, PHS Publication Number 37).

The first Restaurant Sanitation Regulations by the Public Health Service were issued 1934. Prior to that, since 1896 the Public Health Service had been interested in milk as a means of spreading disease, and in 1923 an Office of Milk Investigations was established to help the states in developing an effective milk-control program.

Similarities and Differences between Philadelphia, Pennsylvania, and Columbus, Ohio, Food Protection Programs

The Columbus program and the Philadelphia program are similar regarding regulatory foundation, trained regulatory staff, foodborne illness and emergency response, compliance and enforcement, and program assessment. They differ in that the Columbus program utilizes HACCP principles as part of the inspection process and regarding the potential for food terrorist acts, the close relationship with industry groups in problem solving, the use of training materials and staff members speaking the native languages of the new Americans in our current society, and the color-coded signs project which makes the public aware of the conditions within food service entities. The Columbus program uses the advanced technology of today, whereas the Philadelphia program used the advanced technology of the late 1950s and early 60s. The regulatory foundations in Philadelphia and Columbus were similar. The programs were based on the most current scientific recommendations of the United States Public Health Service through its Sanitarian Engineering Section in the 1950s and the FDA in 2009. Also both programs were based on actual codes, laws, and ordinances of either city or state. In Philadelphia, the sanitarian had to have a 4-year degree in science or environmental health science plus 1 year of progressive field experience and had to take a comprehensive written exam designed by the Public Health Service and the equivalent of today's registration examinations for employment as a Sanitarian I. After employment, the individuals went through intensive field training by existing experienced sanitarians and district environmental health supervisors. They also underwent a minimum of 40–80 hours of in-service training each year, presented by the Public Health Service. In Ohio, the individual has to have the same academic qualifications plus be a State of Ohio Registered Sanitarian to be qualified. Foodborne illness investigations were conducted in Philadelphia by a public health nurse and a sanitarian in a team approach. This worked well because both individuals were part of the same district health office. Additional assistance was available at the central office level, state level, and national level through CDC. In Columbus, the Primary Nurse Investigator or a member of the Communicable Disease Prevention Team does the initial study and then additional support is provided at the local, state, and national level. Where a food establishment is involved, a sanitarian becomes part of the team and investigates the establishment and its past records. Unfortunately, because of severe budget cuts there have been sharp reductions in the various programs of the Philadelphia Department of Public Health at the present time.

In Philadelphia, emergency response was initiated through a call to the Philadelphia Department of Public Health or the Mayor's Office of Complaints. The emergency typically fit into the categories of fire, flood, raw sewage present, inadequate water supply, inadequate hot water supply, unapproved water supply sources, contaminated water supply, loss of electricity or gas, unapproved food handling process, contamination from structural defects, extreme insect and rodent infestation, and gross insanitary conditions. The complaint was handled either immediately depending on the severity of the public health emergency, or at most within 24 hours. The full range of legal authority was available for use by the environmental health practitioners, including immediate closing of facilities and condemnation and disposal of contaminated products. The Columbus program has approximately the same process. A 3-1-1 telephone process is in place and residents may call the number for any city services complaint. The 3-1-1 operator enters the complaint into the electronic system where it is tracked and then routed to the proper agency.

Consistency of inspections was achieved in Philadelphia when the Public Health Service certified the district environmental health supervisors and the supervisors certified the level of consistency in field personnel. In Ohio, the state certified supervisory individuals who also certified field personnel.

Compliance and enforcement are basically similar in both departments. In Philadelphia, the procedure followed included initial inspection or survey. If serious problems related to potential outbreaks of foodborne disease or poor sanitation were found the following procedures were used:

1. A warning letter citing the nature of the problems was sent to the proprietor, and the individual was given a short but definitive time period to make the necessary corrections.
2. A follow-up inspection was made. If considerable problems or severe problems persisted, an administrative hearing was scheduled for the District Health Office, with the health officer acting as presiding officer and the environmental health supervisor identifying the problems and situations.
3. A follow-up inspection was made prior to the hearing.
4. At the hearing, the individual could either be given additional time to complete the necessary corrections or a recommendation would be made for license suspension or revocation.
5. An inspection was made to determine the status of the establishment prior to the court hearing.
6. The judge of the Quarter Sessions Court could fine the individual and/or shut down the establishment. If the judge issued a court injunction and the owner violated it, the owner would be held in contempt of court and go to jail.
7. Depending on the severity of the problem, any or all portions of this enforcement procedure could be suspended and a court injunction could be issued at once. In Columbus, although the enforcement procedures were somewhat different, they followed the same basic principles.

In program assessment, in Philadelphia, the United States Public Health Service determined the effectiveness of the food inspection program. The health department determined through the use of IBM cards filled out by all field sanitarians and supervisors, the number of inspections made, the type of inspections made, the amount of time spent on the inspection, the amount of travel time, and the amount of office time. There was no computerized determination of the effectiveness of the inspections. However, the district environmental health supervisors could quickly review the key points of inspections, those we would call today critical points, and have brief training sessions at the staff meetings to teach or discuss with the field staff the most urgent problems to look for when inspecting. Also, the district environmental health supervisors would at a minimum pick up a day's work, each year, for each sanitarian and go out and redo the inspections the following day. Also, as a minimum, each year the supervisor would spend an entire day with the field sanitarian. The field person and the supervisor would make separate inspections of the various sites including food, insects and rodents, nuisances, water problems, etc. The strengths and weaknesses were then discussed privately with each field person. In reality, the supervisor would spend several days in the field each year with each person to strengthen the field person's observation, inspection, consultation, and educational skills. Further, the supervisor would handle all special problems with the field person. A reduction in repeat inspections and in numbers of warning letters, office hearings, suspensions and revocations of licenses, and court hearings, was a rough way of measuring a successful program. In Philadelphia, it was frowned upon if the field sanitarians spent more than roughly 30 minutes in the morning and 30 minutes in the afternoon in the office. It was also frowned upon if the district environmental health supervisor spent entire days in the office reading reports. This was considered to be a waste of time. The individual was supposed to be the most knowledgeable person in the district office and therefore should be working on special problems in the field as well as assisting field personnel.

Obviously, with today's computer capacities it is possible, as Columbus Public Health does, to gather a substantial amount of very useful information and determine what are the most important issues to be followed closely. The computerized approach helps digest data and reduces the office time needed by the field staff and supervisory staff. The question is always how much data is needed to do a good job, and how much data becomes mind boggling and its analysis a waste of time and money.

Program support in Philadelphia was through the regular budget for 60 field sanitarians, 9 district environmental health supervisors, 7 chiefs of programs, 9 assistant chiefs of programs, and the environmental health director. All licensing fees went directly to the City of Philadelphia and did not affect the environmental health budget. In Columbus, Ohio, the financing of the environmental health division is through licensing fees, grants, and other sources, which affects the numbers of personnel and the type of programming, and may vary as frequently as annually based on funds available. Columbus, Ohio, went beyond the Philadelphia experience by use of HACCP principles, excellent industry and community relations, food defense and preparedness.

Prioritizing of Inspections   In Philadelphia, the first priority for inspections was outbreaks of disease, complaints about contaminated food or improper handling of food, and follow-up on establishments that had serious deficiencies which could potentially cause foodborne disease and those establishments going through the enforcement process. All establishments were inspected at least once a year and the more complex food preparation facilities were inspected more frequently because they would tend to have a larger number of unsatisfactory items related to potential foodborne disease. Less attention was given to items such as the physical facilities, unless they were very dirty or they potentially could contaminate the food.

In Columbus, Ohio, a priority system was set up based on the potential risk for foodborne disease. Risk level 1 has the least risk of spread of foodborne disease. It consists mostly of storage areas and prepackaged food. Risk level II is higher in importance because there is a higher potential risk to the public regarding heating or improper holding temperatures for potentially hazardous food. Risk level III has a higher potential risk because of improper cooking temperatures, improper cooling temperatures, improper holding temperatures, and contamination through preparation of food by handling, cutting, grinding raw meat products, cutting or slicing ready-to-eat meats and cheeses, preparing or cooking potentially hazardous food, and reheating of foods. Risk level IV has the highest potential for the production of foodborne disease because the handling and preparation of food involves several steps and multiple temperature controls. The foods include ready-to-eat raw potentially hazardous meat, poultry, fish, or shellfish or foods containing potentially hazardous items or ingredients. Also, Risk level IV can be reached by supplying food to high-risk individuals including those who are immune-compromised or elderly in either a healthcare facility or assisted living facility. There are two types of inspections: the Standard Inspection Report, which is of a routine nature, and the more in-depth Retail Food Establishment Process Review, which is concerned with the proper temperatures, holding time, and transportation of potentially hazardous foods.

In Philadelphia, each inspection was made using the systems approach and flow from the point of origin of the food entering the establishment through all processing and consumption. The sanitarians were instructed to look for key problems such as those involving running hot and cold water, temperatures of food storage and preparation (both hot and cold), the health of the food handler, insects and rodents, sewage, contamination of food preparation surfaces, etc. These were determined to be major problems and therefore any actions taken by the health department would be to reduce the potential for disease by elimination of these problems. The inspection form was only filled out after the actual inspection was done, therefore the role of walls and floors and ceilings were not considered to be significant compared to the conditions which could lead to disease outbreaks.

In Columbus, Ohio, the seven HACCP principles were followed:

1. Conduct a hazard analysis to determine food safety hazards and locations, and appropriate preventive measures for control.
2. Identify critical control points where a procedure can be provided to prevent, eliminate, or reduce a food hazard.
3. Establish critical limits for each critical control point with maximum or minimum allowable physical, biological, or chemical agents to be controlled to prevent, eliminate, or reduce these agents to acceptable levels to prevent disease or injury.
4. Establish critical control point monitoring requirements, that is, what activities are needed to monitor the control of critical points in the food delivery, storage, preparation, and serving process to prevent, eliminate, or reduce the level of a food hazard.
5. Establish corrective action when there is a deviation from the critical limit of a potential hazard.
6. Establish appropriate record keeping procedures.
7. Establish procedures to verify that the HACCP is working.

Obviously, this process is certainly better and more detailed than the Philadelphia technique of 1961. However, the question becomes the reality of time usage by environmental health field personnel and the amount of time they tie up for the person in charge at a food preparation and serving facility and if this is acceptable to avoid problems in that particular facility.

Food Security   Food defense preparedness and security is obviously a very new concern, especially since the terrorist attack of September 11, 2001. There is now a very serious concern about terrorists using food as a vehicle to inflict serious sickness and death in the population. See "Emergency Handbook for Food Managers," a project of the Twin Cities Metro Advanced Practice Center (APC), funded by NACCHO, September 2005.

Working with Industry and Community Relations   There is a substantial difference between the Philadelphia and Columbus programs in the area of industry and community relations. In Philadelphia, little attention was paid to either group. However, in the Community Rodent Control Program, there was a very close relationship between the Philadelphia Department of Public Health, other city agencies, other community agencies, and various community groups. Unfortunately, at that point this type of cooperation had not been extended to the food service area.

Columbus Public Health has been an active participant in the Ohio Retail Food Safety Advisory Council. Columbus Public Health is a long-time partner with the retail food industry and industry groups such as the Ohio Chinese Restaurant and Business Association, etc., academia and consumer groups. Columbus Public Health works closely with people who are new to America and American customs and language. A bicultural/bilingual field sanitarian is fluent in English and Spanish. The health department contracts out for assistance with the Mandarin language. Although other environmental health personnel are fluent in Somali, Russian, German, Italian, Farsi, and Arabic, they are not necessarily assigned to the food protection section, but could be utilized in the event of an emergency. Numerous workshops and special food service training courses in English, Spanish, Mandarin, and Somali are made available to supervision and management personnel in a variety of restaurants. Further, a highly innovative means of presenting information in a Food Safety Toolbox provides placards in several languages. A toolbox is given to restaurants as needed. The essential areas of food safety most commonly utilized to control foodborne disease are the subjects of detailed description on the placards. They are dishwashing (three-compartment sink); thermometer calibration; advisory on eating raw or undercooked meat, poultry, seafood, and shellfish and eggs; cooking temperatures; cooling of foods; foodborne illness investigation; hand washing sink; no bare hand contact with food; refrigeration safety; and personal hand washing.

The comparison of the two programs is significant because basic environmental health procedures used to protect consumers from food borne infection have not changed significantly in the last 50–60 years. Yet there are now global sources of food and although some organisms that were not apparent years ago are now causing outbreaks of disease in addition to the usual culprits, approaches to protecting the food supply by highly competent individuals has not changed that much. What has changed is the use of HACCP principles, which has been both good and problematic. The good is in the identification of critical points where emphasis is placed on known potential problems which can lead to disease outbreaks. What is problematic is the use of time by local and state government officials to carry out these surveys, when there are such limitations on the number of highly competent environmental health personnel who are available because of very strict budgetary demands. Critical safety elements have to be sacrificed in other program areas in order to meet the needs of the food protection and safety program.

What has also changed is the establishment of working relationships with numerous partners interested in and concerned with the spread of disease through food, and the ability to present information in such a manner that it is possible to reach a diverse population and better inform them of appropriate measures to be used. Also, it is evident that local health departments need excellent laws, appropriate budgets to obtain and retain highly qualified individuals, and constant supervision of field staff. Budgets cannot and should not depend on the whim of legislators or the ability to obtain grants to carry out necessary activities to protect the public. Grants also may distort the priorities of the community because they are addressing a special problem at times at the expense of other major concerns in potential environmental health problem areas.

These budgets should and must be based on the ability of the environmental health division to respond to all environmental concerns. This includes potential emergencies that can affect the health and safety of the community.

In each of these two departments, only those with appropriate educational backgrounds should be hired to carry out the work of the environmental health professional. Both of these departments have had extremely strong, very well-qualified, highly disciplined administrators with a comprehensive vision of the needs of their communities, in order to achieve the ultimate award of excellence, the Samuel J. Crumbine Consumer Protection Award.

ENDNOTES

1. Koren, Herman, Bisesi, Michael. 2003. *Handbook of Environmental Health: Biological, Chemical, Physical Agents of Environmentally Related Disease*, volume 1. CRC Press, Boca Raton, FL.
2. Centers for Disease Control and Prevention. 2014. *Estimating Foodborne Illness: An Overview.* Atlanta, GA.
3. Food and Agricultural Organization of the United Nations. 2002. *Trends in Veterinary Public Health and Food Safety: Problems and Challenges: Chapter 3 in FAO Animal Production and Health Paper*, Number 153. Rome, Italy.
4. Drake, John M., Hassan, Ali N., Beier, John C. 2013. *A Statistical Model of Rift Valley Fever Activity in Egypt.* Journal of Vector Ecology, pp. 251–259.
5. Centers for Disease Control and Prevention. 1998. Foodborne Disease Outbreaks, Five-Year Summary, 1983–1987: Surveillance Summaries. *Morbidity and Mortality Weekly Report*, Atlanta, GA.
6. US Food and Drug Administration. 2004. *Good Manufacturing Practices (GMP) For the 21st Century: Food Processing.* Washington, DC.
7. National Environmental Health Association. 2013. *Assessment of Foodborne Illness Outbreak Response and Investigation Capacity in US Environmental Health Food Safety Regulatory Programs.* Denver, CO.
8. Food and Drug Administration. No Date. *FDA Food Code 2009: Annex 4-Management of Food Safety Practices-Achieving Active Managerial Control Foodborne Illness Risk Factors.* Washington, DC.
9. Egg Safety Center. 2010. *FDA Egg Safety Final Rule.* FDA, Washington DC.
10. Egg Safety Center. 2010. *Food Safety Programs.* Alpharetta, GA. FDA, Washington DC.

11. The United States Department of Agriculture, Food Safety and Inspection Service. 1958. *Key Facts: The Seven HACCP Principles.* Washington, DC.
12. Canadian Food Inspection Agency. 2012. *Hazard Analysis Critical Control Point (HAACP).* Ottawa, ON.
13. Food Standards Australia, New Zealand.2009. *Public Health and Safety of Eggs and Egg Products in Australia, Explanatory Summary of the Risk Assessment.* Canberra, Australia.
14. BC Centre for Disease Control, Food Protection Services. 2011. *Fish Processing Plants: Guidelines for the Application of a Hazardous Analysis Critical Control Point (HACCP) Program.* Vancouver, BC, Canada.
15. Global Aquaculture Alliance. 2014. *Global Aquaculture Alliance Best Aquaculture Practices: Seafood Processing Standards, Food Safety Management Component.* St. Louis, MO.
16. US Food and Drug Administration, Food CGMP Modernization Working Group Center for Food Safety and Applied Nutrition. 2005. *Food CGMP Modernization: A Focus on Food Safety.* Rockville, MD.
17. World Bank Group, International Finance Corporation. 2007. *Environmental Health and Safety Guidelines Fish Processing.* Washington, DC.
18. Halden, Rolf U., Schwab, Kellogg J. 2008. *Environmental Impact of Industrial Farm Animal Production.* Pew Commission on Industrial Farm Animal Production, Washington, DC.
19. Xin, H., Gates, R.S., Green, A.R. 2010. *Emerging Issues: Social Sustainability of Egg Production Symposium: Environmental Impacts and Sustainability of Egg Production Systems.* Poultry Science Association, Champaign, IL.
20. US Food and Drug Administration. 2014. *Potential for Infiltration, Survival, and Growth of Human Pathogens within Fruits and Vegetables.* Silver Spring, MD.
21. US Food and Drug Administration by the Institute of Food Technologists for the FDA. 2013. *Evaluation and Definition of Potentially Hazardous Foods.* Silver Spring, MD.
22. US Food and Drug Administration, Center for Food Safety and Applied Nutrition, 2008. *Guidance for Industry: Guide to Minimize Microbial Food Safety Hazards of Fresh-Cut Fruits and Vegetables.* College Park, MD.
23. Produce Marketing Association. 2014. *Produce Safety Best Practices Guide for Retailers.* Newark, DE.
24. US Food and Drug Administration, Center for Food Safety and Applied Nutrition. 2004. *Guidance for Industry: Juice HACCP Hazards and Controls Guidance First Edition; Final Guidance.* College Park, MD.
25. Grocery Manufacturing Association. 2010. *Industry Handbook for Safe Processing of Nuts.* Washington, DC.
26. US Food and Drug Administration, Office of Safety. 2011. *Guidance for Industry: Evaluating the Safety of Flood-Affected Food Crops for Human Consumption.* College Park, MD.
27. Gretchen Goetz. 2011. Sprouts Pose a Unique Danger, EFSA Says. *Food Safety News.*
28. US Food and Drug Administration, Center for Food Safety and Applied Nutrition. 2014. *Guidance for Industry: Reducing Microbial Food Safety Hazards for Sprouted Seeds.* Silver Spring, MD.
29. Food Standards Agency Mycotoxins Branch, Chemical Safety Division. 2007. *The UK Code of Good Agricultural Practice to Reduce Fulsarium Mycotoxins in Cereals.* London, England.
30. US Environmental Protection Agency. 2012. *Crop Production,* Ag 101. Washington, DC.
31. Harris, Kerri B., Savell, Jeff W., Facilitators. 2009. *Best Practices for Beef Slaughter.* National Meat Association, College Station, TX.
32. Food and Agricultural Office of the United Nations. No Date. *Meat Processing Technology for Small-to Medium-Scale Producers.* Bangkok, Thailand.
33. US Department of Health and Human Services, Public Health Service, Food and Drug Administration. 2011. *Grade "A" Pasteurized Milk Ordinance.* 2011 Revision. Silver Spring, MD.
34. Claudio, Luz. 2012. Our Food: Packaging and Public Health. *Environmental Health Perspectives,* 120(6):a232–a237.
35. Food Packaging: Roles, Materials, and Environmental Issues. 2007. *Journal of Food Science,* 72(3).
36. Ackerley, Nyssa, Sertkaya, Aylin, Lange, Rachel. 2010. Food Transportation Safety: Characterizing Risks and Controls by Use of Expert Opinion. *Food Protection Trends,* 30(4):212–222.
37. US Food and Drug Administration. 2014. *"FSMA" Proposed Rule and Sanitary Transportation of Human and Animal Food.* Silver Spring, MD.
38. National Grocers Association. 2014. *NGA Comments on the FDA Proposed Sanitary Transportation of Human and Animal Food.* Arlington, VA.
39. Association of Nutrition and Food Service Professionals. 2011. *ANFP Practice Standards: Food Storage Guidelines.* St. Charles, IL.

40. Frozen Food Handling and Merchandising Alliance. 2009. *Frozen Food-Handling and Merchandising.* McLean, VA.
41. Garden-Robinson, Julie. 2012. *Food Safety Basics: A Reference Guide for Food Service Operators.* North Dakota State University Extension Service, Fargo, ND.
42. US Department of Health and Human Services, Food and Drug Administration. 2009. *Good Importer Practices: Guidance for Industry(Draft Guidance).* Silver Spring, MD.
43. US Food and Drug Administration. No Date. *Food Safety Modernization Act: Frequently Asked Questions.* Silver Spring, MD.
44. United States Code. 2011. Food and Drugs. In *Egg Products Inspections.* 2011 Edition, FDA. Washington, DC.
45. United States Code. 2011. Food and Drugs. In *Poultry and Poultry Products Inspection.* 2011 Edition, FDA. Washington, DC.
46. United States Code. 2011. Food and Drugs. In *Meat Inspection.* 2011 Edition, FDA. Washington, DC.
47. Public Law 111-353. 2011. *FDA Food Safety Modernization Act*, 21 USC 2201. Washington, DC.
48. US Department of Health and Human Services, Public Health Service, FDA. 2011. *Grade "A" Pasteurized Milk Ordinance.* 2011 Edition. College Park, MD.
49. US Department of Health and Human Services, Public Health Service, Food and Drug Administration. 2013. *Food Code-2013.* College Park, MD.
50. Trust for America's Health and Robert Wood Johnson Foundation. 2014. *Investing in America's Health: A State-by-State Look at Public Health Funding and Key Health Facts.* Washington, DC.
51. Centers for Disease Control and Prevention. 2009. *Strategic Options for CDC Support of the Local, State, and Tribal Environmental Public Health Workforce.* Atlanta, GA.
52. Association of State and Territorial Health Officials. 2013. *Policy and Position Statements: Public Health Workforce Position Statement.* Arlington, VA.
53. National Environmental Health Association. No Date. *Position: Retail Food Protection on the Local, State and Tribal Levels: Left Out of the New Federal Food Protection Initiatives?* Denver, CO.
54. Partnership for Food Protection Laboratory Tests Group. 2013. *Food/Feed Testing Laboratories: Best Practices Manual (Draft).* FDA-Silver Spring, MD.
55. Partnership for Food Protection National Workplan Work Group. 2013. *Model for Local Federal/State Planning and Coordination of Field Operations and Training.* FDA-Silver Spring, MD.
56. US FDA, Center for Food Safety and Applied Nutrition, CFSAN Constituent Update. 2013. *FDA Reports High Enrollment in Voluntary National Retail Food Regulatory Program Standards.* Silver Spring, MD.
57. US FDA. 2015. *FDA Investigates Listeria Monocytogenes in Ice Cream Products from Bluebell Creameries.* Washington, DC.
58. Foodservice Packaging Institute. 2015. *2015 Crumbine Guidelines Released.* Falls Church, VA.
59. US Environmental Protection Agency. 2005. *Water Health Series: Bottled Water Basics.* 816-K-05-003. Washington, DC.
60. U.S. Food and Drug Administration. 2015. *Hazard Analysis Critical Control Pint (HACCP).* http://www.fda.gov/Food/GuidanceRegulation/HACCP/
61. U.S. Food and Drug Administration. 2006. *Managing Food Safety: A Manual for the Voluntary Use of HACCP Principles for Operators of Food Service and Retail Establishments* FDA. College Park, MD.
62. U.S. Food and Drug Administration Managing Food Safety. 2006. *A Regulators Manual for Applying HACCP Principles to Risk-based Retail and Food Service Inspections and Evaluating Voluntary Food Safety Management Systems.* College Park, MD.

BIBLIOGRAPHY

Acheson, David. 2007. Testimony. Subcommittee on Horticulture and Organic Agriculture House Committee on Agriculture. *Fresh Produce and Foodborne Illness Outbreaks.* Washington DC.
Association of Schools of Public Health. 2009. *Report Calls Food Safety System Antiquated, Calls for Reform*, ASPH Friday Letter Number 1562. Washington, DC.
Association of State and Territorial Health Officials. 2008. *State Health Agencies and Food Safety: Responsibilities, Challenges and Opportunities.* Arlington, VA.
Baker-Hegerfeld, Joan. 2008. *Growing and Direct-Marketing Produce: Potential Hazards.* South Dakota State University/College of Agriculture and Biological Sciences Extension Service/USDA, Brookings, SD.
Centers for Disease Control and Prevention, Food Safety Office. 2009. *Overview of CDC Food Safety Activities and Programs.* Atlanta, GA.

Columbus Public Health. 2009. *Retail Food Establishment Process Review, Authority: Chapters 3717.* Ohio Revised Code, Columbus, OH.

Columbus Public Health, Environmental Health Division. 2008. *Food Safety Program: Compliance and Enforcement-Standard Operating Procedures.* Columbus, OH.

Columbus Public Health. 2006. *Foodborne Illness Investigation Manual.* Columbus, OH.

Columbus Public Health, *Food Safety and Security,* personal communications from Keith. Krinn, Environmental Health Administrator, (Columbus, Ohio, August 13, 2009).

Levi, Jeffrey, Siegel, Laura M., Vinter, Serena. 2009. *Keeping America's Food Safe: A Blueprint for Fixing the Food Safety System at the US Department of Health and Human Services.* Robert Wood Johnson Foundation, Princeton, NJ.

Michigan Department of Agriculture. 2008. *Michigan Department of Agriculture and Local Health Department Optional Risk Based Evaluation Schedule.* Lansing, MI.

National Association of County and City Health Officials, Annual Conference. 2009. *2009 Samuel J. Crumbine Consumer Protection Award Recipient, Columbus Public Health.* Orlando, FL.

National Advisory Committee on Microbiological Criteria for Foods. 1977. *Food Security: A Workbook for Food Operators, Columbus Public Health. Hazard Analysis and Critical Control Point Principles and Application Guidelines.* Columbus, OH.

Ohio Department of Agriculture/Ohio Department of Health. 2009. *Standard Inspection Report: Chapters 3717 and 3715 Ohio Revised Code, Columbus Public Health Department.* Columbus, OH.

Patrick, Mary for Centers for Disease Control and Prevention. 2009. An Introduction to FoodNet Sites. *FoodNet News,* 2(4). Atlanta, GA.

Philadelphia Department of Public Health. 2001. *Food Establishment Self Inspection Checklist.* Philadelphia, PA.

Philadelphia Department of Public Health, *Environmental Health Services Food Safety Certification Program.* Philadelphia, PA, 2009.

President's Food Safety Working Group. 2009. *Food Safety Working Group: Key Findings.* Washington, DC.

Public Health Accreditation Board. 2009. *Proposed State Standards and Measures.* Alexandria, VA.

Surveillance for Foodborne Disease Outbreaks: United States, 2006. 2009. *MMWR Weekly,* 58(22).

Taylor, Michael R., David, Stephanie D. 2009. *Stronger Partnerships for Safe Food: An Agenda for Strengthening State and Local Roles in the Nation's Food Safety System.* Department of Health Policy, School of Public Health and Health Services, George Washington University, Washington, DC.

U.S. Department of Agriculture, Food Safety and Inspection Service. 1998. *Introduction to a Risk-Based Food Inspection Program, Office of Food Protection. Philadelphia Department of Public Health: Key Facts: The Seven HACCP Principles.* Washington, DC.

World Health Organization. 2008. *Terrorist Threats to Food: Guidance for Establishing and Strengthening Prevention and Response Systems.* Geneva, Switzerland.

8 Healthcare Environment and Infection Control

STATEMENT OF PROBLEM AND SPECIAL INFORMATION FOR HOSPITALS

The healthcare environment starts with the emergency medical services that may be bringing the patient to the medical facility. The healthcare facilities are a mixture of temporary or permanent housing and health management practices which include primary care, diagnostic care, acute care, complex health care, rehabilitation, and continuing care. These actions take place in hospitals, stand-alone surgical units, nursing homes, facilities for the aged, rehabilitation centers, assisted living facilities, and individual homes and doctors' offices. For the purposes of this discussion, the greatest amount of information will be about the hospital and its environment and then much briefer material will be provided on some of the other facilities (See endnotes 1, 6).

The hospital is a small, very complex community with highly vulnerable individuals (patients) who are the recipients of preventive measures, or seeking diagnosis of conditions or diseases or treatment for a variety of diseases and/or injuries. Patients with infections or healthy carriers of disease frequently move from one area of the hospital to another. Highly susceptible individuals are concentrated in specialized units for newborn infants, burn patients, intensive care patients, and surgical patients. The emergency department is of special concern because of the vulnerability of the patients coming in with traumas, acute diseases, or chronic conditions. The emergency department also may be extremely crowded for short periods of time and therefore disease may spread more easily.

The employees, with their own range of health and safety problems, carry out a vast variety of tasks which require different levels of education and skills. However, their overriding concern is to do no harm while performing their many duties. The presence of visitors and vendors increases concerns for health and safety within the institution. All of these individuals interact with each other and with the environment of the hospital or facility for the elderly. They are exposed to the various environmental health, safety, and infection control problems in their homes, neighborhoods, occupational environments, and in other situations and then bring the result of these exposures to the hospital environment, thereby potentially increasing hospital health hazards. All of these people then become the means of transmission of disease and environmental hazards from the hospital to the community. The health and safety hazards may be microbiological, chemical, or physical in nature.

Medical errors and healthcare-associated infections are among the leading causes of death in the United States. Patients and medical personnel are rightfully very concerned and frustrated because these problems have not been resolved in a positive manner. Peer-reviewed studies have shown that multiple patients in a room, noise, lighting, ventilation, ergonomic design, and poor workplaces and layouts contribute to errors, safety hazards, stress, poor sleep, pain, healthcare-associated infection, and other poor outcomes.

The acute care hospital, ambulatory surgery center, inpatient rehabilitation facility, outpatient dialysis facility, long-term acute care hospital, and long-term care facilities such as nursing homes, assisted living, residential care, and chronic care facilities, and skilled nursing facilities have many responsibilities for the care of the individuals within the particular institution. Problems of an environmental nature include infection control, injury control, medication errors, environmental hazards internal to the facility, and pollution hazards external to the facility.

Various surfaces, equipment, and linens may become rapidly contaminated with a variety of microbiological agents. Problems of indoor air increases substantially. Water utilized in various treatments may become contaminated. A large quantity of highly contaminated materials is produced in the hospital and then may go out to the outer community.

SUB-PROBLEMS INCLUDING LEADING TO IMPAIRMENT AND BEST PRACTICES FOR ACUTE CARE HOSPITALS

HEALTHCARE-ASSOCIATED INFECTIONS

(See endnote 20)

The concept of the prevention of the spread of infection in hospitals goes back to a least 1771 when Thomas Percival wrote *Internal Regulation of Hospitals* and there was a move to provide isolation for patients with infectious diseases and to improve the sanitation of the facilities. In subsequent generations as scientific knowledge improved concerning the spread of disease, new techniques were used to reduce the level of infections occurring in the institutions. Additional concern was shown during the Civil War, Crimean War, World War I, and World War II because of the continuing high level of infections in patients.

In addition, microorganisms constantly change and therefore new problems arise. For example, in 1954 in courses in medical microbiology taught in colleges, *Staphylococcus aureus* was used as an indicator organism. Shortly thereafter it was discovered that one of the serotypes had become resistant to penicillin and then became a hospital problem for infection control. Methicillin resistant *S. aureus* (MRSA) is now a worldwide problem in hospitals and in communities (especially in areas where there is close contact and contaminated surfaces utilized by sports teams, child day care centers, prisons, injection drug users, military personnel living in dormitories, and Native Americans), causing bacterial infections involving the bloodstream, lower respiratory tract, skin, and soft tissues.

Based on data from the National Nosocomial Infection Surveillance System collected by the Centers for Disease Control and Prevention (CDC) in 2002, it was estimated that there were 1.7 million cases of healthcare-associated infections in US hospitals. It was estimated that approximately 99,000 people died from these infections. The yearly direct hospital cost of treating healthcare-associated infections in the United States in 2007 dollars ranges from $28.4 billion to $33.8 billion. (See endnote 2.) (These numbers vary from year to year and increase or decrease according to the level of performance of staff and the availability of infectious agents in communities. These numbers also only represent hospitals and not the total healthcare system.)

Although there are some new organisms that are causing infections in hospitals and in the community today, there are many organisms which were present in the past that are still creating serious problems. Many of the protocols used in the past are still beneficial if applied appropriately and consistently. The CDC has always been involved in advocating high-level standards of performance to prevent and control healthcare-associated infections. However, these protocols are only recommendations to be followed by hospitals and are not mandated by law. The problem is that once we reduce infections, we get complacent and loosen our approach to preventive measures. This leads to increased infections and a new surge of resistant organisms as well as an increase in healthcare-associated infections.

The level of healthcare-associated infections is being exacerbated by improper and too frequent use of antibiotics. Further, when patients do not complete the antibiotic program, they are in effect destroying those microorganisms which are most easily killed but perpetuating those that are most resistant. Antibiotic resistance from microorganisms may also be occurring because of the extensive use of antibiotics in animals.

Minimal discussion will be presented concerning the very substantial area of potential healthcare-acquired infections associated with medical/nursing techniques. Most of the discussion will be about the potential for the spread of microorganisms, which can be the cause of healthcare-acquired infection through contact with equipment, air, water, linens, surfaces of all types, food, and hands.

EMERGENCY MEDICAL SERVICES AS A SOURCE FOR HEALTHCARE-ASSOCIATED INFECTIONS

(See endnote 39)

Emergency medical services systems personnel respond to situations where the individuals may either be highly infectious from a variety of diseases or may be highly susceptible to disease because of existing chronic health conditions, age, medications being taken, being debilitated, or severe trauma from motor vehicle accidents, falls, cuts, burns, gunshot wounds, domestic violence, or other kinds of violence. The personnel arrive at the scene of numerous high-risk situations and may be exposed to a variety of diseases without even knowing it. The individuals responding may vary from paid to unpaid, from highly experienced to people with little experience, and they may be paramedics, emergency medical technicians, police, firefighters, and others. The amount of knowledge about infection prevention may vary from none to fairly intensive training. In any case, the patients typically end up at a hospital and may either be at risk themselves, put the first responders at risk, and/or put the hospital at risk of serious infections.

The personnel not only have exposure to a variety of infectious or contagious diseases but also to body fluids, sharps which are contaminated and may cause immediate problems to personnel if they penetrate their skin, high-risk procedures such as intubation, starting IVs, bandaging injured people, poorly lit work areas, hazardous work areas with fires or hazardous materials, patients who become violent, etc.

The totality of first responders may be exposed to cutaneous anthrax, hepatitis B, hepatitis C, human immunodeficiency virus (HIV), viral hemorrhagic fevers, measles, tuberculosis, chickenpox, diphtheria, influenza, meningitis of various forms, mumps, whooping cough, MRSA, SARS, etc.

Best Practices for Preventing Infections from Emergency Medical Services System Practices
- Vaccinate all emergency medical services personnel with all appropriate vaccines to immunize them against possible contamination and disease from patients they are treating or transporting. The vaccines should include but not be limited to hepatitis B, influenza, measles, mumps, rubella, and varicella zoster.
- Provide to all personnel with intensive and continuous training in infection prevention and control including, but not limited to, proper hand washing (this is of greatest significance), use of gloves, proper use of respiratory devices and outer gowns when necessary, covering of exposed skin, proper use of eye shields if necessary, techniques of disinfection of equipment and all surfaces within the vehicle, etc.
- Conduct all cleaning and disinfecting within the vehicle before leaving the facility to which you have brought a patient.
- Remove all disposable contaminated items in a waterproof bag labeled biohazard.
- Remove all linens carefully and place in a waterproof bag marked contaminated laundry.
- Place all needles and sharps in a special waterproof solid box for later disposal.
- Instruct all patients to cover their mouth and nose when sneezing or coughing and dispose of tissues in a no-touch receptacle. They should wear a surgical mask if they are in an infectious stage and dispose of it properly when it becomes wet. Patients need to wash their hands thoroughly before doing any routine type of lifestyle work.
- Clean and disinfect all high touch areas that are close to the patient after each patient.
- If the patient had a communicable or highly infectious disease, totally decontaminate the vehicle before using it for other patients.

- Report all potential infection problems to healthcare personnel when bringing patients into a facility.
- For large quantities of blood or other bodily fluids, wear proper personal protective equipment, use super absorbent pads and dispose of them in biohazard bags, use disinfectants over the entire area, and if necessary rinse the area with small quantities of water. Bodily fluids include semen, vaginal secretions, cerebrospinal fluid, synovial fluid, pleural fluid, pericardial fluid, peritoneal fluid, amniotic fluid, saliva, and anything contaminated with blood, tissue, or organs.
- If blood or body fluids come in contact with the emergency medical services personnel's intact skin, wash the area thoroughly for at least 15 seconds, examine the skin to make sure there are no breaks or chapped areas, and then disinfect the skin carefully with appropriate skin disinfectants. Non-intact skin should follow the previous care system and then have the appropriate antibiotics applied and supplied orally if directed by medical care personnel.
- Clean and decontaminate or dispose of all equipment used on patients. (See the section on "Equipment" later in this chapter.)

HEALTHCARE-ASSOCIATED INFECTIONS CAUSED BY MEDICAL/NURSING TECHNIQUES

Some of the major areas or people involved where infections occur include central line catheter insertion sites, dialysis units, catheter-associated urinary tract patients, surgical sites, and ventilator-associated patients who develop pneumonia. Also discussed will be *Clostridium difficile* infections and MRSA infections. (See endnote 3 for details.) Other portals of entry for microorganisms may include the eyes, ears, nose, skin, and gastrointestinal system. Normal organisms present in certain parts of the body can be transferred to other parts of the body where they may cause a healthcare-associated infection. Hand washing among staff varies with the type of position the individual holds, the facility, and the level of supervision provided for this highly important preventive activity. Physicians and nursing assistants most frequently wash their hands. However, there needs to be improvement in all staff positions.

Best Practices for Preventing Healthcare-Associated Infections Caused by Medical/Nursing Techniques (See endnote 3 for details on each of the subjects mentioned immediately above) (See endnotes 4, 17)
- Establish an infection control committee which meets regularly and on an emergency basis when needed to: review epidemiological surveillance data from all areas of the institution; review case histories of patients and/or employees with healthcare-associated infections; establish a baseline of data concerning healthcare-associated infections and compare to national CDC data to see where the individual institution actually stands in relation to other institutions and the amount of infections present; determine critical areas for use of intensive preventive and control techniques; develop appropriate training measures for all levels of personnel and ensure that the training results in reduced levels of infections; and provide a comprehensive infection control manual.
- Provide an infection control team of specialized professionals who can rapidly enter any area of the institution, evaluate the problems, and make necessary changes to stop an outbreak of infectious diseases or prevent one from occurring.
- Put special emphasis on catheter-associated urinary tract infections, surgical site infections, ventilator-associated infections, central line-associated bloodstream infections, blood and other bodily fluids procedures, and healthcare personnel exposure.
- Establish comprehensive evaluations of the technical knowledge and skills of all medical/ nursing personnel involved in performing the necessary functions in an aseptic manner for

various catheterizations, use of dialysis units, a variety of surgeries and invasive procedures, and insertion and use of ventilators.

- Provide appropriate supervision especially to new and existing personnel in performing any of the above-mentioned skills when working with these procedures.
- Provide necessary in-service training courses and periodic continuing education to refresh individuals in the above-mentioned skills after specialized training.
- Teach each of the individuals utilizing invasive procedures all necessary techniques used for the prevention of infection in the patient and spread of infections from the patient to other patients and staff.
- Minimize the use of invasive medical devices, enhance and insist upon proper hand hygiene of all personnel, and improve cleaning of all surfaces and equipment to reduce potential for healthcare-associated infections.
- Use only specially trained personnel to insert catheters while utilizing all appropriate aseptic techniques.
- Increase staffing to appropriate levels to avoid undue fatigue in staff which may lead to improper use of aseptic techniques.
- Document on the patient's record the physician's order, the catheter placement, and any problems related to insertion and maintenance. Also record catheter removal and if necessary justify continued use of catheters.

HEALTHCARE-ASSOCIATED INFECTIONS CAUSED BY ENVIRONMENTAL HAZARDS INTERNAL TO THE FACILITY

(See endnotes 9, 14, 16, 18)

BUILDING CONSTRUCTION OR RENOVATION

(See endnotes 5, 8)

Many billions of dollars are being spent annually to renovate and upgrade existing hospital and other healthcare facilities and also to construct new ones. Many of the facilities in the United States were built in the 1950s and 1960s and no longer can provide efficient and safe healthcare delivery or utilize rapidly emerging technologies. Healthcare-associated infections and injuries can increase as a result of the close proximity of renovation areas and the work being done to patient care areas. The activities of construction increase the potential for airborne, waterborne, and contact infections by simply disturbing the existing facilities as repair or remediation occurs. There have been in the past outbreaks of disease when construction has caused microorganisms to become airborne or waterborne. This problem is especially a concern in high-risk patient areas.

However, healthcare-associated injuries and healthcare-associated infections can be reduced by following the evidence-based standards and recommendations established by The Joint Commission for the six phases of the building process including: planning; schematic design of the project including preliminary room layout, structure, and scope; design and development including all details; construction documents requiring a template using which contractors can make bids and also understand all the details of the construction; the actual construction phase of the building which can create numerous hazards to the contractors and existing patients and staff in renovated areas; and the commissioning of the building when actual ownership is taken after all specifications have been met as determined by inspection by the proper authorities in each area. The planning, construction, and design of these facilities impact the health and safety of employees, patients, visitors, and contractors.

There are numerous problems in facilities that need to be addressed. They are: traffic flow in highly sensitive areas including where high-risk patients are being treated; improper spatial separation of patients, especially where potential spread of healthcare-associated infection may occur; inadequate numbers and types of isolation rooms; inadequate access to hand-washing facilities; use of carpets and other types of materials which are difficult to clean in patient treatment and care areas; inadequate, poorly maintained, or inappropriate ventilation for isolation rooms, operating rooms, intensive care units, cancer care units, special nurseries including neonatal intensive care, emergency departments, care areas for other highly susceptible patients; exposure to fungi, dust, asbestos, etc. during renovation projects; inadequate or inappropriate water supplies for critical areas; etc.

Besides the normal hazards associated with renovation and construction, the contractors, staff, and patients are also exposed to numerous hazardous drugs and chemicals which may have a profound health effect on the individuals. Cleaning agents may also result in potentially hazardous situations and affect all individuals who are present.

Best Practices for Building Construction or Renovation (See endnote 7)
- Include environmental and infection control personnel in the team that prepares necessary plans and evaluates all healthcare facility demolition, construction, and renovation.
- Evaluate all potential hazards through a risk assessment process for demolition and construction activities prior to construction or renovation and determine how best to prevent or control these hazards to prevent accidents and resulting injuries and avoid environmentally created disease or healthcare-associated infections.
- Determine if construction will create water leaks, loss of negative air pressure or positive air pressure as needed in sensitive areas, barriers, and their types that need to either come down or be erected before construction, potential utility failures, fires, or other emergencies, potential spread of healthcare-associated infection, or interference with ingress or egress of staff, and make necessary plans for correction prior to construction.
- Observe all National Fire Protection Association standards and American National Standards Institute standards for construction, personal protective equipment, and construction practices.
- Monitor and then record and make necessary corrections on a daily basis of any adverse air flow in isolation rooms and reverse isolation rooms where immunocompromised patients receive care and treatment during all types of renovation and construction.
- Minimize all unnecessary stresses such as noise, vibration, dust, mold, etc.
- Ensure that all exits are unobstructed, all construction areas are secured, and that there is immediate entrance for emergency services including fire, police, and medical personnel in the event of an accident.
- Ensure that all alarm, detection, and suppression systems for fires are in working order.
- Ensure that all temporary partitions used in construction are smoke tight and non-combustible.
- Provide specialized training for all personnel in firefighting and give them additional firefighting equipment as needed.
- Prohibit all smoking within the premises.
- Determine appropriate storage, housekeeping, and debris removal and enforce it rigorously.
- Conduct fire and disaster drills on a routine basis.
- Work with a buddy system in dangerous areas such as the cleaning of large tanks.
- Control traffic flow during all construction and renovation areas to reduce dust levels and do not allow unauthorized people to enter construction sites.
- Create appropriate traffic flow in a new or renovated facility to avoid access problems for people coming into critical areas.
- Evaluate all proposed flooring materials for use of sealants, adhesives, maintenance, and treatment products needed, durability, safety and potential traction and prevention of falls, glare avoidance, acoustic properties, wear ability, and time needed for cleaning and

maintenance, and utilize those products which are best able to protect the health and safety of all people while being cost-effective.

- Establish special separate rooms for infected individuals, highly infected individuals such as Ebola patients, immunocompromised patients, and special needs patients.
- Quickly respond to problems of water damage and determine if the wet material can be dried successfully without growth of mold or whether it needs to be removed and replaced immediately.
- Locate sinks and hand-washing dispensers in appropriate and convenient locations for employee use.
- Locate sharps containers in convenient places.
- Physically separate dirty work areas from clean work areas.
- Clean all work zones daily by using wet wiping techniques and cover all debris before removing from the work zone.
- Flush the water system to clear any sediment which is in the pipes to prevent growth of waterborne microorganisms.

AIR AND AIRBORNE INFECTIONS

(See Chapter 2, "Air Quality (Outdoor [Ambient] and Indoor)")

All populations of patients, employees, visitors, and contractors are subject to disease transmission by organisms which are airborne. The disturbance of soil, water, dust, or decaying organic material during construction or otherwise can release these microorganisms into the air. The microorganisms may also enter the air from oral or nasal secretions from infected patients or healthy carriers, which may then be inhaled by other individuals. The microorganisms may land on surfaces or on various parts of the body of the individuals and lead to infections.

Large patient populations can contribute to the healthcare infection problem because of the sheer numbers of individuals who may be contributing microorganisms to the air, water, and surfaces. Pathogens may be transmitted from the hands, other parts of the skin, respiratory system including nose and throat, urinary tract system, and digestive tract system of individuals, patients, and staff. A variety of bodily fluids may be involved. Pathogens may also be present in the soil, water, dust, or on disturbed surfaces or decaying organic material and can be released into the air and thereby transmitted throughout the healthcare environment. Organisms can be transmitted from person to person by droplets in the air produced during coughs, sneezes, or simply talking. Droplets may be suspended in the air for an extended periods of time and travel long distances. Construction, renovation, and cleaning can disturb pathogens and make them airborne and easily inhaled.

Ventilation systems may malfunction and become contaminated. Numerous factors related to ventilation can affect the amount of microorganisms being transmitted. These factors include cleanliness of air filters, cleanliness of grills, type of air filter, direction of air flow, air pressure, number of air changes per hour in the room, humidity, cleanliness of the ventilation system, maintenance of the ventilation system, and level of contamination within given areas. Accumulations of dust and moisture within high-efficiency particulate air filters (HEPA) increase the risk of the spread of fungi and bacteria. Also if the system is shut down, negative pressure can be created and air from other contaminated areas can be sucked in. This may be due to a malfunction, inadequate, or cleaning, or simply running the system periodically instead of continuously. Filtering systems can become contaminated with pigeon droppings and other bird droppings which contain microorganisms.

Best Practices for Preventing Airborne Infections
- Set up the heating, ventilation, and air-conditioning (HVAC) systems according to the specific guidelines from the American Institute of Architects for special patient and treatment areas to prevent the spread of airborne contaminants.

- Monitor outdoor air intake systems and inspect filters on a regular basis.
- Remove any bird roosting areas or nests near outdoor air intake systems.
- Place air exhaust systems downwind from the building away from air intake systems.
- Establish and maintain appropriate humidity controls in all HVAC systems.
- Use filtration, which is the removal of particulates containing microorganisms, chemicals, dust, and smoke, to remove at the source the substances which may cause disease.
- Use ultraviolet germicidal irradiation as a secondary air cleaning measure in the duct work leaving the area containing the exhaust air. Frequent maintenance and cleaning of the ultraviolet units is necessary.
- Establish an appropriate schedule for frequently evaluating the cleanliness of ventilation systems, grills, and filters. Replace the filters on a routine basis and also when found to be dirty.
- Use a system of 90% efficiency filters, and clean and maintain all ductwork and grills in the ventilation system to prevent the spread of disease.
- Do not use fans in high-risk areas especially where there are immunocompromised patients.
- Use 99.97% efficiency HEPA filters and appropriate maintenance schedules in specialized areas such as surgical suites, burn units, ICU units, areas with immunocompromised patients, neonatal nurseries, etc.
- Use the latest appropriate ventilation guidelines which indicate where negative pressure is mandated and where positive pressure is mandated for all areas of the facilities including patient rooms, labor/delivery areas, isolation rooms, emergency rooms, radiology, procedure rooms, surgery, critical care areas, pharmacies, sterilizing and supply areas, linen areas, etc., as identified by the Facility Guidelines Institute and updated every 4 years. These guidelines are the consensus of a multidisciplinary group of experts from the federal government, state governments, and private sector. They are based on practical evidence-based decision-making processes that are conducted in a scientific manner. (See endnotes 12, 13.)
- Use single bed rooms instead of multiple bed rooms with HEPA filters to prevent transmission of healthcare-associated infections through the air.
- During construction and renovation, use portable HEPA filters, physical barriers between patients and construction areas, negative air pressure in construction/renovation areas, and sealing of patient windows.

EQUIPMENT (MEDICAL AND SURGICAL)

(See endnote 15)

In the United States, annually there are about 46.5 million surgical and other invasive medical procedures including about 5 million gastrointestinal endoscopies. In each of these procedures, pathogenic organisms can be introduced into the body by means of the medical device or surgical instrument, thereby setting up a site of infection leading to potentially serious consequences for the patient and outbreaks of healthcare-associated infections caused by a break in procedure in the proper cleaning, disinfecting and, where necessary, sterilization of the equipment.

Sterilization is a process that destroys all forms of microbial life including spores and it may be accomplished by use of steam under pressure, dry heat, ethylene oxide gas, hydrogen peroxide gas plasma, or liquid chemicals. Disinfection is a process that eliminates all pathogenic microorganisms except bacterial spores on inanimate objects. Cleaning is the visible removal of all soil, either organic or inorganic, from objects and surfaces and can be accomplished either mechanically or by hand utilizing appropriate detergents. Decontamination results in the removal of pathogenic microorganisms from objects. The proper techniques utilized in any case include first meticulous cleaning and thorough rinsing, and then depending on the nature of the equipment, utilizing decontamination, disinfection, or sterilization techniques.

The Spalding technique which was set up as a rational approach to disinfection and sterilization of patient care items and equipment many years ago, although still being used in some areas, is considered to be an oversimplification, however the reader should understand what it is and how it works. The technique consists of dividing all patient care items and equipment into three categories including: critical items which enter sterile tissue or the vascular system and have to be sterile; semi-critical items that come in contact with mucous membranes and non-intact skin including respiratory and anesthesia equipment and certain endoscopes which have to be free from all microorganisms but bacterial spores are allowed; and non-critical items that come in contact with intact skin but not mucous membranes such as bedpans, blood pressure cuffs, etc. and can be decontaminated on site without having to go to central supply for processing. One of the problems with implementing this procedure is the reprocessing of complicated medical equipment that is heat sensitive or the inactivation of certain infectious agents such as prions. Another problem is the processing of an instrument in the semi-critical category such as an endoscope that may be used with a critical instrument that comes in contact with sterile body tissues. The optimal contact time for high-level disinfection has not been defined and therefore there is no one Best Practice established for this type of procedure.

Factors that affect disinfection and sterilization include the number and location of the microorganisms and whether or not they produce spores; the resistance of the microorganisms; the concentration and potency of the disinfectants; the temperature, pH, relative humidity, and water hardness of the solution containing the disinfectants; the type and amount of organic and inorganic matter that is on the surface of medical equipment; the time of exposure of the equipment to the disinfecting solution; and the presence or absence of biofilms which are large groups of microorganisms that are tightly attached to surfaces and may be surrounded by additional materials and are difficult to remove.

Each of the major groups of instruments or pieces of equipment in use will be discussed and the most modern Best Practices will be stated as follows.

1. Endoscopes of all types are used to diagnose and treat various disorders. Endoscopes, which are heat sensitive, become contaminated when they enter body cavities. A prime example is the duodenoscope which has a very complex design and therefore may receive ineffective reprocessing. Multidrug-resistant microorganisms can be very difficult to destroy even if healthcare personnel properly follow the instructions established by the manufacturer. The reprocessing of endoscopes has four steps. They are: mechanically thoroughly clean all internal and external surfaces of the unit including brushing, with a detergent or enzymatic cleaner, thoroughly clean the internal channels and rinse very well with water and then test for leaks; disinfect by immersing the endoscope in an appropriate disinfectant or chemical sterilant, for example, 2% glutaraldehyde for 20 minutes; rinse the endoscope and all channels with sterile water or high-quality potable water; and rinse the insertion tubes with alcohol and dry with forced air after disinfection and before storage. A question that should always be asked prior to a procedure is "Do the benefits of the procedure outweigh potential risk due to the invasive nature of the task?"

2. Other instruments which can transmit a variety of infections from one patient to another include tonometers to measure eye pressure, cervical diaphragm fitting rings, cryosurgical instruments, and endocavity probes. The Food and Drug Administration (FDA) requires that the manufacturer lists at least one validated cleaning and disinfection/sterilization protocol in the labeling of each instrument. However, other than thorough cleaning, there is no consensus or significant research showing the best way to disinfect these types of equipment. It is recommended that further research be carried out in these areas.

3. Dental instruments that penetrate soft tissue or bone must be cleaned thoroughly and either sterilized after each use if heat stable or put in appropriate disposal containers. Covers can be used for other types of equipment and then disposed of after seeing the patients. Surfaces which are frequently touched should be cleaned and disinfected between patients.

4. Hepatitis B virus, hepatitis C virus, HIV, or TB-contaminated devices should be thoroughly cleaned and then treated with 2% glutaraldehyde solution for 20 minutes.
5. Disinfection of the hemodialysis unit includes the machines, water supply, water treatment systems, and distribution systems which have become infected with blood-borne viruses and pathogenic bacteria. Thoroughly clean all surfaces and properly disinfect all water. Use a hypochlorite solution or a disinfectant with a tuberculocidal agent for disinfection of the equipment. Disinfectants include peracetic acid, formaldehyde, glutaraldehyde, heat pasteurization, and chlorine-containing compounds. (See the most current recommendations for the reuse of hemodialyzers issued periodically by the Association for the Advancement of Medical Instrumentation.)

General Best Practices for Cleaning and Disinfecting Medical and Surgical Instruments and Equipment

- General Best Practices for medical and surgical instruments and equipment includes extremely thorough and intensive cleaning of all surfaces, interior and exterior, especially those heavily contaminated with blood, other organic materials, lubricants, or inorganic deposits. Complete rinsing of all of the interior and exterior surfaces to remove all cleaning materials should be done before disinfection is accomplished or sterilization is utilized. It is extremely important that the proper time for cleaning and correct concentration of cleaning materials and disinfectants be used. In the case of sterilization by heat, it is extremely important that not only all of the instruments be thoroughly cleaned and all deposits removed but also that the appropriate time and temperature under pressure be utilized.
- Cleaning, which is the physical removal of all organic material including blood and tissue as well as inorganic salts and various types of soil, is best accomplished by cleaning as soon as possible after use of the equipment and utilizing ultrasonic cleaners, washer disinfectors, and washer sterilizers instead of hand washing which varies in intensity according to the individual performing the work. In some cases where the level of soil is intense, it may be necessary for a person to preclean the instruments and equipment to remove a large amount of the soil prior to using the mechanical equipment for cleaning. In any case, the proper amount of detergents or enzymatic cleaners that are compatible with the metals or other materials of the medical and surgical instruments must be used according to the manufacturer's instructions. Time and temperature are of great importance. All equipment must be inspected after cleaning to make sure that there are no breaks or cracks in it.
- Use disinfectants that are: broad-spectrum antimicrobial agents; fast acting producing a rapid kill; active in the presence of various types of organic matter; non-toxic to people; produce continued killing of microorganisms on surfaces for a period of time; easy-to-use, odorless, soluble in water, good cleaning substances, and stable in the concentrated and diluted forms; not supporting bacterial growth in the cleaning solution; economical; and not damaging to surfaces, water, or the rest of the environment.
- Take into consideration when sterilizing equipment or medical instruments the following issues: is the cleaning complete and thorough to avoid a higher bioburden and salt concentration; are you dealing potentially with spores; type of pathogens and resistance to sterilization techniques; is there a biofilm on the equipment or medical instruments; the length and diameter of various tubes; is there a restricted flow in the tubes because of sharp bends; the types of materials utilized in the equipment and how this affects sterilization; are there screws or hinges in the equipment that may affect the sterilization technique being utilized.
- Despite the best cleaning and decontamination procedures utilized on single use medical equipment or instruments, do not reuse them.

Food

(See Chapter 7, "Food Security and Protection")

Food service at a hospital is similar to food service at extremely large complex facilities housing potentially thousands of people on a daily basis. In hospitals, food has to be provided 24 hours a day to patients with special diets, debilitated people, staff, and visitors. Hospitals in addition to sheer numbers of people, typically have numerous other problems including: improper storage and handling of incoming foods and improper timing and refrigeration of perishables; improper defrosting of chickens and turkeys and other meats outside of refrigeration; improper cleaning and decontamination of a variety of surfaces because of the constant use of the kitchen facilities by workers who may only stay for short periods of time and need constant training and supervision; inadequate lighting in a variety of areas especially the ware washing rooms; sick food handlers who continue to process and serve food; inadequate supply of hand-washing sinks; improper breaking down of the equipment and thorough cleaning procedures; reusing disposable water carafes and glasses; improper cleaning and temperature control, both refrigeration and heat in food carts that go on to the various floors; delaying service of food to patients and thereby compromising temperature controls; potentially serious insect and rodent control problems based on the distribution of substantial amounts of food in various parts of the facility; improper handling and removal of food from isolation units; etc.

Best Practices for Food Preparation, Handling, and Disposal (See Chapter 7, "Food Security and Protection")
- Develop and utilize a self-inspection concept for food service operations. See the Food Service Inspection Report on pages 526–527 of *Handbook of Environmental Health— Biological, Chemical and Physical Agents of Environmentally Related Disease*, Volume 1 (CRC Press, Boca Raton, FL, 2003). (See endnote 61.)
- Develop a food service self-inspection program for a 19-week, 2-hour a week training program for food service supervisors and managers. This program should include the following: Session 1: introduction of course and administrative personnel of hospital; Sessions 2–3: discussion of injuries, infections, and proper hospital practices; Sessions 4–5: discussion of Hazard Analysis and Critical Control Points (HACCP) principles and practices; Session 6: the self-inspection form and how and why to use it; Sessions 7–8: field experience within the institution in use of the form (make actual inspections under supervision); Sessions 9–10: review of actual food service department problems within the institution and start using the inspection form on a weekly basis; Sessions 11–12: discuss general housekeeping principles and practices and results of weekly self-inspections and how to make appropriate corrections; Session 13: discuss general food service equipment and ware washing techniques and continue the discussion of the weekly inspections; Session 14: microbiology and spread of foodborne and waterborne disease as well as hospital infections; Session 15: food preparation, storage, serving, transportation, and disposal as well as results of weekly inspections; Session 16: insect and rodent control as well as results of weekly inspections; Session 17: supervisory problems; Session 18: supervisory techniques; and Session 19: final examination covering the entire course.

Hand Washing

Poor hand-washing compliance may be due to numerous factors including: inconvenience sink locations and shortage of sinks, soap, or paper towels; high patient census or bed occupancy rates; overworked staff members; and indifference or a lack of supervision on the part of managers. Hands are the most frequent means of transmission of microorganisms from contaminated people and surfaces to the staff themselves and to patients who are susceptible to infection.

Best Practices in Hand Washing (See endnote 19)

- Increase the ratio of hand-washing sinks and hand cleaner dispensers to the number of beds within the facility and place them in highly convenient areas close to the patient bed.
- Use mandatory short-term continuing education frequently to teach staff proper hand-washing techniques for the areas they service and make violations of appropriate practice a serious demerit on annual job rating forms. Repeated problems of lack of hand washing or improper hand washing should result in disciplinary action by supervisors and managers.
- Place posters close to hand cleansing equipment stating that "Hand Washing Is Mandatory by All Personnel to Prevent Spread of Healthcare-Associated Infections."

INSECT AND RODENT CONTROL

(See Chapter 5, Insect Control; Chapter 6, Rodent Control; and Chapter 7, Pesticides in *Handbook of Environmental Health—Biological, Chemical, and Physical Agents of Environmentally Related Disease*)

Insects and rodents continue to be a serious concern in all institutions especially those involved in health care. There are numerous entrances into facilities and the constant movement of food and other supplies from vendors and visitors presents a unique opportunity for insects to enter the premises. Food, water, and shelter are abundantly available within the institutions because of the size of the facilities and because food is transported everywhere and stored, there are innumerable sinks that may leak, and there are a huge number of cracks and crevices where insects can harbor. Food crumbs from meals being eaten at the bedside as well as in nurses' stations and other employee areas on the various floors of the facility are attractive to insects and rodents. The various pests found in health-care institutions include German roaches, American roaches, bedbugs, ants, flies, fleas, termites, food storage insects, mice, and rats. Whenever repairs take place within the institution, there is an opportunity for a new natural pathway to be created for the invasion of the various pests.

Best Practices for Insect and Rodent Control

(See Chapter 9, "Insect Control, Rodent Control, and Pesticides")

LAUNDRY

Laundry may be contaminated with various body substances including blood, skin, feces, urine, vomit, etc. Disease transmission may occur if linens are shaken vigorously while they are being removed from beds and other laundry items are mishandled. This may be not only a potential patient care problem but also an occupational problem as the various linens are prepared for washing and then final processing. Dirty and soiled linens may be placed in the same laundry carts as clean linens, resulting in cross-contamination. Linens coming from isolation units are of special concern because of the high level of highly infectious microorganisms that may be present. Linens may also contain sharps or other types of equipment that can injure laundry personnel.

Best Practices for Preventing the Spread of Hospital-Acquired Infections through Laundry

- Use a minimum of handling and shaking when sorting contaminated laundry to prevent microorganisms from becoming airborne.
- Use personal protective equipment including masks, gowns, and gloves when sorting and processing contaminated laundry containing blood or other infectious materials.
- Do not launder at home potentially contaminated personal protective clothing or equipment including laboratory coats, although uniforms or scrubs worn in the same manner as outer clothing and not contaminated with blood or other infectious materials may be laundered at home.

- Change white coats frequently since not appearing dirty is deceiving and microbiological load may be considerable around sleeves and pockets. Have them laundered through the hospital because constant contact with patients may lead to cross-contamination.
- Provide a dirty area in the laundry for receiving and sorting and a totally separated clean area for processing and packaging.
- Do not leave damp laundry overnight in machines.
- Place all highly contaminated laundry in special bags that are leak proof and color-coded to indicate microbiological contamination, close securely, and double bag if necessary.
- Segregate highly contaminated laundry from regular laundry and transport by carts to the laundry facility.
- Do not sort highly contaminated laundry until the washing process has been concluded.
- Steam autoclave all laundry such as surgical drapes and reusable gowns, etc., after laundering and before use in surgical areas or other highly sensitive areas. Transport them separately to the specific areas.

MEDICAL WASTE AND OTHER HAZARDOUS SUBSTANCES INCLUDING CLEANING MATERIALS

(See Chapter 12, "Solid Waste, Hazardous Materials, and Hazardous Waste Management")

Solid waste is a high cost area of hospital and other healthcare facility expenses. The solid waste has to be divided into normal solid waste produced by various businesses including considerable paper and paper board, plastic, glass, food remnants, and medical waste and other hazardous substances. Utilize programs of minimization and pollution prevention including source reduction, recycling, and better use of materials to help reduce the overall solid waste problem and save considerable money for the institution. The majority of this discussion will be on medical waste and other hazardous substances.

Infectious waste is produced in patient care areas, laboratory areas, and research areas and consists of a variety of disposable materials including parts of equipment which have come in contact with the infected patients. Sharps include needles, scalpels, and sharp pieces of broken objects, especially plastics and glass. Major medical waste includes: cultures and stocks of microorganisms from laboratories; bulk blood, blood products, and bloody body fluid; pathology and anatomy wastes such as human tissue, and body parts; and contaminated materials from highly infectious patients.

Pharmaceutical waste includes materials contaminated by pharmaceuticals, expired pharmaceuticals, and liquids which have come in contact with the pharmaceuticals. Pharmaceutical waste is a concern because pharmaceutical residues have been found in drinking water in major metropolitan areas serving large numbers of people. Human and veterinary drugs, both prescription and over-the-counter, even at very low concentrations, may cause a risk to sensitive populations such as the fetus, individuals with chemical sensitivities, and individuals with existing diseases taking large numbers of prescribed pharmaceuticals. Hormone-disrupting chemicals may be a particular challenge to the fetus and young children at very low levels.

Chemical waste includes a vast variety of materials from laboratories, housekeeping cleaners and disinfectants, and chemicals used in swimming pools and hydrotherapy. Research waste includes animal carcasses, various containers, and testing media. Special waste containing heavy metals includes batteries, thermometers, certain chemicals, etc. Pressurized containers include empty gas cylinders, gas cartridges, and aerosol cans. Radioactive waste containing radioactive material from therapy or research laboratories includes contaminated glassware and other containers, urine and feces from patients receiving radioactive therapies, and unused portions of the radioactive material. Chemicals used in cancer treatment can damage the cells or genes.

Hospitals primarily used to take care of their own medical waste and other hazardous substances. Before 1995 there were over 4000 medical waste incinerators in the United States. They produced in addition to many other toxins, dioxin and mercury, which ended up in the air and water.

Today this has been reduced substantially and much of the waste is now sent to specialized treatment companies who can carry out proper shredding and incineration with appropriate air pollution controls and water pollution protections. The generation and transportation of medical waste and other hazardous substances is of greatest concern in the occupational environment, and therefore all employees must take necessary precautions to avoid becoming contaminated and ill.

Best Practices for Medical Waste and Other Hazardous Substances including Cleaning Materials plus General Solid Waste (See endnote 4) (See Chapter 12, "Solid Waste, Hazardous Materials, and Hazardous Waste Management")

- Place all medical waste and other biohazardous substances in special containers which are then double bagged and specially marked as highly infectious material, and collect it separately from other waste materials.
- Put all sharps in puncture-proof and leak-proof containers which are specifically marked as such and marked as highly infectious material.
- Use the international infectious substances symbol on all infectious waste.
- Sterilize by autoclaving all microbiological laboratory waste in a special autoclave used only for contaminated materials prior to disposal.
- Cytotoxic waste must be placed in hardened leak-proof containers, labeled cytotoxic waste, and moved to specialized approved medical incinerators by trained individuals.
- Chemical waste must be packed in special chemical-resistant containers, appropriately labeled, and removed to specialized treatment facilities for hazardous chemicals.
- Use steam sterilization for all sharps, disposable syringes, IV sets, catheters, and various tubing used in the body and then shred them and place them in a landfill in a deep burial secure area above the groundwater table.
- Incinerate all body parts unless there is a religious requirement to bury them.
- Incinerate all animal carcasses used in research areas.
- Exclude all contaminated plastics from incineration and use steam sterilization prior to shredding and land disposal.
- Use steam sterilization for all highly infectious patient wastes contaminated with body fluids, excretions of all types, as well as infectious laboratory waste prior to incineration.
- Determine when to use disinfectants in order to limit chemical exposure by personnel and patients.
- Determine if new cleaning technologies such as peroxide vapor and steam vapor are both effective in disinfecting contaminated areas and also cost-effective.
- Conduct appropriate research to determine at what level pharmaceuticals in water may become a health problem to various subpopulations of people and what levels of pharmaceuticals are typically found in the water supply.
- Determine the amount of waste anesthetic gases released into the air and whether or not they have a negative effect on personnel.
- Develop and enforce stringent air pollution rules and regulations for use of hospital medical incinerators.
- Determine if thermal treatment such as the use of microwave technologies will destroy microorganisms that can cause disease.
- Despite the expiration of the Medical Waste Tracking Act of 1988, record medical waste generation from starting point to final resting place, and determine if all necessary preventive techniques have been utilized to protect workers.
- Inform all workers of the nature of the hazardous waste being transported and proper means of self-protection.
- Train all workers in the proper handling of hazardous medical waste and provide appropriately trained immediate supervisors to evaluate the operations, make necessary corrections, and keep proper records.

NOISE

(See endnote 6)

Noise is excessively high in hospitals. There are numerous noise sources which are allowed and disturbing including public address systems, alarms, telephones, voices of personnel, other patients, equipment being moved back and forth in the hallways, bed rails, etc. The various surfaces which may be excellent for cleaning typically bounce sound off of floors, walls, ceilings, and various pieces of equipment. The sound-reflecting surfaces cause the sounds to echo and overlap each other for long periods of time. Studies have shown that medical equipment and staff voices can produce 70–75 dB of sound at the patient's ears, which is equivalent to the noise level in a busy restaurant. Noises from alarms and certain equipment can produce 90 dB of sound, which is equivalent to the noise produced next to a highway.

High sound levels or noise can affect the health of individuals. Noise can decrease oxygen saturation, elevate blood pressure, increase the heart and respiration rate, and interfere with sleep. Frequent and sudden awakening can be detrimental to the patient. Stress levels especially in adults increase and healing is delayed when noise levels increase.

Best Practices to Reduce Noise
- Install special acoustical tiles to help absorb sound and reduce noise levels.
- Use a noiseless paging system by contacting individuals through vibrating phones.
- Provide single bedrooms rather than multiple bedrooms.
- Train all staff to reduce the volume of their voices especially in corridors.
- Only use medical alarms when absolutely necessary.
- Do not check vital signs unless medically necessary during the sleep hours of patients.
- Monitor levels of noise in different parts of the hospital and determine how best to lower the level.
- Provide sleep mask and earplugs for patients where feasible.
- Program TVs so that the volume cannot exceed acceptable levels and use headsets where appropriate.

SURFACES

(See endnote 10)

The hospital environment may contribute to the transmission of organisms that may cause healthcare-associated infections. Surfaces within the hospital and other healthcare settings may become easily contaminated with a variety of microorganisms, some of which may be pathogenic. Patients, staff, visitors, and others may be the source of the contamination or the surfaces may become contaminated by individuals touching the surface or stirring up contaminated dust on the surface into the air and inhaling it. The ability of the microorganism to persist on the surface depends on several factors including the nature and quantity of the organism, the presence of organic matter, and the presence of moisture. Contamination can be spread from individuals who are sick from a given microorganism or individuals who show no symptoms but still carry the organisms in their nose and throat, on their skin, or in feces or urine and then contaminate surfaces with their hands. The actual amount of contamination that is spread in this manner is not well understood.

Surfaces include all areas such as floors, beds, bedside tables, windowsills, wall areas, doorknobs, light switches, toilets, medical equipment, etc. The surfaces may also become contaminated by poor housekeeping techniques and/or by the sheer number of people in the area. Microorganisms may grow because of availability of substances supporting bacterial growth, type of surface, and whether the surface is horizontal or vertical.

Although carpet in certain hospital areas is effective in reducing noise, making walking easier, and reducing the number of falls and seriousness of the injuries in patient care areas, carpeting may contain fungi and bacteria. Carpeting is harder to clean, especially after spills of body fluids and blood substances. Pushing of carts and beds is more difficult. Vacuum cleaners can easily move microorganisms from the carpet into the air where they become potential airborne infections.

The reduction of surface contamination in good part depends on the people carrying out the necessary cleaning and disinfecting processes, the cleanliness and maintenance of the equipment utilized, the nature of the detergents and disinfectants being used, the frequency of the changing of cleaning solutions, the amount of traffic in a given area, the type of air flow and use of filters, and the presence of infected individuals, either sick or healthy carriers.

Best Practices in Reducing Microorganisms on Surfaces

- Develop a housekeeping program that meets the needs of the particular area to reduce all visible dirt and dust on floors, walls, windowsills, ventilation systems including grills, door ledges and frames, tops of cabinets and bedside tables, equipment, bathrooms, etc., while enhancing the physical environment of the institution.
- Focus cleaning and disinfection especially on high-touch surfaces, especially those that come in contact with the skin or bodily fluids including all bathroom surfaces, bed rails, tray tables, IV poles and pumps, all controls, call boxes, telephones, chairs, sinks, door handles, etc.
- Immediately clean up all spills of fluids of human origin and any other sources of bacteriological contamination and then disinfect with either a phenolic compound or a chlorine compound. Keep all floors, walls, and other surfaces free of blood and other bodily fluids.
- Do not use alcohol to disinfect surfaces. While it is effective on skin, it does not work well on a variety of surfaces.
- Train appropriate housekeeping staff in specific cleaning techniques to be used in areas where isolation and reverse isolation (protecting the patient from the environment and people) is required.
- Immediately clean up all chemical spills to prevent hazardous conditions for patients, employees, and others.
- Utilize a housekeeping evaluation form such as that found in the *Handbook of Environmental Health—Biological, Chemical, and Physical Agents of Environmentally Related Disease*, Fourth edition, Volume 1, pages 528–529 (see endnote 62), to do self-inspections of all areas to determine if the areas are being properly cleaned to remove all visible soil and potentially large quantities of microorganisms which may cause healthcare-associated infections.
- Develop a specific training program for housekeeping supervisors and managers including classroom topics and practical work experiences. This should include some elementary knowledge of the spread of microorganisms and potential means of control, potential hazards from cleaning products and personal protective equipment to be used, use of appropriate cleaning products for the project to be completed, use and care of cleaning equipment, techniques of cleaning and disinfecting, and use and evaluation of the results of the self-inspection program.
- Develop and enforce a hand-washing program specific to the jobs performed by housekeeping personnel.
- Develop a specific training program for housekeeping supervisors and managers including classroom topics and practical work experiences on supervision and management of people including discussions on leadership, planning and organizing work assignments, employee selection, people management, understanding human behavior, how to be firm but fair, the significance of good morale, how to teach adults to carry out assignments accurately and in a timely manner, and how to properly communicate with all levels of individuals.

- Develop a specific training program for new employees in job orientation and various cleaning techniques. Provide immediate close supervision to all new employees.
- Develop short training sessions for existing employees to reinforce skills of cleaning in various situations and introduce new techniques and products as needed.
- Evaluate the work of individuals on a weekly basis and make necessary corrections.
- Use the appropriate detergent and disinfectant, quantities of chemicals needed, and amount of time used in cleaning for the potential microbiological load from potential contamination and the types of surfaces to be cleaned, recognizing that detergents are surface-active chemicals that emulsify soil and then need to be removed from the surfaces by proper rinsing techniques before disinfecting. There are some chemicals which are both detergents and disinfectants to which the above does not apply.
- Make sure that all detergents and disinfectants being used are compatible, that adequate contact time periods are being utilized and water is changed very frequently.
- Do not use the same mop water in areas of potentially high contamination and other areas. Change water and clean equipment or use new equipment when mopping each different isolation or reverse isolation room.
- Recognize that floor finishes and floor strippers are chemical agents that can be hazardous to individuals if inhaled and then take necessary preventive actions.
- Recognize the limitations of disinfectants on certain surfaces and the potential hazards associated with them to people.
- Clean up immediately all spills of blood and other bodily fluids and substances and then disinfect with a chlorine solution.
- Where carpeting is used, vacuum areas with special equipment that contains HEPA filters.
- Privacy curtains in areas where there are large numbers of patients may be contaminated rapidly and therefore should be cleaned appropriately in a timely manner.
- Use swab cultures only to determine if microorganisms are present on a given surface to see how a disease might be potentially spread by contact or become airborne. The results of the swab sample can be used as a teaching tool. A sample size must be so large to get an accurate determination of the reduction of microorganisms by any given type of cleaning procedure, that it is not cost-effective and also very time-consuming to acquire for the potential benefits realized.

WATER

(See Chapter 13, "Water Systems [Drinking Water Quality]")

Treated water from a public water supply enters the healthcare facility through water mains and is then distributed through a series of pipes throughout the institution. The pipes may be of iron, copper, or polyvinyl chloride. Hot water is produced in boilers and may become chemically or physically contaminated by the contents of the boiler. Water becomes stagnant when it stays in the lines for a period of time without forward motion out of a fixture.

Chlorine content may be reduced below acceptable levels and even to zero because of the retention of the water in the hot water tank.

Because of the size of the institution and the different needs of the staff and patients, the temperature and pressure of the water will vary considerably. Too high a temperature will cause burns especially in older people in nursing care facilities. Depending on the nature of the equipment being used, the temperature of the water may differ substantially and there may need to be booster heaters within given areas, especially in food service and laundry facilities.

When water systems are being worked on because of a water failure, it is necessary to make sure that there will not be any back siphonage into the public water supply. The vacuum breakers may not be working effectively and compound the problem.

Waterborne infections can occur from direct contact in various forms of hydrotherapy, baths, potable water which has become contaminated during dialysis, laboratory solutions, and ice, or through the ingestion of drinking water which has been sitting around and is contaminated. Also equipment used for preparing distilled water needs to be evaluated regularly in case of breaks in lines or improper back flows. Sterile water if exposed to air or physical contact after use may become contaminated.

Legionella bacteria have been found in healthcare facilities in aerosols generated by cooling towers, showers, faucets, respiratory therapy equipment, heated potable water systems, and distilled water systems. The vast number of plumbing fixtures within the institution can also harbor a variety of bacteria. Most of the organisms including fungi may accumulate around the various openings of the fixtures being used. A variety of decorative fountains and waterfalls and other uses of water if not maintained properly can create aerosols containing microorganisms that can potentially lead to healthcare-associated infections.

Other Gram-negative bacteria may be found in the potable water and may cause healthcare-associated infections. These include various species of *Pseudomonas*, *Acinetobacter*, *Enterobacter*, etc. These organisms can grow in distilled water, contaminated solutions and disinfectants, dialysis equipment, nebulizers, water baths, contaminated mouthwash, contaminated respiratory therapy solutions, contaminated disinfectants and antiseptics, and a variety of medical equipment that requires a source of moisture for operation.

Best Practices in Water Problems
- Keep water temperatures cool or cold in all decorative uses of water, perform frequent cleaning and maintenance, and do not put near high-risk patients.
- All water sources where water will be standing or new pipes have been installed should be checked frequently and thoroughly cleaned and maintained in proper order. Bacterial testing should be done where appropriate.
- Avoid long runs of pipes and potential dead ends so that water will not collect and microorganisms start to grow.
- Make all valves for shutting off and restarting water flow readily accessible and always serviceable.
- Clean thoroughly all hydrotherapy tanks between uses by patients and then thoroughly sanitize and air dry. Maintain 15 ppm chlorine residual in the water of hydrotherapy tanks. Routinely culture the tanks especially around the faucets and drains to determine if there is bacterial growth.
- Evaluate water used in laboratories and pharmacies on a periodic basis to determine if they are contaminated in any way.
- Determine on a monthly basis if water used in hemodialysis contains toxins or bacteria and if so determine the source and eliminate it.
- Establish contingency plans for obtaining a substantial quantity of water if there is a break in the potable water system.
- Evaluate the pressure and temperature of water fixtures periodically to determine if what is coming out of the unit meets the needs of the special area and if not make necessary corrections.
- After a break in the line, if thermal shock treatment is available where the water temperature can be safely raised to between 160°F and 170°F (71–77°C) and kept at this level through 5 minutes of flushing, do so, if not add chlorine overnight so that there is a residual of 2 ppm throughout the system.
- In the event of flooding, emergency preparedness plans must be used to produce immediate action to move patients to safe areas, provide emergency generators, provide adequate quantities of potable water, correct the flooding situation, and then use intensive cleaning prior to the return of patients to the area.

- Use environmental surveillance and culture potable water on a monthly basis for all types of containers, especially cooling towers, evaporative condensers, showerheads, around faucets and drains, and respiratory therapy equipment, where *Legionella* or other bacteria or fungi may be possibly growing. If found, institute immediate treatment procedures and retest.
- Where microorganisms and chemicals have been removed, use specially treated water in hemodialysis procedures.
- Clean thoroughly and disinfect with chlorine solutions, all ice storage chests and ice making machines on a regular basis to avoid growth of microorganisms and use scoops or direct flow of the ice into cups or containers which are kept very clean and sanitized.
- Do not store samples of bodily fluids, lunches, chemicals, or anything else in ice storage chests.
- Clean thoroughly and disinfect with chlorine all whirlpools and other hydrotherapy tanks, hydrotherapy pools, birthing tanks, and other equipment after each patient usage.
- Where water holding reservoirs are used in equipment, have them checked microbiologically on a periodic basis. Determine when it is necessary to flush the water out, clean the equipment, and replace the water.
- Frequently change the water in plants and cut flowers in patient care areas to prevent the growth of microorganisms which may already be present on the vegetation.

SUB-PROBLEMS INCLUDING LEADING TO IMPAIRMENT AND BEST PRACTICES FOR EMPLOYEE AND PATIENT SAFETY (OCCUPATIONAL HEALTH AND SAFETY)

(See endnote 43, an extremely valuable document) (See Chapter 6, "Environmental and Occupational Injury Control")

Health care is a high-hazard and high-risk industry for patients and staff. The Office of Inspector General of the US Department of Health and Human Services in 2010 stated that poor hospital care contributed to the deaths of an estimated 180,000 patients in Medicare each year. (The report of the numbers of patients dying each year from poor hospital care varies with the group doing the studies. The Institute of Medicine report in 1999 suggests that 98,000 Americans die each year.)

The healthcare industry has the second largest number of injuries and illnesses to employees in the country. The number of non-fatal injuries and illnesses requiring time away from work has increased among healthcare workers.

PATIENT SAFETY

(See endnote 42)

Patients are people who typically are debilitated, may be confused, physically disabled, weak from being in bed, disoriented, and very young or older, and face numerous other problems related to their disease, the medicine they take, and the normal problems faced by all people within a residential environment. There is a considerable concern about falls and resulting injuries as a result of the above conditions. In addition, there are numerous other safety considerations including: events related to medication; events related to patient care; events related to surgery or other procedures; and events related to infections.

Overall, adverse events to Medicare patients in 62% of cases required prolonged hospital stays. Permanent harm was found in 5% of patients. The need for life-sustaining intervention occurred 27% of the time, while the event contributed to the death of the patient 10% of the time.

Medication problems contributed about 31% of the total adverse events that occurred to the patients. Patient care including falls with injuries contributed 28% of the adverse events. Surgery or other procedures contributed 26% of the adverse events. Infections of all types contributed 15%

of the adverse events to patients. In addition, there were temporary harmful events the result of which did not cause any type of permanent damage to the patient. It was felt that 60% of the infections could have been prevented, 50% of the medication problems could have been avoided, 51% of patient care problems could have been prevented, and 17% of surgery and other procedure problems could have been avoided. (See endnote 41.)

There are several categories considered to be serious reportable events including those related to surgery, devices, patient protection, patient care management, and environmental and criminal situations. The environmental events include electric shock while receiving care at a facility, the wrong gas being delivered to a patient or contaminated with toxic substances, burns received at the facility, falls with injuries at the facility, and death or serious disability from the use of restraints or bed rails.

Unfortunately, in many institutions safety is the concern of a special committee rather than senior leadership of the institution. Because of this lack of high-level involvement, the issues related to patient safety do not receive the amount of time and control that is necessary for a successful program.

Best Practices in Patient Safety (See Best Practices in Employee Safety below)
- Senior leadership including chief executive officers, executives reporting to them, senior clinical administrators, and the Board of Trustees must be involved deeply in patient safety and establish goals and objectives for the entire institution and for specific departments in the reduction of medical errors, infection problems, environmental and safety problems, and other areas of patient and employee safety.
- At periodic intervals (such as 3 months, 6 months, and 1 year), each of the results of the work being done to achieve goals and objectives to resolve patient care problems should be evaluated and a determination should be made on how to redirect the program if necessary.
- Develop and implement a hospital incident reporting program and if one exists, evaluate how it is working and the results of the various reports being analyzed individually and en masse.
- Evaluate the hospital communications program to make sure that the proper individuals are being kept informed of all necessary information related to potential patient care problems and existing patient care problems.
- Carefully evaluate the amount of time that various members of staff have to work on a continuous basis since fatigue is one of the contributors to poor patient care and potential accidents.
- Place on the agenda of all senior executive meetings an update and discussion of patient safety problems, measurements, and trends, techniques being utilized to reduce safety problems, and the results.
- Encourage all frontline staff members to offer suggestions on how best to improve patient safety and reward them accordingly.
- Involve physicians to help resolve patient safety problems since most of the work is typically performed by nurses, pharmacists, and other allied health professionals.
- Involve patients and families in the work of the care team to help prevent patient care problems.
- Senior administrative staff should make unannounced rounds on a periodic basis to present a presence to the staff members and help enforce the concept of the necessity for excellence in patient care and avoidance of medical errors, infection problems, and environmental and safety problems.
- Since 80% of medical errors or adverse events are due to the system which has been established for a particular technique or treatment being used, all systems must be examined in depth by experts in that particular field in order to determine weak spots and how to correct them and institute the necessary procedures and tools to make this happen.

- All tests, medications, procedures, and information must be given at the appropriate time and meet quality control standards set up by experts within the particular field.
- Continuously provide short-term continuing education programs to all employees in their respective areas to keep them current on new practices and procedures and advise them of the types of patient safety problems that are occurring and how to resolve them.
- Avoid patient and treatment mix-up by immediately identifying the patient with an identification bracelet, including use of barcodes, with the understanding that staff will check the bracelet at least twice before performing any type of procedure or providing medications or other treatments.
- Design tubing connections so they cannot fit into the wrong ports.
- Patients or caregivers must provide a list of currently used medications and potential adverse reactions at time of entry into the patient care system. Before leaving, the patients or caregivers are given a list of medications given at the hospital and those to take home or prescriptions for them plus instructions on their use.
- Monitor very closely all patients given anticoagulation therapy and make necessary adjustments when shown to be necessary by blood work or symptoms.
- Establish an infection control team with a highly trained nursing specialist, physician, and other essential people including environmental health practitioners to conduct routine and when necessary special studies to determine potential sources of infection, prevention, and control throughout the entire institution. The team should work closely with local, state, and federal health authorities to deal with current infection problems and to prevent potential infection problems.
- Perform appropriate hand-washing techniques to prevent the spread of infection. Alcohol-based gels may be used by the staff and visitors in all patient care areas. All staff members must use the gel before and after taking care of the patient. (See endnote 19 for a comprehensive discussion of hand washing.)
- All staff members must use gloves, masks, eye protection, and gowns when dealing with infections or potential infections.
- Specialized staff members need to be highly trained in utilizing central venous lines, urinary catheters, ventilators, and other techniques which may be needed to prevent healthcare-associated infection.
- Patient falls must be prevented by lowering the beds as much as possible, utilizing bed rails, giving extra care to individuals who are at high risk for falling, and ensuring that the high-risk condition is noted on their wristband.
- Prevent pressure ulcers when patients spend extended amounts of time in bed by frequently changing their position in bed, cleaning the skin and drying it well, and using specialized pressure-relieving medical beds or other devices.
- Manage the most critical patients in the intensive care units by providing highly skilled doctors and nurses 24 hours a day.
- Provide a computerized system for physician orders and prescriptions to be written electronically to improve accuracy and help prevent errors because of handwriting legibility.
- All medications for patients must be double checked to make sure that the correct medication and the proper dosage is given to the correct individual. The time of providing the medication is also very important to avoid an error by giving the patient too little or too much over a specific period of time.
- Move patients safely when either adjusting them in bed or taking them for tests or other procedures. Provide adequate staff to do this to prevent injury to the patient and the staff member.
- As part of proper monitoring of patients, make sure that all treatments and medications are properly identified to the patient or caregiver and presented in a timely manner.

- Managers and administrators should give close supervision to all nursing personnel and other staff members to ensure that they provide adequate and timely care without endangering the patient.
- Ensure that there is a constant flow of effective communications among all caregivers of the patient so that everyone is working toward a common goal.

EMPLOYEE SAFETY (OCCUPATIONAL HEALTH AND SAFETY)

Hospitals are in fact small communities and have all the problems related to this type of environment, including automobile accidents, home accidents, etc. In addition, there are hazards related to the use of: various types of equipment; gases which are toxic, explosive, or carcinogenic, and may cause fires; a variety of chemicals found in various parts of the institution; and various procedures used in patient diagnosis and treatment. Excessive heat may be found in laundries, kitchens, boiler rooms, and furnace rooms. Poor lighting which may lead to accidents may be found in innumerable areas throughout the institution. High levels of noise may be created by a variety of situations. Radioactive material is used in various treatment and diagnostic situations. Research areas may become hazardous from animal contamination and other safety considerations such as slippery floors and specific problems related to disease being studied by research staff. Fires are a constant threat due to potential breaks in procedure in housekeeping as well as the use of electrical appliances throughout the institution, flammable liquids, heating units, explosive materials, paint storage areas, compressed gases, etc.

Employee safety in the healthcare setting is complicated by a wide range of hazards including sharps injuries, back injuries, latex allergies, violence, and stress. Sharps expose healthcare workers to blood-borne pathogens. Reproductive risks may be associated with cleaning and decontaminating materials, laboratories, x-ray rooms, pharmacy, various chemotherapies, and specialized antibiotics.

Musculoskeletal injuries occur in patients and healthcare workers because of the job demands, workload, physical factors, and health conditions of the individuals. Back injuries occur primarily because of improper lifting of patients who may be too heavy, lifted improperly, combative, agitated, totally unable to lift himself/herself because of a lack of upper body strength, and have a physical condition which needs special types of lifting. Slips, trips, and falls which are the second most common cause of employee injuries in hospitals may occur throughout the institution. Problems exist on the exterior grounds, parking lots, walkways, and staircases and may be created either by poor maintenance of surfaces or by weather-related elements such as rain, ice, or snow. Within the facility there are numerous places where floors may become a safety hazard because of water, body fluids, spilled liquids, food, or grease. There are special risk factors in the emergency room, operating rooms, laundry facility, and food facilities because of the concentration of employees working in a limited space with considerable potential unknown hazards. There are a large number of sinks in the facility which are frequently used and therefore are more prone to contribute, because of poor use or overflow, to liquid and cleaning materials ending up on the floors.

There is a serious concern about occupational exposure to infectious diseases because of blood-borne pathogens that are either transmitted to the worker by a sharps injury (needles, scalpels, broken glass or plastic, or other sharp objects) or contaminated blood or body fluids entering a mucus membrane or exposed damaged skin of the worker. Large numbers of healthcare workers experience these problems annually. Sharps may transmit hepatitis B virus, hepatitis C virus, HIV, and many other potential diseases. Under-reporting of these incidents is considerable because of fear of being disciplined, lack of time to report the information, or lack of concern about the risk of disease transmission. The healthcare worker who has been contaminated in this manner now may become a hazard to new patients. The worker is also affected because the exposure may trigger physical problems or emotional problems giving rise to fear and anxiety in the individual and his/her family.

Healthcare workers are exposed to hazardous drugs, chemicals, and other substances. Many treatments and diagnostic techniques used in healthcare areas unfortunately have unintended consequences of employee exposure to hazardous substances. The employees may be subjected to antineoplastic agents and cytotoxic drugs which are used in cancer chemotherapy. Pharmacists who prepare these drugs and nurses who administer them have the highest potential exposure and the greatest opportunity to suffer unwanted consequences. There is also potential exposure by physicians, operating room personnel, custodial workers, laundry workers, and those who handle waste. Chronic effects can include liver and kidney damage, damage to the bone marrow, lungs and heart, infertility, and hazards to a developing fetus in a pregnant woman. Environmental contamination and worker exposure may occur. Exposure may occur through inhalation, skin contact and absorption, ingestion, and needle stick or sharps injury. (See endnote 44.)

Healthcare workers are also exposed to antiviral drugs, hormones, bioengineered drugs, and radiographic diagnostic materials. Cleaning products, disinfectants, sterilants, and anesthesia gases have been previously mentioned.

Radiation of various types is used in the diagnosis and treatment of patients. Healthcare workers are at risk because of unintended exposure to the radioactive material. In the past these workers had a higher than normal risk of leukemia, skin cancer, and breast cancer. The use of lead aprons has reduced this risk substantially. However, there are new technologies that require that the healthcare worker be very close to the patient and therefore close to the source of radiation.

Workforce staffing schedules may create extreme fatigue and result in health and safety problems for the employees as well as for the patients. Employee performance errors increase with shift work, rotating shifts, periodic night shifts, extended work hours and excessive workloads, and inadequate or non-existent work breaks and lunch hours. These create problems of chronic fatigue, sleep deprivation, nervous reactions, effects on the cardiovascular, metabolic and immune systems, burnout, and exhaustion. Mentally and emotionally these people are drained because of the work schedules and in addition all of the health problems and potential deaths that they deal with daily. Work schedules can also affect the reproductive system.

Violence in the healthcare setting is affecting workers' emotional as well as general health and safety. People are being intimidated by threatening language, sexual harassment, bullying (especially through the use of electronic devices), disruptive behavior, and assaults. Hospitals have a large number of parking lots, entrances, and exits to the facility and a substantial number of people of all types who are either delivering vital supplies or are visitors or patients and staff. Many of these individuals are traumatized, may be disoriented or experiencing the side effects of medications, are sleep deprived, or are just plain angry at the current situation and are having difficulty coping with it. Patients may assault healthcare workers or other patients. Waiting rooms, emergency rooms, and mental health units are potentially high areas of conflict. People being treated may be under the influence of alcohol or drugs. Factors that contribute to the potential for violence include professionals working alone; understaffing of vital areas in an attempt to save money; unrealistic productivity demands; cost-containment or reduction demands; transportation of patients; long waiting times to be seen by a healthcare professional; overcrowded, uncomfortable, poorly designed, and poorly administered waiting rooms; badly lit areas throughout the institution and the parking lots; improper training or lack of training of staff and how to deal with a serious situation; and an inadequate security systems and extremely poor response time to emergencies.

Older workers have higher rates of injury than younger workers especially among the direct care staff. They may have preexisting conditions as well as decreased strength and therefore are more susceptible to injury from a variety of factors.

Best Practices in Employee Safety (See Best Practices in Patient Safety above)
• Conduct an intensive field study by department of all hazardous situations through observation, walk-through, screening surveys, incident and record analysis, and medical review of injuries. Assess the scope of the hazards as a baseline for improving existing programs

and establishing new programs, and develop techniques of hazard prevention and control utilizing recognized Best Practices.

- Integrate employee safety and patient safety into one program with one director to avoid duplication of effort and confusion in implementing policies.
- Establish a safety and health policy and program with goals and objectives involving top management leadership as well as employees, and teach managers and supervisors how to work within the program successfully, and periodically conduct evaluations and redirection as necessary to improve employee and patient safety.
- Supervisors and managers as well as employees should analyze routine hazards for each job or process and the necessary steps along the way to complete the work assignment. They should then meet together to resolve the problems and reduce or eliminate the hazards.
- Submit yearly to senior management and the Board of Trustees a report including: all system or process failures; numbers and types of incidents and how handled; whether patients and families were informed of the incidents; and what actions are now being taken to prevent these types of problems in the future.
- Establish the position of Environmental Health, Safety, and Infection Control Specialist at all larger institutions and give this individual high-level administrative authority as well as working staff to assist other departments in carrying out their assigned duties in a manner that will reduce environmental, health, safety, and infection control problems. For smaller institutions, set up a process where the Environmental Health, Safety, and Infection Control Specialists can be housed at one of the institutions and shared by two others.
- All levels of staff should be involved in accident/incident or near accident/incident investigations to get complete input on what occurred and how to resolve the issues. The investigations should be carried out utilizing standardized forms in order to be able to statistically analyze what is occurring by department and types of procedures.
- When conducting an incident investigation, always attempt to determine who was involved, what happened, when did it happen, where did it happen, why did it happen, how did it happen, and what systems and/or procedures were in place that should have prevented it from happening.
- Periodically, highly qualified individuals should conduct a proactive comprehensive industrial hygiene, safety, health, and potential healthcare-associated infections survey throughout the major problem areas of the institution and report results to high-level management, middle management, and the employees working within each of the departments. Excellence should be rewarded and problems should be corrected.
- Maintain a written current inventory of all hazardous materials and hazardous wastes, their storage and disposal.
- When a new chemical or new piece of equipment is introduced into the facility, determine in advance all potential hazards and preventive measures needed to avoid the hazards.
- Provide safety, health, and infection-control training at the orientation programs for all new employees. Also provide brief continuing education programs in all areas on a periodic basis, when new problems arise, when new procedures are being put into effect, and when new research occurs which furthers the knowledge of the various members of staff.
- Develop a nurse staffing plan that has adequate resources for 24/7 staffing coverage and provide active highly skilled managers that will ensure patient safety.
- Enforce all current hand-washing requirements that have been established by CDC or the World Health Organization. Employee failure to comply with these requirements, whether the individual is a physician or housekeeping personnel, should result in the first steps of disciplinary action leading, if necessary, to suspension.
- Provide excellent care and treatment for all staff who become sick or injured while at work and provide for physical examinations before returning to normal duties.
- All employees must receive proper immunizations before starting work assignments.

- Design work processes taking into account potential ergonomic problems and seek to simplify and standardize the work assignment as well as provide sufficient numbers of personnel to carry out the assignment.
- Ergonomic injuries can be reduced by rotating workers through repetitive tasks.
- Machine guards should be provided to protect the operator and other employees from all machines that can injure people while operating.
- Back injuries can be reduced by assessing the situation and determining how best to transfer patients from beds to chairs, chairs to toilets, chairs to chairs, or car to chairs, and how many people are necessary to assist the individual depending on weight and physical condition.
- Slips, trips, and falls can be reduced by: using non-slip surfaces; using non-slip shoes; providing water absorbent mats; immediately cleaning up all spills of liquids and bodily fluids, placing cones on floors that are being washed, and washing only one part of the corridor or room at a time so that there is always a dry surface to walk on; keeping all floors clean and dry; and reporting all hazards immediately and dispensing cleaning crews swiftly.
- Provide prompt ice and snow removal and constant inspection of walkways, staircases, and parking lots.
- Provide umbrella sleeves and bags at all entrances.
- Establish policies and procedures to prevent and if necessary immediately treat those individuals who have been exposed to blood-borne pathogens including HIV virus through sharps injuries.
- Provide a confidential medical examination as part of a total postexposure evaluation for all employees who have been exposed to blood-borne disease as well as any other type of injury or illness.
- Present in-service training including safe injection practices on a routine basis to all employees who may be exposed to sharps injuries and blood-borne disease transmission.
- Substitute safe alternatives for invasive procedures where possible.
- Provide special vaccinations if necessary for individuals who are exposed to hepatitis B virus or other blood-borne pathogens.
- Never use needles or syringes for more than one patient or to draw up additional medication.
- Use safe needle disposal methods in waterproof boxes.
- Conduct environmental sampling techniques consisting of using moistened wipes on work surfaces and extraction and analysis for specific antineoplastic drugs, six to eight of which can be identified at this point.
- Use bleach to decontaminate and deactivate certain classes of antineoplastic agent. Reference should be made to those listed by the US Environmental Protection Agency (EPA) as hazardous waste and dispose of them accordingly.
- Use personal protective equipment when handling antineoplastic agents and other hazardous pharmaceuticals. Always use gloves and gowns. For those with latex allergies there are new materials on the market. Depending on the drug being used, it may be necessary to double glove. Use a mask or respiratory equipment depending on the level of hazard of the drug being utilized. Check with Occupational Safety and Health Administration (OSHA) requirements or CDC requirements.
- Employees involved in the preparation, administration, or disposal of hazardous drugs that they may be exposed to, should report any symptoms of occupational exposure including nervous system effects, rashes, sore throats, dizziness, headaches, nausea, diarrhea, vomiting, or any effects on pregnancy.
- Use a mask, eye protection goggles, or face shields during any procedures where there are likely to be splashes or sprays of blood, body fluids, or secretions. This is especially true during sectioning or endotracheal intubation.

- Use a mouthpiece, resuscitation bag, or other ventilation devices to prevent contact with secretions from the mouth or nose of the patient.
- Soiled patient care equipment is hazardous and must be disposed of in special containers marked biological hazards.
- Bag soiled laundry and remove to laundry washing facilities immediately. Do not shakeout linens or try to wash any of the material in any place other than the laundry facility. In the event of soiled laundry from people with infections, double bag it and then mark it as infectious and hazardous.
- Use ventilated biological safety cabinets and closed system drug transfer devices when preparing antineoplastic drugs. A closed system drug transfer device is a mechanical piece of equipment that does not allow the transfer of environmental contaminants into the system or the escape of the hazardous drug or vapor outside the system. (See endnote 44.)
- Immediately clean up and remove all spills of hazardous drugs and dispose of them in an appropriate manner.
- Use non-toxic drugs or safe procedures when possible instead of toxic drugs on the patient.
- Provide a written workplace violence prevention policy, educate all levels of employees on how to avoid or contain violence, and enforce it rigorously. Any acts of violence by a staff member against another staff member should be severely punished and if necessary the individual's employment should be terminated.
- An in-depth analysis should be made of the entire institution to determine areas where violence might occur and policies and practices should be put into effect to reduce the potential or eliminate it.
- All acts of violence should be recorded as incidents and reviewed by the safety committee to determine how to prevent this from occurring in the future and also whether there a larger problem than just this single incident.
- Install and maintain alarm systems in high-risk areas and check the response time of appropriate levels of protection from security and the local or state police if necessary.
- Provide adequate lighting in all parking lots and an escort service especially for women at night from the hospital to their cars and make sure that the person is secure within the vehicle and leaving the premises in an appropriate manner.
- Control access to various parts of the institution based on the need for the individual to be in any given place.
- When people come into the facility obviously arguing or fighting, have security separate them into different areas and make sure that they do not come together later.
- Develop appropriate work schedules to keep personnel working within the same timeframe over long periods of time. Decrease the mandatory hours of work and only allow several additional hours of overtime in the event of an absolute emergency.
- Evaluate carefully the hand-off time between shifts because this is prime time for fatigued workers to become injured or make poor patient decisions, and patient coverage is reduced.
- Establish continuing education programs which explain to the healthcare workers the relationship between fatigue and occupational health and safety problems as well as work errors resulting in danger to the patients.
- Provide adequate rest and lunch breaks to allow healthcare workers to rest and regroup as well as get nourishment in order to have better performance.
- Provide a realistic staffing pattern and do not overburden the staff with too high a workload while maintaining appropriate quality of care.
- Conduct routinely fire drills, disaster drills, radiological disaster drills, biological triage and decontamination drills, and all hazards emergencies.
- Prepare an evacuation plan for the institution in the event of a fire, flood, or other disaster which may affect the institution. Determine how critical care patients will be removed and where patients will be taken to get them out of the danger zone.

- Prepare, institute, and test frequently a disaster plan which will go into effect if there is a disaster within the community and the institution will be utilized for treatment of the victims and of emergency medical service personnel.
- Install metal detectors at all entrances to reduce the chance of firearms and knives being brought into the institution by employees, patients, visitors, construction people, and suppliers.

SUB-PROBLEMS INCLUDING LEADING TO IMPAIRMENT AND BEST PRACTICES FOR POTENTIAL HAZARDS EXTERNAL TO THE FACILITY

AIR POLLUTION

(See Chapter 2, "Air Quality (Outdoor [Ambient] and Indoor)")

Hospitals produce air pollutants from medical waste incinerators, boilers, sterilization units, paint booths, emergency generators, anesthetic use, air conditioning and refrigeration, laboratory chemicals and fume hoods, motor vehicles, etc. These pollutants consist of volatile organic compounds, nitrogen oxides, carbon dioxide, and hazardous air pollutants. Most of the on-site medical and infectious waste incinerators have been eliminated due to federal rules and regulations. This means that commercial units using remote sites are carrying out this function. Industrial boilers which are common in hospitals may also produce sulfur dioxide, carbon monoxide, and particulates. When hospital waste is burnt either on site or at special incinerators, it can produce additional pollutants such as hydrochloric acid, dioxins, furans, lead, cadmium, mercury, etc.

Best Practices in Controlling Air Emissions
- Service all air conditioning and refrigeration units on a routine basis and prevent leaks of the gases.
- Where feasible, use steam generated from incinerators to partially replace boilers. Use ENERGY STAR certified boilers and a set of standard boilers to conserve energy and reduce air pollutants.
- Minimize the use of medical waste incinerators and use other techniques of sterilization or decontamination.
- In older buildings where asbestos or lead paint needs to be removed, utilize only highly trained, certified companies to avoid creating air pollutants.
- Install a filtering ventilation system in paint booths to collect paint fumes.
- Use scavenger systems to contain anesthesia gases and prevent them escaping to the inside air or outside air.

HAZARDOUS WASTE

(See Chapter 12, "Solid Waste, Hazardous Materials, and Hazardous Waste Management") (See endnote 11)

Hospitals produce municipal solid waste, half of which is paper and cardboard, 17% of which is food and other organic material, 15% of which is plastics, and 15% of which is hazardous waste. It is the hazardous waste which is of great concern. It is made up of pathological waste, radioactive waste, chemicals used in chemotherapy and antineoplastic agents, biomedical waste, pharmaceuticals, waste anesthetic gases, sharps, and pressurized containers.

The biomedical wastes can lead to infections and outbreaks of disease. Hepatitis, HIV infections, and many other potential diseases may become a serious problem when people come in contact with disposable syringes, needles, IV sets, containers that have been contaminated, and other wastes.

Hazardous chemical waste may be generated in laboratories, maintenance areas, grounds keeping, and diagnostic and treatment areas. These chemicals may contaminate the air or water, or be included with solid wastes and cause potential hazards for employees.

Best Practices in Handling and Disposal of Solid and Hazardous Waste and Materials
- Reduce municipal-type solid waste by: requiring that suppliers use less packaging material; recycling as much bulky furniture, carpets, etc. as possible; using composting programs for lawn waste; using double-sided printing for documents; and filing insurance claims and purchase orders electronically.
- Reduce food waste by predetermining appropriate portions for staff and patients and by utilizing a controlled system of purchasing and storage.
- Turn old linens into rags which could be used in various maintenance areas and other areas.
- Use reusable dishes and silverware in areas where the potential spread of infection is not a concern and appropriate dishwashers can properly wash and sanitize them.
- After recycling construction and demolition wastes that can be reused, send the rest to clean landfills if not contaminated with asbestos, lead, mercury, or PCBs.
- For biohazardous waste see above topic on Best Practices for Medical Waste and Other Hazardous Substances.
- Collect, clean, and recycle solvents where feasible.
- Implement strict inventory control for pharmaceuticals, chemical purchases for laboratories, etc., and cleaning materials. Always put the new chemicals behind those that are already in the facility so that wastage will not occur because chemicals pass their use-by date.
- Minimize use of ethylene oxide.
- Identify all lead-containing equipment and supplies and determine if they can be reused, recycled, or must become hazardous waste and handled specially.
- Return to the vendors any pressurized containers or aerosol cans that have been used in the institution.
- Where items containing mercury are used, determine if there are mercury-free alternatives and if not collect and recycle the mercury-containing instruments.

WATER POLLUTION

(See Chapter 14, "Water Quality and Water Pollution")

Healthcare facilities discharge their wastewater directly into publicly owned treatment works. Therefore, if chemicals or highly infectious wastes are dumped into toilets, they end up at the treatment facility and may be partially treated or not treated before they enter the waters of the United States.

Best Practices in Water Pollution
- Wastewater containing oils, hydraulic fluid, or chemical spills must be cleaned up, trapped, and treated instead of being released into the public sewage systems.
- Stormwater which includes runoff from buildings, lawns, parking areas, storage tanks, or disturbed soils, must follow the Best Practices identified in Chapter 14, "Water Quality and Water Pollution."

Best Practices in External Pollution Prevention
- Conduct a baseline study to determine the amount and type of pollutants going into the air, water, and land from the institution.
- Develop an environmental management system to review and apply necessary preventive and corrective actions to solid waste disposal, especially hazardous waste from the institution.

- Develop and enforce a strict waste segregation system where biohazardous waste, chemical waste, and radioactive waste is placed in specially marked containers and handled according to the current rules and regulations of the EPA and OSHA.
- Substitute less toxic or less polluting products for more toxic or more polluting products. Examples would be purchasing non-mercury-containing products for all departments of the institution, and substitution of microfiber mops for usual mops to reduce the amount of water, cleaner, disinfectant, and chemical storage space needed, etc.
- Change the process from imaging to digital imaging thereby avoiding use of x-ray films with silver on them.
- Use recycling for all paper wastes, aluminum and plastic wastes, durable furniture, equipment, pallets, etc.
- Reuse packing materials such as foam peanuts, foam inserts, airbag inserts, or other types of packaging instead of disposal.
- Recycle as much of construction and demolition waste as possible by separating out the metals, bricks, good pieces of wood, old equipment, etc. and determine if these can be used in other areas or sold.

SUB-PROBLEMS INCLUDING LEADING TO IMPAIRMENT AND BEST PRACTICES FOR EMERGING AND RE-EMERGING DISEASES

(See endnote 28)

SPECIAL HIGHLY INFECTIOUS PATHOGENIC ORGANISMS

Travelers are especially at risk for infectious diseases. In 2011, United States citizens made over 85 million visits abroad. They are exposed to diseases endemic in the countries that they visit and then become potential hosts for the diseases and can easily bring them back to the United States. Hepatitis A and typhoid fever are considered high potential problems and therefore the traveler should be vaccinated prior to leaving the United States. Border entry screening for infectious diseases in people during a SARS outbreak was found to be very costly and a diversion of limited public health resources without a measurable effect. (See endnote 26.)

Emerging and re-emerging infectious and communicable diseases are of constant concern because they simply cannot be eradicated. Virtually every time people believe this has occurred the microorganisms seem to adapt to the current environment and new strains appear. Individual organisms may become resistant to drugs and new organisms may emerge from animals.

Pathogenic organisms of concern currently and the problems that they cause include:

- *Acinetobacter*, a group of bacteria usually found in soil and water, causes infections especially in intensive care units and other areas housing very sick patients.
- *Borrelia hermsii* causes a tick-borne relapsing fever usually occurring after sleeping in cabins infested with the ticks.
- *Burkholderia*, a group of bacteria that can be found in soil and water, causes infections in people with weakened immune systems or chronic lung diseases.
- Chikungunya virus is spread by the mosquito *Aedes aegyptii* and causes insomnia, high fever, severe headaches, and joint and muscle pains that last for weeks.
- *Cryptosporidium* is a microscopic parasite that causes a diarrheal disease resulting in dehydration, vomiting, elevated fever, and weight loss. It is found in water including recreational water that has been contaminated with sewage or feces from humans or animals. It also can be contracted by eating contaminated food or touching surfaces that are contaminated by contaminated feces or sewage.

- *C. difficile* is a bacterium that causes inflammation of the colon, diarrhea, and fever in individuals who have overused antibiotics.
- *Clostridium perfringens* is the third leading cause of foodborne illness in the United States and may cause diarrhea and severe dehydration in the very young, the very old, and those debilitated by sickness. (See endnote 30.)
- Dengue fever creates symptoms of sudden high fever, severe headaches, pain behind the eyes, severe joint and muscle pain, vomiting and skin rash, and is a viral infection transmitted by the bite of an *Aedes* mosquito.
- *Escherichia coli* 0157: H7 found in food and also acquired from person-to-person contact, produces Shiga toxins which create diarrhea (some people have no symptoms while others have severe diarrhea), abdominal cramps, and blood in the stools. Children under five can have a complication called hemolytic uremic syndrome which is extremely serious because red blood cells are destroyed leading to kidney failure. (See endnote 40.)
- *Helicobacter pylori* is a bacterium that is transmitted through the oral–fecal route or through the oral–oral route and causes nausea, vomiting, loss of appetite, and bleeding, and is the major cause of peptic ulcers and gastritis.
- Hepatitis A is an oral–fecal route disease and may be transmitted by contaminated water or other sources of consumption of fecal material.
- Hepatitis B and hepatitis C viruses can be transmitted to patients and healthcare workers through injections, improper use of needles and syringes, and breaks in infection control techniques, especially in hemodialysis units, outpatient areas, long-term care facilities, and hospitals.
- Human metapneumovirus is a common but usually undetected cause of human respiratory diseases especially among children.
- *Klebsiella*, a bacterium that has become increasingly resistant to antibiotics, is usually found in human feces in the healthcare setting and causes pneumonia, bloodstream infections, wound or surgical site infections, and meningitis in already sick patients.
- Malaria is caused by the parasite *Plasmodium* and transmitted through the bite of a mosquito. In 2010, there were an estimated 219 million cases with 660,000 deaths.
- Meningitis can be caused by a group of different bacteria and viruses as well as various chemicals. The bacterial form is caused by the bacterium *Neissera meningitides*. Transmission is person-to-person through respiratory secretions.
- *Mycobacterium abscessus*, a bacterium related to those that cause tuberculosis and leprosy, is found in water, soil, and dust and may contaminate medications and medical devices, leading to skin and soft tissue infections as well as lung infections.
- *Mycobacterium tuberculosis*, an organism which causes infectious tuberculosis, is especially readily transmitted in healthcare facilities to patients and healthcare personnel.
- Non-polio enteroviruses, although similar to the common cold, for individuals with weakened immune systems can cause serious consequences.
- Norwalk-like calciviruses frequently cause food and waterborne outbreaks of disease.
- *Norovirus* is a highly infective group of viruses spread from feces to food and water, from contact surfaces, and person-to-person in close contact, and can infect the individual repeatedly. This virus is especially problematic for the very young, debilitated, those with impaired immune systems, and the very old and produces symptoms of nausea, severe vomiting, abdominal pain, diarrhea, malaise, low-grade fever, and muscle pain.
- Polio enteroviruses are once again becoming a source of concern in countries where vaccination may be not only opposed but opposed violently. The spread of the disease to the United States and other countries may happen quite rapidly through travel of individuals who become contaminated.
- Prions are small pieces of protein that cause disease. They are neither bacteria, viruses, nor fungi. They may cause degenerative brain diseases and an unusual form of dementia.

- Rotavirus. (See Respiratory and Enteric Viruses in Pediatric Care Settings below.)
- *Toxoplasma gondii* is transmitted in unfiltered water containing the parasite's oocysts. It also can be transmitted from food or water contaminated with cat feces or soil or by eating undercooked meat that contains the oocysts.
- *Vibrio cholerae*, the causative agent of cholera, has had a resurgence in Dhaka and Bangladesh causing 30,000 cases of the disease.
- Yellow fever, a mosquito-borne viral infection, is still endemic in tropical countries in Africa and South America and can be transmitted to individuals from the bites of a mosquito, *A. aegyptii*, which can be found in the United States.

SPECIAL PROBLEMS RELATED TO SOME OF THE EMERGING PATHOGENS AND BEST PRACTICES

While most of the microorganisms listed above can be controlled with appropriate cleaning techniques and chemical disinfectants or sterilants, there are some pathogens that are resistant to these techniques. These pathogens are *Cryptosporidium* and *H. pylori*.

Cryptosporidium is resistant to chlorine at the normal concentrations found in potable water. It is also resistant to ethyl alcohol, glutaraldehyde, 5.25% hypochlorite, peracetic acid, and phenols. The only thing that works as a disinfectant or sterilant is 6% or 7.5% hydrogen peroxide used after thorough cleaning of the area. Sterilization using steam, ethylene oxide gas, or hydrogen peroxide gas plasma will fully deactivate the *Cryptosporidium* and spores.

H. pylori is resistant to formalin and cleaning with soap and water as well as ethanol. Mechanically washing endoscopes and disinfecting with 2% glutaraldehyde for 20 minutes is effective.

Rotaviruses may be resistant to quaternary ammonium chloride compounds. After proper cleaning and rinsing, 2% glutaraldehyde, 800 ppm free chlorine, and sodium hypochlorite and certain phenol-based compounds can decontaminate surfaces and destroy the rotavirus. On skin, the rotavirus can be destroyed with 95% ethanol.

Antibiotic-Resistant Bacteria
(See endnotes 24, 25, 29)

The CDC estimates on a yearly basis 2 million illnesses and 23,000 deaths are caused by antibiotic-resistant bacteria in the United States. There has been a huge overuse of antibiotics and misuse of antibiotics both in the healthcare fields and in food production. Outbreaks of disease-causing organisms resistant to antibiotics include carbapenem-resistant *Enterobacteriaceae* and particularly *Klebsiella* pneumonia carbapenem-resistant *Enterobacteriaceae* (See endnote 27), and *C. difficile* which is resistant to many drugs. Also vancomycin-resistant *Enterococci* (VRI), methicillin-resistant *S. aureus* (MRSA), and *S. aureus* with intermediate levels of resistance to glycopeptide antibiotics are serious concerns in the spread of hospital-acquired infections. This is especially important in intensive care units and other areas where patients may be immunocompromised.

People are the primary reservoir of the organisms even though they may be found on various types of equipment that come in contact with the patient, floors, frequent touch surfaces, and in hydrotherapy tanks which are of special concern for burn victims. Contamination of the surfaces increases substantially when patients have diarrhea and when there is improper removal and cleaning up of the fecal contamination. Misuse of gloves by healthcare workers can spread the microorganisms to the various surfaces. Also these organisms can be easily spread from patient to patient, from patient to healthcare worker to their families and back to patients and to others who come into direct or indirect contact with the individuals or especially their hands. The organisms can persist on surfaces for periods lasting from a week up to several months. They can persist on the hands or gloves of healthcare workers for up to 60 minutes.

Best Practices for Preventing the Spread of Antibiotic-Resistant Bacteria
- Decrease the amount of antibiotics being used in health care and in food production.
- Use a national One Health surveillance system to reduce the numbers of resistant organisms. This system integrates public health and veterinary disease, food, and environmental control. A regional laboratory network is needed to support the system by using innovative diagnostic tests and rapidly disseminating the results. Best Practices in infection control must be instituted when resistant organisms are identified.
- When transferring patients from one healthcare facility to another, clearly communicate to the new facility the nature, care, and treatment of individuals with infections so that proper infection control procedures may be used.
- Encourage state authorities to make the records of antibiotic-resistant bacteria patients a reportable event so that there can be accurate information concerning possible spread of disease.
- Use specialized teams of trained housekeeping personnel wearing gowns, gloves and if needed masks, and under strict supervision to thoroughly clean the entire area where the patient is treated and is housed, and especially frequent contact surfaces. Use increased frequency over and above normal cleaning of patient units for areas contaminated with infections. Dispose of all contaminated clothing prior to leaving the room.
- Healthcare workers and housekeeping personnel should wash hands thoroughly before entering the area where there are infected patients and after removing gloves and gowns prior to leaving the area.
- All housekeeping equipment should be considered to be contaminated and needs to be decontaminated before use in other areas.
- Make sure that all healthcare employees are aware of the potential for infection within the room where the patient is residing.
- Make sure that bathrooms are especially cleaned and disinfected.
- Teach the patients the necessity for proper hand washing after going to the bathroom, before coming in contact with other people, and also before preparing or consuming food.
- Have patients use chlorhexidine gluconate-impregnated bath cloths to reduce microorganisms when taking baths. (See endnote 27.)

Clostridium difficile

C. difficile most frequently causes healthcare-associated diarrhea today. Although most patients do not have symptoms but have organisms in their stools, it can become a problem with individuals who have had substantial antibiotic therapy, those having gastrointestinal procedures and surgery, and older people. Frequent touch surfaces as well as patient care items may also become contaminated. The most frequent routes of contamination are the hands of patient care employees and patients, and environmental surfaces. Carpeted rooms become heavily contaminated with the organism, while non-carpeted rooms are less contaminated. *C. difficile* produces spores and non-chlorine-based cleaning agents contribute to the problem.

The greater the number of patients present in any given environment where these organisms are found, the greater the opportunity for contamination of the various surfaces, especially the frequent touch ones.

Best Practices for Preventing the Spread of *C. difficile*
- Meticulously clean all frequent contact surfaces and other surfaces, and then disinfect with chlorine compounds following specific instructions for use in this type of situation.
- Use a dilute solution of hypochlorite at 1600 ppm available chlorine for a thorough disinfection of rooms with patients who have *C. difficile*-associated diarrhea. This will kill the spores.
- Utilize standard isolation techniques for enteric infections including a single room, and disposable gowns and gloves.

Respiratory and Enteric Viruses in Pediatric Care Settings

The typical common respiratory viruses found in pediatric areas include rhinoviruses, respiratory syncytial virus (RSV), adenoviruses, and influenza viruses. They are transmitted by direct contact with small particle aerosols or hands contaminated with respiratory secretions to the children's noses and eyes. Frequent touch surfaces which can be contaminated for periods of time from 30 minutes to 10 hours may also be a source of hand contamination. Enteric viruses found in pediatric areas include enteric adenovirus, astroviruses, caliciviruses, and rotavirus. The transmission of these viruses is typically through the fecal–oral route from contaminated surfaces and fomites (an inanimate object which is contaminated which can transmit microorganisms to an individual). Organisms may also be transmitted through the creation of aerosols during vomiting. These viruses may cause a variety of enteric diseases.

Best Practices for Control of Respiratory and Enteric Viruses in Pediatric Care Settings
- Use intensive cleaning procedures to remove all visible soil including fecal material and vomit from frequent touch surfaces and any other surfaces and then dispose of the cleaning solution into a toilet in a very careful manner as not to produce aerosols. Dispose of the cleaning materials in special packaging for biologically active materials.
- Use either fresh quaternary ammonium chloride compound solutions or chlorine solutions to disinfect the area after all visible soil has been removed. Wear disposable protective clothing, gloves, and where appropriate masks, when performing the cleaning and disinfecting operation and remove them to a special contaminated clothing area for special processing.

Severe Acute Respiratory Syndrome (SARS) Virus

This is an atypical pneumonia of unknown etiology which came out of Asia 2002 and caused severe viral upper respiratory infections. It is found at very frequent levels in the respiratory secretions, feces, and blood of infected people. It is spread by infected droplets in the air and person-to-person contact. Any aerosol producing procedures such as intubation, broncoscopy, and sputum production may contaminate surfaces.

Best Practices for Control of Severe Acute Respiratory Syndrome
- Follow Best Practices for Control of Respiratory and Enteric Viruses.

Ebola Virus

Ebola virus is transmitted by direct contact with blood or body fluids such as urine, feces, and vomit or by exposure to objects such as needles and bloody gloves that have become contaminated. The virus is highly infectious and can live on solid surfaces for several days.

Best Practices for Control of the Ebola Virus (See endnotes 22, 58)
- All staff members including housekeeping staff must wear appropriate personal protective equipment to avoid direct skin and mucous membrane contact with Ebola virus. (See the latest guidelines from CDC on the use of personal protective equipment by healthcare workers in endnote 21.)
- All housekeeping personnel must go through intensive training in putting on and taking off personal protective equipment when cleaning and disinfecting areas where Ebola patients have been treated. A highly skilled housekeeping supervisor must supervise not only the use of the personal protective equipment but also the cleanup and disinfecting process as well as the proper disposal of all equipment and fluids.
- All other personnel must follow the strict CDC guidelines for treatment of Ebola patients or suspected Ebola patients.

- If heavy-duty gloves are used for cleaning and disinfecting purposes, they should be disinfected after use and kept in the room or attachment to the room and then disposed of as highly infectious waste in an appropriate manner when the patient is either discharged or dies.
- Use a US EPA registered hospital disinfectant which is approved for non-enveloped viruses.
- Thoroughly clean all surfaces, especially frequent contact surfaces, and remove all visible soil before disinfecting the area. In the event of a large spill, neutralize the material with a very potent disinfectant and remove it before cleaning and disinfecting the area.
- Remove all single use equipment including food-service, linens, etc. and place in an impermeable waste bag in a rigid container marked as highly infectious before removing to processing and then final disposal of the contents.
- Use only mattresses and pillows which are impervious to fluids.
- Do not use carpeted rooms for Ebola patients and remove all unnecessary furniture.
- Discard all linens, non-permeable pillows or mattresses, privacy curtains, other cloth materials, dressings, portable toilets, emesis pans, etc. which have come in contact with the patient as highly infectious waste.
- Clean at least daily and possibly more frequently the area in which clothing changes occur and personal protective equipment is put on. The individual conducting the cleaning must be wearing clean fresh personal protective equipment and after finishing the various tasks dispose of them in an appropriate manner as determined by the CDC.

Tuberculosis

(See endnote 23)

Tuberculosis continues to be one of the most serious communicable diseases in the world. In 2013, an estimated 9 million people developed the disease and 1.5 million died of whom 360,000 were HIV-positive. Tuberculosis has the potential to be a healthcare-associated infection. It has become a very serious secondary infection for HIV patients as well as a primary infection for healthcare workers infected by people being admitted to hospitals and other healthcare institutions who have not been identified as tuberculosis active or carriers.

Best Practices in Control of Tuberculosis
- Screen high-risk patients especially those with productive coughs, and healthcare staff for unrecognized tuberculosis.
- Reduce the quantity of infectious airborne particulate droplets going into the general indoor air by proper early identification of individuals who have the disease, proper care including chemotherapy and treatment of individuals, use of isolation rooms and isolation techniques, and personal protective equipment for healthcare workers.
- Infectious patients must cover all coughs and sneezes with a tissue and place it in a special bag for highly infectious material.
- When the patient is transferred from area to area, he/she should wear a well-fitted impermeable surgical mask which should be disposed of as highly infectious material.
- Use special booths with local exhaust ventilation and HEPA filters, which remove almost 100% of airborne particles, for treatment of individuals with tuberculosis where sputum induction or aerosolized medications are part of the treatment process. These treatments can make the area a greater risk for other individuals. The exhaust fan in the booth should maintain negative pressure to avoid aerosol particles from entering adjacent areas.
- The general air of the room in which the patient is housed should be continuously diluted with fresh air and the exhaust air should go to the outside away from all air intake vents, people, and animals. The air supply should come from the ceiling at higher pressure and contaminated air should be exhausted near the floor at lower pressure, which produces

a constant downward movement of air through the potential contamination zone at the breathing level of the individual.

- Ultraviolet radiation is a controversial means of controlling microorganisms. If the ultraviolet lights are used within ductwork and are properly cleaned, they may be effective in destroying microorganisms in the air passing through. However, ultraviolet light used within patient rooms can be quite dangerous and not necessarily effective.
- All healthcare personnel should wear proper personal protective equipment including a disposable particulate respirator designed to filter out particulates 1–5 μm in diameter.
- Critical items such as surgical instruments, cardiac catheters, etc. should be thoroughly cleaned and then sterilized.
- Semi-critical items such as non-invasive endoscopes, bronchoscopes, and anesthesia breathing circuits should be thoroughly cleaned and then sterilized.
- Non-critical items such as crutches, frequent touch surfaces, blood pressure cuffs, etc. from an infectious patient should be washed thoroughly with a good detergent before reuse.
- Walls, floors, and other surfaces should be washed very thoroughly with a good detergent, rinsed, and then decontaminated.
- Establish a tuberculosis screening and prevention program for all healthcare personnel and periodically test them for the disease.
- Establish a special surveillance and reporting program for tuberculosis among patients in healthcare facilities. Report the results to the local and state health departments

SUB-PROBLEMS INCLUDING LEADING TO IMPAIRMENT AND BEST PRACTICES FOR BIOTERRORISM AND POTENTIAL AGENTS

(See endnote 36)

Note: Only those agents listed in Category A will be discussed and the rest of the section will be about the CDC Office of Public Health Preparedness and Response which is responsible for providing the necessary research and resources for the control of bioterrorism in the United States.

Bioterrorism is the deliberate release of viruses, bacteria, or other microorganisms to cause illness or death in people or animals, or destroy plant life which may be used as a food source. These microorganisms already exist and can cause disease, but by increasing their quantity and altering certain parts of the organism, terrorists can make the attack extremely deadly and cause considerable destruction, death, and fear in a population. These agents, which are very stable in the environment, can be spread through air, water, or food and may be very difficult to detect until a disease outbreak occurs. The organisms in the case of smallpox and pneumonic plague could then be spread further from person to person.

Although the bacteria or toxins from a biological terrorist attack can enter the body many times by all potential routes including inhalation, ingestion, absorption through the skin or eyes, and arthropod-borne transmission from animals or other humans, the inhalation route for many organisms is the most common. The aerosols may contain *Burkholderia pseudomallei*, *Burkholderia maiiei*, *Coxiella burnetii*, the causative agent of Q fever, and select other *Rickettsia*, all of which can be inhaled and cause human disease, and some of which have a mortality rate of 50% and cause severe or chronic disease even if treated with antibiotics.

Arthropod-borne viruses that can be used as a biological weapons include the alphaviruses that are the cause of Venezuela equine encephalitis, Eastern equine encephalitis, and Western equine encephalitis; flaviviruses that are the cause of West Nile virus disease, Japanese encephalitis, Kyasanur Forest disease, tick-borne encephalitis, and yellow fever; and Bunyaviruses that are the causes of California encephalitis, La Cross virus diseases, and Crimean–Congo hemorrhagic fever.

Food- and waterborne viruses include hepatitis A virus and the caliciviruses, most notably Norwalk. They are excellent for bioterrorist threats because they can spread rapidly and are highly infective.

Hepatitis A virus causes about 55% of the cases of hepatitis in the United States each year. Transmission can be by water and food or directly from person to person. Norwalk virus and other caliciviruses cause a diarrheal disease which has been very prominent in outbreaks in camps, hospitals, nursing homes, and cruise ships. They cause about one third of the viral diarrheal diseases in the United States. Enteric protozoa such as *Entamoeba histolytica*, *T. gondii*, *Giardia lamblia*, *Cryptosporidium parvum*, etc. can also be disseminated by terrorists in food or water. In all food and water contamination, those at greatest risk are the very young, very old, immunocompromised, and those with debilitating acute or chronic diseases.

Toxins which can be utilized in bioweapons which include those from *C. perfringens*, Staphylococcal enterotoxin B and *Clostridium botulinum* can be delivered in air, food, or water. Ricin toxin comes from the castor plant. *C. botulinum* is heat stable and available throughout the world. The toxin prevents protein synthesis and leads to cell death.

CATEGORIES OF POTENTIAL ORGANISMS USED AS BIOWEAPONS

The CDC has established three categories of organisms. These categories are as follows:

- Category A which includes organisms or toxins that are of the greatest risk to the public and national security because they: can be spread easily from person to person; cause high death rates; cause panic and disruption of society; and require special precautions by public health officials.
- Category B which holds second priority and includes organisms that spread easily, cause moderate illness and few deaths, and require special CDC laboratory capacity. (See endnote 36 for a list of diseases.)
- Category C which includes emerging pathogens that have been bioengineered to spread in the future, are readily available, are easily produced and spread, and have potential for high morbidity and mortality. (See endnote 36 for a list of diseases.)

Category A Microorganisms

Anthrax

(See endnote 37)

Anthrax is a highly infectious disease caused by the bacterium *Bacillus anthracis* which is normally found in soil and affects domestic and wild animals throughout the world. People are affected when the spores of anthrax get into the body, become bacteria, multiply, and produce toxins that cause fever, cough, chest discomfort, a period of improvement and then respiratory failure plus a collapse of appropriate blood flow through the body. Anthrax makes a good weapon because the spores are found in nature, can be produced in a laboratory, and can be dispersed into the environment and remain there for a long period of time. It can be put into powders, sprays, food, water, and done in such a manner that no one would know what has happened, and has already been used as a weapon. It can cause mass casualties and devastate the economy while creating panic in the public. The spores can be carried by the wind or on people's clothing. Inhalation of anthrax spores will kill quickly if not treated at once.

Botulism

(See endnote 31)

Botulism is caused by a spore-forming organism, the bacterium *C. botulinum*, which is naturally found in soil, and can easily be isolated and therefore concentrated. Three of the five types of botulism are very important because they involve food, infect infants, or get into wounds. In foodborne botulism, the toxin causes illness within several hours to several days after food consumption and results in double vision, slurred speech, difficulty in swallowing, muscle weakness starting in the upper body first and moving down through the feet, and paralysis of the breathing muscles which may result in death. Infant botulism is caused when the infant consumes the spores of the bacteria and then they grow in the intestines and release the toxin. Wound botulism is caused when the toxin is produced in a wound that is infected with the microorganism. Botulism toxin in solution is colorless, odorless, and believed to be tasteless. Once the toxin is absorbed in the body, the bloodstream carries it to various nerve endings and the toxin binds with them blocking the release of acetylcholine.

Best Practices in Preventing Botulism
- Use appropriate therapy for botulism including supportive care, fluid and nutritional help, proper ventilation and treatment of complications, and passive immunization with an equine antitoxin which should be administered as quickly as possible.
- Test all possible food or water consumed by the patient.
- Antibiotics showed no known properties for altering or protecting against the toxin, however they may be used for secondary infections.

Plague
(See endnote 32)

Bubonic plague is caused by the organism *Yersinia pestis* which is carried by the Oriental rat flea. The reservoir of infection is rodents, prairie dogs, squirrels, and rabbits. Either the bite of the infected flea or touching or skinning an infective animal can cause disease transmission.

Pneumonic plague is transmitted from person to person when the bacteria affect the lungs and another person inhales the infected material. Symptoms of the disease include high fevers, chills, cough, difficulty in breathing, bloody mucus, and then death. An aerosolized weapon would cause fever and cough within 1–6 days and then a rapid move to septic shock with a substantial number of people dying.

Smallpox
(See endnote 33)

Smallpox, which has been eradicated worldwide, may still be used as a bioweapon by terrorists if they have access to supplies of the variola virus. Although there are two designated World Health Organization repository laboratories, no one knows if terrorists have access to other sources of the virus. The mode of transmission is from person to person by inhaling infective droplets in close face-to-face contact. The symptoms usually begin in 12–14 days with high fever, malaise, severe headaches, and backaches followed by a rash all over the body. This is a highly infectious disease.

Best Practices for Controlling Smallpox
- Administer smallpox vaccine within 3 days after exposure, if smallpox is suspected, to prevent the disease or decrease the severity and reduce the risk of death.
- Obtain and utilize the most up-to-date version of the CDC Smallpox Response Plan.
- Identify and isolate as quickly as possible all smallpox cases to prevent disease spread.
- Identify, vaccinate, and monitor all known contacts of the individuals who have smallpox.
- If deemed necessary, vaccinate a larger population of people as soon as possible.
- Identify for vaccination and treatment individuals who had close face-to-face contact with the patient, those initially exposed to the virus release, household members of actual cases or potential cases, and healthcare personnel including first responders.
- Treat any individuals rapidly if they show adverse reactions to the vaccination.
- Use strict isolation procedures and quarantine on known or suspected cases.
- Inform the public by all sources of media of what has occurred and the necessity to look for symptoms and to immediately contact the appropriate authorities and be vaccinated.
- CDC should coordinate all activities in investigating the problem, and vaccination and treatment of patients with all state and local public health authorities.

Tularemia
(See endnote 34)

Tularemia is a bacterial zoonosis caused by the microorganism *Francisella tularensis*. The disease can be caught by the bite of an infected tick, deer fly, or other insects, working with infected animal carcasses, eating or drinking contaminated food or water, or breathing in the bacteria especially from an aerosol. It is not spread person to person. It can be fatal if not treated rapidly with antibiotics. Airborne release of the organisms can cause primarily pleuropneumonitis and also possibly ocular tularemia. It can penetrate broken skin and cause glandular disease. People usually become infected 3–5 days after exposure and develop a variety of symptoms including hemorrhagic inflammation of the airways, life-threatening bronchopneumonia, fever, chills, progressive weakness, sore throat, and systemic infection. Since the organism which causes tularemia is highly infectious, a small number of bacteria will cause the disease. The bacteria are widely present in nature and therefore readily accessible. Therefore, it might make an excellent bioweapon used as an infectious aerosol.

Best Practices for Control of Tularemia as a Biological Weapon
- If exposure to tularemia is suspected, individuals should immediately be given vaccinations for the disease, appropriate antibiotics, and supportive therapy where necessary.
- Microbiological sampling should be handled in the appropriate biosafety cabinets.
- Avoid autopsies where possible of victims to prevent aerosols from entering the environment and causing a potential hazard for healthcare workers.
- Decontaminate and disinfect all linens and clothing containing the body fluids of the patient.

Viral Hemorrhagic Fever
(See the section "Ebola Virus" above) (See endnote 35)

Viral hemorrhagic fevers are a group of highly infectious organisms, one of which is the Ebola virus, from several distinct families of viruses. The viral hemorrhagic fever viruses affect multiple organ systems in the body, often accompanied by severe bleeding. Symptoms include fever, headaches, nausea, vomiting, diarrhea, and chest pain. Some of the viruses cause mild symptoms, while others are severe and cause a large number of deaths. Viruses are all RNA viruses and depend on an animal or arthropod host from a specific geographic area. Human cases occur when people come in contact with the animal or arthropod, which may be a mosquito or tick, and then the disease may be spread secondarily from person to person especially in a healthcare setting. Primary concerns are Ebola hemorrhagic fever, Marburg hemorrhagic fever, Lassa fever, and Argentine hemorrhagic fever.

Best Practices for Control of Bioterrorism and Potential Agents

Note: Other than the specific Best Practices listed under Ebola, in order to avoid repetition general Best Practices will be given for all potential Category A, Category B, and Category C agents of bioterrorism, based on the method of distribution of the microorganism or toxin.

- Healthcare providers need to be taught the symptoms of the illnesses and best techniques for diagnosing the effects of agents of bioterrorism when a cluster of patients with similar symptoms appear in either their offices, clinics, or hospitals, and this should be reported immediately to local, state, and federal public health authorities and the FBI. Laboratory samples should be used only for confirmation and identification purposes while immediate precautions need to be taken based on the symptoms.
- Healthcare providers should be looking for unexplained fevers, sepsis, which is a life-threatening complication of an infection where chemicals are released into the bloodstream

to fight an infection and trigger inflammatory responses throughout the body which then causes multiple organ system damage or failure, pneumonia, respiratory failure, or severe rashes especially in apparently very healthy people. Information concerning individuals with these symptoms aids in the identification of potential use of bioterrorism agents.

- Healthcare providers should be looking for what appears to be chickenpox among adults as possible use of smallpox as a weapon.
- Healthcare workers should be looking for cases of acute flaccid paralysis with prominent bulbar palsies (extreme loss of muscle tone that affects the lower cranial nerves) suggesting exposure to botulism toxin.
- Physicians treating individuals contaminated with various biological agents should consider the use of antibiotics, appropriate vaccines, and antitoxins.
- Clinical laboratory personnel should be alert to a large number of samples of blood or stool cultures or a request for analysis of a large number of these samples, as potential outbreaks of a serious disease or a bioterrorism agent.
- Use personal protective equipment when dealing with suspected highly infectious material, handle all samples in the appropriate biological safety cabinets, and immediately inform appropriate public health authorities. After determining the causative agent, a priority immediate report should be given to the public health agencies as well as to the practicing healthcare worker. All laboratory equipment and facilities should be immediately decontaminated after suspicion of a highly infectious agent.
- Infection control personnel should be constantly upgraded in their knowledge and skills concerning potential biological agents and it should be top priority protocol that current phone numbers for notifying appropriate public health officials, the CDC, and the FBI are prominently posted for use by healthcare workers.
- State health departments should have current plans and means of implementing them to quickly respond to any outbreaks of disease including use of biological agents. They should have appropriate educational programs and reminders to all healthcare workers about each of the potential disease outbreaks as well as use of biological agents.
- Since aerosol dissemination of biological agents is extremely hard to detect, if individuals become aware of a suspicious substance in the air, they should immediately cover their mouths and noses with layers of cloth to act as a filter, get away from the area as quickly as possible, wash and rinse thoroughly with soap and water, wash all clothing carefully, notify the proper authorities, and if told to get a specific vaccination, do it immediately. Seek immediate medical treatment since antitoxins exist for biotoxins and supportive therapy may be needed immediately to preserve life because of biological weapon use.
- If the aerosol release occurs in an enclosed environment, shut off the ventilation system to the building, trains, aircraft, or subways where terrorist groups in the past have released anthrax and botulism toxin. Thoroughly clean and disinfect the ventilation system, repeatedly if necessary, to remove all microorganisms or toxins. Chlorine dioxide may be excellent for this purpose.
- Monitor the level of biological agents to determine the quantity and specific type and perform the necessary cleanup and decontamination of the area by properly equipped personnel based on the organisms released.
- Where arthropods including insects are purposely infected with biological agents and an outbreak of disease occurs, there should be a determination of the type and nature of the organism, immediate public warnings for necessary precautions including vaccinations, immediate spraying campaign to destroy the insects, destruction of potential resting places and harborage, and in the case of mosquitoes staying indoors at dawn and dusk.
- If the biological weapon is a form of a contagious disease, immediately institute isolation procedures in healthcare centers, quarantine all suspected cases, have medical personnel

use personal protective equipment of the appropriate kind depending on the disease under close supervision, and in all cases use appropriate hand-washing techniques.

- If the release of the biological substance is into water and cases of a disease start to track to a water source, immediately issue a boil water order to the public for any water used in drinking, eating, preparation of food, brushing teeth, or other essential personal needs, bathing, etc. for people and pets. Bottled water may be used for these purposes or the water may be disinfected by using bleach with 1/8 teaspoon of bleach per gallon of water, shaking the container and letting it stand for 30 minutes before using. After the boil water notice is withdrawn, thoroughly flush all water outlets including those for swimming pools, hot tubs, spas, icemakers, etc. All swimming areas need to be thoroughly cleaned using gloves and masks if necessary, disinfected, and restocked with treated water.
- If the release of the biological substances is into the food, as has been done in salad bars by a terrorist group, as soon as cases of a disease are tracked to the food source, there should be an immediate halting of all activities, thorough evaluation by the appropriate health authorities, and intensive cleaning and decontamination of all surfaces, vessels physical facilities, and equipment. All contaminated and potentially contaminated food should be disposed of in an appropriate manner. (See Chapter 7, "Food Security and Protection.")
- If human carriers are involved in the distribution of biological substances, as soon as the disease symptoms appear, the individual should be put into isolation and treated.
- If infected animals are involved in the distribution of biological substances, as soon as the disease symptoms appear, the animals should be examined by veterinarians and a determination made as to the nature of the disease and then the animal should be destroyed and incinerated. All other animals should be given necessary protection including appropriate vaccinations. All areas where the infected animal lived should be thoroughly cleaned and disinfected, and the bedding and other materials including contaminated food should be incinerated.
- For mail attacks see below.

SUB-PROBLEMS INCLUDING LEADING TO IMPAIRMENT AND BEST PRACTICES FOR USE OF MAIL AS A MEANS OF DISPERSING BIOLOGICAL WEAPONS AND OTHER POTENTIAL THREATS

(See the section above on "Sub-Problems, Factors Leading to Impairment, and Best Practices for Bioterrorism and Potential Agents") (See endnote 38)

There are a wide variety of substances that can create harm to individuals or groups of people and disrupt operations including those of a chemical, biological, radiological, nuclear, or explosive nature or other threats. In addition, white powder hoaxes and threatening letters are sent through the mail.

Chemical agents include nerve agents, blood agents, pulmonary agents, blister agents, and various chemicals which may be hazardous or may be simply irritating to individuals. They can be shipped in solid, liquid, or gaseous/vapor states. Typically, there needs to be a timer included so that it will cause damage to the targets rather than in the general mail room. Besides mailing from United States Postal Service drop boxes, these packages or letters may be delivered by courier or a local mail delivery service.

Biological agents may include microorganisms which cause anthrax, plague, smallpox, and tularemia. They may be distributed throughout the mailroom or other areas readily because of their small size and the large amount of dust and paper dust within the environment, since dust facilitates dispersion of microorganisms. Individuals would not know if they had been exposed to the biohazard and this would hinder treatment and extend the potential for recovery. A prime example is the anthrax letters that were sent out in October 2001, resulting in 11 cases of cutaneous anthrax and 11 cases of inhalation anthrax, with 5 people dying from the inhalation anthrax. Ricin, which is a toxin which comes from castor beans, may be used as a biological weapon and will enter the body through a cut in the skin or through inhalation and cause death.

Radiological/nuclear agents can be dispersed from a dirty bomb inside a package. This can lead to immediate and long-term effects including cancers of various types and death.

Explosives in letter or package bombs can be set off upon opening of the package or by a timer. Various explosives have been used for this purpose.

Hoaxes are mail items that are meant to disrupt the lives of people, but do not cause actual harm from a known hazard. The best examples of this are envelopes with different types of white powder enclosed since the anthrax letters contained a white powder.

Dangerous items may include things that can cause cuts or shocks to the individuals opening a package or letter. Contraband would include packages that may contain illegal drugs, guns, knives, and other sharp objects. Letters with threatening content may include death threats against the President or other elected officials or heads of corporations. Interoffice mail in a large corporation, university, or governmental agency can be subject to many different types of at-risk mail or packages, since they can be introduced into the system flow at a variety of points within the facility.

Most large organizations are very vulnerable to receiving hazardous packages or mail. The mail centers are typically within the facility and therefore the potential for serious problems is enhanced. The size and volume of the amount of mail being processed is a contributing factor.

Although the front entrance to buildings may be well guarded and various protective mechanisms used, a significant weak point is the loading docks, as well as the numerous delivery vehicles entering the building. Further, the mailroom location may be hazardous to the health and welfare of all employees because biological or chemical weapons may be released into the ventilation systems and travel throughout the structure.

Best Practices for Mail Screening and Handling Processes
- Conduct a complete risk analysis study of all mail streams within the facility and determine all the weaknesses and strengths of the system. Enumerate the weaknesses and necessary steps to be taken to correct the situation. The United States Postal Service can be of great value in making this determination.
- Determine based on mail volume the most efficient, economical means of screening mail and the types of facilities and processes needed to do so including mail centers off-site.
- Train all employees of the mail center the necessary techniques and use of equipment for dealing with suspicious mail and packages and how to process them.
- Train all personnel in techniques of decontamination in the event a terrorist action occurs.
- Use appropriate technologies to determine if explosives are present in letters and packages at high-risk facilities.
- For high-risk facilities, establish an off-site screening facility for all deliveries including mail, furniture, food, and other supplies and have security accompany all vehicles with their contents that have been approved to enter into the secure facility.
- Provide separate, isolated HVAC systems in lobbies, loading docks, and mail rooms that are susceptible to intake of letters and packages that may contain biochemical weapons.
- Personal protective equipment for medium- and high-risk environments where mail and packages may be handled should include a Tyvek suit (a specially woven lightweight high density polyethylene abrasion-resistant material which protects against small size particles), nitrile gloves (protects against chemicals and puncture resistant), foot coverings, and a National Institute for Occupational Safety and Health (NIOSH)-approved disposable filtering facepiece respirator. The employees must have annual physicals to determine if they are able to wear the respirators on a regular basis. All personal protective equipment should be put on prior to entering the mail screening area and then removed in the exit chamber before leaving the area.

- All personal protective equipment should be disposed of daily in sealed bags after it is determined that there is no biologically hazardous material present in the facility.
- All mail centers should have air sampling systems with an automatic alert if sensors detect chemicals, radiological material, and/or explosive material.
- All mail centers should be under negative pressure in the event of the presence of biologically hazardous materials.
- All mail and packages should be screened by an x-ray scanner and suspicious items segregated until security personnel can arrive and make appropriate decisions on disposal of the items.
- Personnel who process, sort, and deliver mail should view the CDC video entitled "Protecting Your Health" and be given short-term continuing education courses on a regular basis in order that they might understand the potential hazards involved and know what best to do in the event of an emergency.

SUB-PROBLEMS INCLUDING LEADING TO IMPAIRMENT AND BEST PRACTICES FOR TOXICOLOGICAL, ENVIRONMENTAL, AND OCCUPATIONAL CONCERNS FOR CLEANING MATERIALS AND DISINFECTANTS

There are health hazards associated with the use of various cleaning materials and disinfectants. They may seriously damage mucous membranes, the skin, eyes, and internal organs of all types if inhaled, ingested, or absorbed through direct contact. The potential for health risk depends on the type of chemical being used, the concentration of the chemical, the time of exposure, the route of entry into the body and whether the individual is working with the material on a routine basis, periodically, cleaning up a spill of the chemical, or has limited exposure. The individual involved may have a specific sensitivity to the chemical. The individual's past medical history, previous exposures to a variety of different chemicals, state of health, age, types of pharmaceuticals taken, etc., contribute to the potential effects of the various chemicals being used. Acute toxicity may result from an accidental spill, whereas chronic toxicity may occur from repeated exposure to low levels of the chemical over a period of time.

Best Practices for Avoiding Health Problems Related to Cleaning Materials and Disinfectants
- All employees should be given a physical examination including an extensive history, before working with any types of chemical substances.
- All employees should be taught about the potential hazards of the chemicals that they are using for cleaning and decontamination and should be under close supervision to make sure that they follow appropriate instructions not only for proper cleaning purposes, but also for self-protection.
- All employees should wear the necessary personal protective equipment based on the types of chemicals being used.
- All employees should thoroughly wash their hands frequently during the course of the work shift and then wash their hands thoroughly and change their clothing before leaving work.
- Employers must be constantly aware of the hazards from cleaning materials and are responsible for informing the workers about these hazards and how to protect themselves.
- Employers must follow the necessary rules and regulations published by OSHA and use the information on the material safety data sheets for each chemical or mixture of chemicals and must observe the exposure limits which have been established for a normal 8-hour workday for 40-hour work weeks to protect the employees utilizing cleaning materials and disinfectants. Particular attention must be given to the use and disposal of glutaraldehyde, formaldehyde, and some of the phenols.

STATEMENT OF PROBLEMS AND SPECIAL INFORMATION
FOR OTHER HEALTHCARE OPTIONS

By 2050, the population of older Americans aged 65 and over is projected to grow from 40.2 million in 2010 to 88.5 million in 2050. (See endnote 47.) This creates a huge new problem of housing for the elderly and severe healthcare situations due to infirmity and a wide range of chronic and infectious diseases.

There are several senior housing options which include staying in place; staying in place with outside help to take care of medical needs, physical needs, home maintenance, loneliness, socializing, financial needs, and loss of independence such as no longer driving; an independent living facility; an assisted living facility to take care of personal care and other things but not on a 24-hour basis; and eventually a nursing home to provide medical and personal care which is greater than that which can be handled at home or in less intense facilities. The nursing home or rehabilitation hospital may be used as a temporary recovery area after hospitalization and until the individual is capable of once again resuming a fairly normal life. Each of these options may present problems for the individual and family. This discussion will be primarily about those receiving home healthcare and those in nursing homes. It will also include a brief discussion on stand-alone surgical units, clinics, and physicians' offices.

SUB-PROBLEMS INCLUDING LEADING TO IMPAIRMENT
AND BEST PRACTICES FOR HOME HEALTHCARE

Home health care is the fastest growing area in the healthcare industry. A recent sample study has identified risk factors in the health and safety of people receiving healthcare in the home. (See endnote 48.) Individuals in the homes were exposed to roaches, cigarette smoke, other insects and rodents, irritating chemicals, peeling paint, temperature extremes, unsanitary living conditions in the home and neighborhood, violence, and crime. Poor indoor air quality and lead paint are also concerns because many of the elderly live in older properties that are not well maintained.

The potential for the spread of healthcare-associated infections is considerable because of the surroundings which may include highly unsanitary conditions especially in the bathroom and kitchen, as well as the treatment being given to the individual and the lack of continuous professional nursing care throughout the day and night. There is a serious concern about the mishandling of biomedical waste, especially sharps and bandages. Indwelling catheters are the source of numerous infections in both use and potentially disposal. Home hygiene including disinfection procedures and effective hand washing may be a serious problem. The accumulation of the problems noted can lead to a situation where individuals acquiring infections in the hospitals, return to the communities and spread them to others in the family and community, and then return to the hospital where they become the source of new infections. (See Chapter 3, "Built Environment—Healthy Homes and Healthy Communities," Sub-Problems Including Leading to Impairment for Housing.)

The healthcare providers may be highly experienced, such as visiting nurses, or have limited training or expertise and work under little or no direct supervision. Financial constraints on agencies providing home health care continue to increase with the result of poor pay for many health aides, constant turnover of people, and inadequate training or supervision of these individuals.

The safety of the patient is always a concern because of the many accidents that occur within the home setting. The safety of the healthcare worker revolves around many issues but especially lifting and exposure to contaminated bodily fluids. (See Chapter 6, "Environmental and Occupational Injury Control.")

Best Practices for Home Health Care (This discussion will not include appropriate nursing techniques for patient care which may be found in endnote 59)
- When nurses make visits to patients within the home and they observe numerous problems of housing and other environmental issues, they should contact the local health department and request that an environmental health technician come out to the site and conduct a housing evaluation in order to determine the types of problems which may lead to disease or injury. Part of the care plan for the individual and family should include the necessary correction of these deficiencies in order to provide quality care.
- The care plan established for the patient should help prevent infections, other health problems, and injuries by preventing and where necessary mitigating practices and situations which will avoid medical errors, medication errors, injuries due to accidents, especially falls, and infections.
- The individual and family should be taught how to properly care for the patient concerning the use of: sterile barriers and sterile equipment; appropriate bedding and moving of the patient frequently to prevent pressure ulcers; patient self-management for anticoagulants and other drugs; appropriate knowledge and use of information concerning adequate nutrition; and the types of situations in which the patient may experience falls or other debilitating situations because of safety hazards.
- The patient and family should be taught when to seek immediate health assistance in the event of an emergency, and especially when to contact 911 for emergency medical services.

SUB-PROBLEMS INCLUDING LEADING TO IMPAIRMENT AND BEST PRACTICES FOR PHYSICIANS' OFFICES AND MEDICAL CLINICS

(See endnote 56)

In 2007, there were almost 1 billion visits to physician's offices in the United States. Physician's offices or medical clinics may be located in a variety of facilities ranging from old houses and remodeled stores to modern medical offices. The older facility may contribute numerous environmental concerns to the safe handling of patients including problems of poor indoor air quality, contaminated surfaces, contaminated waste, lead and asbestos prior use, etc. There may be inadequate room for dirty areas and clean areas as is needed in other types of advanced healthcare facilities. The opportunity for infections increases in the physician's office or medical clinic because of a variety of factors. There is no central infection control team or individual who is overseeing medical nursing techniques as well as environmental concerns. The disinfection of surfaces and/or sterilization of equipment may not be supervised to ensure a proper job is done. The waiting room becomes a problem because sick people as well as those who are healthy but still vulnerable and getting blood work or other procedures done, may be sitting together. This is especially a problem in pediatric practices where there are so many sick children and on the other hand well babies who are in for routine examinations and vaccinations. Injection practices within offices may also be a problem, particularly reinsertion of used needles into multiple dose vials or containers of solution. The personnel at the facility may be both paid professionals and also unpaid volunteers who may expose patients to a variety of diseases or become infected by contaminated materials, medical supplies and equipment, environmental surfaces, or air problems. Clerical and housekeeping personnel may also become exposed to infectious agents.

Best Practices for Physician's Office or Medical Clinic
- Develop and maintain an infection control program as well as an occupational health and safety program. Assign one knowledgeable person to be in charge.
- Develop written infection prevention policies and procedures as well as the occupational health and safety policies and procedures appropriate for the practice.

- Train all employees in the previous policies and procedures during new employment orientation and periodic brief reviews.
- Follow all necessary reporting requirements on healthcare-acquired infections and occupational health and safety rules and regulations from all local, state, and federal agencies.
- Use appropriate hand-washing techniques as discussed under acute hospitals.
- Use appropriate personal protective equipment as discussed under acute hospitals.
- Utilize safe injection practices as discussed under acute hospitals.
- Use appropriate cleaning and disinfection of surfaces and cleaning, disinfection, and/or sterilization of medical equipment as discussed under acute hospitals.
- Use and teach patients appropriate respiratory hygiene for coughs and sneezes and how to dispose of the tissues without contaminating others.

SUB-PROBLEMS INCLUDING LEADING TO IMPAIRMENT AND BEST PRACTICES FOR STAND-ALONE SURGICAL UNITS

(See endnotes 54, 55)

Specialized surgical outpatient programs have been developed in stand-alone units where individuals do not have to remain overnight in a hospital for recovery purposes. Up to 80% of operative procedures are now carried out in these day surgery units in the United States. Also, many of the invasive procedures for diagnostic purposes are also carried out in these units. This practice of using off-site facilities has become popular because of a substantial amount of cost-saving and because it reduces the number of inpatient beds needed at hospitals. It has been found that there is a considerable reduction in the risk of cross-infection compared to hospitals if all procedures and standards established by the appropriate authorities including The Joint Commission in the United States are followed rigorously. There is less stress for patients and their families and the patient has a quicker return to normal life. However, it is absolutely essential that each of these units be accredited by The Joint Commission and that the facility follows the accreditation standards for all procedures.

Best Practices for Stand-Alone Surgical Units (contact The Joint Commission for further information, and see endnote 60)
- It is absolutely essential that each stand-alone surgical unit is accredited by The Joint Commission and that the facility follows the accreditation standards in great detail for all procedures and that a single individual who is a high-level administrator supervises the entire program.

SUB-PROBLEMS INCLUDING LEADING TO IMPAIRMENT AND BEST PRACTICES FOR SKILLED NURSING FACILITIES

There are many excellent skilled nursing facilities in the United States. They are well supervised and the patients are treated with great care and love. However, there are others which have fundamental problems of health and safety usually based on financing, poor training for inadequately paid personnel, and poor supervision throughout the 24/7 cycle.

A recent study by the Office of Inspector General, US Department of Health and Human Services indicates that there are numerous problems occurring in skilled nursing facilities or nursing homes used by Medicare patients. (See endnote 49.) An estimated 22% of Medicare recipients were involved: 79% of these individuals had prolonged stays at the facilities because of these problems; 14% required some form of intervention to keep them alive; and 6% were so affected that it contributed to their death. The adverse events are due to medication problems, resident care, and infections. The medication may cause induced delirium or change of mental status, excessive bleeding, and in some cases, contribute to falls. Poor residential care has led to falls, exacerbation of existing conditions, acute kidney injury, electrolyte imbalance, and pulmonary embolisms. Infections have included pneumonia, other respiratory problems, surgical site infection because of poor wound care, urinary tract infection, and infection with *C. difficile*. Patients have suffered low or significant drops in blood glucose, trauma with injuries, allergic reactions, and pressure ulcers. It was determined by physicians reviewing the patient charts that 69% of the adverse events were preventable. Most of the preventable problems were due to substandard treatment and inappropriate monitoring of the patient.

Infections are very common in long-term care facilities. Residents are usually in confined living situations and daily activity takes place in groups which promotes the spread of infections. Since many of the individuals are at least restricted in their cognitive abilities and others are senile, it is difficult to maintain an appropriate level of personal hygiene, especially hand washing, unless the staff works on a one-to-one basis with each individual. This gets to be very difficult when under-staffing is a very common problem and improperly trained and poorly supervised individuals are hired to assist the residents. An increase in drug-resistant microorganisms adds to this complex situation. The most common infections found in nursing homes are respiratory, urinary, skin, soft tissue, and gastrointestinal. The most common diseases are: scabies, which is a mite infestation and is transmitted by person-to-person contact; influenza, which is a major source of illness and death in this high-risk population; *C. difficile*, which is a diarrheal infection ranging from mild to fatal; and MRSA, which is not only serious within the nursing home but also becomes a major risk factor when the individuals are admitted to hospitals. A re-emergence of tuberculosis has become a renewed healthcare problem. A serious communication problem exists between the nursing home and the acute care hospital, and often when the individual is moved to the hospital appropriate records and warnings of existing infections do not accompany him/her. (See endnote 50.)

Patient abuse, physical, sexual, and verbal, is increasing in a portion of the nursing homes in the country. Many of these facilities are understaffed, underfinanced, and improperly supervised, and unfortunately the patients are getting much older and feeble and there are increased levels of senility. In these situations, it is common to see patients with untreated bedsores, malnutrition, dehydration, and poor hygiene. Unsanitary conditions exist and medical care is spotty. Overmedication is a concern, especially with the use of psychoactive drugs. There may be excessive therapy services that are medically unnecessary or even harmful in order to increase the reimbursement from Medicare and Medicaid.

Approximately 1800 older adults who live in nursing homes die each year from fall-related injuries and those that survived the fall may have a permanent disability and have their quality of life affected. About 1.4 million people 65 and older live in nursing homes. This is estimated to increase to 3 million by 2030. Each year a typical 100-bed nursing home reports 100–200 falls. This is probably vastly under-reported. Patients frequently fall more than once. Falls may occur because of medications, the disease process, muscle weakness, balance problems, walking problems, environmental hazards such as wet floors, poor lighting, incorrect bed height, improper shoes or poor foot care, improper fitting, and use of wheelchairs, etc. The falls may also be the result of staff negligence. The routine use of physical restraints does not lower the risk of falls or injuries. (See endnote 51.)

OCCUPATIONAL SAFETY AND HEALTH PROBLEMS

Nursing homes and residential care facilities employ about 2.8 million people at 21,000 worksites. There is a considerable amount of strenuous physical labor involved, especially in the moving of patients from one place to another and in helping them with their routine daily hygienic practices. The injury rates at these types of facilities are double those in other occupations. There are health-care facility hazards which occur in all areas such as potential contamination with blood-borne pathogens, ergonomic problems where the job should be matched to the worker including physical requirements and the physical capacity of the worker, and in the handling, removal, and washing of laundry. There are special occupational health and safety problems: in dietary areas; with all types of maintenance work; within the nurses' station; in the pharmacy; in housekeeping; and in the use of whirlpools or showers for patients. Violence is increasing and may cause serious injuries not only to patients but also to employees. (See endnote 52.)

Best Practices for Other Healthcare Options

Note: There will be several Best Practices for Other Healthcare Options, but the major source of Best Practices by area and situation will be found in each individual section related to the acute hospital situations previously discussed.

- Improve home healthcare personnel by providing proper training, increased salaries, and proper supervision by highly skilled, licensed individuals.
- Reduce the amount of cases that a caseworker has to deal with on a weekly basis in order for the individual to make repeat unannounced visits in situations that are poor or deteriorating to prevent disease and injury and protect the health of the person, family, and community.
- Where necessary, bring in clean-up crews to remove potential environmental hazards from homes and establish a clean and sanitary situation.
- Immediately investigate all adverse events in nursing homes and other healthcare facilities to determine the cause of the event and how best to correct it. Do not assess blame but rather use the situation as a teaching moment to correct the problems now and into the future for all patients.
- Establish a comprehensive medication program supervised by registered nurses and evaluate each patient individually and have a physician determine appropriate quantity, schedule for delivery, and type of medication which will be combined with all other medications taken by the individual. Report side effects immediately to supervisors for action.
- Highlight on the top sheet of the patient chart special needs including fluid and electrolyte maintenance, kidney injury or insufficiency, embolisms, pressure ulcers, skin breakdown of any type, surgical site infections, urinary tract infections, respiratory tract infections, other infections, etc. Train all personnel to check these situations each time they see the person and report any changes immediately to supervisory personnel.
- Train all personnel in methods of fall prevention. Determine if medical treatment, the patient's physical condition, or environmental changes are the cause of the problem. Provide appropriate rehabilitation activities and supervision when the individual might most frequently fall. (See endnote 51.)
- Conduct routine surveillance of all patients for infections including available microbiological data. If an individual has a serious infection problem and it may spread to others, utilize appropriate isolation techniques or remove the individual to an acute care hospital for treatment.
- Teach all personnel preventive health care to keep these workers from becoming infected, or spreading the disease to other patients.
- Conduct a comprehensive survey of patient safety and quality care by using the "Nursing Home Survey on Patient Safety" and utilize the results to make changes at the facility. (See endnote 53.)
- Utilize the Occupational Hazards in Long-Term Care: Nursing Home eTool provided by OSHA to determine: potential hazards and possible solutions for healthcare-wide hazards such as blood-borne pathogens; ergonomic problems facing employees; and safety and health problems in the dietary department, laundry, nurses' station, pharmacy, housekeeping department, and maintenance department. Also utilize the etools established for workplace violence, helping patients in the whirlpool and shower, and in dealing with tuberculosis and Legionnaires' disease. (See endnote 52.)

LAWS, RULES, AND REGULATIONS

NEEDLESTICK SAFETY AND PREVENTION ACT

The Needlestick Safety and Prevention Act revised the OSHA standard regulating occupational exposure to blood-borne pathogens. This includes HIV, hepatitis B virus, and hepatitis C virus. It helps to reduce healthcare workers' exposure to blood-borne pathogens by increasing requirements on employers, especially in hospitals, regarding sharps procedures and potential blood-borne pathogen contamination of the worker.

CLEAN WATER ACT

The function of the Clean Water Act is to restore and maintain the physical, biological, and chemical integrity of the surface waters of the United States, whether the institution discharges directly to the waterways or to publicly owned treatment works. The EPA has established specific standards for healthcare facilities for their effluents, stormwater, and oil pollution prevention requirements. There are different permits required for direct wastewater discharge to surface waters or indirect wastewater discharge to treatment plants.

SAFE WATER DRINKING ACT

The Safe Drinking Water Act mandates that the US EPA establishes regulations to protect human health from contaminants in drinking water. This has resulted in the development of primary and secondary drinking water standards. The hospital is considered to be a non-transient, non-community water system and therefore must comply with the Safe Water Drinking Act.

RESOURCE CONSERVATION AND RECOVERY ACT

The Resource Conservation and Recovery Act manages the disposal of wastes from municipalities and industries. It regulates how facilities generate, transport, treat, store, or dispose of hazardous waste. Under this law, most hospitals are hazardous waste generators and therefore must comply with all rules and regulations. There are different levels of generators depending on the size of the institution and specific requirements for the particular size. There are standards for: identifying solid and hazardous waste; generators of hazardous waste; transporters of hazardous waste; land disposal restrictions for hazardous waste; used oil management standards; underground storage tanks; boilers and industrial furnaces; and imminent hazards.

EMERGENCY PLANNING AND COMMUNITY RIGHT TO KNOW ACT

This act is also known as the Superfund Amendments and Reauthorization Act. It is intended to give citizens, local governments, and local response authorities information about the potential hazards in their community. Title III of this act (better known as EPCRA) requires emergency planning and designates state and local governments to receive information about certain chemicals in the community.

MEDICAL WASTE TRACKING ACT

This act was a 2-year demonstration program for medical waste tracking from June 1989 to June 1991. The program expired and there are no federal tracking requirements in place now. However, many states have developed tracking and management programs.

CLEAN AIR ACT

The Clean Air Act and its amendments are there to protect and enhance the nation's air resources to protect public health and welfare and encourage the production of goods and services in such a manner that it will not be harmful. Under this act many institutions have to obtain permits for their air emissions. Most hospital boilers and medical waste incinerators fall under these standards. Hazardous air pollutants including asbestos and those coming from boilers and process heaters as well as chemical accidents and their prevention, are covered by federal regulations. There is a refrigerant recycling rule for maximum recovery and recycling of refrigerants so they do not escape to the air.

TOXIC SUBSTANCES CONTROL ACT

The Toxic Substances Control Act creates a regulatory framework for chemicals to evaluate, assess, mitigate, and control risk from manufacturing to processing to alternate use. This includes: lead hazards from paint, dust, and soil; polychlorinated biphenyls regulations and how to handle waste; and asbestos regulations for use, removal, and disposal. Hospitals and other healthcare institutions are clearly covered by this law.

FEDERAL INSECTICIDE, FUNGICIDE, AND RODENTICIDE ACT

The Federal Insecticide, Fungicide, and Rodenticide Act and its amendments gives the EPA the authority to enforce rules and regulations concerning the registration, distribution, sale, and use of pesticides. This also includes antimicrobials which are used extensively on surfaces in healthcare settings. The CDC recommends strongly that healthcare institutions use EPA-registered products for disinfecting surfaces or sterilizing or disinfecting medical equipment and medical facilities.

FEDERAL HAZARDOUS MATERIALS TRANSPORTATION LAW

This law requires institutions to meet specifications for hazard communications, packaging requirements, training in the handling of biohazardous wastes, etc.

NUCLEAR REGULATORY COMMISSION

The Nuclear Regulatory Commission under the authority of the Atomic Energy Act has authority in the areas of nuclear medicine, radiation therapy, and research. They also make determinations of how radioactive material as well as of radioactive wastes will be handled.

OCCUPATIONAL SAFETY AND HEALTH ADMINISTRATION

The Occupational Safety and Health Administration has numerous regulations and standards established for prevention of disease and injuries in workers. Hospitals and other healthcare institutions must comply.

RESOURCES

AGENCY FOR HEALTHCARE RESEARCH AND QUALITY, US DEPARTMENT OF HEALTH AND HUMAN SERVICES

This agency works in numerous areas including healthcare-acquired infections. They provide funding for special research projects to determine how best to prevent and then control healthcare-acquired infections. Examples of the grant topics include Barriers and Challenges for Preventing HAIs in 34 Hospitals; Initiative Examines Tools and Interventions to Assist Hospitals in Reducing HAIs; Hand Hygiene is Important for Preventing HAIs; Testing Spread and Implementation of MRSA-Reduction Practices; etc. (See endnote 57.)

CENTERS FOR DISEASE CONTROL AND PREVENTION

Healthcare-associated infections monitoring systems and specialized training materials from the CDC are essential to the work of the modern environmental health and infection control specialists, since they provide rapid information on disease problems. They are as follows:

- Active Bacterial Core Surveillance, ABCs (cdc.gov/abcs/index.html). This surveillance system is used to determine and record invasive bacterial pathogens of public health importance.
- Emerging Infections Program, EIP (cdc.gov/hai/eip/). This surveillance system and research network made up of 10 state health departments and academic units works with CDC in developing innovations in surveillance, evaluating epidemiology techniques, and evaluating practices in all areas of health care.
- The Healthcare Infection Control Practices Advisory Committee is a federal advisory committee whose function is to advise and guide the CDC and the Secretary of the US Department of Health and Human Services on Best Practices for surveillance, prevention, control of healthcare-associated infections, and antimicrobial resistance.
- The National Healthcare Safety Network, NHSN (cdc.gov/nhsn/). This very secure internet public health surveillance system is very frequently used by all types of healthcare facilities in all 50 states, collecting data on patient safety and employee safety including injuries and infections to: help establish appropriate surveillance techniques; determine the amount and type of healthcare-associated infections in the healthcare facility, similar types of healthcare facilities, state or region and the epidemiological pattern of the causative agents; see the facilities healthcare-associated infection data in real time; use appropriate measures to reduce the problems; establish priorities for eliminating these infection problems; comply with the state and federal reporting rules. The CDC through the National Healthcare Safety Network provides a series of training modules relating to Best Practices for various procedures, use and cleaning of medical devices, antimicrobial use and resistance including *C. difficile*. It provides an in-depth series of reports from current and previous data. This network provides information for healthcare personnel on problems related to blood/body fluids exposure and management of body fluids, as well as influenza exposure and vaccination. It provides modules for insertion practices and blood-stream infections as well as discussions on problems of infections related to ventilators, catheters, dialysis treatments, surgery, and anti-microbial use and multidrug resistance.
- Surveillance for Emerging Antimicrobial Resistance Connected to Healthcare (S.E.A.R.C.H.). This surveillance system is made up of a group of voluntary participants from hospitals, state health departments, professional organizations, and clinical microbiology laboratories that are reporting the isolation of *S. aureus* which is suspected of being resistant to vancomycin.
- CDC Healthcare Infection Control Practices Advisory Committee General Guidelines provides internet sites for specific guidelines on: disinfection and sterilization in healthcare facilities; isolation precautions; environmental infection control; hand hygiene; public reporting of healthcare-associated infections; device-associated infection prevention; procedure-associated infection prevention; prevention and control of drug-resistant organisms; a norovirus prevention toolkit; multi-drug-resistant organisms; and healthcare personnel guidelines.
- CDC Healthcare Infection Control Practices Advisory Committee General Guidelines presents resources on: Biological Hazards and Control; Physical Hazards and Controls; Safety; Slips, Trips, and Falls; Violence; Reproductive Health; Dentistry; Emergency Preparedness and Response; CDC Emergency Preparedness and Response; and Surveillance and Statistics.
- The National Health Worksite Program helps employers implement science-based Best Practices in prevention and control of chronic illnesses and disabilities due to employment.

- Emergency Department Ebola Preparedness Training Videos include specialized modules: Considerations for Preparedness; Screening Patients for Ebola Risk Factors and Symptoms; Isolation of a Patient with Ebola Risk Factors and Symptoms; and Evaluate and Briefly Manage Patients: Ebola Assessment Hospitals.

The *Office of Public Health Preparedness and Response*, which is part of the CDC, is the lead agency working with local and state health departments to save lives and protect communities from public health threatening situations. It operates the CDC Emergency Operations Center. It provides strategic direction, teams of experts, support services, coordination of all activities at the local, state, tribal, national, and international levels, funding, and technical assistance. It is involved in the strategic national stockpile preparedness system, and works with select agents and toxins. It helps with emergency operations, education, and training. (See endnote 45.)

The National Institute for Occupational Safety and Health (NIOSH) provides a website (cdc.gov/niosh/) with a variety of topics including biological, chemical, and physical hazards, and controls, etc. The latest information in these areas, and rules and regulations are presented.

The *Occupational Safety and Health Administration* (OSHA) mission is to save lives, prevent injuries, and protect the health of the workers of the United States. It is involved in every workplace within the country. Hospitals have to meet a large number of standards established by OSHA covering: hospital investigations: health hazards; blood-borne pathogens standard; healthcare-wide hazards—electrical; fire prevention plans; use of fire extinguishers; cardiopulmonary resuscitation (CPR) and self-contained breathing apparatus; chemical spill control procedures; search and emergency rescue procedures; hazardous materials emergency response; emergency communications; communication programs for employers using hazardous chemicals; etc.

Along with the standards that are issued are numerous documents in Best Practices in all areas within healthcare institutions. Some of these documents cover the following departments: administration, central supply, clinical services, dietary, emergency, engineering, housekeeping, ICU, laboratory, laundry, pharmacy, and surgical suite. Common topics and potential solutions are given for the following areas: ergonomics (reaching, lifting, repetitive motions, etc.); equipment and machine guards; tire safety; hazardous chemicals; healthcare-associated infections and infectious diseases; slips, trips, and falls; electrical safety; and infectious materials.

The *European Agency for Safety and Health at Work* has produced a document entitled "Occupational Health and Safety Risks in the Healthcare Sector-Guide to Prevention and Good Practice" (see endnote 63) which discusses the following six major topics: management's role in health prevention and promotion; how to perform a risk assessment; biological risks; musculoskeletal risks; psychosocial risks; and chemical risks. This guide presents up-to-date technical and scientific knowledge concerning the prevention and control of various occupational risks.

The *Society for Healthcare Epidemiology of America* provides a compendium of a variety of strategies used to prevent healthcare-associated infections. The document is entitled, "Compendium of Strategies to Prevent Healthcare-Associated Infections in Acute Care Hospitals." (See endnote 17.)

The Joint Commission is a not-for-profit independent agency that accredits more than 20,500 healthcare organizations throughout the country. Without this accreditation hospitals and other healthcare facilities have great difficulty in operating. Their mission is to improve health care for the public by evaluating healthcare organizations and providing a variety of training modules and Best Practices papers to improve quality of care. Healthcare facilities must receive on-site evaluations every 3 years and laboratories every 2 years. Many of the documents and standards established by The Joint Commission are available for use by healthcare facilities to upgrade their existing protocols and practices.

The Joint Commission establishes health and safety standards based on OSHA requirements. These standards are in several categories: management leadership; employee involvement; worksite analysis; hazard prevention and control; safety and health training; and annual evaluation.

The *Institute for Healthcare Improvement* of Cambridge, Massachusetts, has three critical objectives: improving the health of certain populations, enhancing patient care, and reducing or controlling the cost of health care. They work in conjunction with numerous other health organizations including the CDC, Association for Professionals in Infection Control and Epidemiology, public health systems, healthcare delivery systems, the community, the state, universities, and private organizations to improve health care including the control of healthcare-associated infections as is shown in their "How-to Guide: Improving Hand Hygiene." (See endnote 64.)

PROGRAMS

In the 1960s and 1970s, the CDC and the National Environmental Health Association, Hospital Sanitation Committee were leaders in developing new techniques to prevent disease and injury in institutions. The CDC as part of its mission of improving the knowledge of existing professionals in the field of environmental health have presented training to professionals shortly after its founding on July 1, 1946, in a variety of areas including healthcare-associated infection control. The National Committee on Hospital Sanitation brought together a group of environmental health specialists who were working in various areas of the hospital environment. This committee, the forerunner of the technical sections of the National Environmental Health Association, served as a place for individuals throughout the country to share information and work on areas of the environment that needed further improvement. The chairperson of the committee from 1963 to 1967, the author of this book, developed many unique programs to control environmental problems and help prevent disease and injury at Philadelphia General Hospital. The committee acted as a sounding board for many of the ideas and willingly gave input to turn ideas into reality. Many of the self-inspection forms and a discussion on continuing education programs for food service and also housekeeping supervisors may be found in Chapter 9, Institutional Environment in the *Handbook of Environmental Health—Biological, Chemical, and Physical Agents of Environmentally Related Diseases*, Volume 1, Fourth Edition (CRC Press, Boca Raton, FL, 2003). (See endnote 65.)

SUCCESSFUL HOSPITAL HEALTH AND SAFETY PROGRAMS

(See endnote 46)

University Medical Center at Brackenridge, Austin, Texas

A board member of this institution in 2005 asked how safe were employees and patients. After a study and comparison with other similar institutions, University Medical Center at Brackenridge determined that they were not among the best and immediately started to inquire how best to upgrade their program areas to become the best. The Board of Trustees and the chief operating officer developed a new safety culture to provide all employees from managers to field staff levels with the necessary tools, resources, authority, and accountability to integrate safety into every one of the daily activities. All individuals received intense training in safety problems and solutions for patients and employees.

St. Thomas Midtown Hospital, Nashville, Tennessee

The institution developed a program where employees were encouraged by leadership to consider safety to be a core business value. At the beginning of each day the chief operating officer first discusses safety and how to implement information found in case incident reports and solutions to the incidents. She also discusses what is going right and how to encourage it as well as what has gone wrong and how to correct it. As part of orientation for new employees, she stresses the importance and significance of safety and how best to work with your colleagues to prevent injuries and illness to staff and patients. Every other week senior managers make rounds throughout their departments to determine for themselves safe and unsafe practices in the institution.

Because of the profusion of different ethnic groups coming to the hospital and working for the hospital, there was a need to speak 17 different native languages. The hospital hired individuals who could speak each of these languages and then started a buddy system with individuals who spoke English. When new information needs to be dispersed, it is done rapidly and accurately.

Because of The Joint Commission accreditation all hospitals have to conduct annual reviews of programs and develop goals. St. Thomas utilizes this annual process as a way of integrating worker safety into the existing program of self-improvement. Each year the safety manager proposes a set of safety goals which is reviewed by various groups. These goals and objectives which are short-term and long-term, require continuous innovation and application of Best Practices to reduce employee and patient injury and illness.

Blake Medical Center, Bradenton, Florida

The chief executive officer encourages all employees to email him with questions, concerns, and suggestions including areas of safety. He receives more than 500 of these each month and after consulting with his senior team, he answers them personally. Many worthwhile ideas including those about safety are brought to his attention by the staff and he shares these with the management team for implementation. The money which has been saved by this hospital program is utilized to purchase new equipment while making the workplace safer.

The safety committee is made up not only of the safety professionals, and management and supervision personnel but also of 28 general employees from all departments of the hospital. One quarter of the committee members are replaced with new individuals each year, allowing for new ideas and new solutions. This allows every department and major unit to have representation on the committee and therefore their concerns are brought to the attention of the proper individuals and specific recommendations can be made to reduce problems that lead to disease and injury for employees and patients.

Precise data are collected on all incidents, especially on falls, to determine how they occurred and what interventions can be utilized to improve the situation and prevent injuries. This led to a decrease in falls in parking lots, when the bumpers that the car pulls up to were painted yellow. Also they discovered that 40% of the patient management injuries came from moving a patient up in bed. This was corrected by buying beds that can move, lift, and recline in more directions.

Lima Memorial Health System, Lima, Ohio

This institution joined the OSHA initiative called the Voluntary Protection Program. This program is for private industry, federal agencies, and local and state agencies and its function is to prevent injuries and illnesses in the workplace by using worksite analysis, hazard prevention and control, intensive training and retraining where necessary, and excellent cooperation between management and staff. Because of the size of the community many of the individuals working at the institution know each other and each other's families. They participate in many of the same extracurricular activities. This makes it far easier for individuals to discuss their concerns about safety at the institution and to share the passion for excellence in patient care and protection for the staff. All incidents that are investigated are examined with the thought of helping correct problems instead of blaming someone for what has occurred.

All new employees go through an intensive program prior to assignment. Safety is constantly taught and reinforced during this program. Refresher training, videos, demonstrations, handouts, and reminders are utilized to reinforce the message of safety at the institution.

St. Vincent's Medical Center, Bridgeport, Connecticut

This institution trusts their safety staff and workforce to carry out studies of potential hazards and existing hazards and make sound recommendations for correction. The managers understand that the best recommendations come from those working within the areas where the specific procedures and equipment are being used. One such recommendation from a staff member was to use a safe enclosure bed instead of a sitter. This saved the institution a considerable amount of money and also reduced the number of falls of patients who typically wandered, were confused, or were aggressive.

A new program was established where staff members assumed responsibility for their colleagues and others. Now there are many pairs of eyes checking such things as working an excessive number of hours, which can result in poor decisions, health problems and injuries, or individuals not using appropriate hand-washing techniques, etc. This is extremely effective because all employees are looking out for the welfare of everyone else including patients.

St. Vincent's has improved their communication of safety concerns by issuing safety alerts whenever something unusual occurs. For instance, when four sharp injuries were reported in a 6-day period instead of the usual one or two, an immediate alert went out to the entire hospital about the situation, reminding them of proper procedures for the use and disposal of sharps.

Because there is so much electronic equipment being used and cords can result in tripping and other types of injuries, a new program has been started where all excess cords must be tied together. Reaching for supplies on high shelves has caused falls and therefore step stools have now been put in appropriate places for employee use. The current locations of all hand-washing dispensers were evaluated and many of them were changed to new locations or there have been increases in hand dispensers to meet specific needs. Mobile patient equipment requires the use of heavy batteries which were causing injuries when they were being recharged. This was corrected by lowering the outlets to the level of the batteries.

Cincinnati Children's Hospital, Cincinnati, Ohio

In the 1990s, this institution became one of the leaders in workplace safety. The chief executive officer and chairman of the Board of Trustees believed in transparency, sharing data, and long-term improvement especially as related to patient safety and care. Employee safety was integrated into the same program as patient safety, which provided for a much more efficient operation and reduced incidents of injury and illnesses to employees and patients. Each day a safety officer is designated for the day and there are two check-in meetings with representatives from all departments of the hospital to discuss problems of employee and patient safety. The top administrators established ambitious safety goals and everyone participated from physicians and surgeons to all levels of personnel. Outside consultation and information is always sought to try to further improve the practices and procedures of the hospital.

This hospital, which specializes in pediatrics, has special concerns about violent behavior by children. Existing medical problems may be compounded by severe psychological issues and the fact that children do not understand at a very young age the consequences of their actions. The hospital has implemented the use of Kevlar sleeves, which protects the arms, wrists, and palms against cuts, abrasions, and bruising, when working with patients. This has reduced the number of injuries to the staff and at one point the staff went over 200 days without a recorded injury. Sharps injuries due to children squirming have been reduced substantially by utilizing a parent to hold the child in a comfortable position while keeping him/her stable. Slings, lifts, and other equipment are in each of the patient rooms to help the staff prevent back injuries and other injuries.

Tampa General Hospital, Tampa, Florida

The hospital administration supported the development of an injury prevention program in 2000. A full-time physical therapist was hired to evaluate employee injuries and establish a hospital-wide program to prevent and control situations in which injuries occurred. The hospital established a two-person lift team and specialized equipment to lift and transfer patients. This has resulted in a substantial decrease in employee injuries especially among the nursing staff.

The lift teams resulted in a 65% decrease in patient handling injuries and a 90% reduction in lost workdays for employees, with considerable savings to the hospital. Part of their job is to continuously evaluate the equipment to make sure that it is maintained properly and the procedures being used. Team members are permitted to devise new techniques when there is a complex lifting situation.

Washington State Department of Social and Health Services, Olympia, Washington

This department provides numerous services for hospitals and other healthcare situations, one of which is a proactive accident prevention program to reduce the number of accidents, injuries, and illnesses throughout the hospital. The program consists of the following:

- Prompt and thorough investigation of all incidents resulting in injuries and illnesses and determining how to prevent them from occurring in the future
- Continuous evaluation of injuries and healthcare-associated infections by reviewing all incident reports and prevention recommendations and sharing them with all personnel
- Continuous assessments of the potential for environmental problems contributing to injuries and healthcare-associated infections and risk assessments during daily rounds evaluating the care of patients
- An ergonomic assessment program to prevent repetitive injuries or illnesses to employees
- A return to work program for employees who have been injured and checks that the environmental problems have been corrected
- Provision of a fire and evacuation plan for the institution and testing it frequently

ENDNOTES

1. Koren, Herman, Bisesi, Michael. 2003. *Handbook of Environmental Health: Biological, Chemical, and Physical Agents of Environmentally Related Disease*. CRC Press, Boca Raton, FL.
2. Klevens, Monina, Edwards, Jonathan R., Richards, Chelsea L. 2007. Estimating Healthcare-Associated Infections and Deaths in US Hospitals, 2002. *Public Health Reports*, 122(2):160–166.
3. Centers for Disease Control and Prevention. 2012. *Top CDC Recommendations to Prevent Healthcare-Associated Infections*. Atlanta, GA.
4. Ducel, G., Fabry, J., Nicolle, L., editors. 2002. *Prevention of Hospital-Acquired Infections: A Practical Guide*. WHO/CDS/EPH/2002. World Health Organization, Malta.
5. Joint Commission on Accreditation of Healthcare Organizations. 2009. *Planning, Design, and Construction of Healthcare Facilities*, Second Edition. Oakbrook Terrace, IL.
6. Ulrich, Roger, Quan, Xiaobo, Zimring, Craig. 2004. *The Role of the Physical Environment in the Hospital of the 21st Century: A Once-in-a-Lifetime Opportunity*. Report to the Center for Health Design for the Designing the 21st Century Hospital Project. Robert Wood Johnson Foundation, Concord, CA.
7. Methodist Health System. 2014. *Construction, Renovation, Repair, Contractor's Orientation Handbook*. Omaha, NE.
8. Kaplan, Susan, Orris, Peter, Machi, Rachel. 2009. *A Research Agenda for Advancing Patient, Worker and Environmental Health and Safety in the Healthcare Sector*. University of Illinois at Chicago School of Public Health, Chicago, IL.
9. Joseph, Anjali. 2006. *Impact of the Environment on Infections in Healthcare Facilities* The Center for Health Design, Concorde, CA.
10. Steinberg, James P., Denham, Megan E., Zimring, Craig. 2013. The Role of the Hospital Environment in the Prevention of Healthcare-Associated Infections by Contact Transmission. *Health and Environment Research and Design Journal*, 7(1 Suppl):46–73.
11. US Environmental Protection Agency, Office of Compliance, Office of Enforcement and Compliance Assurance. 2005. *Profile of the Healthcare Industry-Sector Notebook Project*. EPA/310-R-05-002. Washington, DC.
12. Facility Guidelines Institute. 2014. *2014 Guidelines for Design and Construction of Hospitals and Outpatient Facilities*. Dallas, TX.
13. Facility Guidelines Institute. 2014. *2014 Guidelines for Construction of Residential Healthcare, and Support Facilities*. Dallas, TX.
14. Centers for Disease Control and Prevention, National Center for Emerging and Zoonotic Infectious Diseases, Division of Healthcare Quality Promotion. 2014. *General Guidelines*. Atlanta, GA.
15. Rutala, William A., Weber, David J. 2008. *Guideline for Disinfection and Sterilization in Healthcare Facilities*. Healthcare Infection Control Practices Advisory Committee, Atlanta, GA.
16. World Health Organization. 2004. *Practical Guidelines for Infection Control in Healthcare Facilities*. Environmental Management Practices, Geneva, Switzerland.

17. The Society for Healthcare Epidemiology of America. 2014. *Strategies to Prevent HAIs: Compendium of Strategies to Prevent Healthcare-Associated Infections in Acute Care Hospitals.* Arlington, VA.

18. US Department of Health and Human Services, Centers for Disease Control and Prevention. 2003. *Guidelines for Environmental Infection Control and Health-Care Facilities: Recommendations of CDC in the Healthcare Infection Control Practices Advisory Committee.* Atlanta, GA.

19. Institute for Healthcare Improvement. No Date. *How-to Guide: Improving Hand Hygiene: A Guide for Improving Practices among Healthcare Workers.* Cambridge, MA.

20. Siegel, JD, Rhinehart, E, Jackson, M. 2007. *2007 Guideline for Isolation Precautions: Preventing Transmission of Infectious Agents in Healthcare Settings.* Healthcare Infection Control Practices Advisory Committee, Centers for Disease Control and Prevention, Atlanta, GA.

21. Centers for Disease Control and Prevention. 2014. *Guidance on Personal Protective Equipment to Be Used by Healthcare Workers during Management of Patients with Ebola Virus Disease in US Hospitals, Including Procedures for Putting on (Donning) and Removing (Doffing).* Atlanta, GA.

22. Centers for Disease Control and Prevention. 2015. *Interim Guidance for Environmental Infection Control in Hospitals for Ebola Virus.* Atlanta, GA.

23. Francis J Curry National Tuberculosis Center. 2007. *Tuberculosis Infection Control: A Practical Manual for Preventing TB.* San Francisco, CA.

24. The White House. 2014. *National Strategy for Combating Antibiotic Resistant Bacteria.* Washington, DC.

25. The White House, Office of the Press Secretary, Executive Order. 2014. *Combating Antibiotic-Resistant Bacteria.* Washington, DC.

26. Selvay, Linda A., Artillo, Catherine, Hall, Robert. 2015. Evaluation of Border Entries Screening for Infectious Diseases in Humans. *Emerging Infectious Disease Journal,* 21(2):197–201.

27. Hayden, Mary K., Lin, Michael Y., Lolens, Karen, Weiner, Shayna, Blom, Donald, Moore, Nicholas M., Fogg, Louis, et al. 2015. Prevention of Colonization and Infection by *Klebsiella pneumoniae* Carbapenemase-Producing Enterobacteriaceae in Long-Term Acute-Care Hospitals. *Clinical Infectious Diseases,* 60(8):1153–1161.

28. Rapose, Alwyn. 2013. Travel to Tropical Countries: A Review of Travel-Related Infectious Diseases. *Tropical Medicine and Surgery,* 1(128).

29. Gupta, Neil, Limbago, Brandi M., Patel Jean B. 2011. Carbapenem-Resistant Enterobacteriaceae: Epidemiology and Prevention. *Clinical Infectious Diseases,* 53(1):60–67.

30. Centers for Disease Control and Prevention. 2012. Fatal Foodborne *Clostridium perfringens* Illness at a State Psychiatric Hospital: Louisiana 2010. *Morbidity and Mortality Weekly Report,* 61(32):605–608.

31. Working Group on Civilian Bio Defense. 2001. Botulism Toxin as a Biological Weapon: Medical and Public Health Management. *Journal American Medical Association,* 285(8):1059–1070.

32. Inglesby, Thomas V., Dennis, David T., Henderson, Donald A., Bartlett, John G., Ascher, Michael S., Eitzen, Edward, Fine, Anne D., et al. 2000. Plague as a Biological Weapon: Medical and Public Health Management. Working Group on Civilian Bio Defense. *Journal American Medical Association,* 283(17):2281–2290.

33. Centers for Disease Control and Prevention. 2003. *CDC-Smallpox Response Plan and Guidelines, for Distribution to State and Local Public Health Bioterrorism Response Planners.* Atlanta, GA.

34. Dennis, David T., Inglesby, Thomas V., Henderson, Donald A., Bartlett, John G., Ascher, Michael S., Eitzen, Edward, Fine, Anne D., et al. Consensus Statement: Tularemia As a Biological Weapon: Medical and Public Health Management. *Journal American Medical Association,* 285(21):2763–2773.

35. Centers for Disease Control and Prevention. 2013. *Virus Families: Viral Hemorrhagic Fevers.* Atlanta, GA.

36. US Department of Health and Human Services, National Institutes of Health, National Institute of Allergy and Infectious Diseases. 2003. *NIAID Biodefense Research Agenda for Category B and C Priority Pathogens.* NIH publication number 03-5315. Bethesda, MD.

37. Iowa State University, College of Veterinary Medicine, the Center for Food Security and Public Health. 2007. *Anthrax.* Ames, IA.

38. Department of Homeland Security, Interagency Security Committee. 2012. *Best Practices for Mail Screening and Handling Processes: A Guide for the Public and Private Sectors,* First Edition. Washington, DC.

39. The Association for Professionals in Infection Control and Epidemiology. 2012. *Guide to Infection Prevention in Emergency Medical Services: APIC Implementation Guide.* Washington, DC.

40. Rangel, Josefa M., Sparling, Phyllis H., Crowe, Collen, Griffin, Patricia M., Swerdlow, David L. 2005. Epidemiology of *Escherichia coli* 0157: H7 Outbreaks, United States, 1982–2002. *Emerging Infectious Disease Journal,* 11(4):603–609.

41. Department of Health and Human Services, Office of Inspector General. 2010. *Adverse Events in Hospitals: National Incidence among Medicare Beneficiaries*. Daniel R Levinson, Inspector General, OEI-06-09-00090. Washington, DC.
42. Institute for Healthcare Improvement. 2005. *Leadership Guide to Patient Safety: Resources and Tools for Establishing and Maintaining Patient Safety*. Cambridge, MA.
43. Miller, Kristine M. 2012. *Improving Patient and Worker Safety: Opportunities for Synergy, Collaboration and Innovation*. Chapter 2-Management Principles, Strategies, and Tools That Advance Patient and Worker Safety and Contribute to High Reliability and Chapter 3-Specific Examples of Activities and Interventions to Improve Safety. The Joint Commission, Oakbrook Terrace, IL.
44. Centers for Disease Control and Prevention. 2014. *Anti-Neoplastic Agents, Occupational Exposure to Anti-Neoplastic Agents and Other Hazardous Drugs*. Atlanta, GA.
45. Centers for Disease Control and Prevention. 2014. *Office of Public Health Preparedness and Response: Overview*. Atlanta, GA.
46. Occupational Safety and Health Administration. 2013. *Safety and Health Management Systems: A Roadmap for Hospitals*. Washington, DC.
47. Vincent, Grayson K., Velkoff, Victoria A. 2010. *The Next Four Decades: The Older Population in the United States: 2010 to 2050*, P25-1138. US Department of Commerce, Economics and Statistics Administration, US Census Bureau, Washington, DC.
48. Gershon, Robin R.M., Pogorzelska, Monica, Qureshi, Kristine A. 2008. Home Healthcare Patients and Safety Hazards in the Home: Preliminary Findings. In *Advances in Patient Safety: New Directions and Alternative Approaches* (Volume 1: Assessment). Battles, Henriksen K., Keys, J.B., editors. National Library of Medicine, National Institutes of Health, NCBI Bookshelf, Rockville, MD.
49. US Department of Health and Human Services, Office of Inspector General. 2014. *Adverse Events and Skilled Nursing Facilities: National Incidence among Medicare Beneficiaries*. Daniel R Levinson Inspector General, OEI-06-11-00370. Rockville MD.
50. Mathei, Catharina, Niclaes, Luc, Suetens, Carl. 2007. Infections in Residents of Nursing Homes. *Infectious Disease Clinics of North America*, 21:761–772.
51. Centers for Disease Control and Prevention. 2015. *Falls in Nursing Homes*. Atlanta, GA.
52. US Department of Labor, Occupational Safety and Health Administration. No Date. *Occupational Hazards in Long Term Care Nursing Homes*. Washington, DC.
53. Agency for Healthcare Research and Quality. 2014. Nursing Home Survey on Patient *Safety*. Rockville, MD.
54. Australian Day Surgery Council of Royal Australian College of Surgeons. 2004. *Day Surgery in Australia: Report and Recommendations of the Australian Day Surgery Council*. Melbourne, VIC.
55. Joint Commission. No Date. *Accredited Ambulatory Healthcare Centers*. Oakbrook Terrace, IL.
56. Centers for Disease Control and Prevention. 2011. *Guide to Infection Prevention for Outpatient Settings: Minimum Expectations for Safe Care*. Atlanta, GA.
57. Agency for Healthcare Research and Quality, US Department of Health and Human Services. 2009. *Fact Sheet: AHRQ's Efforts to Prevent and Reduce Healthcare-Associated Infections*. Rockville, MD.
58. Columbus Public Health. 2014. *Policy and Procedure for Ebola Virus Disease*. Columbus, OH.
59. Huges, Rhonda G., editor. 2008. Patient Safety and Quality in Home Health Care. In *Patient Safety and Quality: An Evidence-Based Handbook for Nurses*. Ellenbecker, Carol Hall, Samia, Linda, Cushman, Margaret J., editors. Agency for Healthcare Research and Quality, Rockville, MD.
60. The Joint Commission. 2015. *Joint Commission FAQ*. Oakbrook Terrace, IL.
61. Koren, Herman, Bisesi, Michael. 2003. *Handbook of Environmental Health–Biological, Chemical and Physical Agents of Environmentally Related Disease*, V1, 4th edition 526–527. Boca Raton, FL: Lewis Publishers–CRC Press.
62. Koren, Herman, Bisesi, Michael. 2003. *Handbook of Environmental Health–Biological, Chemical and Physical Agents of Environmentally Related Disease*, V1, 4th edition. 528–529. Boca Raton, FL: Lewis Publishers–CRC Press.
63. European Commission, Directorate-General for Employment, Social Affairs and Inclusion. 2011. *Occupational Health and Safety Risks in the Healthcare Sector–Guide to Prevention and Good Practice*. European Agency for Safety and Health at Work. Available: https://osha.europa.eu/en/legislation/guidelines/occupational-health-and-safety-risks-in-the-healthcare-sector-guide-to-prevention-and-good-practice.
64. Institute for Healthcare Improvement. No Date. *How-to Guide: Improving Hand Hygiene: A Guide for Institute for Healthcare Workers*. Cambridge, MA.
65. Koren, Herman, Bisesi, Michael. 2003. Handbook of Environmental Health: Biological, Chemical, and Physical Agents of Environmentally Related Disease. CRC Press, Boca Raton, FL.

BIBLIOGRAPHY

Association of State and Territorial Health Officers, Centers for Disease Control and Prevention. 2011. *Eliminating Healthcare Associated Infections: State Policy Options.* Arlington, VA.

Centers for Disease Control and Prevention. 2011. *Monitoring Health Care-Associated Infections.* Atlanta, GA.

Centers for Disease Control and Prevention. 2012. *Diseases and Organisms in Healthcare Settings.* Atlanta, GA.

DeLeo, Frank R., Chambers, Henriette. 2009. Reemergence of Antibiotics-Resistant *Staphylococcus aureus* in the Genomics Era. *Journal of Clinical Investigation,* 119(9):2464–2474.

Diab-Elschahawi, Magda, Assadian, Ojan, Blacky, Alexander. 2010. Microfiber Compared with Other Commonly Used Cleaning Cloths. *American Journal of Infection Control,* 38(4):289–292.

HAIs Declined in 2010, CDC Data Show. 2012. Virgo Publishing Company, Phoenix, AZ.

New York State Department of Health. 2007. *Health Advisory: Prevention and Control of Community Associated Methicillin-Resistant Staphylococcus Aureus.*

Scott, Douglas R. 2009. *The Direct Medical Costs of Health Care-Associated Infections in US Hospitals and the Benefits of Prevention.* Center for Disease Control and Prevention, Atlanta, GA.

US Department of Health and Human Services, Office of Disease Prevention and Health Promotion. 2015. *National Action Plan to Prevent Healthcare-Associated Infections: Roadmap to Elimination.* Rockville, MD.

9 Insect Control, Rodent Control, and Pesticides

STATEMENT OF PROBLEM AND SPECIAL INFORMATION

Insects and rodents are one of the primary means of spreading disease throughout the world. Typically, a microorganism causing a specific disease lives in a wild animal, bird, or human being. This is called a reservoir of infection. A vector (an insect or rodent) then transmits the disease from the reservoir of infection to a new animal, bird, or human being, called the host. Yellow fever, a mosquito-borne disease, evolved in Africa about 3000 years ago. Malaria, a mosquito-borne disease, was described in ancient Chinese medical writings in 2700 BC. It was responsible for the destruction of many of the city states. It was known about 2000 years ago in Roman times and in fact was called Roman fever. Bubonic plague, involving Norway rats and fleas, was first traced to Asia in the 1330s and 1340s. It spread to Sicily and the rest of Europe over a period of time. Bubonic plague, also called the Black Death, was the most devastating pandemic in history. It resulted in the deaths of an estimated 75–200 million people and had its peak years from 1346 to 1353. (See endnote 4.)

Each of the diseases mentioned was devastating to the populations at the time and later in various parts of the world. However, this was not the end of worldwide epidemics. The most recent concern is the Zika virus which is transmitted by *Aedes* species mosquitoes. Although the symptoms for those individuals who become ill are usually mild, there is a serious concern of newborns, who have had the virus transmitted to them from the mother, having microcephaly, which is a serious birth defect causing the baby's head to be smaller than expected. This is creating great fear in areas where the disease is prevalent. (See endnote 31.)

Insects and other arthropods of public health significance include bedbugs, fleas, flies, food product insects, lice, mites, mosquitoes, roaches, and ticks.

SUB-PROBLEMS INCLUDING LEADING TO IMPAIRMENT AND BEST PRACTICES FOR INSECTS AND OTHER ARTHROPODS

BEDBUGS

Bedbugs are not a vector and not known to transmit diseases, but they are pests of public health importance because they suck the blood of people. They can cause physical and mental health problems, as well as economic problems. Recently they have been spreading rapidly through the United States, Canada, the United Kingdom, and other parts of Europe. (See endnote 1.)

Best Practices in Bedbug Control (See endnotes 2, 15)
- Use the integrated pest management approach including necessary inspections with flashlights, magnifying glasses, and probes to determine how best to eliminate bedbugs.
- Do not use bug bombs or foggers, discard beds and bedding not enclosed in plastic, and discard other furniture and infested items since this may only add to the problem by causing the bedbugs to spread.
- Immediately report bedbug problems to the proper individuals for treatment and control.

- When returning home from traveling or purchasing secondhand clothing or furniture, thoroughly check all luggage, clothing, and furniture for possible bedbugs, and then wash all clothing, both dirty and clean, in a hot wash cycle and wipe down all luggage inside and out with a disinfectant that will kill mites and lice.
- Move all furniture away from walls and closely inspect the furniture, remove content from drawers, and launder all clothing and bedding in as hot a wash cycle as your machine allows for laundry. Some items may be dry cleaned and enclosed in plastic.
- Lift up the edges of all wall-to-wall carpeting and inspect carefully.
- Use compressed air or pyrethrin to flush bedbugs out of cracks and crevices.
- Do not use bleach in areas where you have used insecticides because it may cause a chemical reaction and a far more dangerous product.
- Use a vacuum cleaner with a nylon or spandex pantyhose fitted tightly over the vacuum hose to catch them and carefully remove physically all bedbugs and eggs especially in cracks and crevices. After use dispose of the nylon or spandex pantyhose immediately in a tightly sealed plastic bag. Be careful to inspect the vacuum cleaner and make sure it is not transmitting bedbugs to other areas. Dispose of vacuum bags in sealed plastic bags.
- Use steam cleaning or heat and cold treatments of materials as needed.
- Use appropriate pesticides applied by licensed, well-trained pest management personnel, when needed. However, observe the latest Centers for Disease Control and Prevention (CDC) health advisory concerning health issues related to the misuse of pesticides for bedbug control. (See endnote 15.)

FLEAS

There are about 2500 different species of fleas in the world. Fleas transmit the causative agents of bubonic plague, murine typhus fever, salmonellosis, and tularemia. Cat fleas are a host of the dog tapeworm, which infects people, especially young children. The bites of all fleas may be very painful.

Best Practices in Controlling Fleas
- Use topical or oral flea medication on pets for flea treatment or flea prevention. Check with your veterinarian to determine which of the various medications should be used on the individual pet. (See endnote 3.)
- Conduct a comprehensive survey for fleas and flea larvae especially in areas where dogs and cats sleep and away from pedestrian traffic or exposure to sunlight. Adult fleas may have left dried blood and feces in these areas.
- Thoroughly clean all areas where adult fleas, flea larvae, or flea eggs can be found by vacuuming rugs, carpets, upholstered furniture, and all crevices, and then launder all pet bedding at least weekly.
- Use effective insecticides, only if absolutely necessary, containing insect growth regulator (IGR) methoprene or pyriproxyfen in a hand sprayer or aerosol directly to infested areas of carpets and furniture, while continuing to vacuum at least every 2 weeks. Avoid using room foggers because they do not come in contact with all areas where fleas may be harboring. As a precaution, keep all pets away from these areas until the fleas are gone and the areas are properly decontaminated.
- Out of doors if fleas are very concentrated in locations where pets rest and sleep, beneath decks, and next to foundations away from traffic, it will be necessary to use insecticides in these areas. Once again keep all pets away from these areas until all the fleas are gone and the area has been decontaminated.

FLIES

(See endnote 5)

Houseflies are a major vector in the transmission of diarrheal disease through mechanical transmission from the source of the organisms, such as fecal matter, to the food supply for individuals. Houseflies transmit the organisms of typhoid fever, cholera, polio, etc. The deer fly in the Western United States is involved in the transmission of tularemia. The bites of the horsefly and deer fly may be very painful.

Flies breed very quickly and can complete their development in as little as 7 days in warm weather. Houseflies also contaminate surfaces and food by regurgitating part of what they have eaten, through their feces, and through physical contact with them. Their larvae are attracted to fecal material and rotting garbage, while the adult is attracted to human food sources. They can travel up to 10 miles unless it is very windy when they can travel much further. Blowfly larvae develop inside of things such as rodents, animal waste, and dead people. Fruit flies are found near fermented materials in trash cans and floor drains. Drain flies breed in raw sewage or near broken pipes and need standing water to breed. Phorid flies breed in decaying organic matter and are found typically near hospitals, in restaurants, in the bottom of trash cans, under kitchen equipment, and in drains.

Best Practices in Fly Control
- Conduct a survey of the sources of the flies, types of flies, and concentration of flies to determine appropriate control procedures.
- Check all incoming material for signs of fly infestation.
- Evaluate all areas near buildings for breeding places and treat them accordingly.
- Remove all decaying material and clean the area effectively.
- Rotate all produce and remove that which is starting to deteriorate, place in a sealed bag or container and have it removed daily if possible but if not at least twice weekly to an appropriate disposal site.
- Clean all garbage collection sites and floor drains daily and remove any organic material as well as eliminate standing water. Remove all odor-producing materials. Place dumpsters as far away from the building as possible. Ensure dumpsters are equipped with lids and drainage lines are sealed to minimize flies from entering.
- Caulk all cracks and crevices into buildings, screen all openings, and/or use an air curtain of high-speed air to keep flying insects out of entrances and exits to the structure. Do not leave doors open.
- Use insect traps outside of the buildings.
- Use insecticides approved by the US Environmental Protection Agency (EPA) and applied by trained personnel only as a last resort.
- Never contaminate any food contact areas. If inadvertently they do become contaminated, wash them very carefully, rinse the areas very carefully, and wash and rinse again and allow them to air dry.

FOOD PRODUCT INSECTS

Food product insects infest stored foods including grains, flour, nuts, spices, packaged herbs, and dry fruit. They include the Indian meal moth, flour beetles, sawtooth green beetles, and carpet beetles as well as others. They lay their eggs in crevices, in the food supply, or in cabinets or other storage areas.

Best Practices in Food Product Insect Control
- Survey and find all sources of infestation, clean up all spillage of foods, and vacuum the area meticulously.

- Destroy all infested foods in homes and food preparation areas.
- If possible, store all products in the refrigerator because refrigeration slows down the growth of new insects.
- Use of insecticides in pantry areas including the use of pyrethroids is not recommended, since the food may become contaminated and unfit for human consumption. Food in contact with any type of pesticide shall be discarded. Food should never come in contact with any type of pesticide in any place or at any time.

LICE

Lice live, feed, and reproduce on the body of a living host. Lice can transmit, particularly during times of war or disasters, epidemic typhus fever, epidemic relapsing fever, and trench fever. Further, their bites can be painful. A louse infestation on an individual can be a great source of embarrassment.

Best Practices in Lice Control (See endnote 6)
- When head lice or other body lice are suspected because of intense itching and annoyance, individuals, generally children, need to have their heads, scalps, and hair examined as well as other parts of their body if necessary for either the lice or nits, which are the eggs. First a black light should be used in a darkened area to see if there are nits on the scalp or on other areas of the body. This should be done in a private area. Then a well-lit area should be used by the nurse or other individual conducting the search and a special fine-tooth comb should be used to remove the lice.
- Use special pyrethrum-based shampoos or lotions and completely lather the hair and allow it to stay on for several minutes. Rinse the hair thoroughly, comb it out while it is wet with the fine-tooth comb, and do not re-shampoo the head for several days, thereby allowing the chemicals to continue working. Use a fine-tooth comb for several weeks to comb out lice or nits. Repeat the re-shampoo process in 7–10 days to prevent the hatching of remaining nits.
- Keep all children with active head lice infestations out of school until there is satisfactory control of the situation, which is the elimination of all nits and lice from the person. Constant checks and re-checks are necessary on a daily basis.
- Vacuum all floors, rugs, overstuffed furniture, pillows, mattresses, and wash in hot water of at least 130°F all hair brushes, combs, clothing, linens, cloth toys, hats, and other materials that come in contact with the contaminated individual. That which cannot be washed should be put in a sealed plastic bag for a minimum of 14 days. Headphones should be thoroughly cleaned and decontaminated before reuse.
- Pesticides should only be used as a last resort, since they can cause more danger than the persisting louse infestation. If pesticides are used, they should only be applied to surfaces and objects the person will not touch.

MITES

Mites are tiny insect-like arachnids which can cause a variety of health problems such as bacterial infections and allergic reactions. They may be carried into facilities or homes by birds or rodents. Chigger mites can cause itchy red bumps to appear at the site of the bites and are competent vectors of scrub typhus fever. Scabies mites live on and inside the human skin and may cause severe itching and secondary bacterial infections; individuals may also have allergic reactions to their feces which may be within the skin. Dust mites are a primary trigger of allergies and allergic reactions, such as asthma and eczema, from house dust.

Best Practices in Control of Mites

- If you have scabies mites or other mites, see a dermatologist or your primary care physician, take a thorough and complete bath, dry yourself totally and apply prescribed medicine to the entire body from the neck to the toes including the hands and any crevices. Leave it on from 8–14 hours before you wash it off. All people who come in contact with you must follow the same procedure, since they may not see any symptoms for 2–6 weeks.
- Wash all items in a washing machine with the hottest water setting and dry the items in a dryer at the highest heat setting. If an item cannot be washed, take it to a dry cleaner or seal in a plastic bag for at least 1 week.
- Special dust mite sprays should be used on surfaces including carpeting and then after 2–3 hours all surfaces including furniture should be vacuumed very thoroughly. Dispose of the vacuum bag in a sealed plastic bag before putting into the trash.
- Do not treat your pets for the human itch mite since this mite cannot survive on animals.

MOSQUITOES

Of the numerous mosquito groups, only *Aedes*, *Anopheles*, and *Culex* mosquitoes are primarily responsible for infecting humans with disease. *Aedes* mosquitoes lay their eggs above the waterline in tree holes, containers including tires, and even dampened ground or bottle caps. They bite humans especially fiercely and live within 100–300 feet of the target area. *Anopheles* mosquitoes breed in fresh or brackish water and are most active at night when they enter buildings, bite, and then return to forested areas. *Culex* mosquitoes breed in quiet stagnant waters, artificial containers, or large bodies of permanent water, frequently where sewage may be found. Adult females remain inactive during the day, bite during the night, especially at dawn and dusk, and may have a range of 10–12 miles. Mosquitoes may bite fiercely, which can be very annoying, and are the vector of a variety of diseases including: West Nile fever, St. Louis encephalitis, La Cross encephalitis, Eastern equine encephalitis, Western equine encephalitis, yellow fever, malaria, dengue fever, chikungunya virus, and most recently the Zika virus. Mosquitoes from other parts the world may be introduced into a given area in containers and tires, and can be of public health significance as a capable vector of disease.

Misting systems are an effective mechanism for reducing adult mosquito populations identified as a threat to public health, although they may create a hazard for the public. If windows are open in houses when the mist is applied, people within the structure may be subjected to insecticidal exposure through inhalation, ingestion, or absorption. Children on bicycles create hazards for themselves as well as equipment operators when they follow closely behind misting trucks. Wind speed and direction should be taken into account before residential misting occurs, as chemicals can drift to non-target areas subjecting people, animals, or consumable products to unwarranted pesticide exposure. Misting systems may leak and can potentially contaminate recreational areas or surface waters which may be used for drinking water.

Best Practices in Controlling Mosquitoes (See endnotes 8, 16)

- Conduct intensive studies in areas where mosquitoes are most likely to breed. Larval surveillance is accomplished through using a dipstick to remove water containing larva and estimating their density in it. Adult surveillance involves the use of traps including carbon dioxide or light to attract and catch biting and/or egg laying mosquitoes to determine the types and densities of mosquitoes in a given area. Virus samples are acquired by sampling mosquitoes for viruses, sampling bird blood, especially of dead birds, for viruses, and evaluating equine cases of the disease. Repeat all surveillance on a regular basis to evaluate control efforts and redirect the program as necessary.

- Focus on the control of mosquito larvae by eliminating areas of standing water and treating those which are too large to eliminate. Use proper water management practices allowing a free flow of water. Keep all storm ditches clear of plants and debris to allow free flow of stormwater. Local mosquito control programs should work closely with agencies and departments responsible for keeping ditches clear of debris.
- Remove all artificial containers containing water where feasible. If not feasible, empty them out or keep fish in them to feed on the larvae.
- Fill in all holes or depressions in the ground or tree holes where water is retained.
- Construct drainage ditches where appropriate for areas which are flood prone.
- Supplement prevention programs and biological controls with appropriate insecticides when necessary. Use biorational products containing bacteria which produce insecticidal toxins. These bacteria are cultured and packaged and then put into areas where mosquito larvae can ingest them. Another technique is the disruption of mating. Surface agents such as highly refined mineral oils should be applied to stagnant water breeding sites with mosquito pupae where other larvacides will not be effective. IGRs containing the active ingredient methoprene disrupt the physiological development of the larva by stopping them from becoming adults, and are an effective alternative larvicide. Lastly, use very carefully and highly trained people to apply chemicals to control larva without negatively impacting others and the environment.
- Adult mosquitoes can only be controlled with pesticides which target either the resting mosquitoes or adult mosquitoes over large areas. Trucks and aircraft can be used for dispersal of chemicals through misting systems. Use when the adult mosquitoes exceed action levels for nuisances or where there is a disease outbreak. Malathion, naled, pyrethrins, and pyrethroids (manufactured pyrethrins) are common active ingredients used in adult mosquito controlling products. They should be utilized with great care and with an understanding of weather conditions, especially wind.
- Personal protective measures include the use of mosquito repellents, screening of facilities and homes, mosquito netting over beds, long-sleeved clothing for use when outdoors especially at dawn or dusk, and public education to identify the source, potential for disease, and means of protection of the individual and family.
- Above all else, public communication and outreach on how people need to be aware of and protect themselves from being bitten is of greatest significance.

ROACHES

Roaches are long-lived and have been around for at least 145 million years. The four major types found in structures in the United States are the American roach, the brown banded roach, the German roach, and the Oriental roach. The German roach readily becomes resistant to insecticides and is probably the greatest pest in most homes. They can use almost all organic substances for their food supply. They can use food, cereals, baked goods, grease, glue, starch, wallpaper binding, fecal material, garbage, and dead animals. They must have a moist or damp area for breeding and to exist. These conditions are especially found in food service operations. They may be found around cracks and crevices by the wallboard, under and around mechanical dishwashers, in bathrooms, and in other warm and moist areas. The egg sack is immune to insecticides. The gestation period is 21 days and therefore the area should be retreated after this period. Roaches can carry pathogenic bacteria onto food and therefore be responsible for causing outbreaks of food-borne disease. An infestation may be very difficult to get rid of. They may also cause allergic reactions and therefore may be associated with increased levels of asthma, particularly in children.

Best Practices in Controlling Roaches (See endnote 9)
- Conduct an intensive survey of where roaches may be hiding, especially warm places with access to water.
- Reduce or remove all food and water sources as the initial phase of the control effort. Seal all cracks and crevices that are available for roach harborage.
- Thoroughly clean and remove all evidence of roaches including feces and body parts, etc. to prevent susceptible individuals from having asthma attacks.
- Depending on the roach species, prepare bait traps as a primary means of control. Place them in areas where roaches forage for food. Insecticide sprays may be used as a secondary approach after removal of food, water, and harborage, and use of traps.
- Store all foods in insect-proof containers.
- Put all garbage and trash in containers with liners and use tight-fitting lids.
- Remove all trash, newspapers, piles of paper bags, or other clutter and carefully vacuum the area, especially cracks and crevices. Seal all conduits, doors, false bottom cabinets, and hollow walls. Use boric acid powder in these areas as an insecticide.
- Check all new packaging coming into dry storage areas for possible transportation of roaches into the facility.
- Check to see if egg cases are glued to the underside of furniture, or in the motor casing of refrigerators and other appliances.
- Where insecticides are used, they should be applied only by a well-trained certified professional. They should never be put on food contact surfaces or kept stored with any type of foods.
- Establish and utilize a continued maintenance program with frequent inspections to determine if the roaches have been eliminated or if a new infestation has started.

TICKS

Ticks are bloodsucking parasites of humans, pets, livestock, and wild animals. Some ticks may carry the microorganisms that cause babesiosis, human granulocytic ehrlichiosis, Lyme disease, tick-borne relapsing fever, Rocky Mountain spotted fever, or tularemia. There are about 25,000 cases of Lyme disease reported in the United States each year with the highest reported amounts in 5- to 14-year-old children and 45- to 54-year-old adults. Anyone who is outside and unprotected in areas where ticks are prevalent may get the disease if bitten by an infected tick. The longer the tick feeds on the person or pet, the greater the chance of developing Lyme disease. (See endnote 10.)

Best Practices in Controlling Ticks (See endnote 11)
- Use a tick repellent when going into forest areas, especially those with high grasses and heavily wooded that are known for tick infestation.
- Use protective clothing which is light-colored and tight around the ankles and wrists when walking in wooded areas.
- When leaving these areas check the entire body for ticks, and remove them with tweezers and destroy them. Removing ticks within 48 hours of attachment sharply reduces the transmission of Lyme disease.
- Remove brush and leaves from backyard areas and put woodchips around the perimeter to reduce sharply tick populations. Restrict the use of certain types of groundcover near houses. Move piles of firewood and birdfeeders away from the house.
- Exclude deer from areas around the house by using special deer fencing.
- Keep dogs and cats from roaming in wooded areas. Check your pets carefully for ticks when they have been outside and if found, remove immediately and destroy.

- Have licensed trained personnel apply approved pesticides to backyards at least once or twice a year to sharply reduce tick infestation.
- Use cardboard tubes stuffed with permethrin-treated cotton, which can be found in hardware stores, where there are substantial populations of mice. The immature ticks that feed on the mice will be killed off since the mice take the cotton back to their nests.
- Rodent proof homes to keep especially the white-footed mouse from entering.
- Move swing sets and other toys away from the wooded portion of the house lot.
- Use special feeding stations for deer where they have to come in contact with paint rollers containing pesticides. This will help kill off part of the tick population.

SUB-PROBLEMS INCLUDING LEADING TO IMPAIRMENT AND BEST PRACTICES FOR RODENTS

Rodents may be the vector of several diseases which include hantavirus pulmonary syndrome, hemorrhagic fever with renal syndrome, Lassa fever, leptospirosis, lymphocytic choriomeningitis, Omsk hemorrhagic fever, plague, rat-bite fever, salmonellosis, South American arenaviruses (Argentina hemorrhagic fever, Bolivian hemorrhagic fever, Venezuelan hemorrhagic fever), and tularemia.

Best Practices for Controlling Rodents
- Conduct a comprehensive survey to determine the size and scope of rodent infestation and identify causes and conditions contributing to their existence.
- Eliminate all sources of food, harborage, and where possible water.
- Make sure that all trash and organic materials are bagged and are kept in containers with tight-fitting lids and removed at least twice a week for organic material and at least once a week for trash.
- Pick up and dispose of all fruits and nuts which have fallen from trees as well as garden vegetables and berries.
- Remove food after dogs and cats have been fed and make sure that the areas are perfectly clean and free of remnants. Bulk birdseed and pet food should be stored indoors in containers with tight-fitting lids.
- Remove all abandoned vehicles, old furniture, unused appliances, tall grass, dense undergrowth, and remnants of pruning, and prune all bushes to keep them off the ground and separated from each other. Keep trees at least 6 feet away from the house or utility wires. Elevate all firewood at least 18 inches above the ground and keep at least 12 inches away from walls and fences.
- Seal all openings into structures.
- Set snap traps or glue boards instead of poison baits nearby active rodent signs.
- Wear rubber gloves and a respirator when cleaning out an area where there are mice or rats, their droppings, urine, or nesting materials. This also includes the removal of dead rodents. All this should be put into a plastic bag and double bagged before being taken for disposal.
- Ventilate all areas that are contaminated by rodents and decontaminate the area with a 10% solution of bleach. Dispose of all sponges or other cleaning materials along with the contaminated material itself.
- Discard all food or single service items that have come in contact with rodents, their feces, or urine.
- Thoroughly clean all food preparation or storage surfaces and dispose of the cleaning sponges, and then sanitize using a bleach solution.
- For a large rat infestation in a community, see "An Example of a Successful Program: Community Rodent Control, Philadelphia, 1959–1963" in Chapter 1.

STATEMENT OF PROBLEM AND SPECIAL INFORMATION
FOR PESTICIDES AND ANTIMICROBIALS

In our modern world, pesticides make a significant contribution to the elimination of disease, the production of adequate food supplies, and the protection of our homes from damaging insects, for a world population that is continuing to grow. People come into contact with a variety of pesticides from their homes and garden, their pets, their communities, and the food they eat which may be thoroughly contaminated, and through their workplaces. There is an accumulative problem of being exposed to a variety of pesticides in different concentrations over sustained periods of time. Special populations at risk include infants, children, and pregnant women, who have some level of problems with all pesticides and major concerns with others since the chemicals can breach the placenta and enter the fetal blood supply and brain. Also, small children crawl around the floor and may be exposed to chemicals after treatment of cracks and crevices. The elderly and infirm may have a more serious reaction to the chemicals than younger people, as may individuals taking a variety of prescription medications.

Problems with pesticides relate to: the chemical selected, with its ingredients including contaminants; its level of toxicity for non-target plants or animals including people; its level of retention in the environment, especially in the soil, food, water, and air; and the unintended consequences related to its use. The chemicals may be short-term or long-term toxins, endocrine disruptors, carcinogens, etc. Organophosphates and carbamates affect the nervous system, while others may affect the skin, eyes, hormones, or various organs within the body. Occupational and casual exposure to the pesticides by people may lead to acute and chronic disease and/or injury. Other problems include appropriate storage and disposal of the chemicals.

Antimicrobial pesticides which are used to disinfect, sanitize, or reduce or inhibit the growth or development of microorganisms may also cause significant chemical exposure problems for people. These pesticides are frequently used on floors and walls as well as other surfaces in water containers such as cooling towers to prevent the growth of bacteria, viruses, fungi, algae, etc. They also are used to preserve paints, metalworking fluids, wood supports, etc.

SUB-PROBLEMS INCLUDING LEADING TO IMPAIRMENT AND
BEST PRACTICES FOR PESTICIDES AND ANTIMICROBIALS

The use, storage, and disposal of pesticides in a variety of situations and what effect these chemicals have on our health and our environment is of major concern. Pesticides are utilized in urban areas, suburban areas, and rural areas. Each type of usage will be addressed.

Urban usage includes residential areas, public areas, commercial buildings, and industrial areas. In residential areas pesticides are applied by trained individuals as well as by citizens in and around homes. Each time the person is exposed to the pesticides in all the aforementioned settings, there is a potential for increased body burden of the chemical. Individuals can easily purchase a variety of pesticides in grocery stores, hardware stores, general merchandise stores, etc. They buy them and use them because of an existing insect or rodent problem. Their major concern is getting rid of the pests. This is especially true of inner city areas, where there is overcrowding and where food, water, and harborage are readily available. The individuals apply as little or as much pesticide as they feel is necessary to solve the problem. The concern is that the sheer volume of pesticides utilized, approximately 2 billion pounds of active ingredients a year in the United States, and the location of the application can lead to multiple hazards to people. Semi-volatile pesticides leave a residue within the home environment, which when released can be inhaled by the people living there. Pesticides and fertilizers used outside the home add to the total exposure of the individuals. Some of the pesticides bioaccumulate in food as well as within the body of the individual.

Some 50% of the 2 million cases of poisoning in the United States are found in children less than 6 years of age and 90% of these cases occur in the home. Many thousands of calls are made to poison control centers in the United States each year concerning potential exposure to common household pesticides by children. Most people store pesticides in unlocked cabinets and within the reach of children. The children are particularly vulnerable because of physiological and behavioral circumstances. (See Chapter 4, "Children's Environmental Health Issues.")

Suburban areas, although typically less overcrowded and therefore with less opportunity for the presence of insects and rodents, contain additional problems because of increased usage of the chemicals to improve the visual beauty of the outside environment surrounding the home. There is a growing concern about the contamination of the water supply by these chemicals. Suburban areas were typically, at one time, rural areas and therefore the opportunity for increased levels of mosquitoes and other insects exists. Farms on the urban fringe contribute high levels of organic materials which can act as an excellent source of food, water, and harborage for insects and rodents. Agricultural chemicals become a serious concern for the water supply. Pesticide residues are readily spread through this environment.

In rural areas, agricultural chemicals contaminate people and the environment from use, misuse, storage, and disposal. Of special concern is pesticide drift from the application of pesticides through the air and onto the surface of soil. Very small chemical particles can drift for miles before they reach the soil. Another problem is that the pesticides evaporate during and after application and then reform to create hazardous situations. Residues left on food and vegetables can create additional hazards in the food chain. Family members and other individuals living in the rural setting beside field workers and applicators are subjected to the pesticides and can become sick or injured.

Outdoor residential misting systems for mosquito control are poorly regulated since the regulations vary by states and communities and the EPA has not been involved. The frequency and timing of the use may not be coordinated with actual weather conditions and create problems of chemicals being spread within homes and businesses or of extensive chemical drift. Although label instructions should be followed by individuals using such devices, there is no guarantee that improper chemicals are not being used. The effectiveness of this type of system has never been properly evaluated.

It is absolutely essential that all mosquito control program personnel who are conducting adulticiding and larviciding are professionals and are licensed by the appropriate licensing authority. Beyond actually carrying out all survey and spraying activities, these individuals become an important part of the public education and outreach effort of the local health department programs. They explain the program that is in operation and also how and where to eliminate standing water to reduce mosquito populations. There is a constant flow of information to the public about when and where mosquito control spraying will occur and what to do to avoid becoming contaminated by the spray.

In addition to the individuals in the urban, suburban, and rural areas who are exposed to pesticides, there are two other high-risk groups. They are workers in the pesticide industry and farm and migrant workers. Those in the pesticide industry typically, on a daily basis, mix, load, transport, store, apply, and dispose of complex chemicals which may be highly toxic and/or cause serious long-term illnesses. They are exposed to the chemicals through accidental spills, leakages, and poor spraying equipment, which leads to inhalation, skin absorption, and contaminated air, food, and water. Air temperatures and high humidity may contribute to activation of the compounds. These conditions potentially intensify the levels of exposure and the chemicals are added to the normal exposure the individuals experience in their everyday lives. Farm and migrant workers are constantly exposed to high levels of pesticides and fertilizers as they work in the fields and as they live in their homes which are in close proximity to the areas where the pesticides and fertilizers are used in substantial quantities.

Pesticides are used to protect crops and livestock from insects, weeds, and disease. It is estimated that 76% of the total pesticides used in the United States is for agriculture, with 24% used in urban, industrial, forest, and public areas. Even small quantities of these chemicals can contaminate both ground and surface water and cause additional problems with the drinking water supply. Non-point contamination is the application of the chemicals to the land, while point source contamination occurs when there are spills of the concentrated chemicals during production, transportation, storage, and at loading and mixing sites, and at disposal sites. The active ingredients in the products may last for months or even years in the ground and continue to contaminate the water. Additional problems may occur when the chemicals are broken down either by sunlight or microorganisms, or taken up by plants. Soil properties and weather conditions affect the actual problems of use of pesticides over large areas. (See endnote 19.)

Adverse human health effects from pesticides range from headaches, weakness, blurred vision, and irritability to suppression of the immune system, depression, asthma, blood and liver diseases, nerve damage, cancer, and mutagenic and teratogenic effects. Some of these symptoms may be recognized quickly and others not for many years. It appears that the amount of birth defects increases in children who were conceived between April and July of the previous year in rural areas.

Human health risks may be due to the totality of pesticides taken into the body through food, water, air, and direct contact with surfaces that have been treated. Different pesticides may have the same effect on the human body and therefore there is a cumulative effect from being exposed to them. Occupational exposure may add to the previously mentioned exposures and cause additional accumulative problems.

The EPA has evaluated about 490 chemicals to date to determine the potential level of a carcinogenicity hazard. It has not included exposure information. This list is a supplement to the individual full risk assessment of the chemical and it provides the date at which the particular chemical was evaluated. Check the most current chemical evaluation for the one which you are concerned about. (See endnotes 17, 18.)

The cancer risk review is based on: laboratory animal findings; metabolism studies; structural relationships with other carcinogens; and any other information including epidemiological findings in people. The categories of risk include carcinogenic to humans; likely to be carcinogenic to humans; evidence of potential carcinogenicity; inadequate information available; and not likely to be carcinogenic to humans. Currently, the guidelines discuss dose–response, exposure assessment, and risk characterization.

Low-dose chronic exposure to the chemicals can impact the nervous system in a negative way. Long-term, low-dose chronic exposure to the chemicals is not well understood. However, it is thought that the chemicals may contribute to the causation and/or exacerbation of asthma. Low-dose exposure of pregnant women in residential settings may contribute to potential birth defects. They may also contribute to birth and early childhood-related health problems. High-dose exposure can lead to acute poisoning incidents, which has been well documented. This type of problem has been particularly identified in children.

The World Health Organization in 2006 issued a document listing pesticides of public health importance that were safe for the destruction of mosquitoes, flies, fleas, bedbugs, lice, roaches, ticks, mites, and rodents. (See endnote 20.) The state of Oregon trains pesticide operators in the use of aluminum phosphide, *Bacillus thuringiensis*, diphacinone, malathion, methoprene, naled, permethrin, pyrethrins and pyrethroids, resmethrin, temephos, warfarin, and zinc phosphide. However, as scientific knowledge expands concerning the use of various pesticides, what was considered to be safe at one time may now be considered to be hazardous to certain groups of people or the natural environment. An example of this would be the pesticide chlorpyrifos which is an organophosphate used to control insects, ticks, and mites. It has been used in both agricultural and non-agricultural areas since 1965. It can over-stimulate the nervous system in people at high exposures and has become an occupational exposure hazard. Frequently, a reassessment is being carried out by the EPA for protecting workers who mix, load, and apply the chemicals and how to best avoid short-term and long-term risk to their health. This chemical also creates a hazard in vulnerable water supplies.

Pest resistance to certain pesticides and families of pesticides has increasingly become a problem. The pests adapt to the chemicals and develop avoidance systems. Those which are weakest die off and those which are most resistant live on to propagate and create more and more resistant pests.

The rapid reproduction rate can make the time span quite short. Many of the new pesticides that would help eliminate pests may be far more hazardous to people than those that are currently being used.

Best Practices in the Use of Pesticides and Antimicrobials
- Develop and use an integrated pest management program before using any pesticides.
- Review all active ingredients within the pesticide formula and read the instructions for use before selecting the least toxic and most effective chemical to treat the area for specific pests.
- Determine if there is the potential for cumulative risk from the use of different pesticides that act in the body in the same manner before deciding which pesticide to use. See the work of the US EPA in this area and use their most current material as a guide in the selection process. Human health risk assessment is based on: hazard identification from each of the chemicals or components in the pesticide; dose–response assessment; exposure assessment; and risk characterization based on the scientific data available.
- Avoid overuse of pesticide applications which can be modified.
- Understand pesticide persistence in the environment as well as mobility and adsorption before using it.
- Ensure that the pesticide applicator is highly trained, licensed, and knows exactly where to apply the chemicals, the amount of soil moisture, as well as the prevailing weather conditions to avoid over-concentration as well as chemical drift.
- Establish buffer zones of 50–100 feet from wells and surface water when spraying.
- Avoid repetitive use of the same pesticide or the same family of pesticides to help prevent resistance from insects and weeds.
- Calibrate and maintain all pesticide application equipment in excellent condition to get the proper amount of pesticide released for the particular situation.
- Provide backflow prevention devices on irrigation systems which apply insecticides.
- Avoid retreatment of areas by utilizing the proper quantity of pesticides initially.
- Use personal protective equipment including gloves, goggles, respirators, and special clothing when handling, storing, applying, and disposing of pesticides. Wash the chemicals off of the clothing and gloves and allow them to dry before storage. Do not let the chemicals enter bodies of water. Change clothing and shoes before entering the house and wash appropriately.
- Provide emergency eye and skin washing in the event of contamination from the chemicals during storage, mixing, use, or disposal. Have immediately available special phone numbers for emergency assistance.
- Handle and use pesticides only in well-ventilated areas when working indoors.
- Keep all children, pets, sensitive people, the elderly, and the chronically ill from areas where pesticides are being applied. Protect all food, pet bowls, and clothing from the pesticides. In outdoor areas obey all signs, flags, or postings about a pesticide application.
- Never use an outdoor product indoors.
- Use insect repellent that are both safe and effective based on information from the EPA. Check for the length of protection and whether there are any allergic reactions to the chemicals or carriers.

STATEMENT OF PROBLEM AND SPECIAL INFORMATION FOR INTEGRATED PEST MANAGEMENT

(See endnote 7)

There are several approaches to the elimination of pests from a given area. In past times, individuals were involved in removal of food and water (prevention and control), elimination of harborage (prevention and control), and then the safe use of chemicals to complete control efforts. Evaluations were carried out to determine if and when further control was necessary. In the present, the technique of integrated pest management is being utilized. There is still an emphasis on the removal of food, water, and harborage, and evaluation of control efforts, but now the use of actionable thresholds of levels of the pests has been added, and a determination whether there are other cost-effective options to control the pests. The threshold depends on the pest and whether it constitutes a nuisance problem and how much of a nuisance or health problem. When that threshold is reached, a variety of chemicals is utilized in a safe manner, then becomes part of the process for the elimination of the pest. It is recognized that not all insects, weeds, and other plants or living organisms create enough of a threat to compensate for the potential harm to the environment and to people by use of a variety of chemicals.

Best Practices for Pest Control through the Use of Integrated Pest Management

The integrated pest management approach is the most effective way to get rid of insects and rodents without having to use chemicals. It consists of the following steps:

- Perform a comprehensive inspection and survey, using risk assessment techniques in both agricultural and non-agricultural situations, of all infested areas including all surrounding parts of the structure inside and, depending on the pest identified, outside.
- Keep accurate records and maps of all locations where the pests are found and make sure that the dates are included.
- Reduce all clutter and remove all sources of harborage and habitat from areas where the pests are found.
- Thoroughly clean all areas and remove all live or dead insects or rodents.
- Remove all sources of food and water.
- Use preventive measures such as naturally occurring parasites and predators; pruning and raking and use of pest-resistant plants; alternative chemicals instead of chlorinated or brominated organic molecules or those based on phosphoric acid or carbamic acid; and pheromone traps, boiling water for anthills, or diatomaceous earth.
- Use action thresholds (the point at which the pests cause more problems than the potential hazards caused by using chemicals) to determine when to apply appropriate pesticides in proper quantities and under exact environmental conditions.
- Use highly trained, certified technicians to conduct the chemical insect and rodent control work and take into consideration time, temperature, weather conditions, etc.
- Re-clean all areas and determine if the insect or rodents have been eliminated or reduced in numbers.
- Seal all openings where feasible to prevent entrance into the house or business.
- Teach people about the reasons for insect and rodent problems and how to prevent them.
- Re-inspect periodically and if necessary retreat the areas.

LAWS, RULES, AND REGULATIONS

FEDERAL INSECTICIDE, FUNGICIDE, AND RODENTICIDE ACT

The Federal Insecticide, Fungicide, and Rodenticide Act (FIFRA) provides the framework for the federal pesticide program. The EPA is responsible for registering and licensing of pesticide products in the United States. The decision to register or license must be based on a detailed assessment of the potential risks involved to human health and the environment. By law, the individuals utilizing the pesticide must follow the directions and precautions listed on the label exactly as stated. Older pesticides have to be reviewed by the EPA and a determination must be made as to whether or not they should be discontinued or restricted. The EPA and states must establish programs to train and then certify applicators of pesticides.

FEDERAL FOOD, DRUG, AND COSMETIC ACT

This act establishes the program that provides the tolerances or maximum level of pesticide residues for human food and animal feed and makes the EPA responsible for administering the act.

FOOD QUALITY PROTECTION ACT OF 1996

This act amends the previous two acts and establishes a tougher standard for pesticides used on food. There is now only a single health-based standard that is to be used when determining the risks of pesticide residues in food or feed. Additional requirements are as follows:

- The EPA can only establish an acceptable tolerance level for a pesticide if there is reasonable certainty that no harm to humans, especially children and the infirm, will occur from the combined effects of multiple exposures to the chemical or families of chemicals including of different composition that act in the same way in the body (a tolerance can only be established after all issues related to aggregate, non-occupational exposure through diet, drinking water, and use of pesticides in and around the home, and the cumulative effects from exposure to pesticides with common action in the body have been assessed).
- By 2006, the EPA reviewed all old pesticides to make sure that residues on or in food meet the new safety standard. This action is repeated every 15 years.
- All pesticides must be tested to determine if they have the potential for endocrine disruption which may result in sexual, developmental, behavioral, and reproductive problems.
- The EPA must prepare and distribute a brochure to supermarkets and other food outlets to discuss pesticides on food in order to alert the public as to what is proper.
- The EPA must reassess the pesticide registration of every chemical used at least once every 15 years.

PESTICIDE REGISTRATION IMPROVEMENT ACT OF 2003

This act amended all the original acts. It requires that companies pay service fees and that the EPA has to meet decision review time schedules. It provides shorter decision times for reduced risk pesticides. It was renewed in 2007 and 2012.

FREEDOM OF INFORMATION ACT

This act legally requires all agencies to respond to information requests from the public unless the information is confidential and protected by disclosure rules. Trade secrets and financial information from various companies cannot be released.

FEDERAL ADVISORY COMMITTEE ACT

This act establishes a variety of special committees who advise the EPA on various areas of scientific research. The committees include the:

- Tolerance Reassessment Advisory Committee
- Food Safety Advisory Committee
- Endocrine Disruptors Screening and Testing Advisory Committee
- Pesticide Program Dialogue Committee

- FIFRA Scientific Advisory Panel and Scientific Advisory Board
- State FIFRA Issues Research and Evaluation Group.

Safe Drinking Water Act

In 1996 the Congress amended the law for screening and testing programs for chemicals and pesticides that may cause endocrine disruptor effects. The EPA developed and presented a screening program to Congress so that it could be implemented.

RESOURCES

Centers for Disease Control and Prevention

Pesticide-Related Illness and Injury Surveillance: A How-To Guide for State-Based Programs

(See endnote 25)

The manual provides information on how to develop and maintain programs to evaluate acute and sub-acute health effects from pesticide exposure. Pesticide poisoning is very complex to understand since it can produce a variety of symptoms and can occur over various time periods. The poisoning also may be due to single exposures or multiple exposures of the same toxin or a mixture of toxins over a period of time. The guidelines help set standards for conducting the investigation of the incident, the types of technical resources needed, methods of data collection, methods of reporting the information, and outreach to various groups and educational efforts.

Sentinel Event Notification System for Occupational Risk (SENSOR)

(See endnote 26)

SENSOR is an occupational illness and injury surveillance program for the states. NIOSH provides technical support to state health departments on how to gather these data. Financial support may come from NIOSH or EPA. The data are utilized to develop preventive techniques in certain industries in order to eliminate or mitigate pesticide hazards. They are also a timely means of identifying for local and state governments acute occupational or environmental exposure to chemicals and provides immediate data on what is happening and therefore how best to deal with it and the chemical or chemicals which may cause massive health, safety, and environmental problems.

National Poison Data System

This is a compilation of all the data from the phone calls that are received at the poison control centers in all states. This helps NIOSH determine if an acute incident is occurring and immediate action needs to be taken.

Toxic Exposure Surveillance System

The Toxic Exposure Surveillance System is a national, real-time database including all human exposures reported to poison control centers since 1985. About 44% of the initial reports are followed up to get additional information. The database searches for unusual occurrences in the hourly case volume for the poison control centers, and additional information may be sought and interpreted by clinical toxicologists on individual cases or on a pattern of cases. This helps scientists understand what is occurring and when to make a significant decision about what to do concerning prevention, mitigation, and control of the chemical event. (See endnote 29.)

US Environmental Protection Agency

Integrated Pest Management in Buildings

The Office of Pesticide Programs has written a document entitled "Integrated Pest Management in Buildings." It discusses an integrated environmentally friendly pest management program and uses a common sense approach to the control of pests. (See endnote 22.)

Pesticide Fact Sheets

The US EPA has published a list of topical and chemical fact sheets easily accessed on the internet. These fact sheets cover pesticides used in decontamination of anthrax spore bioterrorism attacks, antimicrobial pesticides, the pesticide atrazine, chromated copper arsenate, DDT, diazinon, and many others. It also presents fact sheets entitled Assessing Health Risks from Pesticides, Citizen's Guide to Pest Control and Pesticide Safety, Consumer Products Treated with Pesticides, EPA and Food Security, Mosquito Control in the United States, Laws Affecting EPA's Pesticide Programs, etc. (See endnote 23.)

National Pesticide Information Center

The National Pesticide Information Center located at Oregon State University at Corvallis is a cooperative effort between the university and the US Environmental Protection Agency. It provides unbiased, science-based information on EPA-registered pesticides and pesticides used by people for a variety of purposes. It is involved in a variety of topic areas including pesticides and human health, pesticides and animal health, safe use practices, pest control, pesticide ingredients, occupational safety, etc. (See endnote 24 for topics and phone number of the center, where individuals can receive help and guidance concerning pesticides.)

US Department of Agriculture

EXTOXNET

This is a pesticide information project of the Cooperative Extension offices of Cornell University, Oregon State University, University of Idaho, University of California at Davis, Institute for Environmental Toxicology, and Michigan State University. It is funded by the US Department of Agriculture Extension Service, National Agricultural Pesticide Impact Assessment Program. It provides detailed information for a series of pesticides of public health significance. The information includes: a discussion of the use of the product and toxicological effects including acute toxicity, chronic toxicity, reproductive effects, teratogenic effects, mutagenic effects, carcinogenic effects, organ toxicity, and how it reacts in humans and animals. It also includes information on ecological effects on birds, aquatic species, and other animals that are non-targeted by the chemical. The environmental fate of the chemical in soil, groundwater, surface water, and vegetation is discussed. Physical properties are shown and exposure guidelines are stated. The basic manufacturer of the chemical is listed as well as the phone number for emergency situations. (See endnote 21.)

Pesticide Data Program

(See endnote 27)

This is a national database program of pesticide residues in foods. The US Department of Agriculture gathers data from state agricultural departments and other federal agencies, analyzes it, and determines the pesticide residues in agricultural commodities in the US food supply, especially those consumed in large quantities by infants and children. Drinking water sampling also is essential. Fresh and processed fruit and vegetables constitute 86% of the total samples collected. Special well-equipped laboratories are used to make the determinations, and the data are supplied to the US EPA. A company cannot sell or distribute any pesticides in the United States until the EPA has made a thorough study of the potential for unreasonable risk to human health or the environment. The company then may either receive a license or a restricted license, or be denied use.

PROGRAMS

NATIONAL PESTICIDE PROGRAM

(See endnote 28)

The EPA's National Pesticide Program protects human health and the environment from unreasonable adverse effects from pesticide use, storage, and disposal. It helps protect agricultural workers and individuals who are preparing, using, storing, or disposing of pesticides. The field program is implemented by the various states through investigations, establishing worker protection, certification and training of people, and water quality protection. The program ensures appropriate pesticide use and the use of alternatives where feasible. It also provides funds to make the program work. National decisions on the use of specific pesticides are implemented at the local and state level.

ENDOCRINE DISRUPTOR SCREENING PROGRAM

(See endnotes 12, 13, 14)

This program was mandated by Congress after scientists found in the 1990s that certain chemicals were disrupting the endocrine systems of humans and wildlife which resulted in developmental and reproductive problems. The requirements were embodied in the Food Quality Protection Act and Safe Drinking Water Act Amendments in 1996 and the EPA was given the authority to screen, test, and if necessary mitigate or limit a group of chemicals to determine if they mimicked the effects produced by female hormones and produced other potential endocrine effects. The EPA selected chemicals as tier 1 for first review based on recommendations from an advisory committee. A second list of chemicals was selected as tier 2 for the next review. The EPA also established a universe of chemicals and general validation principles in November 2012. (See endnotes 13, 14.) The EPA developed an Endocrine Disruptor Screening Program Comprehensive Management Plan in 2014. (See endnote 12.)

The program evaluates potential human and ecological effects from chemical exposures, and determines how these chemicals react with estrogen, androgen, and thyroid hormone-related processes. This includes pesticides, non-pesticide chemicals, contaminants, and mixtures of chemicals. The plan provides strategic guidance for all work between the years 2014 and 2019. However, it is evaluated on a yearly basis and redirected as necessary. It utilizes advanced computational toxicological methods, screening assays, and scientific validation for humans as well as animals, and uses a weight of evidence guidance technique to determine what chemicals need to be banned or limited. Risk assessments are done on all chemicals for humans as well as the ecological environment. The chemicals are prioritized so that the most dangerous can be dealt with immediately. The management structure is so set up that there is a constant flow of information and essential data from one group to another of the researchers to avoid duplication of effort and waste of time. Decision-making is done in a timely and highly competent manner based on science. A technical review process is built into all steps of the research on the chemicals. Cross agency communications and training programs have evolved. The end result will be the protection of humans and the ecological environment.

CITY OF BERKELEY, CALIFORNIA, RAT CONTROL PROGRAM

The City of Berkeley not only deals with complaints about rodents but also works with tenants or property owners to remove rodents or insects from their properties. The staff carry out in-depth surveys, and provide consultation and recommendations to keep rodents from entering properties and control them if they already are causing an infestation. They remove overgrown vegetation, trash, garbage, and debris, thereby eliminating rodent harborage. They monitor sanitary sewers throughout the city and use appropriate rodent control techniques. They work with the Public Works Department to find broken sewer lines or laterals and make sure that they are repaired to keep the rats from leaving the sewers and entering properties. They conduct surveys throughout the city especially along the waterfronts, in the parks, and in vacant lots. They then carry out proper rodent control activities.

CALIFORNIA MOSQUITO CONTROL PROGRAMS

(See endnote 30)

The California mosquito control programs are built around the document "Overview of Mosquito Control Practices in California" by the Vector-Borne Disease Section, California Department of Public Health. The document discusses the fundamentals of organized mosquito control programs in California including: the types of mosquitoes; citizen responsibilities; mosquito breeding and harborage reduction; the job of the public agencies; and techniques of surveillance. Prevention and control of mosquitoes includes discussions on: integrated pest management; habitat modification; biological mosquito control; and chemical mosquito control. Pesticide issues include pesticide basics; pesticide labels and the significance of following instructions carefully; and pesticide resistance. Larval growth is controlled by bacterial agents, surface agents, insect growth regulators, and only if necessary, chemical larvacides. Adult mosquito control discusses (ultra-low volume spraying and barrier spraying. The adult control chemicals are examined and discussed including malathion, naled, pyrethrins, and pyrethroids such as resmethrin, Sumithrin, permethrin, and deltamethrin. Relative risk is introduced including the potential for disease from the mosquitoes as well as from the chemicals to try to control the mosquitoes.

CITY OF COLUMBUS PUBLIC HEALTH DIVISION OF ENVIRONMENTAL HEALTH—MOSQUITO CONTROL PROGRAM

The mosquito control program includes: monitoring mosquitoes for microorganisms and human disease monitoring; larval mosquito monitoring and control; adult mosquito monitoring and control; analyzing mosquito monitoring data; education and communications; investigating complaints and enforcement; and preparing for emergency situations.

Since West Nile virus was introduced into the United States over 25 years ago, Columbus Public Health as well as 19 other counties in Ohio and many others across the country have been trapping mosquitoes and testing them for various viruses. In 2014, pools of mosquitoes in Ohio were tested and 369 of them were positive for West Nile virus. The largest contributor of samples tested came from Columbus Public Health. In 2014, there were nine cases of West Nile virus in humans and 29 cases of La Crosse encephalitis virus in humans in Ohio.

Larval mosquito monitoring and control was accomplished through inspections of over 10,800 catch basins, 184 rain barrels, 804 scrap tire generators, 93 stagnant pools, and 1208 stagnant water sites. Larviciding of each of these areas was a major means of reducing the mosquito population. Adult surveillance and control programs start with identifying the areas where actual disease has occurred or adult mosquitoes have been found to carry the viruses which cause disease. Mosquito traps are set in these areas and also set in other areas based on information about past positive findings of either viruses or people with diseases from viruses transmitted by mosquitoes. The type of mosquito, quantity, and whether or not they are infected is analyzed and adulticiding programs are initiated. Throughout the mosquito season, truck-mounted equipment sprays tiny droplets of chemicals into the air. The mosquito control unit also treats all city parks especially at the time of holidays.

Using mapping techniques, analysis is done of the monitoring data from year to year to determine if there are changes which need to be noted in establishing larval and adult mosquito control programs.

Since the citizens in the community are an integral part of all public health control activities, they are kept constantly aware of what the problems are and how best to control them. Also if any citizen finds standing water or is concerned about mosquitoes proliferating, he/she can call the 311 call center. All complaints are handled on a priority basis. (See endnote 32.)

In the event of a serious emergency or anticipated serious emergency, the environmental health division administrator prepares a memorandum for the assistant health commissioner to explain the nature of the emergency, the potential for outbreaks of disease, the activities which will be carried out by the division to prevent, mitigate, and control the mosquito problem, and the necessary emergency budget. An example of this occurred on February 1, 2016, when the World Health Organization declared the Zika virus outbreak to be an extraordinary event that needed a coordinated response which constitutes a public health emergency. Increased public information and the initiation of a source reduction program were instituted at once. This was followed by increased surveillance and larviciding and then adulticiding where necessary. Physicians, clinics, and hospitals were sent notices reminding them that any patients with the diseases under concern must be reported immediately to the state health department. Confirmed cases must be separated from any potential mosquito bites.

It will be necessary to increase the number of vector control aides who are trained for mosquito control work, buy additional mosquito traps, prepare public outreach material, and obtain a supply of an effective water-based adulticide. Appropriate budget amounts must be set aside for these activities. (See endnote 33.)

ENDNOTES

1. Centers for Disease Control and Prevention. 2013. *Bed Bugs FAQs*. Atlanta, GA.
2. Cornell University. No Date. *Guidelines for Prevention and Management of Bedbugs in Shelters and Group Living Facilities*. Ithaca, NY.
3. Rust, M.K., Reierson, D.A. 2014. *Statewide Integrated Management Program: How to Manage Pests-Pests of Homes, Structures, People, and Pets-Fleas*. University of California Agriculture and Natural Resources, Davis, CA.
4. Koren, Herman, Bisesi, Michael. 2003. *Handbook of Environmental Health: Biological, Chemical, and Physical Agents of Environmentally Related Disease*. Volume 1. Fourth Edition. CRC Press, Boca Raton, FL.
5. Baumann, Greg, Harrison, Ron. 2012. Fly Control: Understanding Threats to Your Business. 2012. *American School and Hospital Facility Magazine*. V1 Whitepaper on www.facilitymanagement.com.
6. Harrison County Board of Education, Office of Health Services. 2011. *Lice Management Protocols*. Clarksburg, WV.
7. US Environmental Protection Agency. 2012. *Agriculture-Integrated Pest Management*. Washington, DC.
8. California Department of Public Health, 2008. *Best Management Practices for Mosquito Control in California State Properties*. Sacramento, CA.
9. Rust, M.K., Reierson, D.A. 2014. *Statewide Integrated Pest Management Program: How to Manage Pests-Pests of Homes, Structures, People and Pets-Cockroaches*. University of California Agriculture and Natural Resources, Davis, CA.
10. University of Maryland Medical System. 2015. *Lyme Disease and Related Tick-borne Infections*. Baltimore, MD.
11. Stafford, Kirby C. 2007. *Tick Management Handbook*. The Connecticut Agricultural Experiment Station, New Haven, CT.
12. US Environmental Protection Agency, Office of Chemical Safety and Pollution Prevention, Office of Water. 2014. *US Environmental Protection Agency Endocrine Disruptor Screening Program Comprehensive Management Plan*. Washington, DC.
13. US Environmental Protection Agency, Office of Chemical Safety and Pollution Prevention, Office of Water, Office of Research and Development. 2012. *US Environmental Protection Agency Endocrine Disruptor Screening Program Universe of Chemicals and General Validation Principles*. Washington, DC.
14. Wissem, Minif, Hassine, Aziza Ibn Hadi, Bouazi, Aicha. 2011. Effective Endocrine Disruptor Pesticides: A Review. *International Journal of Environmental Research and Public Health*, 8(6):2265–2303.
15. Centers for Disease Control and Prevention, Health Alert Network. 2012. *Health Concerns about Misuse of Pesticides for Bedbug Control*. Official CDC Health Advisory. Atlanta, GA.

16. Rose, Robert I. 2001. Pesticides in Public Health: Integrated Methods of Mosquito Management. *Emerging Infectious Diseases*, 7(1), 17–23.
17. US Environmental Protection Agency, Office of Pesticide Programs. 2014. *Chemicals Evaluated for Carcinogenic Potential*. Washington, DC.
18. US Environmental Protection Agency. 2012. *Pesticides: Health and Safety—Evaluating Pesticides for Carcinogenic Potential*. Washington, DC.
19. Bauder, Troy, Waskom, Reagan, Pearson, Robert. 2010. *Best Management Practices for Agricultural Pesticide Use to Protect Water Quality*. Colorado State University Extension, Fort Collins, CO.
20. World Health Organization, Department of Control Neglected Tropical Diseases. 2006. *Pesticides and Their Application for the Control of Vectors and Pests of Public Health Importance*. Sixth edition. Geneva, Switzerland.
21. US Department of Agriculture/Extension Service/National Agricultural Pesticide Impact Assessment Program. 2002. *EXTOXNET*. Oregon State University, Corvallis, OR.
22. US Environmental Protection Agency, Office of Pesticide Programs. 2011. *Integrated Pest Management in Buildings*. EPA 731-K-11-001. Washington, DC.
23. US Environmental Protection Agency. 2014. *Pesticides: Topical and Chemical Fact Sheets-Alphabetical List of Pesticide Fact Sheets*. Washington, DC.
24. National Pesticide Information Center, Oregon State University. 2015. *Home Page*. Corvallis, OR.
25. Centers for Disease Control and Prevention. 2006. *Pesticide-Related Illness and Injury Surveillance: A How-To Guide For State-Based Programs*. DHHS (NIOSH) number 2006-102. Atlanta, GA.
26. Centers for Disease Control and Prevention. 2014. *Sentinel Event Notification System for Occupational Risk (SENSOR-Pesticide Illness and Injury Surveillance)*. Atlanta, GA.
27. US Department of Agriculture. 2014. *Pesticide Data Program's Annual Summary-2012*. Washington, DC.
28. Watson, William, Litovitz, A.T., Reuben, C. 2012. *About Pesticides-Pesticide Field Programs Contribution to National Pesticide Program Mission*. US Environmental Protection Agency, Washington, DC.
29. Centers for Disease Control and Prevention. 2004. *MMWR*, 53 (supplement). Atlanta, GA.
30. Vector-Borne Disease Section, California Department of Public Health. 2008. *Overview of Mosquito Control Practices in California*. Sacramento, CA.
31. Centers for Disease Control and Prevention. 2016. *Zika Virus: Questions and Answers*. Atlanta, GA.
32. Columbus Public Health Division of Environmental Health. 2015. *2014 Vector Control Annual Report*. Columbus, OH.
33. Krinn, Keith L., Personal Communication, Environmental Health Division Administrator to Roger Cloern, Assistant Health Commissioner. 2016. *Preparations for Zika Virus Response*. Columbus, OH.

BIBLIOGRAPHY

Centers for Disease Control and Prevention. 2010. *Yellow Fever History, Epidemiology and Vaccination Information*. Atlanta, GA.
Centers for Disease Control and Prevention. 2012. *The History of Malaria, an Ancient Disease*. Atlanta, GA.
Damalas, Christos A., Eleftherohorinos, Ilias G. 2011. Pesticide Exposure, Safety Issues, and Risk Assessment Indicators. *International. Journal of Environmental Research Public Health*.
Purdue University. No Date. *Public Health and Medical Entomology*. West Lafayette, IN.
Rose, Robert I. 2001. Pesticides and Public Health: Integrated Methods of Mosquito Management. *Emerging Infectious Diseases*, 7(1).

10 Recreational Environment and Swimming Areas

STATEMENT OF PROBLEM AND SPECIAL INFORMATION

There are several major components of the recreational environment that can contribute to the spread of disease and increase the amount of injuries within the United States and all other world communities. People travel from one point in the world to another or within a given country. Unless they are walking to the travel destination, they need to use some form of transportation, so they travel by motor vehicle (automobiles and buses), trains, planes, and ships. The individuals need to have available but may be lacking: potable water; safe food; safe and secure housing; disposal of solid and hazardous waste; and disposal of sewage including human waste. Throughout the centuries various microorganisms, insects, and rodents have accompanied the travelers from one area to another which has resulted in innumerable outbreaks of disease, some of them devastating epidemics or pandemics. Animals and plants have been moved from one local ecosystem to another, sometimes resulting in disastrous consequences.

Hundreds of millions of people, United States citizens and visitors from abroad, utilize a variety of recreational areas each year. Many of the recreational areas were not designed to be utilized at this level of intensity. Some of these individuals are subclinical carriers of various diseases which may spread throughout the complex recreational environment. Typical environmental problems include inadequate or improper sewage disposal; poorly drained land; contaminated water supplies; improper solid waste disposal; inadequate or malfunctioning refrigeration; poorly constructed and poorly maintained food preparation areas; insect and rodent hazards; animal problems; noxious weeds; inadequate, overcrowded, or poorly maintained housing; and improperly maintained or contaminated water recreational areas. In addition, there are problems of inappropriate feeding and petting of livestock, mechanical erosion of soils, substantial increase in traffic and resulting safety concerns, disturbance of wildlife, camping which may affect the ecosystem, litter, and forest fires. For specifics in each of these areas, see the relevant chapters of this book by topic area and see *Handbook of Environmental Health Biological, Chemical and Physical Agents of Environmentally Related Disease*, Volume 1, Fourth Edition (CRC Press) and *Handbook of Environmental Health Pollutant Interactions in Air, Water, and Soil*, Volume 2, Fourth Edition (CRC Press). (See endnotes 65, 66.)

There are a huge number of potential safety problems related to the unfamiliar environment, the behavior of the individuals on vacation, and/or the level of physical fitness of the visitors to these areas. Some 24% of all injuries are caused by falls, while 19% of all injuries are related to motor vehicles. Other types of injuries may be related to hiking, swimming, boating, climbing, biking, skiing, horse riding, camping, and snowmobiles. (See endnote 3.)

Many recreational areas are of a seasonal nature. The physical facilities may have deteriorated. The equipment may become outdated, inadequate, or simply not operable. The individuals operating various pieces of equipment are typically either students or transient workers. Because of a lack of experience by the workers, the risk of spreading disease or causing injury increases.

Since it would be much too lengthy to go into a discussion of all types of recreational areas and potential programs that are utilized by a variety of health departments, the means of transportation to and within recreational areas including airlines and cruise ships, summer camps for children, and swimming areas will be the major concern of this chapter. (The transportation issues will be limited to airlines and cruise ships since environmental health as well as safety issues of motor vehicles have been discussed in: Chapter 2, "Air Pollution (Outdoor [Ambient] and Indoor)"; Chapter 4, "Children's Environmental Health Issues"; and Chapter 6, "Environmental and Occupational Injury Control.") Passenger trains especially those which are long distance have some of the same types of problems as cruise ships and therefore will not be discussed. Freight trains are an entirely different concern because of the cargo that they carry and will not be part of this discussion. (See Chapter 5, "Environmental Health Emergencies, Disasters, and Terrorism" for further information.) Transportation problems related to cruise ships will be part of the cruise ship discussion. The problems as well as the Best Practices used to resolve the environmental, and health and safety issues for these four topic areas can be used for all recreational areas and transportation concerns with certain modifications.

SUB-PROBLEMS INCLUDING LEADING TO IMPAIRMENT AND BEST PRACTICES FOR AIRLINE TRANSPORTATION

Approximately 1 billion people travel by commercial aircraft every year and this is expected to increase to 2 billion people within 20 years. The opportunities to acquire a communicable disease or be subjected to unwanted chemical contamination not only includes the point of origin but also the potential contaminants found in buses, trains, taxis, and public waiting areas before boarding the plane. Within the plane are the concerns of potentially contaminated indoor air, close proximity to other passengers who may have subclinical cases of a disease or be clinically sick, lack of hand washing after using toilet facilities, or remnants of disinfectants and insecticides which may have been used in or on the plane. There is a concern with the transmission of: tuberculosis from person to person by airborne respiratory droplet nuclei; meningococcal disease caused by *Neisseria meningitidis* transmitted by direct contact with respiratory droplets and secretions, which may be quickly fatal; measles; influenza; and other upper respiratory or oral–fecal route diseases. (See endnote 1.)

The 2003 SARS (severe acute respiratory syndrome) outbreak started in China and spread to 25 countries and Taiwan, principally along air transport routes. This phenomenon is not new since the pandemic of bubonic plague in the 14th century followed the trade routes across the Mediterranean Sea. SARS is a viral respiratory disease caused by a recently discovered coronavirus which results in flu-like symptoms and commonly leads to pneumonia. (See endnote 4.)

In March and April 2003, a new swine-originated 2009 H1N1 influenza A virus was spreading rapidly through the United States and Mexico and a total of 30 countries worldwide. It was the first pandemic of the 21st century and was extremely fast moving because of a rapid rate of globalization through sharply increased travel and trade, leading to very large outbreaks of the disease which causes severe illness and may cause death especially in the very young, the very old, immune-suppressed people, and those who are debilitated. The disease originally is transmitted from animals or birds to people and then transmitted from person to person by aerosols which contain the virus or by direct contact with the saliva, nasal secretions, and feces of sick individuals. (See endnote 5.)

These two examples of outbreaks of communicable diseases spreading along trade and travel routes, obviously including airlines, are but a foretaste of the potential for the spread of disease which has been an unwanted consequence of globalization. It has always been this way and always will be this way.

Other health problems may not be environmentally related but will still reduce an individual's resistance and may increase the ability to be affected by communicable diseases or chemicals found in the environment. These conditions may include fatigue from walking long distances, carrying baggage, airline delays, noise, vibration on the plane, cramped sitting spaces, inability to exercise muscles, and frustration; low cabin humidity resulting in dehydration; exacerbation of various chronic diseases especially those of a respiratory nature; barotrauma which occurs when the pressure in the middle ear, sinuses, or abdomen is not the same as the air pressure in the aircraft cabin and may cause pain and potentially allergic reactions for individuals who have allergies; deep venous thrombosis (blood clots) due to long-term immobility of the lower limbs; decompression sickness (bends) for scuba divers especially when returning from the sport location; peanut allergies; use of alcohol which can contribute to severe dehydration; motion sickness; eye diseases which can be exacerbated by changes in cabin pressure and the dry cabin air; mental health conditions which can be exacerbated by numerous changes brought on by air flight; heart disease which may be exacerbated by changes in air pressure and increased heart rate; diabetes which can be exacerbated by the long flights and missing necessary meals; and surgical conditions where anesthesia had been used recently. (See endnote 2.)

Best Practices for Preventing Spread of Communicable Disease or Exacerbation of Existing Chronic Conditions in the Use of Airline Transportation (See endnotes 6, 7)
- Advise the Division of Global Migration and Quarantine, part of the Centers for Disease Control and Prevention (CDC) National Center for Emerging and Zoonotic Infectious Diseases in Atlanta, of any actual or potential outbreaks of communicable disease including those which are quarantinable: cholera, diphtheria, infectious tuberculosis, plague, smallpox, yellow fever, SARS, and viral hemorrhagic fevers (Lassa, Marburg, Ebola, Crimean-Congo, South American, and others not yet identified). (See endnote 9.)
- Utilize the quarantine stations supervised by the CDC as needed to help prevent and control outbreaks of communicable diseases at the ports of entry in Anchorage, Atlanta, Boston, Chicago, Dallas, Detroit, El Paso, Honolulu, Houston, Los Angeles, Miami, Minneapolis, New York, Newark, Philadelphia, San Diego, San Francisco, San Juan, Seattle, and Washington, DC. (See endnote 10.)
- Evaluate all sources of information including the media and electronic reports to determine if an outbreak of disease is occurring or likely to occur and report this to the CDC.
- If there is an existing clinical disease outbreak in a specific country or region, provide immediate active surveillance of all individuals, those who are flying, and people in all of the support services which are involved in processing materials or people for the flights throughout the airports. Active surveillance may include taking the temperatures of the individuals and other procedures outlined by the appropriate medical staff. In the case of an active surveillance program, all of the workers need to wear appropriate personal protective equipment. (See endnote 6.)
- Establish a disease-specific response including appropriate epidemiological and medical screening techniques for all individuals having symptoms of communicable diseases or having traveled from an area of severe outbreaks of communicable diseases to prevent a major public health problem and contain the spread of the microorganisms to others. This means that there has to be a constant flow of information from areas of potential or actual outbreaks of disease to all other areas of the world. Medical and epidemiological investigation must be done immediately and extremely thoroughly and the results broadcasted to all appropriate world entities as rapidly as possible, especially to the airports since this mode of travel can result in immediate contamination of areas and people by healthy and sick carriers of the disease. Disease alerts should be issued by all forms of media and electronic devices to all potential passengers and airline workers.
- Follow the Airports Council International document "Airport Preparedness Guidelines for Outbreaks of Communicable Disease." (See endnote 7.) This guideline involves the following: communications especially with the public; screening processes; transportation of infected people to health facilities; equipment which should be used for the prevention and control of disease; entry and exit procedures for all aircrafts, especially the one which is affected, personnel and passengers; and coordination with all local, state, national, and international public health agencies.

- Travelers from areas of known outbreaks of disease such as Ebola should not be permitted to board an aircraft if potentially exposed to the disease and showing symptoms. The individual should undergo medical evaluation prior to departure to determine when it is safe for him/her to travel on a commercial carrier if he/she is asymptomatic and is in a low but not necessarily zero risk category. The individual should still be monitored actively for 21 days. Individuals in the no identifiable risk category do not have to be monitored or have travel restrictions. (See endnote 8.)
- When traveling abroad, obtain all appropriate vaccinations prior to leaving and carry the vaccination certificates which may be required by some countries.
- Have a physician determine if anti-malarial drugs should be taken prior to the flight to help prevent the spread of this disease. A special concern is that some of the drugs may have a profound effect upon pregnancy and therefore should not be used.
- People with uncontrolled or frequent seizures should avoid traveling on airlines.
- People who have had a recent stroke or TIA (transient ischemic attack) should not be flying until the acute phase has passed and a physician has cleared the individual.
- Persons with recent surgery, especially if it relates to the spine and back, should be cleared by their physicians before they fly.
- Individuals with heart disease should be cleared by their physician and carry with them all medications, a copy of their most recent electrocardiogram, a pacemaker card if applicable, and limit walking by requesting wheelchair assistance. They should also request a special diet and medical oxygen if necessary.
- Patients with ear, nose, and throat problems should not fly until the conditions are cleared up unless absolutely necessary and then carry proper medications.
- Passengers with existing chronic conditions which may worsen during flight should receive a medical evaluation prior to departure.
- Individuals requiring special meals because of diabetes, other conditions, or for religious purposes should contact the airlines prior to departure.
- Request wheelchair assistance or trolley service within the airport if needed to reduce levels of fatigue.
- If an individual requires oxygen, determine in advance if the airline provides it in flight and if so make a request for this service. The individual will not be able to bring his/her own oxygen aboard the plane.
- Wash hands thoroughly with soap and water or use an alcohol-based disinfectant after touching surfaces, after using toilet facilities, and before eating or drinking.
- Exercise legs, walk around the cabin periodically to avoid blood clots forming in the legs, and make sure that the individual is well hydrated. Keep seatbelts fastened continuously while seated especially if pregnant and make sure that the belt is snug over the pelvis or upper thighs to prevent injury to the abdomen.
- If the individual has been scuba diving, in order to reduce the risk of severe joint or muscle pain and even a stroke or collapse of the cardiovascular system, avoid flying for at least 12–24 hours after the last scuba dive.
- Worldwide but not in the United States, the World Health Organization recommends that the cabin of an aircraft be sprayed with an aerosol, usually pyrethroids, to kill insects that may carry disease and have been transferred from the point of origin to the destination in the aircraft.
- For individuals with peanut allergies, do not consume them aboard the aircraft and ask for a seat in the peanut-free area.
- Individuals with asthma or emphysema should consider not flying unless absolutely necessary and if necessary carry on board all asthma medications. These individuals should be especially well hydrated during the flight and avoid caffeine and alcohol.
- Individuals with diabetes should carry all medications as well as needles, syringes, blood glucose monitors, snacks etc., in carry-on luggage and not in checked luggage. For security

reasons have a note from the physician that these are necessary. Be careful to eat all meals especially prepared by the airlines to help control blood sugar levels.
- Use alcohol with great care since it will cause dehydration and may interfere with sleep.
- Avoid motion sickness by taking appropriate medication determined by a physician prior to travel, flying in large airplanes, sitting in a window seat and looking at the horizon, having cool air focused on the face, and avoiding alcohol.
- Avoid flying for at least 72 hours after a cast is put on a body part because of a fracture to prevent excess swelling and pain.

OCCUPATIONAL HEALTH AND SAFETY CONCERNS

(The discussion on occupational health and safety concerns will not include information concerning the actual operation of the aircraft since the author has absolutely no experience in this area and the topic is far too complex for a book of this nature, although there are numerous potential health and safety problems related to the crew as well as the passengers. The Commercial Aviation Safety Team is deeply involved in this process and focuses their resources on the following areas of risk: weather conditions, turbulence, loss of control, icing, maintenance, etc. They analyze data, identify hazards, develop safety recommendations, implement cost-effective safety enhancements, evaluate implementation of the regulations, and use Best Practices of specific knowledge and practices to improve the overall aviation system.) (See endnote 14.)

Employees are exposed to severe noise, temperature extremes, fuel vapors, diesel emissions, carbon monoxide, chemical cleaners, solid and hazardous wastes, bodily fluids, blood-borne pathogens, SARS, influenza, etc. They also are exposed to numerous motor vehicles with potential for accidents, falls especially from a height, ergonomic concerns, and physical stresses related to repetitive actions, movement of baggage, and disabled people. Also individuals in each of the specific occupations involved with the aircraft may have occupational hazards related to their work, such as electricians being electrocuted while performing specific tasks.

Best Practices to Reduce or Eliminate Occupational Health and Safety Concerns from Airlines (See endnotes 11, 12, 13, 15)
- Reduce noise exposure and potential hearing loss, ringing in the ears, fatigue, and inattention, by wearing appropriate noise reduction devices and limiting the amount of time that an individual is in an environment that exceeds 85 dBA in an 8-hour work shift. Require that individuals who are in noisy environments be medically tested on a periodic basis to determine any potential hearing loss problems and assign the individual to a quieter work situation.
- Reduce heat exposure and subsequent potential illness, inattention, and fatigue by determining the environmental temperature and humidity, radiant heat sources, amount of contact with hot objects, direct sun exposure, limited air movement, and use of special bulky equipment and remove the individual from this environment periodically and frequently. Require that the individual has frequent physical examinations, is well hydrated, is acclimated to the heat in increasing periods of time, and does not drink alcohol. Instruct all supervisory personnel to immediately provide medical attention for any employees if any symptoms of heat-related conditions occur. These same guidelines apply to situations where the workers are performing in very cold or wet conditions.
- Reduce falls and severity of injuries for maintenance work, inspection work, cleaning and washing of aircraft, food catering, when opening doors to the plane, etc., by using appropriately constructed work platforms, lift equipment, ladders, fall arrest systems, and appropriate restraint systems. Provide a qualified program manager for fall protection who will supervise the individuals at risk for falls and the equipment necessary to prevent this type of incident. If the individual can fall 4 feet or more, he/she should be trained to recognize

the hazards, how to use the fall protection equipment, and wear appropriate shoes and other clothing. All leaks of hydraulic fluids, oils, or other fluids must be immediately and completely cleaned up. Use designated walkways when walking on the wing of the aircraft. Use extra caution when the aircraft surfaces are wet, or covered with snow, frost, or ice. Personnel should never be allowed to walk on wet surfaces unless it is extremely urgent and then he/she must wear proper safety harnesses.

- To prevent musculoskeletal disorders due to ergonomic problems in the workplace, the employers should: establish a program to avoid these types of injuries with clear goals and objectives and the process necessary to protect the employees; identify the problems in advance and determine how best to correct these situations while involving the workers in the survey as well as the solutions; provide appropriate training and continuing education to all employees concerning the hazards of their job and how to avoid serious consequences; encourage the early reporting of potential injuries and muscular skeletal symptoms; and implement solutions to ergonomic problems quickly and evaluate the progress of the program regularly.
- Workers including the individuals moving passengers in wheelchairs as well as aircraft cabin cleaners must not clean up the blood, urine, feces, or vomit of sick passengers without proper personal protective equipment, appropriate cleaning and disinfecting materials and equipment, and appropriate disposal of all hazardous waste. The airline and airport employees should have all necessary vaccinations provided by the employer. Medical treatment and follow-up medical treatment should be provided by the employer to all employees who have been exposed to a variety of diseases. Have the Occupational Safety and Health Administration (OSHA) investigate all instances of the transmission of blood-borne diseases.
- All airport workers should be taught how to protect themselves from the spread of infectious diseases which may be brought into the airports by people from abroad or American citizens returning from areas where there are outbreaks of disease.

Environmental Concerns

Environmental concerns and Best Practices regarding airliners and airports are similar to those for any large institution such as a hospital but with a lesser degree of intensity unless there is an outbreak of a communicable disease. The typical airliner may contain from fifty to hundreds of people in a confined environment and therefore the same types of hazards exist as in any other type of confined environment with numerous people except the traveling public comes from a vast variety of areas and brings with them the viruses and other microorganisms that they have been exposed to at their point of departure and where they live. The airports at Chicago and Atlanta average about 250,000 people a day. That's the size of a medium city in the United States. The airports have all of the support services that a city would normally have and also all of the potential problems. The environmental problems range from indoor air quality to environmental and occupational injuries to food security and protection to health care and infection control to insect and rodent control and use of pesticides to sewage disposal systems, solid waste, hazardous waste, water systems, noise, and on and on. If you read Chapter 8, "Healthcare Environment and Infection Control" and modify it for the concerns of the airport environment you will gain considerable knowledge on how best to prevent and mitigate environmental health, safety, and personal health problems.

Air Pollutants Released by Aircraft

Aircraft engines emit small amounts of nitrogen oxides, organic gases, carbon dioxide, carbon monoxide, sulfur gases, particulate matter including metals, and unburnt hydrocarbons. (See endnote 28.) Follow new standards adopted by the US Environmental Protection Agency (EPA) for nitrogen oxide emissions from aircraft gas turbine engines. (See endnote 29.) New standards are also expected in the future for greenhouse gases, etc., from the US EPA.

SUB-PROBLEMS INCLUDING LEADING TO IMPAIRMENT AND BEST PRACTICES FOR CRUISE SHIPS

Cruise ships have become floating hotels with entertainment, but still must meet the health and safety standards established for all ships at sea. The ship design, equipment, training of personnel, housing, and provision for a large number of different services from food to housekeeping have to be evaluated on each ship to make sure that there is compliance with existing regulations and that the passengers as well as the crew will return to the port in a healthy and safe manner. This may be very difficult to accomplish in this floating city unless there is strict adherence to appropriate supervision and management techniques and all procedures needed to prevent illness and injury are understood by crew and passengers and are followed closely.

The North American cruise industry of over 200 ships carrying about 15 million people annually is the largest in the world. Cruise ships vary in size. Some of them may carry thousands of people including passengers and staff. Passengers tend to be older and therefore are already subject to the numerous health and physical conditions that have been described in other parts of this book. Most passengers come from the United States or Canada, however they may be coming from all parts of the world and bringing with them the microorganisms which are common in their own homelands. Crew members tend to be from an international group and also may be contributing microorganisms from many areas. The cruise ship environment is such that it is crowded with people who may be perfectly healthy, have subclinical symptoms of various diseases, or sick and their microorganisms can be transmitted from person to person or from contaminated food, water, or surfaces. The stress of travel can reduce resistance to disease especially among the elderly, children, individuals with underlying health conditions, and pregnant women who are more susceptible to various levels of contamination. The elderly have also had immunological changes resulting in loss of certain antibodies, a decrease in immunity, and chronic diseases. Drugs that reduce gastric acidity can make the individuals more susceptible to gastrointestinal pathogens. Diuretics, frequently used by heart patients, can increase the severity of the disease because a typically mild diarrhea could cause severe dehydration. (See endnote 1.) Repeated outbreaks especially of noroviruses on the same ship on different cruises with different passengers indicates that the disease is easily transmitted from person to person in a closed environment and that the problem may be with infected members of the crew and/or contaminated surfaces on the ship.

HEALTH RISKS

Food and Waterborne Diseases

Infectious diarrhea, most frequently caused by the norovirus (Norwalk virus) is found more frequently on cruise ships than in the normal population of travelers. The virus can survive on practically all surfaces including door handles, sinks, railings, glassware, etc. and therefore may easily be spread in the closed environment of the ship. Enterotoxigenic *Escherichia coli*, *Shigella*, *Salmonella*, and other organisms have also caused outbreaks of disease, however the most frequent cause in the recent past has been norovirus. (See endnote 17.)

Hepatitis A is spread by the oral–fecal route and may also be transmitted through contaminated food or water. Exposure may occur on board the cruise ship or by eating food or drinking water which is contaminated at the various ports of call.

Respiratory Infections, Rubella, and Tuberculosis

These diseases may be spread because of the places from which the passengers and crew have come, the susceptible elderly passengers, the common ventilation system, the large population in close quarters, and through aerosols generated in spas, pools, and from buffet misting devices. Influenza and pneumonia have both occurred aboard ship. Legionnaires' disease has been linked to ships where several people died and others have become ill. *Rubella* outbreaks have been limited to crew members. There appears to be limited transmission of tuberculosis in public transportation and then only when there is close contact. There is always a potential for SARS if infected individuals are part of the cruise ship passengers or crew.

Mosquito-Borne Diseases

When cruise ships travel to tropical and subtropical countries, there is potential risk of malaria, dengue fever, and yellow fever especially in the ports and in any land-based side trips. (See endnote 16.)

Best Practices in Avoiding Outbreaks of Disease on Cruise Ships

- Passengers who are acutely ill with contagious diseases should not be permitted to board the ship.
- If a passenger becomes ill with a communicable disease, the individual should be treated by the medical staff and then isolated.
- All passengers should be given specific information concerning the spread of disease, symptoms, instructions on what to do in the event the individual becomes ill, how to properly wash your hands, and when to utilize alcohol-based sanitizers in a variety of situations where the person comes in contact with surfaces, food, drinking water, etc.
- All crew members should be given the same instructions as in the item above.
- Passengers and crew should have appropriate vaccinations or protective medications when going into areas such as tropical and subtropical communities for specific types of diseases such as malaria, yellow fever, etc.
- Proper hand washing techniques using soap and water are the most effective means of preventing and controlling outbreaks of disease on cruise ships. Hands should be washed thoroughly after using the bathroom, changing diapers, before eating, preparing or handling food, and at any time before putting hands to the mouth. The use of alcohol-based hand sanitizers on the hands is effective after the soil or potential soil has been removed from the hands and is secondary to thorough hand washing in these situations.
- Wash fruits and vegetables and cook seafood thoroughly before consuming them.
- Discard all foods that may have become contaminated or suspected of being contaminated with noroviruses.
- Keep infants and children who are sick away from areas where food is being handled, served, or prepared.
- When sick, do not prepare food or care for other people who are sick.
- After vomiting or having diarrhea, cleaning staff should clean and disinfect all contaminated surfaces. First use a good detergent and water and then rinse the area. Finally, use a chlorine bleach solution of 5–25 tablespoons of household bleach per gallon of water on the area and then allow it to air dry.
- Handle all soiled laundry gently and put it through a complete washing cycle. Dry thoroughly on maximum cycle. Wear disposable gloves and then wash hands and forearms very carefully with soap and water. (See endnotes 18, 19.)
- If the same ship has more than one outbreak of noroviruses, then it must be taken out of service and thoroughly decontaminated before it can be used again. The crew members should be checked carefully to determine if they have gastrointestinal symptoms or if they have had them in the past several weeks.
- When leaving the ship for short trips, consume only food and water which can be certified as safe. Wash hands frequently during the trip and before returning to the ship.
- Make sure that all passengers are well hydrated at all times.

- Since food is the most common means of transmitting noroviruses, be extra vigilant concerning preparation, storage, refrigeration, and heating, and protection of salads, peeled fruits, deli-type sandwiches, finger foods, hors d'oeuvres, and dips.
- Cleaning staff should clean and decontaminate all surfaces which typically come in contact with passengers and crew including carpets, toilet seats, toilet handles, faucets, sinks, phones, door handles, horizontal surfaces such as railings, toys etc. (See Chapter 8, "Healthcare Environment and Infection Control" for details on cleaning.)

Safety Risks

In the recent past, some cruise ships have had a series of safety problems which could affect the health and well-being of the passengers and crew. Engine room fires have resulted in a loss of electricity, functioning toilets, air-conditioning, and preparation of hot food, and the ships have even had to be towed into port. From January 2009 to March 2013, seven cruise ships ran aground. Propulsion problems, collision with piers, maintenance problems, engine problems, electrical panel problems, plumbing issues, and cabins with toilets that would not flush occurred on 21 cruises from March 2009 through April 2013. There were 353 incidents involving mechanical problems and accidents from January 2009 to June 2013.

Best Practices to Avoid and Mitigate Safety Risks
- The ship design, equipment, and training of the crew must be such that the ship can be abandoned in the event of a major disaster without loss of life or injury. This can be accomplished when the captain or his designee determines to "abandon ship" and the following actions are taken: the crew immediately goes to evacuation duty stations to assist passengers; proper signs are installed for passengers to go to the specific stations; emergency lighting is provided; smoke detectors are provided throughout the ship; all passengers are trained in emergency measures and families are kept together; the crew is trained to deal with the rolling of the ship, darkness, severe weather conditions, fire and smoke, and especially in helping the elderly, disabled, intoxicated, and special needs passengers.
- Provide handholds in all areas for passengers to prevent falls.
- Alcohol consumption, which is a substantial money maker for the cruise ships, must be limited when passengers seem to have imbibed too much.
- Enforce a "no alcohol" policy for all crew members with random testing.
- Design the ships or retrofit older ships so that the lifeboats are readily accessible and the muster station near the lifeboats has adequate space for people to assemble.
- Each muster station must have a comprehensive first-aid pack and a defibrillator plus individuals who have been trained to use the equipment.
- Re-evaluate the space requirements needed for people to be seated in lifeboats and life rafts, and recalibrate the number of individuals who can be put on each of the boats based on current average weight and average height of men, women, and children. Provide adequate space for disabled individuals.
- Advise all passengers that they must bring warm clothing aboard the ship in case they have to abandon ship.
- Lifejackets must be distributed properly, efficiently, and rapidly and help should be available for individuals, especially those who are disabled and are having trouble putting them on.
- Double hulls should be required for all the new large cruise ships.
- Use appropriate navigation lights which are not masked by the light in the cabins and on the decks.
- Provide more than enough lifejackets in conveniently located areas that will properly be sized for each person aboard the cruise ship. The lifejackets must be approved by the US Coast Guard. (See endnote 22.)

- Provide adequate numbers of lifeboats supplemented by appropriately constructed life rafts to accommodate all passengers and crew and piloted by specially trained crew members. There should be at least three crew members, who are not entertainers, on each lifeboat and at least one crew member on each life raft.
- All officers involved in the actual operation of the ship should wear uniforms that easily distinguish them from hotel staff.
- Use acoustic hailing devices that provide communication throughout the ship to help protect the cruise ship from terrorist attacks.
- Secure all heavy objects either permanently when not in use or during heavy weather.
- Cruise ships operating from US ports should have accident investigations conducted by the National Transportation Safety Board as a condition of using the ports.
- The United States Coast Guard without advance warning should conduct safety inspections. (See endnotes 20, 21, 22.)
- Observe new policies established by the Cruise Lines International Association to improve cruise ship safety including: Passenger Muster; Personnel Access to the Bridge; Excess Lifejackets; Common Elements of Musters and Emergency Instructions: Lifeboat Loading for Training Purposes; Harmonization of Bridge Procedures; Location on Lifejackets Stowage; and Securing Heavy Objects. (See endnote 23.)
- In case of passengers falling overboard, provide special systems for all cruise ships which will detect falling objects by their heat temperatures and record their image as well as sound an alarm. The passengers must be brought together at their muster point and the lifeboats be put in proper position so that all individuals know what to do in the event of a disaster aboard the ship.

ENVIRONMENTAL CONCERNS

(See endnotes 30, 40)

Air Pollution from Cruise Ships

(See endnotes 31, 32, 33, and Chapter 2, "Air Quality (Outdoor [Ambient] and Indoor)")

Cruise ships as well as other ships are a prime source of air pollutants in ports and in the bodies of water. Most oceangoing ships are not registered in the United States and may not even buy their fuel here. Although modest controls for pollutants have been established internationally in the MARPOL (International Convention for the Prevention of Pollution from Ships) agreement, they were very modest and still allowed substantial air pollutants to be released. For instance, the bunker fuel used in ships contains 27,000 ppm of sulfur, which is 2000 times as much as that allowed for trucks on US roads. Oceangoing ships are among the largest sources of nitrogen oxides being emitted to the air. Some 70% of the emissions of sulfur oxides typically come from ships. Particulates are also a very serious problem. Cruise ships may produce black carbon, a pollutant from the incomplete burning of fossil fuels, which is totally banned in the United States. MARPOL Annex VI limits the sulfur content of fuel to 45,000 ppm, whereas highway diesel fuel in the United States is limited to 15 ppm. In certain special areas, the level of sulfur dioxide is limited to 15,000 ppm, which is way beyond that which is acceptable. Nitrogen oxide levels for new engines and engines that have been renovated have to limit emissions to a range 20 times as great as that of electrical power plants in the Eastern United States. The agreement does regulate the emissions of volatile organic compounds, prohibits the emissions of ozone-depleting substances, and prohibits the incineration on ships of polychlorinated biphenyls. Even California, which has the most stringent rules in the country, still allows 1000 ppm of sulfur in the fuels. Obviously there is an enormous public health risk involved in breathing in these air pollutants and they contribute to extremely poor health as well as the death of some individuals. They also may affect the amount of ozone that is present near the ground. In the immediate vicinity of ports, there may be residential neighborhoods, schools, and playgrounds which would make individuals in them highly susceptible to the pollutants created by the cruise ships.

Best Practices for Improving Air Quality
- Provide in the port an alternate power source for ships, which would allow them to hook in and then not utilize auxiliary engines to generate electricity and produce air pollutants.
- Clean up all harbor craft including the tugboats used to bring in the ships to the ports. Retrofit all existing engines and use only low polluting engines if new ones are required.
- Limit the idling time of oceangoing vessels and tugboats.
- Require that ships coming into port must use the cleanest grade of diesel fuel available.
- Require emission control devices on cruise ships and other ships.
- Require that all equipment being used to move cargo be no older than 10 years of age and then replace it.
- Require that all ships have air pollution control devices that are functioning and test the exhaust from the ship to determine that it does not exceed appropriate air pollution control limits.
- Ensure that all ships meet engine and fuel standards as developed by the US EPA. (See endnote 34.)
- For shipboard incineration, the flue gas outlet temperature should exceed 850°C.
- Incineration residues should be treated as hazardous waste and not allowed to become airborne.
- Firefighting or refrigeration systems must not contain chlorofluorocarbons.
- All painting and stripping of the ship during maintenance should be done by professional individuals who utilize compounds that will not cause air pollution problems. Do not use methylene chloride as a paint stripping agent.

CHILD ACTIVITY CENTERS

(See Chapter 4, "Children's Environmental Health Issues" and endnote 24)

FOOD SAFETY

(See Chapter 7, "Food Security and Protection")

Food safety aboard ship is a highly complex process which must be examined in great detail and all of the elements including the human elements must be considered in preventing outbreaks of disease. This process consists of: production or growth of the foodstuffs, processing, packaging, transportation, storage, preparation, serving, and ultimately disposal of the remnants of the food. The food must come from approved sources which are inspected by the appropriate public health authorities. Throughout the process it is necessary to evaluate: the use of potable water, food temperatures, holding times, cleanliness and disinfection of equipment and surfaces, health of and use of proper handling techniques by employees, washing of hands, etc. Of special concern are buffets, small eating areas throughout the ship, and any leftover foods from previous meals. The time of the shift change of personnel, as well as changing menus from breakfast to lunch to snack time to dinner is also crucial and needs to be supervised carefully. (See endnotes 24, 25 for details.) A major new Best Practice is to lower the sneeze guard on buffet lines in order that the passengers cannot in any way handle the dispensing utensils or touch the food. Servers should be located behind the buffet tables and serve the guests as they make their choice of food. This is one of the primary techniques used to prevent disease and is called isolation.

Hand Washing and Toilet Facilities
(See Chapter 8, "Healthcare Environment and Infection Control")

HAZARDOUS MATERIALS USE, WASTE STORAGE, AND DISPOSAL

(See endnote 27, section 6)

The hazardous wastes may be liquid, solid, semi-solid, or contained gases. Typically, they are stored aboard ship until they can be offloaded at a port for recycling, treatment, storage, or disposal. The waste may come from photo processing, dry cleaning, equipment cleaning, paints and thinners, aerosol liquid waste from crushing aerosol containers, incinerator ash, florescent and mercury vapor light bulbs, batteries of all types, pharmaceuticals, other chemicals, etc.

Best Practices for Hazardous Materials Use, Waste Storage, and Disposal
- Create an industry-wide cruise ship hazardous materials use, waste segregation, minimization, storage, and disposal program based on federal and state laws including the use of Best Practices and the maintenance of complete records.
- Prepare and implement spill prevention procedures for oil and other substances in ports and at sea in the event of collisions, grounding, fire, explosion, etc.
- Conduct proper techniques of transfer of fuel from shore to ship or from ship to ship when fuel is needed. Follow all existing regulations.
- Adequately secure all hazardous materials in a well-ventilated area accessible only to special personnel.
- Avoid using anti-fouling paint containing tributyltin on the hull of the ship since it may persist in the sea water environment including the sediment as a contaminant.
- Create an education and enforcement program for the cruise industry regarding hazardous wastes generation, separation from other solid wastes, storage, treatment, and disposal.
- Determine the effects on air quality for passengers and crew as well as for the environment from incineration of hazardous waste aboard cruise ships.
- Review, understand, and adhere to the state requirements for disposal of all types of hazardous wastes where ports will be used.
- Prohibit the discharge of hazardous materials into US waters to the 200-mile limit.
- Prohibit incinerating hazardous waste aboard ship while in port.
- Establish a mandatory incinerator ash-testing program including keeping appropriate logs.
- Understand that the cruise industry, whether carrying the flag of the United States or another nation, is subject to all hazardous waste generator requirements and frequent inspections by official agencies including the United States Coast Guard, who will make unannounced visits and examine all logs as well as observe actual processes of minimization, segregation of materials, storage, and disposal.
- Prohibit ships violating hazardous waste requirements as well as sister ships under the same ownership from using the ports of the United States.

(See Chapter 12, "Solid Waste, Hazardous Materials, and Hazardous Waste Management")

HEATING, VENTILATION, AIR CONDITIONING, FOUNTAINS, MISTING SYSTEMS, HUMIDIFIERS, AND SHOWERS

Naturally occurring biological threats such as some bacteria and viruses can enter the cruise ship's heating, ventilation, and air-conditioning system through the air intakes and be spread throughout the ship. These may be organisms causing tuberculosis, influenza, colds, Legionnaires' disease, etc. Also molds, microbiological toxins, dust mites, fungi, and various pests' droppings can cause allergic responses in hypersensitive people. Improper air filtration caused by poor installation, poor maintenance, lack of proper cleaning of the system, etc. can exacerbate the problem. (See endnote 36.)

Best Practices for Heating, Ventilation, Air Conditioning, Fountains, Misting Systems, Humidifiers, and Showers

- Make readily available for inspection, maintenance, and cleaning, as well as self-draining all air handling unit condensate drain pans in collection systems.
- Only use potable water for cleaning the heating, ventilation, and air-conditioning systems as well as the fountains, misting systems, humidifiers, and showerheads.
- Develop and preserve the records of an inspection and maintenance plan based on the manufacturer's recommendations and industry standards concerning the cleaning and maintenance of all of the above-named systems.
- Only use potable water and an automated treatment system for all types of water sprays, decorative fountains, humidifiers, and misting systems.
- Ensure that all nozzles are readily removable for cleaning and disinfection.
- Ensure that all pipes and reservoirs can be drained when any water systems are not being utilized. Clean and disinfect these pipes and reservoirs before putting back into operation.
- Provide shock treatment with either a chemical disinfectant or appropriate temperatures of hot water for an appropriate period of time if there is any potential contamination in any of the systems which come in contact with water.
- Ensure that the supervisor of heating, ventilation, and air-conditioning systems has complete knowledge of these systems and the other ones mentioned above and can demonstrate the knowledge in actual practice to the inspection team coming aboard the ship.

INTEGRATED PEST MANAGEMENT

(See Chapter 9, "Insect Control, Rodent Control, and Pesticides")

LIQUID WASTE (SEWAGE, BLACK WATER) DISPOSAL

(See endnote 27, section 2, and Chapter 11, "Sewage Disposal Systems")

Sewage or blackwater is the human body waste from toilets and urinals and is treated before being released to the surrounding waters. Marine sanitation devices that receive, retain, treat, or discharge sewage should be installed on board the ship. They might also include holding tanks which will be emptied onshore at the port. In certain specified areas there should be no discharge from the ship either of sewage or graywater which has been treated.

- Establish standards for Best Practices for operation and maintenance to decrease or eliminate contaminants and reduce the volume of treated sewage effluent being released to the surrounding waters.
- Require that the effluent meets the national federal water quality criteria and state criteria at point of discharge and at the edge of the mixing zone.
- Require that the effluent contaminants are no greater than those coming from secondary treatment plants on land.
- Use the best available technology economically achievable to treat the sewage for disposal.
- Utilize technologies that will eliminate any pollutants of concern beyond those of a biological nature.
- Require periodic sampling and testing by knowledgeable individuals on board and maintain proper logs.
- Restrict discharge of treated sewage effluent to 12 nautical miles from the shore or any sensitive area which has been identified. Do not permit discharge of untreated sewage.
- Substantially increase penalties for failures to meet established standards of performance and treatment and disposal of sewage. Prohibit all non-complying vessels and their sister ships from using US ports.

LIQUID WASTE (GRAY WATER) DISPOSAL

(See endnote 27, section 3)

Graywater is wastewater from drainage from dishwashers, sinks, baths, showers, laundry, and food preparation areas. It may include contaminants such as bleach, nitrates, oil and grease, sodium, suspended solids, bacteria, organic matter, be turbid, have a high pH, high water temperature, and have both a high chemical oxygen demand and biological oxygen demand. It is estimated that the amount of graywater produced is an average of 67 gallons per day per person (different studies give different estimated averages of graywater produced on cruise ships). This was originally potable water (drinking water quality) that has been used for the purposes identified above. Graywater does not include any drainage from toilets, urinals, hospitals, or animal spaces. The graywater goes through special pipes to holding tanks for different sources of the liquid. It may go through gross particle filters or grease traps prior to the holding tanks, depending on the ship, and may be discharged without treatment into the surrounding waters or it may be treated before discharge. The holding capacity aboard ship varies considerably from as little as 5 hours to as much as 90 hours. Except in Alaska, treatment of graywater is not required before discharge. However, it should not be discharged in port or within 12 nautical miles of shore. Untreated graywater from ships may contaminate the surrounding waters.

Best Practices in Controlling Contamination of Surrounding Waters by Graywater Discharge
- Establish Best Practices for the design, operation, and maintenance of systems for the treatment of graywater aboard ship before discharge.
- Establish or revise existing standards for the discharge of graywater to require: effluent that meets federal water quality criteria at the point of discharge and the mixing zone, and those required as secondary treatment standards by publicly owned treatment works; effluent standards that are attainable by use of the best available technology economically achievable; and effluent standards that eliminate pollutants of special concern.
- Require the use of advanced wastewater treatment systems aboard ship.
- Require periodic sampling and testing of effluent discharges by onboard ship monitors.
- Require that there be no discharge of untreated graywater within 12 nautical miles of shore or other potentially sensitive areas.
- Prohibit any non-compliant vessels or their sister ships from using port facilities under US jurisdiction.

MEDICAL FACILITIES

Inspections performed by the US Public Health Service include an evaluation in the medical facilities of the documentation for gastrointestinal illness surveillance and outbreaks, and the availability of medical logs. If illness has occurred, then proper fecal samples should have been taken for analysis and the results posted. If possible a determination should be made about the etiology of the outbreak and the corrective conditions invoked to halt the outbreak and prevent it from occurring again. Cruise ships are inspected twice a year. They also are inspected if there is an outbreak of disease.

OILY BILGE WATER

(See endnote 27, section 4)

Oily bilge water is a mixture of water, oily fluids, lubricants, degreasers, detergents, cleaning fluids, etc. created by leaks from various pieces of equipment, onboard spills, wash water, and wastewater from various pieces of equipment. Bilge water is the most common source of oil pollution from cruise ships. Bilge water may also contain solid waste including rags, metal shavings, paint, glass, and a variety of chemical substances making it a very complex potential source of ocean pollution. It may also include foreign plants and marine creatures that can cause invasive problems in the waters of the United States and overwhelm the existing fish and shellfish population. Bilge water may damage the propulsion systems and constitute a fire hazard. Cross-contamination may result from the impurities in bilge water being mixed with the impurities in the sludge tank which is used for storage of the wastes from the cleaning of the fuel oil before use. Oil can kill marine organisms, reduce their ability to survive within the given environment, and disrupt the structure and function of their ecosystems. It can also damage coral reefs, kill birds, and sicken marine animals.

Oil water separators are needed to remove the oil from the water before discharge into the sea. The United States Coast Guard has found deliberate bypassing of the system and tampering with the monitoring equipment as well as falsifying of records. There has been an improper use of cleaning chemicals and surfactants to try to conceal the oil discharge sheen.

Best Practices for Treating Oily Bilge Water
- Utilize secure oily water separators that are failsafe and that have oil content monitors which cannot be bypassed or subject to tampering.
- Establish Best Practices for operation, maintenance, and training of personnel to decrease levels of contaminants in bilge water, and teach them how to properly treat the contents without illegal discharge.
- Encourage the cruise ship industry to switch to water-based lubricants wherever possible.
- Encourage states to pass laws prohibiting the discharge of any petroleum product into marine or freshwater.
- Prohibit the inclusion of any hazardous waste into the bilge area.
- Ban the discharge of any untreated or treated oily bilge water into the waters of the United States.
- Ban the discharge of any ballast water into ports or the waters of the United States to prevent alien aquatic organisms from invading the surrounding areas.
- Require that there be periodic inspection on board all ships, whether flying the United States flag or any other flag, and monitor the sampling and treatment of bilge water. This should be done aggressively and without warning by the appropriate authorities. Violators should be fined heavily and if necessary banned from using US ports. This should also apply to sister ships under the same management.

PLUMBING AND CROSS-CONNECTIONS

(See endnote 37)

Where there is a connection of any sort between the potable water supply, which is safe to drink, and any other source of water, there is the potential for the other source of water, usually seriously contaminated with microorganisms or chemicals, to cause disease or injury to people if the contaminated water flows back into the potable water supply. This is called a cross-connection, which is illegal and very dangerous. The contaminant enters the potable water supply when the pressure of the polluted liquid exceeds the pressure of the potable water. This is usually called backsiphonage or back flow. These cross-connections usually exist because of errors made when individuals doing the plumbing inadvertently connect the wrong pipes or hoses to the potable water supply or a hose from an acceptable plumbing fixture is submerged in a body of fluid, causing a submerged inlet of potable water to occur.

Cross-connections have resulted in numerous serious health problems. Because of a cross-connection in a large southern city, human blood coming from a funeral home ended up in water fountains in a building. In another incident, a key water system was contaminated with sodium hydroxide which may have come from a chemical company disposing of it. When a water main broke it caused a pressure differential allowing for the back flow. Antifreeze and other chemicals such as paraquat, chlordane, heptachlor, and other herbicides and insecticides, have been found in the water supply because of reduced pressure problems. Salt water from ports and water from rivers have also entered the potable water supply because of cross-connections.

Cruise ships are extremely vulnerable to problems of cross-connections because of the large number of pipes, hoses, and potential for submerged inlets in a vast number of sinks. It would be a problem to train all of the crew members in avoiding these situations especially when there are so many different languages spoken. Also, the passengers could easily contribute to the problems of submerged inlets in their own cabins.

Best Practices for Prevention of Backflow and Backsiphonage
- During construction of the ship or renovation, there should be strict inspections by highly qualified individuals to determine that cross-connections do not exist.
- When repair work is done, it is essential that highly trained supervisors make sure that cross-connections have not been installed in various piping systems.
- All individuals who hook up hoses must be under close supervision by highly competent people to prevent the potential for cross-connections.
- All potable water pipes and hoses should be painted or striped with an easily seen blue color at least every 15 feet and around partitions, bulkheads, and decks. No blue color should be used on piping downstream of any reduced pressure devices.
- All other pipes and hoses should have distinct colors representing the contents that they are transporting. A color chart should be posted at regular intervals explaining what each of the pipes and hoses may transport.
- Create an air gap between the potable water line and a receiving tank or sink. The air gap must be at least twice the diameter of the potable water line.
- Use mechanical types of backflow preventers such as: atmospheric vacuum breakers, hose bib vacuum breakers, pressure vacuum breakers, and double check valves, and in case of fires, double check detectors. (See endnote 37 for details including illustrations and discussion.)

POTABLE WATER

(See Chapter 13, "Water Systems (Drinking Water Quality)")

There have been numerous outbreaks of waterborne disease associated with ships. From 1970 to 2003, there were over 100 outbreaks of disease, with 21 of them being associated with water and 33 of them of unknown origin, some of which could have also been associated with water. Over 6000 people became ill from the known outbreaks of waterborne disease. Seven outbreaks were due to enterotoxigenic *Escherichia coli*. Three of the outbreaks were due to norovirus. One outbreak was due to *Salmonella typhi*. One outbreak was due to non-typhoidal *Salmonella*. One outbreak was due to *Cryptosporidium* species. One outbreak was due to *Giardia lamblia*. One outbreak was traced to chemical poisoning. Five outbreaks were of unknown origin. These data may be at the low end of what is occurring, since numerous times reporting is not necessarily uniform and accurate. (See endnote 38.)

The number of outbreaks of disease aboard cruise ships from 2004 through May 14, 2015 has been established as well as the causative agent if known, but the number of people affected is not listed nor a determination whether the outbreak was due to food contamination, water contamination, or a combination of both. If the reader wishes to delve further into the next data released, contact the CDC Vessel Sanitation Program. (See endnote 17.)

Outbreaks of waterborne disease can occur from: contaminated water sources where there might be high turbidity, various chemicals present, as well as increased levels of microorganisms; defective filters; contaminated hoses; contaminated water hydrants; cross-connections with non-potable water when loading or aboard the ship; defective backflow preventers at loading or aboard the ship; and sick people especially food service workers who may be spreading the microorganisms to drinks or food.

Contamination of the water supply even though it is safe to start with can occur readily aboard ship because the potable water system is hooked up to provide water supply lines to swimming pools, whirlpools, hot tubs, bathtubs, showers, garbage grinders, hospital and laundry equipment, boiler feed water tanks, and toilets. It also may be hooked up to the salt water ballast systems, bilge or other wastewater, international shore connections, and hospital and laundry equipment.

Best Practices for Preventing Disease Spread by Use of Potable Water Facilities (See endnote 38)

- Develop and implement a water safety plan for each cruise ship and each port. The plan should include an assessment of the entire water system from procurement to treatment to storage, to ensure a safe water source at the point of consumption by people; a hazard analysis of all steps in the water system and means of correcting problem areas; the use of acceptable water treatment technologies; a management plan in writing including appropriate control measures and corrective actions; and a monitoring system in accordance with the plan and meeting the highest level of requirements for chemicals and microorganisms as established by the appropriate official agency, US EPA, CDC, World Health Organization, and the national health agencies of other countries adhering to World Health Organization standards.
- Clean and disinfect all water holding tanks before filling with freshwater and after emptying existing remaining water in tanks.
- Determine the safety and security of all potable water being brought to the ship. Also determine if the means of transportation of the water is safe and secure and not contaminated.
- Test all incoming waters to determine if they meet acceptable microbiological, chemical, physical, and radiological standards as set by the US EPA, CDC, World Health Organization or nation states where the regulations are equal to or better than those of the World Health Organization.
- Have appropriate water treatment processes and equipment aboard all ships to be used for ensuring that the potable water is safe.
- Determine if the water once treated can become contaminated aboard ship during storage and distribution and if so immediately correct the situation and decontaminate the equipment.
- Frequently inspect all back flow prevention devices to ensure that contaminated water will not be merged with potable water. This can especially be a problem in the event of a fire on board the ship.
- Document and maintain the records of all inspections of the water system, the various tests being conducted, the sources of the water, and any outbreaks of waterborne disease.
- When bringing seawater on board ship either for a fresh seawater swimming pool or as a source of potable water after treatment, do not take it from ports or other areas where there is a considerable amount of traffic.
- Where hoses are used to deliver potable water from a municipal or private supply at the dock to the ship, mark the hoses with a distinct color of blue to indicate potable water use only and provide signs at reasonable distances and around corners to alert people not to use them for any other purpose than a clean water connection. Protect the inlet and outlet of the hoses when not in use.
- Do not use water boats or barges for any other purpose except delivery of potable water to the ship and make sure that the tanks and pipes or hoses are kept clean at all times and not used for any other purpose except delivery of potable water.
- Have the appropriate public health authorities test the water source on a frequent as well as emergency basis. This is especially important in the event of any flooding. Keep all hydrants and other attachments above flood stage in ports where the water will be used as a potable water source.

- If a water source is questionable and there is no other available, then the water must be first filtered with high-efficiency filtration systems, treated through the use of water treatment systems, and appropriately disinfected before bringing aboard ship. Retest the water before using it as a potable water supply.
- Use appropriately trained individuals in all aspects of the delivery, storage, treatment, and use of potable water.
- Label all non-potable water outlets as unfit for human consumption.
- It is of greatest significance to have highly trained individuals under close supervision making all repairs to pipes and hoses as well as the operation of the hoses when needed.
- Frequently inspect and test all water systems aboard ship to ensure the safety of the potable water supply.

RECREATIONAL WATER FACILITIES ON BOARD THE SHIP

There are numerous microbiological, chemical, and physical problems that can cause disease and injury aboard cruise ships while recreational water facilities are being used. A brief discussion concerning these issues which relate primarily to cruise ships will appear here and the broader discussion plus Best Practices will be discussed below in the special section on recreational water areas, which applies to all water facilities used for recreational purposes.

Operation of flow-through seawater recreational water facilities at sea and in the port is of concern. This type of seawater supply system must be used only when the ship is at least 12 miles from the nearest land. Before arriving in port or at a harbor, the seawater recreational water facility must be drained and not refilled until the ship reaches the 12-mile mark from land.

The World Health Organization has stated that there have been over 50 outbreaks of legionellosis with hundreds of cases associated with ships between 1977 and 2000. The source of the microorganisms has varied from whirlpools to leaking boilers to ventilation systems, etc. Studies of people on cargo ships have shown a high proportion of antibodies to the organism, suggesting that people on board ships are at increased risk of the disease compared to the general communities. The risk factors aboard ship include the water quality from the original source of the water taken aboard, the residual disinfectant, the water storage and distribution systems, and the temperatures of the water upon loading into the ship. The organism can grow in both hot and cold water piped water systems. It can also grow in whirlpools, spa pools, and any of the equipment associated with these bodies of water. (See endnote 35.)

Best Practices for Recreational Water Facilities Onboard Ship

(See endnote 24, pages 55–69 for a detailed discussion on acceptable Best Practices in this area.)

SOLID WASTE STORAGE AND DISPOSAL

(See endnote 27, Section 5)

Solid waste includes garbage, refuse, sludge, rubbish, trash, and discarded materials and can either be non-hazardous or hazardous. Non-hazardous waste which will be discussed here includes packaging, newspapers, disposables, food waste, plastic, wood, glass, metal cans, incinerator ash, etc. Cruise ships generate large volumes of solid waste. For example, one cruise ship with 2500 passengers and 800 crew members can generate 1 ton of garbage a day. It can also generate thousands of pounds of glass and cans per week as well as a large amount of dunnage, which is packing material to protect and support cargo in the ship's hold, as well as all the packaging material used around the ship. It is estimated that a cruise ship can generate 70 times more solid waste each day than a cargo ship. Solid waste improperly handled and discharged to the ocean, litter thrown into the water by passengers, or things such as towels, clothing, plastic bottles, cans, etc. blown overboard, create a serious environmental impact on the ocean or fresh bodies of water and may have a detrimental effect on sea life. Plastics are especially a problem since they accumulate on the surface of the water and on beaches. Birds, fish, turtles, and even marine animals may be seriously affected. Garbage can increase the biological oxygen demand and chemical oxygen demand, increase turbidity, and increase nutrient levels in the water. The ash from incineration may be toxic in nature to sea life.

Best Practices for Solid Waste Storage and Disposal
- Increase the use of onboard compactors, pulping equipment, shredders, and incinerators to reduce the volume of solid waste aboard ship.
- Use techniques of source reduction to minimize the amount of potential solid waste brought aboard ship.
- Recycle all plastics, paper, metal cans and objects, and wood by first separating them from the solid waste stream and then compressing and packaging them for appropriate disposal on shore.
- Separate all solid waste from hazardous waste.
- Pulp or compress all food waste and remove food liquids during dehydration. Send food liquids to graywater holding tanks. Then incinerate the food waste aboard ship regularly. When arriving in port, offload any remaining solid waste and ash for proper disposal.
- Provide adequate port reception facilities for all solid waste and recyclable materials.
- Require that incinerator ash from the onboard incinerators be tested.
- Prohibit discharge of any solid waste or food into a marine sanctuary or other sensitive area.
- Expand training of crew members in waste reduction and proper waste disposal.
- Require that at all ports of call for cruise ships there are reception facilities for offloading solid waste.
- Prohibit the use of incinerators while in port.
- Request that the EPA and Coast Guard develop plans where polluters will pay for the damage that they cause.
- Require that ports provide receipts for all garbage and other solid wastes being offloaded at their facilities.
- Require that cruise ships maintain certified logbooks documenting various solid waste storage and disposal actions. Require that there be uniform rules consistent with international law on all ships from all nations entering US waters and going into US ports. Refuse port privileges to any ship not obeying these orders.

(See Chapter 12, "Solid Waste, Hazardous Materials, and Hazardous Waste Management")

SUB-PROBLEMS INCLUDING LEADING TO IMPAIRMENT AND BEST PRACTICES FOR OCCUPATIONAL HEALTH AND SAFETY PROBLEMS ON BOARD SHIP AND IN PORTS

Docks and ports can be very hazardous to employees. Workers are seriously injured or even die because of falls from various heights, being crushed by equipment, or being struck by a vehicle or an object. Most of the non-fatal accidents, but still resulting in injuries, may be caused by slipping on wet or greasy surfaces, tripping and falling, or mishandling of equipment. A dock is a highly complex and constantly moving area with people, cargo, and equipment. Various groups of people who do not know each other interact. There may be multiple languages spoken, causing serious communications problems. The skill levels of the individuals vary enormously from unskilled and possibly very low educational levels, to highly skilled. Inattention, a negative attitude toward safety rules, the effect of heat or cold, or fatigue can also contribute to the substantial potential level of injuries which may occur.

The crew of the ship is involved in innumerable occupational risks since a ship has all the complexities faced by a small town but confined to a single structure which is constantly in motion and subjected to different weather conditions. The crew performs electrical work, plumbing work, maintenance and operation of equipment, painting and rehabilitation of areas, food service, provision of water, provision of facilities, and storage, treatment, and disposal of graywater, sewage, oily bilge water, hazardous materials, solid waste, etc. The employees work at heights which could cause serious falls and even death, and during very severe weather they may be on the decks and can fall overboard. They operate very complex equipment many times in limited areas of space. They are subjected to a large number of chemical hazards which may include PCBs, toxic paint, heavy metals, and volatile organic compounds from solvents and solvent-based paints, etc. Hydrogen sulfide is a special risk in confined areas such as engine rooms.

The crew members are also subjected to outbreaks of disease acquired from passengers and other crew members. They are responsible for cleaning up after sick people who have diarrhea and vomiting. These individuals work long hours, typically 7 days a week, while the ship is sailing or in port. Some of the crew members' work is in confined areas which may be very hot. All of these situations may lead over time to physical problems, mental and psychological problems, and extreme fatigue. Any of these conditions can lower the immunity of the individuals to outbreaks of disease or reduce the individual's ability to function properly, resulting in severe injury.

Worker fatigue has enormous consequences. It is driven by the need to stay competitive in the cruise line business. This reduces the number of workers and increases what they have to do. It is also driven by the 24/7 cycle where crew members have to perform a certain set of duties in order to keep the ship operating properly. If there is an outbreak of disease aboard the ship and crew members are affected, it may mean more work for those who are healthy. Fatigue may lead to heart disease, high blood pressure, inattention, suicide, and major problems involving the ship and the protection of the passengers.

Best Practices in Protecting Workers at Ports and Onboard Ships (See endnote 41)
- All work at ports must be planned, organized, and closely supervised by highly competent people who have the communication skills to deal with individuals who speak different languages and are part of the workgroup.
- The supervisors and administrators must interact with other groups of individuals at the port in a comprehensively planned manner with one highly competent, high-ranking individual totally responsible for all port activities concerning the health and safety issues of the workers. This individual makes the final decision in adjudicating situations that can cause health and safety risks.
- Special highly trained individuals should be responsible for carrying out written risk assessments including what to do to eliminate or reduce any risks that are found in different operations and in the interactions between various groups as well as passengers embarking or debarking at the dock.
- All health and safety regulations must be posted clearly and in all appropriate languages, at various points on the dock so that individuals have immediate and quick access to them.
- All individuals working on the dock must be thoroughly trained to carry out their duties safely and then closely supervised.
- All passengers must be kept in safe areas until allowed to board the ship. Boarding and disembarkation should be kept away from all work areas, obstructions, or where cargo or suspended loads are traversing. The area should be extremely well lit and checked for hazards before the individuals are allowed to enter or exit the ship.
- Because of the potential hazards of using the gangway, extra care and sufficient time should be allowed to complete this action, not only for workers but also especially for passengers including those who have disabilities.
- Emergency plans should be in effect, frequently tested, and administered by one high-ranking individual in the event of imminent danger due to flooding, high winds, fires, explosions, leaking containers, fractured pipes, or acts of terrorism. Emergency crews

should be highly trained and respond immediately to any imminent dangers. Passengers and workers should be immediately moved to places of safety.

- Based on the types of incidents that have been documented on various cruise ships, it is important to provide proper equipment, training, preventative maintenance of equipment, and detailed procedures on what to do if an incident occurs, in order to mitigate or prevent injury from occurring.
- Provide alarms and remote sensors in the event a person falls overboard.
- Provide drills for the crew and passengers for a potential series of incidents including fire, extremely rough seas, and catastrophic events.
- In the event of a maritime disaster, determine the resources that are available immediately to the cruise ship and communicate with them with special equipment if the normal communications system has been disrupted.
- Cruise ships have a rapid turnaround time in ports and unnecessary risks may be taken to meet schedules. These actions should be analyzed carefully by using risk analysis techniques and appropriate solutions should be anticipated before problems occur. The entire operation should be closely monitored and supervised by highly competent people.
- On board the ship it is recommended that an individual trained academically with occupational health and safety certification, 5 years of practical experience in occupational/environmental health and safety, knowledge of ergonomics, government regulations, and effective principles of business and management, including cost analysis, be in charge of the occupational health and safety program. This individual will evaluate all problems, incidents, and injuries and will establish appropriate actions to eliminate or minimize these incidents from reoccurring. The individual will also evaluate all data concerning health and safety problems, establish programs including training of crew, and provide draft reports to management and the appropriate health and safety authorities.
- Before performing any work on the ship, thoroughly investigate the area to make sure there are no hidden hazards behind molding or other cosmetic features.
- Maintain all steps properly and on a routine basis aboard ship to avoid slips and falls especially when the ship lurches.
- When conditions change aboard the ship because of the weather or rough seas, re-evaluate all potential hotspots for injuries to the passengers or crew and make necessary changes in procedures.
- Routinely check all non-slip surfaces aboard the ship to make sure that they are still performing as needed.
- Ensure that all crew members wear shoes at all times which have slip resistant soles.
- Establish a regular inspection and maintenance program for all deck areas, railings, catwalks, and other walking areas. Immediately remove all grease and other substances which can cause falls, as well as repair areas that have been weakened or cracked. Remove all tripping and falling hazards.
- When crew members have to work in confined areas, provide appropriate respiratory protection equipment and use the buddy system. This work should be closely monitored by well-trained supervisors and the time allowed in confined areas should be limited. Do not use highly combustible materials in confined areas.
- Implement a total no smoking policy aboard the ship for crew members and passengers.
- Test for the buildup of static electricity before working within a given area.
- Follow all environmental health and safety guidelines as established by the International Labor Organization for all ship maintenance activities.
- Follow the OSHA standard "Permit-Required Confined Spaces." Use a special compact portable multigas monitor to detect levels of oxygen, combustibles, hydrogen sulfide, and carbon monoxide.

SUB-PROBLEMS INCLUDING LEADING TO IMPAIRMENT AND BEST PRACTICES FOR CHILDREN'S CAMPS

Children's camps are typically set up for the summer months and are places of residence and recreation for 1 or more weeks for each of the children. Since this is a seasonal venture, usually during the summer, the people who become employees may either be students or individuals with limited experience working in temporary jobs. The equipment, whether for food service, recreational swimming facilities, on-site water and on-site sewage, playground or athletics, etc. may become dilapidated and in poor shape because of lack of maintenance and use.

The potential for disease varies with the microorganisms that the individuals bring to the camp environment, including influenza, as well as other diseases and the environmental health hazards that are present. This is exacerbated by people coming from different areas bringing their own set of viruses and bacteria and intermingling with others who do not have the same resistance to the organisms.

Problems found in overnight camping situations include:

1. Inadequate or poor supervision of the children who are attending the camp.
2. On-site potable water systems which are contaminated microbiologically and/or chemically.
3. The use of unapproved or contaminated water supplies during at least part of a day.
4. Water treatment systems, which require disinfection and removal of contaminants, not working properly.
5. Inadequate disinfection of water.
6. Presence of cross-connections and other plumbing and water distribution system concerns.
7. Inadequate quantities of water to be used for drinking water purposes and for other purposes.
8. Overflowing on-site sewage systems.
9. Improper transportation of campers on truck beds.
10. Improper supervision of swimming or other water areas.
11. Improper storage of solid or hazardous waste.
12. Improper storage of flammable liquids.
13. Improper storage, preparation, serving, and disposal of food.
14. Overcrowded and dilapidated sleeping areas for the children and staff.
15. Inadequate insect and rodent control.
16. The presence of noxious weeds, such as ragweed, poison ivy, poison oak, and poison sumac, which may contaminate campers and/or staff.
17. Inadequate medical facilities and poorly trained personnel.

An example of another type of problem is the presence of toxigenic fungi and mycotoxins in damp temporary housing and also in the outdoor recreational environment. Since fungi may be found anywhere, especially where there is dampness, they are a potential source of disease, which may be spread by food or inhalation.

There were 299 bat incidents in 109 children's camps in the state of New York from 1998 to 2002. A total of 1429 campers and staff were involved and 46.1% of these individuals received treatment for rabies. (See endnote 59.) It is also essential to be alert to problems with snakes and other wildlife as well as plants that can cause poisoning, as mentioned above.

Children and staff are exposed to disease and injury when they go on field trips away from the camp to animal farms, hiking in strange areas, climbing hills or mountains, caving, etc. There are large numbers of human–animal contacts each year in petting zoos, fairs, educational exhibits, etc. These contacts can lead to the transmission of various diseases from animals to these young people. (See endnote 57.)

More than 3.5 million children aged 14 and under sustain sports and recreation-related injuries every year from all activities including camps, and 21% of all traumatic brain injuries among children occur during sports and recreation. (See endnotes 50, 56.) Some 30% of all injuries are due to tripping, slipping, or falling.

However, children and/or staff illnesses are even more significant because they typically occur at a rate twice that of injuries.

Best Practices to Prevent and Mitigate Disease and Injury in Children's Summer Camps (See endnotes 44, 45, 46, 47, 60, 61)
- Prepare and submit a plan for the maintenance and operation of the summer camp to the local and state health department in keeping with the requirements of the given health authorities. The plan should be submitted at least 60 days prior to the anticipated opening of the facility in order for proper inspections to be performed, problem areas corrected, and appropriate certification or licensing to be granted.
- Prepare the plan to be a complete and intensive review of all aspects of camp life including: reporting of disease and/or injury to the appropriate authorities; the potential for abuse and neglect of the children; protecting children and adults from the sun; the qualifications and experience of all members of the staff including the camp director, water safety director, medical director, counselors, etc.; requirements for safe use of potentially dangerous items such as firearms, archery, playground and athletic equipment, boating equipment, etc.; building construction and sleeping space; fire safety; emergency procedures for weather or other types of emergency situations; transportation; horseback riding; and all environmental issues including food, laundry, solid waste, hazardous wastes, insects, rodents, water quality, plumbing, sewage, chemical and physical hazards, etc.
- Be prepared for outbreaks of disease such as influenza which may be found in the community, and follow all of the recommendations for prevention, mitigation, and control of the disease as noted by the appropriate authorities including the CDC. (See endnote 58.)
- Children and/or staff who become ill should be immediately segregated from the rest of the camp community and treated appropriately by medical personnel. They should not be integrated back into the camp community until the potential for transmission has passed.
- Special emphasis should be placed on the care and cleanliness of feet and the wearing of appropriate footwear within the camp and also on special trips. Flip-flops or sandals must be banned from all activity periods.
- Appropriate protective equipment as approved by the American Camp Association must be used for sports and other activities. Counselors must make sure that the equipment is used and properly put on.
- There should be training for children and staff and constant reminders concerning techniques of good personal hygiene including hand washing and protecting others against coughing and sneezing.
- Evaluations of the health record logs must be completed regularly to determine the causes of injuries and illnesses if possible and how to prevent these from occurring again.
- Knife safety must be taught and enforced since 15% of injuries at camps involve the use of knives.
- Blows to the head and concussions are of special concern because of the long-lasting potential problems involved. When these injuries occur, the individual should be immediately taken to appropriate medical authorities, examined, and the protocol for head injuries should be followed implicitly.
- Reduce the impact of fatigue on individuals by removing them from the activity if needed and re-evaluating how best to offer the activity or when to eliminate it. The temperature and humidity of the area as well as the time involved in conducting the activity should be also analyzed and necessary changes made. Allowances should be made for illness, size and age of the children, previous injury, and other conditions prevailing.

- Evaluate existing free time to determine if individuals are getting adequate periods of rest and are properly hydrated during the course of the day, especially when the weather conditions are extreme.
- All specialty counselors must be certified in their specialties and extremely knowledgeable about appropriate health and safety practices.
- Health authorities should make a minimum of two complete inspections of the camp including all facilities from medical care to all environmental issues. This would include the infirmary with its storage of medicine, appropriate isolation if needed, storage of medical logs, and submission to appropriate authorities of outbreaks of disease or severe injuries reports. This would also include inspection and evaluation of sleeping spaces, bathing, hand washing and toilet facilities, solid waste disposal, hazardous waste storage and disposal plans, potential insect and rodent problems, on-site water systems, on-site sewage disposal systems, food facilities, and all recreational swimming areas.
- Before using the water in the camp, flush the well, chlorinate the water and allow the water to be chlorinated for an appropriate period of time, usually 24 hours, and then run water through all taps until the water runs clear and there is no significant odor of chlorine. Hook up the chlorination equipment and make sure that it is operational. Flush all water lines within the camp until there is a free chlorine residual of at least 2.0 ppm at each of the taps in the distribution system. Shut off all taps and allow the water to remain in the water lines for at least 24 hours. Check the entire system for leaks and losses of pressure which should be a minimum of 20 pounds per square inch. Flush the entire system again and then determine if the water is clear, colorless, and has a free residual chlorine concentration of 0.2 ppm at the taps. (See endnote 48.)
- Wash and decontaminate all walls, floors, ceilings, beds, and bathroom facilities and air out all of the cabins prior to use by the campers. Check underneath the cabins to make sure that wildlife has not set up harborage. Determine if all stairs, railings, etc. have deteriorated and if they have, make sure that they are replaced immediately.
- Completely wash and decontaminate all kitchen equipment, floors, walls, ceilings, windows, dishes, utensils, pots and pans and allow to air dry. Check to see that all equipment is functioning properly and that there is adequate exhaust from areas where natural gas is used as fuel. Particularly check refrigerators and freezers to make sure that they reach appropriate temperatures. Check dishwashers to make sure that they reach proper temperatures. Check the hot water heaters.
- Prepare wholesome meals under proper sanitary conditions always using appropriate temperature and time sequences and excellent sources of food. (See Chapter 7, "Food Security and Protection.")
- Have professional sewage haulers clean out all septic tanks and grease pits before starting any water flow in the camp. Test the on-site sewage systems by allowing the water flow used in testing the wells to enter the septic tank systems and see if there is any sign of overflow onto the ground. Also conduct a complete on-site sewage system evaluation. Check previous years' reports and look for signs of overflow where the distribution system is located.
- Look for insect and rodent harborage and remove it after appropriate insect and rodent control programs. Exterminate the excess rodent population as well as mosquitoes and other insects within the areas where the children will be exposed. Remove all harborage.
- Make sure that all solid waste areas are constantly kept clean and the material is removed rapidly from the site.
- Store hazardous materials safely in a well-ventilated area which is locked. Remove all hazardous waste rapidly from the site to a secure hazardous waste disposal area.
- When campers or staff have arrived at the camp, determine if they are ill or have recently been ill. In either case send them to the infirmary to be evaluated by medical personnel.

If there are multiple individuals ill, immediately contact the local health department to report the event and to request additional help if necessary.

- All staff and campers must be given instructions in proper hand washing techniques and when to utilize them, as soon as they arrive at the campsite. This should be strictly enforced. Alcohol-based sanitizers for hands should only be used if soap and hot and cold running water is not readily available. However, after using the bathroom, tending to someone who was ill, or if an individual is ill, and always before eating, it is absolutely essential to wash hands thoroughly with soap and water for at least 20 seconds before rinsing. Segregate all sick people from the rest of the camp community and use appropriate isolation techniques including a proper gown, gloves, and mask if necessary.
- Exclude all sick food handlers from work until cleared by the medical team.
- In the event of an outbreak of potential food or waterborne disease, discontinue all salad and sandwich bars, buffets, or family-style food serving at tables. Wash all tables, chairs, equipment, utensils, and all parts of the dining area with the appropriate detergent and water and then use 1 ounce of bleach to 50 ounces of water as a disinfectant on all surfaces. Allow to air dry.
- Remove sick animals, especially those with diarrhea, immediately and thoroughly clean and disinfect the area. This includes the removal and destruction of sawdust, straw, and any other type of material that may be in the area where the animals have been present.
- All children and adults coming in contact with animals must immediately wash their hands with soap and water extensively and thoroughly under supervision to make sure there is no transfer of enteric diseases from animals to humans. There must be strict supervision of the children to make sure that they do not put their hands in their mouths or touch their face before hand washing. Strict hand washing procedures also apply if the individuals go into any facility that was used by animals.
- Educate all children and adults about the potential and actual spread of disease occurring when coming in contact with animals or their habitat.
- Emergency drills should be conducted periodically and without warning based on written plans for how to deal with fires, floods, outbreak of contagious diseases, or other emergencies. All staff members need to be trained and to immediately go to their stations to carry out their responsibilities when the alarm is sounded. Communication systems must be such that all individuals can be reached immediately to inform them of the hazardous or potentially hazardous situations occurring. Additional emergency health care must be readily available and easily contacted.
- All medications for campers or staff must be in locked cabinets and supervised by the camp nurse.

SUB-PROBLEMS INCLUDING LEADING TO IMPAIRMENT AND BEST PRACTICES FOR RECREATIONAL SWIMMING AREAS

(See endnotes 63, 64, 65, chapter 4)

Swimming is one of the most popular of sports and/or exercise activities in the United States. There are over 300 million visits to recreational water areas including treated pools each year. The health benefits of water-based exercise are well known. It helps people with chronic diseases such as arthritis. It helps with mental health concerns by improving the mood of both men and women, decreasing anxiety, and helping parents with children with developmental disabilities. It is excellent for older adults to improve their quality of life and decrease disability and pain as well as for pregnant women and their unborn children through this moderate level of exercise. It is just plain fun for others.

Different types of recreational water environments include coastal and freshwaters; and swimming pools, spas, water parks, hot tubs, etc. Each one has its own set of potential hazardous situations which can lead to disease, injury, and/or death.

Disease Outbreaks

The water quality of recreational water areas may be poor because of contamination from feces, other organic matter, or free-living microorganisms introduced by human or animal fecal contamination, sewage spills, or animal waste runoff following rainfall. Human contamination can occur because people have not thoroughly cleaned their rear end before entering the water and especially when someone has diarrhea or defecates into the water. Beach sand may be contaminated, especially from animal fecal material, primarily of dogs and birds, but also of any wild or domesticated animals. Algae, which may be toxic to humans, may also be found in this environment. Recreational water illness may be caused by microorganisms that enter the body most frequently through swallowing the water, breathing mist or aerosols, and being in contact with the contaminated water in any of the constructed or natural swimming areas.

For example, in coastal waters, *Vibrio parahaemolyticus* and *Vibrio vulnificus* can not only cause disease by humans eating raw oysters or other raw shellfish but also can cause disease if the individual has an open cut and comes in contact with the organisms in warm seawater. Whereas the first organism has typically a mild effect on the individual with a self-limiting gastrointestinal illness, the second organism can cause a severe gastrointestinal disease and, especially for immunocompromised individuals, lead to severe life-threatening illness with potential for septic shock and death. (See endnote 62.)

Contamination may also be found on surfaces surrounding the water. The most frequent illness reported is gastroenteritis and the leading cause of disease is *Cryptosporidium*.

Hot tubs and spas have been the sources of rashes for people of all ages because of infection with *Pseudomonas aeruginosa*. This type of rash can also occur in swimming pools which have not been properly maintained. Typically, disinfectant and pH levels are not correct.

Decorative water fountains may not be chlorinated or filtered and when children, especially those of diaper age, play in the water they may contaminate it with fecal matter. When another child or the same child swallows the contaminated water, they can become ill with various gastrointestinal diseases.

Since 1978, the number of outbreaks of recreational waterborne disease has increased substantially. They have been caused by *Cryptosporidium*, *Pseudomonas*, *Legionella*, *Giardia*, *Shigella*, *E. coli*, norovirus, *Campylobacter*, adenovirus, hepatitis A, *Salmonella*, *Staphylococcus*, *Streptococcus*, and disinfecting agents and their byproducts. From January 2007 through December 2008, there were 134 recreational water-associated outbreaks reported in 38 different states, with 116 of the outbreaks occurring in treated recreational water including pools and interactive fountains. This resulted in 13,480 cases with over 60% being acute gastrointestinal illness. Also there were dermatological illnesses and acute respiratory illnesses. This was a substantial increase over the 78 outbreaks reported for 2005–2006. Previously, recreational water outbreaks of Legionnaires' disease had been reported. A total of 62 outbreaks of gastrointestinal disease were associated with parasites especially *Cryptosporidium*. *Giardia* has also been found in pool water. *E. coli* 0157:H7 has been found in water slides and interactive fountains in water parks. *Shigella* has caused outbreaks as has *Pseudomonas*. Norovirus was the etiological agent in five outbreaks of disease. Nine outbreaks of disease were due to chemicals or toxins. Improper levels of chlorine or bromine, improper pH, and contact time contribute to the dissemination of the disease organisms.

Although all individuals are subjected to a variety of organisms and may become sick, those who are most vulnerable are children, pregnant women, and individuals with weakened immune systems. Individuals who have had organ transplants or chemotherapy are also greatly at risk.

Physical Problems Leading to Injuries

Frequently individuals who are on vacation are exposed to unusual amounts of sun and heat which lower their resistance to a variety of diseases and increases their potential for injury and sunburn. Prescription drugs and alcohol may be a contributing factor.

Spinal injuries and concussions are frequently caused by diving into water of unknown depth or with unknown obstructions. Diving in shallow water unless the individual is trained can also cause these types of injuries when coming in contact with the bottom of the pool or other recreational water facility. Poor underwater visibility may contribute to these problems.

Other injuries include impact from slipping or tripping and falling; cuts, lesions, and punctures; and retina tears or dislocations.

From 2005 to 2009 in the United States, there was an average of 3533 fatal unintentional drownings not associated with boating accidents. In addition, 347 people died each year from drowning related to boats. Some 80% of the drowning victims were male with the 1- to 4-year-old age group being most frequently involved. The fatal drowning rate for African-Americans aged between 5 and 14 is about three times as high as for white children in the same age group. Use of alcohol among adolescents and adults was involved in up to 70% of deaths associated with water recreational sports. (See endnote 49.)

In 2009, 3358 people were injured and 766 people died in boating incidents, 90% of whom were not wearing life jackets. (See endnote 51.) In coastal and freshwater areas, drowning associated with watercraft or swimming is a major cause of death. Immersion in cold water can be a shock to the system and contribute to the potential for drowning. Diving accidents result in spinal cord injuries, brain and head injuries, fractures, cuts, punctures, etc.

Drowning and near drowning, which may affect the child forever, in swimming pools and other constructed facilities occurs especially among the young. Special problems occur because pools may not be protected by properly locked fences, there are inadequate measures to prevent trapping of hair and body parts in drains or grills, or improper diving. Slippery decks, chairs, tables and toys out of place, poor maintenance of surfaces and lighting, and high temperatures in hot tubs contribute to injuries.

Chemical Hazards

Chemical hazards found in the swimming pool may come from the source water, disinfectants, sweat, urine, soap residues, cosmetics, suntan oil from the swimmers, and air pollutants. Chemical disinfectants may interact with chemicals present in the water to produce unhealthy byproducts. Chemicals may be inhaled, ingested, or absorbed through dermal contact. A frequent problem is respiratory concerns and eye irritation, especially from pool treatment chemicals.

Pool chemical storage areas and pump rooms, which are typically in confined areas, may create hazards for the staff and individuals using the pool facilities. Spills of hazardous chemicals may occur. Gas tanks may leak. Poor lighting can result in improper use of the chemicals. Outdated chemicals may deteriorate and create potential hazards. Slick surfaces can be the cause of falls. Incompatible chemicals in storage if mixed improperly can lead to fires and explosions especially if stored in areas with flammable materials such as gasoline, oil, grease, and solvents, etc. Improper handling of the chemicals when introducing them into the water can result in serious skin conditions, and problems of the respiratory tract, eyes, nose, and throat.

Aesthetic Problems

The aesthetic nature of recreational waters is such that visible contamination affects the psychological needs of individuals. Chemical and physical agents can enter the surface water and contaminate the beaches from a variety of sources, including runoff from land, contamination from landfill sites and contamination from old mines, working industries and ship disasters at sea.

Swimming Pool Codes and Inspections

Swimming pool codes and inspections vary from community to community and therefore it is difficult to compare problems from one area with those of another one. However, the CDC has compared violation summaries on more than one occasion of thousands of pools from several different states. Approximately 12% of the pools closed immediately because of serious violations, particularly a lack of disinfectant in the water. This has been shown repeatedly through microbiological tests of the water. Tests of the filters used in public pools showed evidence of contamination by many organisms.

Childcare pools had the highest percentage of immediate closures (approximately 17%), followed by hotel/motel and apartment/condominium pools which each had about a 13% immediate closure rate. Interactive fountains had the highest percentage of immediate closures at about 17%. Kiddy/wading pools had the highest percentage of disinfection level violations.

Best Practices in Preventing and Mitigating Illness and Injury Associated with Recreational Water Facilities
- Use the Model Aquatic Health Code as the basis for the construction and operation of all swimming pools and other water venues to reduce injuries and levels of disease transmission. The most up-to-date science and Best Practices are incorporated throughout the entire guide. The guide covers: facility design standards and construction; facility operation and maintenance; policies and management of the facility; and resources. (See "Resources" section below and endnote 54.)
- All approved swimming areas must be supervised by an individual with credentials similar to American Red Cross swimming instructor certification, plus practical experience. Other individuals involved in water safety must have as a minimum credential training which is similar to American Red Cross senior lifesaver certification.
- To prevent drowning teach individuals swimming skills and life-saving skills including cardiopulmonary resuscitation (CPR).
- Fence off all swimming pools to prevent children from falling into the water and drowning.
- Make the use of lifejackets mandatory for all individuals who cannot swim and where swimming or boating occurs in natural bodies of water. Only US Coast Guard-approved lifejackets are acceptable in natural bodies of water.
- Provide appropriate supervision for all swimming areas.
- Insist on the use of the buddy system or no swimming allowed.
- If caught in a rip current, swim parallel to the shore until free of the current. Watch out for dangerous waves and rip currents by looking for discolored, choppy, foamy water filled with debris and moving away from the shore.
- Do not dive into shallow water or water of unknown depths.
- Obtain information on local weather conditions and leave the area in the event of severe weather including thunderstorms.
- Insist upon no alcohol use or very limited alcohol use by individuals involved in water recreational sports.
- Only trained individuals should operate powerboats and then within safe limits.
- Carbon monoxide poisoning can occur aboard a boat and create headaches, confusion, fatigue, dizziness, or loss of consciousness. The engine exhaust builds up within the boat and the carbon monoxide is colorless and odorless. Immediately ventilate the area and remove the individuals to a fresh air situation. Seek immediate medical help.
- Build spill containment, fire suppression equipment, adequate lighting, alarms, check valves in chemical feed lines, and deactivation devices for the chlorine-pH feed pumps when the recirculation system stops, into pump rooms as well as pool chemical use and storage areas. Include emergency showers and eyewash stations as well as chemical safety equipment. Secure the areas against unauthorized individuals. (See endnote 52.)
- Install separate air handling systems for the chemical storage area, pump room, and pool areas from the rest of the structure.
- Use proper scheduled maintenance by experienced people for all recirculation and filtration systems.

- Individuals adding chemicals to the pool, involved in chemical storage, handling, and maintenance and repair of equipment must be specially trained and certified to do the work. These individuals need to be closely supervised because of the potential hazards to the individuals and the people using the recreational water facility.
- Maintain appropriate levels of chlorine or bromine at appropriate pH levels for an adequate contact time in order to destroy organisms that can cause outbreaks of recreational water disease.
- Test all waters frequently during the course of the day to make sure that all chemical levels are maintained at the appropriate amounts.
- Immediately close swimming pools or other water venues if there is a disinfectant level violation.
- Train all operators of water venues including pools and spas in how best to disinfect the water, how frequently to test it, and when to decide to close the operation voluntarily if problems exist.
- Inform all individuals using water venues that if they are having diarrhea they must not enter the water. This is especially true of individuals who are infected with *Cryptosporidium* since the organisms are chlorine tolerant and are spread through the oral–fecal route of transmission. Babies in diapers should not be allowed in the water.
- Where an individual defecates into a recreational water source, all other individuals should be removed immediately and the area should be thoroughly cleaned and disinfected. Free chlorine levels need to be at 1–3 mg/L for an extended period of time.
- All individuals must take a shower with soap and water prior to entering any water venue.
- In spas or hot tubs, the higher water temperature makes it hard to maintain proper disinfectant levels. This allows organisms such as *Pseudomonas* and *Legionella* to grow and cause outbreaks of disease. Greater care should be taken in the testing and the control of chemicals within these areas.
- Interactive fountains, splash pools, and other water play areas readily become contaminated since the water drains into a water reservoir and may be sprayed back onto the individuals. These venues should be regulated in the same way as pools. (See endnote 55.)
- Avoid swimming in oceans, lakes, and rivers after heavy rainfall and runoff or if health departments determine that the water is unsafe for use.
- In children's pools, at least once every hour all the kids should come out of the water and take bathroom breaks which are supervised to make sure that they wash their hands properly and clean themselves properly. Sunscreen should be re-applied and the children should be given ample quantities of fluids to prevent dehydration.

RULES AND REGULATIONS

PUBLIC HEALTH SERVICE ACT OF 1944 WITH AMENDMENTS AND EXECUTIVE ORDERS UPDATED TO JANUARY 4, 2012

(See endnote 26)

The Public Health Service Act established the quarantine authority for the federal government for the first time and gave it to the US Public Health Service to enforce. The Public Health Service is responsible for preventing the introduction, transmission, and spread of communicable diseases from foreign countries into the United States.

CRUISE VESSEL SECURITY AND SAFETY ACT OF 2010

This act was passed by Congress to reduce the number of crimes committed on ocean-going cruise ships that either embark or disembark passengers in the United States. The Coast Guard enforces the various requirements of this act. Some of the items include rail height to prevent passengers from going overboard; peepholes or other visual identification in state rooms; for ships where construction started after July 27, 2010, security latches and sensitive key technology for all passenger and crew cabins; information on location and how to contact US embassies and consulates; current licensure for physicians or registered nurses; equipment to prevent, treat, and collect evidence for sexual assault cases; forensic sexual assault examination training; a means for individuals to contact law enforcement if needed free of charge; provision of data on the website of criminal activity on the company's vessels; ability to detect and capture images of passengers who fall overboard; acoustical handling and warning devices as well as video surveillance; and training of special crew members in how to preserve evidence of criminal actions.

INTERNATIONAL CONVENTION FOR THE PREVENTION OF POLLUTION FROM SHIPS (MARPOL)

This act regulates the disposal of garbage into the sea and prohibits the disposal of plastics into the waters. It requires governments to ensure that there are facilities at ports and terminals for the reception of garbage. It also addresses bilge water, oil pollution, and other discharges from cruise ships.

SOLAS

(See endnote 42)

The International Convention for the Safety of Life at Sea was first adopted in 1914 as a result of the sinking of the *Titanic* in 1912. Cruise ships are put in the same category as cargo ships for this international agreement. It was determined that seafaring was one of the most dangerous occupations with numerous deaths from maritime disasters, illnesses, suicides, homicides, and occupational accidents. Most of the incidents (approximately 80%) are due to human error. The main purpose for this convention is to establish minimum standards of construction for all parts of passenger ships including watertight compartments, safe electrical systems and machinery, fire protection and detection equipment, communications equipment, etc.

ACT TO PREVENT POLLUTION FROM SHIPS

This act and the implementing regulations prohibit the discharge of: all garbage within 3 miles of shore; certain types of garbage 3–25 miles offshore; and plastic everywhere. It also applies to bilge water and other discharges. The United States Coast Guard regularly inspects the discharge records and logbooks of vessels.

CLEAN WATER ACT

The Clean Water Act prohibits any individual from dumping pollutants from any point source, which would be a cruise ship or any other type of ship, into the waters of the United States, including the territorial seas. A special permit is needed if such a discharge of a pollutant does occur.

NATIONAL MARINE SANCTUARIES ACT

This act protects marine resources and ecosystems, such as coral reefs, sunken historical vessels, and unique habitats, from degradation of any type. The National Oceanic and Atmospheric Administration designate specific areas and provide comprehensive conservation management.

Resource Conservation and Recovery Act

This act defines and provides regulation of solid waste and hazardous waste. It minimizes hazards of waste disposal through recycling, recovery, and reduction, which helps conserve resources.

Comprehensive Environmental Response, Compensation, and Liability Act

This act regulates the release of hazardous substances. It insists that any person in charge of a vessel or an offshore or onshore facility handling hazardous wastes contact the National Response Center and give them the appropriate information if a hazardous substance is released, except for those which have a federal permit, for any vessel or facility in quantities exceeding federal regulations.

Marine Protection, Research, and Sanctuaries Act (Ocean Dumping Act)

This act prohibits the transportation of any substances from the United States in any manner for the purpose of dumping in the oceans.

Oil Pollution Act

This comprehensive act is designed to deal with oil spill prevention, preparedness, and the response capabilities of the federal government and industry.

Virginia Graeme Baker Pool and Spa Safety Act

This act became effective on December 19, 2008, and stipulates the types of controls that need to be taken to prevent children from dying or being injured in swimming pools. (See endnote 53.)

Coast Guard Regulations

There are several Coast Guard regulations concerning solid waste and its discharge overboard, to another ship, to a reception facility, and for incineration aboard the ship. Contact the Coast Guard for specific details.

RESOURCES

Centers for Disease Control and Prevention

CDC Health Alert Network

The CDC Health Alert Network provides the public health infrastructure for access to information, communications, and distance learning at the state and community level. If a Health Alert Notice has been issued regarding a specific disease in a particular country, the quarantine station personnel distribute copies of the notice to each arriving adult traveler in that country. Different levels of notices are distributed to professionals, travelers, and airline personnel. At airports a trigger to activate all preventive techniques is a combination of conditions that potentially can cause the dangerous situation to affect people.

CDC's Traveler's Health

CDC's Travelers' Health is a program that provides information on global disease outbreaks and issues travel notices and other information, to help provide information to individuals concerning precautions to be taken and what to do if he/she becomes ill from that disease.

NATIONAL CENTER FOR ENVIRONMENTAL HEALTH'S DIVISION OF
EMERGENCY AND ENVIRONMENTAL HEALTH SERVICES

This center within the CDC uses US Public Health Service officers to protect passenger and crew health by developing and implementing comprehensive sanitation programs to minimize gastrointestinal illness aboard cruise ships, inspect vessels using the Vessel Sanitation Program Operations Manual, collect and monitor outbreaks of gastrointestinal disease, and provide technical assistance to international groups on outbreaks of disease.

UNITED STATES COAST GUARD

The United States Coast Guard has oversight of the safety, security, and environmental stewardship of commercial vessels, facilities, and mariners. US law dictates that all accidents or incidents occurring on cruise ships must be reported to the Coast Guard immediately. The Coast Guard reviews the information and makes in-depth studies of the occurrences to determine the causes and establish corrective actions that need to be taken. The particular cruise line is held accountable for the problems and making necessary corrections to avoid this from occurring again. Passengers as of July 2015 will have to attend mandatory safety briefings immediately after the vessel sails. Because of recent fires on several cruise ships, there has been an intensive effort made to improve equipment and training of crew members. Coast Guard field inspectors witness training programs for crew members as well as information programs for passengers as part of the inspection process.

United States Coast Guard Cruise Ship National Center of Expertise

The United States Coast Guard carries out an annual examination of foreign passenger vessels to determine the competency of crew members in firefighting, life-saving, and emergency systems by actually witnessing and reviewing comprehensive fire and boat drills. Periodically the Coast Guard will board vessels to re-evaluate the skills and training of crew and officers. The Coast Guard is involved in almost all aspects of cruise ship oversight from ship design to construction to operation if the ships are flying the flag of the United States or if the ships are operating in US territorial waters.

WORLD HEALTH ORGANIZATION

Global Outbreak Alert and Response Network

The Global Outbreak Alert and Response Network administered by the World Health Organization is a technical program of existing institutions and networks working together to rapidly identify diseases of international importance. It provides an opportunity for people to share information rapidly.

PUBLIC HEALTH AGENCY OF CANADA GLOBAL PUBLIC HEALTH INTELLIGENCE NETWORK

The Global Public Health Intelligence Network is a secure internet early warning system that gathers all reports of public health significance in seven different languages on a real-time 24-hour basis. Information is filtered for relevancy and then analyzed by the Public Health Agency of Canada which puts it into categories and immediately alerts all public health outlets.

ILO MARITIME LABOR CONVENTION AND ITS GUIDANCE
CONCERNING HEALTH AND SAFETY ON-BOARD SHIPS

(See endnote 43)

This convention establishes the general duties of the competent authorities of the various nations whose flags are flown on cruise ships. It provides for the responsibilities of the ship owners, the general duties of the master (captain), the general duties and responsibilities of the crew, the general duties and responsibilities of the shipboard safety and health committee, and the general duties and responsibilities of the safety officer and the safety representatives. It also discusses structural features of the ship, machinery, loading and unloading equipment, sanitary measures below deck, fires, fire prevention techniques, and how to lower temperatures, noise, and vibration. It provides for reporting of accidents, collection of statistics and investigation, and preventive programs. The objective of the guidance is to prevent disease, accidents, and any other harmful effects on the crew while establishing responsibility for safety and health, promoting consultation, and ensuring proper representation of the crew in onboard safety and health matters.

MODEL AQUATIC HEALTH CODE

(See endnote 54)

The Model Aquatic Health Code was developed by a series of technical committees who worked on 14 different modules to present a document which could be used by local and state health authorities as well as industry to understand and utilize the most up-to-date information and technology to improve the use of pools and spas while preventing disease and injury from occurring. Recognizing the serious problems of injuries, deaths, and spread of various illnesses, and since in the United States there is no federal agency responsible for swimming pools and spas, the CDC worked with the Council for the Model Aquatic Health Code, a non-profit organization created in 2013 to support and improve public health by providing a healthy and safe water environment. The Model Aquatic Health Code is a voluntary guidance document and program which is constantly being updated to utilize the best science available as well as Best Practices. This guide covers: facility design standards and construction; facility operation and maintenance; and policies and management.

PROGRAMS

CDC VESSEL SANITATION PROGRAM 2011 OPERATIONS MANUAL

The CDC Vessel Sanitation Program helps the cruise ship industry prevent and control the introduction, transmission, and spread of gastrointestinal illnesses on cruise ships. This is accomplished by US Public Health Service officers boarding the ships and making intensive evaluation studies and inspections utilizing detailed forms and procedures set forth in the Vessel Sanitation Program 2011 Operations Manual. Subject areas include acute gastroenteritis surveillance reports, specimen collection, and isolation of sick people both crew and passengers; potable water sources, halogenation and pH control, monitoring, storage tanks, water piping, water hoses, potential sources of contamination, chemical treatment, microbiological monitoring, water distribution systems, and back flow prevention devices; recreational water facilities seawater sources, recirculation systems, halogenation, monitoring, and safety; food safety personnel, cleanliness, and hygienic practices; food sources, receiving temperatures, hand contamination by employees, food and ingredient contamination, use of ice, equipment storage and cleanliness, storage and preparation, and display and service; food destruction of pathogens and parasites, reheating, holding temperatures and times, cooling and holding temperatures and times, and removal of adulterated products; food equipment and utensils design and construction for multiple use, and single service use and disposal; ware washing design, construction, use, maintenance, and temperatures; poisonous and toxic materials storage, use, and disposal; hand washing and toilet facilities installation, maintenance, cleanliness, and use of disinfectants; solid waste storage and disposal; liquid waste pipelines and disposal; plumbing and backflow preventers; lighting; ventilation; integrated pest management; housekeeping; child activity centers; heating, ventilation, and air-conditioning systems, fountains, misting systems, humidifiers, and showers. (To review the entire manual, see endnote 24)

CDC Vessel Sanitation Program 2011 Construction Guidelines

The guidelines are a framework for consistent construction and design that helps protect passengers and crew from spread of disease because of poor or inadequate physical facilities. It covers the facilities aboard ship that are related to public health including food, potable water storage and distribution, equipment placing and mounting, lighting, waste management, black and graywater disposal systems, fire hose connections, cross-connections, heat exchangers, recreational water facilities; etc. (See endnote 25.)

United States Coast Guard Cruise Ship National Center of Expertise

(See endnote 21)

The United States Coast Guard is responsible for safety pertaining to passenger ships, both United States flagged and foreign, that come into US ports, and with special emphasis on large ships. Because of the continuing growth in the size, capacity, and complexity of cruise ships, the Coast Guard created the Cruise Ship National Center of Expertise in 2008 to provide highly proficient individuals who evaluate cruise ship safety and environmental security compliance. These individuals focus completely on foreign cruise ships and the Passenger Vessel Safety Program for foreign flagged ships. Consultation is given to the cruise industry on the construction and renovation of ships with environmental health officers doing plan reviews to analyze the ship's design to help eliminate environmental health risks and to create situations where there will be a healthy and safe environment. Fire protection and detection systems are of prime concern. Environmental surveys are conducted aboard ship to determine if the waste stream of oil, non-hazardous material, hazardous material, and graywater which comes from sinks and showers, and black water which is sewage, is handled properly by highly trained crew members using appropriate equipment for storage, treatment, and disposal.

Interagency Marine Debris Coordinating Committee

This group co-chaired by the US EPA and the National Oceanic and Atmospheric Administration work to reduce the impact and source of marine debris of all types.

ENDNOTES

1. Brunette, Gary W. 2015. *CDC Health Information for International Travel-2016*. Centers for Disease Control and Prevention. Oxford University Press, New York.
2. Bagshaw, Michael, DeVoll, James R., Jennings, Richard T. 2002, *Medical Guidelines for Airline Passengers*. Aerospace Medical Association, Alexandria, VA.
3. Tuler, Seth, Golding, Dominic. 2002. *A Comprehensive Study of Visitor Safety in the National Park System: Final Report*. Clark University, George Perkins Marsh Institute, Worchester, MA.
4. Brown, John T., Jr, Laroe, Christian. 2006. Airline Networks and the International Diffusion of Severe Acute Respiratory Syndrome. *Geographical Journal*, 172(2):130–144.
5. Institute of Medicine. 2010. *The Domestic and International Impacts of the 2009-H1N1 Influenza A Pandemic*. National Academies Press, Washington, DC.
6. US Department of Transportation. 2005. *National Aviation Resource Manual for Quarantinable Diseases*. Washington, DC.
7. Airports Council. 2009. *Airport Preparedness Guidelines for Outbreaks of Communicable Disease*. Montréal, Canada.
8. Centers for Disease Control and Prevention. 2014. *Interim US Guidance for Monitoring and Movement of Persons with Potential Ebola Virus Exposure*. Atlanta, GA.
9. Presidential Documents. 2003. *Revised List of Quarantinable Communicable Diseases, Executive Order 13295*. Federal Register 68:68. Washington, DC.
10. Centers for Disease Control and Prevention. 2014. *History of Quarantine*. Atlanta, GA.

11. United States Department of Labor, Occupational Safety and Health Administration. No Date. *Safety and Health Topics: Airline Industry.* Washington, DC.

12. Federal Aviation Authority, Occupational Safety and Health Administration. 2000. *FAA/OSHA Aviation Safety and Health Team (First Report).* Washington, DC.

13. US Department of Transportation, Federal Aviation Administration, Occupational Safety and Health Administration. 2014. *The Memorandum of Understanding Between Federal Aviation Administration, US Department of Transportation and Occupational Safety and Health Administration, US Department of Labor-Occupational Safety and Health Standards for Aircraft Cabin Crew Members.* Washington, DC.

14. Federal Aviation Administration. 2015. *Fact Sheet: Commercial Aviation Safety Team.* Washington, DC.

15. United States Navy. 2006. *Guidance Document: Fall Protection for Aircraft Maintenance and Inspection Work.* Washington, DC.

16. Committee to Advise on Tropical Medicine and Travel. 2005. *Statement on Cruise Ship Travel (archived).* Canada Communicable Disease Report, Volume 31, ACS-8.

17. Centers for Disease Control and Prevention. 2015. *Outbreak Updates for International Cruise Ships.* Atlanta, GA.

18. Centers for Disease Control and Prevention. 2014. *Preventing Noroviruses Infection.* Atlanta, GA.

19. Hall, Aron J., Vinje, Jan, Lopman, Benjamin. 2011. *Updated Noroviruses Outbreak Management and Disease Prevention Guidelines.* Morbidity and Mortality Weekly Report, 60 (RR03). Centers for Disease Control and Prevention, Atlanta, GA.

20. International Cruise Victims Association, Inc. 2014. *National Transportation Safety Board Form on Cruise Ships: Examining Safety, Operations and Oversight.* Phoenix, AZ.

21. United States Coast Guard. 2014. Proceedings of the Marine Safety and Security Council: Passenger Vessel Safety Aiding Cruise Industry Regulatory Compliance. *The Coast Guard Journal of Safety and Security at Sea,* 71(2), 8–66.

22. Boat US Foundation for Clean Water and Boating Safety. 2015. *Lifejackets Are for Everyone.* Alexandria, VA.

23. Cruise Lines International Association. 2012. *Operational Safety Review: Executive Summary.* Washington, DC.

24. US Department of Health and Human Services, US Public Health Service, Centers for Disease Control and Prevention, National Center for Environmental Health. 2011. *Vessel Sanitation Program 2011 Operations Manual.* Atlanta, GA.

25. US Department of Health and Human Services, US Public Health Service, Centers for Disease Control and Prevention, National Center for Environmental Health. 2011. *Vessel Sanitation Program: 2011 Construction Guidelines.* Atlanta, GA.

26. United States Code, Title 42-The Public Health and Welfare, Chapter 6 A-Public Health Service, Subchapter II-General Powers and Duties, Part G-Quarantine and Inspection. 2012. *Regulations to Control Communicable Diseases.* Washington, DC.

27. US Environmental Protection Agency. 2008. *Cruise Ship Discharge Assessment Report.* EPA 420-R-0 9-902. Washington, DC.

28. Knighton, W.B., Herndon, S.C., Miake-Lye, R.C. 2009, *Aircraft Engine Speciated Organic Gases: Speciation of Unburned Organic Gases in Aircraft Exhaust.* EPA-430-F-00-005. US Environmental Protection Agency, Washington, DC.

29. US Environmental Protection Agency. 2012. *EPA Adopts Nitrogen Oxides Emission Standards for Aircraft Gas Turbine Engines.* EPA 420-F-12-027. Washington, DC.

30. US Environmental Protection Agency. 2000. *Cruise Ship White Paper.* Washington, DC.

31. McCarthy, James E. 2008. *CRS Report for Congress: Air Pollution from Ships: MARPOL Annex VI and Other Control Options.* Congressional Research Service, Washington, DC.

32. Natural Resources Defense Council. 2004. *Harboring Pollution: Strategies to Clean up US Ports.* New York.

33. Eckhard, S., Hermansen, O., Grythe, H. 2013. The Influence of a Cruise Ship Emissions on Air Pollution in Svalbard: A Harbinger of a More Polluted Arctic. *Atmospheric Chemistry Physics,* 13:8401–8409.

34. US Environmental Protection Agency. 2015. *Non-Road Engines, Equipment, and Vehicles: Ocean Vessels and Large Ships.* Washington, DC.

35. World Health Organization. 2001. *Sanitation on Ships: Compendium of the Outbreaks of Foodborne and Waterborne Disease and Legionnaires Disease Associated with Ships: 1970–2000.* WHO/SDE/WSH/01.4. Geneva, Switzerland.

36. UPMC Center for Health Security. 2014. *Protecting Building Occupants*. Baltimore, MD.
37. US Environmental Protection Agency, Office of Water. 2003. *Cross-Connection Control Manual*. Washington, DC.
38. World Health Organization. 2011. *Guide to Ship Sanitation*. Third Edition. Geneva, Switzerland.
39. World Bank Group, International Finance Corporation. 2007. *Environmental, Health, and Safety Guidelines for Shipping*. Washington, DC.
40. Health and Safety Authority. 2015. *Management of Health and Safety Imports and Docs: Information Sheet*. Dublin, Ireland.
41. Mileski, Joan P., Wangmr, Grace, Beacham, L. Larnar. 2014. Understanding the Causes of Recent Cruise Ship Mishaps and Disasters. *Research and Transportation Business and Management*, 13:65–70.
42. International Labor Organization. 2006. *ILO, Maritime Labor Convention and Its Guidance about the Health and Safety On Board Ships*. Geneva, Switzerland.
43. Massachusetts Department of Public Health. 2003. *Minimum Standards for Recreational Camps for Children (State Sanitary Code, Chapter IV)*. Boston, MA.
44. Oregon Health Authority, Public Health Division. 2007. *Oregon Administrative Rules, Chapter 333 Division 030-Organizational Camps*. Portland, OR.
45. New York Department of Health. 2011. *Children's Camps in New York State*. Albany, NY.
46. New York State Department of Health. 2015. *Requirements for Children's Camps in New York State*. Albany, NY.
47. New York State Department of Health. 2015. *NYS Children's Camps Acceptable Annual Water Supplies Startup Procedures*. Albany, NY.
48. Centers for Disease Control and Prevention. 2012. *Unintentional Drowning: Get the Facts*. Atlanta, GA.
49. Safe Kids Worldwide. 2014. *Facts about Childhood Recreational Injuries*. Washington, DC.
50. Centers for Disease Control and Prevention. 2012. *Stay Safe While Boating*. Atlanta, GA.
51. Centers for Disease Control and Prevention. 2014. *Recommendations for Preventing Pool Chemical-Associated Injuries*. Atlanta, GA.
52. United States Congress. 2014. *Virginia Graeme Baker Pool and Spa Safety Act*. Washington, DC.
53. Centers for Disease Control and Prevention, Conference for the Model Aquatic Health Code. 2014. *Model Aquatic Health Code*. First Edition. Atlanta, GA.
54. Hlavsa, Michelle C., Roberts, Virginia A., Anderson, Ayana R. 2011. *Surveillance for Waterborne Disease Outbreaks and Other Health Events Associated with Recreational Water: United States, 2007–2008*. Morbidity and Mortality Weekly Report, volume 60 (SS 12). Centers for Disease Control and Prevention, Atlanta, GA.
55. Youth Sports Safety Alliance. 2013. *Youth Sports Safety Statistics*. Carrollton, TX.
56. Centers for Disease Control and Prevention. 2011. *Compendium of Measures to Prevent Disease Associated with Animals in Public Settings, 2011: National Association of State Public Health Veterinarians Incorporated*. Morbidity and Mortality Weekly Report Volume 60 (RR04). Atlanta, GA.
57. Centers for Disease Control and Prevention. 2010. *CDC Guidance for Day and Residential Camp Responses to Influenza during the 2010 Summer Camp Season*. Atlanta, GA.
58. Robbins, Amy, Eidson, Millicent, Keegan, Mary, Sackett, Douglas, Laniewicz, Brian. 2005. Bat Incidents at Children's Camps, New York State, 1998–2002. *Emerging Infectious Disease Journal*, 11(2):302–305.
59. Garst, Barry, Erceg Linda Ebner, 2015. *Ten Ways to Reduce Injuries and Illnesses in Camp*. American Camp Association, Martinsville, IN.
60. American Camp Association. No Date. *Sample of Health-Care Policies and Procedures*. Martinsville, IN.
61. Ohio Department of Health. 2014. *Disease Facts Sheet: Vibriosis*. Columbus, OH.
62. World Health Organization. 2009. *Guidelines for Safe Recreational Water Environments: Coastal and Freshwaters*, Volume 1. Geneva, Switzerland.
63. World Health Organization. 2006. *Guidelines for Safe Recreational Water Environments: Swimming Pools and Similar Environments*, Volume 2. Geneva, Switzerland.
64. Koren, Herman, Bisesi, Michael. 2003. *Handbook of Environmental Health: Pollutant Interactions in Air, Water, and Soil*. Volume 2, Fourth edition. Lewis Publishers, CRC Press, Boca Raton, FL.
65. Koren, Herman, Bisesi, Michael. 2003. *Handbook of Environmental Health: Biological, Chemical and Physical Agents of Environmentally Related Disease*. Volume 1, Fourth edition. Lewis Publishers, CRC Press, Boca Raton, FL.
66. Koren, Herman, Bisesi, Michael. 2003. *Handbook Environmental Health: Pollutant Interactions in Air, Water and Soil*. Volume 2, Fourth Edition. FL: Lewis Publishers, CRC Press. Boca Raton, FL.

BIBLIOGRAPHY

Blake, Rob, PetersJ., 2012. Model Aquatic Health Code (MAHC) and International Swimming Pool and Spa Code (ISPSC). *Journal of Environmental Health*, 74(9).

Centers for Disease Control and Prevention. 2003. Surveillance Data from Swimming Pool Inspections: Selected States and Counties, United States, May–September 2002. *MMWR Weekly*. Atlanta, GA.

Centers for Disease Control and Prevention. 2007. Cryptosporidiosis Outbreaks Associated with Recreational Water Use: Five States, 2006. *MMWR Weekly*.

Centers for Disease Control and Prevention. 2009. Pool Chemical–Associated Health Events in Public and Residential Settings: United States, 1983–2007. *MMWR Weekly*.

Centers for Disease Control and Prevention. 2010. Giardiasis Surveillance: United States, 2006–2008. *MMWR Weekly*.

Centers for Disease Control and Prevention. 2010. Violations Identified from Routine Swimming Pool Inspections: Selected States and Counties, United States, 2008. *MMWR Weekly*.

Centers for Disease Control and Prevention. 2012. Drowning United States, 2005–2009. *MMWR Weekly*.

Griffiths, Tom. 2003. *The Complete Swimming Pool Reference.*Second Edition. Sagamore Publishing, Champagne, IL.

New York State Department of Health, New York Government Regulations. No Date. *Children's Camps.*Albany, NY.

Sudakin, Daniel, Fallah, Payam. 2008. Toxigenic Fungi and Mycotoxins in Outdoor, Recreational Environments. *Clinical Toxicology*, 46.

11 Sewage Disposal Systems

STATEMENT OF PROBLEM AND SPECIAL INFORMATION FOR SEWAGE

Proper sewage disposal is necessary to protect human health, welfare, and the economy. The contaminated liquids and solids from sewers may back flow into streets, properties, businesses, and homes, or overflow into bodies of water, underground or on the surface. Sanitary sewage may contain bacteria, viruses, and other microorganisms, as well as a variety of chemicals, including heavy metals and pharmaceuticals that can cause disease. Industrial waste coming from businesses and industry, mixed into the public sewage in the sewers, can contain a huge variety of chemicals and other substances which can affect the health of individuals. Correction of the sewerage system problems constitutes a huge financial burden on communities throughout the country (See endnotes 3, 28, 38).

Each year, there are over 40,000 sanitary sewer overflows, the majority of which are from combined sewer/sanitary sewer overflows during large wet weather events, with a great potential for disease and environmental degradation. Public wastewater collection in the United States is made up of over 16,000 different sewer systems used by over 190 million people with 740,000 miles of gravity sewers and 60,000 miles of pressure mains. There is an estimated additional 500,000 miles of private sewer laterals. The estimated cost per year over a 20-year period to correct the infrastructure for sewerage gathering systems is between $12 billion and $21 billion, whereas about $3.3 billion are actually being spent annually.

POLLUTANTS FOUND IN SEWAGE

The pollutants found in sewage include oxygen-demanding substances, pathogens, nutrients, and inorganic and synthetic organic chemicals, while thermal conditions can reduce the capacity of water to retain oxygen.

Since dissolved oxygen is necessary for aquatic life, anything, especially the many pollutants and wastewater which reduces the level of dissolved oxygen found in water, may cause disastrous effects to various ecosystems. This process is called the biochemical oxygen demand (BOD) and is also a measure of how effective the sewage treatment plant is operating through its various processes. Especially troubling for the sewage treatment plant is the industrial waste that comes from paper mills, food processing plants, tanning operation, and other manufacturing processes.

There is a multitude of pathogens which may be found in wastewater including those that cause gastroenteritis, leptospirosis, typhoid fever, salmonellosis, shigellosis, cholera, balantidiasis, cryptosporidiosis, amoebic dysentery, giardiasis, infectious hepatitis, etc. Individuals may come in contact with the pathogens either through drinking the water which is contaminated or swimming in contaminated water.

Carbon, nitrogen, and phosphorus are essential for living organisms and are the major nutrients found in natural water. Added to this is the large amount of these nutrients that may be found in sewage effluent, even that which has gone through conventional secondary treatment plants. In addition, there are large numbers of nutrients which come from fertilizing the soil. All of these can over-stimulate plants found in water, and cause serious problems with drinking water, making it toxic and also aesthetically unacceptable. Non-toxic algae blooms in reservoirs caused by septic leachate can contribute to taste and odor issues created in the reservoir. Uncontrolled algae can block out sunlight and cause serious problems to aquatic plants and animals by depleting dissolved oxygen in the water at night.

There are numerous inorganic and synthetic organic chemicals, especially from detergents, household cleaners, pharmaceuticals, synthetic organic pesticides, industrial chemicals, and manufacturing wastes, which cause severe problems in surface and groundwater supplies. Heavy metals can be very toxic to the various ecosystems.

Legal pharmaceuticals and personal care products are frequently deposited in toilets, as a supposedly safe way of disposing of them. In addition, discharge of bioactive metabolites of these drugs may also be found in the water, coming from urine and feces. Unfortunately, these drugs or metabolites may not be destroyed by the treatment process and therefore are found in bodies of water which may be used for drinking water purposes after treatment. Illicit drugs may also end up in the water supply.

SUB-PROBLEMS INCLUDING LEADING TO IMPAIRMENT AND BEST PRACTICES FOR PUBLIC SEWAGE SYSTEMS

An increased temperature of the water is caused by the use of water in power plants and industries for cooling purposes. Also releases from wastewater treatment plants and stormwater retention ponds subjected to the heat of the sun, especially during the summer, can interfere with the surface bodies of water.

INFILTRATION AND INFLOW

Infiltration and inflow of water into the collection system may have a profound effect on the eventual treatment of the wastewater because of surges and too much wastewater to be treated by the existing treatment facilities. The three major components of wastewater in the sanitary sewer system are the actual flow of wastewater from homes, businesses, and industry, and groundwater infiltration and inflow, which come from rainfall or snow melt into the system. A determination must be made of the amount of inflow and infiltration into the system and how best to reduce this in order to prevent sanitary sewer overflows as well as problems in the sewage treatment plants. (See endnote 40.)

Best Practices to Mitigate or Prevent Infiltration and Inflow into Sewer Pipes
* See Best Practices in Inspection, Maintenance, and Repair of Sewer Laterals and Best Practices in Sewer Line Replacement below.

Sewer Laterals

Sewer or building laterals, which are private connections to the sewerage system and therefore not under the control of the public authorities, make up about half the total length of all the pipes connected to the system. There is a substantial amount of infiltration into these building laterals as well. Typically, the pipes are very narrow in diameter with frequent changes in diameter, multiple bends and multiple fittings for cleanout, or minimal or no cleanout or effective access. They may be laid with a minimum slope and put in by local contractors with little or no inspection and supervision. There can be numerous defective connections including to the main sewer line. Pipes that are not aligned properly occur as well as pipe joints that are cracked or displaced. The laterals may pass close to trees which allows for disruption by roots and infiltration of the line, private property, a roadway or driveway, or bodies of water. There is considerable failure because of unsuitable materials or installation. Since these lines are privately owned, this creates a serious problem of private property issues. There are certainly serious financial issues for the property owners when laterals need to be replaced.

Best Practices in Inspection, Maintenance, and Repair of Sewer Laterals
* Locate cleanouts for laterals on each property by visual inspection.
* Locate laterals on each property by smoke testing, use of dye in toilets, use of closed-circuit TV, ground penetrating radar, radar tomography, etc.
* Determine if there is leakage by smoke testing, use of dye in toilets, and visual inspections of uncapped cleanouts.

- Redirect downspouts from structures away from the sewer lateral line while providing an absorption zone for the water through a garden.
- Redirect footing drains from the laterals.
- When absolutely necessary, permanently repair the laterals by using a backhoe to excavate the area and install new larger piping to the mainline. This is usually best when the laterals are in very shallow trenches.
- If the problem is within 2 feet of the mainline, robotic repair can be utilized which also can remove roots from the system.
- Re-grout the lateral chemically from the mainline or through cleanouts.
- Use one of the various types of cured-in-place linings for the laterals. Typically, a repair of 100–200 feet from the cleanout is then possible.
- Consider adding a small insurance premium to the monthly cost of sewage disposal to provide for lateral repair in the event it is malfunctioning.
- Understand the private property issues particular to a given community since a substantial amount of sewer lines and laterals are beneath private property. Typically, the homeowner should be responsible for the entire lateral to the mainline in the street, but this is not always necessarily so.
- Provide a special fund for homeowners to be able to borrow the necessary money at very low interest rates to correct the problems of laterals on their property. This could be very costly and could drive the homeowner into bankruptcy or out of the property and still not fix the problem of the lateral.
- Do not allow businesses to put fats, oils, and grease into drains since this is a major contributor to sewer line blockage.

Collection Systems

Public sewerage systems deteriorate with age causing pipe or equipment failures, blockages, and breaks in sewer pipes. The ground may settle over time and water may infiltrate into the sewer line, increasing the volume going to the sewage treatment plant, thus overwhelming the system. Sewer laterals may cause substantial problems as noted above. Sanitary sewer overflows may also be caused by sewer design defects, improper sewer system operation and maintenance, power failures, and vandalism. The pipes may be constructed of materials which corrode or deteriorate and potentially can collapse. Pipes may be located on shallow slopes, floodplains, or in areas where the depth to groundwater or bedrock is minimal and the soil may be compressible, susceptible to frost, and easily eroded. Hydrogen sulfide frequently corrodes sewers and equipment used in wastewater collection.

Typically, the pipes that are used to transmit sewage to a treatment plant are hidden deep underground and therefore forgotten. In the United States, significant numbers of installations in urban areas can be tracked back 100 years or more. Obviously, these pipes, because of disrepair and dysfunction, need to be replaced with modern piping systems.

Blockages and infiltration of water during wet weather may overburden the hydraulic capacity of the sewers as well as the wastewater treatment plants, and the raw sewage is then discharged unintentionally but illegally to surface bodies of water. Improper sewer system inspection, operation, and maintenance, power failures, and poor sewer design also contributes to the problem.

Best Practices in Reducing Flow of Water and Contaminants to Collection System
- Track and quantify the location, area, and volume of water being treated by the infrastructure and how best to improve upon this by means of reducing the volume to the collection system including the use of permeable parking lots, green roofs, and vacant land stormwater detention ponds. (See Chapter 14, "Water Quality and Water Pollution.")
- When improving roads and various piping systems, incorporate stormwater Best Practices into the plan.

- If large animals such as horses or cattle are on parcels of land serviced by municipal sewerage systems, all floor drains must be screened in the stable buildings to keep straw and sawdust out of the drains.
- Do not dispose of used oil through culverts into the collection system.

ANCILLARY STRUCTURES INCLUDING PUMP/LIFT STATIONS

(See endnote 31)

Wastewater pump/lift stations are facilities which contain a receiving well for the sewage flow from the properties or businesses. The sewage flow is screened or ground up to get rid of course materials and then sent through pipes using pumps to a point where it can be lifted up to a higher level. The sewage flow can once again move by gravity toward the sewage treatment plant. These units are frequently used in the centralized collection system to keep the sewage flow from the houses and businesses moving forward. The advantage of the lift station is that it reduces the cost of sewer system construction because the trenches do not have to be dug more than 10 feet deep. The disadvantage of this system is that power is required to operate the pump/lift station and there need to be individuals trained to maintain the system and pumps properly. There is also substantial diversity between the various components in the different systems including necessary sensors and control mechanisms. Pump stations can age the pipes when surges of sewage occur. The structures for these units are frequently underground and a combination of groundwater leakage and condensation may corrode steel supports and frames as well as other equipment. Also, if there is a failure in the power system, the pump/lift station may shut down.

Drop Shaft

A drop shaft is used to connect a shallow storm or sanitary sewer with a deeper pipe or interceptor tunnel. The flow may either drop freely within the shaft or be piped downward. Sometimes a spiral technique is used to decrease the velocity of the flow. These shafts need to also be maintained properly based on the internal structure and how they have deteriorated.

Valve, Diversion, and Overflow Structures

Valve chambers shut off the flow to a pipe or divert it into a parallel line. These are most frequently used on a pressure pipe system. Overflow structures are susceptible to corrosion and deteriorate faster than the rest of the system. There are few standard shaped parts in these systems and this creates difficulty in maintenance and repair, and because the sewer lines tend to be very deep, the groundwater pressure in the actual groundwater is high and puts considerable stress on the pipes allowing for inflow, causing greater problems with volume at the sewage treatment plant.

Manholes

Manholes are put in sewer systems to assist in maintenance and cleaning of the sewer pipes. They can be spaced between 100 feet and 500 feet apart. There are over 12 million manhole structures in the United States. It is possible to inspect, repair, clean, and maintain pipes or install new ones in a segment between manholes using trenchless methods. The older manhole structures were typically made of brick or concrete, and they deteriorated because of the presence of hydrogen sulfide gas. The leaks from the structure may wash away the surrounding soil and weaken the structure. Manhole structures may be fractured because of the heaving of the soil from frost in cold climates. Manholes are a significant source of unwanted water in the sewerage system because of high water tables and infiltration, or simply pouring into the structure around the cover, frame, or frame seal, or holes in the cover. The ladders then become corroded and hazardous to employees using them.

Best Practices in Repairing Ancillary Structures

Pump/Lift Station
- Establish appropriate pump operations, maintenance, and emergency procedures to operate pump/lift stations in a manner which avoids damage to the adjacent pipes.
- Eliminate groundwater infiltration into pump/lift stations and waterproof as well as possible.
- Make external and internal shapes of pump/lift stations as uniform and as simple as possible, while avoiding creating sharp corners, angles, or bends when constructing them or waterproofing them.
- Slope internal supporting members and floors slightly toward the wall of the structure and provide a drainage space behind the walls and floors to reduce corrosion from standing water.
- Install an internal drain at the floor wall junction to remove leakage and dispose of it in an appropriate manner.
- Where possible when upgrading pump/lift stations, eliminate them by using new technologies involving micro-tunneling or directional drilling. (See endnote 42.)
- Since there is such a variation of structures and equipment in all ancillary units, have on call specialized work crews who can repair or replace whatever equipment structure is encountered.
- Since surface problems can be deceiving especially in concrete, use core samples to determine if an actual problem is occurring and has to be dealt with.
- Always keep complete logs of all problems and repairs. There are a series of techniques which should be used in making determinations of problems in ancillary structures. They include visual inspections, digital photography, tapping with a hammer to find damaged sections, measuring hammer rebound to determine hardness of the surface, testing of the bonding material, ultrasonic measurement of thickness of materials, tilting gauges if the structure appears to be tilting, and installation of crack gauges to monitor openings and shifting of cracks.
- Repairs can vary from patching and sealing to inserting new coats and linings. Also grouting is very important to prevent water leakage.

Drop Shafts
- Use pipes to smooth the velocity of the flow of the sewage and not overwhelm the system with turbulence as well as producing excessive amounts of hydrogen sulfide. Piping also reduces corrosion and erosion in the sanitary sewers.
- See Pump/Lift Stations above for other Best Practices.

Valve, Diversion, and Overflow Structures
- When repairs are made to these structures and the flat surfaces, floors, walls, and roofs have deteriorated and need rebuilding, sealing, or waterproof coating, it is necessary to bind and anchor the new material totally to the structure and have a complete and total waterproof seal applied.

Manholes
- Determine the types of defects in the manhole cover, frames, missing or damaged gaskets, etc., and correct them, sealing off potential entry for external water. Replace the manhole cover and frame if necessary and seal it properly.
- Provide locking lids to keep individuals other than those who are authorized to be in the manhole from entering. This is also very useful to prevent terrorists from planting bombs under the lids.
- Patch and plug all minor leaks in the manhole structure and use appropriate sealants at all joints and rings internally or externally if accessible.

- Where the manhole itself is structurally sound, re-seal all joints with waterproof material to prevent excess entry of water into the structure.
- Depending on the nature of the problem and the cost involved, utilize either chemical grouting, flood grouting spray, or spin coatings and liners including polymer coatings, cured-in-place liners, cast-in-place liners, etc.
- Establish a strong inspection and maintenance program of all manholes on a regular basis.

COMBINED SEWER SYSTEMS

A frequent problem is storm sewers, designed to remove excess water from areas, which use a readily accessible existing system which is also used to transport sanitary sewage away from a community. In 1843, the City of Hamburg, Germany, used the skills of English engineers, combining sanitary sewer discharge with the storm sewer discharge, to make the first combined sewer. Philadelphia is an excellent example of the use of combined sewers. As the city developed in the 19th and 20th centuries, a majority of the streams were diverted into pipes and then the areas were brought up to ground level with fill of millions of yards of dirt. These pipes were then also used for the collection of stormwater runoff and raw sewage from homes as well as wastewater from industries. There was a need to get rid of the sanitary waste because of a sharp increase in the use of water closets. There was also a need to get rid of the wastewater from an ever-expanding industrial economy. Other cities did the same thing. Since, in the United States, many communities had already built storm sewer systems to relieve flooding, it became practical that sanitary sewage wastewater be added to this system. The exception to the combination of the two types of water waste was in Memphis, Tennessee and Pullman, Illinois, where in 1880, engineers separated sanitary waste from runoff into two separate systems. Where wastewater treatment plants were built, they were designed to treat the sanitary sewage and frequently waste from local industries, and not the runoff water. In time of high amounts of rainfall, the sanitary sewerage systems were overwhelmed and the liquid waste had to be diverted to a body of water, thereby contaminating it with raw sewage and chemical and other industrial wastes. Even in properly operating systems, there are times when untreated sewage, stormwater, toxic materials, and industrial waste have to be diverted into surface bodies of water because of unusual volumes which cannot be handled by the treatment plants. This affects swimming, boating, access to beaches, fishing, seafood harvesting, and public drinking water supplies. In 1967, the American Public Works Association conducted a nationwide survey to determine the extent of environmental problems coming from combined sewers, basically found in the Northeast, Great Lakes region, and the Ohio River basin. Over 1300 municipalities and over 36 million people were served by these combined systems; however, the number of industries was not listed. Combined sewers represented about three quarters of all the overflow problems. There was a significant amount of pollution added to surface bodies of water from the raw sewage from people, storm runoff including debris, animal droppings, soil erosion, de-icing compounds, pesticides, PCBs, fertilizers, heavy metals, air pollutant deposits, and industrial waste including hazardous chemicals from factories and other sources. This continues to be a problem today.

After the passage of the Clean Water Act of 1972, numerous new municipal sewerage systems were built and now are approaching the usable time limit for the facilities and pipes. In 2002, the Environmental Protection Agency (EPA) estimated that 23% of sewer pipes were rated as poor, very poor, or no longer acceptable. It is estimated that by the year 2020, 45% of sewer pipes will be in need of immediate repair. It is estimated that 94% of the international market is made up of facilities which need to be refitted or replaced.

Best Practices in Use of Combined Sewers
- The best approach would be to provide separate storm sewers and sanitary sewers for all new developments and where possible for older units. Unfortunately, for older units the cost may be prohibitive.

- Develop a written inspection and maintenance program for the combined sewers to ensure that they are functioning as well as possible because they are being cleaned and repaired on a regular basis. No dry weather discharges of overflow from the system should be allowed.
- Determine the amount of system storage capacity for the combined flow and utilize it when the flow increases because of wet weather.
- Industrial pretreatment of all waste must occur and be verified before the industrial flow is allowed to enter the combined sewer system.
- During times of high water flow due to weather conditions, the industrial flow should be stored at the industry until there is a reduced amount of combined flow within the collection system.
- Reduce the non-point source water flow from land into combined sewers by utilizing special techniques which are discussed in Chapter 14, "Water Quality and Water Pollution."
- Maximize the combined flow to the publicly owned treatment works and store any excess there until it can be processed.
- Develop and enforce laws about street sweeping and litter control to prevent these materials from entering the storm sewers through culverts.
- Educate the public not to throw any items into catch basins and not to dispose of any personal items into toilets, but rather put them in with the solid waste.
- Evaluate all catch basins to determine if mosquitoes are breeding there and if necessary have the health department treat them.
- See the sections on "Sewer Cleaning and Inspection and Sewer Line Repair."

Sewer Cleaning and Inspection

(See endnote 32)

As was discussed above, there are many potential problems in sewer system networks because of their age, potential deterioration, blockages which slow down the flow of the waste material or bring it to a halt, collapsed and leaking pipes, infiltration of water, flooding, and the entire massive concern of combined sewers. Debris and foreign material along with grit, sand, fatty deposits, roots of trees, and accumulations of biosolids help produce the blockages which lead to backups in houses, odors, and/or dangerous hydrogen sulfide gas buildups.

Adequate funding for inspection and maintenance programs is lacking in many communities. Legislative bodies do not like to fund that which they cannot see even though poor infrastructure in these systems can lead to numerous complaints and even serious hazards and eventually cost more than the original repairs.

In order to counteract the serious problems in our sewage system infrastructure, it is necessary to have an appropriate inspection program, sewer cleaning, and replacement of pipes hopefully before they start to leak and collapse. Some 90% of sewer main backups are in pipes with a 12-inch diameter or less.

Serious problems can occur when personnel conducting the inspections improperly dispose of the materials and chemicals which have been gathered together during the cleaning process. Also there are concerns about the handling of chemical dyes, which are powders, and the evaluating instruments that may become coated with petroleum products and pose a fire hazard especially if there is a buildup of gas in the sewer line.

Gravity sewers have gradients downward so that they should be self-cleaning from the velocity of the fluid. However, this is compromised by debris entering the system, pipe settlement causing a loss of gradient, buildup of fats, oils, and grease, pipe collapse, and infiltration of grit and sand through openings in the pipe.

Laterals from properties can cause a serious problem in releasing the sewage from the house into the surrounding ground and possibly the groundwater, especially in high groundwater areas. When the main lines in the street are fixed, this increases the flow of the sewage within the system and if the laterals are in poor shape this may lead to odors or actual leakage on the properties.

Best Practices for Sewer Inspection and Cleaning

- Using the latest technology, map out the entire system of sewer lines, pumping and lift stations, manholes, etc., and record all known information from previous studies and complaints about problem areas throughout the system.
- Determine if mosquitoes are breeding in the various catch basins of the combined sewers or storm sewers and if so immediately contact the local health department for appropriate control measures.
- Sewer line inspections should be conducted at low-flow times, preferably between midnight and 5 AM or if necessary when the sewers are temporarily plugged to reduce the flow. Inspections can be conducted with: closed-circuit television which, because of the late night hours when the operators are less efficient and may possibly lead to errors due to inattention or fatigue; cameras where the pictures of the inside of the pipes may be done haphazardly and therefore not provide a comprehensive view of the problems; visual inspection and recording of the problems, which is fine for large sewer pipes but in smaller sewer pipes only allows the individual to see what is happening close to the manhole; lamping used in low priority pipes laid 20 years ago or less, which is the lowering of a camera into a maintenance hole and then positioning it in the center of the manhole frame and the sewer and taking pictures of the surrounding area; and the use of mirrors to reflect images of the inside of the pipes. Innovative acoustic technology using portable equipment can be used for smaller diameter pipes of 12 inches or less. This can give you information within a few minutes about blockages and save substantial amounts of money and time in the testing process. (See endnote 37.)
- Use grease traps from businesses and in some cases homes to trap the grease before it can get into the sewer line. It is essential to clean the grease traps regularly and dispose of the material in an appropriate manner to avoid contaminating the land or water.
- Insist that industries trap and treat most of the effluent before it is allowed to enter the sewer line.
- Use mechanical techniques for breaking up grease deposits, cutting roots, and loosening debris and removing it from the inside of the sewer pipe. This can be accomplished either through rodding, which consists of using an engine, rods, and blades to break up and cut debris and remove grease and other substances, or using a bucket machine with open hinge jaws to scrape the material off of the pipe into the bucket and disposing of the contaminated contents of silt, sand, gravel, and biosolids.
- Use hydraulic techniques which scrub the interior of the pipe and increase the flow in the sewer line. This helps remove floatables and some sand and grit. It can be accomplished by using a threaded rubber cleaning ball that scrubs the inside of the pipe, by introducing a heavy flow of water into the sewer line through the manhole, or directing high velocities of water against pipe walls.
- Use chemicals only as a last resort in the sewer lines, and be certain that the material safety data sheets have been reviewed in order to not cause additional problems when trying to control roots, grease, odors, corrosion, insects, and rodents.
- Monitor the flow of the fluid within the sewer lines at various points in the collection system to determine if there are obstructions and deposits which may be slowing down the fluid.
- Determine if there are frequent sewage overflows from the system and record them along with the weather conditions at that time. The maintenance of good records is essential to aid in establishing a proper maintenance schedule and to avoid violating environmental laws.
- Establish emergency repair crews with all the necessary available equipment and materials as well as procedures to be followed in the event of an emergency.
- When sewer lines are repaired, the laterals should be repaired if feasible at the same time.

Best Practices in Sewer Line Replacement (See endnotes 33, 34)

The traditional method of digging up old, deteriorating, and cracked sewer lines and replacing them is an extremely expensive and time-consuming process. The latest methods are called trenchless sewer rehabilitation which uses: pipe bursting, or in-line expansion; slip lining; cured-in-place pipes; and modified cross-section liners.

- Pipe bursting occurs when a special tool forces the pipe to burst and thereby opens it up. An expansion head is pulled by a special machine through the old pipe and replaces the old pipe with the new pipe, which is immediately behind the expansion head. It is necessary after the pipe bursting to reroute the flow and reconnect laterals to the new pipe.
- Slip lining is the placement of a liner of smaller diameter inside the existing pipe. It is necessary to grout between the liner and the old pipe to prevent leaks and provide a seal for the space between the new liner and the old pipe.
- Cured in-place pipe is a process where a flexible fabric liner coated with a special heat setting resin is inserted into the existing pipe and cured under pressure to form a permanent liner.
- Modified cross-section lining consists of using a new flexible pipe deformed in shape and inserted into the original pipe. The liner is pulled through the existing pipe and is heated and pressurized to conform to the original pipe shape.

All of these techniques may be useful in replacing the damaged and deteriorated existing pipes.

WASTEWATER TREATMENT SYSTEMS

(See endnote 3, and Chapter 11, "Sewage Disposal Systems" for specifics)

Primary Treatment

Primary sewage treatment involves the removal of biosolids and other settleable material in domestic wastewater. Prior to this, the wastewater entering the facility goes through a screen to remove large floating objects that can clog pipes and pumps. The screens can vary from removal of coarse material to the removal of very fine material while the rest of the wastewater continues on to the primary treatment process. Grinding mechanisms known as comminutors can cut and shred material which can then be prepared for further processing. The wastewater then flows into a grit chamber where sand, grit, cinders, and small stones can settle out, and then into a primary sedimentation tank where the wastewater is slowed down and suspended solids can be settled to the bottom of the tank by use of gravity or also with the addition of a chemical coagulant. The settled materials or primary sludge is then removed either manually or continuously depending on the age of the plant. The sludge is removed to a sludge digester where the process is carried out by anaerobic bacteria, occurring in the absence of air, that are present in the solids. The fluid that is left over should be introduced into secondary treatment. The material that settles should be considered to be biosolids and dealt with accordingly.

Secondary Treatment

The wastewater effluent from primary treatment flows into secondary treatment which uses natural biological treatment processes to reduce the organic matter by 90%. Microorganisms found in the wastewater use the organic matter as a food supply in the presence of oxygen. This process can be accelerated by adding oxygen to the wastewater beyond that which is already present. Typically, two different processes are most commonly used. They are called the attached growth process and the suspended growth process.

The attached growth or a fixed film process takes place on the surface of stones or plastic. The wastewater is spread over the media along with air and trickles down slowly allowing the biomass containing microorganisms to consume the organic matter and produces carbon dioxide. These systems though they may vary in type are very successful and they include trickling filters, rotating biological contactors, and bio towers. Problems occur when there are sharp temperature differentials which may slow the process down, the storage space between the different types of media become filled, there is a reduction in the air present within the system, groups of organisms, their waste products, and byproducts crack off the stones or other parts of the media, and the under-drainage system to capture the effluent becomes clogged or deteriorates.

In the suspended growth process, the biodegradable organic material and organic nitrogen-containing material through the use of pumped-in oxygen or air and agitation of the fluid allows an oxygen transfer to convert ammonia nitrogen to nitrates, which then may be processed further. This creates large numbers of microorganisms and speeds up the process. As the activated sludge accumulates, part of it is removed and part of it is returned to the aeration tank to be mixed with the incoming wastewater and this activated sludge must be kept in suspension to make the system work. Most activated sludge processes are more expensive to operate than trickling filters because of the cost of high-energy use to run the aeration system. Toxic chemicals from industry also impact the process.

Membrane bioreactors may be used for secondary treatment. This process combines the activated sludge treatment with the use of a membrane that separates the solids from the liquid, leading to effective removal of the soluble and biodegradable materials at higher loading rates. The problems are that they cost more than the usual secondary treatment and the membranes can become less effective over time.

Wastewater lagoons, which are scientifically constructed 3–5 feet deep ponds that allow sunlight, algae, bacteria, and oxygen to interact, especially in smaller communities where land size is not a problem, serve as a secondary treatment method for about 25% of the municipal secondary treatment facilities. It typically works as well as secondary treatment of other types in removing biodegradable organic material and some of the nitrogen if the lagoon is constructed properly and operated properly. Cold temperatures have a significant effect on the operation of the lagoon. Of great concern is the breaching of the lagoon, especially those containing animal waste during storms, which will have a major impact on the environment and health of people.

Advanced Wastewater Treatment (Tertiary Treatment)

Tertiary treatment can be an extension of the secondary treatment and get rid of more of the biological material including a variety of microorganisms and/or it may be utilized specifically to get rid of certain nutrients which can affect bodies of water.

Physical/chemical treatment processes include adsorption using activated alumina, granular activated carbon, granular iron, and powdered activated carbon; disinfection using ozone, chlorine, bromine, and ultraviolet light; nutrient removal by air stripping, denitrification filters, ion-exchange, chemical precipitation with alum, iron salts, and zeolite, and using a solids contact clarifier to remove phosphorus; chemical oxidation with chlorine, hydrogen peroxide, oxygen, and ozone; and advanced oxidation using catalysts, hydrogen peroxide plus ferrous ions, and ultraviolet light plus titanium dioxide. (See endnote 44, Chapter 2.)

Biological treatment processes innovative technologies for tertiary treatment include anaerobic processes such as anaerobic attached growth systems, the anaerobic contact process, anaerobic sequencing batch reactors, and anaerobic sludge blanket upflow; BOD removal and nitrification including aerated lagoons, complete mix activated sludge, contact stabilization, conventional extended aeration high purity oxygen, sequencing batch reactors, etc.; biofilm processes using, for instance, biological aerated filters, fluidized bed bioreactors, trickling filters, etc.; nitrogen removal using, for instance, a denitrification filter, a simultaneous nitrification denitrification process, etc.; nitrogen and phosphorus removal; membrane process; and bioaugmentation. (See endnote 44, Chapter 3.)

Land Treatment of Wastewater

Where there is adequate land available and the soil is able to readily process the wastewater as it passes down through it by means of physical, chemical, and biological processes, the wastewater from the treatment plant can be utilized as a means of irrigating crops. This can be accomplished through slow-rate infiltration where the disinfected wastewater is applied either by spraying, flooding, or irrigating furrows and part of the liquid evaporates and the plants utilize the other part. Rapid infiltration is used to polish and recover the wastewater effluent after pretreatment when large quantities percolate down through permeable soils into the water table. Wells can be used to extract the water and utilize it again. Overland flow is a process where the wastewater effluent passes over gently sloped, compact soil and the soil and its microorganisms form a gelatinous slime similar to a trickling filter, which removes the solids, pathogens, and nutrients. The water at the bottom of the slope if it meets appropriate standards can then be discharged or further treated.

Wetlands

Most wetlands systems are considered to be free water surfaced where the surface of the water is exposed to the atmosphere. These include the natural wetlands, such as bogs which are primarily vegetation mosses, swamps that are primarily vegetation and trees, and marshes that are primarily vegetation grasses and emergent macrophytes. (Macrophytes are plants that are rooted in shallow water with their vegetative parts above the surface of the water. They are extremely productive because their roots are in the sediments and their photosynthetic parts are in the air.) These natural systems cannot be used for disposal of effluent from sewage treatment plants. However, if a system is specially constructed for the use of an advanced wastewater treatment plant for tertiary polishing of the effluent, it must be used safely.

Constructed Free Water Surface Wetlands

(See endnote 36)

This type of system mimics the systems that are naturally occurring in marshes. The effluent flows over a vegetated soil surface from a point of inlet to a point of outlet. The liquid may be removed by evapotranspiration or seepage downward through the constructed material. To discharge to a natural source requires a permit which satisfies the National Pollutant Discharge Elimination System limits. Therefore, very few natural systems exist. The size of the units can vary from that which would be used on a single property to treat septic tank effluents to something as large as 40,000 acres. The system usually has one or two shallow basins or channels with barriers so that the effluent will not seep into groundwater and a submerged soil layer to support the roots of the plants. There is a uniform distribution under very low velocity that resembles laminar flow of the effluent which allows the accumulated materials including the live plants, dead plants, and litter to help process the fluid. In the United States, commonly, the effluent gets some sort of treatment before it is released to the constructed free water surface wetlands. The advantages of the system are that: the wetlands system produces no residual biosolids, BOD, chemical oxygen demand (COD), total suspended solids, metals, or persistent organics; the system works effectively with a minimum of equipment and energy use; and it provides green spaces for the community. The disadvantages of the system are: the need for very large areas to process the effluent; cold climate problems; mosquito and other insect vector concerns; increased bird populations which may affect airports that are nearby; and potentially incomplete removal of fecal coliform.

Subsurface Flow

(See endnote 35)

A subsurface flow wetland is specially designed to treat some types of wastewater, and it consists of a bed or channel with coarse rock or gravel and planted with vegetation usually found in marshes. The water surface is kept below the top surface of the medium which helps prevent the growth of mosquitoes as well as odors. Since it had been noted that in natural wetlands the quality of the water improves due to physical, chemical, and biochemical reactions, this type of system has been designed and installed to work at even higher rates of success than the natural wetlands. The advantages of this system are that it provides effective treatment without requiring the use of mechanical equipment, sources of energy, or a skilled operator. This helps contain cost. It can be used year-round, has no residual biosolids, and removes BOD, COD, total suspended solids, and metals effectively. The disadvantages of this system are: the requirement for large land areas; in cold climates the amount of BOD, ammonia, and nitrogen is decreased because of the low winter water temperatures; and the amount of *Escherichia coli* removed may not be sufficient to meet standards of water for discharge.

DISPOSAL OF BIOSOLIDS AND OTHER WASTEWATER RESIDUALS

(See endnote 45)

Biosolids

Sewage sludge is a combination of solids and liquid which comes from the treatment of domestic sewage. It contains beneficial nutrients for plants, as well as microorganisms which can cause disease. Untreated sewage sludge when applied to the land becomes a potential cause of diseases in people as well as animals. It contains bacteria, viruses, protozoa, parasites, and other organisms which may cause disease. This problem can be mitigated by various forms of sewage sludge treatment including digestion, drying, composting, lime stabilization, heat treatment, etc. Sewage sludge as well as biosolids can attract vectors, which are a variety of insects and/or rodents who may carry viable microorganisms that can carry disease from the site of disposal to people and animals. Disease can also spread by direct contact with the sewage sludge, walking through the area where it has been deposited, and handling the soil from the fields. It also can be spread through ingestion of food, drinking water, or recreational water which has become contaminated with runoff from the application areas. Airborne contact with the microorganisms can occur when the liquids are either sprayed or splashed on the ground.

Biosolids are treated sewage sludge which after treatment and processing can be recycled by farmers and gardeners to be used as fertilizer containing essential crop nutrients, especially nitrogen and phosphorus, to stimulate plant growth. This material must adhere to the standards of the federal biosolids rule found in 40 CFR, part 503, which establishes numerical limits for metals and pathogens, and other standards for site restrictions, crop harvesting restrictions, and monitoring, record-keeping, and reporting requirements. This helps avoid the use of chemical fertilizers which may readily contaminate the environment. Biosolids are also used in mine site reclamation projects to improve the land, and forestry where it promotes rapid timber growth. Biosolids may also be used in plant nurseries, cemeteries, parks, gardens, and for home and business lawns. If there are too many biosolids to be used in the above ways, then it may also be disposed of in landfills and by incineration at high temperatures.

Biosolids may contain bacteria, viruses, protozoa, and helminths, depending on the amount of pretreatment, which may be pathogenic to people and to animals. Care should be taken not to expose workers or people in surrounding areas to airborne microorganisms at land application or biosolids storage sites.

Dewatering and thickening of the biosolids is important to make the transportation and eventually use of this material easier. This can be accomplished through several innovative processes involving membrane thickening, electro-dewatering, and geotextile dewatering.

Best Practices in Turning Sewage Sludge into Safe Biosolids (See endnotes 46, 47)
- Utilize an aerobic digestion process for the sewage sludge after secondary treatment for 40 days at 68°F and 60 days at 59°F.

- Utilize an air drying process where the sewage sludge is dried on beds of sand for at least 3 months and the temperature is above freezing.
- Utilize an anaerobic digestion process where the sewage sludge is treated in the absence of air for 15 days at 131°F or 60 days at 68°F.
- Utilize a composting technique where the temperature of the sewage sludge is 104°F or higher and stays this way for 5 days. For 4 hours during this time the temperature must be raised to 131°F.
- Utilize a lime stabilization technique where there is sufficient lime added to the sewage sludge to raise the pH to 12 for at least 2 hours of contact time.
- Utilize the innovative supercritical water oxidation process for thermal oxidation of the organic material within the waste stream by increasing the temperature and pressure in the equipment until the organic compounds are oxidized.
- Utilize an innovative sludge-to-fuel process in which a carbon-rich fuel is produced from the sludge under high pressure and heat which after dewatering can be used as a fuel.
- Utilize an innovative vitrification process where the minerals in the biosolids are melted and create a glass-like product that can be used as an aggregate.

Septage Treatment/Disposal: Land Application

(See endnote 13)

Septage includes liquids, solids, concentrated BOD, nutrients, toxics, inorganic materials (such as sand) pathogens, oil, grease, hair, debris, and odorous materials which have accumulated over a period of time and have been pumped out of septic tanks and cesspools to be treated and for eventual appropriate disposal. This can be accomplished through either surface or subsurface land application, or added upstream to a sewage treatment plant at a sewer manhole or at the beginning of the process. Surface application can be accomplished through spreading the material on the ground or using spray irrigation. Subsurface application can occur through plowing the septage into furrows or injecting it below the surface. It can also be buried in sanitary landfills. Surface disposal can result in odor problems, and fly and mosquito problems, and provide an opportunity for aerosols to be created containing viruses and bacteria.

Best Practices for Septage Treatment/Disposal: Land Application
- The septage should be removed from the septic tanks' or cesspools by licensed trained sewage haulers who must remove it in a safe and effective manner and take it to the point of disposal.
- Buffers should be provided and setbacks observed between the areas of application of the septage and any water sources. The slope of the area, type of soil, climate, weather conditions, and application rates are very significant. Food used for human consumption should not be grown in this area.
- Stabilization of the septage or pretreatment should be done if feasible prior to the land application. This can be accomplished by using stabilization lagoons, chlorine oxidation, aerobic digestion, and biological and chemical treatment.
- In the spray irrigation process, the septage must be screened and pretreated and applied to the ground at a rate of 80–100 pounds per square inch through nozzles. The material can also be spread out in furrows and on the land in thin layers. This is then covered up by a plow.
- Septage or sludge must never be applied to frozen agricultural land.
- Use alternate means of disposal, if available, such as taking the septage to a sewage treatment plant for processing which is preferable if the septage is properly controlled and released into the treatment process without affecting what is occurring.

Wet Weather Concerns at Combined Sewer Wastewater Treatment Facilities

(See endnote 43)

The treatment plants that serve combined sewer systems are readily impacted by storm factors. Wet weather impacts are not only due to the amount of rain or snow melt but also due to: the age and physical condition of the collection system; the level of groundwater around the sewer lines and manholes; the presence of footing drains and drainpipes from the roofs of structures leading into the sewer pipes; how the storm and sanitary sewers are joined together; the storage capacity in the collection system; the storage capacity of each unit of the treatment plant; and the plan put into operation to prevent hydraulic overloading in each step of the treatment process.

When the volume of sewage flow is greater than the capacity of the treatment plant, whether it be for combined sewers or sanitary sewers, it may be diverted into surface bodies of water which will contaminate these waters with physical, chemical, and biological agents leading to potential disease or injury to people, destruction of shellfish areas, lowered dissolved oxygen ratios, permanent or temporary closing of bathing beaches, and severe aesthetic concerns about floating debris, oil slicks, etc.

Best Practices for Wet Weather Concerns at Combined Sewer Wastewater Treatment Facilities
- Develop a comprehensive wet weather operating plan for combined sewers to prevent or minimize the discharge of combined sewer overflows into bodies of water without proper treatment and disinfection. Include in the plan, the nature and condition of the collection system, typical flows and pollutant loadings at the wastewater treatment plant, what occurs during exceptional wet weather flows, and how it affects the operation of the plant. Monitor the typical amount of grit, screenings, BOD, and total suspended solids in primary sedimentation, operation of the aeration tanks, final settling of the solids, and the disinfection system, and predict the increase due to extreme weather conditions.
- Determine the amount of space available for storage of access sewage flow in unused equipment and tanks for later treatment as well as maximum hydraulic capacity, maximum treatment capacity, and how to improve these to avoid sewage overflows.
- Determine how to bypass a portion of the treatment process when hydraulic overload will occur and yet not release the flow into surface bodies of water.
- Reduce plant recycling such as the supernatant fluid from the sludge digesters and store until the flow rate through the plant is reduced and then treat it.
- Equally distribute the flow into multiple units in any given process by using special flow splitting devices.
- Determine if the secondary clarifiers in the treatment plant are working effectively during wet weather and if not alter the process.
- Determine prior to wet weather if the pumping stations and the collection system are working properly and any unused storage of fluid will be able to be used.
- Determine before wet weather if the bar screens and aeration tanks are working appropriately and if cleaning and maintenance needs to be carried out immediately.
- Place all units of the sewage treatment system into active service and set the controls for continuous operation during wet weather. Cleanout all channels and screening containers, as well as grit and other substances frequently.
- Constantly evaluate the velocity of the sewage flow as it moves into the plant and through the process to prevent hydraulic overloading and make necessary corrections rapidly and efficiently.
- Avoid high solids loading in the primary treatment process by removing them more frequently during wet weather. Check all baffles and weirs and make sure they are operating properly.
- Discontinue sending secondary sludge to the primary clarifiers until the extreme wet weather is over.

- Use chemical coagulants to help remove the total suspended solids in the primary clarifiers during primary treatment.
- Activated sludge processes may be severely affected by large flows which will cause a loss of the biomass in the aeration tanks and in the secondary clarifiers. Utilize the step feed and context stabilization mode which handles higher hydraulic loadings, but will not be as efficient as the normal mode in BOD removal, in order to prevent this from occurring. Do not allow excess solid storage in this process.
- Determine what amount of chlorine and contact time is necessary to effectively reduce any solids and the amount of *E. coli* present before discharge of the effluent.

FLOODING A SEVERE DANGER TO WATER AND WASTEWATER TREATMENT FACILITIES

(See endnote 41)

Flooding causes more damage in the United States than any other problems caused by weather conditions due to overwhelming amounts of water and/or debris. Drinking water and wastewater treatment facilities can become inundated by the floodwaters causing loss of power and damage to equipment and buildings, and creating extremely hazardous conditions for the employees. Flooding can happen extremely rapidly and with little or no warning. It will have a profound effect on individuals in the community by eliminating the public potable water supply and/or send wastewater back into homes, onto the streets and highways, and into the surface bodies of water creating enormous potential for disease and injury. Typically, water and wastewater treatment plants are located close to large bodies of water which may be prone to flooding.

Best Practices to Prevent or Mitigate Damage and/or Destruction of Water and Wastewater Treatment Facilities
- Determine the threat of flooding by studying the maps prepared by the Federal Emergency Management Agency for a 100-year flood event, meaning that there is a 1% probability of this occurring in a given year. Use other information sources that are available including records of previous floods.
- Identify the vulnerable parts of the treatment system including elevations of the equipment, instrumentation, and electrical controls, power supply, etc., and determine the potential of damage or destruction.
- Evaluate the possible mitigation procedures, determine which are cost-effective and of highest priority, and develop a plan to implement them prior to the flooding conditions.
- For water treatment plants, prior to flooding, top off the water storage tanks, bolt to the floor all chemical tanks and ensure that waterproof covers are tightly in place, elevate all equipment especially generators, provide emergency back-up generators in a safe location, and put flood barriers around equipment. Relocate equipment if needed.
- Relocate or elevate pump houses that are in flood zones. Install waterproof pump motors or submersible pumps.
- Protect all surface intakes of water against floating debris, silt, or other contaminants by installing jetties or breakwaters to divert these substances away from the intake.
- For groundwater supplies prior to flooding, evaluate and correct the seal of well casings and protect the well areas above potential flooding heights.
- Utilize substantially reinforced screens at the intake to prevent debris from entering the system.
- Top off all fuel tanks prior to the flooding event.
- Where feasible install solar panels or wind turbines to produce electricity for the plant. Also capture waste heat and use it to produce electricity.
- Keep in a secure flood-proof area, which is easily accessible, a collection of all necessary parts needed to operate equipment in the event of an emergency.

- Develop a program to keep all culverts and drains clean and clear of debris to reduce flooding potential. Have sand bags available to place in vulnerable parts of the facility.
- Install backflow preventers for the potable water supply and on sewers and drains.
- Provide in-depth and periodic training to staff on all aspects of their role in the event of a flooding situation, when to turn power on and off, and when to remove all computers from the area.
- Provide an access plan in the event normal openings to the structure are blocked to secure the safe movement of personnel from the buildings.
- Maintain all records not only at the treatment plant but also at a distant area outside of the flood zone to prevent the destruction of invaluable information as well as customer records.
- During flooding, it is assumed that the water is heavily contaminated whether the source be from a surface body of water or a groundwater supply, so immediately test the intake supply both microbiologically and chemically to determine additional treatment procedures and advisories to the public.
- Provide permanent physical barriers around all wastewater collection system lift stations to prevent floodwaters from inundating the unit.
- For wastewater collection system lift stations' protection, make sure that all vent lines are above potential flooding elevation. If the station becomes unusable, obtain and utilize a list of vendors who can divert the wastewater flow to an alternate system.
- Install non-electrical back-up controls where possible.
- Install external connected compressed air systems to keep air flowing to the sewage treatment process.
- Install physical barriers to prevent flooding around all wastewater treatment plants.
- Install green infrastructure around all treatment works, water, and wastewater, to help divert or mitigate floodwater surges.
- Establish secure communications with the emergency management agency to keep them advised of the situation at the treatment facilities and when to issue certain orders related to the health of the public.
- See the section above on "Wet Weather Concerns on Combined Sewers and Wastewater Treatment Facilities for additional information."

PACKAGE TREATMENT PLANTS

(See endnote 53)

A package treatment plant is a premanufactured facility used to treat wastewater in small communities, mobile home parks, and summer camps, or for use by schools, businesses, or industry. They are typically extended aeration plants, sequencing batch reactors, oxidation ditches, or rotating biological compactors.

The extended aeration plant is made of steel or concrete and is a modification of the activated sludge process where air may be supplied by mechanical means or diffused aeration. Mixing by mechanical means or diffused aeration is essential for the process as well as extended contact time. For the microorganisms to work on the dissolved organics, the pH must be controlled. The remaining sludge is removed to a digester. Clarification of the effluent and disinfection occurs before disposal. This type of plant is used for small municipalities, subdivisions, apartment complexes, highway rest areas, trailer parks, etc. These units are easy to install and operate, odor free, produce very little sludge, and can handle fluctuations in the organic load and flow of the wastewater. It is best if the wastewater collection goes to the system by gravity flow. The system cannot be used for denitrification or removal of phosphorus, and requires considerable energy and a large amount of space and larger tanks than the other units.

A sequencing batch reactor is a variation of the activated sludge process where all biological treatment occurs in a single tank. The sludge is removed to a digester and the effluent is disinfected.

This type of system is used where there is little suitable land available for sewage waste processing and disposal and minimal operator attendance. It is good for industries with high BOD levels in the wastewater and requiring nitrification, denitrification, and removal of phosphorus. These units optimize the treatment efficiency and do so in a single tank which lowers the requirement for energy and space. They have few operational or maintenance problems and can be operated remotely. Sludge must be disposed of frequently and it is difficult to adjust the cycling times for a small community. Small units can have problems with operation, maintenance, and hydraulic loading.

An oxidation ditch is a modification of the activated sludge process where there is an aerated, long-term mixing of the sewage waste in a continuous manner typically in a concrete tank, although a steel and concrete tank may be used. This system can be used to treat any kind of wastewater that works by aerobic degradation and can be used for denitrification and removal of phosphorus. It provides high quality effluent with the removal of total suspended solids, BOD, and ammonia, and produces a small amount of sludge. It can be noisy, large in size, and unable to treat toxic waste. The system is expensive to operate and maintain, but it can be used if people understand its limitations.

All package treatment plants have the potential for problems caused by: a substantial and sudden temperature change; the removal efficiency of grease and scum; very small flows of wastewater; fluctuations in flow of wastewater and in the amount of BOD; hydraulic shock from overload of wastewater; and problems of the air supply.

Best Practices in Use of Package Treatment Plants
- Where public sewage systems are not available and where feasible, utilize a cost-effective package treatment plant, located in a flood-free zone, which meets the needs of the community, business, or industry, for the present and for the foreseeable future, instead of numerous on-site sewage disposal systems to prevent land and water contamination and provide a satisfactory effluent for disposal.
- Provide effective and continuous maintenance as well as operational capabilities; especially monitor wastewater flow, aeration, electronics, and equipment for the system.
- Dispose of biosolids in an appropriate manner.
- Frequently conduct tests to determine levels of BOD, total suspended solids, ammonia, nitrate levels, phosphorus, and *E. coli* and determine if they meet the standards established by the appropriate regulatory agency.
- Provide a standby generator in the event that the electrical source is disrupted.

DISINFECTION OF EFFLUENT

(See endnotes 54, 55)

A disinfectant must be used on the effluent from the wastewater treatment plant, from on-site sewage disposal systems where the effluent flows onto the ground and into a body of water, and if there are combined sewer overflows. The disinfectant must be able to penetrate and destroy the infectious organisms under normal conditions, be safe for use, storage, and shipping, and not leave toxic or mutagenic residuals.

Chlorine

Chlorine is very effective in many situations in disinfecting effluent from wastewater treatment plants. Chlorine's mode of action is well understood. It works well on numerous organisms, can reduce odors, is cost-effective, and leaves a residual that continues to kill the organisms. The problem is that chlorine residual can be toxic to aquatic life, highly corrosive and toxic to people, oxidizes organic chemicals at times into more dangerous substances, and does not work well in high concentrations of chlorine-demanding substances such as combined sewer overflows. Some parasites are resistant to low doses of chlorine including *Cryptosporidium parvum*, *Entamoeba histolytica*, and *Giardia lamblia*.

Chlorine Dioxide

Chlorine dioxide is an effective disinfectant of wastewater. It is used as a gas which is generated from excess chlorine on-site since it is unstable, reactive, and difficult to transport. It oxidizes phenols but does not produce trihalomethanes unless there is an excess of chlorine remaining after the generation of the chlorine dioxide, which means that the individuals generating the chlorine dioxide must be very careful with the amount of chlorine being used. The chlorine dioxide will not form chloramines but it can produce toxic byproducts including chlorite and chlorate.

Ozonation

Ozone is a strong oxidizing agent which is applied to wastewater as a gas. It is equal to or better than chlorine in destroying organisms and it does not form halogenated organics. However, it does not produce a residual to prevent bacterial growth. It increases coagulation, removes iron and manganese, needs a short contact time, and can be used for taste and odor control. It is costlier than chlorine, and forms nitric oxides and nitric acid which are corrosive. Ozone is chemically unstable as a gas and difficult to transport. Low doses may not inactivate some viruses, spores, and cysts.

Ultraviolet Radiation

Ultraviolet disinfection occurs during application of light with a wavelength of 40–400 nm with the germicidal range being between 200 nm and 300 nm and 260 nm being most lethal. The disinfection occurs when the ultraviolet light penetrates the cell walls of the organisms and alters the DNA. Ultraviolet light does not alter the physical or chemical properties of water, but it is affected by the amount of suspended solids present which can reduce the effectiveness of the treatment. It is necessary to closely monitor the flow rate, level of suspended solids, initial and final coliform count, number of lamps and average output of the ultraviolet lamps, and the average transmissibility of the transmitting system.

USE OF ENERGY IN WATER AND WASTEWATER TREATMENT PLANTS

(See endnotes 51, 52)

The cost of energy is 10% of a typical local government's operating budget and a good portion of this goes to the operation of water and wastewater treatment facilities because the pumps, motors, and other equipment have to operate 24 hours a day and 7 days a week. Energy efficiency is a very practical consideration.

Best Practices in Use of Energy in Water and Wastewater Treatment Plants
- Use automatic controls when possible to monitor and control the potable water processing system.
- Incorporate Best Practices in use of energy in all changes and additions to the potable water processing system.
- Utilize a computer model to determine changes in pipe size, pumping grades, system pressures, and use and location of booster pumps and storage facilities, and to determine where variable flow rates should be put into operation.
- Develop and implement a system-wide leak detection and repair program.
- Install, where possible, energy-efficient variable speed drives to control the flow rate from pumps.
- When using groundwater supplies, monitor well production and drawdown to determine if there are any major mechanical problems and if the well is becoming inefficient because of overuse. Use the well sequentially to keep from over pumping.
- Promote water conservation which also will have a profound effect on reducing the amount of wastewater which needs to be treated.

- Reduce lawn sprinkling and utilize where possible treated effluent from the sewage treatment plants. It is absolutely essential to keep both systems separate and not allow any cross-connections.
- Charge a premium for high-volume users and establish conservation programs in their facilities.
- Use variable frequency speed drives in wastewater facilities based on the load requirements at a given time when there are minimum, average, and peak flows.
- Establish operational flexibility with all equipment in order to store wastewater at high peak flows, including seasonal adjustments and during times of tourists visiting the area, and process it during times of low peak flows, especially during the night.
- Recover excess heat from wastewater and use it to produce in-house electricity.
- Especially in the northern part of the country, cover the basins to avoid heat loss and freezing during the colder weather.
- Evaluate the aeration system and determine if it is working at maximum efficiency and, if not, add additional compressed air as needed. Use variable speed devices in introducing the air supply.
- Determine if the aerobic digester will work better with a smaller blower that would control the airflow with a fine bubble diffuser.
- Determine if anaerobic digestion would be effective for biosolids, since aerobic digestion is very energy intensive.

STATEMENT OF PROBLEM AND SPECIAL INFORMATION FOR PRIVATE SEWAGE SYSTEMS (ON-SITE SEWAGE DISPOSAL)

In 1955, with only a degree in biology and chemistry and having taken the exam for sanitarian for the state of Pennsylvania, the author achieved an extremely high grade and was appointed to the position of a field practitioner in Montgomery County, Pennsylvania. At that time, a vast number of houses were being built very quickly, on small pieces of land, sometimes only 10,000 ft^2, and having to use well water and septic tank systems to take care of the water and sewage disposal needs of the family that had been reared in most cases in the City of Philadelphia. Mortgages were easily obtained by using either the Veterans Administration program or the Federal Housing Authority program. In either case, the well and the sewage system had to be approved by the county health department which was an arm of the state. Many times, the sanitarian was called out to evaluate several septic tank systems before they were covered and the time allowed was very brief. This led to a quick visual look at the drainage field or seepage pit and almost universal approval.

The author's training initially was to go out into the field with a practicing sanitarian and squat down beside a hole and be shown that there were 3 feet of shale, and therefore, a seepage pit could be used after the septic tank for disposal of the effluent instead of a tile field system. Tile field systems were also a problem because, although it was required that a professional engineer sign off on a percolation test for a tile field system, they were never conducted by the health department personnel or ever verified. It was only 7 months later during a 9-week, 40-hour-a-week training program conducted by the US Public Health Service at the field training center in Pittsburgh, that the author learned how to properly assess land in order to approve tile field systems and utilize seepage pit systems as a last resort. The reason for this story is so the reader will understand that there are a huge number of non-performing on-site sewage disposal systems throughout the country that are outdated and probably have been overflowing illegally into bodies of water for years. The land site of the house is so small typically that in many cases, it is virtually impossible to replace the system.

What happened here was typical in many areas of the country because of a lack of properly trained sanitarians with the skills and ability to approve a new septic tank system. Speed was essential to meet the demands of the construction companies. Typically, the entire project of hundreds of homes was in an area where a single home had existed for the farmer and his/her family.

The land worked well as a farm because the water wells and sewage disposal were on a vast area for the single house. Now this area was converted into all of these new homes, each of which had to have its own well and septic tank system. The effluent load even if the soil was good was well beyond the capacity of the land.

In 2007, it was estimated that 20% of the housing units (over 26 million housing units) in the United States used on-site septic systems to remove domestic wastewater from the homes in areas where municipal or public sewage systems did not exist. This was an increase of over 1.5 million septic systems since 1985 and is still growing. Some 50% of the systems were in rural areas and 47% of the systems were in suburban areas. (See endnote 1.)

The systems have a typical life expectancy of 20 years. They function by providing a place for solids to settle and then for the effluent to be distributed evenly in the ground. They may be an individual unit or in a cluster. A septic tank system consists of basically:

1. A large watertight, underground concrete, fiberglass, or polyethylene/plastic container, with one or more chambers, is used to collect domestic wastewater from the house for primary treatment and holding, for at least a 24-hour period, allowing solids to go to the bottom of the container and floating scum and grease to the top.
2. Pipes which come from the house or other structure go to the underground container and from the container are plumbed directly to a header or to an underground distribution box and then to the actual effluent absorption system.
3. Install baffles at each of the inlets and outlets of the septic tanks and make them an integral part of the tank. The inlet baffle must extend at least 6 inches above the total liquid depth and at least 1 inch above the inlet sewer pipe. The outlet baffle must extend above the surface of the liquid and depending on the jurisdiction, 40% below the total liquid depth. The system should allow the liquid to flow from the inlet, retain the sewage for 24 hours for settling, prevent floating solids from leaving the tank, and then let it out through the outlet to the header or distribution box.
4. The header and the distribution box serve the same purpose to evenly distribute the effluent. (In some jurisdictions headers are primarily used and in others distribution boxes.)
5. The area of soil is used to absorb the liquid and also may allow it to go into the air through evapotranspiration.

The soil effluent absorption systems may have different configurations based on the on-site soil conditions, however they all have the task of efficiently removing the effluent from the septic tank in such a way that it does not contaminate ground or surface water, come to the surface and contaminate the land, or back up into the house or other structure. Some of the systems include tile leaching fields, mound systems, leaching chambers, recirculating sand filters, low pressure pipe systems, home aerobic treatment systems, etc.

The number of these systems continues to grow because of several factors including: people moving to fringe areas surrounding metropolitan communities; people moving into rural areas; people utilizing waterfront seasonal recreational areas on a year-round basis; and people, especially retirees, seeking remote areas for permanent housing.

There are numerous problems related to on-site systems. The most significant problem is the amount of water that is utilized in the property by the people and then the amount of effluent put into the system for disposal. Other problems include the size of the house lot, the location of the septic tank system and relationship to the well and the house or a neighbor's system, well, or house, the type and permeability of the soil which is used for drainage of liquid effluent, clogged drain fields, the nature and type of soil testing, the mechanical failure of the system, removal and disposal of solids from the septic tanks, the size of the septic tanks, the age of the system, the presence of bushes and trees in the liquid disposal field, the slope of the land, the site of the system which may not have been laid out properly to start with, the presence of bedrock close to the surface, the concentration of systems in a given area, the functioning of the system in various types of weather conditions especially frequent heavy rains or very heavy snows that last for long periods of time, the maintenance of the system, and the management of decentralized systems.

On-site septic systems should not: cause problems with community or private drinking water supplies by adding nutrients, chemicals, or organic matter; create direct human exposure to fecal material and disease; create harborage, water, or food for insects, rodents, birds and so on; or create odors or aesthetic nuisances for the neighbors or communities.

A Survey of Household Sewage Treatment System Failures in Ohio

(See endnote 27)

Approximate 31% of the household sewage treatment systems in the state of Ohio have some degree of failure. This is an increase from the 23% failure rate reported by the Ohio Department of Health in its 2008 survey of household sewage systems. This is probably not unique and may apply at different levels to all other states. All types of systems were involved starting with the septic tanks and utilizing tile systems, mound systems, sand filters, seepage pits, etc. With almost 194,000 systems failing, this constitutes a very serious environmental and public health hazard and indicates the need for considerably more work to be done in this area to rectify existing problems and to determine how best to deal with situations that seem to be without resolution.

SUB-PROBLEMS INCLUDING LEADING TO IMPAIRMENT AND BEST PRACTICES FOR PRIVATE SEWAGE SYSTEMS (ON-SITE SEWAGE DISPOSAL)

(See endnote 5)

Size and Nature of Land for Houses

The size and nature of the land used for new houses is of considerable importance since there has to be adequate space for disposal of the effluent and the soil must be able to allow it to be dispersed in a safe and sanitary manner. As has been indicated previously, there are many structures that have been built on improperly sized pieces of land and in many cases the soil cannot accept more than a small amount of effluent without the liquid rising to the top or back flowing into the structure.

Best Practices for Determining the Size and Nature of Land to Be Used for New Houses (See endnote 2)
- Before issuance of a septic tank permit, house plans including layouts of the structures must be submitted to the licensing authority and reviewed in depth for: size of building lot; location of structure compared to existing structures, driveways, graded cuts and concrete pads; number of potential bedrooms; depth of house sewer as it exits the structure to make sure that the disposal system will not be too deep; elevations of various concrete pads to ensure that the layout of the system can actually be implemented; removal of all trees and bushes within the absorption field; placement of water wells on the lot; and location of wells and septic systems on adjoining properties.
- Have a professional design a septic system according to the soil permeability, peak daily flow of sewage in gallons per day based on the number of bedrooms, and total usable land area for the septic system utilizing all restrictions for easements, watercourses, other wells, and septic systems. Submit this plan along with the house plan to the appropriate public health authority for approval. Include all soil testing results.
- The bottom of the proposed septic system must be above the highest anticipated groundwater level in that area with the distance between the two varying from jurisdiction to jurisdiction depending on the type of soil and the local and state regulations.

SOILS AND SOIL TESTING

(A complete discussion of soils and appropriate graphics can be found in endnote 3.)

Soil is any mixture of particles that come from disintegrated rock and can support the growth of vascular plants, which are those that can internally transport water and food. Soil particles range in size from sand, which is the largest particle, to silt which is of medium size or smaller, to clay which is the very smallest size and most easily compacted. The inherent properties of soil and the depth of it are highly significant because it influences the movement of water and the decomposition of organic material. The function of the soil is to move the water through the soil pores away from the area and to allow microorganisms to operate in the aerated soil to renovate the wastewater.

Soil varies greatly from property to property in a given area and may vary within the individual property from its natural deposition to the removal of the soil when preparing the land for the building of a structure to the depositing of material in order to build up the slope of the given piece of land. This fill material may be of various compositions. Climate and the presence of organic matter can cause the soil to be altered.

The soil profile of the given area provides information on the ability of the soil to aid in the removal of certain chemicals and minerals, and the decomposition of the organic material within the effluent from the septic tank as the liquid flows into the ground. The soil profile depending on the depth of the soil at any given point may contain clay, a mixture of sand and clay or silt and clay, a mixture with loam, or be predominantly silt or sand. The soil profile whether it be the original ground or fill can vary enormously from place to place and therefore soil that appeared to be good for drainage at any given point where a test is conducted may be just a few feet away from soil that may be poor for drainage. When the soil is examined and mottling is encountered, it indicates that there is a seasonally high groundwater table problem and this area may be unfit for an on-site sewage disposal system.

Soil drainage is the ability of the soil pores to avoid saturation and the speed at which the water will move through the soil. Soil texture is related to the various types of soils mentioned in the soil profile. The soil best able to deal with effluent is sand. However, along with the quick flow of the liquid is the problem of microorganisms and chemicals moving into the groundwater supply or, depending on the slope of the land, into surface bodies of water. The problem with silt is that it can flow like water into spaces within the soil and cause clogging. The problem with clay is that the clay will swell and thereby reduce the ability for the effluent to flow away from the septic tank and reduce the organic load. Rarely do you find simply one type of soil.

The color of the soil can indicate if the soil is saturated or unsaturated. When the soil is saturated, it is gray instead of brown from iron oxide minerals. Iron oxides are a major pigment in the soil. Soils with a uniform bright red, brown, or yellow color may be able to drain normal rainfall but have problems with sewage effluent.

The texture of the soil has a significant influence on the percolation rate of effluent from the septic tank. With sand, water percolates at less than 10 minutes per inch. With sandy loam, it percolates at 3–30 minutes per inch. With loam, water percolates at 10–45 minutes per inch. With silt loam, water percolates at 30–90 minutes per inch. With clay loam, water percolates at over 45 minutes per inch. And with clay, water percolates at over 60 minutes per inch. It is the rate of percolation which may help determine the size and nature of the septic tank system.

The slope of the land is very important because if it is greater than 10% then there are severe limitations to subsurface disposal fields for the effluent. Also erosion may remove some of the better soils that could have been used for the disposal of the septic tank effluent.

Soil in a well-designed well-aerated properly functioning subsurface disposal system of proper depth which is not overloaded can reduce the sewage effluents significantly and allow the remaining liquid to disperse appropriately. Part of the dispersion is through evapotranspiration in climatic conditions which allow this to occur.

Soils in the subsurface trenches and beds may become clogged because of the physical, biological, and chemical agents present in the effluent from the septic tanks. Silt, as has been previously mentioned, can flow like water and contribute to this clogging. This may lead to a drastic reduction in the percolation rate of fluid into the ground.

Soil Tests

The three basic types of tests used to determine whether or not soil is suitable for a septic tank system are: the percolation test (perc test), soil core analysis, and backhoe cuts to a maximum depth of 12 feet. The percolation test consists of digging a series of holes in the area of the effluent distribution field, prewetting the inside of the holes, adding a measured amount of new water to the holes, and then measuring how fast the water flows out. (See endnote 1 for total details of an acceptable procedure.) Core sample analysis consists of digging a hole to remove a core sample of the various soil layers and taking it back to a laboratory for analysis or digging a deep hole with a machine typically 7–10 feet deep or more and looking at the sides of the hole for variations in color and types of soils. This will indicate the drainage characteristics of that particular piece of land at the hole. The backhoe cut is made into good holes 25 feet apart. It is necessary in order to get a permit to demonstrate at least 4 feet of sand, sandy loam, or loam which is 4 feet above the seasonally high water table as determined by soil conditions. Where clay is encountered first and then sand still 4 feet above the seasonally high water table, 50% of the clay could be removed and replaced with clean sand which creates a conduit to the virgin sand below. In this type of test especially it is very important for the individuals conducting the tests, whether it be the local environmental health practitioner or others who are certified, to wear hard hats and special boots, and stay away from the perimeter of the hole to prevent a cave in and significant damage or death to the individual.

The problems associated with percolation tests are numerous. There is a huge variation in the qualifications of individuals who perform these tests depending on local and/or state regulations. Tests that are conducted during dry weather give false readings. There are times when the test holes are dug under the house and under the driveway according to the plans instead of in the distribution field. Where fill dirt has been used in a distribution field, the tests will be inaccurate. The number of test holes depends on the average percolation rate of the soil and where that is greater than 60 minutes per inch, there will probably be future problems of liquid dispersal and this will restrict the use of the land. Since soil conditions may vary considerably just a few feet away from the test hole, the results of the percolation test may not be accurate for the entire area being represented. The slope of the land may affect the test if it is very steep.

The problems associated with core samples and core sample analyses are numerous. Once again, the qualifications of the individuals making a decision on permeability by looking at the type of soil in a core analysis or a hole dug by a machine, lead to different interpretations of color and soil type. Thus there may be improper decisions made on soil usability for a septic tank system. A single core sample will not give a complete picture of the soil percolation capabilities of the entire potential system. Multiple soil core samples must be taken within the future distribution field site.

Best Practices for Soil Analysis for On-Site Sewage Distribution Systems
- Only use highly qualified certified individuals with local experience who are closely supervised to conduct percolation tests and core soil analysis tests.
- Follow all local and state rules and regulations concerning the tests and how to conduct them properly.
- Mark on the house plans and sewer plans the places where the percolation tests and soil core analysis tests are conducted and make sure that the places chosen are actually used.
- Use both a soil core analysis test and a percolation test to determine if a septic tank system can be installed, the size of the system, and the type of system.

- Document the seasonal high water table depth, soil types and depth of each different type of soil stratum and soil texture going down into the hole, and depth of bedrock or hardpan (a layer of tight clay or tightly packed soil).
- Presoak all holes used for percolation tests and maintain 12–14 inches of clear water for at least 4 hours. Begin the test 15–30 hours after the presoak. Adjust the water to 6 inches above gravel in the hole and take two readings at 30-minute intervals. Use the slower of the two readings as your percolation rate.
- Determine the size of the effluent distribution field by the percolation rate, the type of soil, and the number of bedrooms within the house, which gives a rough estimate of the number of people living there.
- If percolation rates are greater than 60 minutes per inch, then a much larger lot size will be needed for a house providing the slope of the land is not too great and the depth below the drain field to an impervious formation is at least 10 feet of usable soil for disposal purposes. The extra lot size will provide for reserve areas if the effluent distribution field fails and a new one has to be built.

ON-SITE WASTEWATER SYSTEMS

Septic Tanks and Distribution Boxes

A septic system includes the septic tank, drain field, and soil beneath the drain field. A pipe coming from the house carrying the contents of the toilets and sinks goes into a watertight container in the ground outside of the house. It is made of a durable material which is resistant to corrosion or decay. The material of the tank may be concrete, some form of reinforced fiber, polyethylene, or coated metal. There should be inspection ports which are secure in the top of the tank to be able to determine what is occurring within the tank and to allow the removal of accumulated sludge and scum. The length to width ratio of the tank and the liquid depth are very important considerations in the construction of the septic tank. (In the 1950s, the minimum size septic tank for a three-bedroom home was 500 gallons and for four-bedroom home was 750 gallons. By today's standards these are far too small, since one of the main purposes of the tank is to allow for as much retention time as possible to allow the separation and movement of the oil and grease upward and the settleable solids to the bottom of the tank.) The tank may have two or three sections which are used to better remove suspended solids from the liquid. Inlet baffles force the incoming wastewater downward and the outlet baffles, unless they are overwhelmed by the amount of liquid flow, should keep the scum and solids from leaving the tank thereby not allowing these substances to move out into the disposal field along with the effluent. The settleable solids typically settle to the bottom of the tank while the grease and scum rises to the surface. The biological process has begun and there is decomposition of the material by the bacteria which are using anaerobic digestion on the solids to create liquids and gases.

A solid pipe leads from the septic tank to the header which then sends the effluent to perforated pipes within the field, or the septic tank is connected to a distribution box which has several openings leading out to the effluent disposal areas, which may consist of various types of systems, all with the purpose of using the surrounding soil for dispersal of the liquid.

The effluent in the soil is further treated by microorganisms in an aerobic manner. (See below for different types of effluent disposal systems.) This is typically a gravity flow system from the house to the effluent disposal area. Problems related to septic tanks and distribution boxes are usually caused by improperly installing them causing them to malfunction, not enough depth in the ground to place the septic tank because of bedrock or high seasonal water tables, and lack of proper maintenance on a regular basis. Other problems are related to dumping unnecessary items into the toilets including grease, wash water from floors, unwanted chemicals, cleaners, disposable diapers, sand and grit, old pharmaceuticals, remnants of garbage, cigarettes, etc. Garbage grinders cause substantial problems because the remaining ground material helps fill up the septic tank. The worst problem of all is the overuse of water by the residents of the property, whether this occurs through extended shower time, plumbing leaks, or running the water unnecessarily in various parts of the house into the sinks and drains.

Sewage system failure can be easily seen in sewage backup on the ground, and sewage backflow into the drains of showers, sinks, and toilets. Odors coming from the drain field or from the water supply of the properties downhill or the well on this property may indicate system failure.

Best Practices in Installation and Maintenance of Septic Tanks and Distribution Boxes
- Obtain information about the seasonal high water table and how close bedrock is to the surface on the housing site to predetermine where the house and septic tank system can be placed to avoid problems of design and installation of the system.
- Make sure that all regulations are observed concerning appropriate setbacks from water supply, bodies of water, property lines, houses, and drainage lines in planning the septic tank system including the location of the septic tank and distribution box or header.
- Provide a properly sized, minimum two-compartment septic tank, based on the usage of the system and the projected volume of liquid entering the unit at peak times. Hydraulic residence time should be between 6 and 24 hours. Typically, the minimum size is 1000 gallons or more.
- In order to protect groundwater, it is necessary to have a sufficiently large lot exceeding one half acre per house for a large subdivision utilizing septic tank systems.
- Do not install the septic tank or effluent distribution system when the soil is wet.
- Provide immediate access to the septic tank.
- Determine if all of the pipes leading from the house to the septic tank to the distribution box and the effluent disposal system are downhill in an appropriate slope in a gravity system.
- Determine if the septic tank, header, or distribution box is on level original ground and not fill material.
- Use effluent screens, which are a physical device placed on the outlet pipe of the septic tank to help remove solids from the effluent before it goes on to the distribution box, to assist in preventing solids and non-biodegradable material from entering the drain field and clogging it. (See endnote 4.)
- Never enter a septic tank or cesspool that has been used or work alone on these units because of the many dangers which can cause illness or death.
- Check tank covers and make sure they are safe and put on securely.
- Do not bend or lean over septic tanks or cesspools.
- Determine if there are sewage backups, odors, or seepage onto the ground or into nearby water sources when inspecting the septic tank system.
- Determine if the sludge is accumulating rapidly and have the tank cleaned out by a licensed sewage hauler on a periodic basis.
- If a system does fail, have the septic tank and distribution box pumped out by a licensed hauler and allow the system to rest for several days before use again.
- Do not wear contaminated clothing from working with the septic tank system in the home and make sure that the individual's tetanus inoculations are current.
- If the septic tank is no longer being used, the material inside called septage, which consists of an odorous slurry of organic and inorganic material containing hair, nutrients, pathogenic microorganisms, oil, grease, chemicals, and whatever else has been thrown down the toilet over the years, should be pumped out by a licensed individual who will dispose of the contents in an appropriate manner. Then crush the tank in place and backfill the area. Grade the area and plant grass over it.

Septic Tank Systems for Large Flow Applications
(See endnote 7)

Large flow applications are modifications of the traditional septic tank system treatment technology for several homes and/or commercial discharges. This can be accomplished by having individual septic tanks for each of the structures with pipes leading to a single treatment system. For commercial establishments, the wastewater goes through a special oil and/or grease removal system prior to going to the septic tank which should contain the fluid for at least 24 hours to allow solids to settle. The biggest problem of this type of system is the carryover of solids, oil, and grease because of poor design and poor maintenance. (See the discussions on septic tank systems above and perforated pipe or tile field below.)

Septic Tank Effluents, Treatment, and Disposal Systems
Perforated Pipe or Tile Field (Leach Field or Drain Field)
(See endnote 8)

The distribution box empties its contents, in roughly equivalent amounts, into a series of underground drainage trenches which have had the sides and bottoms scored to allow proper movement of fluid. Gravel is placed in the bottom of the trenches, which have to be at least 3 feet above a nonpermeable layer of soil. Perforated pipe is placed on the gravel, and more gravel is backfilled over the tile or pipe. This is covered by untreated building paper or straw and finally soil is backfilled on top of it and grass is planted. The straw and untreated building paper disintegrate over time. In the absorption field, a biomat forms which helps renovate the wastewater as it percolates through the soil. This usually requires between 2 and 4 feet of unsaturated soil below the drain field. The effluent as it percolates down through the ground goes through physical, chemical, and biochemical processes which help clean the effluent.

The size of the drain field is determined by the nature of the land and soil percolation rate times, and the number of bedrooms which it is assumed will have at least two people in them. Numerous configurations of the perforated pipe or tile lines may be used as long as there is an equal distribution of the effluent in the drain field and one portion of it is not overburdened and another portion underused. A bed of gravel may be used in place of the drain field because of limited space. However, it must create the same level of cleaning of the effluent before it moves down to the groundwater supply. Another technique used is the dry well, which is a container without a bottom, with slit sides and surrounded by 3 feet of stone on all four sides resting on 3 feet of stone. There are usually two to three of these in tandem connected to the septic tank and utilized in areas where there are severe space limitations and extremely sandy soil. The distance above the seasonally high water table applies to this system as well as the others. (See Seepage Pits section below.)

Two to five feet of unsaturated, aerobic soil can do a good job in reducing the BOD and suspended solids. Phosphorus and metals are removed. The ability of the soil to continue doing this effectively varies with the soil mineralogy, organic content, pH, etc. Not much is known about viruses and toxic organic compounds going into the groundwater supply. Nitrates and chlorides leach readily down to the groundwater supply.

Older tile field systems and perforated pipe systems fail for a number of reasons. During installation heavy rains occurred and sealed part of the pores in the soil where the field tile or perforated pipe was being installed. Surface water crosses the system and creates conditions where there is more liquid to dispose of than the system can handle effectively. Too much water is used in the house. A garbage disposal unit is being used. All types of litter and inorganic materials are dumped down the toilets. Chemicals are dumped down the toilets. The septic tanks and/or distribution boxes are not level. The lines are installed uphill instead of downhill. Part of the system may have been crushed by heavy vehicles. The neighbors' systems are overflowing and put more pressure on this property's system. Flooding has occurred and inundated the ground. Trees and bushes are planted on the drain field or close to it and the roots invade the laterals. Poor maintenance of the septic tank leads to malfunctions of the system.

Best Practices for a Perforated Pipe or Tile Field Effluent Removal System
- After excavating the trenches, score the bottom and the sides of each one to prevent sealing of the pores by the use of the equipment.
- Recheck the header system or distribution box to make sure that it is level and that the inlet from the septic tank is above the outlets of each of the pipes going to the tile lines or perforated pipes.
- Install, inspect, and properly maintain oil and grease traps where this may become a problem due to the use of the septic tank system by businesses.
- Install the perforated pipe system according to the approved plans and make sure that the perforated pipes are gradually going downhill and not uphill.
- Inspection of the system must be made before backfill is put on top of the straw or untreated building paper. The individual making the inspection must be wearing appropriate personal protective equipment, be highly trained, have an excellent understanding of the soil conditions of that particular site, and be a member of the appropriate health department. This individual must have the builder or contractor at random move aside the material to check the depth of the gravel underneath the pipe and above the pipe, the width and length of each of the trenches, and whether or not the septic tank and distribution box are level, and check the baffles and screens within the septic tank before approval. This should be completed prior to any heavy rains and the contractor should be ready to backfill as soon as approval is given. If the system is red tagged, then the contractor must make all necessary corrections and call the environmental health practitioner out for a re-inspection before closing the system.
- All downspouts of the house must be redirected from the septic system and distribution field.
- Surface water should not be allowed to cross the field from adjoining properties. If necessary, curtain drains should be placed around the entire property to prevent this or to lower high water tables. The water collected from the perimeter of the lot is discharged further down into a ditch, creek, or stream. This should not contain any contaminants from the septic tank system. (See endnote 6.)
- Do not allow heavy equipment or trucks to be parked or cross drain fields since they may crush the lines.
- Remove all bushes and trees from the drain field to prevent the roots from invading the lines.
- Prevent flooding of systems by using proper flood control procedures on bodies of water and physical barriers. Large amounts of rain in short periods of time could still inundate the system. If so, it would be necessary for the inhabitants of the property to leave if possible for a period of time and allow the system to drain naturally.
- If the system has failed because it is not well aerated due to high groundwater, heavy rains, or excess wastewater overloading the system, pump air into the septic tank and air into the leaching field which will allow aerobic bacteria to resume their role in cleaning the effluent and reducing biological clogs in the soil pores.

Evapotranspiration
(See endnote 10)

In addition to the treatment and drainage of the effluent through the ground, evapotranspiration occurs during certain seasons of the year and allows the liquid to move to the surface and be used by grasses or transpiration by plants, or evaporate into the atmosphere from the soil. The problems of depending on evapotranspiration in climates other than those which are very dry includes the amount of precipitation, the wind speed, the humidity, solar radiation, and temperature. Also as the effluent evaporates salts are left behind which can end up clogging the surface of the unit. In wet and cold climates, there is very little or no evapotranspiration occurring.

Pressure System (Pumping and Dosing)

Where the existing drain field may not be operating properly, a pressure distribution system may be necessary. A pump pressurizes the effluent from the septic tank and sends it to the drain field lines through a small PVC line. This system distributes the effluent through the entire length of the trench when the pump is working and therefore a smaller amount of the effective soil for removal of the water is needed for processing the effluent. Typically, 2 feet of clearance is needed instead of 3 feet below the lines. Pumps fail because of poor maintenance and electrical failure.

Best Practices for Pressure Systems
- Same as listed under perforated pipes or tile field systems.
- All pressurized systems should have at least two drainage beds with a switch that can move the flow from one field to the other, giving each one an opportunity to rest.
- The pump must be serviced and maintained properly in order for the system to work and not have sewage backup or pooling above the ground.
- The pump must have a back-up electrical supply in the event of electrical failure.

Seepage Pits

A seepage pit is a big hole in the ground with soil on the vertical walls which will absorb effluent. It is lined with concrete brick, block or rings with loopholes or notches. The joints should be staggered and allow passage of fluid through them into the surrounding soil. The seepage pit follows the septic tank and is used in place of a distribution field where the upper layers of soil are not conducive to the movement of effluent. Typically, new seepage pit systems are not approved by health departments or the local environmental protection agency. Where an existing seepage pit was approved in the past, under proper conditions of adequate amounts of vertical soil in the hole for removal of effluent and the percolation rate is under 30 minutes per inch, then a new seepage pit may be approved. The depth of the pit and the width of the hole plus the percolation rate determine the amount of absorption area available in the seepage pit for appropriate disposal of the effluent. There are numerous problems with seepage pits because of the concentration of the effluent which must be processed through the sidewalls of the unit. If the septic tank is not properly maintained, it becomes more difficult for the unit to work properly. The use of evapotranspiration is virtually eliminated as a means of getting rid of the liquid. The installation of the pit is absolutely crucial and frequently not done as planned. Maintenance of the pit is another factor. The greater concentration of the septic effluent in a small area creates the need to separate the seepage pit by much greater distances from wells and other bodies of water.

Best Practices for Seepage Pits or Dry Wells
- Recommended percolation test procedures for seepage pits include dig a test hole 12–48 inches in diameter and at least 10 feet below the proposed bottom of the seepage pit and check after 72 hours to determine the level of groundwater that may be present in the hole; backfill the test hole to 10 feet above the bottom of the hole or groundwater level whichever is shallower; presoak the test hole with at least 5000 gallons of water; fill the pit with water up to the cap and then record at 15-minute intervals the drop in the water level. This will help determine whether or not a seepage pit will work under normal conditions. Also have a professional conduct a soil analysis of the sides of the proposed pit.
- The bottom of the pit must be at least 5 feet above the seasonal high water table.

- Conduct construction work on seepage pits only in dry soil and after excavation scratch all surfaces to make sure they have not been sealed by the use of the equipment.
- The depth of the sidewalls must be at least 10 feet below the cap.
- Clean gravel should be backfilled around the porous inner lining of the pit up to the top of the concrete cap.
- If more than one pit is used, all pits shall be 20 feet apart from the edge of any other.
- A riser with a 9-inch diameter opening with a cap on it should be installed in the pit so that the pit can be cleaned out properly by a professional hauler. Frequent inspection of the pit is necessary.

Aerobic Treatment
(See endnote 11)

In an aerobic treatment system, a mechanism is used to inject air throughout the treatment tank thereby helping aerobic bacteria degrade the solids. This helps achieve a superior effluent quality. Typically, a system of this type would be chosen where the regular septic systems are failing. Problems are: the unit is more expensive to operate than a septic tank; electricity is required and a back-up system is mandatory in areas where there is frequent electrical failure; mechanical parts can break down; more frequent routine maintenance is needed than for a septic tank; sudden surges of heavy loads of solids can upset the system; more nitrates may be released to groundwater than from a septic system; and pretreatment steps are needed to remove grease, trash, garbage grindings, or other materials which are not biodegradable.

Best Practices for Aerobic Treatment Systems
- Use a manufacturer that carries the NSF International class I certification for its aerobic treatment systems for equipment purchased for the effluent disposal site.
- Use a pretreatment device of a trash trap, septic tank, or comminutor to remove grease, trash, garbage, and non-biodegradable materials prior to the settling chamber for primary treatment and settling of sludge. The effluent then goes into the aerated chamber and aerobic decomposition occurs. The amount of effluent flowing must be controlled to prevent overwhelming the system.
- Frequent maintenance must be provided to keep the system effectively working.
- Provide alternate source of electricity if the area is subject to electrical failure.
- Do not diminish the size of the effluent disposal system since the liquid flow is still the same as if it were coming from a septic tank although the effluent is far cleaner.

Sand Filters
(See endnote 9)

Sand filters are used to clean up the effluent from the septic tank by using physical, chemical, and biological processes. Suspended solids are removed by mechanical straining and sedimentation. Bacteria on the sand grains help remove suspended solids. At the upper part of the sand filter, there are aerobic conditions and the microorganisms assist in removing the BOD and in the conversion of ammonia to nitrates. The nitrates in the lower part of the sand filter are converted by anaerobic bacteria to nitrogen. The effluent which has percolated down through the sand goes into an underdrain and is dispersed into the effluent dispersal field. The sand filter usually has two sections and each section is dosed individually allowing the other one to rest and process the effluent. There are three types of sand filters used after the septic tank: buried, intermittent, and recirculating. The advantages of these systems are that they reduce the contaminants in the effluent considerably in a consistent well-understood manner and have less problems with clogging of soil, they are relatively compact, and they can be used on soils that have high groundwater and poor permeability where other systems have failed. Problems with these systems include: there may be too little land available to add the system; there has to be a sufficient amount of pressure or head to make the filters work properly; odors can occur from the anaerobic parts of the single-pass filtering system; power outages; pumping and distribution unit failures; poor maintenance; and allowing water from other areas to cross the sand filter area. Aesthetically especially in suburban communities, the sand filter looks like an Indian burial mound and may not be very appealing to the neighbors. The recirculating sand filter sends part of the effluent back into the new effluent which and are then processed together thereby reducing potential odor problems.

Best Practices in Use and Maintenance of Sand Filters (See endnotes 21, 22)
- When installing the system make sure that the design recommendations for the specific size of sand are followed closely.
- Use a professional maintenance group to clean the filters every 6–12 months or earlier if required.
- Make sure that the septic tank is watertight.
- Install appropriate electricity and a back-up electricity source in the event of electrical failure, as well as an alarm system if the filter shuts down.
- If the effluent is released to the surface under the condition it is allowed by local and state laws, it must be chlorinated and regularly tested for contaminants including *E. coli.*
- If problems occur, the system must be taken off-line until the sand is cleaned or replaced.
- If a filter is exposed to sunlight it may develop algae mats. This can be corrected by shading the surface.
- Remove all weeds at the surface above ground filters.
- Use recirculating sand filters instead of the single-pass sand filter to eliminate odors and increase the oxygen content of the filter bed. The advantages are that a very good effluent is produced with a 95% removal of BOD in total suspended solids as well as a significant reduction in nitrogen. Less land is needed than for the single-pass system. Weekly maintenance is required for the media, pumps, and controls. The system may not function very well in extremely cold weather.
- Use intermittent sand filters instead of the single-pass system in order for one bed of sand to rest while the other one is being dosed and working. This also provides a high quality effluent and requires minimal maintenance but more land for treatment and disposal.

Alternate Filter Material

Alternate filter material for the above systems could include filters of peat which is very permeable and can absorb materials readily, loam, crushed glass, and textiles. The advantages of the alternate media filters are they are moderately inexpensive, have low energy requirements, and do not require very skilled people to produce a high quality effluent. These systems may be used if an existing tile system has failed. They can operate over long periods of time. The disadvantages are that the technologies have not been fully approved, cost and maintenance has not been standardized, the filter medium may not be readily available, and there is a possibility of odors. (See endnote 19.)

Mound Systems

A mound system is an above-ground leaching field consisting of a mound of various sizes of aggregate including sand and gravel as well as fill material. The effluent coming from the septic tank either flows down through the mound by gravity or is pumped to the top of the mound and then flows down to drains where the effluent has to still be absorbed in the ground. All of the processing of the effluent must take place in the mound. People object to this type of system because it might not work as described and it makes the property have a funny look. Mounds only work in certain types of conditions. They are especially used if soil permeability is extremely slow or extremely fast, if there is shallow soil cover over cracked or porous bedrock, and if there is a high water table. Construction costs are typically much higher than for the typical septic tank system, the permeable topsoil available is limited and is damaged easily, the mound may affect drainage patterns, and all systems have to have pumps or siphons. (See endnote 18.) A raised bed system is similar to a mound because it is constructed on the top of the soil and a certain amount of treatment of the septic tank effluent happens.

Best Practices for Mound Systems
- Use mound systems only when there is a high water table, shallow soil cover over cracked or porous bedrock, or soil permeability is either extremely high or extremely low.
- Determine how best to design the mound system by how the effluent moves away, the direction of flow, and how fast it travels away from the system.
- There should be at a minimum 1 foot of highly permeable soil between the mound system and the inhibiting ground conditions unless local or state ordinances direct otherwise. However, it is recommended that there be at least 4 feet of highly permeable soil between the mound and the inhibiting ground conditions.
- Mounds appear to work best when they are long and narrow and if the conditions of the soil are very poor, the system should be longer and narrower.
- Mounds can be placed on slopes of up to 25%, however steeper slopes are not advised because of the danger to the equipment operator in working in this type of environment.
- Where a site has been filled with soil, in the case of the mound system, the soil should be placed on top of the natural soil and allowed a time to settle before utilizing a mound system.
- Mounds must not be installed in flood plains. They must not be installed in drainage ditches or depressions unless there are specific flood protections for the most severe type of floods. They should be separated from water supplies, surface waters, springs, the boundary of the property, and the house according to the local or state ordinances.
- Install a mound system that will have sufficient reserve for expansion, and have an upper limit of the number of people allowed to use it, in order to prevent hydraulic overload and biological clogging of the system.
- On a periodic basis, check the septic tank and dosing chamber to make sure that the sludge and scum buildup is not too great to avoid overloading the mound system. Have the septic tank and dosing chamber cleaned out as needed by professional sewage haulers. (See endnote 18.)

Other Advanced Treatment Systems

(See endnote 12)

The problem with these systems is that they need a lot more maintenance than the typical septic tank system and if not done properly, they will not work well. Studies have shown design and construction flaws, loss of electricity, and difficulties caused by lightning, cold, and most significantly, homeowner abuse and neglect. Harsh cleaners, pharmaceuticals, solvents, paints, and other materials may disrupt the systems.

Fixed Film Method

In the fixed film method, the effluent from the septic tank is further treated in a special chamber that has permeable materials, either natural or synthetic. The surface area forms a biomat with aerobic organisms that processes the effluent and reduces the biosolids. At the bottom of the treatment chamber, the effluent is dispersed through the leaching field. In a recirculating system, the effluent may contain excess biomass which includes microorganisms which have fallen off of the filter medium and therefore is recycled through the process to clarify it further. By distributing the original effluent from the tank in special time doses, it allows the part of the system that is not dosed to recover and be more productive. Finally, accumulated sludge at the bottom has to be removed for safe disposal.

Rotating Biological Contractor Systems

These systems have rotating plastic discs on a horizontal shaft that are submerged about half way within the effluent. This allows the aerobic bacteria on the discs to be exposed to air and wastewater to help improve the effluent.

Suspended Growth Method

In this system, the effluent is circulated rapidly and agitated within an anaerobic chamber to encourage the digestion of the biodegradable substances. Changes in flow or the solids affect the system. This may result in the incomplete settling of the solids.

Activated Sludge Technology

This is a variation of these suspended growth systems which combines aerobic and anaerobic processing in a single unit. They operate in alternate sequences.

Membrane Bioreactor Method

This method improves the performance of the suspended growth method by using membranes as super clarifiers filtering the microscopic solids and even some bacteria.

The membranes can be damaged by air or stringy or abrasive materials which have a fine diameter. The membrane filtration works faster in clarifying the effluent than the settling process. This should leave salt and relatively less sludge for disposal. The membranes must be cleaned regularly to prevent problems that could cause loss of permeability.

Best Practices in the Use of Other Advanced Treatment Systems
- Determine flow to the watertight septic tank, risers, and pipe connections, condition of baffles and outlets screens from the tank, wastewater level, and sludge and scum levels in the septic tanks to make sure they will not affect the rest of the process.
- Check the control panel and controls to ensure that all pumps, floats, valves, electrical controls, and alarms are working properly as well as the dosing volume being accurately calculated and set. (See endnote 20.)
- Inspect the surface of the filters for ponding of the effluent. Ponding is the accumulation of water on top of a trickling filter because the organisms which form a slime to utilize the incoming wastewater have sloughed off and have filled the gaps between the rocks in the filter, not allowing the wastewater to penetrate and be treated biologically.
- Determine if there are unusual odors or insect infestations.
- Take samples of the filter effluent to determine the level of clarity and odor as well as the amount of dissolved oxygen and BOD.
- Protect the systems against freezing.
- Provide electrical backup where needed in the event of electrical failure.
- Establish frequent maintenance checks by trained individuals to avoid future breakdowns.
- Have local or state health department or EPA personnel frequently monitor the systems to determine if they are working appropriately.

Low Pressure Pipe Systems
(See endnote 29)

This is a soil absorption system, leading from a septic tank or aerobic unit, of a shallow, pressure-dosed series of small-diameter perforated pipes placed in narrow trenches. The pressure which comes from a pumping tank causes uniform distribution of the effluent in the system. The advantages of this system are that: the shallow placement in the trenches helps promote transpiration and furthers the growth of aerobic bacteria; absorption fields may be placed on sloping ground: there is improved distribution of the effluent because of the pressure unit; the periodic dosing and resting periods encourage aerobic conditions in the soil; and less land is needed for the absorption system. The disadvantages of the system include: the potential for clogging of the holes in the laterals by either solids or the roots of trees and bushes; the lack of storage capacity around the laterals; wastewater may accumulate in the trenches and saturate the soil; and a substantial need for continued maintenance of the system. Large systems have problems with hydraulic overloading because of infiltration of water, poor hydraulic design, improper installation, and orifice and lateral clogging.

Small-Diameter Gravity Sewers
(See endnote 30)

Where there are numerous problems of disposal of the effluent on-site, a small-diameter gravity sewer may be considered for collecting the effluent from several properties and allowing it to go to a central treatment location or pump station to be moved to a treatment facility. The pipes of the system are smaller in diameter than normal collection lines and are placed along the contours of the land always leading downward toward the collection station and pumping system. Most of the suspended solids have been removed by the septic tanks, thereby reducing clogging. Cleanouts and air release risers must be furnished and utilized. Odor control is a problem with the system. The advantages of this system are that: construction is rapid; unskilled people can operate and maintain the system; instead of manholes, cleanouts are used; there are reduced costs of excavation and materials (plastic pipes can be used); and final treatment is less because the organic load has been reduced. The disadvantages are that: these systems cannot handle commercial wastewater with lots of grit or settleable solids; the collected septage still has to go somewhere for treatment and disposal; and the system must be buried deep enough to keep from freezing.

Disinfection for Small Systems
(See endnote 26)

A small system can be a septic system, sand filter, or a system that serves an individual house or group of houses. If wastewater enters any body of water, it must be disinfected. Wastewater can readily contaminate water used for drinking or recreation and may contain bacteria, protozoa, helminths, and viruses, all of which can cause disease in humans. Disinfecting the wastewater will only be effective if the fluid had been adequately treated prior to disinfection to reduce the suspended solids and the BOD.

Although ultraviolet radiation can and is being utilized in some areas, there are safety considerations to be concerned about, frequent maintenance is absolutely necessary, and the amount of contaminants in the effluent can affect the final product. The advantage of ultraviolet radiation is that if it is working properly it can effectively inactivate most viruses, bacteria, and spores, and even some cysts. The disadvantages are that the low dosage may not effectively inactivate some of the viruses, spores, and cysts, may not work effectively with increased levels of turbidity and total suspended solids, and may require large numbers of ultraviolet lamps.

Chlorine is most usually used and most practical for disinfecting wastewater effluent. Chlorine oxidizes the cellular material of the bacteria. Chlorine can come in many forms. The advantage of chlorine is that it is reliable and effective against a large number of pathogenic organisms, more cost-effective than ultraviolet or ozone disinfection, leaves a chlorine residual that can keep on killing the microorganisms that are still existing in the effluent (depending on the dosage), and dosing rates are easily controlled. The disadvantage of the chlorine residual is that it can be toxic to aquatic life, highly corrosive and toxic to people, create hazardous compounds when reacting with some organic matter in the effluent, and may be unstable under certain conditions. Some organisms are quite chlorine resistant and it may require more than normal contact time to destroy them.

Special Toilets to Reduce Water Flow

High-Efficiency Toilets
(See endnote 23)

New toilets have to meet a high-efficiency water standard of 1.6 gallons per flush. This reduces the toilet water per flush by over 50% and total average usage within the property by 16%. This is accomplished by increasing the velocity of the flush water. Most of these toilets work very well.

Incineration Toilets
(See endnote 24)

The incineration toilet is a self-contained unit with the usual toilet seat connected to a holding tank and a gas-fired or electric unit to incinerate the waste products in the holding tank. This is used especially in rural areas with extremely poor ground conditions, where the weather is extremely cold, at roadside rest areas, for work crews, in ships, in areas where water is scarce, and where the level of seepage of contaminants into local water supplies is potentially very hazardous.

Composting Toilets
(See endnote 25)

Composting toilets may be used in rural and suburban areas which do not have sewers. This system which processes feces, urine, toilet paper, and sometimes garbage, depends on aerobic bacteria to break down the waste in unsaturated conditions. The remnants must be removed by a professional sewage hauler or buried. These units are especially good for reducing the quantity and strength of the wastewater and the remnants can be buried. Problems include the maintenance of the units, disposal of the end product, lack of maintenance, and creation of odors, and most of the units require some source of power.

Personal Habits in Use of Water and the Disposal of Waste

One of the greatest problems in living with a septic tank system is the overuse of water which individuals living in cities typically do. Leaking faucets and toilets, long showers, running washing machines and dishwashers with only a few items, allowing water to run while doing other things, all create an additional water burden on the system. Garbage disposal units, water softener waste, and swimming pool filter cleaning waste going into the septic system add more mass to the septic tank and water for disposal. People also typically throw down their toilets chemicals, tissues, cigarette butts, grease, pesticides, coffee grounds, medicines, etc., and all of these items go into the septic tank which is no more than a box where the effluent goes through some transformation. Septic tank additives or cleaners when released with the effluent may be toxic and therefore should not be used.

Problems with the septic system can be determined by: the slow flushing of toilets; surface flow of wastewater; very green grass over the absorption field indicating upward flow instead of downward flow where the soil may be clogged; nitrates and bacteria found in well water; increase in aquatic weeds or algae in ponds or other bodies of water adjacent to the home; and bad odors around the house. The extreme occurs when the sewage backs up into showers, sinks, floor drains, and/or toilets, which then overflow.

Best Practices in Use of Water and the Disposal of Waste
- Institute and enforce a water conservation program within the home by all household members. This may be especially difficult with teenage children, however they have to understand that excess water use may cause the septic system to back up and make their home barely livable.

- Frequently check for leaks throughout the inside and outside of the house and immediately correct them.
- Install water restriction valves where appropriate to reduce the amount of water being used in the structure.
- Do not throw anything down the toilets or other drains other than human excrement.
- Do not attach a garbage grinder where the contents go into the septic tank system.
- Do not empty water softener materials into the septic tank system.
- Use a separate system for the disposal of swimming pool waste from cleaning filters.
- Maintain adequate vegetative cover over the absorption field, but do not allow trees or shrubs to be there because the roots will grow toward the effluent disposal lines and can block them.
- Do not allow surface water from other properties or this property to flow over the effluent disposal field.
- Do not allow heavy equipment, trucks, or automobiles to travel across or be parked on the effluent disposal field.
- Do not do any kind of digging unless it is done by a professional to correct the problem in the disposal field.
- Install a manhole over the septic tank which cannot be easily moved by children and animals and is used for inspection of the septic tank.

HOLDING TANKS

A holding tank is a large tank which is used to receive the flow of a house sewer or other sewer structures on a temporary basis until the sewer line is available for hookup to the treatment plant. This time period has to be guaranteed to be no more than 6 months. The tank has to be periodically pumped out by licensed sewage haulers and the contents taken to the sewage treatment plant for disposal. This is a very expensive process and frequently the haulers are not paid, thereby leading to secession of service and a potentially serious environmental hazard. Many states do not allow the use of holding tanks.

LAWS, RULES, AND REGULATIONS

The Federal Water Pollution Control Act Amendments of 1972, known as the Clean Water Act, made water pollution control a national concern instead of only being the individual concerns of the various states. The federal government set standards and helped state governments financially in basic water research and water quality maintenance of interstate waters. People discharging pollutants into the water had to obtain permits for the discharges and therefore had to follow rigorous rules based on the best practical technology available. This led to the awarding of numerous federal grants for the construction of sewerage systems. Subsequently, it was determined that the pollutants coming from non-point sources, such as runoff from agricultural lands, construction sites, urban areas, etc., were substantial in nature.

STORMWATER PHASE II FINAL RULE

For further improvement of the Stormwater Program which used the permit rule under the Federal Water Pollution Control Act Amendment of 1972, a phase II rule was enacted. This rule included several additional steps in improving water quality such as: a public education program to reduce pollutants in stormwater; public involvement in the development, implementation, and review of stormwater programs; the elimination of illegal discharge into storm drains; construction sites and contamination of stormwater runoff; post construction and contaminated runoff; and establishment of techniques to prevent contamination of stormwater runoff. (See endnote 50.)

PLANNING A NEW OR EXPANSION OF A WASTEWATER TREATMENT PLANT

The following steps should be used: determine the quantity of flow during different weather conditions and composition of the wastewater; establish appropriate objectives and goals for the operation of the wastewater treatment plant; develop appropriate financial resources for building and maintaining the wastewater treatment plant; utilize national or international standards for equipment, facilities, and means of disposing of the effluent and biosolids; determine the quantity and quality of sludge for disposal and how this will be dealt with; establish appropriate industrial pretreatment processes before allowing their effluent to enter the wastewater treatment plant for further treatment; and establish emergency measures for hydraulic overload due to infiltration, high water tables, and/or flooding.

RESOURCES

The US EPA has developed a program and guide for communities to help them manage on-site wastewater disposal. It is entitled "Voluntary National Guidelines for Management of On-Site Clustered (Decentralized) Wastewater Treatment Systems" and was published in March 2003. (See endnote 16.) These guidelines are still being used.

The program covers the many areas which must be managed effectively in order to decrease the risk of disease and contamination of the environment from disposal of household wastewater into the ground and that which flows on the surface of the ground and into bodies of water. These areas include site evaluation, construction, operation, and maintenance, training and certification of personnel, appropriate record keeping, public education, and financial assistance. The purpose of the program is to increase the performance level on the on-site and clustered wastewater treatment systems by using improved management concepts and techniques.

The US EPA *Design Manual On-Site Wastewater Treatment and Disposal Systems* provides information on how to select a site, and design, install, and replace on-site wastewater treatment systems. (See endnote 17.)

The US EPA *Guide to Field Storage of Biosolids and Other Organic By-products Used in Agriculture and for Soil Resource Management* manual discusses problems of water quality, pathogens, and recommended management practices and community relations. (See endnote 48.)

The US EPA website entitled "Contact Us about Biosolids-Frequently Asked Questions" discusses biosolids, how they are produced and used, how they are applied to the land, environmental concerns, regulations, how to manage their disposal, and an Environmental Management System. (See endnote 49.)

PROGRAMS

(See endnote 14)

US ENVIRONMENTAL PROTECTION AGENCY

1. *Clean Water State Revolving Fund*—This fund has provided over $100 billion for more than 33,320 low interest loans to build wastewater treatment plants, control non-point source pollution, and protect estuaries.
2. *WaterSense*—This program teaches and encourages society to protect water, our most valuable resource, by creating efficiencies for utilities, manufacturers, retailers, and consumers. In November 2012, the WaterSense program released information to help facility owners, operators, designers, and managers better understand effective water management. This document was entitled "WaterSense at Work: Best Management Practices for Commercial and Institutional Facilities." (See endnote 56.)
3. *Sustainable Water Infrastructure program*—This program promotes adoption of better utility management practices regarding water efficiency and protection of the watershed.

4. *Sustainable Communities Program*—This program provides technical and financial assistance to partner groups who support improving the water infrastructure while reducing health risk. This is done for small, underserved communities who lack access to safe water.

5. *National Pollutant Discharge Elimination System program*—This program controls water pollution in the waters of the United States by regulating point source discharge of pollutants into the surface waters. Permits have to be issued for any discharge as part of the Clean Water Act from industrial sources, municipal treatment plants, and stormwater management groups. The permit ensures that the discharge will not contaminate the US waterways as they have done in the past contributing an estimated 223 billion pounds of pollutants to the US waters. (See endnote 50.)

6. *Green Infrastructure*—This program identifies stormwater as a valuable resource and promotes the use of techniques that will allow it to percolate to the groundwater supply or go cleanly to surface bodies of water.

7. *State and tribal water pollution control programs*—These programs use federal assistance to states, territories, interstate agencies, and tribes to help establish ongoing water pollution control programs to provide better water quality.

8. *Decentralized Wastewater Management Program*—This program promotes the proper management of the septic systems and other decentralized wastewater treatment systems which treat more than 4 billion gallons of sewage every day. It is a formal partnership between the federal, state, and local governments as well as academia and industry. It encourages the proper design, operation, and maintenance of the decentralized systems. In the program, the federal government works with partners of all types to help local decision-makers understand the economic, environmental, and health benefits of using on-site systems properly. These benefits include mitigating the risk of disease and human exposure to pathogens which may be found in surface water, drinking water, or shellfish beds when they are contaminated with unnecessary overflow from the decentralized wastewater system; and using the decentralized wastewater system appropriately to reduce the cost of the substantial infrastructure and amount of energy needed in typical wastewater collection and treatment for small numbers of people spread out over large areas.

9. Examples of demonstration projects of decentralized wastewater programs include La Pine, Oregon, where an innovative nitrogen removal technology was developed and used; Ephesus, Virginia, where a model program was developed for the education and training of citizens in the use of septic tank systems; Humboldt State University, Arcata, California, which developed a special tool for helping plan water and wastewater treatment systems including reuse of wastewater effluent; and the National Environmental Health Association which has developed credentialing programs to test the knowledge, skills, and abilities of individuals who will be approving the installation of or installing on-site wastewater treatment systems. (See endnote 15.)

10. SepticSmart Week which is a program to help homeowners learn how to use and maintain their septic systems since malfunctioning systems are the second greatest threat to groundwater quality in the United States and may cause homeowners to spend, individually, many thousands of dollars to replace the septic systems if they are not maintained properly.

11. The National Community Decentralized Wastewater Demonstration Project funded by Congress to rapidly move ahead the technology transfer of improved treatment methods and management approaches. The Rodale Institute, Kutztown, Pennsylvania, is a model for the effective use and treatment of water resources including rainwater used for toilet flushing, and constructed wetland treatment for wastewater. This center which is now called the Water Purification Eco-Center is used to help educate large numbers of different types of individuals including officials of local agencies, watershed management groups, and the general public in how best to manage and utilize decentralized wastewater treatment systems.

12. A Model Program for On-Site Management in the Chesapeake Bay Watershed helps states to develop and implement a model program to manage on-site sewage systems and minimize the impact of nitrogen on the Chesapeake Bay.
13. The Wastewater Collection System Toolbox provides a series of free sources on approaches of how best to manage wastewater collection systems including: financial and regulatory needs; preventive maintenance programs; managing the existing infrastructure assets; using information systems including GIS; improving system capacity; and dealing with system overflows. (See endnote 39.)

FAIRFAX COUNTY, VIRGINIA, SEWER MAINTENANCE PROGRAM

(See endnote 32)

This program covers over 3000 miles of sewer lines and consists of visual inspections, scheduled sewer cleanings based on past history, unscheduled sewer cleanings as needed, utilizing mechanical and hydraulic cleaning procedures, appropriate record-keeping and analysis to determine problem areas, and scheduling cleaning prior to problems actually occurring, increasing the efficiency and productivity of the staff.

CITY OF LOS ANGELES, CALIFORNIA, SEWER MAINTENANCE PROGRAM

(See endnote 32)

This program covers over 6500 miles of sewer lines with diameters ranging from 6 inches to 12.5 feet. About half of the system is more than 50 years old. A computerized maintenance management plan has been put into effect and it emphasizes prevention and correction quickly of potential problems as well as actual problems. Inspection, cleaning, and rehabilitation of the sewer system is prioritized based on a scoring system that uses the age of the pipes, size, and construction materials that were used. GIS, computer, and logic programs are utilized to effectively determine where inspections and cleaning should be done most frequently. Corrosion abatement is considered to be of extreme importance.

ENDNOTES

1. US Environmental Protection Agency. 2008. *Septic Systems Fact Sheet*. EPA 832-F-08-057. Washington, DC.
2. County of San Diego, Department of Environmental Health, Land and Water Quality Division. 2013. *Design Manual for On-Site Wastewater Treatment Systems*. San Diego, CA.
3. Koren, Herman, Bisesi, Michael. *Handbook of Environmental Health-Pollutant Interactions in Air, Water, and Soil*. Volume 2, Fourth Edition. Lewis Publishers, CRC Press, Boca Raton, FL.
4. US Environmental Protection Agency. 2003. *Decentralized Systems Technology Facts Sheet-Septic Tank Effluent Screens*. EPA 832-F-03-023. Washington, DC.
5. US Environmental Protection Agency. 2002. *On-Site Wastewater Treatment Systems Manual*. EPA/625/R-00-008. Washington, DC.
6. Lee, Brad, Franzmeier, Don. No Date. *Residential Wastewater-High Water Tables and Septic System Perimeter Drains*. Purdue University Cooperative Extension Service, West Lafayette, IN.
7. US Environmental Protection Agency. 2000. *Decentralized Systems Technology Fact Sheet-Septic Tank Systems for Large Flow Applications*. EPA 832-F-00-079. Washington, DC.
8. US Environmental Protection Agency. 1999. *Decentralized Systems Technology Fact Sheet-Septic Tank-Soil Absorption Systems*. EPA 932-F-99-075. Washington, DC.
9. US Environmental Protection Agency. 2002. *Decentralized Systems Technology Fact Sheet-Septic Tank Polishing*. Washington, DC.
10. US Environmental Protection Agency. 2000. *Decentralized Systems Technology Fact Sheet-Evapotranspiration*. EPA 832-F-00-033. Washington, DC.

11. US Environmental Protection Agency. 2000. *Decentralized Systems Technology Fact Sheet-Aerobic Treatment*. EPA 832-F-00-031. Washington, DC.
12. Miner, Margaret, Annes, Eric, Gager, Sarah. 2011. *On-Site Wastewater Management in Connecticut: The Role of the Advanced Treatment Systems (ATS)-A Guide to the Technology, Science, Law, and Policy of Advanced Treatment Systems*. (financed through a grant from the Jeffrey C Hughes Foundation) Rivers Alliance of Connecticut, Litchfield, CT.
13. US Environmental Protection Agency. 1999. *Decentralized Systems Technology Fact Sheet-Septage Treatment/Disposal*. EPA 932-F-99-068. Washington, DC.
14. US Environmental Protection Agency, Office of Wastewater Management. 2014. *2013 Annual Report*. EPA 832-R-14-002. Washington, DC.
15. US Environmental Protection Agency. 2014. *Decentralized Wastewater Management Program-Annual Report 2013*. EPA-832-R-140006. Washington, DC.
16. US Environmental Protection Agency, Office of Water. 2003. *Voluntary National Guidelines for Management of On-Site and Clustered (Decentralized) Wastewater Treatment Systems*. EPA 832-B-03-001. Washington, DC.
17. US Environmental Protection Agency. 2002. *Design Manual On-Site Wastewater Treatment and Disposal Systems*. EPA/625/R-00-008. Washington, DC.
18. US Environmental Protection Agency, Office of Water. 1999. *Decentralized Systems Technology Fact Sheet-Mound Systems*. EPA 832-F99-074. Washington, DC.
19. US Environmental Protection Agency, Office of Water. 2000. *Decentralized Systems Technology Fact Sheet-Types of Filters*. EPA 832-F-00-034. Washington, DC.
20. US Environmental Protection Agency. 2002. *Decentralized Systems Technology Fact Sheet-Control Panels*. EPA 832-F-02-011. Washington, DC.
21. US Environmental Protection Agency, Office of Water. 1999. *Decentralized Systems Technology Fact Sheet-Recirculating Sand Filters*. EPA 832-F-99-07. Washington, DC.
22. US Environmental Protection Agency, Office of Water. 1999, *Wastewater Technology Fact Sheet-Intermittent Sand Filters*. EPA 932-F-99-067. Washington, DC.
23. US Environmental Protection Agency, Office of Water. 2000. *Wastewater Technology Fact Sheet-High-Efficiency Toilets*. EPA 832-F-00-047. Washington, DC.
24. US Environmental Protection Agency, Office of Water. 1999. *Water Efficiency Technology Fact Sheet-Incinerating Toilets*. EPA 832-F-99-072. Washington, DC.
25. US Environmental Protection Agency, Office of Water. 1999. *Water Efficiency Technology Fact Sheet-Composting Toilets*. EPA 832-F-99-066. Washington, DC.
26. US Environmental Protection Agency, Office of Water. 2003. *Wastewater Technology Fact Sheet-Disinfection for Small Systems*. Washington, DC.
27. Ohio Department of Health. 2013. *Household Sewage Treatment System Failures in Ohio-A Report on Local Health Departments Survey Responses for the 2012 Clean Watersheds Needs Survey*. Columbus, OH.
28. US Environmental Protection Agency, Office of Water. 2004. *Primer for Municipal Wastewater Treatment Systems*. EPA 832-R-04-001. Washington, DC.
29. US Environmental Protection Agency, Office of Water. 1999. *Decentralized Systems Technology Fact Sheet-Low Pressure Pipe Systems*. EPA 832-F-99-076. Washington, DC.
30. US Environmental Protection Agency, Office of Water. 2000. *Decentralized Systems Technology Fact Sheet-Small-Diameter Gravity Sewers*. EPA 832-F-00-038. Washington, DC.
31. US Environmental Protection Agency, Office of Water. 2000. *Collections Systems Technology Fact Sheet-Sewers, Lift Station*. EPA 832-F-00-073. Washington, DC.
32. US Environmental Protection Agency, Office of Water. 1999. *Collections Systems O and M Fact Sheet-Sewer Cleaning and Inspection*. EPA 832-F-99-031. Washington, DC.
33. US Environmental Protection Agency, Office of Water. 2006. *Water Technology Fact Sheet-Pipe Bursting*. EPA 832-F-06-030. Washington, DC.
34. US Environmental Protection Agency, Office of Water. 1999. *Collections Systems O and M Fact Sheet-Trench List Sewer Rehabilitation*. EPA 832-F-99-032. Washington, DC.
35. US Environmental Protection Agency. 2000. *Wastewater Technology Fact Sheet-Wetlands: Subsurface Flow*. EPA 832-F-00-023. Washington, DC.
36. US Environmental Protection Agency. 2000. *Wastewater Technology Fact Sheet-Free Water Surface Wetlands*. EPA 832-F-00-024. Washington, DC.
37. Panguluri, Srinivas, Skipper, Gary, Donovan, Steve. 2014. *Demonstration of Innovative Sewer System Inspection Technology: SL-Rat™, by* EPA/600/R-14-031. National Risk Management Research Laboratory, US Environmental Protection Agency, Cincinnati, OH.

38. Stirling, Ray, Simicevic, Jadranka, Allouche, Erez. 2010. *State of Technology for Rehabilitation of Wastewater Collection Systems*. US Environmental Protection Agency, Urban Watershed Branch, National Risk Management Research Laboratory, Edison, NJ.
39. US Environmental Protection Agency. 2015. *Wastewater Collection System Toolbox-Eliminating Sanitary Sewer Overflows in New England*. Boston, MA.
40. US Environmental Protection Agency, Water Infrastructure Outreach. 2014. *Guide for Estimating Infiltration and Inflow*. Boston, MA.
41. US Environmental Protection Agency, Office of Water. 2014. *Flood Resilience: A Basic Guide for Water and Wastewater Utilities*. EPA 817-B-14-006. Washington, DC.
42. Oregon Department of Transportation. 2005. *Hydraulics Manual*. Salem, OR.
43. New York State Department of Environmental Conservation. No Date. *Wet Weather Operating Practices for POTW's with Combined Sewers-Technology Transfer Document*. Albany, NY.
44. US Environmental Protection Agency. 2013. *Emerging Technology for Wastewater Treatment and in-Plant Wet Weather Management*. Washington, DC.
45. US Environmental Protection Agency. 2003. *Environmental Regulations and Technology-Control of Pathogens and Vector Attraction in Sewage Sludge-(Including Domestic Septage)*. Under 40 CFR Part 503. EPA/625/R-92/013. Cincinnati, OH.
46. New York State Department of Environmental Conservation. 2011. *Biosolids Management in New York State*. Albany, NY.
47. Water Environment Federation, National Biosolids Partnership. No Date. *Emerging Biosolids Treatment Technologies Fact Sheet*. Alexandria, VA.
48. US Environmental Protection Agency. 2014. *Guide to Field Storage of Biosolids and Other Organic By-Products Used in Agriculture and for Soil Resource Management*. San Francisco, CA.
49. US Environmental Protection Agency. November 1, 2016. *Contact Us about Biosolids-Frequently Asked Questions*. Washington, DC.
50. Emanuel, Elizabeth. 2010. *International Overview of Best Practices and Wastewater Management*. International Institute for Sustainable Development, New York.
51. Science Applications International. 2006. *Water and Wastewater Energy Best Practice Guidebook*. Corporation State of Wisconsin, Department of Administration, Division of Energy, Madison, WI.
52. US Environmental Protection Agency. 2013. *Energy Efficiency in Water and Wastewater Facilities*. Washington, DC.
53. US Environmental Protection Agency, Office of Water. 2000. *Wastewater Technology Fact Sheet-Package Plants*. EPA 832-F-00-016. Washington, DC.
54. US Environmental Protection Agency, Office of Water. 1999. *Combined Sewer Overflow Technology Fact Sheet-Alternative Disinfection Methods*. EPA 832-F-99-033. Washington, DC.
55. US Environmental Protection Agency, Office of Water. 1999. *Wastewater Technology Fact Sheet-Chlorine Disinfection*. EPA 832-F-99-062. Washington, DC.
56. U.S. Environmental Protection Agency. 2012. WaterSense at Work: Best Management Practices for Commercial and Institutional Facilities. USEPA. Washington.

BIBLIOGRAPHY

American Society of Civil Engineers. 2010. *2010 Report Card for Pennsylvania's Infrastructure*. Camp Hill, PA.
City of Philadelphia. 2012. *Combined Sewer Flow Program*. Philadelphia, PA.
Johnson County Wastewater Authority. No Date. *Private Inflow/Infiltration Source Control Program Helps Reduce SSOs*. Johnson County, KS.
US Environmental Protection Agency. 1991. *Handbook: Sewer System Infrastructure Analysis and Rehabilitation*. Washington, DC.
US Environmental Protection Agency. 2006. EPA Enforcement: Preventing Backup of Municipal Sewage into Basement. *Enforcement Alert*, 8(1).
US Environmental Protection Agency. 2012. *Combined Sewer Overflows, National Pollutant Discharge Elimination System*. Washington, DC.
US Environmental Protection Agency. 2012. *National Enforcement Initiatives for Fiscal Years 2008–2010: Clean Water Act: Municipal Sewer Overflows (Combined Sewer Overflows and Sanitary Sewer Overflows)*. Washington, DC.
US Environmental Protection Agency. 2012. *Sanitary Sewer Overflows and Peak Flows, National Pollutant Discharge Elimination System*. Washington, DC.

US Environmental Protection Agency. 2012. *Septic (On-Site) Systems-Guidance, Manuals and Policies.* Washington, DC.

US Environmental Protection Agency. 2012. *Septic (On-Site) Systems-Handbook for Managing On-Site and Clustered (Decentralized) Wastewater Treatment Systems.* Washington, DC.

US Environmental Protection Agency. No Date. *Wastewater Collections Systems, Aging Water Infrastructure Research.* Washington, DC.

US Environmental Protection Agency National Exposure Research Laboratory. 2011. *Pharmaceuticals and Personal Care Products in the Environment: Scientific and Regulatory Issues.* Washington, DC.

US Environmental Protection Agency, Office of Water. 2002. *Collections Systems Technology Fact Sheet-Sewers, Conventional Gravity.* EPA 832-F-02-007. Washington, DC.

US Environmental Protection Agency, Office of Water. 2003. *Voluntary National Guidelines for Management of On-Site Clustered (Decentralized) Wastewater Treatment Systems.* Washington, DC.

Washington State Department of Health, Wastewater Management Program. 2002. *Disposal Treatment Options: Highly Pretreated Effluent, Rule Development Committee Issue Research Report-Draft.* Olympia, WA.

12 Solid Waste, Hazardous Materials, and Hazardous Waste Management

There will be three major but related topics discussed in this chapter. They are solid wastes, hazardous materials, and hazardous waste management. Hazardous materials including chemicals are included here because eventually either by accidental release or purposeful disposal the substances become hazardous wastes.

STATEMENT OF PROBLEM AND SPECIAL INFORMATION FOR SOLID WASTE

(See endnotes 3, 10, 97; Chapter 2, "Air Quality (Outdoor [Ambient] and Indoor)"; Chapter 3, "Built Environment—Healthy Homes and Healthy Communities"; Chapter 4, "Children's Environmental Health Issues"; Chapter 5, "Environmental Health Emergencies, Disasters, and Terrorism"; Chapter 8, "Healthcare Environment and Infection Control"; Chapter 9, "Insect Control, Rodent Control, and Pesticides"; and Chapter 14, "Water Quality and Water Pollution")

GLOBAL WARMING

The manufacture, use, and disposal of materials as well as the production of the raw materials needed to make a given product will also produce gases which can alter the ambient atmosphere and cause unnecessary changes in weather patterns. The disposal or processing of solid wastes in incinerators or landfills produces an unusually large amount of methane gas and also carbon dioxide which affect the weather and may add to global warming.

SUB-PROBLEMS INCLUDING LEADING TO IMPAIRMENT AND BEST PRACTICES FOR SOLID WASTE COLLECTION, STORAGE, AND TRANSPORTATION

There are different modes of collection, storage, and disposal of each of these waste streams. Therefore, different problems are associated with each of these systems.

NON-HAZARDOUS SOLID WASTE STREAMS

There are six basic types of non-hazardous waste streams which can cause health and safety problems to humans and have a detrimental effect on the environment.

Municipal Solid Waste

In the United States in 2013, people produced 254 million tons of non-hazardous municipal solid waste, and recycled and composted over 87 million tons of this material. Schools, hospitals, and businesses created a mass of material which needed disposal, including packaging from various products, furniture and appliances, clothing, newspaper, glass, etc. Then add to the waste stream paperboard, plastics, electronics, metals, road debris, yard trimmings, and food waste. Also produced were paint and solvents, other corrosive or volatile materials, used oil, batteries, household cleaners and pesticides, etc., which become part of the hazardous waste stream.

Per capita consumption of goods and products worldwide is projected to increase, and therefore both the non-hazardous waste stream and the hazardous waste stream will increase with it. (See endnote 1.)

Municipal waste from residences is pretty uniform across the country, with the exception of probably higher concentrations of organic materials in warm climates than in colder climates. Municipal waste from business is usually the largest amount of material that needs to be disposed of from cities. The type of waste depends on the type of commercial establishments that are within any given area. Restaurants generate much larger quantities of food wastes and food packaging than other types of materials. There is also a substantial amount of paper, cardboard and packaging, and process waste. Other municipal waste include industrial waste, institutional waste, construction and demolition wastes, wastes from municipal services including waste treatment plants, industrial process waste such as scrap.

Collection
(See endnote 11)

There are fixed and variable factors for the solid waste collection system for individuals and the community. The fixed factors are related to the climate conditions, seasonal factors, topography, and lay-out of the various areas including the width of the streets and whether parking is allowed on one side or both sides, population density, traffic patterns and density, and the type and quantity of solid waste produced. The variable factors include type of storage, recycling of materials, frequency of collection, crew size, type of collection equipment, distance from collection point to disposal point, and political factors. Consideration has to be given for the collection of yard waste, special waste such as large objects, commercial solid waste, occupational health and safety concerns for the workers, and the collection of data necessary for determining proper planning, organization, and management of the solid waste collection program. The collection process accounts for about 75% of the cost of the solid waste management program.

The collection system and the type of equipment used vary with the community and the type of solid waste needed to be collected. The size of crews and the daily routes are established based on the data collected concerning the volume and type of solid waste and the distance to the disposal point.

There are several means of financing solid waste programs, each with advantages and disadvantages. Property taxes may be utilized with the advantage of having funds always available to carry out the program. Disadvantages include: there are no incentives for the individuals to reduce the amount of solid waste being produced; this may be inequitable because different people produce different quantities of solid wastes; and the actual cost of the collection system may not be transparent because it may be part of overall governmental funding. A flat fee system is easy to administer, it is easier to adjust fees when needed, and it can be done without government if needed. The disadvantages include: this fee-for-service may not be used for solid waste collection but for other purposes; the fees may be difficult to collect; residents may dump their solid waste to avoid paying the fee; and it does not provide an incentive for reduction of quantities of solid waste. The variable-rate system gives a direct economic incentive to produce less solid waste to reduce the cost of collection and makes people more aware of what they are putting in the disposal system. The disadvantages include: it is difficult to administer because it is complicated; it requires enforcement programs; and it is difficult to project anticipated revenue to keep the system working properly.

The operation of the solid waste program may be conducted by municipal departments of sanitation, contracts with waste management companies, or private collection between the company being serviced and the waste management company. In any case, they must obey all appropriate governmental regulations and eventually could come under the supervision of the governmental agency.

Best Practices in Collection Systems for Solid Waste

- Conduct an in-depth survey of the specific community to be serviced concerning all the fixed and variable factors as well as the distance to disposal in order to establish a plan for how best to collect solid waste. After a reasonable period of time, re-evaluate the plan and its implementation to make readjustments where necessary to be more cost-efficient and produce the highest level of service.
- In residential areas typically only allow the use of 30-gallon weatherproof cans of metal or plastic with tight sealing lids for temporary storage and removal purposes, or tightly tied plastic bags filled with the solid waste and placed at the curb or in the alley at the time of collection. Never allow the use of 55-gallon drums, which may be too heavy for the worker to lift without injury, or cardboard boxes which may break through from wet solid waste.
- If dumpsters are used in some areas, they must be of such size and specifications that the equipment being used to collect the solid waste can lift it and drop the contents into a compactor truck. The lids on the dumpsters must fit securely and they must be kept clean as well as the area around the dumpster where possible leakage can attract insects and rodents.
- Communities may want to consider solid waste separation by the public prior to the collection by placing all recyclables in special containers to be picked up by a special truck, thereby reducing the amount of solid waste being disposed of in landfills or through incinerators.
- Bulky solid waste should be put out separately for special trucks to remove.
- Hazardous solid wastes should be removed to specific areas for disposal by the citizenry and not mixed in with the non-hazardous solid waste.
- In some communities, garbage is removed separately twice a week and all other solid wastes are removed once a week by different collection vehicles. In any case, where there is organic material present in the non-hazardous solid waste, the removal must be at least twice a week to avoid the growth and distribution of fly infestations.
- In areas where there are sufficient numbers of homes close together, it is most efficient to collect solid wastes at the curb or in the alley. This would involve specific collection days. Where there are insufficient numbers of homes that are close together, then it may be better to have the citizen bring the solid waste to an off-site collection point where it will be placed in dumpsters and removed on a frequent schedule which will be made available to the individuals. These areas need to be closely watched to make sure that the areas are kept clean, the dumpster lids are kept down, any drainage is promptly cleaned up, and hazardous solid waste is not placed with the non-hazardous solid waste.
- Preventive maintenance programs should include inspection and servicing of all parts of the vehicle being used for transportation of the solid waste every thousand miles or every 30 days by highly competent mechanics. Accumulate written reports of types of breakdowns and other service problems and evaluate them periodically to enhance the preventive maintenance program.
- Establish a program for the purchase of new equipment on a periodic basis before the old equipment breaks down and becomes very costly to operate. Allow for those cases where a piece of equipment may be defective and has to be replaced before the usual schedule.
- In busy communities, collect solid waste on one side of the street one day and on the other side of the street the other day. Erect signs that ban parking on the side of the street where collection will be done on the day of collection. Enforce the ban by asking local law enforcement to hand out tickets if the instructions are not obeyed. In quieter communities, especially if the streets are fairly narrow, collect solid waste on both sides of the street at the same time to be more efficient and cost the community less money.

Transfer Stations
(See endnote 12)

- Waste transfer stations which provide temporary storage and, depending on the size of the facility, appropriate recycling, compacting of waste, and bailing of waste to reduce space needed for transportation and disposal, can be very efficient and cost-effective. The location of the site is highly significant because considerable cost can be saved by having it close to the waste collection area, easily accessible for transfer trucks, and close to utilities which will be needed to operate it.
- There are several environmental problems associated with these units including noise, odors, dust, vectors, high traffic, and litter. In urban areas, typically there are not substantial buffer zones between the transfer station and the community, and therefore the people may be subjected at different levels of intensity to the aforementioned problems.

Best Practices for Transfer Stations
- Establish transfer stations in well-drained land away from population centers and in flood-free areas. Design the site such that the buildings fit in with the surroundings and the area is landscaped to disguise its true purpose. The transfer station must be readily accessible to good roads including superhighways, rail lines, and waterways if the waste will be transmitted by boat to the disposal facility.
- Establish a facility noise level limit of 55 dB, test the noise level frequently, and enforce the appropriate standards.
- Use buildings made of concrete with double-glazed windows and sound-absorbing materials at the transfer building site to reduce noise levels.
- Surround the facility with trees, berms, or concrete walls to absorb sound. Also, use wing concrete walls which will extend beyond the length of the longest vehicle at all transfer buildings to reduce sound.
- Keep the doors of the facility closed during operating hours except when the vehicles are entering or exiting.
- Either use visual warning devices for vehicle backup or the lowest setting of the back-up alarms.
- Establish operating hours away from early morning and late evening times.
- Remove all wastes from the site before nightfall and frequently clean the area, especially the tipping pit, carefully.
- Install a ventilation system with air filters or scrubbers and a misting system with deodorizers to mask or neutralize odors. Also use biofilters, where the odor in the air has to pass through organic materials such as wood chips, mulch, or soil to capture the odor molecules.
- Pave or use gravel all-purpose roads leading to the transfer station site to reduce dust levels. Keep the facility roads clean and washed and wash all waste collection vehicles before they leave the site.
- Make sure that there is little space between the sides of the vehicle and the dump area and use plastic curtains at that space to help prevent dust from entering the air of the community.
- Use a professional pest control company with an appropriate contract to make sure that all insects and rodents are eliminated and that harborage, food, and water are not available. Have this supervised closely to ensure compliance.
- Pretreat all areas of suspected or anticipated insect or rodent harborage.
- Provide a special fund for assisting the neighbors in insect and rodent control if the source is potentially the transfer facility.

- Restrict collection trucks from using residential streets unless they are making actual pick-ups of solid waste.
- Create acceleration, deceleration, and turning lanes at site entrances to avoid tying up traffic.
- Constantly maintain and upgrade all roads leading to the facility and within the facility.
- Ensure that all transport vehicles comply with a plan which has been developed by the facility and the community to cause the least upset in traffic patterns and least annoyance to the community.
- Do not allow incoming trucks to line up on community streets.
- Require that all incoming and outgoing loads of solid waste are fully covered and the trucks are leak proof.
- Use a team of individuals to scout the area around the facility to spot litter and remove it promptly.
- Use a perimeter fence to prevent litter blown by the wind from entering surrounding neighborhoods.
- Establish a facility and complaint operating log to record all incidents to be analyzed later in order to determine how best to prevent the problem leading to the complaint from occurring again. Also record any worker accidents, health incidents, and accidental releases to the air, water, or land.
- Set up a community phone line so that the installation can respond immediately to community complaints.

Agricultural and Animal Wastes

Agricultural waste management is both economically important and also environmentally necessary. The reuse of animal wastes reduces the cost of hauling the material away and substitutes for the cost of a commercial fertilizer. Agricultural waste management consists of nutrient management at a given farm and site-specific waste management. The agricultural waste that is not used immediately should be stored in a building or in holding ponds away from bodies of water. Other strategies include avoiding over-application of waste to land; avoiding waste application before or after heavy rain; excluding livestock from sensitive areas; minimizing runoff and erosion of fields; and confining the waste from cattle to a given area.

Agricultural wastes from farms and feedlots which can degrade water quality and cause aesthetic problems may be solid waste or hazardous waste, or a combination of both. It is estimated that over 1 billion tons of agricultural wastes are produced annually. Besides this, domestic animals produce over 1.5 billion tons of fecal material and over 600 million tons of liquid waste yearly. (See endnote 3.) The GRACE Communications Foundation using US Department of Agriculture data estimates that there are over 335 million tons of dry matter waste, after water has been removed, coming from farms and that animal feeding operations create each year about 100 times more manure than the amount of human sewage sludge processed in sewage treatment plants. (See endnote 4.) Feeding 800,000 pigs can create one and a half times as much sanitary waste as the City of Philadelphia on a yearly basis. (See endnote 6.) These wastes may make the water unfit for drinking by humans and animals and help destroy aquatic life. Nitrates can enter surface bodies of water and seep into the groundwater from agricultural waste runoff and fertilizers used on the land.

Agricultural and animal wastes consist of the residues from crops, residues from processing of foods, decomposing vegetables and fruits, animal feed, animal urine, feces, bedding, dead animals, etc. These wastes come from crops, orchards, vineyards, dairies, feedlots, farms, etc. And then of course there is available a substantial amount of hazardous waste including used oil, pesticides, herbicides, etc., which will be discussed later in this chapter.

Animal feeding operations as well as concentrated animal feeding operations are increasing in size, although the numbers of them have decreased, and producing considerably more waste material which may contaminate the air, land, and ground and surface water. These wastes if used in proper quantities would be an excellent source of nutrients for plants, but also contain pathogens, heavy metals, antibiotics, and hormones. Reactive nitrogen as a nitrate or ammonium stresses water quality. Hormones are the most potent endocrine-disrupting chemicals that can impact ecosystems and human beings. Pathogens may contaminate all environmental media and food and lead to outbreaks of disease in humans. Antibiotics may create in humans a serious problem of dealing with antibiotic-resistant organisms. Other environmental problems include odors and providing food and harborage for insects and rodents. (See endnotes 5, 6.)

Groundwater and surface water may be contaminated through runoff from applying manure to the land as well as leaching from that which has been applied and from leaks or breaks in the manure storage units. Veterinary antibiotics have been found in well water from areas where manure has been spread. The rate of application and prevailing weather conditions are highly significant in the potential for the spread of microorganisms to food and water. Pathogenic organisms and substantial amounts of nitrates have been found in both groundwater and surface water.

Air quality in close proximity to industrial farms has been found to be poor and at times hazardous. Gases from the decomposition of the manure and particulate matter from feed and feathers (in poultry processing areas) are present in the air. There is a volatilization of ammonia from the manure when it is applied to the land at application and the production of nitrogen oxide when the nitrogen applied to the land undergoes nitrification and then denitrification. Children are most susceptible to the problem of breathing in these gases. Also present is methane, which is not a health risk but is highly flammable and is a greenhouse gas contributing to global warming, as well as hydrogen sulfide which comes from anaerobic bacterial decomposition of organic matter.

Odors are one of the most common complaints of citizens living near these industrial farms. The odors are a mix of ammonia and hydrogen sulfide as well as other volatile and semi-volatile organic compounds. The anaerobic reaction leading to many of the odors occurs when the manure is stored in pits or lagoons for long periods of time.

Where there is manure, there will be various types of flies, mosquitoes, and also rodents looking for food that has not been digested but is still in the manure. This can create substantial insect and rodent problems and also lead to outbreaks of disease.

Pathogens that may be found in animal manure include *Bacillus anthracis*, causing anthrax; *Escherichia coli*, certain strains of which will cause diarrhea; *Leptospira pomona*, causing leptospirosis resulting in abdominal pain, muscle pain, vomiting, and fever; *Listeria monocytogenes*, causing listerosis resulting in fever, nausea, vomiting, and diarrhea; *Salmonella* species, causing salmonellosis resulting in abdominal pain, diarrhea, nausea, and fever; *Clostridium tetani*, causing tetanus resulting in violent muscle spasms, difficulty in breathing, and lockjaw; *Histoplasma capsulatum*, causing histoplasmosis resulting in fever, muscle ache, joint pain, and stiffness: *Microsporum* and *Trichophyton*, causing ringworm resulting in itching and a rash; *Giardia lamblia*, causing giardiasis resulting in diarrhea, abdominal pain, vomiting, and fever; and *Cryptosporidium* species, causing cryptosporidiosis resulting in diarrhea, dehydration, and weakness. All these organisms may lead to serious outbreaks of food- and waterborne diseases.

Antibiotics are usually used in animal feed in the United States. They help reduce the chance for infection and reduce the need for animals to fight off bacteria. Because animals are now being so closely confined in the industrial farms, the chance for infection spreading has increased substantially. There is serious concern that these antibiotics are carried over in the food supply and are making the antibiotics taken by humans far less efficient and also creating a potential for allergic responses.

Best Practices in Disposal of Agricultural and Animal Wastes (See endnotes 7, 8, 9; Chapter 2, "Air Quality (Outdoor [Ambient] and Indoor)"; and Chapter 14, "Water Quality and Water Pollution")

An agricultural waste management program includes the areas of production, collection, storage, treatment, transfer, and utilization. The Best Practices listed below will fall into these categories.

Along with the variation in the amount of production at different seasons of the year, there will be a variation in the amount of waste which is produced.

- Project the amount and type of waste for disposal on a daily, weekly, monthly, and annual basis in order to determine how best to set up a waste management program.
- Reduce contaminated runoff from open holding areas, which contributes a large amount of the waste and pathogens coming from livestock operations. This can be accomplished by limiting the size of the open holding area, putting a roof over part of it, and installing gutters and water diversions away from the waste. Also reduce heavy rainfall from coming across the surface of the contaminated areas by installing necessary drainage from the area.
- Prevent heavy rainfall from flowing into bodies of water from areas where manure is being stored or spread by use of diversion ditches, catch basins, and vegetative filter strips which remove sediment, nutrients, and bacteria. Apply the water from the catch basins to the land.
- Place permanent stockpiles of manure on a concrete pad or clay base with at least 2 feet of separation between the base of the manure and the seasonal high-water table.
- Place fences around access to streams, rivers, lakes, or ponds to prevent livestock from entering these areas and contaminating them.
- Establish maintenance programs to correct leaking water facilities and clean up spilled feed.
- Provide a written record of the type, consistency, volume, location, and timing of waste which has been produced in order to determine if this can be reduced and how quickly it can be removed in an effective manner.
- Collect all waste on a regular basis as quickly as possible from the point of origin and remove it to the point of collection and disposal.
- Provide proper leak-proof facilities for storage of wastes which will allow management to distribute it in an appropriate manner based on weather conditions and the availability of resources to carry out the work.
- Transport the waste as a solid, liquid, or slurry in a manner which will not contaminate environmental media.
- Recycle all reusable waste products as a source of energy, bedding, animal feed, mulch, or plant nutrients as feasible.
- Since cattle and sheep are the largest source of methane emissions from human activity and because their normal digestive process produces methane, supplement the animals' diet with urea which increases their ability to digest their food and thereby reduces gas production by 25–75%.
- Utilize the AgSTAR program to develop various biogas systems and recovery technologies at concentrated animal feeding operations.
- Federal agencies should use a regional approach to the development of nutrient water quality criteria, establish the regional criteria, and provide technical guidance documents to individuals and groups on how to protect various bodies of surface water and groundwater supplies in particular areas, as well as provide technical assistance where needed.
- Dispose of animal carcasses to prevent transmission of livestock disease and protect the environment through rendering, burial, incineration, and composting.
- Use appropriate vaccinations to prevent illness. Remove sick or stressed animals from the rest of the animals and also quickly clean up their wastes, including all manure, and dispose of it properly.
- Use organic acids in feed for animals where appropriate to reduce levels of certain microorganisms.
- Use anaerobic lagoons in southern climates and deep pits in northern climates for temporary storage and treatment of manure by means of anaerobic processes, especially to destroy bacteria.

- Use, where appropriate, composting of animal manure and livestock carcasses while maintaining a temperature of 150°F to destroy most pathogens. The compost must be turned and mixed on a regular basis so that the correct amount of carbon and nitrogen will be available to the microorganisms to substantially reduce the material and change its nature.
- Use, where appropriate, aeration techniques such as large shallow, under 5 feet deep, storage containers where oxygen can naturally reach the bacteria to reduce the material and change its nature. Mechanical aeration may be used by pumping air into the storage containers.
- Use lime stabilization of the animal waste to reduce odor and pathogens before applying to the land. Chlorine stabilization does not work well because of the large amount of organics and the chemical reactions between chlorine and the organic matter which may produce toxic and carcinogenic byproducts. Ozone is safe but also does not work well.

Industrial Wastes
(See endnote 2)

Annually, industrial facilities generate and manage about 7600 million tons of non-hazardous industrial waste, 97% of which is wastewater. In addition, there are large quantities of hazardous waste produced. City sewage and industrial waste contaminate rivers and lakes. (See endnote 2 and Chapter 14, "Water Quality and Water Pollution," for information on wastewater treatment and disposal.)

Industrial wastes consist of a wide range of: packaging material, paper, cardboard, and straw; chemicals including oils, solvents, resins, paints, adhesives, raw chemical materials, finished products, and byproducts; metals of all types including heavy metals; plastics of all types and compositions both contaminated and non-contaminated; glass; wood; ashes; sludge; street sweepings; animal remains; used carbon filters; non-combustible inert materials; dirt and gravel; treated infectious waste; food waste from cafeterias, etc.; related to the specific industry which is producing the given products, byproducts, and waste. Industrial wastes may be solids, liquids, or gases. The wastes may range from extremely hazardous to non-hazardous. (See endnote 22.)

(For hazardous waste disposal, see the special section on "Hazardous Wastes" below.) (See endnote 2)

Best Practices in Disposal of Industrial Wastes

Because the large numbers of different industries to evaluate would be beyond the scope of this book, the author has chosen the brewery industry and pharmaceuticals in hospitals as examples of industrial waste disposal. Also see the "Source Reduction/Waste Minimization" section below. (See endnotes 23, 24.)

- Review purchasing practices and purchase quantities of substances which will be needed immediately and a small amount for backup until new orders can arrive. Use an ongoing inventory control system and stock rotation to avoid substances from becoming outdated and unusable.
- Change the composition of the product if possible to reduce the amount of waste, eliminate hazardous materials, and use new technology for measuring and cutting to use only the raw materials that are absolutely necessary.
- Return waste materials to the original process if possible instead of new raw materials.
- Process waste material as a byproduct which can be sold or used by another industry or within the existing industry.
- Provide for a cleaner process in a clean environment to reduce potential levels of pollution caused by poor housekeeping and poor operation procedures.
- Segregate non-hazardous waste from hazardous waste as it is produced to save time and not contaminate the non-hazardous waste.
- Determine if certain substances are expiring before they are used and change the purchasing process to obtain smaller quantities.

- Make a waste audit to determine waste from plastics, processing, packaging, and storage, the quantity of the waste, frequency of pickup, and means of disposal.
- Determine from the audit the amount and type of solid waste generated and what portion of it is hazardous and non-hazardous. Use reduction of raw materials, reuse of materials, and recycling where possible.
- Sort and store wastes on-site for treatment or disposal and train personnel how best to do this in a cost-effective manner. Determine how much can be recycled or reused to avoid disposal.
- Create appropriate waste treatment programs on-site to reduce the amount of waste going to public landfills or incinerators and thereby protect the environment as well as save substantial costs for hauling and disposal.
- Utilize case studies of other companies in the same industry in order to learn the Best Practices to use in all facets of production, storage, transportation, and disposal.
- Gather data within the industry and the company concerning all facets of operation, delivery, on-site treatment programs, and disposal, and determine if the company is meeting the industry norms and if not why not, and then make necessary changes in procedures.
- Determine if increased production can be accomplished with less waste.
- Ask for input from employees and reward them for innovative ideas that may result in savings to the company and reduction of waste. Then establish Best Practices and implement them in all facets of the operation of the company.
- Establish a special communications and education program for all employees to keep them aware of what is going on in the company and how best to implement Best Practices in various operations.
- Evaluate the ongoing process periodically and use specific tests which may be voluntary or directed by the environmental and public health authorities.
- Determine if the employees are wearing appropriate protective clothing for all facets of the operation including waste disposal.
- Spent grains are byproducts of the brewing process which are still high in protein, fiber, and other nutrients. This can be used as animal feed, in bakeries for cookies and dog biscuits, etc. This is an example of other industries using the unwanted byproducts of a different industrial group. It helps protect the environment and reduces the quantity of waste for disposal. Spent grains can also be composted.
- Diatomaceous earth is used for filtering in the brewing process and then is usually sent to a landfill. This material can be reused as an additive to concrete and brick.
- Packaging materials have the greatest potential for recycling. This would include corrugated cardboard, aluminum cans, glass bottles, and paper.
- Pallets are frequently used for storage of products as well as raw materials. They can be reused numerous times if they are sturdy, and if they need to be sent for disposal they can be converted into wood chips that can be used for landscaping.
- Develop a green team that evaluates all portions of the operation to determine where savings can be made and the waste can be reduced.
- Food wastes from the various food service operations can be reduced by sorting out all recyclables before placing in appropriate containers for removal.
- Composting can be done on the property if it is cost-effective and there is room to do it properly. Balers can be used for reducing the volume of solid waste to be recycled or for disposal. Depending on the size and nature of the industry, a waste to energy project could be instituted.
- Never dispose of pharmaceutical waste in sink drains or toilets to avoid contaminating bodies of water.
- Replace pharmaceutical samples with vouchers for free medication for a 30-day test.

- Empty containers that have been used for hazardous materials which can be considered to be non-hazardous waste and dispose of normally if the inner liner has been specially treated to remove all toxic substances.
- Non-hazardous pharmaceuticals which may be both controlled and non-controlled substances can be mixed with liquid or solid substances such as coffee grounds, kitty litter, or other absorbent material and then placed in a sealed bag or other container to prevent leakage and sent with normal solid waste to landfill or an incinerator.

Construction and Demolition Wastes
(See endnotes 25, 26, 27)

Construction activities and demolition of structures or parts of structures produce a significant amount of waste, which contributes a substantial amount of materials to the solid waste stream in a given area. Reducing the waste and recycling materials can help conserve landfill space, reduce the impact of producing new materials, and decrease the impact of these wastes on the environment.

Construction and landfill waste includes many different categories each with its own potential concerns, some of which may include the production of hazardous materials. These categories include asphalt from paving and shingles; earth of various types and levels of contamination; electrical fixtures and wiring; insulation consisting of asbestos, different types of plastics, and fiberglass; masonry and rubble consisting of bricks, cinderblocks, concrete, excess mortar, rock, stone, and tile; metal of various types consisting of aluminum siding and ducts, iron, lead, mercury from electrical switches, steel, copper, and cast-iron pipes, copper wiring, etc.; paint containers and paint products; petroleum products consisting of brake fluid, oil, contaminated fuel tanks, petroleum distillates, and waste oil and grease; roofing materials consisting of asbestos shingles, cement cans, other roofing shingles, and tarpaper; vinyl consisting of siding, flooring, doors, and windows: drywall and plaster; wood from numerous sources within the structure, trees, and bushes; wood contaminants consisting of adhesives and resins, laminates, paints and coatings, preservatives, stains, and varnishes, chemical additives; large and small appliances; furniture of all types; linoleum and carpeting: pesticide, herbicide, and other chemical containers; a variety of different types of light bulbs, etc.

As an example of the distribution by weight of these construction and demolition wastes in categories in the state of Vermont in 1 year, wood accounted for 26% of the waste, concrete 14%, metals 5%, asphalt 46%, and other wastes 9%.

Best Practices in Disposal of Construction and Demolition Wastes
- Consider if the structure or facility can be reused for another purpose instead of tearing it down.
- Develop a waste management and recovery plan for all facets of the project. Determine which individuals will be responsible for each part of the project in eliminating waste and recycling materials. By contract make it the responsibility of the subcontractors to remove their own waste in an acceptable fashion and take it to an approved landfill for disposal.
- Reuse materials, either old ones from the structure or portions of new ones which would have gone to waste, where feasible.
- Put all materials which are safe to reuse but no longer usable for this project, in a special site preferably at the curb where salvage people might take them for free for their own use, before committing them to approved disposal areas.
- Have co-mingled materials which may include recyclables taken to a special separation center where the usable items will be removed and the remainder taken to an approved disposal area.
- Remove all leftover supplies to an appropriate storage area to be used for other projects.
- Computerize inventory management to allow construction crews to find what they are looking for rapidly and cut down on oversupply.

- Establish a take-back policy as part of the contract to purchase materials, with manufacturers, especially for carpet, padding, drywall, and vinyl, for these companies to either recycle these items or dispose of them in an appropriate manner.
- Hire an architect who is known to utilize techniques of waste reduction, reuse of materials on-site, recycling, and using the most innovative new practices to reduce all forms of waste.
- Create waste reduction through an efficient structure-framing program using a modular approach where sections of the framing are brought on to the construction site instead of building it there.
- Before demolition of a structure, strip it of all materials that can be reused or recycled and sell the material or donate it to organizations that will use it to provide better housing for the poor and homeless. An example would be the reuse center of Habitat for Humanity, Raleigh, North Carolina.
- Consider utilizing the program of the Department of Defense decision matrix to help evaluate alternatives to the traditional demolition of structures, which has been recommended by the US Environmental Protection Agency (EPA) to reduce levels of solid waste while being cost-effective. (See endnote 27.)

Sewage Treatment Wastes

Sewage treatment wastes consist of sewage sludge in liquid, semi-solid, or solid form.

Disposal of sludge in either landfills or through incineration typically would not be cost-effective, and it may be problematic because of various environmental rules and regulations. (See Chapter 11, "Sewage Disposal Systems.")

Special Wastes

(See endnote 28 plus attachments accessible through the US EPA Special Wastes (A category established by the US EPA) as noted in endnote 28)

Special waste consists of materials from cement kilns, mining, oil and gas drilling, combustion on a commercial basis, and air pollution control residues, etc. This is a category established by Congress for certain wastes generated in large volumes and believed to have less risk for human health and the environment than hazardous waste. The EPA had to do in-depth studies of each of these wastes to determine if there were hazards and issue a formal report to Congress of its findings. The EPA then had to issue a new final regulatory determination within 6 months on whether the special waste should be regulated as a hazardous waste. (See endnote 28 and its attachments for each of the special wastes.)

Cement Kiln Dust Waste

This is the fine-grained, solid, very alkaline waste that is removed from the exhaust gases by air pollution devices on cement kilns. Much of this is unreacted raw material and can be recycled back into the production process. That which cannot be reused goes to a landfill. The Best Practices listed below only apply if the dust comes from the burning of non-hazardous waste.

Best Practices in Preventing Air or Water Pollution from Cement Kiln Dust
- Compact and periodically wet down cement kiln dust in landfills.
- At the site of the cement kilns, transfer the dust in enclosed covered vehicles and conveyance devices from enclosed tanks and containers within the buildings to the site of disposal and use means of transfer to the landfill that will not create a dust problem in the air. All such landfills must have liners to prevent leachate from moving down to the groundwater supply.
- If the cement kiln dust is used as part of an agricultural application, it must meet the concentration limits for arsenic, cadmium, lead, thallium, chlorinated dibenzodioxins, and dibenzofurans.

Crude Oil, Coal, and Natural Gas Waste

(See endnote 29)

This includes all wastes that come from the exploration, development, and production of crude oil, natural gas, and geothermal energy. Oil and gas come from formations that contain naturally occurring radioactive materials such as uranium and its decay products, thorium and its decay products, radium and its decay products, and lead-210. This radioactive material may contaminate pipes, equipment or their components, waters that have been produced along with the energy source, and sludge. The produced water may not only be contaminated but also have a high brine content. The petroleum industry produces about 260,000 metric tons of waste yearly.

Coal ash is made up of: fine fly ash which is mostly silica; bottom ash which is coarse and can be too large to be carried out of the smokestacks; boiler slag which is made up of the molten bottom ash; and flue gas desulfurization material which is left over from reducing the sulfur dioxide emissions. Coal ash also contains, but is not limited to contaminants such as mercury, cadmium, and arsenic.

Coal ash is one of the largest sources of waste created in the United States, especially from coal-fired electric utilities. In 2012, 110 million tons of coal ash were created. Coal ash may be put in on-site or off-site landfills or surface impoundments with dams.

Best Practices in Disposal of Crude Oil, Coal, and Natural Gas Waste

- Test all scrap metals for radioactive materials and then remove those items to a safe place for disposal, before releasing the metal into the waste metal stream for recycling.
- Clean all piping and equipment and test for radioactive material before releasing into the recycling waste metal stream for smelting.
- Install pollution control devices such as filters and bubblers on smelter stacks to reduce airborne radiation.
- Sludge with elevated radiation levels should be dewatered and held in storage tanks for proper disposal.
- Re-inject waters that have come along with the oil and natural gas production back into the underground areas of exploration using state-of-the-art protected deep wells which shall not allow any of the material to go into the groundwater supply.
- Clean out all contaminated pipes at special areas by sandblasting them with high pressure water or scraping and remove the contaminated scale to drums for later disposal.
- If contaminated equipment cannot be decontaminated efficiently by cleaning and then reused by the petroleum industry, it must be sent to special licensed landfills that are able to handle this type of material.
- All impoundments and landfills used to contain coal ash should have liners which are impermeable and will not permit contamination of the groundwater supply. (The Final Rule from the US EPA only requires that those impoundments that contaminate groundwater or fail to meet location requirements install liners or close. The problem with the rule is that no one knows what will happen in the future because of leaking of contaminants, and therefore the groundwater supply is being put at risk unnecessarily.)
- Install groundwater monitoring devices around surface impoundments and landfills to determine if leaking contaminants are entering the groundwater supply.
- Install liners for all new surface impoundments and landfills and only place these disposal sites in areas which meet engineering and structural standards. Do not build them in sensitive areas including wetlands and areas where earthquakes may occur, to prevent contamination of groundwater.
- Utilize the coal ash Final Rule structural integrity criteria to determine if existing disposal sites are safe and then conduct periodic inspections to determine further structural stability.

- Develop a fugitive dust plan for landfills or surface impoundments and operate a water spray or fogging system to contain the dust. Also use wind barriers, compaction of the material, and vegetative cover as means of reducing potential dust.
- Coal ash can and should be recycled as an ingredient in the manufacture of concrete and wallboard.

Mineral Processing Waste

(See endnote 32)

Mineral processing waste is typically different from the original mineral because of the change in chemical content caused by the process of smelting, electrolytic refining, use of acids, or a digestion process of extracting and concentrating the minerals from the surrounding material. The toxicity range of the waste is based on the mineral involved and the means of removing it from the surrounding material during processing. The types of mineral commodities are vast and therefore cannot be addressed as a single entity. Some of the minerals include aluminum, antimony, arsenic, beryllium, cadmium, chromium, lead, magnesium, manganese, mercury, selenium, strontium, uranium, etc. Unfortunately, 133 waste streams from 30 different commodity sectors out of 553 waste streams from 48 commodity sectors have been determined to contain hazardous waste as defined in the Resource Conservation and Recovery Act (RCRA).

Best Practices in Handling Mineral Processing Waste (See endnote 33)

Each mineral processing system has its own special technical process and identified waste. The US EPA has an online document entitled "Mineral Processing Waste." Clicking on "technical documents" at the bottom of the page will bring up another page listing an important group of individual mineral commodity reviews. Each review discusses the nature and use of the commodity; names of companies producing it; production process and related diagrams; and what material goes to waste disposal. This information along with the EPA's compliance assistance website will help the reader gain specific information about a specific commodity. Reviewing the entire field of mineral processing waste is beyond the scope of this book. However, there are some general Best Practices which should be followed by all mineral processing companies:

- Complete a full set of tests to determine what minerals are present in the material and waste and in what concentrations in order to evaluate potential hazards to people and the environment.
- Establish a mineral waste management plan based on the specific needs of the industry and have it enforced at the highest levels of management.
- Minimize the mass of material to be stored and determine the safest means of storage based on existing regulations and potential future problems.
- Make sure that all storage facilities are physically and chemically safe and will not contaminate the environment or cause health problems to workers.
- Rehabilitate all materials that can be recycled safely to aid in the minimization of waste.
- Use dust control procedures during the sorting, crushing, and washing of rocks and in raw materials storage, loading, and processing.
- Develop good housekeeping practices for the entire facility in order to minimize dust and other potential problems from the mining material.
- Employees should wear appropriate personal protective devices as needed based on the potential hazards from the minerals and mineral waste.
- Install permanent diversions of surface water from areas where mineral waste is stored or raw materials are kept.
- Build roads from the construction site and the facility which are very limited in producing dust problems and can withstand the weight of the equipment as well as the different weather conditions.

- Maintain all equipment in such a manner that it does not produce environmental contaminants.
- Do not allow any floor drains to be connected to storm or sanitary sewers.
- Drain all fluids from materials prior to disposal and treat the fluids before releasing to any body of water or on the land.
- Immediately clean up all leaks and spills which could contaminate surfaces, soil, water, and the air.
- Use a concrete pad for all fueling operations and provide check valves for the fueling hoses and spill and overflow protection devices. (See endnote 37.)

Mining Waste
(See endnotes 34, 35, 36)

There are approximately 500,000 abandoned mines and related or processing facilities across the United States. A total of 130 of these are on the National Priorities List which covers over 1 million acres of contaminated land from past mining activities. The contamination is from waste rock and beneficiation waste including mill-tailing piles which have been scattered in many surface impoundments. Contaminated mining waste has caused problems in surface bodies of water and groundwater through seepage, and potential air pollutants especially from heavy metals and other contaminants from mining and processing. Also there is waste present in the form of slurries and sludge in ponds with questionable retaining walls.

Mining wastes include those wastes which come from extraction, beneficiation, and processing. Beneficiation is a value-added process involving transformation from a primary material to a more finished material of greater value through smelting or an extraction process of the metallic ore from the already extracted material from the ground. The original extraction removes the mass of material including the ore from the earth and then makes it ready for the next step in the process. The mineral processing usually changes the chemical composition or the physical structure of the ore or mineral.

Mining wastes include waste rock which is the material including the overburden (the earth above the deposit) that had to be moved in order to access the ore or mineral; waste rock which is the material that contains the minerals in concentrations that are too low for economic removal; non-metallic byproducts of the metal smelting process; water treatment sludge which comes from the water treatment plants at the mine sites and contains solids which are removed from the water as well as any chemicals that were used to improve the efficiency of the process; gaseous wastes which include particulates from dust, sulfur oxides, and other gases produced during the reactions that occur during processing; tailings which are the residuals usually generated in a slurry form during the beneficiation process; mine water which is groundwater or precipitation that infiltrates during extraction and becomes contaminated; and processing wastes which are the residuals from the processing and beneficiation process. Most processing wastes are water related and are usually placed in surface impoundments.

The mining and mineral processing waste is responsible for a large amount of damage to both human health and the environment. Typically, the wastes are put on the land in unlined units that may leach hazardous materials onto the ground and into surface and groundwater supplies. It can also cause problems of air pollution. The wastes may also be put in heaps which can at some point in time create a landslide with catastrophic results. There are also leaks and spills from pipes that may contribute to the problem. A large variety of potential diseases of either an acute or chronic nature may be a result of these hazards.

Best Practices in the Storage and Disposal of Mine Waste (See endnotes 38, 39)
- Identify the key environmental, health, and risk/safety problems for the given industry and mine in order to be able to establish an appropriate plan for the safe storage and disposal of mine waste.

- Establish a baseline of the quantity and characteristics of the tailings and waste rock which will be or are now being produced and include all information on site selection; assess environmental impacts to air, water, land, and people; develop and carry out a risk assessment study for each potential risk; prepare a plan for routine maintenance and appropriate management of the site; prepare an emergency plan for spills, accidents, and disasters; prepare a disposal plan for treatment, storage, and ultimate disposal of all mine wastes and tailings: determine how best to utilize and dispose of water at the mining site; and develop a decommissioning and closure plan plus lifetime maintenance. Choose drought-resistant plants as part of the cover material.
- Review the best available technologies including emerging technologies for control of the potential hazards from the mine waste and tailings and determine which technologies best suit the specific mine in a reasonably economic manner.
- Develop and implement an environmental management plan that includes: adhering to all environmental laws, rules, and regulations; and establishing necessary structures and responsibilities for implementation of the plan including training, communications, employee involvement, documentation of efforts, establishing maintenance programs, and meeting objectives and goals. Have the program evaluated externally periodically.
- Manage acid rock drainage by placing the material in an impermeable area without earthquake potential, such as solid rock without fissures. If this is not possible, then the material needs to be put in a lined pit that will not corrode, tear, or leak and or be subject to flooding. Choose a liner which is made with low environmental impact. Apply a covering material to this area and supply drainage ditches that will prevent surface water from crossing the area.
- Obtain organic materials from companies that are close to the site to minimize fuel consumption and air emissions from the heavy equipment.
- Determine if there are industrial sources of materials close by that can be used as an appropriate fuel for operations without causing environmental contamination.
- Install remote sensing devices for groundwater supplies, surface water supplies, and air.
- Use solar energy or wind energy where possible for all operations in the mine and processing unit.
- Reuse process water where possible to avoid overconsumption of the potable water supply and release of contaminants to the surface or groundwater.
- Use water treatment systems in which metals or other chemicals may be recovered and reused.
- Mix process water with other substances to neutralize it and make it less hazardous or non-hazardous.
- Use sedimentation ponds to capture the fine materials and then treat the effluent for suspended solids and dissolved metals before releasing into bodies of water.
- Neutralize alkaline waters or contaminants with sulfuric acid or carbon dioxide.
- Remove arsenic from the effluent by adding and mixing with ferric salts.
- Neutralize acid mine water with limestone, hydrated lime, or quicklime.
- Add sodium hydroxide to acid rock drainage and that which has a high manganese content.
- Reduce noise to avoid creating local issues by using continuous working systems, putting the belt drives in an enclosed structure, and creating a barrier around the working face of the mine which will block and absorb the sound.
- Where a dam is needed for retention of the mine waste and tailings, it must be designed to withstand a potential 100-year flood and if there is highly hazardous material present, it must be designed to withstand a 5000- to 10,000-year flood. When building a dam, the natural ground must be stripped of all vegetation and humus soil and be replaced with material that will not weaken under any operational condition or changing climate. All dams for all mine waste and tailings must be inspected on a regular basis by competent engineers, and all recommendations must be implemented as quickly as possible.

- Remove water from tailings where possible and thicken the tailings before disposal.
- In the case of the use of a heap as the disposal method for the mine waste, conduct frequent visual inspections and periodically at least once a year have in-depth studies made by highly competent geotechnical engineers, of the stability and safety of the heap and what needs to be done to stabilize it further. Exercise extreme caution that nothing is permitted below the pathway of the heap in the event of collapse.
- Since mining is an industry with great accident potential, it is essential to develop appropriate plans for accidents and emergencies and test them periodically. Highly competent individuals should evaluate all accidents and continue follow-up for extended periods of time.
- Monitor all pipelines for potential weakness and leaks.
- Prevent the generation of mineral wastes and tailings by conducting in-depth studies by engineers on the best approach to use to remove the material sought with the least amount of waste material being produced.
- Backfill tailings and where the tailings waste rock into mines which are no longer being utilized and may cause damage to the environment.
- Maximize control of soil erosion by wind, rain, or construction activities.

Medical Waste

(See endnotes 40, 42, 43, 44)

Medical waste is material or bodily fluids which are potentially contaminated and may be involved in the diagnosis, treatment, immunization, or research of human beings or animals. Infectious medical waste includes: blood-saturated material; pathological waste including tissues, organs, body parts, and body fluids which may be the result of surgery or autopsies; human blood and blood products; cultures and stocks of a variety of infectious agents coming from various tests or used for research purposes; sharps; waste from different types of isolation units especially where there is a potential for spread of communicable disease; and contaminated animal carcasses, body parts, and bedding from research work.

Medical waste may also be produced during the manufacturing or use of various biological compounds including pharmaceuticals. Pharmaceutical waste includes prescription drugs or over-the-counter human or veterinary drugs which are partially used or expired.

Medical waste may be generated at home in the treatment of patients or the use of sharps. It may be found at the scene of traumas and become a hazard for cleanup. Medical waste may be solid, liquid, or sharps. The waste may also contain traces of chemotherapy agents which may be found on gloves, disposable gowns, towels, and intravenous setups. Medical waste must be separated from the normal waste streams of an institution during collection, containment, labeling, handling, storage, transport, treatment, and disposal.

Solid biohazardous waste includes any solid that will not puncture the skin. This material could be plastic, pipettes, syringes without needles, petri dishes, flasks with cultures in them, animal bedding which may contain hazardous substances, gloves, personal protective equipment, etc.

Sharps are responsible for 385,000 reported injuries each year with an unknown additional number of unreported injuries. The major cause of injuries is the use of disposable syringes and suture needles. They account for 51% of the total injuries. Injuries are also caused by wing steel needles, scalpel blades, and other sharp objects. (See endnote 41.)

Chemotherapeutic agents kill or prevent reproduction of malignant cells. The containers, personal protective equipment, empty tubes, syringes, animal bedding, and treated animals as well as anything coming from treated humans are hazardous to other individuals and must be contained.

Depending on the type of infectious waste, treatment techniques might include incineration, which can also contribute to air pollution; sterilization in special steam sterilizers used for this purpose only; and chemical disinfection, thermal inactivation, and irradiation. In any case after treatment, well-selected representative samples should be taken and analyzed to determine if all infectious agents and spores have been killed.

Best Practices for Disposal of Medical Waste

- Conduct a survey of the entire institution to determine the type and quantity of anticipated non-hazardous and hazardous medical waste to understand the need for an appropriate budget, people, and programs to safely dispose of this material.
- Establish a medical waste plan which involves all phases including the generating, segregating, packaging, treatment, transporting, and disposal of both the non-hazardous and hazardous portions of this special waste.
- Train all personnel in safe operational methods and Best Practices for all phases of the medical waste plan and update the training in the event of an emergency and on an annual basis.
- Track all medical wastes from origin to disposal and keep complete records of all activities involved in its storage, transportation, treatment, and disposal.
- Collect all liquid biohazardous wastes including culture media, supernatant fluids, and human fluids in a sealed container clearly marked with the word "BIOHAZARD" and the universal biohazard symbol. Store in the biosafety cabinet and use an appropriate chemical disinfectant to destroy all organisms and spores. If treated with bleach at proper levels and no organisms or spores can be found on testing, it can go down the sink drain.
- If the liquid biohazardous waste is mixed with either chemicals or radioactive agents, first treat as a biohazardous waste as described above, then treat as either a chemical or radioactive waste, and then use disposal methods which are acceptable for that type of waste stream.
- For non-sharp biohazardous solid waste, place in a red biohazard bag which is kept within a rigid, resistant container with a tight-fitting lid. The biohazard bags must be at least 1.5 ml thick.
- Biohazardous solid waste may be put into a special autoclave and sterilized or chemically disinfected with household bleach and then removed with the solid waste stream. Testing should always be conducted periodically to determine if the processes being used are effective.
- Restrict the use of sharps to areas and procedures where they must be used and provide disposable gloves, and an approved sharps disposal container as well as a first-aid kit.
- Report immediately, verbally and in writing as a priority, all injuries from sharps and the exact circumstances as well as the potential contaminants that might have a health effect on the individual. Seek immediate medical attention if the contaminants may cause a long-term or short-term disease.
- All employees using or exposed to sharps must receive appropriate vaccinations prior to starting work and follow-ups as needed.
- All employees using or exposed to sharps must receive appropriate training in the use and disposal of these items.
- Disposable sharps should be used whenever or wherever possible and placed immediately after use into a leak-proof, puncture-resistant sharps disposal container at the point of use and the container must show the universal biohazard symbol in red or orange.
- When using sharps, be aware of other individuals within the vicinity of your work and do not injure them.
- Have specially trained individuals remove sharps to the ultimate disposal areas.
- When dealing with broken glass whether contaminated or not, never use the hands to pick up the glass but rather forceps, tongs, scoops, or other mechanical means. Contaminated glass must be discarded into a special sharps container. Uncontaminated glass may be put in cardboard glass containers but make sure that sharp edges do not protrude through the container.

- Use incineration as the method of choice for final disposal of sharps.
- Remove all human remains including cadavers, body parts, or recognizable tissues or organs to the proper department for cremation or burial. Small pieces of solid unrecognizable tissue can go with the pathological waste to the incinerator.
- Double bag all animals and animal tissues from research laboratories in red biohazard bags at the site of generation, transport to biohazard freezers, and dispose of as pathological waste which will be incinerated.
- For the body parts from an individual with certain religious beliefs, the body parts must be buried according to religious law.
- Place all chemotherapeutic waste materials in biohazard bags inside of a solid container with a tight-fitting lid and add a label showing that the waste has trace chemotherapy materials. Use yellow color coding on the outside container.
- Conduct a study of the pharmaceuticals and partially used pharmaceuticals in an institution and determine how much of this becomes waste. Reduce the amount of unused pharmaceuticals with better inventory control and obtain where possible special medicines as needed instead of stocking them routinely. Segregate all pharmaceutical wastes from non-hazardous material and dispose of it through incineration at a hazardous waste incinerator or place it in a hazardous waste landfill after it has been treated.
- All equipment used to work with infectious materials must be completely decontaminated before disposal. One of the best techniques is to use a 10% bleach solution which has been freshly made and has been in contact with the equipment for a minimum of 10 minutes.

REDUCTION OF NON-HAZARDOUS SOLID WASTE PRIOR TO GENERATION

General Best Practices for All Forms of Non-Hazardous Solid Waste

Source Reduction/Waste Minimization

Solid waste management consists of materials management, which is the use of the least amount of raw materials for producing products and packaging; reuse of materials as much as possible; resource management of air, water, and land; recycling of materials into new products; recovery of energy; waste reduction through use of composting and combustion to reduce volume; and ultimately waste disposal in sanitary landfills.

Materials management attempts to limit the quantities of material to be ultimately put into a landfill or incinerator by use of source reduction, reuse of materials, and recycling of materials.

Source reduction occurs through a change in the design, manufacture, and use of various items to reduce the quantity and toxicity of the product and ultimately the wastes, while making the same product. Source reduction or waste prevention is an economically realistic means of eliminating or reducing the amount of material used which ultimately becomes waste or hazardous waste. It involves use of minimal or reusable packaging; use of durable equipment and supplies; reuse of products and supplies; reduction of hazardous constituents; using materials more efficiently; reduction of energy usage, water usage, and raw material usage; and eliminating unnecessary items. The best technique utilized for waste prevention to achieve source reduction varies by category of waste and by industry.

Resource management involves limiting the amount of electricity, water, and land needed for production and packaging of products and the use and recapture of energy which can then be made into electricity from the remnants of the products, byproducts, and packaging materials.

Best Practices for Source Reduction/Waste Minimization (See endnote 13)
- Conduct a study of the business or industry to determine the quantity and type of solid waste being created for disposal. Based on this information, establish a program of source reduction to conserve resources and reduce the amount of solid waste being produced at

the source, adding to the bottom line by reducing the amounts of money needed for removing the solid waste and for purchasing additional raw materials, office supplies, wear on equipment, shipping costs, and time spent by personnel.

- Use or manufacture minimal size or reusable packaging.
- Use and maintain properly durable equipment that can be repaired easily and inexpensively, and supplies.
- Reuse products that can be washed and/or disinfected in place of disposables.
- Reuse file folders, interoffice envelopes, outdated stationary, etc.
- Substitute less hazardous or non-hazardous material for hazardous material allowing for less expensive disposal.
- Use supplies and materials more efficiently by purchasing only that which is necessary and not having to throw out outdated substances.
- Review the use of all materials and determine which ones are unnecessary and can therefore be eliminated from purchase.
- Use fewer raw materials in the products when feasible.
- Make products more durable and easier to repair thereby increasing their lifespan.
- Make product packaging recyclable.
- Keep excellent records in order to evaluate the amount and type of solid waste sent to disposal versus the original study and determine areas of strength and reinforce them and determine areas of weakness and change them.

Recycling

(See endnote 17)

Recycling is the removal of useful materials such as paper, glass, plastic, and metals from the solid waste and reuse of the materials in such a manner that they do not contribute to an increase in insect and rodent problems, air pollution problems, water pollution problems, land pollution problems, or odors. Recycling prevents pollution caused by the manufacture of new products from the original materials, saves natural resources, saves the energy needed for production, and decreases the level of greenhouse gases by not adding new ones to the atmosphere. In 2013, about 87 million tons of municipal solid waste was either recycled or composted.

Special types of recycling include e-cycling, scrap tires, and used oil. E-cycling is the reuse of electronic products by various agencies which can extend the life of the products and reduce the level of waste for disposal. It also saves a substantial amount of electricity which was needed to produce the products and large amounts of copper, silver, gold, and palladium.

Scrap tires (there are more than 275 million of them in stockpiles) are used as fuel, used in civil engineering projects, converted to ground rubber, used in rubber-modified asphalt, put into special products, and used in agriculture. Used oil can be re-refined into new oil products such as lubricating oil instead of using crude oil as the base.

Recycling programs are effective where the materials can be economically processed into a form that is competitive with the virgin material. The recycled materials need to be free of contaminants and can be sold on the private market. The use of disposal fees based on the amount of solid waste being produced can help make recycling more cost-effective for the citizens doing it.

Best Practices in Recycling (See endnote 14)
- Local governments should establish mandatory recycling programs utilizing appropriate enforcement procedures when people, businesses, or industries do not participate and reduce their solid waste stream for disposal. Goals should be set for each category of solid waste and results should be measured periodically.
- Laws should be passed to enforce the recycling program, and funds should be provided for adequate numbers of trained people to oversee the programs and utilize appropriate public

education techniques to encourage voluntary participation and only use enforcement when necessary.

- Variable rates should be charged to the citizens for the amount of solid waste and garbage that they produce. There should be no charge for recycling collections.
- Since commercial waste typically makes up 50% of the urban community waste, a variable rate fee should also be charged for this solid waste stream with free removal of recyclables.
- Special fees should be applied to the removal of construction and demolition waste to landfills. Once again, quantity and type of material is of greatest significance and reuse of the materials can save the contractor considerable sums of money. These might include carpets, ceiling tiles, ceramic tiles, recycled exterior paint, building insulation, salvaged materials such as copper wire or pipes, wood, bricks, and concrete that can be used as a filler. Obviously, all hazardous wastes will be banned from landfills or the typical municipal incinerators.
- Institute a ban on styrofoam use by restaurants and use products which are biodegradable or can be recycled.
- Dispose of significant amounts of grass, plant debris, trees, bushes, leaves, etc. by turning them into mulch or composting. Do not allow them to be placed into the landfills.
- Establish in the school systems, programs teaching the value of recycling and the use of green constituents in all aspects of life.
- Establish special facilities for the collection, storage, and disposal of household hazardous waste which can be brought by citizens to the units, and there will be no charge for the disposal.
- Establish a recycling unit for used motor oil without charge to the citizens.
- Donate all usable clothing, small appliances, surplus items, and other usable things to the Goodwill stores, Salvation Army, and other charities.

SUB-PROBLEMS INCLUDING LEADING TO IMPAIRMENT AND BEST PRACTICES FOR SOLID WASTE DISPOSAL FOR MUNICIPAL LANDFILLS

MUNICIPAL LANDFILLS

(See endnotes 73, 74, 75)

There always will be a need for landfills that are properly constructed, operated, maintained, and closed in an appropriate manner when the landfill is full. With growing populations, there is an increase in waste production and necessary waste disposal. The use of waste minimization techniques reduces the immediate need for additional large amounts of land, but over time the land areas will be filled up. The function of the sanitary landfill is to completely degrade biologically, chemically, and physically the material within it. The composition of the wastes varies with the type of waste deposited. Domestic waste which comes from household activities includes food wastes, plastics, paper, ash, broken pottery, metal, glass, rubber, textiles, etc., and unfortunately may contain some of the household hazardous waste. Commercial waste comes from stores, offices, restaurants, hotels, etc., and is composed mainly of packing materials, office supplies, food wastes, and then some of the same waste as domestic waste. Street sweepings are primarily dust and soil and may also contain paper, metal, and other types of street litter including the residues of drains and culverts. Waste from sewage treatment plants may include dried biosolids, and chemical wrappers and containers. Industrial non-hazardous waste, depending on the industry, includes packaging, plastics, paper, metal, and food waste from the cafeterias, etc.

There are six major site restrictions. They are airports where a landfill cannot be located within 10,000 feet of the end of the runway; 100-year floodplains where the solid waste would reduce temporary water storage or allow a wash-out of the material; wetlands with a few exceptions for states or tribes in certain situations; fault areas where landfills are prohibited within 200 feet; seismic impact zones where a landfill can exist if the liners and leachate collection systems can resist ground motion; and unstable areas such as sinkholes, rock falls, or where the soil may become liquefied during extreme wetting and drying cycles.

A special problem for landfills and metal recyclers is the household appliances that contain refrigerants. These units include window air conditioners, motor vehicle air conditioners, water coolers, vending machines, icemakers, and refrigerators. The common refrigerants which are ozone-depleting chemicals are typically chlorofluorocarbons and hydrochlorofluorocarbons.

Landfill gas is a mixture of many different kinds of gases. The gases are produced by bacterial decomposition of the material deposited, volatilization of certain organic compounds, and chemical reactions between different compounds present in the waste. The gases produced include typically 45–60% methane, 40–60% carbon dioxide, 2–5% nitrogen, 0.1–1% oxygen, 0.1–1% ammonia, 0.1–0.6% non-methane organic compounds, 0–1% sulfides, 0–0.2% hydrogen, and 0–0.2% carbon monoxide. The amount and type of gas production is affected by the composition of the waste, age of the waste, presence or absence of oxygen in the landfill, moisture content, and the temperature of the landfill. The landfill gases can move from under the surface through spaces in the refuse and soils depending on the ability of the gas to defuse in a given area, the pressure of the gases, and the permeability of the waste and soil.

Landfill gases can be captured, treated, and converted into a source of energy. Solid waste can be turned into cellulosic gasoline and diesel fuels.

Reducing the use of energy and recovery of energy from the solid waste helps control the cost of operating the landfill, helps reduce pollutants of air and water, and assists in controlling global warming.

Modern landfills are properly engineered facilities where the remains of the solid wastes are deposited after recycling and further processing of the solid waste has been utilized. The landfills are designed, operated, and monitored to meet the requirements of federal regulations and therefore the following essential practices are utilized: restriction to locations in areas where the waste will not contaminate the groundwater or essential land areas; proper use of landfill liners; collection and removal of leachate for treatment; modern operating practices; groundwater monitoring wells; proper closure of landfills; and long-term evaluation and correction of problems in closed landfills.

Bioreactor landfills help to reduce the amount of landfill space needed and make the waste less hazardous by rapidly transforming and degrading organic waste. This is accomplished by injecting liquid, which may be leachate from the landfill which is recycled, wastewater, or sludge from sewage treatment plants and thereby maintains the moisture content at approximately 35–65%. The landfill may be aerobic, anaerobic, or a hybrid of both. The bioreactor produces additional landfill gas such as methane, which when captured and treated may be used as a source of energy.

Landfills are the least expensive method of solid waste disposal. Major controversies have occurred in local areas because there were increasing quantities of solid wastes and decreasing areas for disposal of this material. Existing landfills were filling up, and the sites of new landfills, incinerators, and recycling centers were complicated by citizen complaints and the typical response of not placing a new landfill, incinerator, and/or recycling center in my backyard. The cost of taking the solid waste to areas away from municipalities has grown considerably and the amount of funds available for doing this has decreased sharply. The choosing of a facility site is complicated by a variety of issues: environmental and health risk to groundwater, air, and land; economic concerns related to the effect on property values and cost to the taxpayer; social issues related to aesthetics, community image, and current and future land use patterns; and political issues related to special interests, community groups, local elections, and who profits from the facility.

Landfill facilities, which are utilized for the concentration and containment of wastes and waste byproducts, are rapidly running out of room for the placement of new materials. Major concerns include insect and rodent problems, odor problems, leachate potentially surfacing or entering the groundwater supply, hazardous gases, heat-trapping gases with methane being 20 times as potent a greenhouse gas as carbon dioxide, and injuries and potential illness to the people working with the solid waste.

Best Practices in Disposal of Solid Waste in Municipal Landfills
- Involve the community and various community groups in the entire process of establishing a new landfill. Develop a detailed plan.
- Hire a competent, skilled landfill manager as a full-time position to oversee all activities and ensure all health and safety rules as well as environmental regulations are followed explicitly. This individual should have the authority to either institute immediate fines for anyone violating the rules or contact the proper governmental authorities to issue fines.
- Obtain and/or utilize a site that is isolated from all types of groundwater or surface water to keep any escaping leachate from entering these areas. Site selection should also include ensuring that there are no faults, wetlands, or floodplains which could have an effect on the landfill or the reverse.
- Establish a program to detect and prevent hazardous waste from being incorporated in the landfill site.
- Lead-based paints removed from residences in order to reduce exposure to children and adults can be put into a municipal solid waste landfill if the landfill is properly constructed and operated.
- Establish a vector control program to eliminate flies, mosquitoes, other insects, and rodents. Also control wild or domestic animals.
- Divert rainwater from crossing landfill sites. Provide storage for excess water and manage it according to the rules and regulations of the authorizing authority.
- Do not allow bulk or liquid waste which is not in a container to be deposited in the landfill. This will increase the amount of leachate and methane gas.
- Grade all landfill sites appropriately to prevent pooling.
- Waste tipping from trucks should be done in small areas only, and immediately worked on to compact it and cover it.
- Never deposit any type of waste, especially that which is biodegradable, in water.
- All open burning is absolutely banned. If fires occur in the solid waste, they must be put out because they will cause air contamination and may create voids in the soil which can become hazardous to the landfill operators.
- Geosynthetic clay liners must be used in all landfills. These liners are easy to install, have a low permeability, and self-repair when rips or holes occur. The liner is made of a thin layer of process clay, usually bentonite bonded to a geomembrane between two sheets of geotextile, which is a sheet material less pervious to liquid than a geomembrane. The material of the liner is long-term reliable, will not be affected by freeze and thaw cycles, and is usable for long-term containment of leachate.
- Utilize a leachate collection and removal system and treat the leachate before releasing it to a body of water. Leachate may be treated at a municipal wastewater treatment plant if it is close by and cost-effective.
- Capture gases produced from the decomposition of the solid waste, which may go on for many years after the landfill is closed. These gases include methane and carbon dioxide. The methane can be used as a source of energy. Establish effective gas testing programs.
- Compact and cover wastes as the landfill operation proceeds during the day and put on a final cover at night to prevent problems of insects and rodents as well as odors and aesthetic problems. Ensure that there are no cavities in the ground which will cause heavy equipment to turn over and injure workers.
- Establish groundwater testing wells to determine if contaminants are getting into the water supply and in what concentrations. These wells will be used for many years after the landfill has been covered up and closed.

- Immediately clean up all spills or improper dumping of material.
- Provide evidence of ability to fund all environmental protection efforts during the operation of the landfill and after closure. It may be necessary to post a bond to ensure that this will occur.
- Determine the quantity of waste materials of each category, its composition, its moisture content, and amount of biodegradable material to help the landfill operator choose the most appropriate means of disposal.
- Choose a landfill site that is easily accessible to the road system but away from dense population areas. Use land that has very little value for other purposes.
- Keep accurate records of all waste disposed including information about content, who delivered it, where it came from, and any potentially hazardous waste which may have been mixed in with the non-hazardous waste.
- Prevent illegal dumping, unauthorized use of the facility, and illegal vehicular traffic. Control access to the site.
- Provide a final cover when closing the landfill of at least 18 inches of earth compacted. The closing must be certified by an independent certified engineer.
- After closure there will be settling and additional dirt will be needed to level out the property. It is preferable if gentle slopes are introduced to the surface. Over time the land may be reused for numerous purposes.
- The authorizing authority should make frequent inspections of the landfill site while in operation and then when necessary after closure. They should obtain the results of testing in the groundwater wells and make whatever necessary recommendations should be followed.

Best Practices for Disposal of Household Appliances Using Refrigerants
- All refrigerants must be safely removed from appliances prior to disposal.
- The appliances must be accompanied by signed statements that the refrigerants have been removed, and these records must be kept to verify that these items will not cause problems of air pollution. If the refrigerant has not been removed, then the landfill manager or the recycling manager must oversee proper removal and disposal of the refrigerant by experienced technicians and the records must be kept of these actions.
- Utilize the list of EPA-certified refrigerant reclaimers to assist in the proper removal of the refrigerant.

CONSTRUCTION AND DEMOLITION WASTE LANDFILLS

(See endnotes 26, 77)

Construction and demolition waste comes from the construction, renovation, repair, and demolition of all types of structures, roads, and bridges. It is mostly composed of various types of wood, asphalt, drywall, masonry, metals, earth, shingles, plastics, insulation, paper, and cardboard. It may also contain excess construction materials and containers such as adhesives, paint, and roofing cement cans. It may contain waste oils, grease, other fluids, batteries, appliances, carpet, and treated wood. The bulky material of construction and demolition waste allows moisture to easily infiltrate and therefore be available for production of hydrogen sulfide gas. Where the gypsum drywall is reduced to smaller pieces, there is an increased surface which creates greater amounts of hydrogen sulfide gas.

The composition of any given project varies tremendously from other projects. Most of the construction and demolition wastes are sent to either municipal solid waste landfills or specialized construction and demolition landfills. Some are dumped illegally or go to incinerators.

Two very serious problems are leachate and the production of landfill gases, especially hydrogen sulfide. The leachate may contain hazardous substances such as 1,2-dichloroethane, methylene chloride, cadmium, iron, lead, manganese, and excess total dissolved solids. Hydrogen sulfide can be produced in substantial quantities from landfills that accept gypsum drywall. Hydrogen sulfide is a poisonous, irritating, flammable, colorless gas which may cause substantial odor problems or damage the landfill gas collection system. It is also very harmful to the health of humans. The scope of the problem depends on the duration of exposure and concentration, and whether or not the individual already has upper respiratory problems of either an acute or chronic nature. It also may depend on the concentration. It can paralyze the olfactory system, cause respiratory distress, and at high enough levels cause death. The gypsum is a source of reducible sulfur in the landfill and combined with moisture, organic matter, anaerobic conditions, a pH of 6–9, and a variety of temperatures will produce hydrogen sulfide gas. (See endnote 78.)

Another serious problem in both construction and demolition landfills and municipal solid waste landfills is subsurface heating events such as chemical reactions, including spontaneous combustion, hot waste being deposited, or oxidation of cellulose and plastics which form peroxides, etc. This can lead to fires, explosions from gases, odors, smoke, toxic emissions, contamination of ground or surface water, and damage to various parts of the landfill management system.

Best Practices for Construction and Demolition Waste Landfills
- Use construction and demolition waste minimization practices. (See Best Practices Construction and Demolition Waste Disposal above.)
- See Solid Waste Disposal for Municipal Landfills above for information on siting, design and construction, operation, groundwater problems, leachate, gases, etc.
- Use appropriate soil covering techniques on a daily basis to reduce the potential for stormwater infiltration into the waste material.
- Collect and remove leachate to be treated before release into any body of water.
- Conduct frequent tests in the air and in the leachate to determine levels of hydrogen sulfide or reducible sulfates. The leachate can be a substantial source of hydrogen sulfide.
- Divert drywall from construction and demolition waste landfills. Recycle into either agricultural soil amendments or new drywall.
- Use moisture control techniques which include minimizing the landfill working face, grading it properly, and covering it at the completion of each operation as well as at the end of the day and weekly.
- Dispose of materials containing drywall at the highest level of the landfill, and immediately cover with other waste and soil cover.
- Use various soil amendments in the cover material such as ammonium nitrate fertilizer, coal ash, compost, concrete fines (less than 2.5 cm), Fuller's earth which is a clay-like substance, lime, steel tire shreds, and metallic filter materials.
- Finally, the landfill site should have a low permeability layer of material. Make it erosion resistant with a substantial vegetative thick growth on top of the low permeability area.
- Do not allow any hot wastes to be placed within the landfill.
- Have a written evaluation of all wastes being deposited and look for those which are incompatible and may lead to hot events. Separate or ban them from the landfill.
- Check the temperature of the landfill in various places and make sure that it is not increasing substantially.
- Determine where oxygen intrusion is occurring and limit it.
- If a fire breaks out, immediately put it out in an appropriate manner.
- Conduct regular inspections and look for unusual or rapid settlement, vegetative cover which is stressed, smoke and steam, new odors, and excessive amounts of liquid in gas extraction wells.
- Analyze the landfill gas for methane, levels of oxygen, carbon monoxide, volatile organic compounds, hydrogen, gas pressure, temperature, and whether or not there is hot leachate. These could indicate a hot environment.
- Use a thick layer of low permeable soil as a cover to help suppress hot events. It may also be necessary to use a foam, liquid, or inert gas suppressant. (See endnote 79.)

SUB-PROBLEMS INCLUDING LEADING TO IMPAIRMENT AND BEST PRACTICES FOR SOLID WASTE DISPOSAL USING MUNICIPAL INCINERATORS

(See endnotes 80, 81, 82, 83)

Incinerators reduce the volume of materials in a controlled manner and reduce toxicity and potential sources of infection by burning waste at extremely high temperatures. Major problems relate to noise at the facility, dust and dirt, contamination of the ambient air, land, and water, and potential for injuries and disease to the people working with the solid waste. Energy can be created as a byproduct of the incinerator process.

There are three different types of technologies used in the burning of municipal solid waste. They are mass burn facilities, modular systems, and refuse-derived fuel systems. Mass burn facilities are the most common ones used in the United States, even though presorting and recycling are highly recommended to reduce waste quantities and reuse materials to save the use of raw materials and protect the environment. The burning occurs in a single combustion chamber with excess air which helps mix the materials and the turbulence allows the air and burning to reach all parts of the waste. The grate vibrates to help create further turbulence. The heat within the process converts water to steam and the steam is used to help generate electricity. Flue gases and ash are produced. Flue gas may contain carbon dioxide, nitrogen oxides, sulfur oxides, and a variety of gases which may be produced from reactions with chlorine as well as organic compounds and inorganic compounds.

The ash is taken to a landfill. The air leaving the system carries numerous particulates and must go through a high-efficiency baghouse filtering system. Some 99% of the particulates are captured there. The fly ash particles fall into hoppers and are transported through closed conveyor systems, wetted down and mixed with the bottom ash from the grate. Scrap metals are removed at this end of the process also. The ash is then hauled in leak-proof trucks to a landfill which is designed to protect the groundwater from the landfill contents.

The ash is about 15–25% of the weight of the municipal solid waste that was burnt and 5–15% of the volume. The fly ash, which is the fine particles, is about 10–20% by weight (the numbers vary with the source) of the total ash and may contain numerous contaminants. The bottom ash, which is 80–90% of the weight, contains basically, silica, calcium, iron oxide, and aluminum oxide.

Modular systems burn unprocessed, mixed municipal solid waste in the same manner as the mass burn facilities but are much smaller and can be moved from site to site.

Refuse-derived fuel systems best operate through an air-fed gasification system which has been identified by the US EPA and Department of Energy. In this process, the materials that contain organic carbons are broken up into compounds by their constituent parts. This occurs under very high temperatures of 600–800°C and in an oxygen-starved environment. It can also be carried out under air-fed gasification systems at 800–1800°C, or in plasma or plasma arc systems at 2000–2800°C. Syngas is produced and it is composed mainly of carbon monoxide, carbon dioxide, methane, and water vapor. Where an air-fed system is used, nitrogen gas will also be produced. The syngas is then burnt in a heat recovery boiler which makes steam and then turns a turbine to make electricity. (See endnote 84.)

Best Practices in Disposal of Solid Waste in Incinerators
- See various waste minimization practices above.
- Utilize a variety of pretreatment sorting processes to remove recyclables and to reduce the size of the waste materials. Homogenize the waste to get better burning.
- Obtain a permit for the incinerator facility from the authorizing authority which will cover design review, safe operating procedures, maintenance procedures, operator certification, and inventory and record-keeping.

- Obey all rules and regulations concerning air quality and potential contamination of land and water established by state and local authorities.
- Locate the municipal solid waste incineration plant in areas where there is typically medium or heavy industry, but close enough to collection areas to be economical and reduce pollutants of vehicles traveling long distances to dump the waste. Keep away from residential areas, athletic fields, shopping centers, etc.
- Determine the amount, composition, moisture level, etc. of the waste on a daily, weekly, and seasonal basis and allow room for expansion in the determination of the size of the incineration plant.
- The design and operation of the incinerator should be such that it always achieves desired temperatures and time necessary for burning of all wastes, destroys all pathogens and hazards within the waste, reduces all types of ash, avoids internal damage from the heat and sudden bursts of additional heat from substances being destroyed, and uses a minimum amount of fuel.
- Waste destruction efficiency should be greater than 90% by weight.
- Fully preheat the incinerator system before introducing solid wastes for destruction.
- Separate out from the normal solid waste very wet materials which must then have moisture reduction, and high heat fuels such as plastics, paper, and textiles which may spike the temperature within the unit. Supplemental fuel may be necessary to reduce moisture.
- Provide continuous combustion and emission monitoring through stack tests, in-stack monitoring, and environmental monitoring of the air, soil, food, and water, and also use observational techniques.
- Use appropriate maintenance procedures such as hourly removal of ash; daily evaluation of the temperature within the incinerator as well as the pollution control equipment; weekly lubrication of all latches, hinges, wheels, and other movable parts of equipment; monthly inspections of all parts of the incinerator and chimney and make repairs as needed; and every 6 months inspect the hot external surfaces and paint with a high-temperature paint as needed.
- Provide appropriate personal protective equipment such as a facemask, heavy-duty gloves, proper footwear, and an apron.
- Allow ash to have a cool-down period of 3–5 hours before removal.
- Provide facility inspection by highly trained technicians on a regular basis to determine any potential problems which may be corrected immediately.
- Keep complete and accurate records of all activities, waste delivery, and composition, as well as information concerning any hazardous events which have occurred, results of the problem, and how it was handled and by whom.
- All incinerator facilities must be administered and supervised by operators with professional certification and training.
- Provide necessary roads for the heavy vehicles which will be coming to and going from the incinerator.
- Make the incinerator facility accessible to the part of the electrical grid where it can contribute electricity from the refuse burning process.
- Utilize the heat produced by the incineration plant within the facility and where feasible in other facilities.
- Have an appropriately established landfill for final disposal of all ashes and other unburned materials if not salvageable.
- Use air pollution control devices such as cyclones, baghouse filters, electrostatic precipitators, dry scrubbers, and wet scrubbers to remove air pollutants from the dirty gas stream. High-efficiency particle filters are used within the system to remove tiny ash particles.

SUB-PROBLEMS INCLUDING LEADING TO IMPAIRMENT AND BEST PRACTICES FOR SOLID WASTE DISPOSAL USING COMPOSTING

(See endnotes 18, 19, 20, 21)

Composting is the enhancement of a natural process used to break down organic wastes from raw, putrescible, organic material to a stabilized product, carbon dioxide, and water. Carbon and nitrogen are the two most important elements of the process. Carbon is the basic energy source for the microorganisms and nitrogen is necessary for microbial population growth. Too much nitrogen turns into ammonia gas and other nitrogen compounds and may cause odors. Moisture is essential to the process because most of the microbiological activities take place in a thin film on the surface of the material. However, too little moisture (under 40–45%) slows down the process and too much moisture (over 55%) reduces the oxygen level. Oxygen is essential for proper decomposition by aerobic organisms, whereas too little oxygen can lead to growth of anaerobic microorganisms in part of the mix and cause odors. Heat is a byproduct of the decomposition of the organic matter and needs to be kept in the range of 113–138°F. The compost pile must be turned periodically. The factors above lead to different types of composting systems.

Composting is the means of disposal for about 27% of the US municipal solid waste. When done properly, the volume of the material is reduced substantially and the remainder may be used for numerous purposes including erosion control, improving topsoil, biofiltering, etc. The organic material may consist of leaves, grass, other yard trimmings, soiled paper, garbage, and unfortunately other wastes including tennis balls, plastic bags, and street sweeping material which may contain used oil, asbestos, lead, cans, etc. These items cannot be composted and may interfere with the process. Further, people tend to throw hazardous items and materials into the solid waste and may readily do so into the waste which is to be composted. This creates further problems and may make the compost hazardous and therefore not able to be used. Municipal solid waste is made up of materials of various sizes, quantities of moisture, and nutrient content. Major concerns are that the compost may be a source of odors in the community and a potential source of food and harborage for insects and rodents. Worker safety is a concern since the physical contaminants may contain inorganic and organic chemicals which can cause short-term or long-term health problems.

BEST PRACTICES IN COMPOSTING

- Organic materials to be composted must be separated from non-organics which may be recyclable; reduced in size to increase the surface area which allows for greater biological activity in the material: and mixed thoroughly to get an even distribution of nutrients, moisture, and oxygen. The first separation should be done by the citizens putting out the waste.
- The organic materials for composting should be taken to a centralized separation system which will further recover recyclable or combustible materials, and the visible inert materials such as plastic and glass, and remove potentially hazardous materials that are visible. Screening mechanisms, magnetic separation, and manual separation may be necessary to accomplish this task. This is followed by size reduction of all materials before the compost process proceeds.
- Use windrow composting when a system of natural convection and diffusion of oxygen works best for the community. The pile size, porosity, and frequency of turning to manage temperature control and oxygen levels can be altered as necessary to prevent anaerobic digestion. The pile is shaped like a haystack and is usually 300 feet or more long, 4.5–9 feet high, and 9–18 feet wide. This system must be placed on a layer of impermeable soil to prevent contamination of the groundwater. Forced aeration can be used if necessary, but this adds to the cost. This process is usually conducted outside but cannot be covered by a roof.

- Use an aerated static pile when it is best for the community. It is shaped like the windrow, but it is not mechanically agitated. Process control occurs by use of pressure and/or vacuum-induced aeration using temperature or oxygen as the control variable. Often a layer of wood chips is placed below the pile and also on top of the pile. This process is usually conducted outside but can be covered by a roof.
- Use an aerated static pile when it is best for the community. The pile is usually 12 feet high and can be placed in a silo or other large building. The original material is fed into the reactor at the top and is distributed evenly. It flows by gravity to the unloading mechanism at the bottom. Pressure aeration causing the airflow to be opposite to the downward material flow is the means of process control.
- Use a horizontal reactor when it is best for the community. The horizontal reactor avoids the potential problems of high temperature, oxygen levels, and moisture gradients of the vertical reactor by having a short airflow pathway. The agitation system to turn the material is in continuous mode with materials being shredded to expose new surfaces for decomposition.
- Develop a quality assurance program for the compost by utilizing a proper random sample collection system using multiple samples of sufficient numbers to determine the mean concentration of a specific contaminant in the feedstock and the finished compost product. Metal concentrations, hazardous chemical concentrations, plastic concentrations, and microorganism concentrations and types should be noted. The frequency of sampling and the timing of sampling are essential because different feedstock may be found at different times of the day, week, season, and year. These samples should be mixed thoroughly and homogenized prior to the removal of subsamples for evaluation of different contaminants. Samples should be collected only in clean plastic containers, refrigerated but not frozen, and delivered promptly to the analytic laboratory. Record keeping is absolutely essential at the processing plant and at the laboratory.
- Establish a market to use the compost after it has been created.

SUB-PROBLEMS INCLUDING LEADING TO IMPAIRMENT AND BEST PRACTICES FOR NON-HAZARDOUS WASTE DISPOSAL AND HAZARDOUS WASTE DISPOSAL USING UNDERGROUND INJECTION WELLS

(See endnotes 61, 64)

An underground injection well is a dug or mechanically constructed hole in the ground that is deeper than it is wide and is used for depositing fluid including hazardous wastes and also carbon dioxide into isolated areas typically far below the Earth's surface. This means of disposal started in the 1930s and was used to remove salt water which was present during the drilling for crude oil. This process became even more common in the 1950s when chemical companies began depositing industrial waste into deep wells. In 2010, the EPA finished issuing regulations for geological sequestration of carbon dioxide.

There are at least six ways in which fluids injected underground in wells can migrate into the underground source of drinking water. They are:

1. Migration through the injection well casing
2. Migration through the annulus (void between concentric cylinders) between the casing and the drilled hole
3. Migration from the injection horizon through the confining zone
4. Vertical migration through wells which have been improperly abandoned or completed
5. Lateral migration from an injection zone into a protected area
6. Injection of the fluids above the underground source of drinking water or directly into it

Over half of the liquid hazardous waste and a large amount of the non-hazardous industrial liquid waste in the United States are put into underground injection wells. There are six types of underground injection wells that are regulated to protect the groundwater supply. They are:

- *Class I*—Industrial and municipal waste disposal wells (See endnote 60)
- *Class II*—Oil and gas-related wells
- *Class III*—Mining wells
- *Class IV*—Shallow hazardous and radioactive injection wells
- *Class V*—Shallow injection wells which do not fit in the other four categories and may either be hazardous or non-hazardous
- *Class VI*—Geological sequestration wells

DESCRIPTION OF UNDERGROUND INJECTION WELLS

Class I Underground Injection Wells

Class I underground injection wells, of which there are 523 in the United States, are exempt from any bans related to the Hazardous and Solid Waste Amendments to the RCRA if it can be demonstrated that the hazardous constituents of the waste will not migrate underground from the disposal site for 10,000 years or longer or until the wastewater becomes non-hazardous. The fluids for disposal are injected into brine-saturated formations thousands of feet below the surface. However, the Class I wells may range from 1700 to 10,000 feet beneath the surface of the Earth. The area must be geologically stable and free of fractures or faults which would allow the contaminated fluid to enter the groundwater supply. These wells have a multilayer construction with redundant features and are constructed of corrosion-resistant material. The wells are under constant pressure and continuously monitored to determine if the system is working. According to the EPA, the current Class I wells are protected adequately and will not cause health problems for the human population or damage the environment.

Best Practices for Class I Underground Injection Wells
- Requires a permit prior to construction from the proper authorizing agency including information on casing and cementing, financial assurances, maximum operating pressure to avoid causing or increasing fractures that would allow the fluids to enter the underground safe drinking water supply, monitoring and reporting requirements, demonstration of mechanical integrity, pressure test when the system has been disturbed, and plans for plugging and abandonment of the well.
- All plugs for the well must be made of concrete.
- The area of review for non-hazardous injection wells is a minimum of 0.25 mile while that for hazardous waste injection wells is 2 miles.
- Do not inject substances between the outermost casing and the well bore.
- Casing in all hazardous waste injection wells must be cemented at the surface and extend from the surface through the underground safe drinking water supply.
- Mechanical integrity must be tested before the well goes into operation by using pressure tests, temperature variations, oxygen activation, or noise logs. Annually, a radioactive tracer survey must be conducted to show the integrity of the bottom of the well cement.
- There must be continuous monitoring of annulus pressure, injection pressure, flow rate, and volume of substances.
- Provide a hazardous waste injection well-written waste analysis plan.
- Use groundwater monitoring wells to make sure that the waste is not entering the groundwater supply.
- Maintain accurate logs of all actions and incidents and report problems to the US EPA.

Class II Underground Injection Wells

Class II underground injection wells, of which there are about 168,000 in the United States, are regulated either by the state or the US EPA. A technical review is mandatory to make sure that there will be no chance of contaminating drinking water sources. These wells are used for disposing of salt water, enhanced oil recovery, and hydrocarbon storage. Typically, 10 barrels of salt water are produced with each barrel of crude oil. Enhanced oil recovery wells use the salt water to push the oil into the oil-producing wells. Hydrocarbon storage wells are used to store crude oil and liquid hydrocarbons underground. The wells have a casing made of concrete and it is cemented from below the lowest drinking water strata to the surface. The amount of pressure applied in the process is limited by the US EPA to avoid any potential of groundwater contamination.

The Government Accountability Office report on the protection of underground sources of drinking water from injection of fluids associated with oil and gas production concludes that there are problems that need to be addressed to protect the groundwater supply. They found that the safeguards against contamination of underground sources of drinking water do not include emerging problems such as seismic activity and too much pressure in geological formations causing outbreaks of fluids on the surface. Part of this is because of a lack of financial resources to carry out the necessary evaluations, and also the rulemaking process is extremely difficult and time-consuming. Even though considerable data are collected on the Class II program, the data are not complete or comparable from site to site. (See endnote 68.)

Seismic Activity Caused by Humans (Induced Seismicity)
(See endnotes 69, 70)

Fracking or horizontal drilling and a high volume hydraulic fracturing not only has helped produce more oil and natural gas but also has created a substantial problem of how to dispose of the wastewater including injected fluids deep into the ground. There is a concern that the high volume fracking process and the injection of these large amounts of wastewater and other fluids are causing or increasing seismic activity. Some of the earthquakes have been at a magnitude of 4.0–4.8 and have caused damage to structures on the surface of the ground. Recent research indicates that the majority of wells used for hydraulic fracturing cause micro-earthquakes. Contributing to this seismicity are three factors: a buildup of enough pressure from disposal activities to cause the problem to occur; a fault of concern which is set to create problems in a critically stressed area; and a means for the pressure to move from the injection area to the fault area. Pressure can be transmitted for many miles through fractures. It appears that in certain instances the magnitude of the earthquakes increases with increased exposure to the disposal of wastewater. Unfortunately, many of these areas have natural fractures.

Since the 1920s, it has been known that pumping of any kind of fluids in or out of the earth, as well as the filling of a reservoir, can cause seismic activity. Although there is a need for considerably more research concerning the amount and rate of waste fluid injected into the deep wells as well as other operational aspects, there is a potential for creating fractures in the rock formations and increasing seismic activity.

Best Practices to Prevent or Mitigate Seismic Activity Caused by Humans
- Do not wait for seismic activity to start before taking action. Prepare the well and its operation as if seismic activity will become a problem.
- Use basic petroleum engineering practices and geology and geophysical information to provide a better understanding of the presence and nature of faults as part of site assessment. Evaluate all existing data from the region and locality to find active subsurface stresses and reservoir pathways.
- Determine the potential pressure buildup in any proposed storage disposal area underground.

- Fill in holes in the data with additional geoscience experts conducting whatever they consider to be appropriate studies.
- Measure the initial pressure at the bottom hole and determine if additional pressures will cause seismic activity.
- Review the data from operating wells in the area to determine the amount of pressure that is present, any informational flow behavior in the injection zone, and any reports of seismic activity. Where there have been suspected problems to frequent pressure tests, determine if higher pressure causes seismic activity.
- Inject wastewater intermittently instead of continuously to allow pressure buildup to be relieved.
- Separate multiple injection wells by substantial distances to allow pressure distribution.
- Increase monitoring where problems exist, so that pressure and quantity of injected material can be identified quickly and action can be taken to reduce them.

Best Practices for Class II Underground Injection Wells
- Requires a permit prior to construction from the proper authorizing agency including information on casing and cementing, financial assurances, maximum operating pressure to avoid causing or increasing fractures that would allow the fluids to enter the underground safe drinking water supply, monitoring and reporting requirements, demonstration of mechanical integrity, pressure test when the system has been disturbed, and appropriate plans for plugging and abandonment of the well.
- Use mechanical integrity tests at least once every 5 years for the Class II injection well. Advise the US EPA at least 30 days prior to the test.
- Use a review area which is a minimum of 0.25-mile radius from the well.
- Conduct monitoring of injection pressure, flow rate, and volume regularly and as needed on a continuous basis.
- Sample the injection fluid and determine if it meets the specifications established for the operation.
- Provide the appropriate authority with a record of plugging and abandonment of wells within 60 days of the operation.

Class III Underground Injection Wells

Class III underground injection wells are regulated by an authorized agency typically through the state. The wells are used for mineral extraction of salts, sulfur, and extracting copper, gold, and uranium. The well only allows injected fluids to go into an authorized injection zone and does not allow it to migrate to other areas. Monitoring wells must be placed around the body of ore to detect any horizontal migration of the solutions used in the mining process.

Best Practices for Class III Underground Injection Wells
- Requires a permit prior to construction from the proper authorizing agency including information on casing and cementing, financial assurances, maximum operating pressure to avoid causing or increasing fractures that would allow the fluids to enter the underground safe drinking water supply, monitoring and reporting requirements, demonstration of mechanical integrity, pressure test when the system has been disturbed, and plans for plugging and abandonment of the well.
- Sample the injection fluid and make sure that it meets the appropriate specifications.
- Observe the injection pressure, flow rate, or volume at least twice a month and record the information.
- Perform a mechanical integrity test on the well at least once every 5 years. Notify the authorizing agency 30 days prior to the event.

- Determine the results of the testing of required groundwater monitoring wells and advise the US EPA if problems are occurring.
- Notify the authorizing agency 45 days before starting the plugging and abandonment of a well.

Class IV Underground Injection Wells

Class IV underground injection wells have been determined to be a significant threat to human health and the environment by the US EPA and as such have been banned for use. These were wells where dangerous waste had been introduced either into or above the drinking water strata. They are still being used where it is necessary to clean up existing contamination, but new wells constructed and operated for the purpose of disposal of hazardous waste are prohibited.

Best Practices for Class IV Underground Injection Wells
- Class IV injection wells that are part of an official cleanup need to be approved by the US EPA or the state.
- Use where appropriate in situ bioremediation of groundwater techniques. See information below concerning how these techniques work and their effectiveness.

Class V Underground Injection Wells

Class V underground injection wells have been utilized in various areas of the country, especially where public wastewater treatment has not existed. These wells may range from cesspools to complex injection wells. There may be as many as 1.5 million Class V wells in the United States, some of which are of little consequence and others which can cause serious potential problems depending on the well construction, the local geology, closeness to local water supplies, and the types of waste fluids being injected. Some of the worst shallow injection wells are those used for motor vehicle waste disposal, large size cesspools for multiple dwellings or a business, and stormwater drainage wells. Currently, new large capacity cesspools and new motor vehicle waste disposal wells have been banned by the US EPA. The old ones have to be closed down appropriately in some states and in other states the automobile business can get a permit providing the waste is not contaminating the groundwater supply. (See endnote 62.) For the large capacity cesspool, the individual cannot contaminate the groundwater supply if the state allows a special permit for the cesspool use. (See endnote 63.)

Federal funding for the Underground Injection Control program has been diminished and as a result less attention is being focused on this serious means of contaminating the groundwater supply. Much of the program is carried out by certain states and they are given federal mandates to do things without much funding, which is approximately $11 million yearly for the entire country, to carry out the necessary functions.

There are 32 different types of Class V injection wells in seven different categories. The categories are:

1. Drainage wells which can range from agricultural drainage to stormwater drainage to industrial drainage or other special purposes. Risk for people and the environment varies from low to high, with the high level relating to pesticides, nutrients, pathogens, and heavy metals.
2. Geothermal reinjection wells of several types which have potential moderate hazards from pH problems, hot geothermal problems, and the materials within the solution.
3. Domestic wastewater disposal wells which range from untreated sewage to septic systems to the effluent of public wastewater treatment plants which has gone through secondary or tertiary treatment. Risk ranges from low to high depending on the amount of raw sewage and treatment.
4. Mineral and fuel recovery wells used for mining, fuel recovery, cleaning of air scrubbers, and water softener regeneration. Hazards rank from low to moderately high

depending on the chemicals that are present in the fluid and whether or not the water is acidic.

5. Industrial, commercial, and utility waste wells. Hazards rank from low to high depending on the amount of heat, suspended solids, heavy metals, solvents, and cleaning agents.

6. Recharge wells which are used for recharging an aquifer by injecting fresh water to keep salt water intrusion from occurring and controlling settling of the ground. Aquifer storage and recovery wells are used to store water underground for some use in the future. Hazards rank from low to high depending on the quality of water used on the recharging of the aquifer.

7. Miscellaneous wells which include those for experimenting on unproven technologies and for remediation of aquifers, abandoned drinking water wells now used for sewage disposal, and any other unspecified Class V wells. The hazards rank from low to high depending on the nutrients which are being introduced as well as levels of chemicals and microorganisms.

Best Practices for Class V Underground Injection Wells (See endnote 66)
- Do not discharge any wastewater into stormwater infiltration systems.
- Remove sediment from stormwater prior to the injection well. The sediment may clog the infiltration system and include metals, oil and grease, pesticides, phosphorus, and other contaminants.
- Separate potential contaminants from stormwater by using curbing, containment dikes, or covering materials to prevent leaching of the waste material or raw or finished products from rain which would cause contaminated runoff.
- Fuel vehicles and equipment in areas where a spill or overfill will not be washed into runoff areas from heavy rains. Conduct routine and required maintenance of these vehicles and equipment in these areas.
- Wash all equipment and vehicles in areas where the wash water is self-contained and then treated before release.
- Do not discharge any wastewater directly into the groundwater or below the highest seasonal groundwater table.
- Use spill control procedures to prevent wastewater from going into the groundwater.
- Separate all Class V wells from water supplies, wetlands, surface or subsurface drains, etc.
- The minimum depth between the Class V well measured from the base of the filter sand and stone beneath it to the seasonal high groundwater table is 2 feet for all stormwater wells, 4 feet for other types of Class V wells, and where percolation rates are 2 minutes or less per inch then the well should not be used at all because of potential contamination of the groundwater.
- The minimum distance between the base of the Class V well and bedrock should be 5 feet.
- All closures of these wells must be done in accordance with the regulations of the authorizing authority.

Class VI Underground Injection Wells
(See endnotes 65, 67)

Class VI underground injection wells are used to inject carbon dioxide into areas where oil and gas have been removed, deep aquifers containing saline, and other deep formations for the storage and disposal of carbon dioxide to mitigate its effects on global warming. There are numerous concerns that need to be answered before Class VI injection wells can actually be put into operation in a meaningful way. These technical and regulatory issues include but are not limited to: who owns the injection zones; the cost of injection; the problem of carbon dioxide migrating underground; the long-term effects that carbon dioxide may have on the area in which it has been injected; etc.

Best Practices for Class VI Underground Injection Wells

- Apply for a permit from the authorizing authority to establish a Class VI injection well or modify an existing injection well to become a Class VI. Include a map with cross-sections, list location and description of all injection wells whether producing or closed within the review area, and information on geological structure, hydrogeological properties of the site, suspected faults and fractures, seismic activity and history, depth of well, confining zone, and injection zone, and other data required by the authorizing authority. Include information on the average and maximum injection pressure, sources, and analysis of the chemical and physical characteristics of the carbon dioxide stream.
- Install the injection well on a site that has an injection zone of adequate (areal) space or surface with a proper amount of thickness, porosity, and permeability to accept the entire volume of carbon dioxide over the course of the life of the well. The confining zone must be free of any faults or fractures or permeability that would allow the carbon dioxide to move through it to another area. There will be an area of review and corrective action around the well in which testing will be necessary to make sure that no problems are created by the injection well and its contents.
- All Class VI wells must be constructed to: prevent the movement of fluids into or between drinking water strata; have structural strength for the cement and other materials which is compatible with the material being used in construction; be compatible with the carbon dioxide and any other associated fluids; and last the life of the geological sequestration.
- Provide to the authorizing authority the following information: depth to the injection zone as well as injection pressure, external pressure, and internal pressure; hole size and size and grade of all casing and cement as well as the extent of the casing above the ground and under the drinking water strata; corrosiveness of the carbon dioxide and associated fluids; quantity, chemical composition, and temperature of the carbon dioxide stream as well as the down hole temperatures; the tubing, its size and composition, packer materials and their compatibility with the carbon dioxide stream and other fluids.
- Analyze the carbon dioxide stream frequently to determine chemical composition and physical characteristics.
- Monitor continuously potential corrosion of injection well materials, injection pressure, rate, volume, and the annulus pressure.
- Monitor groundwater quality and any geochemical changes above the confining zone.
- Establish a well plugging plan and have it approved before the well is plugged.
- Prepare a plan for the postinjection care of the site including postinjection monitoring on a regular basis.
- Provide the authorizing authority with an emergency and remedial response plan in the event that the movement of the injected fluids or associated fluids will endanger the drinking water strata. Immediately cease all injection.

IN SITU BIOREMEDIATION OF GROUNDWATER

(See endnotes 71, 72)

Bioremediation is a technology which modifies physical, chemical, biochemical, or microbiological environmental conditions to help them destroy or detoxify organic and inorganic matter in various environmental media. The process works above ground, underground, and in materials or water. Indigenous bacterial groups are helped by stimulating them (biostimulation) through various amendments or certain subgroups which are introduced (bioaugmentation) to treat hazardous materials or waste, to make them less hazardous or non-hazardous. A variety of amendments may be added depending on the contaminant, but the problem appears to be the distribution of the amendments in order to clean up and remove the contaminant.

Aerobic bioremediation effectively treats non-halogenated organic compounds. Additional oxygen can be added to help move the process forward more rapidly. Anaerobic oxidative bioremediation uses electron receptors such as nitrates or sulfates for the microbial metabolic oxidation of the substance. It works well on halogenated organic compounds, especially chlorinated solvents. Hexavalent chromium is reduced to trivalent chromium.

Cometabolism is a process where microorganisms use a compound as an energy source while also producing an enzyme that can chemically transform another compound. The microorganisms can then degrade certain contaminants. This is particularly effective in low contaminant concentrations.

General Best Practices for Underground Injection Wells
- Consider all geological information prior to applying for a permit for an underground injection well. This includes: the geological and hydrological characteristics of the zone of confinement (this is the area that prevents the material injected into the ground from migrating to other areas); the potential for the movement of the fluid from the injection zone to other areas; the number of other wells penetrating the injection zone and adding more waste material to this area; the existence of fractures and faults and the potential for fracturing to occur in the confining zone; and the corrosive characteristics of the fluid and the potential for penetrating the confining zone.
- Construct and maintain the integrity of the well casing to prevent migration of fluids into the ground and into the drinking water strata. Cement the casing properly to the ground and sidewalls of the hole which the well goes through. The casing must go into the impermeable area beneath the drinking water strata and the cement must be attached to the casing below the drinking water strata. Determine the casing size, weight, composition, and protective coating.
- Determine the type of tubing as well as its composition, coating, and ability to withstand bursting from pressure or the waste material disposal.
- Predetermine the mechanical integrity of the well and evaluate it periodically to make sure that the injection pressure and the fluid volumes have not damaged the system and interfered with the mechanical integrity of the unit.
- Continuously monitor injection pressure, volume of material being injected, type of material, rate of injection, and the annulus pressure.
- Where applicable, continuously monitor drinking water wells to determine if they are becoming contaminated.
- Never use a drinking water well for disposal of any type of waste even if the well has run dry. The drinking water well is in an aquifer which may be renewed from various sources and the waste will contaminate any other drinking water wells which will be utilizing this aquifer.
- Develop in advance a plugging and closing plan which meets the specifications of the authorizing authority.

STATEMENT OF PROBLEM AND SPECIAL INFORMATION FOR HAZARDOUS MATERIALS

(See endnotes 45, 48, 50, 56; Chapter 2, "Air Quality (Outdoor [Ambient] and Indoor)"; Chapter 3, "Built Environment—Healthy Homes and Healthy Communities"; Chapter 5, "Environmental Health Emergencies, Disasters, and Terrorism"; and Chapter 9, "Insect Control, Rodent Control, and Pesticides")

The chemical industry was founded in most countries in small towns close to bodies of water especially rivers or the various coastlines. This gave them access to raw materials, a substantial amount of water needed for processing, cheap transportation, needed energy supplies, crude oil and gas, and necessary labor. The body of water also became the source of inexpensive waste disposal which contributed unusual levels of toxins to the air, water, and land. The mineral industry became more significant in the production of appropriate minerals for the production of chemicals, which also led to substantial contamination of air, land, and water.

Chemical industry products are potentially hazardous at each stage of manufacturing, storage, transportation, use, and disposal. The chemicals may be solids, liquids, or gases and be flammable, explosive, corrosive, or toxic. The major task of industry, regulatory agencies, academic institutions, communities, and others is to prevent hazardous materials from becoming hazardous waste and prevent or mitigate the effects of accidental release of the hazardous materials as well as the protection of workers, local communities, air, land, and water.

Depending on the quantity released for human and ecosystem exposure and the type of chemical, any accidental release or planned disposal may cause catastrophic results. The chance for hazardous chemical release is compounded by the very nature of the manufacturing of chemicals because it involves the conversion of organic and inorganic raw materials or intermediaries to produce wanted chemical substances and this takes a variety of different processes to achieve. In addition, there are many additional substances produced, some of which may be useful but still hazardous, and others which may simply constitute additional hazards.

Each time there is a disaster because of a chemical accident of some form which may affect large numbers of people causing illness or death, the question is why we allow such hazardous materials to be produced and transported through our country. The answer is quite simple. The chemical industry in this country manufactures over 70,000 different products which are absolutely necessary for the economic well-being and security of the United States and the world as well as to protect the health of the population. The chemical industry has nearly 1 million employees and generates over $700 billion each year. Although the chemical industry is very important to our modern lifestyle and economy, the production, storage, transportation, and use of these hazardous chemicals present a serious risk to the health and safety of people as well as animals, and potential destruction of various ecosystems in the environment.

Hazardous materials are chemicals used in agriculture, manufacturing, consumer goods, industries, weapons, medicine, and research. The chemicals fit into five major categories. They are basic chemicals, specialty chemicals, agricultural chemicals, pharmaceuticals, and consumer products. The raw materials used to make the chemicals are in part fossil fuels, metals, minerals, etc. Water used in the process becomes readily contaminated. The chemicals and products which are produced may include explosives, flammable materials, poisons, radioactive materials, carcinogens, teratogens, and endocrine disruptors. The raw materials, finished products, byproducts, and waste materials may cause serious injuries, long-lasting health effects, and death as well as damage to property. The chemicals may be found in various industries, households, schools, hospitals and other institutions, waste storage areas, and waste disposal areas, or released into the air, water, and land. In previous years, there has been substantial improper disposal of hazardous materials, which has led to a series of extremely dangerous sites where people became exposed to the chemicals and the interactions of the chemicals. Schools and other public buildings have been built on some of these sites.

SUB-PROBLEMS INCLUDING LEADING TO IMPAIRMENT AND BEST PRACTICES FOR PRODUCTION OF HAZARDOUS MATERIALS

(See endnotes 49, 53)

BASIC CHEMICALS

Companies make basic chemicals by utilizing natural resources such as natural gas, crude oil, minerals, and metals producing both organic and inorganic chemicals, polymers, petrochemicals, dyes and pigments, etc. About 70% of these chemicals are usually sold in large quantities to the chemical industry and are used to create materials for other industries such as the automotive, construction, and packaging sectors.

Petrochemicals can actually be made from coal and biomass as well as crude oil, with methanol as an example, being produced from oil and natural gas in the United States and Europe but from coal in China. Large quantities of ethene are transported in gaseous form through a pipeline in Europe to companies that make polyethene and other polymers. Ammonia is made from natural gases and some of the ammonia is used to make nitric acid. The ammonia and nitric acid are then turned into the fertilizer, ammonium nitrate. Hydrocarbons in crude oil are converted into more useful forms by the chemical industry.

Basic inorganic compounds are used for manufacturing as well as agriculture. They include chlorine, sodium hydroxide, sulfuric acid, nitric acid, hydrochloric acid, and phosphoric acid, and a variety of metals.

This industry has the potential for producing a considerable amount of environmental pollutants in a vast number of areas. The pollutants include:

- Air pollutants, typically nitrogen oxides, sulfur oxides, ammonia, acid mist, carbon monoxide, and carbon dioxide which come from the actual processing of the materials or fugitive emissions, heating devices, pipes with valves, pumps, and compressors, and during the transfer and storage process of raw materials.
- Liquid effluents, typically process and cooling water, stormwater, water from cleaning activities and purges, water from scrubbing gases, water involved in accidental releases or leaks from product storage tanks, condensates, and special effluents depending on the type of manufacturing operation.
- Solid and hazardous waste, which should be kept to an absolute minimum in a well-run and well-managed industrial operation. Typical waste from the manufacturing plants includes: waste oils which can be recycled; used catalysts; wastewater treatment sludge; dust from baghouses used for air pollution control; ash from boilers; and materials from filtration systems.

Best Practices in Control of Production and Use of Basic Chemicals
- Conduct a comprehensive survey of all production, storage, and use of feedstock and basic chemicals to determine potential sources of air pollution, types of releases to the air, and quantities.
- For highly volatile substances, provide a second set of containment vessels to capture release of the substances before the contaminants are released into the air.
- Use good housekeeping practices when transferring products from one system to another or one container to another and clean up all spills immediately.
- Enclose all areas where feedstock is delivered and processed and use either adsorption or absorption techniques in high-efficiency ventilation systems with appropriate bag filters or electrostatic precipitators to trap particulate matter and then pass the air through HEPA filters before allowing it to go to the outside.
- Process all air from energy consumption (this industry uses high levels of energy and therefore produces substantial pollutants) in the same manner as feedstock use stated above.
- Do not use any open vents on tank roofs to prevent escape of fugitive emissions.
- Establish a rigorous maintenance program including use of leak detection devices for all facets of the production and storage program including valves, phalanges, and fittings and carry out necessary work on a routine basis to prevent leakage.
- Use the latest technology and Best Practices and establish a flaring program and record the volumes of gas flared as well as the content. Conduct air pollution surveys downwind from the factory to determine if contaminants are still present in the air after flaring. Wherever possible, try to reuse and recycle the substances instead of burning them off.

- Minimize the use of water in processing and cleaning operations.
- Use primary feedstock with low sulfur content.
- Use closed-loop reactors and evaporators to eliminate process wastewater and to capture chemicals for recycling and reuse.
- Establish a water treatment system to precipitate solids and pretreat effluent to acceptable levels before discharge into other treatment systems.
- Determine if any of the impurities found in the solid and hazardous waste can be put through special reclaiming processes and then can be recycled for further use instead of disposal.
- Determine the impurities and contaminants found in the solid and hazardous waste and ensure that the waste is placed in special landfills which have appropriate linings to prevent leaching to water.

SPECIALTY CHEMICALS

(See endnotes 54, 55)

These chemicals are used for special customer needs and for different types of technical service. They include adhesives, sealants, plastics, additives, catalysts, paints, sealants, etc. These chemicals are used in a variety of industries such as textiles, paper, etc., and for direct consumer use such as in paints, varnishes, sealants, etc. An EPA report of chemical companies producing chemicals in the United States showed that in 2012 there were 1677 companies reporting from 4785 different sites on their production of about 7700 different chemicals. Further, in the document there is a discussion about the use of 82,000 different chemicals worldwide. The latter is a worldwide estimation of chemicals produced, while the former is a survey of a group of companies responding to the survey above a certain level of production of a given chemical in the United States. As can be seen, the numbers vary considerably based on the source.

The special chemicals industry has several very difficult problems to resolve. These include stringent safety and compliance mandates from governmental agencies: high cost of energy and feedstock; the aging of equipment and plants; and the global market. The industry keeps growing, which increases the risk of storing, processing, and using a variety of hazardous or toxic materials. It is necessary to manage very closely a variety of physical properties while keeping the process working on a continuous basis to avoid causing damage to sensitive equipment. This requires the proper type and numbers of experienced professionals to oversee the operation.

Best Practices in Control of Production and Use of Specialty Chemicals

Because of the vast number of companies at different sites making large numbers of different chemicals, many of which are specialty chemicals, it is beyond the scope of this book to get into each of the processes for discussion and the establishment of Best Practices for each of the specialty chemicals. However, all of the Best Practices listed in the general category and all other categories apply even if specifics are not presented.

AGRICULTURAL CHEMICALS

(See endnotes 58, 59, 73)

Agricultural chemicals include fertilizers and pesticides including herbicides. Agricultural chemicals present an immediate and long-term risk for poisoning and serious health effects to workers, the immediate community around the agricultural production, and consumers of agricultural products. The agricultural chemicals utilized may be specific for different crops and each type of pest that needs to be controlled. The agricultural chemicals are used in very large quantities. In fact, worldwide over $125 billion is spent on these chemicals. However, without them there would be a huge decrease in food production and worldwide hunger.

Fertilizers are nutrients needed by plants to grow and thrive. The raw materials used to make fertilizers come from nitrogen, phosphorus, and depending on the type of fertilizer, other substances. Ammonia is a primary nitrogen source for the production of fertilizers. Ammonia in high concentrations is extremely hazardous to people. To produce the most usable form of fertilizer, ammonium nitrate, potassium chloride, and ammonium phosphate are put together and blended. This process may produce substantial amounts of dust. Fertilizer is then supplied to farmers usually in large bags. To fill these bags, the substances needed are dumped in appropriate quantities into a large hopper and this then also becomes a means of contributing dust to the environment. Much of the nitrogen is not used in the soil but contributes to problems when runoff carries it into bodies of water.

Pesticides contain active ingredients that kill the pest and inert ingredients which are used as a binder to enhance the ability of the toxin to be sprayed or coated on the surface of either an unwanted plant or an insect or make the toxin available for consumption by insects or rodents. Many of the chemicals used are hazardous and the end product is hazardous to people and the environment. When the concentrated pesticide is diluted for use, there is potential hazard to the workers and also to the environment. The end-use of the pesticide if not properly applied in appropriate quantities to resolve the problem will contaminate the environment and may cause short-term or long-term health effects in people. Persistence of the pesticide in the environment may be very harmful for the long-term health of ecosystems and people.

Chemicals used for agriculture as well as the fertilizers and pesticides used on lawns are a serious potential water quality issue. Stormwater runoff causes not only erosion and the movement of sediment to bodies of water but also movement of hazardous chemicals into surface bodies of water or downward into the groundwater supply. These chemicals contaminate the air and the land. They may also contaminate raw food. The loss of raw chemical products during production, preparation, storage, and use is not only a hazardous materials problem but also an economic problem for the manufacturer and person applying the substance in the agricultural setting.

Best Practices in Control of Production and Use of Agricultural Chemicals

- Use various measuring instruments to determine the type, amount, and source of basic chemicals, other raw materials, and finished products which are escaping from the delivery and shipping areas, the storage areas and equipment, and the production areas, as well as the equipment being used to apply agricultural chemicals, fertilizers, and pesticides. The application of chemicals may contaminate the air, water, and land.
- Examine past records of the industrial plant and also of the industry in general to determine the major areas where these fugitive emissions or dust occur and where spills frequently occur.
- Do not store insecticides and herbicides together. Do not allow pesticides in any event to contaminate the production or storage areas for fertilizers.
- Clearly mark all pesticide storage areas with appropriate signs exhibited for: poisoning; no smoking; no eating; authorized personnel only; and limit access to this area to only specially trained people wearing appropriate personal protective equipment. Have a special team always on call for immediate cleanup of pesticide spills of any amount. Use appropriate fire extinguishers in the event of fires in pesticide areas used by highly trained individuals and immediately contact the fire department with all necessary facts and data concerning the hazard. Have a working plan with the fire department on how to handle this problem and with the local hospital on how to manage individuals who have been exposed to these chemicals. If any materials are exposed to pesticides, they must be discarded in an appropriate manner as a hazardous waste.
- See the "Agriculture" section in Chapter 2, "Air Quality (Outdoor [Ambient] and Indoor)" and Chapter 9, "Insect Control, Rodent Control, and Pesticides," for additional Best Practices in the use and storage of the chemicals.

Pharmaceuticals

(See endnotes 51, 52)

Pharmaceuticals provide active ingredients, some of which are hazardous, and also binders which are used to allow the active ingredients to work properly for specific diseases. This category includes diagnostic agents, prescription drugs, vaccines, vitamins, and over-the-counter medicines for both human and veterinary use. It is important in the design and operation of the facility to ensure the quality of the product while protecting the workers from potential health effects.

The United States pharmaceutical industry is one of the most highly regulated in the world. The Food and Drug Administration (FDA) enforces stringent government mandates and inspects all facilities on a periodic basis. They make determinations of the efficacy of drugs and the unwanted dangers involved in taking them. System failure at pharmaceutical production facilities must be dealt with in order to avoid serious consequences from the production and use of drugs. Many times there is an incomplete set of data presented due to isolation of parts of the process, redundancy, and confusion due to different reports with different conclusions, missing data, and no centralized control, and complaints assume a low priority at the company.

Best Practices in Control of Production and Use of Pharmaceuticals
- Conduct a risk assessment of all products or materials being handled by the workers and include toxicological data and permissible levels of exposure by quantity of the material, time exposed, and frequency.
- Establish an operational risk-based system at the manufacturer to identify hazards, prioritize seriousness, record accurately all data, determine the causes of the hazard, isolate the problem, and take immediate corrective action. The FDA has to approve all actions taken to eliminate the risk.
- Corporations should establish as a priority, policies and procedures supporting a risk management system with leadership at the top of the company evaluating and enforcing all necessary actions. Always use industry standard procedures and those approved by the FDA.
- There must be complete transparency in everything done to avoid hazards. This allows the regulators to thoroughly understand the conditions occurring and give appropriate advice on corrective actions. Regulators by law must protect business secrets involving processes and products.
- Utilize clean rooms with air locks and with controlled air volume, air cleanliness, and controlled temperature and humidity to help prevent contamination or cross-contamination of the pharmaceuticals which could either reduce the effectiveness of the medication or cause it to become toxic to the individual. The workers should also be wearing special apparel to protect the product as well as to protect themselves from the product.
- For extremely hazardous pharmaceuticals, all preparation work should take place in special biosafety cabinets and if necessary use an isolation system of glove boxes to protect the worker.
- All employees must utilize some form of self-contained breathing device for self-protection when working with pharmaceuticals during production. The type of device used depends on the chemicals involved.
- Never allow the final product or raw materials to escape to the atmosphere or to be dumped down the drain to go to a body of water.
- There should be very limited access to all pharmaceutical production facilities.
- Pharmaceutical production facilities should be extremely well sealed and kept under negative pressure to prevent air leakage carrying dangerous substances to the public.
- Air-handling systems should not vent directly to the outside air until they pass through properly operating HEPA filters. Air pressure alarm systems should be utilized and immediately responded to when there is an unwanted change in air pressure in the system.

- All employees leaving the work area must go through showers and change of clothing before going to other parts of the facility or going home.
- When changing filters for the ventilation system, ensure that the area is enclosed and dust cannot get on the worker or be inhaled. Check all filters on a regular basis to make sure they are functioning correctly and not getting too dirty before replacing.
- When using portable vacuum cleaners or dust collectors, make sure that an appropriate HEPA filter is being used and then when cleaning the equipment, do it in the area under negative pressure.

CONSUMER PRODUCTS

Consumer products include a variety of soaps, detergents, cleaning agents, disinfectants, cosmetics, and other substances used to beautify the body or make it more acceptable to touch or smell. They are also used to clean various environmental areas and equipment.

Cleaning chemicals create physical and chemical risks depending on the material used, the quantity, and corrosiveness to eyes, skin, and/or mucous membranes. There may also be acute or long-term reactions to inhaling the solvents in the cleaning or disinfecting material or the specific chemicals being used. Examples of chemicals used in cleaning and/or protection of surfaces include air fresheners and other types of aerosols; substances used for dusting; fabric protectors; floor polishes and waxes; general purpose cleaners; toilet bowl cleaners; glass cleaners; chlorine-based products; etc. At times, a cancer-causing agent may be present in the material. Quaternary ammonium compounds although excellent disinfectants, over time and with long-term exposure can lead to occupational asthma and hypersensitivity to the material. Floor strippers and polishing compounds are particularly bad because of their chemical composition and may cause a series of symptoms from eye irritation and dizziness to respiratory infections, asthma attacks, and fatigue.

Hazardous chemicals may also be found in materials used for clothing and other purposes, carpets, televisions and computer equipment, plastic containers, flame retardants, toys, furniture, etc.

Best Practices in Control of Production and Use of Consumer Products

(See Best Practices in Control of Production and Use of Specialty Chemicals)

OTHER CONCERNS ABOUT CHEMICAL PLANTS

Noise

There is a considerable amount of noise in the manufacturing of chemicals because of the use of large machines, compressors, turbines, and other types of heating and cooling equipment. If the pressure within an operation has to be suddenly released, it causes considerable additional noise.

Best Practices to Mitigate Noise
- Place the buildings, especially the larger ones, in such a way that they will act as noise barriers from known sources of noise.
- Use low noise generation equipment.
- Use acoustic insulating barriers.
- Use trees as a shield against noise being released into the community.
- Develop and implement a stringent maintenance program for equipment and facilities.
- Replace older equipment on a periodic basis and utilize a new generation of lower noise and lower vibration systems.

Odors

Odors typically come from fugitive vapor releases, wastewater treatment plants, transportation sources, and release of dust during unloading and transfer of raw materials and fuel.

Best Practices to Mitigate Odors
- Determine the major source of leaks and correct them immediately.
- Clean up all spills as rapidly as possible.
- Eliminate leaks from equipment by use of proper maintenance on a scheduled basis and as needed.
- Use dust suppression techniques in all areas where materials are being transferred.

Decommissioning a Chemical Manufacturing Facility

When a chemical manufacturing facility is no longer going to be used, there is an extremely serious concern about the remnants of hazardous materials that may be present in or on the equipment, in the facility, or on the surrounding grounds, as well as potential leakage into ground or surface water and the air.

Best Practices in Decommissioning a Chemical Manufacturing Facility
- Conduct a complete survey of the entire facility, grounds, equipment, ground and surface water supply, and air and record the current levels of chemicals present as well as quantities in storage and/or waste.
- Establish a complete plan of decommissioning with all steps involved and all necessary checks and balances including testing programs.
- Determine the most effective means of cleaning all equipment and facilities, and establish a program with highly trained individuals using proper personal protective equipment to carry out the task.
- Determine with the help of the regulatory agencies and using best industry practices how to remove the hazardous chemicals including spent catalysts in a safe and secure manner and reuse them wherever possible.
- Determine if the facility can be used for other purposes in some form of manufacturing process.
- Determine if the facility should be torn down after decontamination and the equipment be removed for other purposes.
- Determine the safety of the soil and the levels of contamination found there. If contaminated, follow the techniques stated later on how to process contaminated soil and turn the site from one which is hazardous to one which can be reused.

SUB-PROBLEMS INCLUDING LEADING TO IMPAIRMENT AND BEST PRACTICES FOR HAZARDOUS MATERIALS USE, COLLECTION, STORAGE, AND TRANSPORTATION

The hazardous properties inherent in the production, storage, use, and transportation of chemicals have led to serious problems impacting workers, the community, and the environment. Major explosions and/or fires in a variety of chemical plants have contaminated land, water, and air. The release of hazardous substances to the air impacts human health and the local environment. Explosive atmospheres have been created in facilities by concentrations of dust mixed with oxygen and then a source of ignition sets off an explosion. Groundwater, surface water, and the soil have been contaminated from spills or leakage of chemicals from a series of minor leaks which have accumulated over time or a larger spill. Transportation sources have overturned and exposed the area to substantial amounts of chemicals. All these situations have had a profound effect upon communities, the personal health of individuals, and various ecosystems.

There have been recent catastrophic chemical facility incidents in the United States. President Obama, as a result of these very serious chemical exposures, issued Executive Order 13650 entitled "Improving Chemical Facility Safety and Security" on August 1, 2013. The Executive Order sets forth several tasks for the federal government agencies and their numerous partners plus industry to reduce hazards and improve community health and safety by: strengthening community planning and preparation efforts; enhancing the coordination of all operating units of the federal government; improving the collection of data, data management, and data presentation to all significant parties; modernizing all policies and regulations; and incorporating feedback from all stakeholders in the development of Best Practices to prevent and mitigate chemical accidents.

In addition, other major hazardous materials include petroleum and petroleum products and natural gas, as well as radioactive substances used for research, treatment, and nuclear explosives. An extensive discussion of petroleum and petroleum products as well as natural gases is the subject of other books. However, there are limited discussions in this area in Chapter 2, "Air Quality (Outdoor [Ambient] and Indoor)" and Chapter 5, "Environmental Health Emergencies, Disasters, and Terrorism." An extensive discussion on radioactive substances is also the subject of other books. Some references are made to the use of radioactive tracers or material for treatment in Chapter 8, "Healthcare Environment and Infection Control" and very brief mention of this topic in the section discussing disposal of medical waste.

Another major category of hazardous materials includes a variety of biological materials. (See Chapter 8, "Healthcare Environment and Infection Control" for a discussion in this area as well as Best Practices.)

A hazardous material is any substance which could adversely affect the health and safety of the public and the individuals who produce, store, use, transport, or dispose of it.

The Department of Transportation established nine classes of hazardous materials for transportation purposes including:

- *Class 1*—Explosives of all types
- *Class 2*—Gases that are both flammable and non-flammable and those that are toxic through inhalation
- *Class 3*—Flammable liquids and combustible liquids with a flashpoint between 140°F and 200°F
- *Class 4*—Flammable solids, those that combust spontaneously, and those that combust when exposed to water and give off a dangerous gas
- *Class 5*—Oxidizers and organic peroxides
- *Class 6*—Toxic materials and infectious substances
- *Class 7*—Radioactive materials
- *Class 8*—Corrosive materials
- *Class 9*—Miscellaneous dangerous goods including dry ice, asbestos, engines, materials with elevated temperatures, etc.

Each one of these classifications has its own emblem which must be displayed on the transporting vehicle.

General Best Practices for Use, Collection, Storage, and Transportation of Hazardous Materials (See endnote 57)
- Industries involved in the processing, use, collection, storage, and transportation of hazardous materials should use all necessary techniques, procedures, and technologies to avoid an accidental release of hazardous materials in the industry facilities as well as contamination of the air, water, and land. This can be accomplished by: determining through ongoing studies and hazard analysis, those operations which are most risky and therefore provide the

greatest opportunity for accidental releases or spills; using non-hazardous or less hazardous materials in a process where possible; training all workers in the specific skills needed to manage the hazardous materials; providing good design and installation practices; totally enclosing all processing and handling systems; using a separate ventilation system including if necessary safety cabinets; and separating the hazardous materials process, storage, and use from all other processes within the industry facilities. Excellent supervision and management by highly skilled individuals is mandatory to keep a system of this nature working properly. Spills and accidental releases must be kept at an absolute minimum and if they occur, they should be completely cleaned up immediately by highly capable individuals using appropriate equipment and wearing the proper personal protective gear.

- Establish the position and office of Director of Environmental Health and Safety with the authority to cross all departmental lines to make necessary inspections, studies, and recommendations, and carry out enforcement actions concerning all events, spills, occupational safety and health hazards, emergency situations, etc., with the direct authority of the chief executive officer of the facility.
- Develop a working plan with the authority of the chief executive officer and the participation of trained professional consultants, all department heads, middle-management, and representatives of the workers to prevent, mitigate, and if necessary clean up any spills or other hazardous material incidents that may occur at the facility, which may affect those present and/or the community.
- There should be a daily visual inspection of the facility to determine if there are obvious problems that need to be immediately corrected by department managers or supervisors. Especially inspect areas where, according to records of past problems in the plant or in the industry in general, serious incidents have occurred. Be aware of the potential for fires and explosive events occurring.
- Establish an inspection program of all pipes, valves, and containers to ensure that there is not any leakage and correct all problems immediately.
- Use special fittings, pipes, and hoses for transfer of materials from arrival to storage, and from storage to use within the facility.
- Use secondary containment units in the event of frequent problems of escape of materials from storage units.
- Use flame-resistant substances on vents in areas of flammable storage.
- Where underground storage tanks are used, they should be double-walled and specially coated to prevent corrosion and release of hazardous materials to the environment.
- All outside contractors must follow the rules and regulations of the facility and must be overseen by facility workers with appropriate knowledge and skills.
- Carefully measure the quantities of active ingredients needed in a formula and only utilize the amount which is necessary to make the appropriate agricultural chemical. Avoid wastage, which not only saves money but also preserves the environment.
- Use the byproducts of the original process in an appropriate manner including selling them to other industries.
- Reuse any raw materials that are left over from the original process instead of sending them to waste.
- Use an automated filling system with proper alarms and shut-off systems to avoid spillage of materials.
- Return all toxic materials packaging to the supplier for recycling or dispose of in an environmentally safe manner.
- Minimize storage time for all products and send them out to customers as rapidly as possible to prevent deterioration of the chemicals.
- Store all raw materials, byproducts, waste products, and final products in areas totally separate from potential stormwater contamination.

- All facilities used to wash trucks and other equipment must have self-contained areas for the wash water, away from any potential stormwater flow. The wash water must then be treated before it can be released into any body of water or into the groundwater supply.
- Where feasible, reuse the water from the manufacturing process.
- Ensure that containment areas for stormwater are properly designed and constructed prior to putting into use.
- Evaluate frequently and especially prior to heavy storms, all stormwater control structures and containment areas to determine if problems exist which may potentially cause a rupture of the containment system, affecting areas where hazardous chemicals are being stored, transferred, or used.
- Evaluate for potential leakage or fugitive dust loading and unloading areas, elevator shafts, means of movement of the raw or finished product from one area to another, mixers, temporary holding hoppers, and wash areas.
- Utilize dust control agents on the exterior such as oil, sprays of water, and special chemical sprays. The best techniques are those involved in preventing dust production by enclosing those plant areas where dust is frequently produced such as in loading and unloading and moving the material into other areas. These enclosed spaces need to have specially powered dust collection systems and be specially ventilated with excellent filters, and the employees need to have the proper respiratory gear for the situation. Frequent daily dust removal from surfaces using dustless techniques is essential.
- Never store together chemicals with different hazard symbols.
- Store all chemicals in a special enclosed facility with impermeable concrete floors and appropriate ventilation systems.
- Prevent tanks from being overfilled and use alarm systems to warn workers of the problem and automatically shut off the filling device.
- Install groundwater monitoring devices to determine if chemicals are seeping down to the strata in the ground.
- Use explosion-proof, spark-proof equipment and make sure that all equipment is grounded.
- Provide separate storage areas for pesticides from fertilizer storage and blending areas.
- Ensure that all deliveries of chemicals are accompanied by the proper paperwork and that a responsible person receives the delivery and verifies all items listed, the expiration date of the chemicals, as well as those containers that may be leaking or damaged.
- All communities and their leaders should be advised of the facilities involved in the production and storage of hazardous chemicals, an assessment of the potential risk involved for an accident or disaster, and how best to respond to this situation while protecting the health and safety of the citizens. Special concerns include geographic and socioeconomic issues as well as potential weather conditions and the needs of special sensitive populations.
- Establish a local emergency planning commission as a response to potential chemical and other material hazard accidents and disasters. The committee should represent all stakeholders, while utilizing appropriate technical personnel to help in the decision-making process. These committees are most effective at the local level but must have immediate and constant support from the state and federal governments. When a plan has been made operational, it has to be tested by the governmental agencies as well as the numerous stakeholders in the community.
- The local emergency planning commission can obtain from the Centers for Disease Control and Prevention the necessary material safety data sheet for each chemical produced or raw material used. These sheets include product identification information including the manufacturer's name, a list of all ingredients on the label, and regular and emergency telephone numbers; a list of hazardous ingredients and the type of hazard such as toxicity, flammability, reactivity, etc.; the chemical substance using recognized nomenclature as well as common names; the approximate percentage by weight or volume of the hazardous

ingredients in any mixture; hazard data such as parts per million, milligrams per kilogram, permissible exposure levels, and flashpoints; physical data such as melting points, vapor density, evaporation rate, etc.; fire and explosion data; health hazard information including permissible levels of exposure and routes of exposure such as skin contact, eye, inhalation, and ingestion; reactivity data; emergency and first aid procedures; spill or leak procedures; personal protective equipment required; and special precautions as stated on labels and warning placards. Further, information should be provided on toxic, acute, and chronic health effects, allergic and sensitizing effects, and carcinogenic, teratogenic, and mutagenic effects created by the hazardous materials.

- Ensure that firefighting agencies have specific information on the requirements for putting out fires related to specific chemical compounds. These requirements include appropriate extinguishing agents; extinguishing agents which will not create additional hazards; special protective equipment for the firefighters; and special training for all individuals involved in the firefighting and necessary cleanup.
- Establish accidental hazardous materials' measures to be utilized including: immediate information to the appropriate authorities on what has been released, quantities, specific hazards to people and the environment and how to counteract these hazards; accurate structural plans showing all areas of handling, storage, and use of the hazardous materials; provision of local and general ventilation; and appropriate means of cleanup without endangering the workers or community.
- Determine in advance who will do the clean-up work, and how there will be immediate provision of personal protective equipment to deal with the specific hazard and what will be the means of ultimate disposal of the material in a safe manner. Provide specialized training for these individuals on a continuing basis. Have very knowledgeable supervisors in charge of all operations.
- Develop a written hazard identification and communications program involving all agencies, industries, universities, community leaders, and the general public.
- Ensure that all information from all individuals will be funneled to a single operational headquarters where decisions can be made and responsible individuals issue necessary directives to carry out specific tasks during the emergency or disaster.
- Integrate the industry rapid response team with the local rapid response team under a single command structure utilizing the same communications network.
- Establish a rapid communication system between the industries which may be involved in a chemical crisis and the communities which may be the recipients of the hazards in various forms of contamination of air, water, land, shelter, and food. Provide alternate means of disseminating information in the event that the system goes down. Test the system periodically to determine how rapidly information will be disseminated and then make the plan operational including the use of appropriate equipment and trained people.
- Improve basic training and specialized training depending on the type of emergency that is expected for all first responders and emergency management preparedness and response personnel.
- Establish a single set of standard operating procedures for all individuals working to resolve a hazardous materials incident.
- Develop the specifications for personal protective equipment and make it such that all individuals involved will be able to interchange equipment as needed to eliminate surpluses in some area and lack of equipment in other areas.
- Develop a single data service that can be used by all individuals to rapidly disseminate necessary information and how best to respond to specific types of hazardous material incidents. This data service will integrate information from all federal and state governmental agencies and applicable industries.

- Develop a team of experts from all working agencies at the state and federal level as well as industry to standardize the data collection and interpretation.
- Develop a team of experts from all working agencies at the state and federal level to be able to respond rapidly to a situation and provide high levels of specific knowledge and experience to any given hazardous materials event.
- Review all existing policies and regulations of all governmental agencies. Eliminate all confusing and misunderstood policies or conflicts established by different policies at different agencies on the same issue.
- Request and utilize feedback from the stakeholders such as the facility owners and operators, the workers, people in the community and their organizations, unions, environmental organizations, universities, etc., in order to establish Best Practices in the gathering and use of data and the resolving of problems in a uniform manner. Expand and enhance training of all governmental employees, especially those working in the field responsible for chemical facilities and potential emergencies.
- Develop online as well as other types of training sessions for all interested people.
- Update all response teams regularly on lessons learned from actual experiences and new technologies available to prevent and mitigate the exposures from accidental chemical or other hazardous materials releases.
- Work with Congress to develop adequate budgets that will be provided as grants to various communities to consistently upgrade the skills, knowledge, equipment, and reserves of first responders and emergency management personnel.
- Develop a chemical facility safety and security executive committee at the national level to be responsible for maintaining high level competency in all areas, coordinating the efforts of all federal agencies, and immediately inserting appropriate staff and equipment when requested by state and local committees to help resolve specific emergency situations.
- Prioritize potential hazards by using existing data to determine the types of chemical and other hazardous materials accidents, releases, and disasters which are most frequent and develop a timely response for the prevention of these problems. Carry out frequent inspections to enforce necessary actions to eliminate sources of problems before they occur.
- Develop an evacuation plan for employees and the community which will be put into operation under specific conditions. Determine the lines of authority necessary to make this operational. Test the plan periodically to make sure that it works.
- Use chemicals that are less hazardous where possible within the operation.
- Do not allow unauthorized individuals to access oil and gas storage facilities as well as other potential hazardous materials sites.
- Improve local and state Safe Drinking Water Act measures to prevent and mitigate chemical spills into the drinking water supply.

TRANSPORTATION

(See endnote 86)

There have been several very serious incidents in the transportation of hazardous materials which has affected communities and the health and safety of their residents. There are approximately 1 million shipments each day in the United States of hazardous materials by railroad, truck, pipeline, aircraft, and waterways with 99% arriving safely. Even though the rate of problems has decreased, there still are numerous accidents with potential release of hazardous materials to the air, water, or land. The sharp increase in crude oil production and ethanol production as well as movement of other bulk hazardous materials, which may be flammable, by rail and truck has led to some serious incidents affecting communities. The hazardous materials incident may cause not only a huge economic problem to the community but also social and psychological impacts that last for considerable periods of time.

Flammable liquids exceed all other categories of hazardous materials being transported. Highway traffic for total tonnage of hazardous materials exceeds all other categories put together.

There are numerous economic effects from transportation incidents with hazardous materials. They include injuries and fatalities, cost of cleanup, property damage, damage to the carrier, costs of evacuation, product loss, delay in delivery, and environmental damage.

Hazardous materials are regulated by the US Department of Transportation through use of the Hazardous Materials Regulations. These regulations apply to any commercial activity by rail, aircraft, ships, and motor vehicles. They also apply to shippers, brokers, warehouses, packaging, reconditioning centers, and independent testing agencies. The Hazardous Materials Regulations of the federal government define a hazmat employer as one who uses employees who work in various capacities in the transportation or handling of hazardous materials. A hazmat employee is a person who is directly involved in hazardous materials transportation safety. Transportation of hazardous materials is a highly regulated industry and for good cause since any type of accidental release may be catastrophic.

On January 16, 2014, the Secretary of the US Department of Transportation issued a call to action resulting in meetings with rail company executives, petroleum producers, and others to find immediate ways to reduce the potential for the unintended releases of hazardous substances.

Best Practices in Transportation of Hazardous Materials (See endnotes 46, 47)
- Determine the type and amount of hazardous materials that will be transported and means of transportation, determine the classification of the substances, use appropriate placards, and follow all rules and regulations established by the US Department of Transportation for that particular class of hazardous materials.
- Provide initial training to all individuals who will be involved in any aspect of transporting hazardous materials including all new rules, regulations, and Best Practices, before they are allowed to carry out their work assignments.
- Use US Department of Transportation training modules for all employees.
- Use frequent in-service training to bring the individuals up to date and provide complete retraining every 3 years.
- The employer is responsible for all training, testing, maintenance of records of training, and certification of all individuals and must have the records available at any time for evaluation by governmental agencies.
- Never ship materials that are so hazardous that they are designated as forbidden. This may apply to all forms of transportation or only special forms of transportation. This typically includes: materials that will explode in a fire; a combination of materials that will react and produce poisonous gases or vapors or cause a corrosive effect; substances which will react with air to cause a flammable reaction; and substances that will react to cause intense heat.
- Packaging for hazardous materials is based on the level of the hazards and standards set by the US Department of Transportation. There must be no release of the contents of the package under any condition.
- Packaging of liquids for aircraft transportation must withstand a variety of barometric pressures and rapid changes. The inner packing must prevent breakage and be leak proof.
- Substances that may react with each other must never be placed in the same package.
- Hazard warning information must be on all packages being transported and all shipping documents. The packages must be accompanied by the shipper's certification and signed by an individual trained in the appropriate areas of hazardous materials transportation. This provides for emergency responders immediate information on the types of hazards encountered in the event of an accident.
- Display all hazard warning placards and identification numbers on the outside of motor vehicles, freight containers, and bulk packaging of hazardous materials.
- Provide all information concerning the hazardous materials to train crews, motor vehicle drivers, flight crew members, and appropriate personnel on vessels.

- All incidents involving hazardous materials must be immediately reported on the automated Hazardous Materials Information System database. Information would include inadequate or improper packaging, problems during loading, unloading, or handling of packages; and inadequate securing of packages within transport vehicles.
- If a hazardous material is shipped by air and a discrepancy is found in the shipment, it must be reported immediately to the closest federal aviation civil aviation security office by telephone and then in writing.
- Hazardous materials must be properly segregated within the transportation vehicle.
- Visually inspect all shipments to determine if they are properly packaged, labeled, and secured.

Best Practices in Local Community Recovery from Disastrous Hazardous Materials Transportation Incidents
- Local communities need to establish a special plan concerning hazardous materials transportation incidents including the necessity and means for mass evacuation, shelter, mass care, restoration of the road system and bridges, and environmental response to contamination of air, land, shelter, water, and food supplies. (Much of this information can be found in Chapter 5, "Environmental Health Emergencies, Disasters, and Terrorism.") The plan should include a federal role, state role, community resources, and Best Practices learned from other communities who have experienced the same type of problem. It should also cover a short-term recovery phase, intermediate recovery phase, and a long-term recovery phase which may take months or years.
- Establish a single source of information on hazardous materials, potential health and environmental problems in the short-term and long-term, and how to deal with them through various means of decontamination.
- Develop a public education program concerning the various types of hazardous materials that are transported through the community and how best to deal with serious incidents.
- Determine in advance all federal and state funding sources in the event of an emergency and be prepared to instantly request assistance for the community.
- Establish a system of instant communications of various forms to advise the citizens of the hazardous situation and what to do immediately.
- Conduct damage evaluation studies including the safety of public and private structures.
- Establish appropriate shelters and advise endangered residents to evacuate to the shelters including special needs shelters. Use a central system to track the evacuees.
- Restore all essential services including electrical, water, and sewage utilities, means of communications, and transportation.
- Medically evaluate residents who have been exposed as well as emergency management service personnel. Determine various levels of psychological problems and treat accordingly.
- Determine if it is appropriate to close beaches and commercial fisheries as well as any farmlands or other food sources, and advise the public not to use the areas or consume the food.
- Determine when it is safe for individuals to return to their homes, businesses, sources of recreation, and industry.

STATEMENT OF PROBLEM AND SPECIAL INFORMATION FOR HAZARDOUS WASTE DISPOSAL

(See endnotes 85, 97)

There are a large number of toxic chemicals being used and therefore a substantial amount of hazardous waste produced during chemical production, storage, use, and disposal. Chemical interactions and byproducts can be of greater concern at times than the original product.

The top 10 hazardous waste problems include failure to properly mark containers; open containers of hazardous materials; improper disposal of hazardous waste; improper storage in accumulation areas; keeping chemicals in storage too long; not inspecting or properly documenting the waste; improper labeling and dating of containers holding hazardous chemicals; lack of or an inadequate contingency plan for spills; inadequate maintenance of personnel training records; and management of used oil.

There are four major categories of hazardous waste:

1. *Ignitability*—These wastes can create fires under certain conditions, may be spontaneously combustible, or have a flashpoint of less than 140°F.
2. *Corrosivity*—These wastes which are acids or bases have a pH of either less than or equal to 2 or greater than or equal to 12.5. They can corrode metal containers including storage tanks, drums, and barrels.
3. *Reactivity*—These wastes are unstable under normal conditions. They can cause explosions, toxic fumes, gases, or vapors when they are compressed or mixed with water.
4. *Toxicity*—These wastes are harmful or can cause death when ingested, inhaled, or absorbed.

Universal wastes include batteries (See endnote 101), pesticides (see endnote 102), mercury-containing equipment, and lamp bulbs (See endnote 103). They are items that are typically generated by households and small businesses and thrown into the municipal solid waste for disposal. Many of these wastes can be recycled.

Improper disposal of the hazardous wastes allows the chemicals to seep from the containers and penetrate the ground, resulting in potential contamination of the earth and water. These chemicals may also become a source of air pollution.

It is estimated that there are approximately 82,000 chemicals (estimates vary with sources of data), which are used in our society today. Fewer than 200 chemicals are required to be tested under current law. Reports from the National Pollutant Release Inventory and the Toxics Release Inventory indicate that 0.5 million tons of cancer-causing chemicals, 0.5 million tons of developmental and reproductive damage-inducing chemicals, and 2 million tons of suspected reproductive and/or neurological toxicants were released into the environment and transferred to other locations, in the United States and Canada. The most recent report of North American pollution for 2005, issued by the Commission for Environmental Cooperation in 2009, reaffirms these massive amounts.

Hazardous wastes are not only produced by large industries and businesses but are also generated by a large number of small businesses. Hazardous wastes from small businesses may be harder to control because there are so many businesses in operation. These businesses include dry-cleaning, laundries, furniture manufacturers and refinishers, construction companies, laboratories, vehicle garages, printing and allied industries, equipment repair industries, pesticide user industries, vocational parts of schools, photo processing, leather manufacturing, etc.

Agricultural chemicals, so needed for the growth and protection of the food crop, can be deadly for human exposure. It is estimated that 355,000 people each year around the world die from agricultural chemicals, especially those living in Third World countries where the chemicals directly contaminate the air, water, and soil.

There are a variety of building materials that release toxic vapors such as insulation, plastics, sealants, paints and finishes, particle board, carpet, vinyl, furniture, and treated lumber. Recently, wallboard which came from China was heavily contaminated and created health problems in homes where it had been used.

Pharmaceuticals enter the environment through dumping old pharmaceuticals into toilets and flushing them down the drain; drugs and personal care products being washed down the sink or shower drain; medication residues and byproducts passing out of the body in urine and feces;

residues on food which has been washed at home, and the contaminated water goes down the drain; residues from food processing plants; residues from veterinarians clinics; residues from hospitals and nursing homes; residues from the production and use of illegal drugs; and residues from agribusiness.

The risk of a toxic chemical, during human exposure, causing an unwanted reaction, may vary with the individual based on route of entry, genetic differences, age, gender, pregnancy status, dietary or nutritional status, use of alcohol, smoking, and existing disease or health status. Gastrointestinal, respiratory, cardiac, renal, liver, and thyroid disorders can increase the toxicity of environmental chemicals since the body does not efficiently remove the toxicants from tissues and organs. Also, the action of the toxins in the body may be increased or decreased by the routes of entry, mechanisms of absorption and excretion, how quickly the toxic action begins, the biotransformation of the substance in the body, the metabolites produced, and the duration of the effect.

Typically, the fetus or child is exposed to more than one chemical at a time. There is a poor understanding of the synergistic effect of a mixture of chemicals on the human body.

The chemicals may add another complication if they have immunosuppressive properties. In addition, the concentration of the chemical, the length of exposure, the temperature during the exposure, the amount of moisture, the existence of other chemical contaminants as well as other environmental pressures contribute to the ultimate effect on the organs and tissues. Bioconcentration of the chemical may occur if it is in the food chain, therefore causing small amounts of the chemical which may be non-toxic to increase to substantial amounts of the chemical, which may now be toxic or hazardous. In 1990s, it was found that exceedingly low levels of environmental toxicants could be associated with lower intelligence, diminish school performance, and increase rates of behavioral problems, asthma, etc. Some of these chemicals have low-level toxicity for adults, but apparently in combination with ambient air pollutants such as nitrogen dioxide, acid vapor, and fine particulate matter can produce profound effects in children.

There is an ever-evolving concern about children being exposed to low-level pesticides. These pesticides which are commonly used in or around homes or on pets, or may be found as residues on fruits and vegetables consumed by the children, may interfere with immune, thyroid, or neurological and respiratory processes in children. The pesticides include organophosphates, organochlorines, and pyrethroids.

Note: See hazardous chemicals in the section above for further information on each of the sources, storage, use, and disposal of hazardous chemicals which impact many of the streams of hazardous waste that need appropriate disposal. These waste streams include industrial waste, basic chemicals waste, specialized chemicals waste, oil and gas waste, mining waste, radioactive waste, agricultural chemical waste, pharmaceutical waste, and consumer product waste. There will be some supplementary information below for a few of these waste streams. The topics of hazardous chemicals and hazardous wastes are very much inter-related but were kept separate to establish a continuity of thinking in each of the topic areas.

SUB-PROBLEMS INCLUDING LEADING TO IMPAIRMENT AND BEST PRACTICES FOR HOUSEHOLD HAZARDOUS WASTE MANAGEMENT

(See endnotes 87, 104, 105)

Household hazardous waste consists of used oil, antifreeze, oven and drain cleaners, other cleaners, pesticides, batteries, home improvement substances such as solvents, preservatives, strippers, paints, and building materials which have been torn out for repair purposes or replacement, etc. Typically, many of the household hazardous waste items are either poured down a drain, toilet, storm sewer, on the ground, or mixed in with the normal solid waste. This waste includes anything that can be corrosive, toxic, ignitable, or reactive. Universal waste which is discussed in this chapter is also mixed in with the normal solid waste. The improper disposal of these hazardous substances may affect the health and safety of people and damage the environment.

Best Practices in Collection, Storage, and Disposal of Household Hazardous Waste
• Practice substitution of less hazardous for more hazardous materials, and reduction, reuse, and recycling of existing materials to keep them out of the solid waste stream.
• Develop and utilize special household hazardous waste disposal sites in easy to get to and clearly marked places within the community on a monthly basis and use all means of communication to reach various segments of the community. Where individuals do not have appropriate transportation to do this, then there should be a special hotline phone number that they can call and trained volunteers will come to their house to remove the substances. A trained staff member of a governmental agency should be directing the operation with volunteer help from the area. This individual should not only be concerned about safety and security but about sorting the incoming hazardous materials and determining if any can be recycled or reused.
• Teach consumers how to use less hazardous substances in place of more hazardous substances and to fully use the container and not overbuy because of sales.
• Teach consumers how to use, store, and then dispose of hazardous substances in an appropriate manner.
• Never mix any type of leftover chemicals.
• Always open and use chemicals in well-ventilated areas and read the specific instructions on how to use it best, which appear on the label.
• If in doubt about the use, storage, or disposal of hazardous materials, contact your local environmental health agency for assistance.
• Determine if the community has a permanent collection system and facility for dropping off any used chemicals.
• Determine if any businesses will take back used chemicals related to their businesses and if special community collection days and a permanent facility for receiving and processing the household hazardous waste exist.

ELECTRONIC WASTES

(See endnotes 108, 109)

Electronic waste includes any equipment that has a circuit board or cathode-ray tube. It may include such equipment as television sets, answering machines, cameras, computers, video display systems, radios, telephones, etc. It may contain lead, cadmium, mercury, and other hazardous materials. Best Practices include recycling of the materials in special collection centers.

SUB-PROBLEMS INCLUDING LEADING TO IMPAIRMENT AND BEST PRACTICES FOR SCHOOL HAZARDOUS WASTE MANAGEMENT

School hazardous wastes consist of the previous items plus materials found in laboratories, shops, etc. Laboratory chemicals include acids, bases, solvents, metals, etc. Industrial Arts shops use a variety of chemicals but especially degreasers and other solvents. Art classes use paints, solvents, rags which have paints and solvents on them, and paper which may become the source of a rapidly moving fire. Maintenance departments use pesticides, fertilizers, deicers, cleaning and sanitizing chemicals, paints, oils, oven cleaners, and various types of fuels, and equipment which may contain mercury or other chemicals. The infirmary may contain a variety of cleaning and sanitizing chemicals, alcohol, and human infectious materials.

Best Practices in Collection, Storage, and Disposal of School Hazardous Waste
- Establish a team of individuals representing all areas in which hazardous materials are being used and hazardous waste is generated. Put at the top of the team an administrator who has the authority to cross all departmental lines to get problems investigated and resolved.
- Conduct a survey of the entire campus to determine the types and quantity of hazardous materials being used and hazardous waste being generated, how it is used and stored, and techniques of disposal.
- Develop a plan of action to reduce hazardous materials use and reduce and mitigate problems of hazardous waste use, storage, and disposal.
- Evaluate all purchasing orders to make sure that large quantities of hazardous materials are not being ordered, that there are adequate quantities on hand for needs, and that the supply system will furnish the school with emergency supplies when needed.
- Periodically inventory all hazardous materials to make sure that the first one in is the first one used and that there are not any oversupplies or undersupplies.
- Periodically inspect all hazardous materials waste disposal storage areas for cleanliness, orderliness, and incompatible chemicals kept away from each other, etc.
- Establish appropriate training programs and periodic continuing education programs for all individuals involved in the use, storage, and disposal of hazardous materials and hazardous waste.
- Establish a hazard communications plan and system to immediately report any spills or other problems involved with the use, storage, and disposal of hazardous materials and hazardous waste.
- Create a well-trained emergency response and spill clean-up team to go into action as soon as an event occurs.
- All individuals must wear appropriate personal protective equipment.

SUB-PROBLEMS INCLUDING LEADING TO IMPAIRMENT AND BEST PRACTICES FOR MEDICAL WASTE MANAGEMENT

(See endnotes 99, 100) (Also see the section on "Medical Waste" in Chapter 8, "Healthcare Environment and Infection Control")

Medical waste consists of materials including substances contaminated with blood or other bodily fluids from nursing homes, hospitals, clinics, doctors' offices, dentists, veterinarians, blood banks, funeral homes, etc. Sharps are a particular problem because they are usually contaminated and can easily penetrate the skin of an individual handling the material.

In 1988, because syringes and other medical materials were washing up on the beaches on the Atlantic seaboard, Congress passed a 2-year demonstration Medical Waste Tracking Act for certain covered states on the East Coast. The program has expired and there is now no federal control in this area. However, many states have put into effect their own programs by law. Typically, medical waste includes: cultures and stocks of infectious materials; human pathological waste including blood and blood products; sharps; animal wastes from research and sick animals; isolation wastes from people with communicable diseases; unused sharps; and any other contaminated waste material from individuals with communicable diseases. The program covers those who generate the waste, those who transport the waste, and the treatment, destruction, and disposal facilities including incinerators and special steam sterilizers.

Although there is a distinction made in medical waste programs between what is called non-infectious medical waste including IV bags, tubing, non-bloody gloves, packaging, urine-soaked waste, feces, vomit and blood-tainted waste materials, and infectious medical waste including blood-saturated waste, pathological and anatomical waste, cultures of infectious agents, sharps, isolation wastes, contaminated animal carcasses and bedding, this is really just a matter of degree of the level of contamination. Even though the first group appears to be relatively safe and is put into the normal solid waste stream, it is still a potential source of contamination and should be handled with great care. Obviously, the second group is highly infectious and must be considered as infectious medical waste and disposed of with the greatest of care to avoid contamination of the workers, other individuals, and the environment of the institution and the community at large.

Best Practices for Disposal of Medical Waste
- Inventory all waste generated by hospital operating and research departments as well as all other facilities generating hazardous medical waste and determine how this material is collected, stored, and put through proper disposal techniques.
- Clean up all waste spills immediately and remove all personal protective equipment, and gather up materials as well as cleaning equipment to a special contained area and place in hazardous waste containers. Where the cleaning equipment can be reused, it should be completely decontaminated before moving to another area.
- Determine if any of the material is treated and recycled and conduct appropriate tests to make sure that it does not constitute a hazard of any type.
- Do not put any hazardous waste into the sewage system unless it has been previously treated and it is no longer hazardous.
- The facility generating the hazardous chemical waste may use various treatment methods prior to release for disposal including: solidification or stabilization where free liquids in the wastes are eliminated; neutralization to raise or lower the pH to an acceptable level of between 6 and 9; carbon adsorption to bind the chemical contaminants to carbon which helps remove metals, organic solvents, other inorganic, and organic contaminants; separation through using a centrifuge or coagulation or flocculation; filtration to remove solids from liquids; and evaporation which reduces or removes water from the waste and decreases volume.
- Destroy all microorganisms before sending infectious waste to an incinerator. This can be accomplished by the use of a special steam sterilizer, chemical disinfection, irradiation, and gas/vapor sterilization on-site before the biohazardous material is released to the incineration facility.
- When the medical facility uses special private infectious waste haulers and disposal companies, all containers must be sealed, and all sharps must be placed in waterproof and piercing-proof containers which are sealed. All bags containing infectious medical waste must be double bagged in leak-proof material. It is essential that each of the disposal containers be labeled appropriately in order for all workers and others potentially exposed to the materials to be advised of the nature of the hazardous contents being transported. After use of the transport vehicle, the vehicle must be thoroughly cleaned and decontaminated by experienced professionals wearing appropriate personal protective equipment.

LOW-LEVEL RADIOACTIVE WASTE

These wastes may be produced at nuclear power plants, in certain industries, in hospitals, and in research institutions.

Best Practices for Treatment of Low-Level Radioactive Waste (See endnote 107)
- Convert all low-level radioactive wastes to a solid and structurally stable mass before transportation to a disposal facility. Most of this waste is cardboard, paper, plastic, cloth, and glass.

- Shred and compact the waste for easier movement. Incineration may also be used. Contaminated equipment should go to appropriate disposal.
- Where short-lived low-level radioactive waste is involved, keep in special containers until the decay process eliminates the hazard.

STATEMENT OF PROBLEM AND SPECIAL INFORMATION FOR LINERS FOR NEW SURFACE IMPOUNDMENTS AND LANDFILLS

(See endnote 94)

This discussion and information applies to all landfills whether they are hazardous or non-hazardous, waste piles, and surface impoundments. The discussion below on leak detection and alarm systems also applies to all landfills whether hazardous or non-hazardous, waste piles, and surface impoundments.

The *natural soil* in a given excavated area or impoundment can slow the transport and reduce the concentration of various waste streams if they are either low in hazardous wastes or the waste is inert. This is fine as long as components of the waste will not enter the groundwater supply or a surface body of water. The soil must be stable, compatible with the waste, and not in a zone of fracture within the ground.

Single liners may be made of compacted clay, geomembrane for geosynthetic clay. Compacted clay can either be used by itself or as part of a double liner system. Bentonite may be part of the clay liner. The compacted clay liner must be at least 2 feet thick and have a specific maximum hydraulic conductivity, which would be at a very low level, as measured by a soil scientist. This will prevent the migration of leachate. Thicker clay liners also reduce the potential flow. The problem is that depending on the waste, it might interact with the clay and increase permeability and hydraulic conductivity. It is necessary to develop a test pad to demonstrate that the compacted clay liner will work within the parameters established for a given type of waste. The liner must also be protected against desiccation and cracking.

Geomembrane or flexible membrane liners prevent the constituents of the waste or leachate from leaving the contained area. They are made of different plastic polymers and contain carbon black, pigments, fillers, plasticizers, different chemicals, anti-degradant, and biocides. The minimum thickness of the membrane should be 30 mm. It should have significant tensile strength, be puncture and tear resistant, not susceptible to environmental stressing which might cause cracks and tears, ultraviolet resistant, and not interact with a substantial group of hazardous waste substances. It may be necessary to test for chemical resistance to the substances that will be placed within the liner. The liner may be subjected to additional stresses from the foundation soil and its potential settlement, and the design of the trenches. Proper shipment, handling, and storage of the membrane prior to installation by highly skilled workers must be considered. The area in which the membrane will be installed must be properly prepared under the supervision of highly qualified individuals including engineers. Various tests should be conducted to determine if elevated temperatures, wind, or water within the vicinity will affect the integrity of the liner. The seaming process must be overseen and expertly conducted. Exposure to weather conditions must be avoided where possible and the backfilling of the liner must be done in such a way as not to damage it. Conduct tests during the use of the liner in the construction of the unit. Determine whether the material can resist the waste being placed within it and the tensile strength of the liner.

Geosynthetic clay liners are a combination of a factory manufactured geomembrane liner inside of a bentonite clay liner. The internal liner must adhere to the external liner. All of the information above concerning geomembrane liners applies to geosynthetic clay liners. In addition, the type of binding materials to be used between the two liners should be determined to ensure the greatest amount of strength of the unit and its effectiveness in resisting migration of liquids to other areas.

Composite liners are geomembrane liners and the natural soil in place which is made into a compacted clay liner. The geothermal liner is the upper part of the system and the natural soil is the lower part. The geomembrane maximizes leachate collection and removal. The geomembrane is typically 30 mm thick except in some instances where 60 mm is required. The compacted clay ranges from 2 to 5 feet thick.

Double liners consist of two totally distinct liners with one being the primary liner system and the second one being the secondary liner system. Each liner may be compacted clay, a geomembrane, or a composite. A leachate collection and removal system must be constructed above the primary liner system. An additional leachate collection and removal system plus leak detection and warning system must be placed between the primary and secondary liner. The geomembrane liner is typically 30 mm thick except in some instances where 60 mm is required. The compacted clay ranges from 2 to 5 feet thick.

Although the materials used in the various liners have gone through manufacturing quality control procedures and are rated highly, it is also extremely important that the installation be carried out by highly skilled people under very close supervision. The contractor to ensure that the material meets all standards of construction quality and the projected plans and specifications should make necessary inspections throughout the installation process and test the liner material and seams. It is necessary to have frequent on-site meetings with everyone involved to make sure that all parts of the project are working as required. The handling of all liner panels must be done with great care and the seaming must be done impeccably. Everything should be documented in writing.

The contractor should conduct various tests on soil moisture content, density, thickness, and hydraulic conductivity when constructing the soil liners. A test pad should be constructed before the entire project moves forward to see if the soil liner will work properly. The foundation material must have the appropriate bearing strength and be uniform in nature.

STATEMENT OF PROBLEM AND SPECIAL INFORMATION FOR LEACHATE COLLECTION SYSTEMS

(See endnote 94)

Of extreme importance in any of the hazardous waste management units is preventing contamination of the underlying groundwater by leachate. The primary function of the leachate collection and removal system is to ensure that leachate does not leave the system and the depth of the leachate above the liner is controlled. The primary function of the leak detection system is to make sure that the leachate has not escaped from the primary liner.

A leachate collection system consists of a drainage area, collection pipes, a removal system, and the protective filter layer underlain by a thick clay liner. The leachate is collected in the perforated pipe and moved upward by a sump pump through the filter layer of sand and geotextiles through vertical standpipes which extend through the liner to an external pumping station. All parts of the collection system must be constantly checked to prevent clogging.

The leachate may be pumped or hauled to a municipal sewage treatment facility. It should be pretreated with aeration, sedimentation, adjustment of pH, and removal of metals. The activated sludge process is typically used for pretreatment as well as carbon absorption and reverse osmosis.

Leachate collection system tests should be conducted by the contractor and then verified by an independent professional certified by the authorizing authority in this type of service. After the construction of this unit has been completed, it should be inspected closely for any potential damage and necessary corrections should be made.

LEAK DETECTION SYSTEM

(See endnote 94)

The leak detection system should cover both the bottom and side walls of the waste management area, collect any leakage through the primary layer, and move it to a sump pump within 24 hours. The system must be monitored regularly and if the volume of leachate is increasing, then it may be necessary to take further action based on the problem.

SUB-PROBLEMS INCLUDING LEADING TO IMPAIRMENT AND BEST PRACTICES FOR HAZARDOUS WASTE TANK SYSTEMS

(See endnote 93)

Hazardous waste tank systems are typically used on site by the generators of the hazardous waste material. The hazardous waste tank systems and regulations cover not only the tanks but also all ancillary equipment including the leak detection system and secondary containment which includes a liner external to the tank, a vault, or a double-walled tank. There is a system to determine problems at small quantity generators of hazardous waste and large quantity generators, with each of them having their own requirements. The total discussion of each is beyond the scope of this book except for a general discussion on the design and installation of new tank systems which will be included under Best Practices.

Best Practices for Hazardous Waste Tank Systems
- Use the best available technology in designing the tanks and ancillary equipment for the specific types of hazardous materials to be stored and treated.
- Determine the potential for corrosion of tanks by measuring: soil moisture content and pH; soil sulfide levels; soil resistivity and structure; stray electrical currents and underground metal structures such as piping; and techniques used to prevent corrosion.
- Ensure that the tanks are anchored to foundations and that the foundations will be able to support a fully loaded tank.
- Describe and show the evaluation of any leak detection equipment and corrosion protection equipment.
- Describe and evaluate spill prevention or overfill equipment as well as secondary containment systems which must contain 100% of all waste materials in tanks, pipes, and equipment and has its own leak detection system and alarms.
- Use visual inspections on a daily basis to determine any cracks, imperfections in the protective coating, corrosion, etc. and make necessary corrections. Determine if everything is in good working order.
- For each existing tank system that does not have secondary containment, it must be determined by a qualified professional engineer that the system is designed properly, has adequate structural strength and compatibility with the wastes being stored and treated, and is not leaking. If the system is found to be leaking or unfit for use, it must be put out of operation and a new tank system installed which meets all current requirements including a secondary containment system with proper detection for spills and leaks and an alarm system.
- Underground tank system components must not be affected by vehicle traffic. The tanks have to be anchored to the tank foundation which has to maintain the load of a full tank plus anything within the lines or equipment. The system must be able to handle potential ground movement due to frost. Everything around this underground system has to be non-corrosive and not conduct electricity.
- All spills or leaks of wastes must be immediately neutralized and cleaned up.
- Frequently check all pumps, plumbing, piping, valves, and check valves as well as supports.
- Pressurized above-ground piping systems must have automatic shut-off devices if the system loses pressure and an alarm system to notify the operator.
- All underground systems must be established in such a way that ignitable or explosive vapors are not produced.
- Rainwater must be kept away from the systems for at least a 25-year period.
- Utilize all Best Practices for hazardous waste landfills.
- Receive appropriate certifications from the governmental operating authorities.
- Keep an accurate log of all inspections and events which have occurred.

SUB-PROBLEMS INCLUDING LEADING TO IMPAIRMENT AND BEST PRACTICES FOR SURFACE IMPOUNDMENTS USED FOR MANAGING HAZARDOUS WASTES

(See endnote 92)

A surface hazardous waste impoundment is like a hazardous waste landfill cell. It is placed in a natural depression or an excavation in the soil with a dike area. It has to be lined with a special double liner, which will be discussed separately. It must also have a leachate collection and removal system which is located between the liners immediately above the bottom composite liner, along with a leak detection system. The bottom slope of the impoundment must be made of materials that are chemically resistant to the waste and flow downward to a sump pump to pump out the liquids. With permission of the authorizing authority and no leaks in the existing surface of the hazardous waste impoundment, the use of an existing single liner may be accepted.

Best Practices for Surface Impoundments Used for Managing Hazardous Waste
- Follow all information and Best Practices utilized for hazardous waste landfills.
- All earthen dikes must have a protective cover such as grass, shale, or rock to reduce problems from wind and water erosion and maintain structural integrity.
- Waste analysis and trial tests must be run on all waste being brought to the surface impoundment. If necessary, the hazardous waste needs to be treated chemically before it can be put within the unit.
- At closure, all waste residues must be decontaminated including the containment system, contaminated subsoils, equipment, and supplies.
- All free liquids must be solidified and then removed as waste or waste residues.
- Establish a long-term program for minimizing migration of any liquids that may be left in the closed impoundment.
- Maintain the integrity of the unit, as well as monitor the leak detection system.

SUB-PROBLEMS INCLUDING LEADING TO IMPAIRMENT AND BEST PRACTICES FOR HAZARDOUS WASTES

(See endnote 97)

Hazardous waste piles are piles of solid hazardous waste which are not in any type of container, do not contain liquid, and are therefore considered temporary storage.

Best Practices for Hazardous Waste Piles
- Install a double liner and leachate collection and removal system.
- Install a second leachate collection and removal system above the top liner.
- Install run-on and runoff controls for water and prevent wind dispersal of waste.
- Upon closing a waste pile, the entire residue, contaminated soils, and equipment must be removed or decontaminated.

LAND TREATMENT UNITS

(See endnote 97)

Hazardous waste is applied directly to the soil surface or mixed in with the upper layers to promote degradation, transformation, or immobilization of the hazardous materials. The physical, chemical, and biological processes in the top layers of soil denature and decompose the hazardous waste. Before waste can be placed here, the owner and operator have to show to the authorizing authority that the unit will work in the treatment of the hazardous waste.

Best Practices for Land Treatment Units
- Once the unit is in operation, the operator must monitor the unsaturated zone using procedures which have already been identified.
- The operator must demonstrate that all hazardous materials are receiving adequate treatment.
- When the unit is being closed, a vegetative cover must be put around and above it.
- The operator must continue to monitor the unit and show that the hazardous materials in the treatment zone are not exceeding background levels.

STATEMENT OF PROBLEM AND SPECIAL INFORMATION FOR HAZARDOUS WASTE LANDFILLS

(See endnote 91)

The establishment of a hazardous waste disposal landfill site is a complex action. There must be a determination made of: the exact location, its size, and boundaries; the hazardous materials and duration of the work that will be performed; seasonal and daily meteorological data including weather conditions, amount and direction of prevailing wind, amount of precipitation and humidity, as well as temperatures at various times of the year; terrain and geological and hydrological information; population centers and population at risk as well as ecosystems at risk; and accessibility by roads, trains, and planes; and an initial evaluation of the air, soil, surface water, and groundwater, as well as the land being used, for existing types and levels of pollutants.

All the problems stated for the landfills for disposal of non-hazardous waste also apply to hazardous waste landfills. In addition, there are numerous land disposal limitations since the hazardous waste landfill is designed as a means of treating the waste and not allowing any type of seepage to leave the area and move on to the groundwater supply or surface water supply. In addition, there should be no potential for air contamination.

SUB-PROBLEMS INCLUDING LEADING TO IMPAIRMENT AND BEST PRACTICES FOR HAZARDOUS WASTE LANDFILLS

There are restrictions on substances that can be put into the hazardous waste landfill. Under no condition would incompatible substances be put in the same area. This could result in explosions and chemical interactions which would create highly toxic byproducts that could destroy the system and become water pollutants or air pollutants. There are specific restrictions on: dioxin-containing wastes; soils that have the toxic characteristics of metals and PCBs; chlorinated aliphatic wastes; toxic metal wastes; petroleum-refining materials; inorganic chemicals; ignitable and corrosive wastes; and newly identified organic wastes with considerable toxicity.

Hazardous waste must not be put into any land treatment facility unless the waste can be made less hazardous or non-hazardous by degradation, transformation, or immobilization in the soil. The hazardous waste landfill must be designed, constructed, operated, and maintained in such a manner that there cannot be any flow of any type out of the given area. Water volume due to storms must be kept totally away from the disposal area. Collection and holding facilities of water must be such that it can be treated if necessary and discharged after the storm is over into other areas which are not being used for hazardous waste disposal.

Best Practices for Hazardous Waste Landfills
- All new hazardous waste landfills or lateral extensions of old ones must have two or more liners, and a leachate collection and removal system above them and between each of the liners. A single liner is only used in very special situations and then only with the written permission of the authorizing authority after reviewing the situation on site.

- Do not place any bulk or non-containerized liquid hazardous wastes in the landfill.
- The concentration of the hazardous waste, quantity, and type must be clearly identified and recorded before placing in a land area.
- Incompatible wastes must never be stored together in the same land treatment area.
- No food crops must ever be grown in the area where hazardous waste has been put into a landfill.
- No animal grazing or anything else associated with food production can be done in this area.
- Future owners of the property must be notified in the property deed that the site was used for hazardous waste disposal.
- Prepare a plan for a system of monitoring in the zone of aeration of the groundwater supply. The plan must detect any vertical migration of the hazardous waste or any of its constituents; provide information concerning the background concentration of the hazardous waste or any of its constituents in the soil close to the source of water; and incorporate the use of appropriate soil testing techniques at depths below where the waste was placed in the soil and should consist of soil cores and appropriate instruments to monitor the soil pore water with a frequency of testing based on the frequency, time, and rate of waste application. All this information must be kept in a permanent logbook for review by the authorizing authority. Along with this must be a plan for closure and postclosure of that area which will include means of control of migration of hazardous waste and its constituents into the groundwater; means of control of release of contaminated runoff from the facility into surface water; means of control of the release of airborne particulate contaminants because of wind erosion; and means of preventing growth of food or use by cattle of the contaminated area.
- After closure of a given area, there must be a plan for removal and treatment of contaminated soils, and placement of final covering material including appropriate use of vegetation; continued monitoring of groundwater and runoff management of the area; and control of the migration of landfill gas and leachate and all necessary treatment.
- Observe all current rules and regulations of state and federal authorities. These tend to change over time and it is the responsibility of the owner and operator of the landfill to know about the changes and carry out the necessary actions.

SUB-PROBLEMS INCLUDING LEADING TO IMPAIRMENT AND BEST PRACTICES FOR HAZARDOUS WASTE INCINERATORS

Hazardous waste incineration is the burning of hazardous materials at temperatures which are high enough to destroy all of the contaminants. These wastes include soil, sludge, liquids, and gases which may be contaminated with a vast number of chemicals in substantial quantities. About 7% of the approximately 44 million tons of hazardous waste typically generated yearly in the United States is destroyed through combustion. A considerable amount of energy is used in the process. There is a potential for odors, smoke, dust, excessive noise, and hazardous substances getting into the air because of improperly performing air pollution systems.

There are two types of hazardous waste combustion units. They are incinerators, and boilers and industrial furnaces. Incinerators include rotary kilns, fluidized bed units, liquid injection units, and fixed hearth units. Industrial furnaces include cement, lime, aggregate, and phosphate kilns, as well as coke ovens, blast furnaces, etc. The EPA has established National Emission Standards for Hazardous Air Pollutants for maximum achievable control technology for an industry group or special source. (See endnote 90.)

Best Practices for Hazardous Waste Incinerators (See endnote 89)
- Obtain a permit from the authorizing authority by providing information on siting, design, waste streams, transportation, storage, air pollution devices, state-of-the-art technologies used, public education program, and potential hazards, types, and quantities. Also provide

any other information required by the individuals granting permits. Prepare an environmental and public health impact statement.
- Establish a comprehensive monitoring program to detect all types of potential toxic emissions and the quantities released.
- Observe all rules and regulations of the state and federal government. These rules and regulations tend to change over time, and it is the obligation of the owner and operator of the incinerator facility to make all necessary changes as directed.
- Analyze all waste streams that have not been previously submitted for incineration at the facility. As a minimum requirement the analysis must include the heating value of the waste; halogen and sulfur content; and lead and mercury content.
- Monitor all instruments at least every 15 minutes to determine if various parts of the system are operating appropriately.
- Prior to the burning of hazardous waste, the thermal treatment process must be brought up to its normal operating condition and temperature.
- Use a temperature of 1600–2500°F for hazardous materials destruction depending on the type of material, quantity of material, and the contaminants. Use a time of treatment in the initial combustion chamber of 30–90 minutes also depending on the type of material, quantity of material, and the contaminants. Mixing of the waste helps make the process work better and faster.
- Most of the gases that are created during incineration are destroyed; however, for those that are not, they should go through a secondary combustion chamber for further heating and processing. If hazardous gases are still present, they should be trapped by the air pollution devices and then destroyed. Incineration of hazardous materials must result in the removal of 99.9999% of the contaminants.
- The stack plume must be observed visually at least once an hour for color and opacity to see if there are changes due to the substances being burnt and other conditions. Changes would indicate an immediate need to evaluate the system for problems.
- The entire thermal treatment process and all equipment must be evaluated on a daily basis as a minimum for leaks, spills, fugitive emissions, and whether or not emergency shutdown controls and system alarms are working.

SUB-PROBLEMS INCLUDING LEADING TO IMPAIRMENT AND BEST PRACTICES FOR ENVIRONMENTAL IMPACTS OF HAZARDOUS MATERIALS AND HAZARDOUS WASTE

(See endnote 2)

AIR

(also see Chapter 2, "Air Quality (Outdoor [Ambient] and Indoor)")

The basic information on air quality, health effects, National Ambient Air Quality Standards, New Source Performance Standards, and National Emission Standards for Hazardous Air Pollutants have already been discussed in Chapter 2, "Air Quality (Outdoor [Ambient] and Indoor)." Regulations have been established for hazardous waste airborne emissions, municipal solid waste landfill airborne emissions, and offsite waste and recovery operations. There are different operating permits based on the amount of emissions from the unit.

A Title V permit, issued under the Clean Air Act regulations, is required for actual or potential major sources of emission that either meet or exceed the major source threshold for their own location. The threshold is typically 100 tons per year unless the waste incineration unit is located in a non-attainment area for pollutants and then it may be lower. The threshold for hazardous air pollutants is 10 tons per year for a given pollutant and 25 tons per year for total hazardous air pollutants.

Municipal solid waste incinerators, hospital/medical infectious wastes incinerators, commercial and industrial solid waste incinerators, sewage sludge incinerators, and other types of incinerators especially the burning of hazardous waste of all kinds are subjected to the rules, regulations, and specific actions required under Clean Air Act, Title V permits. (See endnote 112.)

Sanitary landfills and hazardous waste landfills may produce a variety of air pollutants. These range from carbon dioxide to carbon monoxide to methane to hydrogen sulfide to a variety of other very hazardous gases. They may also release considerable amounts of particulate matter. (See the section on "Landfills and Hazardous Landfills," above for further details.)

Best Practices to Prevent or Mitigate Air Quality Problems
- Conduct a risk assessment study of the area affected by the placement of central collections systems, incinerators, and sanitary and hazardous waste landfills. Determine the source of the risk, the quantity of the materials and gases being introduced into the ambient air, the nature of the risk, and potential effects on human health, ecosystems, and the general environment. Utilize the data in establishing the plan of operation, control, and how to deal with spills and emergencies.
- Identify specific chemicals being released, specific exposure assessment for humans and the environment, stability of the atmosphere, and prevailing wind patterns, and determine how best to protect workers, the community, food, land, and water.
- Use the Industrial Waste Air Model (IWAIR), which is an interactive computer program with three main components including: an emission model which estimates the release of constituents from waste management units; a dispersion model which estimates the fate and transport of the constituents in the atmosphere and determines ambient air concentrations at specific places; and a risk model that calculates the risk to exposed individuals or the concentrations of contaminants which may be managed successfully. All this information helps establish an appropriate plan for mitigation and control of potential air pollutants from a given unit. (See endnote 113.) The model is user-friendly and flexible but has certain limitations because it does not discuss the potential risk from a particular release or indirect routes of exposure. Other site-specific risk analysis models include AP-42 from the US EPA; Exposure Related Dose Estimating Module (ERDEM) from the US EPA; Landfill Air Emissions Estimation Model from the US EPA; etc.
- Control particulate matter, known as fugitive dust, from becoming airborne by: keeping unpaved roads wetted down with water or a chemical dust suppressant; limiting the vehicles within the landfill to a given area to avoid traveling substantial distances and creating more dust; washing the equipment in a special area with concrete floors and drains to trap the dust and fluid for treatment purposes; using experienced operators of equipment to avoid extra work which would create more dust; and using cover material at the end of each operation within the day and at the end of the day to prevent wind erosion.
- Control volatile organic compound emissions by: using pollution prevention and pretreatment procedures to reduce the risk of the chemicals; choose a site which is sheltered from wind by trees and make sure that the prevailing winds are such that the site is downwind from communities and businesses; enclose the units where possible with a flexible membrane, and capture the air and send it through air pollution devices to neutralize the volatile organic compound emissions by using techniques of carbon adsorption, biofiltration, condensation, absorption, and vapor combustion.
- Do not spray wastes on land if there are volatile organic compounds present.

LAND AND SUPERFUND SITES

(also see Chapter 3, "Built Environment—Healthy Homes and Healthy Communities")

Land has become contaminated from the illegal disposal of a large number of chemicals over many years from factories and businesses. There have been numerous examples of this action as exemplified by the Love Canal episode where chemical waste was buried in drums in an area that eventually became the yards and basements of homes and other structures. The drums corroded and leakage occurred. A total of 89 chemicals were identified, several of which were suspected carcinogens. It was not uncommon for various industries to bury their wastes on their land and then when the industry went out of business and the structures were torn down, the land was reused for other purposes. The land and the water on these sites are typically contaminated and this has led to contamination of other areas to which the chemicals have migrated. In some areas, specific land sites are set aside for hazardous waste land disposal facilities. These facilities must receive specific special permits to operate and for postclosure actions.

Best Practices for Land and Superfund Sites (See the sections on "Hazardous Waste Landfills" and "Land Treatment of Wastewater" earlier in this chapter) (See endnote 114)
* Establish a Superfund site on the National Priorities List if an area of land has had uncontrolled hazardous waste placed there and it has the potential to damage human health or the environment or both. When the land has been abandoned, there are accidental spills or illegally dumped hazardous wastes, and there is a current or future threat to health or the environment, the EPA under the Superfund program will work closely with the communities and other individuals to test the conditions at the site, develop clean-up plans, and then clean up the site. The site once it has had all hazardous wastes neutralized or removed, can be reused for a variety of purposes that all enhance the community. Superfund sites are the country's worst hazardous waste sites in comparison to brownfield sites which are also contaminated but are much smaller and easier to control.
* Obtain permits from the operating authority for the operation and closure of all land disposal facilities which include surface impoundments, waste piles, hazardous waste landfills, and both ground and below-ground tanks which have stored hazardous waste and have leaked into the surrounding soil and groundwater. The land disposal permits require information on the design and performance standards for the liner, the leachate collection and removal system, and the leak detection system; restrictions on the type and amount of waste; a waste analysis in writing; a facility inspection and maintenance program; a groundwater and surface water monitoring program; control systems for water from storms that either run onto the land site or run off of it; a plan for emergencies; and closure and postclosure plans.

WATER

See discussions on hazardous landfills, land treatment of waste, municipal solid waste landfills, and underground injection wells above. Also see: Chapter 11, "Sewage Disposal Systems"; Chapter 13, "Water Systems (Drinking Water Quality)"; and Chapter 14, "Water Quality and Water Pollution."

More than 80% of the most serious hazardous waste sites in the United States have contaminated groundwater. Rainwater has percolated through the soil where hazardous waste has been buried and has carried the chemicals into the groundwater supply. Chemicals in the ground have corroded storage barrels and seep through openings in the soil and rock into the groundwater supply. Different chemicals which should never have been stored together have reacted, and the chemicals and byproducts have seeped into the groundwater supply. This is of great importance since the drinking water supply for half of the country comes from groundwater. Unfortunately, the cleanup of the groundwater supply is very difficult and may take many years to accomplish.

Best Practices in Preventing and Mitigating the Effects of Chemicals on the Groundwater Supply
- Establish a groundwater risk evaluation study to determine the areas of contamination, the source of contamination, and the level of problems. An effective means of doing this is by utilizing the Industrial Waste Management Evaluation Model (IWEM). This is a three-tiered approach with the first tier being based on national data and if it satisfies the need for establishing an appropriate system, this is fine. The second tier is used when the first one does not give enough data for the specific site. This includes in the model the most important site-specific factors in a simple manner and can be used by industry and various governmental agencies. If this does not provide enough data, then third tier is used where experts using groundwater modeling study the specific area and come up with the appropriate set of controls to mitigate the hazardous waste problems related to contamination of the groundwater supply.
- Make a determination of the leachate concentrations, their source and composition, and why they are moving through the ground into the groundwater supply. Various models are available for this purpose.
- Based on the previous data, determine the liner and the type and nature of the hazardous waste site, construct the necessary facility, and install the proper liners and various pollution control devices.
- Establish a leachate collection and treatment system which has appropriate warning systems.

Best Practices in Preventing and Mitigating the Effects of Chemicals on the Surface Water Supply
- Determine the quality as well as the level of microbiological and chemical contamination of surface bodies of water by conducting various tests.
- Do not permit discharge of pollutants to any waters of the United States unless specifically granted under special permits from the authorizing authority.
- Separate all wastewater from stormwater before it enters a surface body of water.
- Reduce stormwater discharges by a large number of barriers which help allow the water to seep into the groundwater supply. (See Chapter 14, "Water Quality and Water Pollution" for specific techniques.)

STATEMENT OF PROBLEM AND SPECIAL INFORMATION FOR OCCUPATIONAL HEALTH AND SAFETY

(See endnotes 95, 106)

Diseases and injuries are a constant problem for individuals who are creating and getting rid of waste material, who are collecting waste material and moving it to other locations, who are treating or recycling the material, and who are responsible for final disposal. Individuals working with hazardous materials have the same sort of problems. In the chemical industry and probably other industries, workers are subjected to additional hazards from the base chemicals which are being produced or being used in the production of other substances in large quantities. The active ingredients of special chemicals in the pharmaceutical industry and other industries include large numbers of solvents and chemicals which may be extremely hazardous to the health and welfare of workers even in minute quantities and may cause numerous long-term chronic diseases and conditions. Consumer chemicals are produced typically in smaller batches and often have to be processed at high or low temperatures and pressures, creating additional problems. Maintenance workers may come into contact with a variety of chemicals which are remnants of the product and byproducts but in large enough concentrations to cause disease and injury. Further, the workers are subjected to the health hazards associated with the cleaning materials. All workers including maintenance workers are subjected to diesel engine exhaust with its complex mixture of polycyclic aromatic hydrocarbons, sanding or blasting operations with the production of dust, toxic gases such as those produced by welding including phosgene, and residues of chlorinated solvents. Injuries are a constant concern not only for the typical worker but especially for the maintenance people. Crude mineral oil is used in various areas to help prevent friction. Confined space injuries and deaths may occur from the production of hydrogen sulfide gas, a low oxygen environment, and other chemical fumes. The catalyst used may be more dangerous than the actual chemicals.

When a unit or plant shuts down for maintenance activities, the employees working on maintenance may come into closer contact than normal with equipment, piping systems, residues of hazardous substances, pipes, storage tanks, and processing areas. The cleaning and washing of pipes, tanks, equipment, and the interior of the facility creates wastewater which may easily be very hazardous to the health and safety of the individuals. Manual brushing to get rid of rust and other residues or sanding creates high concentrations of hazardous dusts. The use of high-pressure hoses to clean the dust off of equipment, pipes, storage vessels, and floors may create contaminated aerosols.

SUB-PROBLEMS INCLUDING LEADING TO IMPAIRMENT AND BEST PRACTICES FOR OCCUPATIONAL HEALTH AND SAFETY ISSUES

The hazardous waste environment is typically an uncontrolled situation where virtually any kind of problem affecting people or the environment can occur. Improper control and handling of the hazardous substances can result in very dangerous results. The situation is further complicated by the fact that there may be hundreds or even thousands of chemicals at the site and they may interact and become even more dangerous to the workers. These individuals are not only stressed by direct exposure but also by the disorderly physical environment and having to wear protective clothing for long periods of time. The combination of all these factors may result in immediate and obvious health and safety problems or long-term problems.

Workers at hazardous waste sites may be exposed to: toxic chemicals, fire and explosions, states of oxygen deficiency, ionizing radiation, biological hazards, safety hazards, electrical hazards, heat stress, cold exposure, and extreme noise. The exposures may be to individual hazards or multiple hazards at the same time, which enhances the potential for disease and injury and possibly even death.

Toxic chemicals may be present in gaseous, liquid, or solid states. They may enter the body through inhalation, ingestion, a puncture wound, or skin absorption. They may cause acute effects which will result in almost immediate symptoms, or chronic effects which may be an accumulation of low concentrations of the chemicals over time. Some chemical effects may be temporary and reversible and others permanent depending on the chemical, time and concentration of exposure, and the individual. Cancer or respiratory diseases are some of the serious side-effects which may occur and may not be seen for many years. Disease potential may depend on the chemical, its concentration, its route of entry, length of exposure, and personal factors such as smoking, consumption of alcohol, medications taken, nutrition, age, and sex. Inhalation is an extremely serious route of entry into the body as well as direct contact of the skin and eyes with the hazardous material.

Explosions and fires can occur spontaneously. However, it is typically the movement of containers of the chemicals or the mixing of incompatible chemicals that causes explosions and fires to occur. The hazards created in addition to those already noted above concerning exposure to toxic chemicals include intense heat, smoke inhalation, open flames, and flying objects.

Diseases caused may include infections, allergies, respiratory conditions, cancers, communicable diseases, and chronic disabilities. Diseases may be spread by food, water, air, animals, vectors, skin contact, and inhalation.

Injuries may occur as a result of damage to joints, fractures, puncture wounds, damage to eyes and ears, sprains, strains, and damage to the skin. Unstable wastes, heavy wastes, cave-ins at disposal sites, fires, explosions, improper processing equipment, and inadequate training of operators may be the cause of injuries. Other concerns are sharp objects, slippery surfaces, steep grades, uneven terrain, and unstable surfaces such as walls or flooring within the disposal area.

Oxygen deficiency occurs when there is less than 16% oxygen in the environment. This may affect the level of attention to surroundings and conditions, judgment, and coordination, and causes nausea, vomiting, brain damage, heart damage, unconsciousness, and death. Typically, this occurs when another gas replaces oxygen within the environment or the individual is working in confined or low-lying spaces.

Radioactive materials may produce alpha particles, beta particles, and gamma rays. Alpha particles can be hazardous if ingested or inhaled. Alpha particles have limited penetration and are stopped by clothing and the outer part of the skin. Beta particles can cause burns to the skin and damage the subsurface blood system. They are also dangerous if inhaled or ingested. Gamma rays pass through clothing and human tissue and cause serious permanent damage to the body. Protective clothing is not effective. (See endnote 94.)

There are numerous electrical hazards from overhead power lines to downed electrical lines, and buried cables which may be cut by heavy equipment when digging up land.

Heat stress is especially significant for people wearing protective clothing. The body is unable to get rid of heat and moisture in an effective manner when enclosed in the personal protective equipment that is used to protect the worker from various hazards at the disposal site. The problem can occur in as little as 15 minutes. Heat stress may cause rashes, cramps, and drowsiness, and impair the individual's ability to function and think clearly. Cold exposure may occur when the individuals are exposed to low temperatures and a significant wind which results in a low wind-chill factor.

Noise is created by the large equipment being used and can easily startle the workers or distract them. Physical damage may occur to the ear and result in pain and temporary or permanent hearing loss. The interference with communications can prevent the individual from being alerted to a dangerous situation.

Best Practices in Mitigation of Occupational Health and Safety Problems
- Establish a risk assessment process within the facility based on good industry knowledge and experience of actual problems within the overall field of endeavor. Pay special attention to the problems and needs of: young workers and older workers; pregnant and nursing women; migrant workers; untrained or inexperienced workers; and individuals involved in maintenance, members of the community, and special contractors.
- Develop and implement a comprehensive occupational health and safety plan for all workers in all departments based on specific hazards, means of prevention and mitigation, employee training, environmental controls, and proper supervision and management of personnel.
- Eliminate known hazards when possible and mitigate the effects of those that cannot be eliminated.
- Develop a special permit-to-work system for those types of high risk jobs which need special personnel who are physically fit, highly trained and experienced, have excellent judgment, and are good workers. There may be a permit system for hot work, cold work, electrical work, confined spaces, complex equipment, isolation areas, radiation, etc.
- Conduct job safety analyses to determine potential critical situations which could lead to disease and injury for all positions within the facilities in order to be able to establish appropriate standard operating procedures, use of personal protective equipment, necessary training, and appropriate supervision.
- Determine the concentrations of the different types of hazards as well as the amount of time of exposure that the workers encounter in any operation whether it is hazardous materials handling, hazardous waste handling, or non-hazardous waste handling.

- Use special safety lockout procedures to make sure that equipment cannot start operating when a maintenance worker is performing an assignment on it.
- Especially during shutdowns and maintenance work, ensure that all vessels are emptied and the workers, using a buddy system, are wearing appropriate personal protective equipment including respiratory devices.
- Have all testing equipment fully calibrated and checked and then utilized during the clean-up process to make sure that hazardous chemicals are not entering the air and causing potential serious injury to workers.
  - Determine if the equipment is cool enough to work on before providing any type of maintenance.
  - Recognize the potential for occupational hazards due to continuous hot work or working in confined areas where hazardous substances have been stored. Use special precautions to prevent worker fatigue, disease, and injury.
  - Minimize the amount of highly volatile and poisonous substances that is kept in storage in the facility.
  - Reduce the amount and frequency of transfer of highly hazardous substances within the facility.
  - Reduce and minimize the levels of toxicity or infectivity of the materials and increase the level of safety of the workers by use of appropriate personal protective equipment, necessary ventilation devices, and limitations of time of exposure to the substances.
  - Utilize rigorous workplace monitoring and competent supervisory control of all activities and personnel.
  - Install gas detectors in hazardous areas which will give immediate indication of leaks and potential disastrous results.
  - Use select barrier creams especially when exposed to aromatic hydrocarbons.
  - Remove all potential sources of ignition where flammable or explosive materials are being processed, used, or stored.
  - Produce, store, use, and transport acids with the greatest care and continuous monitoring as well as an alarm detection system because of the extremely corrosive and hazardous materials involved. Only use specially trained workers.
  - Conduct a noise and vibration survey to determine where the problems occur and provide necessary personal protective equipment as well as limit the time of exposure of workers.
  - Provide automatic alarms and shut-off systems where problems occur with machinery and if safety devices have been altered or removed.
  - Establish areas for workers to walk away from moving vehicles. Ensure that all walkways are made of non-slip materials.
  - Control entry into all enclosed areas and areas of production, storage, and transportation of hazardous chemicals. All individuals entering these areas must have appropriate training and personal protective equipment. They must be closely supervised and use the buddy system.
  - Provide first-aid equipment and emergency showers in the event of spills or other accidents. Make sure that water usage by the individuals will not contribute to the problem.
  - All workers should have physical examinations before working with hazardous materials and periodic follow-ups. The examinations should include all necessary laboratory tests.
  - If a worker is exposed to a hazardous material, the individual should be immediately decontaminated and examined by appropriate medical authorities as well as given necessary tests to determine levels of contamination. Proper treatment should start immediately and be closely followed by medical personnel.

- Employees working with hazardous materials should be kept to a minimum and special precautions should be taken including: limited time exposure by individuals, regular cleaning of contaminated surfaces, proper maintenance of all equipment by specialists, use of personal protective equipment, and prohibiting all eating, drinking, and smoking in hazardous materials areas. Further, facilities should be provided for showers and changes of clothing before the individual leaves the industrial complex.
- Special precautions should be taken around all moving parts of machinery to avoid contact with clothing or the individual.
- Do not wear contact lenses in contaminated areas.
- Never allow a worker to go into a confined space or low-lying area alone and ensure that the individual has proper personal protective equipment as well as a secure communications system within the equipment. These individuals need to be closely supervised and only allowed to work within the confined environment for brief periods of time.
- Determine if radioactive material is emitting alpha, beta, or gamma rays. Alpha radiation can be stopped by protective clothing. It must never be inhaled or ingested. Beta radiation can cause serious burns to the skin and damage the blood system. It must never be inhaled or ingested. Use protective clothing and complete and total personal hygiene and decontamination efforts to prevent problems. Gamma radiation will pass through clothing and human tissue and can cause serious damage to the body. Chemical protective clothing will not help. Use respiratory equipment and other protective equipment to keep the individual from inhalation, ingestion, injection by accident, or skin absorption of alpha and beta radiation. If the radiation level is above 2 mrem per hour, shut down the site and have a health physicist evaluate the potential hazards to workers.
- All biological hazards should be destroyed before sending the material to any landfill site. However, workers are still subject to contact with poisonous plants, insects, or animals which may spread disease to them.
- Employees should be constantly checking for potential safety hazards on-site, avoiding them, and immediately notifying their supervisors concerning the situation.
- Employees should be aware of electrical hazards and have information concerning all buried cables to avoid cutting them during any type of excavation or other work.
- All work should cease during electrical storms and weather conditions should be monitored very closely.
- All workers wearing personal protective equipment must be closely supervised to make sure that they are not suffering from heat stress. They should be immediately removed from the environment and taken to a cool place and allowed to rest before resuming work. During the work there should be frequent breaks for these individuals.
- All workers should be protected against cold injury by removing them to a proper place to warm up, especially when the wind-chill factor is very low. Proper warm clothing in layers should be worn.
- Employees should not be subjected to more than 90 dB on the A-weighted scale without proper means of ear protection and should not be allowed to stay within the very noisy environment for long periods of time.
- Since each portion of the hazardous waste site varies from other portions depending on the chemicals and other hazardous materials being stored and processed, always evaluate temperature at the site which if high may increase the vapor pressure of most chemicals; wind speed and direction; amount of rainfall, diversion tactics, and levels of water in a given area; amount of moisture present which could affect the process or, if limited, increase the amount of dust; actual vapor emissions; and the types of work activities being carried out.

LAWS, RULES, AND REGULATIONS

NATIONAL ENVIRONMENTAL POLICY ACT

(See endnote 124)

This act, one of the first laws written to broadly protect the environment, establishes the basic policy for all branches of government concerning construction of new facilities and alteration of the environment. An environmental impact statement is required for all new facilities and alterations of the environment and it makes assessments of the likely impacts on the environment of the program or project being set forth. A determination can then be made as to whether or not the impacts are greater than the good created by the new project.

RESOURCE CONSERVATION AND RECOVERY ACT

(See endnote 132)

Both non-hazardous solid waste and hazardous solid waste is regulated by the US EPA under the RCRA and its amendments. The goals of the act are to: protect human health and the environment; conserve energy and natural resources; reduce the quantity of waste being generated; and ensure that all solid and hazardous wastes are managed appropriately to protect the environment. Hazardous waste is covered from the cradle to the grave and includes generation, transportation, treatment, storage, and disposal. This act also regulates underground storage tanks. Many problems exist at petroleum refineries, chemical plants, gas stations, and agricultural settings. The act establishes general guidelines for waste management programs and gives the EPA the power to develop and enforce criteria for identifying hazardous waste and its disposal. These regulations and rules were mandated by Congress. The EPA also provides guidance documents on how programs should work.

The major laws will be discussed briefly. However, the rules and regulations are so numerous that it would be impossible to list all of them in a work of this nature. Further, they are in a constant state of change because of legal challenges and new technologies.

COMPREHENSIVE ENVIRONMENTAL RESPONSE, COMPENSATION, AND LIABILITY ACT

(See endnote 116)

This act, which is also known as the Superfund, provides federal funds to clean up hazardous waste sites, accidental spills, and abandoned hazardous waste sites where responsible parties cannot be identified or located or when people fail to act to carry out their responsibilities to clean up all potential hazardous materials that can cause disease and injury to people or harm the environment. This applies to the entire country.

OIL POLLUTION ACT

(See endnote 117)

This act amended the Clean Water Act by: establishing new requirements for construction of vessels and licensing of crews; mandating the establishment of contingency plans in the event of accidents; improving the response of federal agencies; increasing enforcement authority, potential liability, and penalties of shipping companies; and creating new research and development programs.

SUPERFUND AMENDMENTS AND REAUTHORIZATION ACT

(See endnote 131)

This act amended the Comprehensive Environmental Response, Compensation, and Liability Act or Superfund to provide more funding for hazardous waste sites. It also reinforced the significance of human health, community involvement, cooperation with state and local laws, and finding a permanent solution to the hazardous waste site problem.

EMERGENCY PLANNING AND COMMUNITY RIGHT-TO-KNOW ACT

(See endnote 111)

This act required that each governor of a state appoints a State Emergency Response Commission that will report directly to him/her or serve himself/herself as the commission chairperson of a select committee. The commission then establishes emergency planning districts in order to prepare and implement emergency plans and also establishes local emergency planning committees. It is then required by Executive Order that a list of extremely hazardous substances be published with their thresholds, which if exceeded may cause disease and injury. Facilities that manufacture, store, and use substances that exceed the threshold of those on the list are subject to all the rules and regulations of the Emergency Planning and Community Right-to-Know Act.

Each local planning committee must develop an emergency plan which includes resources to respond to problems, identification of facilities that may cause problems, means of determining releases, and how to effectively deal with them.

HAZARDOUS MATERIALS TRANSPORTATION UNIFORM SAFETY ACT OF 1990

(See endnote 119)

This act requires the Secretary of the Department of Transportation to establish standards for the states and Native American tribes that will be used when establishing highway routes, and the limitations and requirements for the movement of hazardous materials within their own jurisdictions. If the states or tribes are in dispute concerning the routing of the hazardous materials, it may come under judicial review to resolve the issue. The Secretary has been granted major powers in all areas of hazardous waste transportation to avoid disastrous consequences from improper packaging, and the inappropriate or misrepresenting of information concerning the movement of hazardous waste. Individuals carrying out these types of activities have strict requirements including reporting of incidents.

CLEAN WATER ACT

(See endnotes 2, 120)

This act sets the basic structure for the regulation of pollutant discharges into the waters of the United States and for establishing water quality standards for surface waters. It gives the US EPA the authority to establish wastewater standards for industry as well as standards for all contaminants in surface waters. It establishes the National Pollutant Discharge Elimination System permits and program. This permit program sets site-specific effluent discharge limitations; both standard and site-specific compliance monitoring and reporting requirements; and schedules to be used for monitoring, reporting, and compliance with all rules and regulations. There are 11 specific categories of stormwater discharges that are involved with industrial discharges, which require permits. They are: manufacturers of certain pesticides; large manufacturing facilities; mining and oil and gas exploration; hazardous waste treatment, storage, and disposal facilities; landfills; recycling plants; steam electric generating plants; transportation facilities; sewage treatment plants; construction operations; and other types of industrial facilities exposed to various types of contamination.

CLEAN AIR ACT

(See endnote 121)

This act is the comprehensive federal law regulating air emissions from both stationary and mobile sources. It gives the US EPA the authority to establish the National Ambient Air Quality Standards and regulates hazardous air pollutant emissions. The law applies to human health as well as the environment and ecosystems.

SAFE DRINKING WATER ACT

(See endnote 122)

This act is the comprehensive federal law that determines the quality of drinking water in the United States. It gives the US EPA the authority to set standards for drinking water quality and to implement the standards through the states, localities, and water suppliers.

TOXIC SUBSTANCES CONTROL ACT

(See endnote 134)

This act gives the US EPA the authority to require information on reporting, record keeping, testing, and type of use, and may restrict certain chemical substances and/or mixtures. Food, drugs, cosmetics, and pesticides are excluded since they are covered under other federal laws.

MIGRATORY BIRD TREATY ACT

(See endnote 125)

This act makes it illegal to be involved in any purchase, importing, exporting, transporting, or sale of any migratory bird or any of its parts or nest without a valid permit from federal authorities. This act protects birds from people, which may create problems when there are large groups of birds at landfills and they potentially can spread disease.

FEDERAL HAZARDOUS SUBSTANCES ACT

(See endnote 133)

This act requires that all containers of household hazardous products be labeled as such in order for the consumer to know how to safely store, use, and dispose of the product. It also provides information on what to do in the event of an accident. The Consumer Product Safety Commission can make a determination if certain products are very dangerous and if the labeling inadequately reflects the potential concerns. This may lead to the banning of certain products.

FISH AND WILDLIFE COORDINATION ACT

(See endnote 127)

This act protects fish and wildlife resources when federal programs change the natural stream or body of water. This includes the movement of pollutants from industrial, mining, and municipal waste or dredged materials into a body of water or wetlands as well as projects for construction of dams, levees, impoundments, and any other type of water diversion structures.

FEDERAL INSECTICIDE, FUNGICIDE, AND RODENTICIDE ACT

(See endnote 128)

This act provides federal control of all pesticide distribution, sale, and use. All pesticides must be registered with the US EPA and they must contain accurate and proper labeling. There are special requirements for agriculture, and commercial and residential use as well as the tolerance levels, worker protection, and determination of certain chemicals which have restricted use or cannot be used. All new pesticides must be registered and the old ones receive a reregistering review every 15 years.

ENDANGERED SPECIES ACT

(See endnote 129)

Congress passed this act to protect native plants and animals that they felt would become extinct. They recognized that the United States has a rich natural heritage which provides beauty and ecological, educational, recreational, and scientific value to the country.

ROBERT T STAFFORD DISASTER RELIEF AND EMERGENCY ASSISTANCE ACT

(See endnote 130)

This act provides the statutory authority for most of the federal disaster programs of FEMA.

RESOURCES

- The National Oil and Hazardous Substances Pollution Contingency Plan, which is called the National Contingency Plan, or NCP, is the way in which the federal government responds to oil spills and hazardous substances releases. The original plan was developed in 1968 and has been upgraded periodically with the latest revisions finalized in 1994. (See endnotes 115, 118, 123, 126.)
- The US Environmental Protection Agency provides expertise on the effects to human health and ecosystems from various pollutants and how best to control them.
- The United States Coast Guard provides physical facilities and expertise in all areas of port safety and security, maritime law, construction, and the handling of hazardous chemicals or waste aboard ships and in the ports.
- The Federal Emergency Management Agency provides technical assistance in the area of hazardous materials, emergency planning, and operational assistance to communities when a serious disastrous event occurs.
- The Department of Defense provides complete information, control, and operational assistance in any type of hazardous materials release or hazardous waste problems in any of their facilities.
- The Department of Energy provides assistance in the control of hazardous materials, especially radiological materials, covered by their mandate.
- The Department of Agriculture has five specific agencies which are involved in providing scientific and technical assistance in a variety of areas impacted by hazardous materials or hazardous wastes as well as fires. These agencies are the Forest Service; Agricultural Research Service; Natural Resources Conservation Service; Animal and Plant Health Inspection Service; and Food Safety and Inspection Service.
- The Department of Commerce's National Oceanic and Atmospheric Administration provides scientific assistance and helps in resolving hazardous situations in coastal environments as well as cleanup and mitigation of the problem.

- The Department of Health and Human Services through its numerous agencies provides technical assistance and professional personnel where needed in the assessment, preservation, and protection of human health. They are the lead department of government in many areas related to hazardous materials and hazardous wastes. The specialized agencies are the Agency for Toxic Substances and Disease Registry (See endnote 110); Centers for Disease Control and Prevention; Food and Drug Administration; Health Resources and Services Administration; Indian Health Service; National Institutes of Health; National Institute for Environmental Health Sciences; and National Institute of Occupational Safety and Health.
- The Department of the Interior through its numerous agencies provides technical expertise and professional personnel for resolving hazardous materials and hazardous waste problems within its jurisdictional areas. These agencies are the US Fish and Wildlife Service; National Biological Service; US Geological Survey; Bureau of Land Management; Bureau of Mines; National Park Service; and Bureau of Indian Affairs.
- The Department of Justice provides necessary advice on a variety of legal issues which may occur as a result of hazardous material spills and incidents or hazardous waste problems.
- The Department of Labor's Occupational Safety and Health Administration provides technical assistance and personnel when there are hazards that affect workers or industries.
- The Department of Transportation provides expert advice on the transportation of all types of hazardous materials in all types of vehicles.
- The Nuclear Regulatory Commission provides specific advice on protecting people and the environment as well as appropriate recovery operations involving radioactive materials.

PROGRAMS

OCCUPATIONAL HEALTH AND SAFETY PROGRAM FOR HAZARDOUS WASTE SITES

(See endnotes 95, 106)

Note: The program described below may be specific for hazardous waste sites but all of its elements constitute basically what an occupational health and safety program should include. Therefore, these elements and program are usable in virtually all of the other chapters in this book as the material relates to occupational health and safety and employee welfare.

Planning and Operation

Planning is not only the first of the elements of a good program but also the most critical one when dealing with any type of activity involving people and/or the natural environment. This is especially true of hazardous waste sites with their many unique problems and opportunities for people, air, land, and water to be exposed to innumerable dangerous substances. Planning consists of: establishing an organizational structure for all working operations; developing a comprehensive work plan including every single facet of the operation, and its potential dangers and means of prevention and mitigation if necessary; developing and implementing a comprehensive health and safety plan which includes all aspects of occupational prevention, treatment, and healthcare, starting with pre-employment physical examinations, necessary immediate treatment, and long-term care as needed; risk analysis of all aspects of the work environment; and risk communications to all individuals concerning what might happen and how to deal with it.

The plan also has a special section on the contamination of air, land, and water and how to prevent, mitigate, and control the various hazardous substances which may degrade these environmental media and the incorporated ecosystems. From its inception, the plan proposal should include the thoughts, knowledge, and experience of all interested parties and partners. A local and state contingency plan identifying all appropriate personnel, policies, procedures, communications systems, and other equipment should be established, and this response team should carry out all necessary functions when accidental releases or disasters occur.

The organizational structure should include a high-level administrator with direct authority to carry out all necessary actions in the implementation of the plan and to order industrial and other operations shut down and emergency actions taken. This authority should come directly from the chief administrative officer and the board of trustees. It should cross all departmental lines. Other personnel needed to carry out program elements should be identified, and specific detailed criteria should be established for hiring them as well as providing the job specifications for the position. Lines of authority, responsibility, and communications should be established clearly. An individual should be identified who will interface with the community and the media. There should be a scientific advisor to the project to clarify all problems where contamination may be released into the environment or individuals may become affected by the hazardous substances. A comprehensive work plan should include the input of all levels of supervision and management and the individuals actually carrying out the various tasks. Appropriate training and retraining should be established.

A written site safety plan based on well-thought-out policies and procedures based on industry experience and best available technologies to protect personnel and the public must be put into action before any work occurs at the site. The site safety plan should list the key personnel, methods of contact, and responsibilities; describe the risks from each portion of the disposal system and how to deal with them; describe the various types of personal protective clothing and equipment to be used; describe any specific medical surveillance needed; list the types of monitoring including air, water, land, and personnel; describe means of mitigating potential and existing hazards; list decontamination procedures for personnel, equipment, and the environment; establish emergency response to spills, fires, explosions, worker injuries, and catastrophic events; and establish all standard operating procedures.

Hazard Assessment

An on-site survey including proper testing of air, water, and land should be conducted at any given site to determine the actual and potential problems that may be occurring that could cause serious health and safety issues for individuals and for the environment in general. The physical survey is an observation of types of materials, types of storage tanks and other areas, compressed gas cylinders, rusted, corroded, leaking, and bulging waste containers, the physical condition of all containers and substances, and the potential pathways of dispersion of hazardous materials. All health and safety hazards must be listed. Air samples should be taken of various parts of the facility for the presence of explosive atmospheres, reduced oxygen levels, inorganic and organic gases and vapors, hazardous airborne chemicals, dust, and other pollutants. Water samples from groundwater supplies and surface water should be taken to determine if there are any levels of hazardous chemicals or materials in the water. Radiological hazards should be quantified and it should be determined if they are alpha particles, beta particles, or gamma rays. This information is necessary not only for mitigation and control but also for determining the appropriate standard operating procedures to be used, the necessary medical supervision, and the type of personal protective equipment to be used and its limitations. In addition, a determination should be made by industrial hygienists as to the implementation of threshold limit values for chemical substances and physical agents that workers can be exposed to over a period of time compared to the permissible exposure limits which have been established as standards by the Occupational Safety and Health Administration.

Training

All workers must receive initial comprehensive training before being allowed to enter any work site. They should have a complete understanding of the potential hazards, and understand and be competent in using knowledge and skills to minimize risk to the individual and others, how to use their equipment properly and when the equipment will fail, and how to leave a hazardous area in the event of an emergency. They should have full knowledge of: the site safety plan and safe work practices; handling, storage, and transportation of hazardous materials; the individuals' rights and responsibilities as employees; and how to use testing instruments and conduct various tests safely and efficiently. Short follow-up sessions should be conducted on a periodic basis to refresh their understanding of potential hazards and how to mitigate these hazards by using actual Best Practices for the individual facility. It is of utmost importance to have available to the individuals during training and then in actual practice highly competent managers and supervisors who work on-site with the individuals.

Medical Program

Develop a comprehensive medical program for the facility which should include a pre-employment medical evaluation and screening program; regular periodic medical examinations; special emergency medical evaluation and treatment in the event of an exposure to hazardous wastes, symptoms of illnesses, or injuries; necessary blood, urine, and tissue tests; and follow-up treatments and examination upon termination of work. Complete records should be kept and be available for review upon request. Employee confidentiality must be maintained during this process. All medical people conducting examinations must consider the person's previous medical history, work history, specific health problems, and chronic conditions. All injuries and illnesses must be reported to the appropriate authority. Individuals must be cleared before assuming any work responsibilities based on medical conditions, ability to work when wearing personal protective equipment, and lifestyle habits such as smoking and use of alcohol and drugs as well as the pharmaceuticals used. Highly skilled medical and ancillary personnel must be available on a 24-hour basis and an on-site emergency facility must be staffed appropriately. A local general hospital must have a complete understanding of the potential hazards, medical symptoms, and how to treat individuals who have been exposed to various hazards and then are brought in to the hospital emergency room from the facility. The entire program should be subject to review by an outside professional group.

Evaluation

Evaluations of all parts of the preceding program should be conducted when specific emergency situations occur and also on a regular planned schedule, and necessary changes should be made immediately as needed to avoid hazardous situations for employees and damage to the environment. An in-depth study should be made of the entire program at least once a year by outside professionals. Results of the study and recommendations for improved operations and the necessary budget should be presented in writing to the chief executive officer and the board of trustees of the company for appropriate action

Integrated Solid Waste Management

Integrated solid waste management is a comprehensive program of waste prevention, recycling, composting, and disposal through the use of incinerators and finally landfills. The program consists of a thorough survey and study of the existing solid waste problems, solid waste programs, and disposal facilities. It incorporates all of the elements of source reduction or waste prevention, collection, transportation and storage of the waste, recycling and composting, and disposal by combustion and use of a landfill. After the completion of the appropriate surveys and studies and the nature of the waste stream is known, a plan is developed for a given area and the planners take into consideration the type of community; the social problems that might exist or be created; the financial cost of the proposal; the economic loss or gain due to implementation of the proposal; the technical problems which need to be resolved in the collection process, transportation process, and storage process; the ultimate environmental degradation which may occur as a result of the solid waste disposal; the potential for causing disease and injury in the citizenry and the workers; and the development and adherence to all federal state and local statutes.

US PROGRAM ON RECYCLING OF TIRES

(See endnote 15)

Used tires in the past were stockpiled, dumped illegally, or taken to landfills. In 1994, it was estimated that in the United States there were 700–800 million scrap tires. By 2003, because of aggressive clean-up and disposal efforts, there were an estimated 275 million tires in stockpiles. The markets for scrap tires were absorbing 233 million, with 130 million used as fuel, 56 million recycled in civil engineering projects, 18 million converted into ground-up rubber for recycling, 12 million converted into ground-up rubber for modified asphalt, 9 million exported, etc. Some 16.5 million were re-treaded in foreign countries and reused there. US EPA Region 5 and the Illinois EPA have developed a Scrap Tire Cleanup Guidebook for use by communities. This recycling of tires is a prime example of how communities can work together with governmental guidance to recycle various types of materials and keep them out of landfills and incinerators. This saves money, raw materials, and energy and helps reduce mosquito problems and aesthetic problems.

BATTERY RECYCLING PROGRAM

(See endnote 16)

Because of federal and state laws on the recycling of lead-acid batteries and strict enforcement, there has been a substantial drop in the amount of these units not being recycled. A study of recycling of these batteries for the years 2009–2013 indicates that 99% of the batteries are recycled. This is well above the rate of recycling for aluminum soft drink and beer cans, newspapers, and glass bottles. This is done only at approved and closely supervised re-processors. The plastic is removed, washed thoroughly and blown dry. It is then melted together and is ready to be reused in battery cases. The lead parts are thoroughly cleaned and then melted together in smelting furnaces and the lead is then poured as needed for the batteries. The sulfuric acid is neutralized and the water is treated, cleaned, and tested to make sure it does not violate clean water standards. The typical new battery contains 60–80% recycled materials.

MODEL SOLID WASTE MANAGEMENT PROGRAM

(See endnote 30)

A model solid waste management program consists of many factors including the industry partners who will be working with the regulatory agencies to follow the appropriate laws, rules, and regulations and utilize the various resources which are provided to them. The program consists of source reduction, recycling/reuse, treatment, and ultimate disposal. When the disposal area if it is a landfill is full, then it is necessary to have an appropriate closure with no opportunity for leakage to the surface or groundwater supply and the prevention of gases contaminating the air.

Source reduction prevents actual waste generation by volume and reduces potential toxicity. It consists of: elimination of certain materials; inventory control and on-time delivery of supplies as well as excellent management; material substitution of less toxic raw materials instead of more toxic ones and materials of less volume instead of those of more volume; modification of the process of making the finished product to avoid producing contaminants and reducing energy use; improved housekeeping by following appropriate techniques on a regular scheduled basis of cleaning and waste removal as well as using special scheduling for certain types of equipment and facilities; and putting into the purchase contract that all unused materials will be returned to the supplier for appropriate credits to avoid creating additional waste and controlling costs.

Recycling/reuse is the collection and use of waste materials free of contaminants that can be put back into the process instead of using raw materials or are sold to other industries as a byproduct. It consists of: reusing existing materials where feasible; reprocessing and reclaiming materials for either internal use or sale as a byproduct to other companies; and using the materials as a fuel to reduce waste and reduce cost.

Treatment techniques vary with the industry and the product and may be physical, chemical, or biological in nature. These techniques consist of: filtration through a variety of different filters which may also produce recycled or reused material; various types of chemical, biological, or thermal treatments to make hazardous material less hazardous and to reduce volume; extraction of the usable materials from the unusable and the non-hazardous from the hazardous; chemical stabilization to make the products inert instead of active; and spreading on the land for agricultural purposes if it is not hazardous.

Disposal may be in lined landfills, through incineration, by discharge to bodies of water with acceptable National Pollutant Discharge Elimination System permits allowing it, turning the materials into solids for later disposal in landfills, burial on-site, and underground injection of the materials.

WASTE MANAGEMENT IN THE OIL AND GAS EXPLORATION AND PRODUCTION PROCESS

(See endnote 31)

This waste management program consists of a series of steps.

1. An overall plan needs to be established at the highest levels of the organization and includes identifying the content of the program based on specific needs, goals, and timed objectives, titles of supervisory and management personnel involved and their roles, resources needed and supplied including the proper budget, means of measuring results, and appropriate record-keeping.
2. Determine the geological area and the businesses and industrial activities to help understand how this program fits in with their existing programs and what type of market would be available for byproducts or recycled material from the company.
3. Conduct an in-depth study to identify the type and quantity of the waste to be removed or recycled and if it is biological, chemical, or physical in nature and the potential level of toxicity.
4. Evaluate all rules, regulations, and laws pertaining to environmental and public health issues and determine what special precautions must be taken to prevent the occurrence of additional environmental problems when waste is being removed from the facility or special processes are being used to reduce hazards.
5. Utilize waste reduction toxicity techniques where possible.
6. Determine the state-of-the-art techniques for waste treatment and waste management, and utilize those that are cost-effective in all parts of the storage, transport, and final disposal of the waste.
7. Evaluate the operational procedures of the plan during actual operation of the program, and determine that which is weak and ineffective and make necessary changes.

US ENVIRONMENTAL PROTECTION AGENCY HOTLINE AND SERVICE LINE PROGRAM

(See endnote 98)

This program provides for individuals immediate information and assistance in a large variety of program areas of the US EPA.

ENDNOTES

1. US Environmental Protection Agency. 2015. *Wastes-Non-Hazardous Wastes-Municipal Solid Waste.* Washington, DC.
2. US Environmental Protection Agency. 2012. *EPA's Guide for Industrial Waste Management.* Washington, DC.
3. Koren, Herman, Bisesi, Michael. 2003. Solid and Hazardous Waste Management. In *Handbook of Environmental Health-Pollutant Interactions in Air, Water, and Soil*, Volume 2, Fourth Edition. Lewis Publishers, CRC Press, Boca Raton, FL.
4. GRACE Communications Foundation. 2015. *Sustainable Table-Waste Management.* New York.
5. US Environmental Protection Agency. 2014. *Groundwater and Ecosystems Restoration Research.* Washington, DC.
6. Heibar, Carrie, Schultz, Mark (editor). 2010. *Understanding Concentrated Animal Feeding Operations and Their Impact on Communities.* National Association of Local Boards of Health, Bowling Green, OH.
7. US Department of Agriculture. 2009. *Agricultural Waste Management Field Handbook.* 210-AWMFH, 4/92. Washington, DC.
8. US Environmental Protection Agency. 2014. *Animal Feeding Operations-Best Management Practices.* Washington, DC.
9. Spiehs, Mindy, Goyal, Sagar. 2015. *Manure Management and Air Quality-Best Management Practices for Pathogen Control in Manure Management Systems.* University of Minnesota Extension Service, St. Paul, MN.
10. World Bank. 1999. *What a Waste: Solid Waste Management in Asia.* Washington, DC.
11. O'Leary, Philip R., Walsh, Patrick W. 1995. *Decision-Makers Guide to Solid Waste Management.* Volume 2. Cooperative Agreement Number CX-817119-01. US Environmental Protection Agency, Office of Solid Waste, Municipal and Industrial Solid Waste Division, Washington, DC.
12. US Environmental Protection Agency, Office of Solid Waste. No Date. *Waste Transfer Stations: A Manual for Decision-Making.* Washington, DC.
13. US Environmental Protection Agency. 1993. *Business Guide for Reducing Solid Waste.* Washington, DC.
14. Mecklenburg County, North Carolina, Land Use and Environmental Services Agency, Solid Waste Division. 2011. *Best Practices for Local Government Solid Waste Recycling, Diversion from Landfill and Waste Reduction.* Charlotte, NC.
15. US Environmental Protection Agency, Wastes, Resource Conservation. 2012. *Basic Information (Tires).* Washington, DC.
16. Smith Bucklin Statistics Group. 2014. *National Recycling Rate Study.* Battery Council International, Chicago, IL.
17. US Environmental Protection Agency. 2015. *Advancing Sustainable Materials Management: 2013 Fact Sheet-Assessing Trends in Material Generation, Recycling and Disposal in the United States.* Washington, DC.
18. Richard, Tom L. 1996. *Municipal Solid Waste Composting: Physical Processing.* Cornell University Waste Management Institute, Ithaca, NY.
19. Richard, Tom L., Woodbury, Peter B. 1993. *Strategies for Separating Contaminants from Municipal Solid Waste Compost.* Cornell University Waste Management Institute, Ithaca, NY.
20. Richard, Tom L. 2000. *Municipal Solid Waste Composting: Biological Processing.* Cornell University Waste Management Institute, Ithaca, NY.
21. Woodbury, Peter B., Breslin, Vincent T. 1993. *Key Aspects of Compost Quality Assurance.* Cornell University Waste Management Institute, Ithaca, NY.
22. Olmsted County, Department of Environmental Resources. 2015. *Industrial Solid Waste.* Rochester, MN.
23. Brewers Association. No Date. *Solid Waste Reduction Manual.* Boulder, CO.
24. US Environmental Protection Agency, Office of Water. 2010. *Guidance Document: Best Management Practices for Unused Pharmaceuticals in Healthcare Facilities.* Draft, EPA-821-R-10-006. Washington, DC.
25. US Environmental Protection Agency, Office of Solid Waste. 1995. *Construction and Demolition Waste Landfills.* Contract Number 68-W3-0008. ICF Incorporated, Washington, DC.
26. National Association of Homebuilders Research Center. No Date. *Waste Management and Recovery-A Field Guide for Residential Remodelers.* Upper Marlboro, MD.

27. US Department of Defense. 2002. *Unified Facilities Criteria-Selection of Methods for the Reduction, Reuse and Recycling of Demolition Waste*, UFC 1-900-01. Washington, DC.

28. US Environmental Protection Agency. 2014. *Special Wastes*. Washington, DC.

29. US Environmental Protection Agency. 2015. *Final Rule-Frequent Questions about the Coal Ash Disposal Rule*. Washington, DC.

30. Gretches-Wilkinson Center for Natural Resources, Energy and Environment. No Date. *Intermountain Oil and Gas BMP Project-Solid Waste*. Boulder, CO.

31. Karami, Sepher, Torkashvand, Porya, Mozafari, Sajad. 2013. The Basic Steps in Waste Management in the Oil and Gas Exploration and Production Process. *Annals of Biological Research*, 4:19–22.

32. US Environmental Protection Agency, Office of Solid Waste. 2015. *Identification and Description of Mineral Processing Sectors and Waste Streams-Final Technical Background Document*. Washington, DC.

33. US Environmental Protection Agency. 2015. *Mineral Processing Waste*. Washington, DC.

34. Congress of the United States, Office of Technical Assessment. 1992. *Managing Industrial Solid Wastes from Manufacturing, Mining, Oil and Gas Production, Utility Coal Combustion*. OTA-BP-O-82. Washington, DC.

35. US Environmental Protection Agency. 2015. *Mining Waste*. Washington, DC.

36. US Environmental Protection Agency, Office of Solid Waste. 2015. *Human Health and Environmental Damages from Mining and Mineral Processing Waste*. Washington, DC.

37. US Environmental Protection Agency, Office of Water. 2006. *Industrial Stormwater-Mineral Mining and Processing Facilities-Fact Sheet Series*. EPA-833-F-06-025. Washington, DC.

38. European Commission. 2009. *Reference Document on Best Available Techniques for Management of Tailings and Waste Rock and Mining Activities*. Brussels, Belgium.

39. US Environmental Protection Agency, Office of Solid Waste and Emergency Response. 2012. *Green Remediation Best Management Practices: Mining Sites*. EPA 542-F-12-028. Washington, DC.

40. University of California, Los Angeles Office of Environment, Health and Safety. No Date. *Medical Waste Management*. Los Angeles, CA.

41. Oregon State University. 2010. *Sharps Safety Plan*. Corvallis, OR.

42. National Center for Manufacturing Compliance Healthcare Environment Resource Center. 2015. *Pollution Prevention and Compliance Assistance Information for the Healthcare Industry-Treatment and Disposal of Regulated Medical Waste*. Ann Arbor, MI.

43. Idaho Department of Environmental Quality. No Date. *Medical Waste Best Management Practices*. Boise, ID.

44. US Environmental Protection Agency, Office of Water. 2010. *Best Management Practices for Unused Pharmaceuticals of Healthcare Facilities*. Washington, DC.

45. Department of Homeland Security, Department of Labor, US Environmental Protection Agency. 2014. *Executive Order 13650-Actions to Improve Chemical Facilities Safety and Security A Shared Commitment-Report for the President*. Washington, DC.

46. US Department of Transportation, Pipeline and Hazardous Materials Safety Administration. No Date. *Hazardous Materials Transportation Training Modules-Version 5.1-Student*. Washington, DC.

47. Oklahoma State University, EHS Safety Training. 2011. *Hazmat Transportation*. Tulsa, OK.

48. Oklahoma State University, EHS Safety Training. No Date. *Hazmat Transportation-Department of Transportation, Dangerous Goods Classifications*. Tulsa, OK.

49. University of York, Promoting Science at the University of York. 2013. *The Chemical Industry*. York, United Kingdom.

50. Stellman, JM, editor. *Encyclopedia of Occupational Health & Safety*, Fourth Edition. International Labor Office, Geneva, Switzerland.

51. World Health Organization. 2009. *Good Manufacturing Practices for Pharmaceutical Products Containing Hazardous Substances*, Working Document QAS/08 .256 Rev. 1. Geneva, Switzerland.

52. Adis, Warren. 2007. *A Risk Modeling Framework for the Pharmaceutical Industry*. Communications of the IIMA, 7:1. California State University, San Bernardino, CA.

53. World Bank Group, International Finance Corporation. 2007. *Environmental Health and Safety Guidelines for Large Volume Inorganic Compounds Manufacturing in Coal Tar Distillation*. Washington, DC.

54. US Environmental Protection Agency. 2014. *2012 Data Reporting Results*. Washington, DC.

55. Hazard, Andrea, Beardsley, Padilla Speer. 2008. *Process Safety and Specialty Chemicals: Turning Industry Challenges into Opportunities*. Rockwell Automation, Milwaukee, WI.

56. European Bank for Reconstruction and Development. 2014. *Sub-Sectoral Environmental and Social Guideline: Manufacture of Chemicals*. London, England.
57. Virginia Tech University-Environmental Health and Safety Services. 2011. *Chemical Storage and Management*. Blacksburg, VA.
58. World Bank Group, International Finance Corporation. 2007. *Environmental, Health, And Safety General Guidelines*. Washington, DC.
59. Florida Fertilizer and Agricultural Association, Florida Department of Agriculture and Consumer Services, Florida Department of Environmental Protection. 1997. *Best Management Practices for Blended Fertilizer Plants in Florida*. Tallahassee, FL.
60. US Environmental Protection Agency, Office of Water. 2001. *Class I Underground Injection Control Program: Study of the Risk Associated with Class I Underground Injection Wells*. EPA 816-R-01-007. Washington, DC.
61. Groundwater Protection Council. 2013. *Injection Wells: An Introduction to Their Use, Operation and Regulation*. Oklahoma City, OK.
62. US Environmental Protection Agency. 2000. *Small Entity Compliance Guide-How the New Motor Vehicle Waste Disposal Well Rule Affects Your Business*. EPA 816-R-00-018. Washington, DC.
63. US Environmental Protection Agency. 2013. *Large-Capacity Septic Systems*. Washington, DC.
64. US Environmental Protection Agency, Office of Water. 2001. *Technical Program Overview: Underground Injection Control Regulations*. EPA 816-R-02-025. Washington, DC.
65. US Environmental Protection Agency, Office of Water. 2013. *Geologic Sequestration of Carbon Dioxide-Underground Injection Control Program Class VI Well Testing and Monitoring Guidance*. EPA 816-R-13-001. Washington, DC.
66. Massachusetts Executive Office of Energy and Environmental Affairs. 2015. *Standard Design Guidelines for Shallow UIC Class V Injection Wells*. Boston, MA.
67. Environmental Protection Agency Part III, 40 CFR parts 124, 144, 145, etc., Federal Requirements under the Underground Injection Control Program for Carbon Dioxide Geological Sequestration. 2010. *Federal Register*, 75:237.
68. Government Accountability Office. 2014. *Report to Congressional Requesters-Drinking Water-EPA Program to Protect Underground Sources from Injection of Fluids Associated with Oil and Gas Production Needs Improvement*. GAO-14-555. Washington, DC.
69. US Environmental Protection Agency. 2014. *Minimizing and Managing Potential Impacts of Injection–Induced Acidity from Class 2 Disposal Wells: Practical Approaches*. Washington, DC.
70. Fulcher, Peter, Tieman, Mary. 2015. *Human, Induced Earthquakes from Deep Well Injection: A Brief Overview*. 7-5700, R 43836. Congressional Research Service, Washington, DC.
71. US Environmental Protection Agency, Office of Solid Waste and Emergency Response. 2013. *Introduction to In Situ Bioremediation of Groundwater*. EPA 542-R-13-018. Washington, DC.
72. Naval Facilities Engineering Command. 2013. *Best Practices for Injection and Distribution of Amendments*. Technical Report, TR-NAV FAC-EXWC-EV-1303. Columbus, OH.
73. US Environmental Protection Agency. 2014. *Landfills*. Washington, DC.
74. US Environmental Protection Agency. 2006. *Safe Disposal Procedures for Household Appliances That Use Refrigerants*. EPA 530-F-06-020. Washington, DC.
75. US Environmental Protection Agency, Solid Waste and Emergency Response. 2001. *Geosynthetic Clay Liners Used in Municipal Solid Waste Landfills*. EPA 530-F-97-002. Washington, DC.
76. US Environmental Protection Agency, Office of Solid Waste. 1995. *Construction and Demolition Waste Landfills*. Washington, DC.
77. US Environmental Protection Agency. 2012. *C&D Landfills*. Washington, DC.
78. Innovative Ways Consulting Services. 2014. *Best Management Practices to Prevent and Control Hydrogen Sulfide and Reduce Sulfur Compound Emissions at Landfills that Dispose of Gypsum Drywall*. EPA/600/R-14/039. US Environmental Protection Agency Office of Research and Development, Cincinnati, OH.
79. Ohio Environmental Protection Agency. 2011. *Subsurface Heating Events at Solid Waste and Construction and Demolition Debris Landfills: Best Management Practices*. Guidance Document 1009. Columbus, OH.
80. US Environmental Protection Agency. 2014. *Wastes-Non-Hazardous Wastes-Municipal Solid-Incinerators-Basic Information*. Washington, DC.
81. US Environmental Protection Agency. 2014. *Air Emissions from MSW Combustion Facilities*. Washington, DC.

82. US Environmental Protection Agency. 2014. *Energy Recovery from Waste*. Washington, DC.
83. World Bank. 1999. *Technical Guidance Report-Municipal Solid Waste Incineration*. Washington, DC.
84. Wilson, Barry, Williams, Neil, Liss Barry. 2013. *A Comparative Assessment of Commercial Technologies for Conversion of Solid Waste to Energy*. EnviroPower Renewable, Inc., Boca Raton, FL.
85. US Environmental Protection Agency. 2014. *Combustion*. Washington, DC.
86. Ranous, Richard A. 2012. *HMCRP Report 9: A Compendium of Best Practices and Lessons Learned for Improving Local Community Recovery From Disastrous Hazardous Materials Transportation Incidents*. Pipeline and Hazardous Materials Safety Administration, Transportation Research Board of the National Academies, Washington, DC.
87. US Environmental Protection Agency. 2012. *Household Hazardous Waste*. Washington, DC.
88. US Environmental Protection Agency, Office of Pollution Prevention and Toxics. 2006. *Chemical Management Resource Guide for School Administrators*. EPA 747-R-06-002. Washington, DC.
89. Sierra Club. 1989. *Hazardous Waste Management*. San Francisco, CA.
90. US Environmental Protection Agency. 2014. *Wastes-Hazardous Waste-Treatment and Disposal-Combustion*. Washington, DC.
91. Interim Status Standards for Owners and Operators of Hazardous Waste Treatment, Storage and Disposal Facilities, CFR-2012. 2012. *Federal Register*, 27(Part 264.1–264.1202).
92. Eagleson, Kyley. 2014. *Surface Impoundments: Design and Operation*. Heritage Environmental Services-Environmentally Speaking, San Francisco, CA.
93. Cal Cupa Forum. No Date. *Requirements for Hazardous Waste Tank Systems*. Sacramento, CA.
94. US Environmental Protection Agency. 2012. *EPA's Guide for Industrial Waste Management-Designing and Installing Liners-Technical Considerations for New Surface Impoundments, Landfills, and Waste Piles*. Washington, DC.
95. US Department of Health and Human Services, Public Health Service, Centers for Disease Control and Prevention, National Institute for Occupational Safety and Health. 1985. *Occupational Safety and Health Guidance Manual for Hazardous Waste Site Activities*. Washington, DC.
96. Shuckrow, Alan J., Pajak, Andrew P., Touhill, C.J. 1980. *Management of Hazardous Waste Leachate*. Municipal Environmental Research Laboratory, Solid and Hazardous Waste Research Division, US Environmental Protection Agency, Cincinnati, OH.
97. US Environmental Protection Agency. 2014. *RCRA Orientation Manual 2014-Resource Conservation and Recovery Act*. Washington, DC.
98. US Environmental Protection Agency. 2015. *EPA Hotlines*. Washington, DC.
99. Washington State Department of Ecology, Spokane Aquifer Joint Board. 2005. *Best Management Practices for Hospital Waste*, 05-04-013. Spokane, WA.
100. Idaho Department of Environmental Quality. 2008. *Medical Waste Best Management Practices*. Boise, ID.
101. US Environmental Protection Agency. 2012. *Waste-Hazardous Waste-Universal Wastes-Batteries*. Washington, DC.
102. US Environmental Protection Agency. 2012. *Waste-Hazardous Wastes-Universal Waste-Pesticides*. Washington, DC.
103. US Environmental Protection Agency. 2015. *Compact Fluorescent Light Bulbs*. Washington, DC.
104. University of Missouri, Extension Service. 2015. *Household Hazardous Products*. Columbia, MI.
105. US Environmental Protection Agency. 2014. *Water: Best Management Practices-Proper Disposal of Household Hazardous Wastes*. Washington, DC.
106. European Agency for Safety and Health at Work. 2012. *E-Fact: 67-Maintenance and Hazardous Substances-Maintenance in the Chemical Industry*. Bilbao, Spain.
107. Fentiman, Audeen W., Jorat, Matthew E., Meredith, Joyce E. No Date. Ohio State University Extension Research Service by RER-40. Columbus, OH.
108. Olmsted County Department of Environmental Resources. 2013. *Industrial Solid Waste Fact Sheet-Electronic Waste*. Rochester, MI.
109. US Environmental Protection Agency. 2012. *Wastes-Resource Conservation-Common Waste-e Cycling*. Washington, DC.
110. Agency for Toxic Substances and Disease Registry. 2013. *Agency for Toxic Substances and Disease Registry Program Overview*. Atlanta, GA.
111. United States Code Title 42. 2011. *Chapter 116-Emergency Planning and Community Right-to-Know*. Washington, DC.
112. US Environmental Protection Agency. 2015. *Operating Permits-Who Has to Obtain a Title V Permit*. Washington, DC.

113. US Environmental Protection Agency. 2012. *Wastes-Non-Hazardous Waste-Industrial Waste Air Model.* Washington, DC.
114. New York State Department of Environmental Conservation. 2015. *Hazardous Waste Land Disposal Facility Operating Permits/Post-Closure Permits.* Albany, NY.
115. US Environmental Protection Agency. 2015. *National Oil and Hazardous Substances Pollution Contingency Plan (NCP) Overview.* Washington, DC.
116. US Environmental Protection Agency. 2015. *Laws and Regulations-Summary of the Comprehensive Environmental Response, Compensation, and Liability Act (Superfund).* Washington, DC.
117. United States Coast Guard, National Pollution Funds Center. No Date. *Oil Pollution Act of 1990 (OPA).* Arlington, VA.
118. US Environmental Protection Agency. 2015. *Superfund Regulations and Enforcement-National Oil and Hazardous Substances Pollution Contingency Plan.* Washington, DC.
119. Library of Congress. No Date. *Summaries for the Hazardous Materials Transportation Uniform Safety Act of 1990.* Washington, DC.
120. US Environmental Protection Agency. 2015. *Laws and Regulations-Summary of the Clean Water Act.* Washington, DC.
121. US Environmental Protection Agency. 2015. *Laws and Regulations-Summary of the Clean Air Act.* Washington, DC.
122. US Environmental Protection Agency. 2015. *Safe Drinking Water Act (SDWA).* Washington, DC.
123. US Environmental Protection Agency. 2015. *Laws and Regulations-Summary of the Toxic Substances Control Act.* Washington, DC.
124. US Environmental Protection Agency. 2015. *Laws and Regulations-Summary of the National Environmental Policy Act.* Washington, DC.
125. US Fish and Wildlife Service. 2013. *Migratory Bird Program-Birds Protected by the Migratory Bird Treaty Act.* Washington, DC.
126. US Consumer Product Safety Commission. No Date. *Federal Hazardous Substances Act (FHSA) Requirements.* 1960 Washington, DC.
127. Federal Environmental Management Agency. 2015. *Fish and Wildlife Coordination Act, 1956.* Washington, DC.
128. US Environmental Protection Agency. 2012. *Federal Insecticide, Fungicide, and Rodenticide Act (FIFRA).* Washington, DC.
129. US Fish and Wildlife Service. 2013. *Endangered Species Act.* Washington, DC.
130. Federal Environmental Management Agency. 2015. *Robert T Stafford Disaster Relief and Emergency Assistance Act (Public Law 93-288) as amended.* Washington, DC.
131. University of Georgia. 2013. *Superfund Amendments and Reauthorization Act (SARA).* New Georgia Encyclopedia. Athens, GA.
132. US Environmental Protection Agency. 2015. *Laws and Regulations-Summary of the Resource Conservation and Recovery Act.* Washington, DC.
133. Consumer Product Safety Commission. 2012. *Federal Hazardous Substances Act Requirements (FHSA).* Washington, DC.
134. US Environmental Protection Agency. 2015. *Laws and Regulations-Summary of the Toxic Substances Control Act.* Washington, DC.

BIBLIOGRAPHY

Centers for Disease Control and Prevention. 2013. *Chemical Safety.* Atlanta, GA.
Contra Costa Health Services Hazardous Materials Program. 2013. *California Accidental Release Prevention Program.* Richmond, CA.
Orum, Paul. No Date. *Environmental-Community-Labor Process Safety-Promising Practice-Best Practices Examples: Safer and More Secure Chemicals and Processes.* Federal Environmental Management Agency. Washington, DC.
Healthcare Environmental Resource Center. 2015. *Pollution Prevention and Compliance Assistance Information for the Healthcare Industry-Hazard Communication Standard.* Ann Arbor, MI.
Michigan Department of Environmental Quality, Office of Waste Management and Radiological Protection. 2015. *Nonhazardous Liquid Waste Generator Requirements: Guidance.* Lansing, MI.
New Hampshire Department of Environmental Services. 2014. Best Management Practices for New Hampshire Solid Waste Facilities. *WMD-13-01.* Concorde, NH.

Olmsted County, Minnesota, Department of Environmental Resources. No Date. *Asbestos Containing Material Disposal Fact Sheet*. Rochester, MI.

Olmsted County, Minnesota, Department of Environmental Resources. 2013. *Industrial Solid Waste Fact Sheet—Ashes*. Rochester, MI.

Olmsted County, Minnesota, Department of Environmental Resources. 2013. *Industrial Solid Waste Fact Sheet—Contaminated Soils*. Rochester, MI.

Olmsted County, Minnesota, Department of Environmental Resources. 2013. *Industrial Solid Waste Fact Sheet—Foundry-Related Wastes*. Rochester, MI.

Olmsted County, Minnesota, Department of Environmental Resources. 2013. *Industrial Solid Waste Fact Sheet—Infectious Wastes*. Rochester, MI.

Olmsted County, Minnesota, Department of Environmental Resources. 2013. *Industrial Solid Waste Fact Sheet—Ink Sludges and Solvents*. Rochester, MI.

Olmsted County, Minnesota, Department of Environmental Resources. 2013. *Industrial Solid Waste Fact Sheet—Machining Waste*. Rochester, MI.

Olmsted County, Minnesota, Department of Environmental Resources. 2013. *Industrial Solid Waste Fact Sheet—Oil Contaminated Wastes*. Rochester, MI.

Olmsted County, Minnesota, Department of Environmental Resources. 2013. *Industrial Solid Waste Fact Sheet—Paint Related Waste*. Rochester, MI.

Olmsted County, Minnesota, Department of Environmental Resources. 2013. *Industrial Solid Waste Fact Sheet—Polychlorinated Biphenyls Contaminated Waste*. Rochester, MI.

Olmsted County, Minnesota, Department of Environmental Resources. 2013. *Industrial Solid Waste Fact Sheet—Resins: Epoxy, Fiberglass, Urethane, and Polyurethane*. Rochester, MI.

Olmsted County, Minnesota, Department of Environmental Resources. 2013. *Industrial Solid Waste Fact Sheet—Sludges*. Rochester, MI.

Olmsted County, Minnesota, Department of Environmental Resources. 2013. *Industrial Solid Waste Fact Sheet—Spent Carbon Filters*. Rochester, MI.

Olmsted County, Minnesota, Department of Environmental Resources. 2013. *Industrial Solid Waste Fact Sheet—Tires*. Rochester, MI.

Safe Drinking Water Foundation. No Date. *Industrial Waste*, Saskatoon, Canada.

US Department of Transportation, Federal Railroad Administration, Office of Research and Development. 2015. *Risk Evaluation Framework and Selected Metrics for Tank Cars Carrying Hazardous Materials*. DOT/FRA/ORD-15-07. Washington, DC.

US Environmental Protection Agency. 2012. *Wastes-Hazardous Wastes-Universal Wastes-Batteries*. Washington, DC.

US Environmental Protection Agency. 2012. *Wastes-Hazardous Waste-Universal Wastes-Pesticides*. Washington, DC.

US Environmental Protection Agency. 2012. *Wastes-Non-Hazardous Wastes-Municipal Solid-Bioreactors*. Washington, DC.

US Environmental Protection Agency. 2014. *Wastes-Hazardous Waste-Universal Waste-Mercury—Containing Equipment*. Washington, DC.

US Environmental Protection Agency. 2014. *Wastes-Non-Hazardous Waste-Industrial Waste-Cement Kiln Dust Waste*. Washington, DC.

US Environmental Protection Agency. 2014. *Wastes-Nonhazardous Waste-Industrial Waste-Crude Oil and Natural Gas Waste*. Washington, DC.

US Environmental Protection Agency. 2014. *Wastes-Nonhazardous Waste-Industrial Waste-Fossil Fuel Combustion Waste*. Washington, DC.

US Environmental Protection Agency. 2014. *Wastes-Resource Conservation-Common Wastes-Used Oil Management Program*. Washington, DC.

US Environmental Protection Agency. 2015. *Lead-Protect Your Family*. Washington, DC.

US Environmental Protection Agency. 2015. *Wastes-Non-Hazardous Waste-Industrial Waste-Mining Waste*. Washington, DC.

US Environmental Protection Agency, Solid Waste and Emergency Response. 2009. *Municipal Solid Waste Generation, Recycling, and Disposal in the United States: Facts and Figures for 2008*. Washington, DC.

US Environmental Protection Agency, Office of Solid Waste. 1998. *Final Technical Background Document-Identification and Description of Mineral Processing Sectors and Waste Streams*. Washington, DC.

US Environmental Protection Agency, Office of Solid Waste, Office of Policy, Planning, and Evaluation. No Date. *Sites for Our Solid Waste: A Guidebook for Effective Public Involvement.* EPA-530-F-009-021. Washington DC.

United States Government Accountability Office. 2015. *Report to the Ranking Member, Subcommittee on Environment and the Economy, Committee on Energy and Commerce, House of Representatives-Hazardous Waste-Agencies Should Take Steps to Improve Information on USDA's and Interiors Potentially Contaminated Sites.* GAO-15-35. Washington, DC.

13 Water Systems (Drinking Water Quality)

STATEMENT OF PROBLEM AND SPECIAL INFORMATION

Historically, waterborne disease was one of the great problems of all societies, and had a profound effect on life and health within communities. The advances made in water treatment and disinfecting of water supplies helped our society, especially the young, since they were typically most affected. Globally, today 88% of diarrhea cases are linked to unsafe water or poor hygiene techniques. This results in 1.5 million deaths each year, mostly in young children. (See endnote 1, Chapter 3; 40.)

About 1% of the fresh water in the world is available for human use, and this has stayed constant over many centuries despite the fact that there is an enormous demand for water created by huge increasing populations, which will continue to grow in the foreseeable future. There is a vast difference in the distribution of fresh water in various parts of the world as well as in the United States. Climate change, increasingly more powerful storms, and droughts are making the situation worse in certain areas. This is compounded by substantial population pressure in coastal areas and the growth of large cities and communities. Certain water sources, such as fossil water which has been trapped for a very substantial amount of time, may be used but not replenished. While the demand for safe drinking quality water is expanding rapidly, the supplies are diminishing.

Disposal of sewage, solid and hazardous wastes, air pollutants, and use and misuse of the many chemicals that make the modern world more livable are also challenging existing water supplies. Modern agriculture consumes a considerable amount of the fresh water which is available. There is a substantial need for protecting watersheds, using existing water more efficiently, preventing environmental pollutants from contaminating fresh water supplies, and cleaning up the results of the carefree attitude of waste disposal in many areas which has occurred since the Industrial Revolution and certainly in much greater quantities since World War II. Water delivery services through pipes which are extremely old result in reduced water at the point of need and water which may become severely contaminated by the pipes or the surrounding ground. A perfect example of the contamination of water by pipes has occurred in Flint, Michigan, where thousands of children were put at severe risk of lead poisoning when the governor's appointee to manage the city because of financial problems made a non-scientific highly questionable decision to stop water purchase from the City of Detroit and instead use the severely contaminated water from the Flint River. In addition, there have been multiple cases of Legionnaires' disease from this drinking water source. (See endnote 47.) Water scarcity exists globally, including the United States and is expected to increase sharply through the coming years. (See endnote 2.)

Fresh water may be found in the form of ice, especially in glaciers, which is not readily accessible; precipitation in the form of rain, snow, sleet, fog, and hail which is readily accessible; in streams, lakes, and rivers which is readily accessible; and in underground aquifers where the water may have been deposited hundreds or thousands of years ago or very recently. The groundwater may be found in spaces between particles of rock and soil, in cracks in the rock, between rock formations (artesian well source), as soil moisture, or at times in underground rivers or underground lakes. The groundwater typically flows slowly through the aquifers. Where water fills an area, it is called the water table and is just below the unsaturated zone, which contains both air and water. The moving groundwater depending on the minerals that it crosses or goes through may become naturally contaminated with a series of substances.

Groundwater is a major component of the hydrologic cycle where precipitation falls and a portion of it evaporates back to the atmosphere, a portion of it is used by plants, trees, or other green materials and goes back to the clouds through transpiration, and a portion of it collects in reservoirs, lakes and ponds, especially in arid areas, and evaporates back to the clouds. The clouds then become the source of more precipitation and this water may be used for irrigation for agriculture where a portion of it re-evaporates back to the clouds, a portion of it runs off to surface bodies of water such as rivers, lakes, streams, and oceans, and a portion of it percolates down through the ground to become groundwater, which can be used at a later date. (See endnote 3.)

The US Environmental Protection Agency (EPA) has promulgated mandatory National Primary Drinking Water Regulations. These are legally enforceable standards which apply to public water systems. The standards provide a list of contaminants including microorganisms, disinfection byproducts, inorganic chemicals, organic chemicals, and radionuclides, their maximum contamination levels, and potential health effects from long-term exposure above the maximum contamination levels or from short-term contamination. The Maximum Contaminant Level (MCL) is the highest level of contaminant that is allowed in drinking water using the best available treatment technology at reasonable cost. (For specifics see endnote 4.)

This chapter will be dedicated to protecting the water supply for human, and agricultural and industrial use, as well as for ecosystems and environmental sustainability. Since water is used and reused constantly and water treatment and delivery systems are essential to the process, these will also be discussed.

Many of the topics which will be discussed in this chapter have already been developed in depth in Chapter 2, "Air Quality (Outdoor [Ambient] and Indoor)"; Chapter 3, "Built Environment—Healthy Homes and Healthy Communities"; Chapter 4, "Children's Environmental Health Issues"; Chapter 5, "Environmental Health Emergencies, Disasters, and Terrorism"; Chapter 7, "Food Security and Protection"; Chapter 10, "Recreational Environment and Swimming Areas"; Chapter 11, "Sewage Disposal Systems"; Chapter 12, "Solid Waste, Hazardous Materials, and Hazardous Waste Management"; and Chapter 14, "Water Quality and Water Pollution." As a result of this and the inter-relationship between air, land, and water environments, and the movement of pollutants from one medium to another, the reader will be referred to other chapters to avoid duplication of written materials and elongating this book. However, when there are additional problems and Best Practices, or the author wants to emphasize certain points which are very significant to the reader, additional information will be provided.

SUB-PROBLEMS INCLUDING LEADING TO IMPAIRMENT AND BEST PRACTICES FOR GENERAL AND SPECIFIC SOURCES OF WATER CONTAMINATION, PREVENTION, MITIGATION, AND CONTROL

As can be seen in the list of potential contaminating circumstances shown below, there are innumerable other problems that can cause drinking water quality to deteriorate in such a manner that it can become undrinkable, cause short-term and long-term disease, and deprive industry, people, and other ecosystems of the necessary potable water to continue to thrive and exist. It is only through prevention, mitigation, control, and appropriate water treatment that life can exist and thrive in our modern society. A quick way to gather the necessary information concerning potential pollutants and how to avoid them or mitigate them is by using the index to find the subjects below.

- *Acid Mine Drainage*—See Chapter 14, "Water Quality and Water Pollution"
- *Acid Rain*—See Chapter 2, "Air Quality (Outdoor [Ambient] and Indoor)"
- *Agriculture*—See Chapter 2, "Air Quality (Outdoor [Ambient] and Indoor)"; Chapter 12, "Solid Waste, Hazardous Materials, and Hazardous Waste Management"; and Chapter 14, "Water Quality and Water Pollution"

- *Air Pollution Deposits*—See Chapter 14, "Water Quality and Water Pollution"
- *Air Toxics*—See Chapter 2, "Air Quality (Outdoor [Ambient] and Indoor)"; Chapter 12, "Solid Waste, Hazardous Materials, and Hazardous Waste Management"; and Chapter 14, "Water Quality and Water Pollution"
- *Businesses*—See Chapter 2, "Air Quality (Outdoor [Ambient] and Indoor)"
- *Cement Kilns*—See Chapter 2, "Air Quality (Outdoor [Ambient] and Indoor)"; and Chapter 12, "Solid Waste, Hazardous Materials, and Hazardous Waste Management"
- *Chemical Industry*—See Chapter 2, "Air Quality (Outdoor [Ambient] and Indoor)"; and Chapter 12, "Solid Waste, Hazardous Materials, and Hazardous Waste Management"
- *Climate Change*—See Chapter 2, "Air Quality (Outdoor [Ambient] and Indoor)"; Chapter 12, "Solid Waste, Hazardous Materials, and Hazardous Waste Management"; and Chapter 14, "Water Quality and Water Pollution"
- *Combined Stormwater and Sanitary Sewers*—See Chapter 11, "Sewage Disposal Systems"; and Chapter 14, "Water Quality and Water Pollution"
- *Composting*—See Chapter 12, "Solid Waste, Hazardous Materials, and Hazardous Waste Management"
- *Construction and Demolition Industry*—See Chapter 2, "Air Quality (Outdoor [Ambient] and Indoor)"; Chapter 12, "Solid Waste, Hazardous Materials, and Hazardous Waste Management"; and Chapter 14, "Water Quality and Water Pollution"
- *Controlled Burns*—See Chapter 2, "Air Quality (Outdoor [Ambient] and Indoor)"
- *Electric Power Plants*—See Chapter 2, "Air Quality (Outdoor [Ambient] and Indoor)"
- *Erosion and Sediment*—See Chapter 3, "Built Environment—Healthy Homes and Healthy Communities"; and Chapter 14, "Water Quality and Water Pollution"
- *Environmental Emergencies, Disasters, and Terrorism*—See Chapter 5, "Environmental Health Emergencies, Disasters, and Terrorism"
- *Floods*—See Chapter 5, "Environmental Health Emergencies, Disasters, and Terrorism"; Chapter 11, "Sewage Disposal Systems"; and Chapter 14, "Water Quality and Water Pollution"
- *Hazardous Wastes*—See Chapter 12, "Solid Waste, Hazardous Materials, and Hazardous Waste Management"
- *Hospital and Medical Waste*—See Chapter 8, "Healthcare Environment and Infection Control"; and Chapter 12, "Solid Waste, Hazardous Materials, and Hazardous Waste Management"
- *Household Products*—See Chapter 12, "Solid Waste, Hazardous Materials, and Hazardous Waste Management"
- *Incinerator Wastes*—See Chapter 12, "Solid Wastes, Hazardous Materials, and Hazardous Waste Management"
- *Illicit Discharges*—See Chapter 14, "Water Quality and Water Pollution"
- *Industrial Wastes*—See Chapter 12, "Solid Waste, Hazardous Materials, and Hazardous Waste Management"
- *Injection Wells*—See Chapter 12, "Solid Waste, Hazardous Materials, and Hazardous Waste Management"
- *Invasive Species*—See Chapter 14, "Water Quality and Water Pollution"
- *Landfills*—See Chapter 12, "Solid Waste, Hazardous Materials, and Hazardous Waste Management"
- *Mining and Extraction of Resources Wastes*—See Chapter 12, "Solid Waste, Hazardous Materials, and Hazardous Waste Management"; and Chapter 14, "Water Quality and Water Pollution"
- *Municipal Waste*—See Chapter 12, "Solid Waste, Hazardous Materials, and Hazardous Waste Management"; and Chapter 14, "Water Quality and Water Pollution"
- *Natural Sources of Pollution*—See Chapter 14, "Water Quality and Water Pollution"

- *Nutrients*—See Chapter 2, "Air Quality (Outdoor [Ambient] and Indoor)"; Chapter 11, "Sewage Disposal Systems"; Chapter 12, "Solid Waste, Hazardous Materials, and Hazardous Waste Management"; and Chapter 14, "Water Quality and Water Pollution"
- *On-Site Sewage Disposal*—See Chapter 11, "Sewage Disposal Systems"
- *Particulate Matter*—See Chapter 2, "Air Quality (Outdoor [Ambient] and Indoor)"
- *Pesticides*—See Chapter 9, "Insect Control, Rodent Control, and Pesticides"; Chapter 12, "Solid Waste, Hazardous Materials, and Hazardous Waste Management"; and Chapter 14, "Water Quality and Water Pollution"
- *Pet Waste*—See Chapter 14, "Water Quality and Water Pollution"
- *Radioactive Waste*—See Chapter 12, "Solid Waste, Hazardous Materials, and Hazardous Waste Management"
- *Runoff and Non-Point Source Pollution*—See Chapter 14, "Water Quality and Water Pollution"
- *Runoff from Hazardous Chemicals Spills*—See Chapter 5, "Environmental Health Emergencies, Disasters, and Terrorism"; and Chapter 12, "Solid Waste, Hazardous Materials, and Hazardous Waste Management"
- *Sand and Salt Storage*—See Chapter 14, "Water Quality and Water Pollution"
- *Schools*—See Chapter 4, "Children's Environmental Health Issues"
- *Sediment*—See Chapter 14, "Water Quality and Water Pollution"
- *Sewage Treatment Plants*—See Chapter 11, "Sewage Disposal Systems"; and Chapter 12, "Solid Waste, Hazardous Materials, and Hazardous Waste"
- *Surface Impoundments*—See Chapter 12, "Solid Waste, Hazardous Materials, and Hazardous Waste Management"; and Chapter 14, "Water Quality and Water Pollution"
- *Temperature (Excessive)*—See Chapter 14, "Water Quality and Water Pollution"
- *Toxic Waste Storage and Handling Areas*—See Chapter 12, "Solid Waste, Hazardous Materials, and Hazardous Waste Management"
- *Wildfires*—See Chapter 5, "Environmental Health Emergencies, Disasters, and Terrorism"
- *Wildlife*—See Chapter 14, "Water Quality and Water Pollution"

General Best Practices for Drinking Water Quality
- Recognize as a top national and international priority the need for drinking water quality sustainability to meet basic human needs, agricultural needs, and industrial needs. Establish all appropriate laws, rules, and regulations and strictly enforce all existing laws, rules, and regulations as a necessary means of protecting the homeland.
- Frequently test groundwater and surface water, especially if there is a concern about contamination, notify the public immediately of the potential hazard, and utilize corrective and preventive measures. Determine if the water meets the maximum contamination levels for microorganisms, disinfection byproducts, inorganic chemicals, organic chemicals, and radionuclides and if not, develop and implement a corrective action plan.
- Use agricultural water much more efficiently since as much as 60% of it is wasted and never reaches the intended crop. Research and develop the latest technologies for this purpose and maintain the system with great care and appropriate supervision.
- Develop and utilize appropriate filtration and disinfection techniques to remove all organisms that can cause waterborne disease.
- Immediately investigate all outbreaks of waterborne disease, notify the public of the potential hazard, and utilize corrective and preventive measures.
- Immediately investigate all outbreaks of inorganic and organic chemical problems, notify the public of the potential hazard, and utilize corrective and preventive measures.
- Establish a system where water-related conflicts will be resolved through formal negotiations thereby avoiding expensive and time-consuming court cases.

- Determine if lead is present in water. Homes, especially those built before 1986, may have lead pipes, fixtures, and solder and therefore create a potential problem of lead in the drinking water which is especially problematic for babies and young children. This source plus the consumption of lead-based paint, dust, etc. can cause lead blood levels which cause lead poisoning.
- Dug wells are not recommended for use for providing a potable water supply because of the large number of potential contaminants that may enter the water table.
- Specific Best Practices will be discussed below under the private drinking water system and public drinking water system categories.

SUB-PROBLEMS INCLUDING LEADING TO IMPAIRMENT AND BEST PRACTICES FOR PRIVATE DRINKING WATER SYSTEMS

Approximately 15% of the people of the United States utilize private drinking water wells as their water supply. Local and state rules and regulations are utilized to try to control the potential for disease. Wells are basically holes in the ground that are dug or drilled in different manners, protected against outside contamination where possible, and contain some mechanism, either a pump or in the case of artesian wells the pressure between two formations of rock, which forces the water to the top for use. There are three basic types of wells: dug wells, which are typically 10–30 feet deep and, therefore, are contaminated by surface water readily; driven wells, which are typically 30–50 feet deep and have a significant chance to become contaminated; and drilled wells, which are typically 100–400 feet deep and penetrate the bedrock and if properly sealed are least likely to become contaminated. Underground sources of water may also come to the surface through springs, which may or may not be contaminated depending on the source of the water. Artesian springs are typically safer than other types of springs.

Other means of collecting and using water include cisterns, which are big concrete boxes in the ground and water collected on the roof of a house which goes through some filters and then into the box for storage and use; infiltration galleries which are bottom drained sand filter trenches that are parallel to stream beds which filter the water as it passes through the filtering material so it becomes usable; and ponds or lakes.

There are numerous types of pumps which are used, based on the type of water well being installed, the depth to the water table, the type of soils encountered, and the quantity of water needed at the surface. All of them have the same function basically of moving the water from the aquifer to the surface so it can be utilized. Besides problems in the well which may lead to contamination, another problem historically for contamination has been the well pit. This has been eliminated by the use of a pitless well adapter that allows the water to go through the wall of the well casing or extension while protecting it from surface water or surface contamination. It is a connection between the underground horizontal discharge pipe and the vertical casing pipe or watertight casing. There are many different types of pitless well adapters that are usable depending on the circumstances.

To have a stable source of continuous water within a structure, there needs to be a pressure water storage tank to allow an adequate quantity of water to be available at all times when needed. About 10–30% of the volume of the tank contains water and the rest of the tank contains air to create the necessary pressure to move it through the system.

Water may be contaminated through naturally occurring sources of pollution. Natural sources of pollution include microorganisms such as bacteria, viruses, and parasites; radionuclides, such as radium and uranium found in rock or groundwater supplies; radon, a breakdown product of radium, which may be found in soil; nitrates and nitrites, found in water from the breakdown products of nitrogen compounds in the soil; heavy metals, such as arsenic, cadmium, chromium, lead, and selenium, which are found in rocks and soils; and fluoride which may be present in excessive amounts in soil and water.

Human activities can pollute the groundwater supply. Bacteria, viruses, and parasites may be found in human and animal waste. Overflowing septic tanks and waste from birds and farm animals create this problem. Concentrated animal feeding operations or factory farms put a large number of animals in a small area and therefore produce large amounts of waste material including nutrients and microorganisms, which when not handled properly can readily contaminate the water supply and the food supply. Heavy metals from mining and construction cause serious groundwater problems. Fertilizers and pesticides used on golf courses, lawns and gardens, and agricultural land pollute the groundwater as well as the surface water. Industrial products used in local businesses, factories, and industrial plants are common sources of hazardous chemicals in the water supply. Leaking underground tanks and pipes allow petroleum products, chemicals, and wastes to leak from corroded storage equipment and infiltrate the groundwater supply. Runoff and seepage from landfills and waste dumps contribute to the chemical wastewater infiltration problem. Household wastes including cleaning solvents, used motor oil, paints, and a variety of other chemicals are frequently mishandled or stored improperly and therefore can contaminate the groundwater. Lead and copper are found in drinking water pipes and therefore can contaminate the potable water used in the homes.

Another potential source of contamination of wells is the use of fracking in the exploration and production of domestic gas and oil. Fracking can lead to the contamination of groundwater because of spills, poor well construction, and the chemicals and materials used to separate the sources of energy from rocks under the ground. In several instances, fracking chemicals have been found in groundwater which has entered private drinking water supplies. The fracking process utilizes large amounts of surface and groundwater and therefore may endanger the amount of water available for human consumption. In some areas, there has been a concern about the potential for earthquakes where known geological faults exist.

As more wells are built in a given area, the cone of depression, which is like a small valley in the water, created by the pumping action underground extends the water table boundaries further out and therefore points of potential pollution which would have never entered this particular water table now do so, and create problems for all wells using this particular water table.

Once the water well has been put into operation, maintenance problems may occur because of: initial poor well design and construction; not developing the well system completely; borehole instability; a crust forming on the well screen from too much calcium, manganese, or iron in the water; biofouling due to iron bacteria forming a gelatin-like mass; corrosion of the metal well casing; over-pumping of the well which depletes the aquifer and also pulls sediment into the well; dissolved gases such as methane and hydrogen sulfide; and problems of the use of the aquifer since the aquifer does not recharge fast enough. (See endnote 8.)

Best Practices for Utilizing Private Drinking Water Sources (See endnotes 6, 7, 9)
- Review the information above on General and Specific Sources of Contamination and Best Practices for Prevention, Mitigation, and Control and implement the Best Practices which relate to the potential problems found in a given locality.
- Bored wells may be used providing the grouting or casing between the well and the soil goes down a minimum of 25 feet to impervious material. The casing should extend 12 inches above the ground and be at least 6 inches thick.
- Drilled wells are the best for producing quality safe water. The drill bit goes down through the ground until the hole is completed and then it is withdrawn and in its place casing which does not allow outside contaminants to go into the well along with the screen at the end of it goes down into the ground. It is grouted to a least 25 feet below the surface or to a solid formation.
- Use an experienced, licensed local well driller who knows the area. Determine if he is certified by the National Groundwater Association, has adequate equipment, and will submit well logs and furnish you a written contract which should cover compliance with local and

state regulations; liability insurance; casing specifications to the bedrock; use of a sanitary well cap and grout seal; development of the well to its maximum yield of water; type of pump and use of a pitless well adapter; disinfection after drilling and installing pump; an estimate of well yield including well logs and pump test results; date of completion; and itemized cost estimate. The contractor must obtain a permit from the authorizing authority before starting any work on the well.

- When building a new home or drilling a new well, understand the topography of the property, the direction of groundwater flow, the direction of surface water flow, and the potential pollution risks in the surrounding area as identified above under General and Specific Sources of Contamination and Best Practices for Prevention, Mitigation and Control.
- All wells need to be uphill from immediate potential sources of contamination and have a minimum number of feet of separation from them. See endnote 6, page 5 for details for typical sources of contamination which may be in close proximity to the well including wetlands, storm drains, sewage disposal sites, lakes, barnyards, etc.
- See endnote 6 for all required construction specifications and inspection requirements for a well as directed by the state of Pennsylvania (used as an example for all states and localities). These specifications include those for: casing; all materials and methods of use of these materials including grout; grout placement; drilling inspection; well casing and grouting inspection; pitless adapter and well seal inspection; well disinfection; and water quality analysis.
- The design of the pump head must include all means of prevention of pollution of the water supply by lubricants, maintenance materials, dusts, grain, birds, flies, animals, etc.
- Install a pressure tank with adequate storage for the water needs of the structure and in the event there is a low yielding well, increase the capacity of storage by adding an intermediate storage tank system. Set the pump pressure switch correctly so that when the water pressure falls below a certain level in the pressure tank, the pump will go on and recharge the pressure. (See endnote 10.)
- Reduce peak water use of well water by purchasing washing machines, dishwashers, toilets, and faucets or showerheads which provide mechanisms for reducing the amount of water utilized during the operation of these units.
- To correct reduced well yield because of pump or water system problems, have a licensed well water contractor or plumber determine if there is a leak in the system or worn pump parts and make necessary corrections.
- To correct reduced well yield because of biofilm buildup, inspect the system by use of a camera and shock chlorinate the well and water system.
- To correct reduced well yield because of mineral crust buildup, determine the problem by using a camera and have a licensed water well specialist treat the system.
- To correct reduced well yield because of sediment plugging the screen or casing as determined by sediment found in the water, have a licensed well contractor redevelop the well.
- To correct reduced well yield because of a collapse of the well casing or borehole usually due to age, depending on the cost, either recondition the entire well or plug it with concrete and construct a new well.
- Always determine if new wells have been added to those in existence using the aquifer and have professionals determine if this is reducing well yield. If so this becomes a governmental problem, where the authorizing agency must make a determination as to the amount of water that may be withdrawn from the aquifer by each well and also whether or not to grant new well permits.
- If sediment appears in a short period of time after the completion of the well, have the licensed well contractor repair all construction problems and if necessary construct a new well. Determine if this sediment will stop flowing into the well with continuous pumping

over time. Establish a level of pumping which should be the maximum allowed in any given period of time. Make sure that the well seal has not been compromised.

- If water quality changes, immediately test the water microbiologically and chemically to determine if there are problems in the well or in the water-bearing strata. Use a licensed well contractor to make all determinations and necessary corrections.
- Carry out appropriate maintenance control procedures on a regular basis.
- Obtain a permit before abandoning the well and follow all local and state specifications including an inspection by the authorizing authority. No surface water or other contaminants must ever be able to enter an abandoned well, which must never be able to be used for disposal of any kinds of substances including hazardous wastes.
- Water taken from a pond or lake as its source must be fenced against livestock, free of weeds, algae and floating debris, must have an intake for the water 12–18 inches below the surface, go to a settling basin, through filtration, and finally be disinfected before use.
- Spring water, unless it is artesian in nature, should be considered to be contaminated and therefore must be disinfected. However, since it may be difficult to determine if it is an artesian spring, always test the water supply microbiologically and chemically before using it.
- Cistern water should go through a filtration process and be disinfected.
- All water should be tested before starting to use it and then at least annually for wells, at the time of major potential contaminating events, and more frequently based on the source.
- After a flood, boil all water for human consumption or other purposes before using. Remember that boiling will not be effective for chemical contamination.
- After a flood, determine the conditions of the well, electrical system, and pump operation by professionals before using the well.
- After a flood which may have contaminated the well, have the local or state health department take samples of the water, test the water both microbiologically and chemically, and if contaminated, have a licensed well contractor thoroughly disinfect the well and if necessary, the aquifer, before using as a potable water supply. Test again after the chlorine is gone before using the water. Wastewater from septic tank systems or chemicals seeping into the ground can contaminate the groundwater supply. In case of chemical pollution, experts in this area will need to make a determination on how to make the water safe and if necessary put in a new well to a different aquifer.
- Emergency disinfection of water includes: straining the water through cheesecloth and then boiling for 10 minutes followed by cooling and transferring from one clean container to another to re-aerate it; or using household bleach which contains 5.25% available chlorine at a rate of two drops per quart of clear water after straining, shaking thoroughly, allowing to stand for 30 minutes, and pouring off the clear liquid from the top while not allowing residue to enter the new container. This can also be accomplished by using chlorine tablets or iodine.

SUB-PROBLEMS INCLUDING LEADING TO IMPAIRMENT AND BEST PRACTICES FOR PUBLIC DRINKING WATER SUPPLIES

(See endnotes 1, 11)

In the United States, public drinking water systems provide drinking water for about 90% of the citizens. There are approximately 155,000 public water systems in the United States. Community water systems, which serve the same population year-round, utilize either surface water or groundwater as their source. The water, which is for human consumption, is then treated and distributed through pipes to at least 15 service connections, or an average of at least 25 people, for at least 60 days each year. A second type of water system is called non-transient, where 25 of the same people are served for at least 6 months a year, but not year-round, such as in schools, hospitals, and factories which have their own water systems. A third type of water system is a transient non-community water system, which provides water for campgrounds, etc. Systems are also classified by their sizes. The community water systems provide water for over 300,000. The other two systems provide for far fewer people.

All the potential pollutants discussed under General and Specific Sources of Contamination and Best Practices for Prevention, Mitigation and Control in the section above potentially apply to the surface water and even possibly the groundwater used in public drinking water supplies as the raw water source. The problems and potential contaminants found in private drinking water supplies are also found in public drinking water supplies.

In addition, there is an impact from increased urban development on water resources. The quantity and quality of the water may be affected by: the hydrologic cycle which may be disrupted; high levels of soil erosion and sediment which may be present; a reduced level of groundwater recharge; an increase in need for water for concentrated populations; and an increase in contamination from higher levels of sewage and waste disposal from these areas. Surface water used as the raw water source may also become easily contaminated by atmospheric deposition, flooding, intentional and unintentional releases of contaminants into the body of water, and municipal, agricultural or industrial wastes which are either treated or not treated, etc.

LIVESTOCK AND POULTRY MANURE CONTAMINANTS AND RAW WATER QUALITY

(See endnote 37)

Although the topic of agricultural waste has been discussed earlier in other chapters and referred to in this chapter, it needs further discussion here. The potential problems created by this type of disposal are significant and can have a profound effect on the raw water which is going to be used at the water treatment plant as its source of supply. Since the 1950s until current times, the production of livestock and poultry, especially in concentrated areas, has increased by 80% and with it has created an extremely large problem of disposal of fecal material and bedding. It is estimated that the 2.2 billion head of livestock and poultry produced about 1.1 billion tons of manure in 2007. The livestock and poultry not only create a massive nutrient problem but also produce large numbers of microorganisms which can cause disease in people. These organisms include *Escherichia coli* 0157:H7, *Campylobacter*, *Salmonella*, *Cryptosporidium parvum*, *Giardia lamblia*, and numerous viruses including rotaviruses. There has also been an extremely large increase in antimicrobial use, which in turn can result in antimicrobial resistance in animals and people. Hormones are produced naturally and are also given to the livestock and poultry to improve meat quality, promote animal growth, and increase milk production. These hormones and their byproducts may be found in their fecal material in very large quantities.

Manure must not get into the raw water by means of runoff from rain, spills, problems with storage lagoons, equipment failures, and improper application of the manure to land for disposal. Pathogens can contaminate food during meat and milk processing.

Best Practices in Managing Livestock and Poultry Manure Contaminants to Avoid Raw Water Contamination (See endnote 38)
- Do not slaughter sick or stressed animals since they shed more pathogens in their manure than healthy animals and therefore create greater opportunity for contamination.
- Make sure that the animals have proper access to food and water, adequate space, temperature, and ventilation, and a clean environment to reduce the stress on the animal.
- Use slotted floors in the animal housing facility to decrease levels of *Salmonella*. Animals housed on dirt lots or solid concrete are continuously exposed to contaminated feces, whereas with slotted lots the feces fall through the slots into an underground pit. This decreases the amount of potential pathogen contact as well as the time of contact. Cattle raised on a pasture do not have higher concentrations of potential pathogens.

- Use appropriate pest control measures to reduce flies, other insects, and rodents, which may act as vectors for the transmission of disease. Birds must also be kept under control.
- Modify the diet of the animals with the consent of the veterinarian in charge, to decrease the amount of pathogens being excreted in the manure. Do not use antimicrobials as growth promoters in the food for the livestock.
- Consider using organic acids in the diet of poultry to reduce the levels of *Campylobacter* and *Salmonella*.
- Reduce acid-resistant *E. coli* and *E. coli* 0157:H7 by abruptly switching from a high grain diet to a high quality hay-based diet.
- Feed pigs coarsely ground diets instead of fine-grained diets to reduce *Salmonella* concentrations. Feed the pigs a meal diet instead of pelleted feed to reduce *Salmonella*.
- Use vegetative filter strips as a means of removing sediment, nutrients, and microorganisms. Direct all runoff and erosion from feedlots across these areas. Be sure to have adequate fly, other insect, and rodent control measures in place.
- Place permanent stockpiles of manure on a concrete pad or clay base with at least a 2-foot distance between the bottom of the stockpile and the seasonally high water table. Catch basins should be used to prevent runoff from this stockpile into any body of water. The contents of the catch basins must be applied to the land or treated before the release of the liquid. Divert the rainwater above the areas of the open lots or manure piles so the water will not carry contaminants to other places.
- Fence in all open water from animals and provide another drinking water source for them.
- Biologically treat manure by using anaerobic lagoons in southern climates or deep pits in northern climates, where pathogens will be reduced within a 30-day period; composting where the organic matter will be broken down by microorganisms and create heat, carbon dioxide, and water vapor, and most pathogens will be destroyed at 131°F; aeration systems, using mechanical aeration devices, in long shallow storage facilities, where the heat generated and the aeration helps reduce pathogens especially viruses; and an anaerobic digester which stabilizes the manure and controls odors as well as reduces *E. coli*, *Salmonella typhimurium*, and *Yersinia enterocolitica*.
- Chemically treat manure with chlorine which is very effective for bacteria but less so for viruses and protozoa; lime stabilization which reduces odor and pathogens before spreading the manure on the land; ozone which is a powerful oxidizing agent to kill bacteria, especially *E. coli*, and will reduce total coliform counts; and pasteurization of the manure which requires temperatures of 158°F for 30 minutes and will reduce all pathogens but may be extremely expensive.
- When applying manure to the land to reduce the potential for disease in people by contaminated runoff, consider the rate of application, the seasonal conditions, the weather conditions, and provide means of capturing the runoff. When the land is dry, then the sunlight containing ultraviolet rays and natural drying will reduce potential pathogens.
- Do not graze animals on land which has had a recent application of manure.

SUB-PROBLEMS INCLUDING LEADING TO IMPAIRMENT AND BEST PRACTICES FOR PUBLIC DRINKING WATER TREATMENT PLANTS

LOCATION, SOURCE, AND PRELIMINARY PRETREATMENT FOR RAW WATER

Source water has to be evaluated in the following way. Determine the source water assessment area for all types of drinking water supplies whether they are groundwater or surface water. For groundwater, use available information about the flow and recharge area. Remember that as the use of groundwater increases because of increased wells, the flow and recharge area will broaden considerably. It is from this area typically that contaminants can readily enter the groundwater supply.

However, fractured limestone or other types of fractured underground areas can create situations where contaminants may flow for many miles underground into the typical groundwater supply. For surface water, determine the watershed upstream from the intake valve and plot it on a topographical map showing all potential sources of pollution and the nature of the pollution within the watershed area. Establish an inventory of all the potential sources of pollutants and quantity that can contaminate the water supply, whether it is groundwater or surface water. These pollutants may come from landfills, fuel or chemical storage tanks, septic systems, runoff from streets, lawns, farms, etc. of fertilizer and pesticides, sludge disposal, businesses and industry, etc. Determine the likelihood of contamination of the water supply and how best to prevent or mitigate the situation. Provide all information to the public as well as the appropriate agencies in order for them to make proper decisions on the best way to utilize the water source and its treatment process. This may involve multiple governmental jurisdictions and should be determined on a regional basis. (See endnotes 5 and 12.)

NITRIFICATION

(See endnote 22)

Best Practices for Location, Source, and Preliminary Pretreatment for Raw Water
- Review the information above on General and Specific Sources of Contamination and Best Practices for Prevention, Mitigation and Control and implement the Best Practices which relate to the potential problems found in a given locality.
- Utilize the Best Practices from the private drinking water systems that apply to the public drinking water systems, especially those related to the use of groundwater as the raw water source.
- Conduct a survey to determine the degree of pollution from all sources entering the watershed and evaluate whether or not the raw water source can be utilized and if so what treatment would be necessary to make it safe and meet all of the minimum specifications of the US EPA and other authorizing agencies.
- Pesticides or herbicides used within the watershed must be used with strict adherence to all posted directions on labels and label restrictions.
- The water quality microbiologically, chemically, and physically must be established from various tests as well as the quantity of water which will be taken on a daily, weekly, monthly, and yearly basis and how this quantity will be replenished must be supplied to the authorizing authority when requesting a new raw water source.
- The water must be continuously monitored at the surface water intakes microbiologically, chemically, and physically and appropriate action must be taken immediately if the tests show the water exceeds any of the established regulations by the US EPA or local authorizing authority.
- Intakes must be located away from swamps and other areas where contamination could readily enter the raw water supply. The intakes must be protected against objects coming down through the water which could damage them.
- Establish a 200-foot radius restricted zone, which has appropriate signs warning people engaged in all recreational activities to stay outside of the special area.
- Keep all raw water inlets at least 1000 feet away from boat launching, marinas, docks, or floating fishing piers.
- The raw water should be taken from several different depths of the surface body of water, which can reduce turbidity caused by storms and also give a different quality of water from each depth.
- Put the raw water intake valve for the water treatment plant away from the source of water feeding the surface body of water. This will also reduce the level of turbidity in the raw water.

- If the raw water source is a river, make sure that there are no pipes within a reasonable distance discharging pollutants into the river upstream from the water intake area.
- Use screens or grates to reduce the amount of debris entering the water treatment plant.
- Keep all water intakes upstream from a sewage treatment plant or any other industrial plant and at least 500 feet away.
- Use a settling pond or concrete basin for presedimentation of the raw water especially if the source has a lot of sand and gravel in the water. Allow a detention time of at least 20 minutes and clean out the pond or concrete basin frequently. Organic polymers may be added to increase the amount of presedimentation.
- Develop a raw water pump station facility in a well-drained area and above any possible flood stage. It should stay in operation during flooding events. The building should be locked and enclosed by a fence which cannot be tampered with by intruders.
- An all-weather road should be constructed leading from main roads and highways to the raw water pumping station.
- The raw water enters the treatment plant and has added to it alum, iron salts, or synthetic organic polymers which bond with metals and salts to cause the small particles present in the raw water to become large particles and then settle to the bottom of the tank. Different types of filtration media as well as ion exchange for inorganic contaminants and absorption for organic contaminants follow this. The water is then either disinfected with chlorine or ozone, with chlorine being by far the most common disinfectant used.
- Develop a comprehensive cyanotoxin management plan. (See "Cyanotoxin Management Plan in the Programs" section below for a model program.)
- All marinas and docks must provide special facilities for proper sewage, and solid and hazardous waste disposal facilities.
- All on-site sewage systems and sewer manholes must be a minimum of 75 feet from the highest level the water source will reach on land and must be protected against flooding at the anticipated 50-year flood level.

DRINKING WATER TREATMENT FACILITY

(See endnotes 13, 14, 15)

A public drinking water system consists of several parts: securing raw water, storing it when necessary, and treating it chemically for flocculation and sedimentation; the water treatment plant itself which contains units for coagulation, sedimentation, filtration, disinfection, and storage; a distribution system; another storage system of potable water if necessary; and final end use by people, businesses, industry, etc. Water systems and water distribution systems have the same types of problems as sewage treatment systems and sewage collection systems. The treatment facilities have aged and the pipes have deteriorated with the potential for inflow of water or sewage or other contaminants into the potable water supply.

The chemicals used in processing and treating the water and the chemical byproducts may become hazardous to people and the environment. Chemicals are used for coagulation, removal of phosphorus, removal of bad taste, odors, and color, reducing inorganic compounds, reducing organic chemicals, reducing hardness, and final treatment to destroy microorganisms with chlorine, bromine, iodine, ozone, and other chemicals.

Filtration systems may include:

- Conventional filtration for drinking water treatment utilizing rapidly mixed alum and a polymer in the raw water (used to help precipitate dissolved solids), followed by actual filtration with a slow sand filter and then disinfection;
- Direct filtration, where the water source is much better and cleaner and therefore the water can go straight to filtration and disinfection;

- Lime softening, where the raw water has high concentrations of calcium and magnesium and the lime helps remove them as well as other toxic minerals, followed by a final step that neutralizes the access alkalinity.

In all cases, filters become dirty and need to be properly maintained and cleaned. (See discussion on sewage treatment plant in chapter 11, pages 411–412.)

Special problems for water treatment plants include *Cryptosporidium*, cyanotoxins, and turbidity. (See endnotes 16, 17, 18, 19, 20.)

Cryptosporidium

The oocysts (the thick-walled stage of a one-celled protozoan parasite) of *Cryptosporidium* are very common and persistent in water, resistant to chlorine disinfection, and can cause massive outbreaks of disease as seen in Milwaukee, Wisconsin in 1993 when 400,000 people became sick from this parasite. *Cryptosporidium* can be transmitted from person to person, livestock to humans, through the oral-fecal route, by inanimate objects or by recreational waters. It causes profuse diarrhea, dehydration, fever, and weight loss especially in individuals who are immuno-compromised, in renal failure, and have liver disease. It may cause an acute episode and also a chronic illness of recurring intermittent episodes. A low dose of the oocysts can cause the disease in healthy people.

Cyanotoxins

Cyanotoxins which are produced by cyanobacteria blooms in freshwater are a health problem both to people worldwide and to the various ecosystems. The sharp increase of nutrients in water including phosphorus and nitrogen has caused eutrophication and this is favorable for the growth of large quantities of harmful algae blooms, especially when there is a high nutrient concentration and high light intensity. The toxins, microcystins, cylindrospermopsin and other cyanotoxins, can cause gastroenteritis, liver damage, and kidney damage.

The algae bloom can cause increased taste and odor problems; increased pH; increased turbidity; shortened filter runtimes; need for increased coagulant; and increased chlorine demand or decreased chlorine residual.

Turbidity

(See endnote 18)

Turbidity is caused by a large amount of small suspended materials. These materials are easily stirred up in water and are very slow to settle. They can be caused by the growth of algae and plankton, sediment and particles found in runoff from land in stormwater, disturbance due to storms, etc. High levels of turbidity interfere with the water treatment process and also are an indicator that a portion of the process is not working properly. The level of turbidity can be affected by the processes of raw water screening, presedimentation, coagulation, flocculation, sedimentation, and filtration. Although the intake screens have no effect on turbidity, the location of the raw water intake can have a significant effect on the amount of turbidity entering the water treatment system. Presedimentation helps to remove silt and other fine suspended solids. The lower the velocity flow of the water, the greater the sedimentation. Since the particles carry electrical charges, they can be brought together if a material of the opposite electrical charge is introduced into the water. Coagulation by itself will not reduce turbidity but may increase it. The flocculation of the larger particles helps reduce the turbidity. Once again, detention time is very important. Sedimentation and clarification can remove 50–90% of the suspended solids. Filtration if working properly can reduce the turbidity to below the acceptable safe drinking water standards. Membrane systems work the best at reducing turbidity.

Best Practices for Drinking Water Treatment Facilities

- The water treatment plant location and its conduits, basins, other structures, storage areas, and distribution systems must be secure against the infiltration of sewage, surface contaminants, and runoff from surrounding lands.
- The water treatment plant must have special extra reserve power sources and pumping equipment in the event that the initial sources are made inoperable. They must be able to be switched in to operation quickly.
- Filter washings and other types of wastes must be removed safely and quickly to points of disposal on a regular basis, except in extreme weather where adequate storage should be available to be used until the extreme weather passes.
- Separate reservoirs are used for presettling of contaminants prior to water treatment and a finished reservoir which must be covered when the water is ready for distribution to the community.
- Develop a disinfection profile for the facility by keeping a log of *G. lamblia* and virus inactivation for at least a 12-month period of time and the amount and type of disinfectant used as well as the time of application. This will help establish the guidelines for disinfecting the water to destroy these organisms and if additional treatment is necessary.
- The maximum contaminant level goal for the protozoan *Cryptosporidium* is zero and since it is chlorine-resistant, the membrane processes using micro- and ultrafiltration as well as the use of ozone or the second-best disinfectant, chlorine dioxide, will reduce the potential for contamination to its lowest level.
- Continuously monitor turbidity in the effluent from the individual filters to determine the level of performance of that filter and whether or not it needs cleaning, maintenance, or replacement.
- All filter backwash water, thickener supernatant, or liquids from dewatering must be recycled to remove contaminants from the effluent.
- The water is tested for 83 different contaminants including volatile organic compounds, synthetic organic compounds, inorganic compounds, and microorganisms. If the water does not meet state and federal standards, it has to be reprocessed and a determination needs to be made of why, after treatment, the water was still not considered to be of drinking water quality.
- All water should be fluoridated before it can be utilized for human consumption.
- All water treatment plant operators must be properly trained and licensed by the appropriate authorizing authority. They must take continuing education courses either yearly or every 2 years to be recertified before being able to be relicensed.
- All water treatment plant equipment must be in excellent operating condition and all problems, breakdowns, incidents, and accidents must be put on a permanent log along with appropriate numbers of samples of the water depending on the size of the plant.
- Graphs must be made of water flow, turbidity, dosage rates of chemicals, etc. and be evaluated monthly by the supervisor and cosigned by him/her.
- Raw water variability by quality or quantity must be charted, problems determined, reports made to supervising personnel, and problems corrected as quickly as possible.
- Incidents of no disinfectant present in the finished water or varying amounts must be brought immediately to the attention of the supervisor and all necessary precautions taken to avoid the potential for creating a health risk.
- All equipment and standby equipment must be evaluated periodically and kept in proper operating condition for immediate use.
- All hazardous chemicals on site must be secured in a properly ventilated storage area and used only by individuals appropriately trained and if necessary using personal protective equipment.

- There should be an immediately usable communication systems and a properly tested and operating back-up system if necessary in the event of an emergency.
- Cleaning and maintenance of all facilities is extremely urgent to prevent contamination of the water.
- Develop a specific safety plan for each facility, communicate its contents to all people working, and enforce all measures necessary prevent injury or illness.
- Conduct sanitary surveys using appropriately qualified professionals on a regular basis to determine the potential for problems in the water treatment plant and the water distribution system before they occur.

DISINFECTION

(See endnotes 22, 23, 24, 25, 26, 28)

After the potable water goes through the various phases of the drinking water treatment plant, the water is then disinfected to ensure that all disease-causing organisms or other pathogens are destroyed. The disinfectants being used are chloramine, chlorine, and chlorine dioxide.

Disinfectants

Chloramine is especially effective in the distribution system pipes and is formed by adding ammonia to the water which contains free chlorine. If used in excess of its maximum residual disinfectant level, it may cause irritation of the eyes, nose, and stomach, or anemia in some people.

Chlorine is added either as a gas or liquid. It is the cheapest of the disinfectants and has a long history of being effective on most microorganisms. If used in excess of its maximum residual disinfectant level, it will cause irritation of the eyes, nose, and stomach for some people.

Chlorine dioxide is very effective in treating the potable water supply and disappears from the water rapidly. The four main ways of producing chlorine dioxide are electrochemical, the acid-chlorite method, the chlorine-chlorite method, and three chemical processes. (See endnote 26 for specific reactions during the production of the chlorine dioxide.) If used in excess of the maximum residual disinfectant level, some infants, young children, and fetuses will suffer from nervous system problems. Drinking excessive amounts of the chlorine dioxide in water over several years may cause anemia. The National Primary Drinking Water Standards for the three disinfectants are chloramine 4 ppm, chlorine 4 ppm, and chlorine dioxide 800 ppb.

Ozone is an extremely strong oxidant and disinfectant utilized in water treatment systems. It works much faster than chlorine, inactivates viruses, oxidizes organic and inorganic compounds, removes iron and manganese, and controls taste and odors. It is used in many water treatment facilities. It is used for control of MS-2, poliovirus, *Giardia* cysts, and *Cryptosporidium*. The disadvantages are the higher cost of equipment and operation, water hardness may affect the process, potential fire hazards, and toxicity problems for people. Also there is no carryover of a disinfectant residual and therefore chlorine has to be used afterwards to prevent growth of microorganisms. Some of the ozone byproducts and reactions with chemicals in the water may include brominated compounds, aldehydes, ketones, and carboxylic acids and these may have serious health effects. (See endnote 27.)

Ultraviolet light may be used for disinfection at the proper wavelength which has germicidal properties (200–300 nm). Some of the problems include breakage of the lamp which could release mercury into the water; the use of a UV reactor which may alter the dose delivery rate and time, and therefore not affect all microorganisms; UV light inactivates the organism but does not destroy it; the proximity of the water to the UV light and how clean the light is kept affects the process; and the amount of turbidity present.

Disinfectant Byproducts

Disinfectant byproducts may be harmful to the health of certain groups of people. Trihalomethanes are produced when chlorine and chloramine react with naturally occurring organic and inorganic materials. Some people who drink the water containing these trihalomethanes over many years may have liver, kidney, and central nervous system problems, or an increased risk of cancer.

Haloacetic acids are produced when chlorine and chloramine react with naturally occurring organic and inorganic materials. Some people who drink this water over many years have an increased risk of cancer.

Bromate is produced when there are bromides in the water and they react with ozone. Some people who drink this water over many years have an increased risk of cancer.

Chlorites appear when chlorine dioxide breaks down. Some infants and young children who drink water with excessive amounts of chlorites have nervous system effects. These effects may also occur in the fetuses of pregnant women. Some people get anemia.

Nitrification

(See endnote 22)

Nitrification is caused by microbes and is a process whereby reduced nitrogen compounds, usually ammonia, are oxidized to become nitrites and nitrates. Nitrification, which can degrade water quality, can reduce alkalinity, pH, dissolved oxygen, and chloramine residuals while helping bacteria grow. Ammonia can also be produced by this process. Nitrifying bacteria grow very slowly and typically nitrification usually occurs in large reservoirs and in areas of the distribution system where the water flows very slowly.

NATIONAL PRIMARY DRINKING WATER REGULATIONS

National Primary Drinking Water Regulations or NPDWRs have been established for microorganisms, disinfectants, disinfectant byproducts, inorganic chemicals, organic chemicals, and radionuclides. These are legally enforceable standards for public water systems. They limit the amount of contaminants allowed in drinking water in each of the above categories. The MCLs are provided in a list of contaminants. (See endnote 25 for the most recent list of contaminants and their MCLs.)

Best Practices for Use of Disinfectants in Water Treatment Plants (See endnote 29)
- To determine the best disinfectant to use for a site-specific water treatment plant, consider the following: determine the watershed that will be utilized for the raw water and any special risks created by different types of pollutants entering the watershed; predetermine by testing the amount and type of pathogens to be removed as well as the quantity, quality, types, and concentrations of raw water to be processed; determine the amount of contact time needed for proper disinfection; determine the efficiency of the disinfectant; determine the level of residual disinfectant which is needed in the distribution system; determine the costs of operating each of the potential disinfection systems; and determine the possibility of disinfectant byproducts which could be harmful to people.
- Determine the efficiency of all parts of the water treatment plant process and how this affects the need for specific quantities of disinfecting materials and specific contact time.
- Determine the effectiveness of the disinfecting process for bacteria, viruses, protozoa (*Entamoeba histolytica, Cryptosporidium, Giardia*) and helminths (parasitic worms).
- Recognize that the diameter of the oocysts of *Cryptosporidium* are too small to be effectively removed by rapid sand filters and therefore there needs to be an additional filtration system, plus appropriate disinfection techniques, to prevent the disease process from occurring. High fecal contamination sources upstream of the water intake valves are a particular problem regarding *Cryptosporidium*. Natural flooding of these areas also is a

serious problem. Where groundwater is used as the raw water source, flooding can also contaminate this source and create problems of potential outbreaks of disease. Other problems include inadequate treatment of the surface water, poor monitoring techniques, improper backwashing of filters, and bypassing the filtration system when the demand for water exceeds the supply available.

- Utilize the physical removal and chemical oxidation of organic and inorganic compounds in the water to also reduce pathogenic organisms.
- Limit the quantity of disinfection byproducts by controlling the residual organic and/or inorganic compounds in the treated water.
- Test frequently during the course of each day to determine if the disinfectant residual is at its proper level in the water to be distributed and in the distribution system and also if there are organisms that could cause disease.
- Consider the potential for health and safety issues among personnel when handling disinfectants which may cause serious concern if spills occur or excessive amounts are used.

FINISHED POTABLE WATER STORAGE FACILITIES

(See endnote 21)

Finished water storage facilities are used to equalize water demands, reduce fluctuations in pressure, and make available a reserve for firefighting, power outages, and other emergencies. The nature of the holding facilities is such that old water may be in the tanks and under certain conditions enter the distribution system and thereby cause problems. There are both ground storage facilities and elevated storage facilities. The ground facilities can be either covered or uncovered. There are numerous potential water quality problems in the storage facilities which may be microbiological, chemical, or physical in nature. Long detention times or old water can be a greater problem for microbial growth and chemical changes than freshwater moving in and out of the storage facility. Stratification in reservoirs can create zones where the water will age more rapidly than in other zones. This can result in disinfectant decay, regrowth of microorganisms, and nitrification. Sediment can accumulate in the storage facility and affect the water quality. Birds or insects create a major water quality problem in storage tanks. Storage facilities have been shown to be the source of several waterborne outbreaks of disease. Coating materials used in steel storage tanks and in concrete tanks may leach into the water and contaminate it. Metal tanks may become corroded. Disinfectant byproducts are more prominent because of the higher water temperatures in steel tanks during the summer.

Best Practices for Finished Potable Water Storage Facilities
- There should be a continuous flow of the water from the distribution system into the top of the storage tanks and out of the bottom of the tanks to help prevent stagnant water from accumulating.
- Note all potential aesthetic indicators as concerns for water storage facilities including: poor taste and odor; accumulation of sediment; and water temperature which may indicate temperature stratification within a reservoir. These indicators may also be caused by the age of the water being stored, inadequate treatment process, the pipe materials, and the condition or age of the distribution system.
- Test the water from the storage tanks or facilities frequently to determine lower disinfectant residuals, higher bacterial counts, and elevated nitrate/nitrate levels indicating nitrification as a means of determining if there are problems in the potable water storage facility.
- Inspect the tanks in both the interior and exterior for physical integrity, security, visible pollutants, vandalism, cleanliness, coating failures, maintenance, and repair problems.

- Follow the appropriate American Water Works Association standards for disinfection of storage facilities and coating systems, and/or the NSF International/American National Standards Institute standard, NSF/ANSI 61.
- Use booster disinfection when needed to restore disinfectant residuals at a water storage facility. Be aware of additional disinfectant byproducts.
- Establish a comprehensive routine maintenance, repair, and replacement program for water storage facilities.

SUB-PROBLEMS INCLUDING LEADING TO IMPAIRMENT AND BEST PRACTICES FOR PUBLIC DRINKING WATER DISTRIBUTION SYSTEMS

(See endnote 39)

EFFECTS OF AGE OF WATER

(See endnote 30)

The deterioration of water quality in the distribution system has two main causes. They are deterioration in the water caused by the wall of the pipe, and reactions in the water itself. Water traveling through the pipe goes through different chemical, physical, and aesthetic changes which affect the water quality. Deposited materials such as sand, iron, and manganese also create water problems in the pipe.

The proper rate of water flow is extremely significant in maintaining water quality. When a new system or an extension of a system is built, the authorities have to construct the distribution facility to last at least 20 years or more. This results in excess capacity and tends to slow down the current water flow. Water demand varies during the 24-hour period, seasonally, and if large users of the system are either added or taken away. There is also a need to have adequate water flow in the event of fires, which means that if there is an unusual demand then the water is used up more rapidly, and if there is little demand the water starts to accumulate and slow down. There must be adequate storage for the water and if not used frequently, the problems noted above can occur and the water becomes contaminated. After 3 days in the distribution system water begins to age.

Disinfection byproducts form more readily as the water ages. Nitrification also occurs and microbial regrowth is present. Phosphate inhibitors and the management of pH deteriorate.

DETERIORATING BURIED INFRASTRUCTURE

(See endnote 31)

The rate of deterioration of a buried water system has little to do with the actual age of the material used but rather with external forces which have contributed to an increased amount of leaks, main breaks, taste, odor and color problems, reduction of water flow, internal pipe corrosion, need for more disinfectants, growth of biofilms, and regrowth of bacteria. Three different older types of cast-iron pipes installed mostly before the 1960s are about at the end of their service life. The problem is not only related to potential health risks and poor customer service, but also the enormous cost involved in replacing the buried water system. The estimated cost of replacement varies with the professional organization or governmental organization making the estimate, and the range of cost from 2002 to 2020 (in 2002 dollars) is $151–$220 billion for the United States. One of the great difficulties is determining when to replace a given section of pipe and the main.

There are numerous reasons why pipes fail. They are manufacturing defects; improper design or installation; geological instability; higher operating pressures; hydraulic transients (a sudden change in velocity of the water); change in water temperature (especially freezing temperatures); excessive external loads; damage from digging; internal corrosion; external corrosion from soil and other sources; leadite (a pipe-joining material) corrosion; incompatibility of materials; deterioration of gaskets; and material fatigue.

PERMEATION AND LEACHING

(See endnote 32)

The distribution system infrastructure and piping, lining, fixtures, and solder can react with the water being distributed and the external environment. These interactions may create problems of water quality. Permeation is the movement of contaminants external to the pipe, especially contaminated soil, through the porous non-metallic joints of the pipe into the water. Leaching is the dissolving of metals, solids, and chemicals from the pipes into the drinking water.

Stagnation of water in the pipes intensifies permeation and leaching. Most permeation problems occur at the service connections because of the small diameter of the pipes and frequent stagnation there.

NEW OR REPAIRED WATER MAINS

(See endnote 33)

The construction and repair of water mains can create potential situations which will contribute to the contamination of the water by microorganisms, which cause disease. Other problems are turbidity and unusual colors, chemical contamination from the surrounding area, excess chlorine use, loss of residual of disinfectants, and pH instability. These problems can be the precursor to disease from microorganisms and chemicals as well as the beginning of problems in the pipes which will lead to leaching and poor water quality. The type and amount of contaminant entering the system plus its distribution will influence the potential for small or large outbreaks of disease or other health conditions, short-term or long-term. The disinfectant needed to be used after repair or construction may contribute to above-normal levels of disinfectant byproducts which may also be harmful to people.

The three times when contamination may occur for water main construction or repair activities are prior to the construction or repair; during construction or repair because of microbial and/or chemical contamination; and contamination after construction or repair from leaking pipe joints, stagnant water in the adjacent piping sections, cross-connections, or variations in pressure of the water. Prior to construction, the pipes may be exposed to soils, sediments, trash, stormwater runoff, harmful chemicals and chemically contaminated soil, and waste from animals and humans. This can create biological and/or chemical problems. During construction or repair which is usually done in open trenches or excavations, the interiors of the pipes and the fittings may come into contact with soil and water in the trench and the soil and water may be contaminated with microorganisms and/or chemicals. This is a substantial concern for contamination of the pipes and fittings. Cross-connections may also occur because of the construction. (See discussion on cross-connections below for more in-depth information.) Contamination of the pipes may occur from leaching during the in-place process which may be utilized for putting coatings inside the pipes and curing in place. Also, bacterial growth resulting in nitrification in reservoirs may occur during each of these phases because of the slowing down of processed water entering the distribution system.

SERVICE LINES

In the United States, there are approximately 880,000 miles of pipe being used as water service lines, which are 0.75–2 inches in diameter and usually made of copper, polyethylene, polyvinyl chloride, polybutylene, and in some older systems lead, galvanized steel, and brass. Obviously lead lines must be removed and replaced. The three problems related to these lines include finding the service connection, re-establishing the service opening, and connecting the service to the carrier pipe. Many of the water problems existing in homes or facilities occur because of the service lines. The service line from the main line to the water meter is taken care of by the water company. The service line from the water meter to the house or other facility is the responsibility of the owner or operator of that home or facility.

HEALTH RISKS ASSOCIATED WITH PRESSURE TRANSIENTS IN WATER DISTRIBUTION SYSTEMS

(See endnote 34)

Flooding is of special concern because if the air valve is open the contaminated water can enter the distribution system. Negative pressures can be created through power failure, main breaks, and flushing of the system. During any negative pressure event, chemicals of all types including pesticides, petroleum products, fertilizers, solvents, etc. can go through the soil and into the water distribution system. Microorganisms from various contaminating sources can do the same thing. This is compounded by leakage from the distribution system pipelines.

HEALTH RISKS RELATED TO MICROBIAL GROWTH AND BIOFILMS IN WATER DISTRIBUTION SYSTEMS

(See endnote 36)

A biofilm is a slimy, glue-like substance excreted by bacteria which are adhering to a surface in a watery environment. These organisms including viruses, protozoa, invertebrates, algae, and fungi, which can survive and grow in the water distribution system, may cause disease in healthy people or in individuals who are sick or immunocompromised. Their metabolic products such as toxins produced may also cause disease.

Microorganisms can also induce corrosion of the pipes, reduce the effectiveness of indicator organisms which would show where there is contamination present, produce taste, color and odor problems, react with the disinfectants to reduce the available dose for control of microorganisms, and produce disinfectant byproducts.

Microorganisms enter the water distribution system because of: raw water problems; breaks in treatment plant techniques; cross-connections and backflow problems; contamination of finished water storage; trespassing and lack of proper security of the water system; etc. The microorganisms can grow because of the presence of nutrients and the slowing down of the water through the system, which also allows for disinfectant residual to be below standard. The accumulation of sediment in the lines also contributes to bacterial growth.

Best Practices for Water Distribution System
- Use tracer studies to determine water quality and water age by measuring the amount of chlorine residual or trihalomethanes. The tracers can be injected chemicals like fluoride or calcium chloride.
- Use the appropriate American Water Works Association standards for designing a water distribution system for the community. The distribution system must also meet the fire protection needs and all state regulations.
- Do not close off distribution system valves because this will result in a dead end which may contribute to contamination of the water when the valve is reopened.
- Maintain a minimum pressure of 20 pounds per square inch at all locations at all times in the water distribution system.
- Ensure that all water which is stored will be mixed to avoid stratification.
- Use water system flushing as a means of removing sediment and stagnant water from the distribution system.
- Make sure that the drinking water treatment plant is working effectively to remove organic matter and organic chemicals to avoid later biochemical stability problems in the distribution system.

- Use frequent testing and monitoring techniques to determine elevated disinfection byproducts (DBPs), lower disinfectant residuals, increased bacterial counts, and increased nitrate/nitrate levels. All of these indicate aged water and other problems.
- Assess the potential for infrastructure problems by accumulating information from the past and mapping the entire water delivery system. Determine the frequency of complaints by location, frequency of breaks in the lines by location, the adequacy of provision of proper quantities of water of high quality, and the adequacy of water needed to fight fires. Place on the map those areas which are most prone to problems and immediately provide a maintenance and replacement program to keep from having larger problems later.
- Develop a present and future pipe replacement program based on facts and determine in advance necessary means of financing.
- Pipe failures can be reduced or eliminated based on the specific types of problems encountered by: higher operating pressures for pressure problems; surge control and operator training for hydraulic transients; cleaning and lining the pipe for internal corrosion; cathodic protection (a technique to control corrosion of a metal surface underground) for external corrosion; replacing the joint only for leadite expansion or corrosion; installing dielectrics for material incompatibilities; and replacing the joint only for gasket deterioration.
- If a water main fails, the least expensive way to replace it is to use an open trench system where the new main is in a trench parallel to the old main and the old main is disconnected but left in place.
- Renew underground pipes by using non-structural, semi-structural, or structural lining techniques within the interior surface of the pipe depending on the particular situation, to restore the pipe's integrity and ability to move the potable water through it without contamination. At times the pipes will have to be replaced, especially if the pipe has had structural loss of strength, lack of adequate flow capacity, and joint leaks that cannot be solved by other techniques, and the water quality problem cannot be resolved with insertions in the pipes.
- Detect suspected permeation and leaching problems by monitoring the water especially if newer rehabilitated mains, for total coliform bacteria, disinfectant byproducts, pH, disinfectant residuals, turbidity, and odor.
- Prevent permeation and leaching by using proper materials and installation practices when designing and installing replacement systems. Utilize the appropriate American Water Works Association manuals to help make decisions.
- Water mains must be separated from potential contamination including sewage pipes, stormwater pipes, or reclaimed wastewater pipes. There must be a 10-foot horizontal separation between the water mains and sanitary sewers and an 18-inch vertical separation for a water main crossing above or below a sewer. Where possible, all water mains should be above sewer lines.
- There must be on the potable water system adequate numbers and types of fire hydrants and valves, and the valves must provide for complete isolation of the potable water.
- There must be a minimum number and lengths of tie-ins to the water main.
- All existing water mains and service connections must have watertight caps or covers.
- All fittings, joints, and valves that have been exposed to the environment and hand tools, tapping machines, and other equipment that come in contact with pipes and fittings must be disinfected before being used.
- Remove any water from a trench so that it will not contaminate the pipes.
- Follow the appropriate American Water Works Association standards for installation and repair of pipes.
- Use the existing data concerning pressure differentials in different parts of the water distribution system to determine which portions are most likely to have low-pressure

events and when. Establish a maintenance program that will deal with these problems and especially where leaks are occurring.

- Train personnel in the use of hydrant and valve operations, the proper use of hydrants, and what the potential problem may be in causing low-pressure events resulting in contamination and potential disease outbreaks.
- Controlling the growth of biofilms in pipes can be accomplished by: reducing nutrients; reducing contamination of materials and equipment used for repair and replacement; maintaining proper disinfectant residuals and corrosion control techniques; and preventing cross-connections and backflow problems.
- Flush the water delivery system on a periodic basis to keep the lines clean and the pressure of the water constant.
- Properly maintain all water storage systems and standpipes to prevent microbial growth and biofilms.

SUB-PROBLEMS INCLUDING LEADING TO IMPAIRMENT AND BEST PRACTICES FOR CROSS-CONNECTIONS AND SUBMERGED INLETS

(See endnote 35)

A cross-connection is a place in the system where non-potable water is connected to potable water sources. This may result in a backflow of the non-potable water into the potable water because of reduced pressure, called backsiphonage, which may occur when the pressure of the non-potable water is greater than the pressure of the potable water. This condition may occur because of main breaks, flushing of the mains, pump failure, loss of electricity, unusual increase in customer demand, booster pumps in high-rise buildings, and emergency firefighting needs. This type of illegal connection can be found in industrial plants, heating and cooling units, waste disposal systems, etc. A submerged inlet occurs when a source of potable water is below a container which has non-potable water in it. For example, if someone puts a hose on a faucet and allows it to fall below the accumulated non-potable water in the container, then a break in pressure in the potable water supply would suck the non-potable water back up through the pipe and into the potable water. Cross-connections and submerged inlets have caused numerous outbreaks of waterborne diseases in people.

Best Practices for Avoiding and Correcting Cross-Connections and Submerged Inlets
- Maintain an operating pressure of 20 pounds per square inch at all locations in a water distribution system at all times.
- To avoid intentional contamination of the water distribution system, homeland security recommends physical security measures be taken to prevent someone from introducing chemical or biological contaminants through cross-connections and backflow.
- Create a comprehensive public education program for communities teaching them how to prevent cross-connections and submerged inlet problems from occurring.
- Create or, if the program already exists, enhance a major cross-connection inspection program for all facilities where chemical or biological contaminants can enter the water distribution system. This program should be enforced by law and penalties should be applied if situations are not corrected immediately upon their discovery.
- Create a comprehensive training program for all individuals involved in new construction, maintenance, or repairs related to the water distribution system, which may cause possible cross-connections, in residential, industrial, and commercial facilities. Make this action part of continuing education requirements for all plumbers and other individuals involved in the connection of pipes in relationship to the water distribution system.
- Never submerge a garden hose in a chemical mixing tank since this constitutes a submerged inlet and in the event of a pressure change in the fresh water line, there may be serious backsiphonage problems.

- Never establish a cross-connection between an irrigation system and the potable water supply because a vacuum breaker valve used to protect the potable water supply could fail and create substantial problems with microorganisms or chemicals.
- Evaluate all connections to cooling towers between the potable water supply and storage facility. This is a common area where contamination may occur and the potable water supply may become contaminated.
- When cleaning out any type of medical equipment, air conditioning systems, etc., do not insert a hose containing potable water into the equipment because of a potential break in pressure resulting in backflow of the contents into the drinking water supply.
- Report all incidents of reduced pressure in the potable water distribution system to the water companies and the health department in the event that this may be the reason for an outbreak of waterborne disease.
- Use mechanical backflow prevention devices in water distribution systems at the service connection to a facility. These devices may include pressure vacuum breakers, double check valve assemblies, and spill-resistant vacuum breakers. They also include air gaps.
- Use physical separation as in air gaps to prevent backflow from occurring between the potable water system and the non-potable source. This means there needs to be a separation of the supply pipe from the overflow rim of a receptacle which is twice the diameter of the incoming supply pipe. In that way, water cannot possibly flow back through the supply pipe.

LAWS, RULES, AND REGULATIONS

SAFE DRINKING WATER ACT

(See endnote 43)

This act and its amendments is the basic law for protecting public water systems and producing potable water in the United States. The law requires numerous actions to ensure the safety of the drinking water supplies. It allows states to implement the program and enforce the regulations. It includes special programs for: restricting the injection of waste in the ground water (Underground Injection Control Program) (See endnote 44); designating a principal groundwater source as the sole source aquifer (Sole Source Aquifer Program) (See endnote 45); prevention of contamination of public wells (Wellhead Protection Program); and determining the susceptibility of public drinking water sources to different types of contamination (Source Water Assessment Program).

The Safe Drinking Water Act amendments of 1996 created the Drinking Water State Revolving Fund Program which provides federal funds to help public water systems finance the cost of improving the infrastructure to either achieve or maintain compliance with the Safe Drinking Water Act requirements. These are grants from the US EPA to the states, which then provide low interest loans to localities to improve the water supply to protect public health. (See endnote 46.)

CLEAN WATER ACT

(See endnote 47)

The Clean Water Act establishes the basic system for regulating any discharges of pollutants into the waters of the United States. The basis for this legal action was the Federal Water Pollution Control Act of 1948, which then served as the basis for the Clean Water Act. Wastewater standards have been established for industry and water quality standards have been established for the contaminants that can be found in surface water. It is unlawful to discharge any type of pollutant from a point source into any navigable water unless a special permit is obtained from the authorizing authority.

The EPA's National Pollutant Discharge Elimination System permit program controls water pollution from pipes, ditches made by people, and any other means of dumping of pollutants into the surface waters. Septic tank systems are not covered by this law. By enforcing the law through the permit system, there has been a substantial decline in pollutants which affect the raw water supply for public water treatment plants and also water used for recreational purposes. (See endnote 48.)

RESOURCES

AMERICAN WATER WORKS ASSOCIATION

This organization, established in 1881, is the largest non-profit, scientific, and educational group working toward managing and treating water properly. It helps protect public health, protect the environment, and improve quality of life. Its technical experts cooperating with other groups of technical experts provide standards for all aspects of water quality and the material and equipment utilized to treat and distribute it. They provide manuals which can be utilized by water professionals and the general public. They provide a variety of educational efforts.

AMERICAN NATIONAL STANDARDS INSTITUTE

This organization for more than 90 years has acted as an administrator and coordinator of voluntary standardization systems for the private sector in the United States. It is the accrediting agency for procedures and standards which are adopted for use in various industries utilizing voluntary standards of performance and equipment. In this role, it works along with other groups to determine the best approach to the handling, treatment, and distribution of water as one of its subspecialties.

NSF INTERNATIONAL

This organization is involved in public health and safety. It has a professional staff of engineers, microbiologists, toxicologist, chemists, and experts in public health in various locations worldwide. Its laboratories conduct a large amount of testing, certification, and technical services. They provide a large amount of services for the water industry to ensure the quality and safety of products. They write standards, conduct tests, and certify products for drinking water, pools and spas, plumbing, and wastewater.

CENTERS FOR DISEASE CONTROL AND PREVENTION

This governmental agency, as part of its mission, provides scientific information on surface water and groundwater which will be used for the raw water source for a drinking water supply. It also provides substantial studies on outbreaks of waterborne disease and will provide experts to help determine causes of disease and how to quickly resolve the problems. It provides information for the prevention of waterborne disease.

UNIVERSITY EXTENSION SERVICES

Agricultural extension agents are excellent resources for individuals wanting to install wells or upgrade those which already exist. These services also provide in-depth documents to help understand various water problems and how best to resolve them.

LOCAL AND STATE HEALTH DEPARTMENTS

These governmental agencies provide all necessary information, rules and regulations, permits for installation, inspection services, and emergency testing and evaluation of groundwater supply delivery systems when problems are suspected. They also provide necessary documents to help with the understanding of the water wells, proper installation, and maintenance.

AMERICAN GROUND WATER TRUST

This organization's mission is to provide information on groundwater and the environment, groundwater management, groundwater science, and solutions to technological problems. It also works in partnership with local, state, and other national organizations as well as various governmental agencies to discuss, determine, and manage water systems, using Best Practices in groundwater supplies.

WATER SYSTEMS COUNCIL

This organization through its wellcare® program helps well owners maintain their wells and the quality of their well water. This program works exclusively with individual water wells and small water well systems.

STATE GOVERNMENTAL AGENCIES

These agencies may be found in every state. The programs of the agencies in Wisconsin cover laws and rules; publications; water quality data; consumer confidence reports; contaminants; health effects; and laboratories. Their material and personnel discuss water protection, wellhead protection, water problems, drinking water issues, and laboratory analysis results. In Maryland and Pennsylvania, they do the same as in Wisconsin and also include information on droughts.

THE US ENVIRONMENTAL PROTECTION AGENCY

This agency has developed a Private Well Initiative under the Safe Drinking Water Act to create a resource that provides data for water quantity and quality in private wells for unregulated drinking water sources. This resource also helps identify, evaluate, and recommend to owners of private drinking water sources techniques to prevent and control drinking water contaminants.

PROGRAMS

FEDERAL SUPPORT FOR STATE AND LOCAL RESPONSE OPERATIONS IN WATER CONTAMINATION CRISES

The federal government, as shown in the Flint, Michigan water crisis, will come together as a team of federal agencies and provide assistance to communities where there are serious water problems that affect the public's health. Typically, the US Department of Health and Human Services' Assistant Secretary for Preparedness and Response will lead the team of agencies which includes the US Department of Agriculture, the US Department of Housing and Urban Development, and the US EPA in providing direct services and various types of aid. US Public Health Service officers are the first ones at the site and make necessary evaluations and rapid recommendations to correct the existing and potential health problems. They involve the Centers for Disease Control and Prevention and other agencies as needed. All federal agencies work closely with state and local entities. (See endnote 48.)

JOAQUIN RIVER DELTA

(See endnote 41)

The delta is more than 50 miles inland from the Golden Gate Bridge but the waters rise and fall with ocean tides. It provides most of the fresh water of the San Francisco estuary. Before California became populated, the delta had a far different configuration than it has today. Dams and aqueducts have interrupted the natural flow of the rivers and the potential for pollution and contamination has increased substantially. The waters from this area are used by a very large number of Californians as far away as San Diego. Two thirds of the people in California and 4.5 million acres of farmland receive part of its waters. However, there is a serious concern today that the delta will sustain damage which will not be able to be repaired as areas will not be replenished by sediments in rivers flooding from the east or will be affected by salinity from the west if rivers flow too slowly. The constant removal of huge quantities of water from this area is causing these problems.

The *CALFED Water Quality* and *Ecosystem Restoration* programs were established to improve water quality for drinking water and environmental and agricultural purposes, with a major focus of providing quality raw water to be used for drinking water. The water in this area already has above national normal levels of organic carbon and bromide, and the fish contain higher levels of metals like mercury and selenium. In 2009, the previous work was recognized by state elected officials and it was directed by law that the Delta Stewardship Council, a California state agency, be created to develop a master plan for the entire Sacramento–San Joaquin River Delta.

Considerable funds have already been invested in a series of projects for the cost-effective improvement of the raw source water, use of Best Practices, and water management and treatment. This new strategic plan was developed with a major component of improving or keeping water quality from deteriorating while controlling sediment in this aquatic ecosystem. The plan includes reducing the amount and concentration of toxic substances, reducing the amount of oxygen-depriving materials brought about by human activities, and reducing fine sediment from human activities along the rivers and streams.

The plan includes working with a variety of agencies together to: reduce the amount of water taken from the delta; improve statewide water quality and supply through investments in local and regional water facilities; teaching consumers how to use water more efficiently and effectively; capturing and storing excess water from rainfall; protecting, restoring, and enhancing the delta ecosystem and protecting endangered species; detecting and destroying invading species, while reducing pollution; preserving rural lands for agriculture and similar use while limiting new residential, commercial, or industrial development; prohibiting encroachment on flood ways and water plains; flood proofing the delta including repairing and replacing where necessary, dams and levees; integrating all government action from all departments while using the best available science and Best Practices in making regulatory decisions; and rapid completion of the Bay Delta Conservation Plan. The two major goals of this plan are to provide a more reliable water source for California for drinking water and agriculture while protecting, restoring, and enhancing the delta ecosystem.

The plan implementation includes: establishing a strong leadership position with open lines of communication to all departments and consumer groups to coordinate all activities effectively and in a cost-effective manner; conducting frequent reviews of all activities and programs as well as in-depth inspections of all areas by professionals to determine if plan objectives are being met and pollutants are being removed from the delta area; and stopping all damaging activities that might affect the delta in an extremely rapid manner. Use court orders where necessary if no other type of action will work while using the latest science and Best Practices available in all decision-making concerning the delta and its waters. Develop means of storing floodwaters, which could be very destructive, until needed for use when droughts occur. Create new wetlands and other places for wildlife where they have been destroyed.

CYANOTOXIN MANAGEMENT PLAN

(See endnote 17)

This plan consists of several steps including performing a specific evaluation for vulnerability to the blooms for a specific watershed and water treatment facility; establish means of preparing and observing for potential blooms; testing for cyanotoxins in the raw water and communicating this information to all individuals in the area, and carrying out necessary treatment to destroy the toxins; determining if there are cyanotoxins in the finished water and advising all individuals in the area to carry out necessary treatment to destroy the toxins; and continuing to monitor and treat the finished water to make sure that cyanotoxins do not exceed acceptable levels.

- *Step 1*—Develop the necessary information to make decisions on how to best eliminate, mitigate, or treat cyanotoxins by evaluation of the watershed used for obtaining the raw water for the specific water treatment plant. This is done by understanding the characteristics of the water including bacteriological, chemical, and physical qualities; climate and weather conditions during different seasons of the year; and the use of the land that is draining into the watershed and potential nutrient types and levels going into the water from the land.
- *Step 2*—If the watershed and raw water are vulnerable to cyanotoxins, determine if there are long-term mitigation strategies which can prevent the blooms from occurring. Conduct constant visual inspections to determine if and when the blooms will occur. There may well be a seasonal variation. Evaluate the existing treatment process of the water from raw water to finished product. Determine if the usual coagulation, clarification, and filtration are effective in removing the cells of the blooms and if frequent backwashing of the filters will make them more effective. Can preoxidation of the raw water cause more cells to disrupt and therefore limit the production of toxins? Can the use of powdered activated carbon at appropriate levels be effective? Can the use of ozone or granular activated carbon in advanced water treatment plants remove the dissolved toxins?
- *Step 3*—Consider several techniques of mitigation of the toxins as follows: relocate the intake source of raw water; use an alternate source of raw water; bypass the increased sedimentation from ponds or reservoirs and have the raw water flow directly to the water treatment plant; use ultrasonic treatment to help prevent the blooms from forming; use an algaecide such as hydrogen peroxide which will not contaminate the water itself; add alum to the water which will help precipitate the phosphorus and coagulate the cells; use aeration pumps at the bottom of the water source which will help mix the water and disrupt the algae blooms; and use mechanical mixing to disrupt the blooms.
- *Step 4*—Test all finished water before distribution, and if cyanotoxins are present in above recommended levels, retreat the water before use.

Texas Commission on Environmental Quality—Public Water Supply Supervision Program

This program on public drinking water supply is part of the Texas Public Water Supply Supervision Program. It includes several actions: adopt, implement, and use drinking water rules at least as stringent as federal law; deliver all necessary data to the EPA; ensure that all water quality meets minimum standards, chemically and microbiologically; determine potential problems in source water and help public water systems counter these problems; review and approve all engineering plans for development or extension of water plant facilities; help communities increase their water available for public consumption; provide all technical assistance for homeland security issues; conduct all sanitary surveys of sources, treatment techniques, equipment, distribution systems, storage facilities, and pump facilities, and verify all data which has been provided by the facility and its operators; make sure that enforcement measures will be utilized if there are violations of the Clean Water Act standards; provide a licensing program for all public water system operators and necessary continuing education; provide a laboratory certification program for analysis of drinking water samples; and provide technical assistance where needed to help resolve critical problems of prevention, mitigation, and control relating to drinking water supplies, treatment, and distribution.

ENDNOTES

1. Koren, Herman, Bisesi, Michael. 2003. *Handbook of Environmental Health-Pollutant Interactions in Air, Water, and Soil,* Volume 2. Lewis Publishers, CRC Press, Boca Raton, FL.
2. The World Bank, Independent Evaluation Group. 2010. *An Evaluation of World Bank Support, 1997–2007, Water and Development,* Volume 1. Washington, DC.
3. Environment Canada. 2013. *Groundwater.* Gatineau, Québec, Canada.
4. US Environmental Protection Agency. 2014. *Water: Drinking Water Contaminants: National Primary Drinking Water Regulations.* Washington, DC.
5. US Environmental Protection Agency. 2004. *Protecting Drinking Water Sources,* EPA 816-F-04-032. Washington, DC.
6. Bucks County, Pennsylvania, Department of Health. 2007. *Rules and Regulations Governing Individual Residential Water Supply Systems and Construction Specifications.* Doylestown, PA.
7. Penn State University. 2015. *Drilling a New Well.* College Park, PA.
8. Penn State Extension Service. 2015. *Water Well Maintenance and Rehabilitation.* College Park, PA.
9. Penn State Extension Service. 2015. *Using Low-Yielding Wells.* College Park, PA.
10. US Environmental Protection Agency. 2004. *Drinking Water Treatment: Public Water Systems.* Washington, DC.
11. Texas Commission of Environmental Quality, Water Supply Division. *2012. Rules and Regulations for Public Water Systems,* RG-195. Austin, TX.
12. Swarts CD, Rajagopaul R, Charles. K. No Date. *Management Guidelines Water Treatment Plants.* Water Research Commission, Gezina, South Africa.
13. US Environmental Protection Agency, Office of Water. 2011. *Surface Water Treatment Rules: What Do They Mean to You,* EPA 816-R-11-009. Washington, DC.
14. US Environmental Protection Agency. 2004. *Comprehensive Surface Water Treatment Rules Quick Reference Guide: Systems Using Conventional or Direct Filtration.* Washington, DC.
15. US Environmental Protection Agency. 2001. *Cryptosporidium: Drinking Water Health Advisory,* EPA-822-R-01-009. Washington, DC.
16. US Environmental Protection Agency Office of Water. 2015. *Recommendations for Public Water Systems to Manage Cyanotoxins in Drinking Water,* EPA 815-R-15-010. Washington, DC.
17. US Environmental Protection Agency. 1999. *EPA Guidance Manual/Turbidity Provisions,* Chapter 10. Washington, DC.
18. US Environmental Protection Agency Office of Water. 2015. *2015 Drinking Water Health Advisories for Two Cyanobacterial Toxins,* 820F15003. Washington, DC.
19. Blaha, Ludek, Babica, Pavel, Marsalek, Blahoslav. 2009. Toxins Produced in Cyanobacteria Water Blooms: Toxicity and Risks. *Interdisciplinary Toxicology.* 2(2), 36–41.
20. US Environmental Protection Agency Office of Groundwater Drinking Water Standards, Risk Management Division. 2002. *Finished Water Storage Facilities.* Washington, DC.
21. US Environmental Protection Agency Office of Water. 2002. *Nitrification.* Washington, DC.
22. US Environmental Protection Agency. 2013. *Basic Information about Disinfectants in Drinking Water: Chloramine, Chlorine and Chlorine Dioxide.* Washington, DC.
23. US Environmental Protection Agency. 2013. *Basic Information about Disinfectant Byproducts in Drinking Water: Total Trihalomethanes, Haloacetic Acids, Bromate, and Chlorite.* Washington, DC.
24. US Environmental Protection Agency. 2009. *Drinking Water Contaminants-National Primary Drinking Water Regulations.* Washington, DC.
25. US Environmental Protection Agency. May 2004. *National Primary Drinking Water Regulations,* 816F09004. Washington, DC.
26. Oram, Brian. No Date. *Ozonation and Water Treatment.* Water Research Center, Dallas, PA.
27. US Army Public Health Command. Updated 2011. *Ultraviolet Light Disinfection in the Use of Individual Water Purification Devices,* 31-006-0211. Washington, DC.
28. Environmental Protection Agency. 2011. *Water Treatment Manual: Disinfection.* Wexford, Ireland.
29. US Environmental Protection Agency Office of Water. 2002. *Effects of Water Age Distribution System Water Quality.* Washington, DC.
30. US Environmental Protection Agency Office of Water. 2002. *Deteriorating Buried Infrastructure Management Challenges and Strategies.* Washington, DC.
31. US Environmental Protection Agency Office of Water. 2002. *Permeation and Leaching.* Washington, DC.

32. US Environmental Protection Agency Office of Water. 2002. *New or Repaired Water Mains.* Washington DC.
33. LeChevallier, Mark W., Gullick, Richard W., Karim, Mohammad. No Date. *The Potential for Health Risks from Intrusion of Contaminants into the Distribution System from Pressure Transient.* US Environmental Protection Agency Office of Water, Washington, DC.
34. US Environmental Protection Agency Office of Water. 2011. *Potential Contamination Due To Cross-Connections and Back Flow and the Associated Health Risks.* Washington, DC.
35. US Environmental Protection Agency Office of Water. 2002. *Health Risks from Microbial Growth and Biofilm in Drinking Water Distribution Systems.* Washington, DC.
36. US Environmental Protection Agency Office of Water. 2013. *Literature Review of Contaminants in Livestock and Poultry Manure and Implications for Water Quality,* EPA 820-R-13-002. Washington, DC.
37. Spiehs, Mindy, Goyal, Sagar. 2007. *Best Management Practices for Pathogen Control in Manure Management Systems,* M1211. University of Minnesota Extension Service, Minneapolis, MN.
38. Morrison, Robert, Sangster, Tom, Downey, Dec. 2013. *State of Technology for Rehabilitation of Water Distribution Systems,* EPA/600/R-13-036. US Environmental Protection Agency, Edison, NJ.
39. Centers for Disease Control and Prevention. 2012. *Global Wash (Program)–Related Diseases and Contaminants: Waterborne Diseases.* Atlanta, GA.
40. State of California Delta Stewardship Council. 2013. *The Delta Plan Ensuring a Reliable Water Supply for California: A Healthy Delta Ecosystem, and a Place of Enduring Value.* Sacramento, CA.
41. Texas Commission of Environmental Quality. 2015. *Public Water Supply Supervision Program.* Austin, TX.
42. US Environmental Protection Agency. 2015. *Groundwater Discharges (EPA's Underground Injection Control Program).* Washington, DC.
43. US Environmental Protection Agency. 2015. *Sole-Source Aquifer Program.* Washington, DC.
44. US Environmental Protection Agency. 2015. *Drinking Water State Revolving Fund: EPA Funding for Drinking Water Activities and Infrastructure Projects.* Washington, DC.
45. US Environmental Protection Agency. 2015. *Laws and Regulations: Summary of the Clean Water Act.* Washington, DC.
46. US Environmental Protection Agency. 2014. *Water: Permitting–NPDES Home.* Washington, DC.
47. American Association for Clinical Chemistry. 2016. Lead Testing Uncovered Flint, Michigan's Water Contamination Crisis. *Lab Tests Online* (registered trademark). Washington, DC.
48. US Department of Health and Human Services, Assistant Secretary for Preparedness and Response. 2016. *Federal Support for State and Local Response Operations: Flint, Michigan, Water Contamination Crisis.* Washington, DC.

BIBLIOGRAPHY

American Waterworks Service Company Inc. 2009. *The Potential for Health Risks from Intrusion of Contaminants into the Distribution System from Pressure Transients, Distribution System White Paper.* Voorhess, NJ.
Centers for Disease Control and Prevention. 2012. *Private Well Initiative.* Atlanta, GA.
Centers for Disease Control and Prevention. 2012. *Water-Related Technical Assistance and Outbreak Response.* Atlanta, GA.
LeChevallier, Mark W., Gullic, Richard W., Karim, Hammond L. 2015. *Troubleshooting Water Well Problems.* Ag-Info Centre, Alberta, Canada.
US Environmental Protection Agency. 2012. *Drinking Water Contaminants: National Primary Drinking Water Regulations.* Washington, DC.
US Environmental Protection Agency. 2012. *Public Drinking Water Systems: Facts and Figures.* Washington, DC.
US Environmental Protection Agency. 2012. *Public Drinking Water Systems Programs* Washington, DC.
US Environmental Protection Agency. 2015. *Drinking Water Laws and New Rules: Safe Drinking Water Act.* Washington, DC.
Watt, Edgar W., Lefrancois, Liz, Boots, Ben F. 2012. *Threats to Water Availability in Canada.* Environment Canada, Gatineau, QC, Canada.

14 Water Quality and Water Pollution

INTRODUCTION

This chapter depends upon and ties together the information, sub-problems, and Best Practices that have been presented in the previous chapters including: Chapter 2, "Air Quality (Outdoor [Ambient] and Indoor)"; Chapter 3, "Built Environment—Healthy Homes and Healthy Communities"; Chapter 4, "Children's Environmental Health Issues"; Chapter 5, "Environmental Health Emergencies, Disasters, and Terrorism"; Chapter 8, "Healthcare Environment and Infection Control"; Chapter 9, "Insect Control, Rodent Control, and Pesticides"; Chapter 10, "Recreational Environment and Swimming Areas"; Chapter 11, "Sewage Disposal Systems"; Chapter 12, "Solid Wastes, Hazardous Materials, and Hazardous Waste Management"; and Chapter 13, "Water Systems (Drinking Water Quality)" (See endnotes 1, 13, 14, 15, 17).

It is clearly seen from actual practice that the various environmental media of air, water, and land are very much inter-related and interdependent regarding the good and necessary things that we have in life and also the problems of environmental pollutants that move back and forth through each of these media. This also reminds the reader that all individuals, no matter what their specialty, whether that be environmental health and its subdivisions or environmental protection, sustainability, and its subdivisions, have an important task to perform by working together to make the world a better place to live in by carrying out the necessary task collectively to protect the health of people and protect the finite environment that supports us as a society. Thus, we now have this highly detailed book with excellent resources, which will appeal to all individuals in all sections of the environmental field.

STATEMENT OF PROBLEM AND SPECIAL INFORMATION

Prevention of water quality deterioration is more effective and less costly than trying to restore the damage to land and waterways after the problems occur. Water quality deterioration may be caused by any natural, physical, biological, or chemical change but most frequently is caused by contaminants made by people. The waters of the United States are supposed to support fish, shellfish, wildlife protection and propagation, recreational activities, agricultural activities, harvesting of animals and/or plants from bodies of water for personal use or sale, public water supply, and industrial purposes. When bodies of water are either threatened (that is, are exhibiting a deteriorating trend to support the above uses) or impaired (that is, are unable to support one of the aforementioned uses), the United States has lost a valuable resource. Impairment may be caused by the presence of pathogens (most coming from fecal contamination from people, animals, and birds), alteration of the habitat, organic enrichment, oxygen depletion, impaired plant and animal life from unknown causes in a given location, nutrients, metals, sediment, mercury, alteration of flow patterns, and the turbidity of the water.

The US Environmental Protection Agency (EPA) in its National Water Quality Inventory Report to Congress in 2004, stated that 44% of all streams and rivers in the United States were impaired while 3% were threatened, 64% of all lakes, ponds, and reservoirs were impaired while 1% were threatened, 30% of all estuaries and bays were impaired and about 1% were threatened, and 93% of the shorelines of the Great Lakes were impaired because of polychlorinated biphenyls (PCBs), toxic organics, pesticides, and dioxins, with the impairment primarily from contaminated sediment. There was inadequate information about wetlands, which are critically important to all forms of life and the environment, and there was inadequate information about the beaches and shorelines of the oceans and the Gulf of Mexico.

The Great Lakes are and have been at risk for many years. Every form of contamination mentioned in this chapter has occurred in the Great Lakes. Despite numerous programs, the level of contamination remains unacceptable in various portions of this huge waterway. The chemical integrity of the Great Lakes is threatened and 29 areas are impacted by toxic contaminants and degraded habitats, and definitely are still in need of cleanup in the United States. Nonpoint source pollution, including runoff of material contributing to the production of sediment and excess nutrients, has impaired water quality. Invasive species have contributed to the water problems. There has been crucial habitat loss due to degradation from development, competition from invasive species, and alteration of the natural lake levels. The loss of wetlands not only affects various ecosystems but also removes a natural source of the removal of harmful substances from water. Harmful algae blooms have seriously affected the potential health and economic well-being of a variety of communities, and this problem has worsened over the past 10 years.

The Grand Calumet River is a perfect example of 100 years of neglect of the waterway as the river has been used as an open sewer for industrial and municipal waste disposal. This waterway is one of the contributors to the Great Lakes dilemma. The river flows for 13 miles through Gary, East Chicago, and Hammond, Indiana, into the Indiana Harbor and Ship Canal and then drains about 1 billion gallons of water each day into Lake Michigan. Typically, 90% of the flow of the river is municipal and/or industrial effluent, cooling and process water, and stormwater overflows. Most of the impairment affects everything from drinking water consumption to fish consumption to degradation of local ecosystems and is due to the huge number of contaminants dumped into this waterway over many decades. The contaminants include PCBs; polycyclic aromatic hydrocarbons (PAHs); and heavy metals, such as mercury, cadmium, chromium, and lead. In addition, there are high levels of fecal coliform, high biochemical oxygen demand (BOD), suspended solids, oil, grease, phosphorus, nitrogen, iron, magnesium, and volatile solids. This area contains from 5 million to 10 million yd.^{3} (3.9 million to 7.7 million m^3) of contaminated sediment up to 20 feet deep (approximately 6 m) from point source and non-point source contributors. In addition, contaminants come from:

1. Industrial waste site runoff of stormwater and leachate from 11 of 38 different waste disposal and storage sites
2. A total of 52 sites on the federal Comprehensive Environmental Response, Compensation, and Liability System list with five of these designated as Superfund sites
3. A total of 423 hazardous waste sites regulated under the Resource Conservation and Recovery Act
4. More than 460 underground storage tanks with 150 leaking tank reports made
5. Atmospheric deposition of toxic substances from burning of fossil fuel, waste incineration, evaporation and direct contact with water, surface water runoff, and leaching of materials from land
6. Urban runoff from rainwater going across paved areas that contain grease, oil, toxic organics, etc.
7. Groundwater contamination by organic compounds, heavy metals, and at least 16.8 million gallons (63.6 million liters) of oil floating on top of the groundwater
8. Three steel manufacturers contributing 90% of the industrial point source discharges
9. Combined sewer overflows from 15 overflows with untreated municipal waste, while conventional and toxic pollutants account for many millions of gallons of waste material being dumped into the harbor and river, and this is still occurring

MEASURES OF WATER QUALITY

The most common measures of water quality are:

1. pH which measures the water's acidity
2. Water temperature which affects fish and plants as well as humans
3. Turbidity which is the amount of cloudiness of the water due to suspended particles
4. Dissolved oxygen which is necessary for fish life
5. Amount of nutrients in the water, including nitrates, phosphorus, nitrogen, and ammonia
6. Enteric bacteria which can indicate contamination of the water and potentially cause disease
7. Toxic substances which may be found in the water, fish, or sediment

SUB-PROBLEMS INCLUDING LEADING TO IMPAIRMENT AND BEST PRACTICES FOR POTENTIAL SOURCES OF WATER CONTAMINATION

Some of the factors leading to impairment and their associated Best Practices are discussed in this chapter.

ACID MINE DRAINAGE

Acidic drainage from mines is caused by sulfide minerals which have been exposed to air or water and then seep into groundwater or surface water. The acidic materials are harmful to aquatic ecosystems and humans through direct contact and ingestion of water. There are an estimated 500,000 abandoned mines in the United States with 130 of them either on the National Priorities List or capable of being on it. This contamination covers more than 1 million acres of land. Surface water and groundwater is contaminated from seepage from various mine openings, waste rock, mill tailings, waste slurries, and waste sludges which have been placed in unlined lagoons.

Best Practices for Acid Mine Drainage (See endnote 5)
1. Since remediation of acid mine waste sites is energy intensive, the following should be observed for testing purposes use field kits, low flow samplers, remote sensing, and existing boreholes for sampling; the use of sonic rotary drilling instead of conventional rotary drilling or hammer techniques; the use of phosphate-free detergents instead of organic solvents or acids; and the use of safe drilling fluid or water in a closed loop system. Utilize renewable sources of energy, such as wind energy, and capture and use any energy produced by the consumption of the agricultural and forestry waste products or gases from the site.
2. Use passive treating systems for the acid water, containing considerable metals, especially existing chemical and biological processes, such as oxidation/reaeration ponds, limestone beds, and biochemical reactors with agricultural and forestry waste products.
3. Capture and sell metals recovered from the biological process.
4. Develop extensive stormwater systems to reroute the water from rain or snow crossing the contaminated site.
5. Only remove the vegetation which is necessary to gain access to the site and use existing roads where possible.
6. When establishing a biochemical reactor, use a geomembrane liner to prevent leakage to the groundwater supply or surface water.
7. Install an appropriate soil cover to stabilize the soil and waste piles.
8. Integrate the cleanup of the land with restoration and reuse of the site.

AESTHETICS

The use of water can be limited by the presence of scum, foam, and unnatural water color and taste. Trash including plastics, litter, debris, and other solid wastes from people also cause degradation.

Best Practices for Aesthetics
1. Develop a periodic clean-up program to remove trash and other litter close to bodies of water, conducted by civic groups, Boy Scouts, Girl Scouts, church and synagogue groups, etc.
2. Sample the water including scum, foam, and color to determine what contaminants are present and where the contamination is coming from. This is typically point source contamination and needs to be either stopped on a voluntary basis or by direct order of a governing body and if necessary the use of further legal efforts.

AGRICULTURE

Manure and other animal wastes are sources of ground and surface water pollution, especially from microorganisms and oxygen-demanding waste which can affect life in streams. About 95% of the fecal coliform found in urban stormwater comes from non-human sources. Feedlots used for animals or poultry produce large amounts of concentrated animal waste which readily contaminates ground and surface water supplies because of the quantities of waste available, through direct surface runoff or seepage into the groundwater, and because of over-application of the waste materials to the land. These contaminants add high levels of nitrogen nutrients to the soil. The increased nutrients create substantial growth of algae in surface waters, thereby killing fish and other life forms because of reduced levels of oxygen. The higher levels of pathogenic bacteria may also cause disease. Medical waste can be generated daily on farms. These wastes include needles, syringes, scalpels, drug or vaccine vials, outdated drugs, etc. If improperly handled they may injure livestock, people, and waste handlers, and pollute the environment or increase the potential of infection.

Best Practices for Agriculture (See endnotes 2, 3, 4, 29)
1. Use proper grazing management measures in sensitive areas such as streams, wetlands, ponds, and other bodies of water. Exclude livestock and use alternate drinking water locations.
2. Leave harvested plant material on the ground to reduce soil erosion and runoff.
3. Reduce nutrient runoff by determining the actual needs for nutrients by type of crop, location of the farm, prevailing weather conditions, and time of year.
4. Use appropriate pest management procedures while reducing use of chemicals.
5. Use strips of vegetation to provide barriers against runoff.
6. Use runoff control and proper waste storage on feedlots.
7. Use proper feed formulations to reduce nitrogen and phosphorus in the feces.
8. Use appropriate erosion and sediment control procedures (see appropriate section).
9. Use irrigation water in quantities needed and not in excess.
10. Use soil efficiently by: testing for quantities of nutrients naturally available and using the results to determine the amount of nutrients to be added; use a proper nutrient (nitrogen and phosphorus) source; use proper timing of nutrient application; and use manure whenever possible as a source of nutrients.

AIR POLLUTION DEPOSITS

Four major groups of air pollutants affect water quality. They are:

1. Organic chemicals, which stay in the environment for long periods of times, bioaccumulate (the uptake, retention, and concentration of environmental substances by an organism)

from low concentrations in water to high concentrations in animal tissue, and are highly toxic at low levels

2. Mercury, which is found in air, water, soils, and sediments, is transformed into a very toxic compound for fish, wildlife, and humans, known as monomethyl mercury
3. Nutrients which are carried through the air into bodies of water and can accelerate eutrophication (a process by which pollutants increase organic and mineral nutrients in a body of water and affect it adversely)
4. Deposits of heavy metals and other contaminants from processing ore

These pollutants along with others, travel long distances rapidly, and may either be deposited on land or water, dry or wet. They may interact with each other, form new even more complex pollutants, be highly bioaccumulative going from low levels of concentration in the water to high levels of concentration in animal tissue, be highly toxic at very low doses, and persist for long periods of time. They may interact with the environment, stick to surfaces to become available at later dates, or re-volatilize into the air and start the process all over again. Typically, pollutants are evaluated individually; however, they are usually present in various combinations and therefore the results of their presence and the potential damage they do may be hard to predict.

Best Practices for Air Pollution (See Chapter 2, "Air Quality (Outdoor [Ambient] and Indoor)")

ALTERATION OF HABITATS

The alteration and destruction of habitat for fish and other marine life may occur from new techniques of fishing including bottom trawling; the cruise ship industry; coastal development; and hydromodification.

There are four major causes of habitat alteration:

1. Bottom trawling uses large bag-shaped nets which are pulled along the seafloor and dig into natural habitats, destroying the ecosystem for the targeted as well as non-targeted marine life.
2. The cruise ship industry destroys habitat through accidental groundings of ships, but most frequently by the dragging of anchors, sometimes weighing as much as 5 tons, along with the chains, through coral reefs, sea grass beds, and sea floors.
3. Coastal development, since 60% of the world's population lives within 60 miles of a coast, has threatened the oceans and the land in the event of storms by destroying or changing the coastal marshes and estuaries (a shallow body of water where a river meets the ocean and fresh water and ocean water mix).
4. Hydromodification (physical modification of water systems) occurs when there is a channeling of water, building of dams, and use of techniques to prevent erosion of stream banks and shorelines of lakes and oceans. This can alter the habitat of marine life, change the pattern of water temperature, produce different types of sediments than that which is normally found, and increase the amount and speed at which non-point source pollutants move from the upper portions of the watershed into coastal waters.

Best Practices for Alteration of Habitats (See endnote 6)

Since the various estuaries throughout the country differ so much, it is difficult to incorporate specific Best Practices. However, the US EPA lists seven general themes for watershed management. They are:

1. Collect necessary information about watershed conditions through monitoring and research.
2. Increase public knowledge about sediments, nutrients, thermal modification, and chemical pollutants.

3. Form local partnerships for watershed protection with governmental agencies, activist groups, business and industry, landowners, and agricultural organizations.
4. Develop a plan for improvement and establish priorities.
5. Obtain and coordinate funding and technical assistance from the local, state, and federal level.
6. Implement the recommended solutions to the problems.
7. Evaluate the results and redirect the efforts were needed.

Climate Change

The water cycle which goes from the evaporation of water from the land, vegetation, and bodies of water to deposition of the water, is affected by climate change. The three major causes of climate change are:

1. The quantity of rain or snow that falls, the intensity of the deposition, and the location may be altered by the ambient temperatures. The amount and nature of pollutants found in the rain or snow is also affected. The quantity of water available to people and the various eco-systems to sustain good health, agriculture, production of energy and manufacturing, flow through the waterways for proper navigation, etc., varies with the amount which is readily available and the amount of usage which increases sharply with higher temperatures.
2. Water quality may be affected by increased runoff, substantial production of sediment, rise of sea levels, incursion of salt water into fresh water areas, and flooding. Large amounts of rain can overwhelm the water treatment plants, sewer systems, and waste treatment plants.
3. The oceans and other bodies of water can become more acid because of the uptake of carbon dioxide from the atmosphere. This can affect all forms of marine life.

Best Practices for Climate Change

Best Practices for climate change are too complex for this portion of this chapter. (See Chapter 2, "Air Quality (Outdoor [Ambient] and Indoor)" for more information.)

Combined Storm and Sanitary Sewer

Overflows in urban areas occur from substantial amounts of rainfall in short periods of time falling in a given locale, which is typically upstream from recreational or other use areas. (See Chapter 11, "Sewage Disposal Systems.")

Best Practices for Combined Storm and Sanitary Sewer (See Chapter 11, "Sewage Disposal Systems")

Erosion

Erosion is the removal of soil and rock fragments by wind, water, snowmelt, flooding, rain, organisms, and gravity. It is also caused by fires, construction, agriculture, tree removal, drought, etc. It may lead to increased sediment in bodies of water.

Best Practices for Erosion
(Note: This topic is not readily itemized)

Control soil erosion and prevent loss of nutrients by: maintaining a soil cover preferably with crop residue; allowing for maximum water infiltration and storage; providing vegetative banks of ditches and channels; sloping roads appropriately; using grass areas wherever possible near roads; and using windbreaks to control wind erosion. Slow the flow of water where possible.

EXTRACTION OF RESOURCES

Hydraulic fracturing or fracking is a process in which a well is drilled and a steel pipe is inserted into the well bore with holes in the bottom and then a liquid under pressure is inserted to overwhelm the natural pressure and fracture or crack rock underground to release gas or oil. This process, which is currently producing a large amount of natural gas, has several potential environmental concerns. They are:

1. Air quality issues, including the release of volatile organic compounds, other hazardous air pollutants, and greenhouse gases into the ambient air. This occurs when organic compounds go into the air from the wastewater, from spills, from the gas or oil being recovered, and from the chemicals being used in the fracking process.
2. Contamination of ground and surface water occurs from spills, poor well construction, etc.
3. Disposal of large quantities of contaminated water is costly and overwhelms existing facilities.
4. Improper use of sewage treatment plants for disposal of the wastewater may affect the efficiency of the plant and allow raw sewage to go into the receiving stream.
5. Potential destruction of habitats and marine life from exposure to toxic chemicals.
6. Potential illness of humans from exposure to toxic chemicals may result from ingestion, inhalation, or direct contact with the skin or eyes. There is a potential for acute or chronic disease of the various body organs.
7. Sand and proppants (agents used to hold open hydraulic cracks in a rock formation) which also contain their own contaminants are used in the fracking process and therefore become a disposal problem.
8. Shale gas wells wastewater from extraction that contains high levels of total dissolved solids and naturally occurring radioactive materials may become part of stormwater disposal and entering surface bodies of water.
9. Surface and groundwater issues where the substantial withdrawal of water for drilling and hydraulic fracturing affects other water uses.
10. Toxic chemicals, some of which are potential carcinogens, are either found in the hydraulic fracturing fluids or may be part of the resulting oil or natural gas produced in the well.
11. Use of surface pits or ponds for storage of the wastewater.
12. Use of underground injection wells and possible contamination of aquifers used for drinking water supplies.

Best Practices for Extraction of Resources (See endnote 7)
Reduce problems associated with extraction of resources by:
1. Improving the scientific knowledge of the effect of hydraulic fracturing on water supplies and the potential for health hazards short-term and long-term in people by:
   - Analyzing existing data of over 25,000 wells in the oil and gas industry
   - Analyzing 12,000 specific wells for chemical use and water use in a registry operated by the Groundwater Protection Council and the Interstate Oil and Gas Compact Commission
   - Analyzing the well operators' records of hydraulic fracturing in 333 oil and gas wells in the United States
   - Developing realistic computer models on drinking water supplies from surface and groundwater sources
   - Utilizing the information from laboratory results of drinking water supplies, toxicity assessments, and case studies of existing problems, to complete the data-gathering process and analyze all this material to determine Best Practices
2. Determining if there are clusters of disease or predisease conditions in areas where hydraulic fracturing has been previously used
3. Ensuring that there is a proper permit for use of diesel fuels in hydraulic fracturing

4. Ensuring that stormwater does not cross the area where hydraulic fracturing is being utilized
5. Ensuring that wastewater is properly disposed of either through underground injection, surface impoundments for storage or disposal, recycling of the wastewater, or least of all the use of municipal treatment plants

GOVERNMENTAL AGENCIES

Government agencies add to the contamination through the use of substances to de-ice roads, through minor road repairs, through major road work, through maintenance activities on automobiles and trucks, through landscaping, and through maintenance of parks and buildings.

Best Practices for Governmental Agencies

Pollution prevention of stormwater from municipal activities by use of proper street sweeping and storm drain system cleaning reduces the potential for ground and surface water contamination. Specific training, inspection, and maintenance procedures help municipal workers carry out this function. The entire stormwater discharge system from municipal areas needs to be evaluated to determine if pipes from sources of contamination are feeding into the stormwater discharge system. Typically, the water flow through the system is not treated before it is allowed to go into a body of water. Special inspection and survey programs are established to prevent illicit discharges from homes, businesses, and factories into storm sewer systems. The programs also provide techniques for handling spills of chemicals and other substances that might flow into the stormwater system.

ILLICIT DISCHARGES

Illicit or illegal discharges are indicated by measurable amounts of pathogens and/or pollutants in a fluid in storm drains during dry periods, or flows across the ground into surface water. Each discharge has a unique source and possible frequency. These discharges may be from direct illegal connections from sewage systems, overflow from on-site sewage, illegal connections from car washes, restaurants, gas stations, other businesses, factories, homes, and apartments, or due to spills or overflows of contaminants which penetrate the ground or run into surface water. They also come from the leachate from landfills, both active and closed.

Best Practices for Illicit Discharges (See endnote 8)

Illicit or illegal discharges can be reduced substantially if the community becomes involved in an illicit discharge detection and elimination program as recommended by the US EPA. The four major steps of the program are also the Best Practices. They are:

1. Determine the existing resources, programs, staff, and legal authority, and their effectiveness, which are available to resolve the problems. Determine the gaps, financially, technically, legislatively, and people-wise and make provisions for correcting these.
2. Determine who will be responsible for administering programs, has legal authority, and how discharges will be tracked and eliminated.
3. Complete a comprehensive study of all potential sources of pollutants in the watershed area and establish priorities for first removing the worst ones that can be done rapidly, the lesser ones that can be removed rapidly, the worst ones that may take extended time, and then the lesser ones that may take extended time.
4. Develop measurable program goals and means of implementation, and provide for redirection of the program as needed.

Also, develop a comprehensive educational program for the public and an approach to working with the many citizens' groups and others who are interested in better water quality in the community.

IMPROPER DISPOSAL OF HAZARDOUS WASTE

(See Chapter 12, "Solid Waste, Hazardous Materials, and Hazardous Waste Management")

IMPROPERLY CONSTRUCTED OR MAINTAINED WELLS

(See Chapter 13, "Water Systems (Drinking Water Quality)")

INVASIVE SPECIES

(See endnotes 22, 23)

Non-indigenous invasive species is a huge topic which is beyond the scope of this book. However, a brief description of the problem follows. The invasion of non-indigenous invasive species of marine life and plants is one of the most challenging environmental issues. They disturb the balance of the natural ecosystems and may cause significant economic impacts, and in fact affect the health and safety of people by disrupting food sources, fiber sources, and drinking water supplies. The single largest source of the unintentional introduction of these organisms into many areas including the Great Lakes basin is through maritime commerce. These organisms attach themselves to the hulls of ships and are also found in ballast tanks in the water that had been taken on in other locations. Several factors facilitate the spread of aquatic invasive species including: moving of boats and ships from one waterway to another; clear cutting of forests; various practices that increase sedimentation and therefore water turbidity; pollution from various industries; many of the contaminants mentioned in other parts of this chapter; and over-fishing. As an example of an invasive species, the zebra mussel may cause severe problems at power plants and municipal water supplies by clogging intake screens, pipes, and cooling systems.

Best Practices for Invasive Species (This will be limited because of the nature of the topic.) Best Practices include but are not limited to:
- An educational program to teach people to wash their boats when moving them from one waterway to another
- A ballast water regulation introduced and enforced by the United States Coast Guard requiring ships to exchange their ballast water or seal their ballast tanks while in a given waterway
- Dredging of contaminated sediments from harbors and other areas

LAND DEVELOPMENT AND BUILDING CONSTRUCTION

Development and construction disrupt the natural habitat and may cause a substantial impact on the water quality of surface and groundwater supplies. Stormwater runoff from construction sites has a significant impact on bodies of water, especially through the depositing of sediments and construction waste into the water. The runoff of sediment from construction sites is typically 10–20 times greater than those from agricultural lands and 1000–2000 times greater than from forest lands. The prevention of the erosion of soil is of primary importance. Postconstruction stormwater management is also of great significance. In the last 20 years, the rate of land development has been more than twice as great as the rate of population growth.

Best Practices for Land Development and Building Construction
- To reduce the impact of contaminated stormwater flow from these areas, it is necessary to treat, store, and ground infiltration of the runoff before it enters bodies of water.
- In order to implement a program of control of construction site runoff, it is necessary to have proper regulatory mechanisms in place, site plan reviews, appropriate

inspections and penalties when the regulations are compromised, and information supplied to the public as to what is an appropriate plan to prevent pollution of bodies of water. (See endnote 12.)

LANDFILLS

(See Chapter 12, "Solid Wastes, Hazardous Materials, and Hazardous Waste Management")

METALS

Metals may have a toxic effect on aquatic life and tend to bioaccumulate in plants, fish, and then in people. The metals become attached to the fine particles of sediment. Mercury is a prime example. Metals typically come from chemical and equipment suppliers and improper waste disposal whether liquid or solid.

Best Practices for Metals
- Modify metal processing storage and handling to reduce potential for exposure and release of raw materials, products, or byproducts to the environment.
- Decrease the amount of suspended solids which have attached metals entering bodies of water.
- Decrease the amount of metals being released into the air as a pollutant.

LOW DISSOLVED OXYGEN

Dissolved oxygen in water is necessary for fish to live. Organic material from excess nutrients, effluent from sewage treatment plants, materials creating colors, and organic material from industrial plants or nature use up the dissolved oxygen, thereby causing an oxygen depleted water environment resulting in plants and fish dying. This is highly detrimental to water quality.

Best Practices for Low Dissolved Oxygen

(See Best Practices for Oxygen-Demanding Substances)

NON-POINT SOURCES OF CONTAMINATION

(See endnote 21)

A pollutant that comes from a non-point source of contamination is any pollutant that does not come from a specific and discrete source of contamination, such as a ship or factory, including stormwater and runoff from urban and agricultural areas. Stormwater from heavy rains falls on a variety of surfaces including sidewalks, yards, driveways, roofs, parking lots, etc., and carries pollutants from littering by individuals, trash and recyclables, pet waste, lawn fertilizers and pesticides, residue from washing cars, residue from motor oil and other contaminants from cars, and leftover hazardous chemicals and paint which are exposed to the rain. Industrial stormwater typically introduces into surface bodies of water substantial quantities of total suspended solids, oxygen-demanding materials, nutrients, metals, hazardous chemicals, and other common pollutants. Runoff is now the most common source of water pollution. Agriculture contributes almost half of the water quality contaminants to rivers and streams and over 40% to lakes, ponds, and reservoirs. Municipal point sources contribute about 37% to the contamination of estuaries. Hydrologic modifications contribute to 20% of the water quality problems in rivers and streams and 18% in lakes, ponds, and reservoirs. Urban runoff and contaminated storm sewer water contribute 32% of the water quality problems to estuaries, while contributing 18% to lakes, ponds, and reservoirs. Industrial discharges contribute 26% of the water quality problems of estuaries and atmospheric deposition 24%.

Runoff from golf courses which contains substantial amounts of fertilizer and pesticides can contaminate groundwater or surface water. Non-point source pollution intensifies with the amount of rain or melting snow in a given area and in a given period of time. The contaminants on the surface of the ground carried into bodies of water may include sediments, pathogens, fertilizers, nutrients, hydrocarbons, metals, oil, grease, toxic chemicals, and acid drainage.

Best Practices for Non-Point Sources of Contamination
- Before digging up the land for any purpose, develop and implement an approved erosion and sediment control plan. Evaluate the plan as the construction proceeds to determine if the goals and objectives are being met and if not modify the plan to do so.
- Identify and prioritize using sound scientific techniques and data, the potential threats to human and ecosystem health from non-point source contamination in a given area.
- Reduce erosion in any type of construction or farming and retain sediment on site.
- Avoid off-site transportation of waste material and chemicals where appropriate.
- Minimize the use, storage, and disposal of all types of chemicals which may then become water pollutants.
- Use small amounts of fertilizer on properties and sweep up driveways, sidewalks, and gutters.
- Do not dump anything down storm drains or in streams.
- Plant vegetation where there are bare spots in the yard.
- Compost the yard waste.
- Use the least toxic pesticides on the property.
- Direct downspouts to grassy areas instead of paved areas.
- Do not wash the car at home.
- Check the car to make sure there are no types of leaks of oil or other substances.
- Inspect the septic tank system on a regular basis and have the sludge and other substances removed by a registered sewage hauler before there is an overflow in the system onto the ground.

Some specific techniques for controlling stormwater flow, which at high levels promotes erosion and spread of pollutants, are bioretention cells, elimination of curbs and gutters, grassy swales, green parking lot design, infiltration trenches, inlet protection devices, permeable pavement, permeable pavers, rain barrels and cisterns, riparian buffers, sand and organic filters, soil amendments, stormwater planters, tree box filters, vegetated filter strips, and vegetated roofs. These are explained below.

- A bioretention cell, also called a rain garden, is a depressed area in which forest material has been placed and is covered by a surface with vegetation. Underdrains are used in these areas to help with filtration and infiltration. The bioretention cell is useful in groundwater recharge, pollutant removal, and containing runoff. It is very useful in parking lots or urban sites where there are very few grassy areas.
- The elimination of curbs and gutters helps reduce the speed of the flow of rainwater, and allows for infiltration into the ground and also the removal of pollutants.
- Grassy swales are shallow indentations in the ground that help capture runoff water and increase infiltration into the soil.
- A green parking lot design utilizes appropriate amounts of parking spaces and grassy areas to allow for infiltration of rainwater flowing off of the concrete or macadam surface.
- Infiltration trenches are ditches filled with rocks that have no outlets. The runoff from the surface flows into the ditches and infiltrates the soil.

- Inlet protection devices, also called hydrodynamic separators, are separation units utilized to remove sediments, oil, grease, trash, and other pollutants.
- Permeable pavement is a porous surface over stone in a hole in the ground, instead of asphalt or concrete surfaces, to allow water to accumulate and disperse through the soil.
- Permeable pavers are interlocking concrete blocks which have voids to allow water to penetrate the ground instead of running off.
- Rain barrels and cisterns are containers which are used to capture rainwater to be utilized at another time for household purposes.
- Riparian buffers are areas near shorelines, wetlands, or streams where development is either restricted or prohibited and the buffer helps manage stormwater.
- Sand and organic filters are used to remove specific types of pollutants from water before the water penetrates to the groundwater supply or flows to a surface body.
- Soil amendments, such as soil conditioners and fertilizers, are utilized to increase water retention capabilities in soil. The soil amendments change the physical, chemical, and biological characteristics of the soil and make the soil more suitable for growth of plants and water retention.
- Stormwater planters are small devices which can be placed either above or below ground and increase the efficiency of the infiltration of stormwater and the filtering of contaminants.
- In-ground tree box filters contain trees, vegetation, and soil that help filter runoff before it enters a catch basin. Aesthetically they are very pleasing.
- Vegetated filter strips are bands of dense vegetation for treating runoff from roads and highways, parking lots, and other impervious surfaces before they get to a body of water.
- Vegetated roofs or green roofs are impermeable roof membranes over which plantings are placed to reduce runoff volume and improve water quality.

NUTRIENTS

Nutrients, including fertilizers, are needed by all vegetative matter for growth. However, excessive amounts enter bodies of water increasing the levels of organic and mineral nutrients allowing algae to grow rapidly and deplete the oxygen supply, causing organic matter to die and producing unpleasant odors. The major components of nutrients include nitrogen and phosphorus.

Nitrogen compounds can be dissipated by uptake by the plants, leaching into surface or groundwater, surface runoff, and losses to the atmosphere when the fertilizers are exposed to sunlight and air, or are stored in the soil. The nitrogen is replaced by humans and if the quantities are too large or if there is a problem of runoff, the nitrogen compounds can enter the bodies of water.

Phosphorus may be found in several different forms in the soil and can be transported by various means to surface bodies of water. Phosphorus may also be found in a variety of chemicals and fertilizers, animal wastes, food wastes, wood and sawdust waste, and anything else that can decompose. It is usually the amount of excess nutrients that controls algae growth in freshwater lakes and leads to oxygen depletion and then the accumulation of sediment as the algae die.

Although ammonia is not a nutrient, it is a nitrogen compound which can have an immediate and deadly effect on humans as well as various ecosystems. Ammonia-based compounds may be used to wash and rinse different types of equipment.

Best Practices for Nutrients (Also see Best Practices for Agriculture and endnote 9)

The use of nutrients for lawns, landscape, golf courses, etc. should be closely supervised by trained people to prevent substantial runoff into bodies of water. This can be accomplished by:

- Conducting soil, nutrient, and plant tissue tests, to determine the best soil productivity, acidity, and the type and amount of nutrients available. The tests measure the amount of phosphorus and potassium available, the pH of the soil to determine acidity, and the soluble

salts to determine the amount of fertilizer already present. With this information and knowledge of the weather conditions and the amount of soil moisture, it is possible to determine the nature, quantity, and type of fertilizer which should be used and how frequently to use it
- Providing diverse healthy eco-friendly trees, grass, ornamentals, and groundcover for the area involved
- Using the diversity of the eco-friendly plants to help prevent a specific disease from destroying the landscape, help prevent serious damage due to drought or poor weather conditions, and make the area more aesthetically pleasing

OXYGEN-DEMANDING SUBSTANCES

Oxygen-demanding substances, coming from organic sources which decompose, such as grass clippings, sugar-containing substances, carbon-based chemicals, and animal wastes, remove dissolved oxygen from the water which is needed to sustain life for aquatic organisms. Stormwater runoff is a prime source of providing large quantities of these wastes to the surface bodies of water. When aquatic microorganisms consume organic matter, they also consume dissolved oxygen. All organic chemicals, especially dyes, utilize dissolved oxygen.

Best Practices for Oxygen-Demanding Substances
- Oxygen-demanding substances can be reduced by providing stormwater detention ponds and filtration devices for the effluent before it goes into the body of water.
- All accidental spills need to be corrected immediately and grass and other organic materials should be stored or removed appropriately.

PESTICIDES

Pesticides which are used on properties for control of insects, rodents, unwanted plants, and fungi, can contaminate groundwater or surface water. (See Chapter 9, "Insect Control, Rodent Control, and Pesticides.")

Best Practices for Pesticides (See Chapter 9, "Insect Control, Rodent Control, and Pesticides")

SAND AND SALT STORAGE

Sand and salt storage areas used to provide the materials to make roads safe during severe winter weather can leach mixtures of salt and other chemicals into the groundwater supply or into surface bodies of water. This raises the chloride level and also introduces other chemicals into the water supply.

Best Practices for Sand and Salt Storage
- Keep the storage areas away from bodies of water or wetlands on high flat ground near a road.
- Enclose the storage area with a high impermeable fence to prevent rain and snow from carrying the salt and sand away from the area and contaminating water.
- Put an impermeable liner under the sand and salt and cover it with another heavy plastic liner.
- Remove all of the remaining salt and sand mixture as well as the liner when abandoning the site.

SEDIMENT

Sediment, composed of clay, silt, and sand, is the material created by the weathering of rocks, soils, and erosion caused by agriculture, urban development, and natural factors. The sediment may block stormwater systems, thereby creating flooding which may create more sediment. A variety of pollutants, such as toxic chemicals, metals, pathogens, phosphorus, etc., become attached to the particles in the sediment. The contaminants may come from industrial plants, municipal sewage treatment plants, septic tank systems, fertilizer and pesticides, other polluted runoff from urban and agricultural areas, air pollution, etc. Excess amounts of sediment degrade water quality by decreasing the clarity of the water and decreasing the light that can penetrate to submerged vegetation. When the sediment is disturbed as through dredging, all marine life and the food sources they feed on can become contaminated and the pollutants can bioaccumulate in the fish and shellfish consumed by people. (See endnote 16.)

Best Practices for Sediment (See Best Practices for Erosion, Best Practices for Non-Point Sources of Contamination, and endnotes 10, 11)

There are a huge number of Best Practices for erosion and sediment control. They may be found in endnotes 10 and 11. Some of the areas covered include soil stabilization, sediment control, wind erosion, snow control, tracking control, non-stormwater, waste management, and the post-construction phase. Fact sheets and training modules are available.
Previously mentioned Best Practices include infiltration trenches, detention and settling ponds, biofiltration, filtration, and flow-through separation process.
Use the following techniques for construction sites.

• Clear the vegetation only from areas that will be immediately used during construction.
• Establish a single point which is stabilized for exit and entry of all vehicles and equipment.
• Protect the perimeter of the site using sediment or silt fences.
• Roughen up the exposed soils.
• Temporarily stabilize all exposed soils with vegetation, sand, straw, compost, or wood.
• Divert stormwater or the water from snow melting from flowing across the construction site.

Best Practices for Sediment Removal from Bodies of Water

Sediment removal from bodies of water can be accomplished through dredging, which may be either done mechanically or hydraulically. Where the sediment does not contain various contaminants, it can be used as clean fill in other areas. Where the sediment is contaminated, it can be contained by a cap of impervious material covered by sand, it can be confined in a special impervious facility, or it can be treated, thereby immobilizing the contaminants, destroying them, or extracting them from the sediment.

SEPTIC SYSTEMS

(See Chapter 11, "Sewage Disposal Systems")

SEWAGE DISCHARGE FROM MUNICIPAL SOURCES

(See Chapter 11, "Sewage Disposal Systems")

SPILLS AND RUNOFF FROM STORED CHEMICALS AND STORAGE TANKS

Ground or surface water is readily contaminated when petroleum products and chemicals are mishandled and spills occur, resulting in the chemicals seeping into the ground and into the water supply or becoming part of runoff and going into surface bodies of water. A single spill or leakage can spread out into the ground and become a source of contamination for many years. In many areas, there are a substantial number of underground storage tanks which have been used over many years for holding a variety of products including substances which are very hazardous. These tanks may be found in gas stations (most common place), airports, dry cleaners, homes, agricultural areas, etc. An immediate problem occurs when there are leaks and spills because of poor housekeeping, overfilling of the tanks, sloppiness in loading and unloading the product, and poor maintenance and inspection of the facility. A long-term problem occurs when the tanks become corroded and start to leak product into the ground.

Best Practices for Spills and Runoff from Stored Chemicals and Storage Tanks
• Inspect all pumps, hoses, and connections between pipes for leaks monthly.
• Check for loose fittings, worn gaskets, or damaged rubber nozzles monthly.
• Check underground storage tank equipment and dispensers for leaks and structural problems monthly.
• Inspect all storage tanks and facilities, more frequently than normally scheduled, during very cold weather and very hot weather.
• Inspect above-ground storage tanks weekly for leaks and monthly for deterioration.
• Check secondary containment areas for any sheen, which would indicate spillage.
• Keep an inspection log with the results of the inspections, dated and signed.
• Contain in a special area, any contaminated stormwater and treat it prior to releasing onto the land or into a body of water.

SURFACE IMPOUNDMENTS FOR FARM WASTE OR OTHER WASTES

The most significant concerns are leakage from the ponds into the groundwater or a body of surface water, overflow due to heavy rain, potential breeding areas for mosquitoes, and odors.

Best Practices for Surface Impoundments for Farm Waste or Other Wastes
• Containment ponds should have impermeable liners to prevent leakage.
• Oxygen should be introduced into the organic mass to enhance the biodegradation of the material.
• Runoff from surrounding areas should be redirected from the containment area.
• The containment area should be built in such a manner that major rainfall will not cause it to overflow.

TEMPERATURE

Temperature has a major influence on cold bodied aquatic organisms causing them to rapidly react to external temperature change. Increases in temperature because of thermal pollution due to hot runoff from urban areas or discharge from industrial plants or municipal sewage systems can cause problems in small bodies of water.

Best Practices for Temperature Problems
• Determine if fish, other aquatic life, or plants are being affected by increased temperature in a given area.
• Determine if there are industrial plants or municipal treatment plants with discharge pipes going into the waterway near the problem area and take the temperatures of the outfall liquid.
• Use voluntary and then enforcement techniques to reduce the outfall liquid to appropriate temperatures.

TOTAL SUSPENDED SOLIDS (SEE THE SECTION ABOVE ON "SEDIMENT")

Total suspended solids include inorganic materials such as sediment, metals, and organic material such as animal and vegetative wastes and debris that have been washed or blown into the bodies of water. They smother fish eggs and larva, make the water turbid, clog fish gills, affect growth of vegetation, increase the cost of water treatment, transport pollutants, and may increase the effects of toxic chemicals which may bioaccumulate in fish and in humans. They are frequently one of the most damaging pollutants found in water.

Best Practices for Total Suspended Solids

Total suspended solids in bodies of water can be reduced by avoiding washing of soil into normal runoff and preventing erosion. During construction activities and other activities, all clearing, grading, filling, logging, and mining should be conducted in such a manner that the least amount of soil is disturbed. Grassy areas do an excellent job preventing the solids from entering water.

Waste from Pets

Waste from pets are sources of ground and surface water pollution. About 20% of the bacteria found in bodies of water come from dogs.

Best Practices for Waste from Pets

Community and individual participation is of greatest importance. Pet waste should be picked up and either flushed down a toilet or put into solid waste.

Wildlife or Other Natural Sources

Rats or other animals contaminate small bodies of water with their feces and urine. Contamination from birds varies with the species of the birds, bird population density, feeding habits, the amount of dilution created by the size of the water body, and the time of the year. Birds that carry highly infectious microorganisms may reside for long periods of time in limited water areas and when they defecate cause substantial water quality problems leading to disease.

Best Practices for Wildlife or Other Natural Sources of Contamination

Because of the vast variety of animals and birds which can contaminate bodies of water, the Canada goose has been chosen as an example of Best Practices.

- Provide barriers to landing and habitation by using 4-inch gauge netting above the water surface, a life-sized replica of an alligator head, or electric fencing that will be annoying but not deadly.
- Utilize a biodegradable food that tastes like sour grapes and is not toxic to humans, dogs, cats, or birds to discourage wild bird feeding.

LAWS, RULES, AND REGULATIONS

(See endnotes 18, 19, 20)

Rivers and Harbors Appropriation Act of 1899

The Rivers and Harbors Appropriation Act of 1899 is the oldest federal environmental law in the United States. It made it a misdemeanor to discharge any kind of refuse material into the navigable waters of the United States without a permit. It also controlled the alteration of any body of water in any manner without a permit. Although this act is still in force, the Federal Water Pollution Control Act of 1948 and its subsequent amendments authorized the Surgeon General of the US Public Health

Service to work with other federal agencies, and state and local entities to prepare comprehensive programs for eliminating or reducing pollution in interstate waters for improving the sanitary conditions of surface and ground waters.

FEDERAL WATER POLLUTION CONTROL ACT (CLEAN WATER ACT)

The Federal Water Pollution Control Act (Clean Water Act) of 1948 and its many amendments including the major ones of 1961, 1966, 1970, 1972, 1977, and 1987 (Water Quality Act) are the basis for water quality and water pollution control programs in the United States. In 1977, a provision was made for the development of a Best Practices program as part of state area-wide planning. In 1987, a provision was made requiring states to develop strategies for cleanup of toxics in waters when the best available technology discharge standards did not provide appropriate water quality and endangered public health. In 1996, the Safe Drinking Water Act amendments were passed. Provisions of this law have been updated through November 2012.

Immediate compliance and enforcement challenges made worse by wet weather are municipal combined sewer and sanitary sewer overflows, concentrated animal feeding operations, industrial liquid waste disposal, industrial stormwater, and runoff from different types of land. It is unlawful to discharge any pollutant from a point source into the navigable waters of the country. It is also necessary to control non-point source contamination.

The US EPA has been given the authority to:

- Monitor compliance with the Clean Water Act, including inspections
- Conduct civil enforcement
- Grant permits to control discharges through the National Pollutant Discharge Elimination System (NPDES)
- Develop water quality criteria and standards to protect human health and the environment
- Develop water quality models
- Explain different approaches for watershed protection
- Act as a resource for animal feeding operations
- Act as a resource for agriculture concerning surface and groundwater pollution
- Conduct periodic inspections of industrial users' pretreatment of wastewater
- Conduct periodic inspections of publicly owned treatment works (municipal sewage)
- Conduct periodic inspections of publicly owned treatment works and industrial facilities that generate, store, transport, and dispose of biosolids
- Inspect facilities that store oil to prevent oil spills
- Conduct inspections of industrial stormwater, construction sites, industrial sites, and municipal stormwater systems
- Conduct inspections of combined sewers and overflows
- Conduct inspections of wetland areas
- Provide financing for specific water quality projects which have been approved by the US EPA

RESOURCES

FARM*A*SYST

Farm*A*Syst is a program offered by many state universities including North Carolina State University and the North Carolina A&T State University Cooperative Extension Services which provide help in improving storage, handling, and disposal of livestock waste. North Carolina extension services also provide documents on protecting water supplies, improving fuel storage, improving storage and handling of hazardous wastes, improving septic systems, improving storage and handling of pesticides, improving storage and handling of fertilizer, grazing livestock and water quality, and managing pests. Land-grant colleges and universities conduct research, teach, and provide professional expertise to various groups in ways of protecting the environment. They typically operate cooperative extension services to work with people in their home communities. These colleges work cooperatively with the National Institute of Food and Agriculture, an agency of the US Department of Agriculture, and the National Integrated Water Quality Program. They attempt to provide innovative research to reduce the problems created by livestock and poultry-feeding operations on water quality.

US DEPARTMENT OF COMMERCE—TRANSPORTATION RESEARCH BOARD

The US Department of Commerce—Transportation Research Board—National Cooperative Highway Research Program has produced a document entitled "NCHRP Synthesis 272—Best Management Practices for Environmental Issues Related to Highway and Street Maintenance" (Springfield, Virginia, 1999). It is a synthesis of highway practice. (See endnote 30.)

US ENVIRONMENTAL PROTECTION AGENCY

The US EPA produces numerous manuals and studies on various phases of water pollution control and protection of water quality. Some of these can be used significantly as resources as follows:

1. The National Agricultural Center was created by the US EPA with the help of the US Department of Agriculture to provide information to the agricultural community through news releases, reports, documents, and other publications to help the individuals understand how to work with various environmental requirements.
2. The document entitled "National Management Measures to Control Nonpoint Source Pollution from Agriculture" (EPA 841-B-03-004-July 2003) is a technical and reference document utilized by state, local, and tribal managers to implement non-point source pollution management programs. (See endnote 31.)
3. The material entitled "Managing Your Environmental Responsibilities: A Planning Guide for Construction and Development" (EPA/305-Be-04-003, April, 2005) is provided to contractors, governmental agencies, and other interested parties by the Office of Compliance, US EPA. (See endnote 32.)
4. The report entitled "National Management Measures to Control Nonpoint Source Pollution from Forestry" (EPA 841-B-05-001, May, 2005) helps people who own forests protect their lakes and streams from polluted runoff from working in the forest using best scientific practices, and helps states implement non-point source control programs. (See endnote 33.)
5. The material entitled "Shipshape Shores and Waters—A Handbook for Marina Operators and Recreational Boaters" (EPA-841-The-03-001, January, 2003) is provided by the US EPA, Office of Wetlands, Oceans and Watersheds. (See endnote 34.)
6. The guidance document entitled "National Management Measures to Control Nonpoint Source Pollution from Hydro Modification" (EPA 841-B-07-002, July, 2007) is a set of voluntary requirements at the federal level but may be mandatory at some state levels for managing hydromodification of streams, lakes, and other bodies of water. (See endnote 35.)
7. Low impact development or redevelopment is a means to manage stormwater close to its source to avoid non-point source pollution. A good reference is "Stormwater to Street Trees—Engineering Urban Forests for Stormwater Management" (EPA 841 B 13001 6-20-2013). Also, see the US EPA sheet "Water: Low Impact Development" on the internet at http://www3.epa.gov/region9/water/lid/, Pacific Southwest, Region 9, Low Impact Development for a substantial amount of additional references.

8. The document entitled "Abandoned Mine Site Characterization and Cleanup Handbook" (EPA 910-B-00-001, August, 2000) is available for people interested in acid mine drainage. (See endnote 36.)

9. SUSTAIN (System for Urban Storm Water Treatment and Analysis Integration Model) is a decision support system developed by the US EPA to assist stormwater management professionals and watershed stormwater practitioners to implement pollution control technologies over stormwater flowing across land, and to select optimal Best Practice combinations that are cost-effective (January 2013).

10. The document entitled "Handbook for Developing Watershed Plans to Restore and Protect Our Waters" (EPA 841-B-08-002, March, 2008) helps communities, watershed organizations, and state, local, tribal, and federal environmental agencies develop and implement watershed plans. (See endnote 37.)

PROGRAMS

The goal of communities at the local, state, and federal level is to develop appropriate programs using Best Practices in all areas to resolve the problems related to controlling water pollution and protecting water quality. Some of the successful programs are included below to give individuals an opportunity to see what actually happens in practice to improve and protect people and the environment.

GREAT LAKES RESTORATION INITIATIVE

The Great Lakes Restoration Initiative is the ultimate response to mitigating, controlling, and remediating a multitude of water quality problems, as well as preventing new sources of contamination and restoring the bodies of water so they can be utilized for all of the purposes intended. The Great Lakes are the largest group of freshwater lakes on Earth. They contain 84% of North America's fresh surface water and are shared by the United States and Canada. (See endnote 24.)

The initiative utilizes an adaptive science-based framework for restoration of the Great Lakes. It is a cost-effective and strategic approach using the best available science and the results of programs from the past and present. It is meant to implement the work of the 16 federal agencies and build on the many years of planning and work done by the non-federal partners. The initiative has already funded over 1500 projects and programs and will plan to do considerably more in the future. This adaptive approach to restoration recognizes that the work needed to solve the many inter-related problems in the environment of the Great Lakes must be accomplished in small pieces, which may have to be repeated numerous times. This is very similar to how this book has been written and should be utilized to respond to environmental problems. This technique includes:

1. Gathering of necessary data in a scientific manner to define the problems in a given area and to identify priorities in prevention, protection, and restoration
2. Establishing goals, objectives, and means of determining progress in work done
3. Developing and implementing projects and programs
4. Monitoring changes due to the project and measuring if they are acceptable
5. Transferring of knowledge to others
6. Making decisions about new projects based on the success of other projects
7. Adapting Best Practices from successful programs to new areas
8. Reassessing major goals, objectives, and techniques utilized annually and every 5 years for redirection of projects or techniques used (See endnote 25)

Since 1970, there have been citizens' groups who have been involved in trying to clean up the huge amount of pollutants that have been introduced into the Great Lakes. These stakeholders have been joined by local, state, tribal, and national agencies to create a means of reducing the numerous pollutants which are the sources of the problems. Considerable funding has been coming from the federal government through various laws including the Great Lakes Legacy Act which is primarily for the remediation of contaminated sediment. The Great Lakes Advisory Board is one of many groups of stakeholders who have important input in the restoration of the Great Lakes. They provide advice and recommendations to the US EPA administrator who serves as the chairperson of the federal interagency task force. Members of this advisory board come from business, agriculture, foundations, environmental justice groups, educational organizations, environmental groups, academia, local, state, and tribal governments, etc. This type of advisory board on a grand scale is similar to what the author used in the community rodent control project in Philadelphia in 1959–1963, on a much lesser scale. There also was an interagency group working on the problems of rodent infestation and rat bites and was chaired by the district health officer as well as the district environmental health supervisor. (See endnotes 26, 27.)

President Obama also, through an Executive Order, established the Great Lakes Interagency Task Force along with the regional working group of 16 federal agencies with the administrator of the US EPA as the coordinator of the group. The group works with the multitude of other interested sources to achieve the rebuilding of the Great Lakes. There are four major focus areas:

1. *Toxic substances in areas of concern* to deal with contaminants and toxic hotspots including the cleanup and restoration of 21 impaired areas at 12 different locations and the remediation of about 3 million yd.³ of contaminated sediments
2. *Invasive species* to deal with disruption of ecosystems by these non-native species including the prevention of self-sustaining populations of Asian carp from invading the Great Lakes
3. *Nearshore health and non-point source pollution* to deal with degradation caused by non-point source pollution and subsequent damage including reducing swimming bans and advisories for the Chicago beaches
4. *Habitat and wildlife protection and restoration* to deal with destruction of fish and wildlife habitats including reopening more than 800 miles of river for fish passage

Grand Calumet River—Great Lakes Areas of Concern

The Federal Water Pollution Act of 1972 set broad national objectives to restore and maintain the chemical, physical, and biological integrity of the waters of the country. It regulated discharges of pollutants including limiting point sources and establishing a clean lakes program. It established the NPDES. This law motivated citizens' groups and various governmental agencies to insist on greater accountability by everyone who had created pollutants in the past and was creating contamination of the waterways in the present. A great deal of attention was given to the Great Lakes, which were dying and had at least 41 toxic hotspots, thus leading to the Great Lakes Restoration Initiative.

The Grand Calumet River projects are examples of the Great Lakes Restoration Initiative. Some of the restoration projects include:

1. A $52 million project to get rid of the contaminated sediment from Roxana Marsh was completed in 2012. It consisted of the removal of 385,000 yd.³ of sediment contaminated with PCBs, PAHs, heavy metals, pesticides, etc., and placement of a cap of 6 inches of organoclay (a chemically modified clay used in water and wastewater treatment) covered by 12 inches of sand, which covered the remaining sediment in this area.
2. The Fish and Wildlife Service helped restore fish, wildlife, and habitat to these cleaned areas.
3. In 2011, Indiana state officials were working with the cities of East Chicago, Hammond, and Gary to help develop plans to separate the combined sewers and thereby help reduce this type of contamination to the waterway.
4. State and federal people are working on restoring 100 areas of wetland to be used for migratory birds and fish and help reduce nutrient pollution.

5. Additional monitoring is being done in a variety of areas to determine the effectiveness of existing programs.
6. A steel company completed the dredging of 5 miles of river on a branch of the Grand Calumet River.
7. The development of a plan for toxic pollution prevention in the beginning of the implementation of the plan.
8. The development and implementation of a household hazardous waste collection project.
9. The development and implementation by the steel industry of a waste minimization program and implementation.
10. The use of Consent Decrees (an order issued by a judge for individuals, companies, or governmental entity to voluntarily comply with certain laws, rules, or regulations) with the three sanitary districts to start the process of eliminating combined sewer overflows.
11. The development of a non-point source pollution control program to reduce contaminants in the watershed. (See endnote 28.)

US ENVIRONMENTAL PROTECTION AGENCY

The US EPA, in its Illicit Discharge Detection and Elimination Program, has a national menu of stormwater Best Practices on special topics. They include public education, public involvement, illicit discharge detection and elimination, construction, postconstruction, and pollution prevention good housekeeping. Educational programs consist of public service announcements in the media, pamphlets, booklets and inserts in utility bills, and programs for educating homeowners and businesses. Public involvement includes volunteer monitoring programs, which need to produce usable data to help determine whether programs are working and what changes need to be made, storm drain marking programs, stream clean-up programs, and developing and implementing new stormwater programs by local people. Appropriate ordinances are approved for all facets of stormwater discharge and all associated concerns at the local and state level. The basic steps in developing the program are:

1. Make a study of the watershed including all bodies of water, pipes leading to water, and stormwater receptacles.
2. Make a study to determine the illicit discharges into all bodies of water, containers, or pipes.
3. Monitor these discharges and determine the composition of the effluent and where it is coming from. This is easiest to do in dry weather.
4. Request voluntary compliance in eliminating the source of contamination.
5. Determine which legal entities have control over illicit discharges in various parts of the selected area.
6. If the source continues to illegally dispose of effluent into pipes, containers, and bodies of water, use appropriate legal enforcement.
7. Resurvey the area to make sure that the problem has not started over again.
8. Re-evaluate the watershed program and make readjustments as necessary.
9. Public education is of greatest significance because the citizens, community organizations, and others can find the cause of problems, report them to the proper people, help in the evaluation of the results, help gain budget for successful programs, and teach our children to protect our air, water, and land for the future.

MINNESOTA POLLUTION CONTROL AGENCY

The Minnesota Pollution Control Agency has developed a cost-effective recommended plan for industrial stormwater control which satisfies the rules related to necessary permits for discharge and other regulations. The plan includes the use of non-structural Best Practices, which is a change of behavior and management to reduce the source of the pollutants. The plan also includes structural

Best Practices, which are physical means of controlling or managing stormwater runoff and drainage including the diversion of surface rainwater from the area, the covering of exposed materials, the prevention of the infiltration of materials into the ground, and the retention and treatment of contaminated water.

The Stormwater Pollution Prevention Plan includes:

1. Planning and organization of the disposal of stormwater by making necessary inspections, gathering data, assessing and reassessing existing problems, and assessing and reassessing existing Best Practices and making changes where necessary
2. Assigning a management team with appropriate oversight responsibility and the ability to make changes in processes when needed
3. Preparing material safety data sheets for all potentially hazardous materials
4. Developing a description map and drainage map of the facility including the industrial activities and potential for discharge of contaminants and at which sites
5. Preparing a list including quantities of significant materials and their associated risks including the raw materials, byproducts, finished products, and waste
6. Knowledge and use of specific Best Practices by individual industries
7. A list of Best Practices and where they are utilized for:
   a. Source reduction and prevention
   b. Diversion of water and pollutants
   c. Treatment
8. Evaluation of all illicit or avoidable discharges, which should be done primarily in dry weather
9. Preventive maintenance program including physical maintenance of equipment, inspection of equipment, and inspection of potential pollutant sources
10. Specialized employee training programs

STATE OF WASHINGTON DEPARTMENT OF ECOLOGY

The State of Washington Department of Ecology has a program modeled after the US EPA guidelines. They utilize citizens, governmental employees, and business people to call their local Stormwater Hotline if there are unusual discharges or unusual odors or colors in the stormwater. The various jurisdictions in the state share information and Best Practices for eliminating the problems. As examples:

- The City of Yakima found sewage discharges from pipes leading to homes, warehouses, wastewater areas and a local hospital. They injected smoke into each of the pipes and found that they were illegally emptying into the Yakima River. They were corrected.
- The City of Seattle Public Utilities employees found while doing routine dry weather screening of stormwater systems, sewage coming from a housing development in Seattle. The effluent was sampled and the city worked with the owners of the development to correct the problem.
- The City of Seattle Public Utilities employees found an illegal connection from an industry which was going into a drain and discharging into the Duwamish River.

PHILADELPHIA SUBURBAN WATER COMPANY

The Philadelphia Suburban Water Company works in partnership with the Bucks County Conservation District and the Penn State Cooperative Extension Service to reduce levels of herbicides used in the spring during peak runoff times. The water company collects samples of the water at different places in the watershed and the other two groups share this information with local farmers and other people to try to get them to reduce the level of chemicals in use. This has shown some reduction of chemicals in the water.

US DEPARTMENT OF AGRICULTURE FOREST SERVICE

The US Department of Agriculture Forest Service has a program to help reduce heavy use of quality water from national forests and grasslands. About 124 million people utilize this primary source of water as their drinking water supply. The Forest Service promotes Aquatic Management Zones, an area adjacent to the stream channels and other water bodies which help protect them from contamination. The width of the zone is determined by administrative decisions based on potential problems that would occur in the event of the pollution of the body of water. Specific Best Practices include the following:

- Reduction of the use of pesticides and fertilizers which lowers the amount of chemicals entering bodies of water. This can be accomplished in parks and golf courses by the use of integrated pest management control programs;
- Reduction of various agricultural chemicals through better practices, thereby lowering the amount of potential pollutants of ground and surface water;
- Proper spreading of appropriate amounts of nutrients and organic waste on land, lowering the quantity of potential runoff;
- Properly enforced pet waste removal programs by the owners helps reduce the level of microorganisms found in water and the level of eutrophication found in bodies of water. All animal wastes from agriculture need to be handled and disposed of in an appropriate manner to prevent pollution of the land and bodies of water. (See *Handbook of Environmental Health, Pollutant Interactions in Air, Water, and Soil*, Volume 2, Fourth Edition (CRC Press), for technical details on how to handle these issues.) Use of a nutrient management plan is a technique to identify nutrient needs, in both timing and quantity, utilized to fertilize crops properly. Excessive amounts of nutrients could contaminate bodies of water, while inadequate amounts of nutrients would result in poor crop production. This allows for appropriate use and disposal of animal manure, commercial fertilizers, irrigation water and wastewater, and naturally occurring nutrients.

US GEOLOGICAL SERVICE

Since 1991, the US Geological Service has been operating the National Water Quality Assessment Program—Source Water-Quality Assessments. The purpose of the program is to help determine whether 280 mostly unregulated organic compounds made by people are found in source water used for community water systems and provide this information to citizens, various groups, industries, and regulatory agencies to help them develop the original data needed for a water quality program. In 2013, the programs were expanded to include an additional 30 surface water and 30 groundwater evaluations. Monthly samples of water are taken and tested and then a determination is made about the quantity and quality of the chemical. In order to eliminate these chemicals, it is necessary to know what they are and their concentrations, which will help lead to the source and hopefully the elimination of the chemical from the water supply. These chemicals may be:

- Byproducts of disinfection
- Pesticides including fumigants, fungicides, herbicides, insecticides, rodenticides, and their products of disintegration

- Gasoline, oil, and their byproducts
- Refrigerants, propellants, solvents, and other organic compounds
- Chemicals used in manufacturing and personal care products

ENDNOTES

1. Koren, Herman, Bisesi, Michael. 2003. *Handbook of Environmental Health-Pollutant Interactions in Air, Water, and Soil,* Volume 2, Fourth Edition, Chapter 7, Water Pollution and Water Quality Controls. Lewis Publishers, CRC Press, Boca Raton, FL.
2. Utah State University Cooperative Extension. 2010. *Manure Best Management Practices.* Logan, UT.
3. U.S. Environmental Protection Agency. 2012. *Agriculture, Pasture, Rangeland, and Grazing Operations: Best Management Practices (BMPs).* Washington, DC.
4. Lilly, J.P. 1997. *Best Management Practices for Agricultural Nutrients.* North Carolina Cooperative Extension Service, Raleigh, NC.
5. U.S. Environmental Protection Agency, Office of Solid Waste and Emergency Response, Office of Superfund Remediation and Technology Innovation. 2012. *Green Remediation Best Management Practices: Mining Sites.* Washington, DC.
6. U.S. Environmental Protection Agency, Office of Water. 2001. *Protecting and Restoring America's Watersheds: Status, Trends and Initiatives in Watershed Management.* Washington, DC.
7. U.S. Environmental Protection Agency, Office of Research and Development. 2012. *Study of the Potential Impacts of Hydraulic Fracturing on Drinking Water Resources-Progress Report.* Washington, DC.
8. Pitt, Robert. 2004. *Illicit Discharge Detection and Elimination: A Guidance Manual for Program Development and Technical Assessments.* U.S. Environmental Protection Agency, Office of Water, Center for Watershed Protection, Washington, DC.
9. Delaware Nutrient Management Commission. 2006. *Water Quality Best Management Practices: Nutrients, Irrigation and Pesticides for Golf Course, Athletic Turf, Lawn Care and Landscape Industries.*
10. Stordahl, Darrell M., Jones, Jeffrey W. 2003. *Erosion and Sediment Control Best Management Practices: Reference Manual, Final Report, State of Montana Department of Transportation.* U.S. Department of Transportation Federal Highway Administration, Helena, MT.
11. Stordahl Darrell M., Jones, Jeffrey W. 2003. *Erosion and Sediment Control Best Management Practices: Field Manual, Final Report, State of Montana Department of Transportation.* U.S. Department of Transportation Federal Highway Administration, Helena, MT.
12. U.S. Environmental Protection Agency. 2007. *Developing Your Storm Water Pollution Prevention Plan: A Guide for Construction Sites.* Washington, DC.
13. Minnesota Pollution Control Agency. 2010. *Industrial Stormwater Best Management Practices Guidebook.* St. Paul, MN.
14. Minnesota Pollution Control Agency. 2011. *Stormwater Best Management Practices Manual.* St. Paul, MN.
15. University of California San Diego. 2006. *Storm Water Pollution Prevention Best Management Practices Handbook,* San Diego, CA.
16. Chesapeake Bay Program. 2006. *Best Management Practices for Sediment Control and Water Clarity Enhancement.* Annapolis, MA.
17. US Department of the Interior, US Geological Service, Science for a Changing World. 2013. *National Water Quality Assessment (NAWQA) Program.* Washington, DC.
18. US Fish and Wildlife Service-Congressional and Legislative Affairs, External Affairs, Resource Laws. 1978. *Digest of the Federal Resource Laws of Interest: The US Fish And Wildlife Service-Federal Water Pollution Control Act (Clean Water Act).* Washington, DC.
19. US Environmental Protection Agency, Compliance Assistance. 2012. *Clean Water Act Compliance Assistance.* Washington, DC.
20. US Environmental Protection Agency, Compliance Monitoring. 2012. *Clean Water Act Compliance Monitoring.* Washington, DC.
21. US Environmental Protection Agency, Office of Water. 2005. *National Management Measures to Control Nonpoint Source Pollution from Urban Areas,* EPA-841-Be-05-004. Washington, DC.
22. US Environmental Protection Agency, Great Lakes. 2011. *Invasive Species.* Washington, DC.
23. Glassner-Schwayder, Katherine. 2000, *Briefing Paper: Great Lakes Nonindigenous Invasive Species.* A Product of the Great Lakes Nonindigenous Invasive Species Workshop of October 1999. US Environmental Protection Agency, Ann Arbor, MI.

24. Alexander, Jeff. 2013. *Great Lakes Restoration Projects Producing Results for People, Communities.* Great Lakes Restoration Initiative, Healing Our Waters, Great Lakes Coalition.
25. Great Lakes Restoration Initiative. 2013. *Adaptive Science-Based Framework for Great Lakes Restoration.* Science Subgroup of the Great Lakes Regional Working Group, Draft.
26. Great Lakes Restoration Initiative. 2013. *Great Lakes Advisory Board.* EPA.
27. Great Lakes Commission. 2008. *Great Lakes Legacy Act-Appropriations Update.*
28. US Environmental Protection Agency, Great Lakes Areas of Concern. 2013. *Grand Calumet River.* Washington, DC.
29. Burkholder, JoAnn, Libra, Bob, Weyer, Peter, Heathcote, Susan. 2007. Impacts of Waste from Concentrated Animal Feeding Operations on Water Quality. *Environmental Health Perspectives*, 115:2.
30. Hyman, William A., Vary, Donald. 1999. *Synthesis of Highway Practice 272—Best Management Practices for Environmental Issues Related to Highway and Street Maintenance.* NCHRP. Washington DC.
31. U.S. Environmental Protection Agency. 2003. National Management Measures to Control Nonpoint Source Pollution. *Agriculture.* USEPA. Washington DC.
32. U.S. Environmental Protection Agency. 2005. *Managing Your Environmental Responsibilities: A Planning Guide for Construction and Development.* USEPA. Washington DC.
33. U.S. Environmental Protection Agency. 2005. National Management Measures to Control Nonpoint Source Pollution. *Forestry.* USEPA. Washington DC.
34. U.S. Environmental Protection Agency. 2003. *Shipshape Shores and Waters—A Handbook for Marina Operators and Recreational Boaters.* USEPA. Washington DC.
35. U.S. Environmental Protection Agency. 2007. National Management Measures to Control Nonpoint Source Pollution. *Hydro Modification.* USEPA. Washington DC.
36. U.S. Environmental Protection Agency. 2000. *Abandoned Mine Site Characterization and Cleanup Handbook.* USEPA. Washington DC.
37. U.S. Environmental Protection Agency.2008. *Handbook for Developing Watershed Plans to Restore and Protect Our Waters.* USEPA.Washington DC.

BIBLIOGRAPHY

Ecological Society of America. 2004. Impacts of Atmospheric Pollution on Aquatic Ecosystems. *Issues in Ecology*, (12).

Erosion Sediment Control (ESC) Bylaw: Best Management Practices, City of Abbotsford, British Columbia, Canada. 2010.

Hill, Lee. 2006. *Best Management Practice Summaries-Riparian Buffers.* Virginia Department of Conservation and Recreation, Annapolis, MD.

Integrated Pest Management Program, Portland Parks and Recreation. 2012. Portland, OR.

Mesner, Nancy, Paige, Ginger. 2011. *Best Management Practices Monitoring Guide for Stream Systems.* University of Wyoming, Laramie, WY.

McCormick, Frank H., Contreras, Glenn C., Johnson, Sherri L. 2010. *Effects of Nonindigenous Invasive Species on Water Quality and Quantity.* US Department of Agriculture, Forest Service, Washington, DC.

Municipal Research and Services Center of Washington. 2012. *Erosion and Sediment Control: Land Clearing and Grading.* Seattle, WA.

New Hampshire Department of Agriculture, Markets, and Food. 2011. *Manual of Best Management Practices (BMPs) for Agriculture in New Hampshire.* Concorde, NH.

U.S. Department of Agriculture, Forest Service. 2012. *National Best Management Practices for Water Quality Management on National Forest System Lands.* Volume 1: National Core BMP Technical Guide, Washington, DC.

U.S. Environmental Protection Agency. 2001. Managing Pet and Wildlife Waste to Prevent Contamination of Drinking Water. *Source Water Protection Practices Bulletin.* Washington, DC.

U.S. Environmental Protection Agency. 2001. *Summary of the Air Water Interface Work Plan.* Washington, DC.

U.S. Environmental Protection Agency. 2010. *Illicit Discharge Detection and Elimination, National Pollutant Discharge Elimination System.* Washington, DC.

U.S. Environmental Protection Agency. 2010. *Post-Construction Storm Water Management in New Development and Redevelopment, National Pollutant Discharge Elimination System.* Washington, DC.

U.S. Environmental Protection Agency. 2012. *Construction Site Storm Water Runoff Control, National Pollutant Discharge Elimination System.* Washington, DC.

U.S. Environmental Protection Agency. 2012. *Hydraulic Fracturing in the Safe Drinking Water Act: Outreach.* Washington, DC.

U.S. Environmental Protection Agency. 2012. *National Menu of Stormwater Best Management Practices, National Pollutant Discharge Elimination System.* Washington, DC.

U.S. Environmental Protection Agency. 2012. *National Water Quality Inventory: Report to Congress 2004 Reporting Cycle.* Washington, DC.

US Environmental Protection Agency. 2012. *Pet Waste Management, National Pollutant Discharge Elimination System.* Washington, DC.

U.S. Environmental Protection Agency. 2012. *Pollution Prevention/Good Housekeeping for Municipal Operations, National Pollutant Discharge Elimination System.* Washington, DC.

U.S. Environmental Protection Agency. 2012. *Public Education and Outreach on Storm Water Impacts, National Pollutant Discharge Elimination System.* Washington, DC.

U.S. Environmental Protection Agency. 2012. *Public Involvement/Participation, National Pollutant Discharge Elimination System.* Washington, DC.

U.S. Environmental Protection Agency. 2012. *Storm Water Discharges from Municipal Separate Storm Sewer Systems, National Pollutant Discharge Elimination System.* Washington, DC.

U.S. Environmental Protection Agency. 2012. *Storm Water Management Best Practices, Greening EPA.* Washington, DC.

US Environmental Protection Agency. 2012. *The Safe Drinking Water Act Amendments of 1996, Strengthening Protection for America's Drinking Water.* Washington, DC.

U.S. Environmental Protection Agency. 2012. *Water: Estuaries and Coastal Watersheds: Challenges and Approaches.* Washington, DC.

U.S. Environmental Protection Agency. 2012. *Water Impacts of Climate Change.* Washington, DC.

U.S. Environmental Protection Agency. 2013. *Natural Gas Extraction: Hydraulic Fracturing.* Washington, DC.

US Environmental Protection Agency, Great Lakes Legacy Act. 2012. *Roxanna Marsh Legacy Act Cleanup–Grand Calumet River Area of Concern.* Washington, DC.

U.S. Environmental Protection Agency, Office of Air and Radiation and Office of Water. 2001. *Frequently Asked Questions about Atmospheric Deposition: A Handbook for Watershed Managers.* Washington, DC.

U.S. Environmental Protection Agency, Office of Water. 2001. Managing Pet and Wildlife Waste to Prevent Contamination of Drinking Water. *Source Water Protection Practices Bulletin.* Washington, DC.

U.S. Environmental Protection Agency, Office of Water. 2005. *National Pollutant Discharge Elimination System (NPDES) Illict Discharge Detection and Elimination.* Washington, DC.

U.S. Environmental Protection Agency, Office of Water. 2005. *Storm Water Phase II Final Rule-Construction Site Runoff Control Minimum Control Measure.* Washington, DC.

U.S. Environmental Protection Agency, Office of Water. 2010. *Source Water Protection Practices Bulletin: Managing Agricultural Fertilizer Application to Prevent Contamination of Drinking Water.* Washington, DC.

U.S. Environmental Protection Agency, Office of Water. 2012. *National Water Program Best Practices and End of Year Performance Report-Fiscal Year 2011.* Washington, DC.

Utah State University Cooperative Extension. 2010. *Manure Best Management Practices.* Logan, UT.

Vermont Surface Water Management Strategy, Watershed Management Division. 2003. *Stressor-Flow Alteration.* Vermont.

Index

1,3 Butadiene, 28, 50, 72

A

acanthamebiasis, 8
Acid Mine Drainage, 6, 530, 561, 577
Acinetobacter, 304, 315
Acrolein, 50, 72
Act to Prevent Pollution from Ships, 394
Activated Sludge, 47, 412, 417–419, 434, 500
Aedes mosquito, 315–316, 345, 349
Aerobic bioremediation, 479
aerobic treatment system, 431
Agency for Toxic Substances and Disease Registry, 2, 95,
    150, 517
Agenda 21, 14
Aging, vii, 2, 9, 19–20, 83, 89, 98, 232, 443, 482
Agricultural and Animal Wastes, 449–450
Agricultural chemicals, 36, 64, 108–109, 482–483
AgSTAR, 34, 71, 451
air pollution, 21–26, 31–33, 66–72, 92–93, 313, 374–375,
    562–563
Air Pollution Control Act of 1955, 66
Air Quality, 21, 25–27, 59–61, 160–163, 171–173
Air Quality Index, 26, 69, 122
air quality standards, 23, 25, 66, 162, 505, 515
Aircraft, 28, 32, 57, 98, 205, 325, 491–492
Alameda, California, 159, 220
algae, 5, 22, 151, 390, 436, 548, 562
algae blooms, 403, 541, 555, 560
America's Children and the Environment, 151
American Academy of Pediatrics, 143–144, 153
    American Academy of Pediatrics Council on
        Environmental Health, 153
American Association of Poison Control Centers,
    133, 234
American National Standards Institute, 96, 264, 292,
    546, 552
American Public Health Association, xxxvi, 3, 153, 274
    American Public Health Association—Public Health
        Nursing Section, 153
American Public Works Association, 408
American Red Cross, 211, 392
American Water Works Association, 546, 548–549, 552
amoebic dysentery, 403
Anopheles Mosquito, 349
Antarctica, 5
Anthrax, 9, 216–217, 239, 289, 325–327, 360, 450
Arctic, 5, 24
Arctic sea ice, 24
Argentine hemorrhagic fever, 324
Asbestos, 60–61, 93, 126, 172–173
Asbestos Hazard Emergency Response Act, 146
Association of Maternal and Child Health Programs, 151
Asthma, 3, 24–25, 107, 121, 146–149, 151–152
Atomic Energy Act, 11, 335
ATSDR, 150, 156
Aurora Corridor, Shoreline, Washington, 102

B

balantidiasis, 403
Baltimore, x, xiii, 17, 103, 105, 158, 165, 363, 400
    Baltimore Maryland Housing Authority, 103
Basic Environmental Health and Radiation Preparedness
        Programs, 3
Battery Recycling Program, 520
Bay Delta Conservation Plan, 554
Baygon, 64
Bearer, Dr. Cynthia, 145
Belmar, Lakewood, Colorado, x, 102
Benzene, 7, 33, 35, 48–50
Berkeley California Rat Control Program, 361
Best Practices
    Accidental Chemical Releases for, 51
    Acid Mine Drainage for, 561
    Acid Rain for, 33
    Aerobic Treatment Systems for, 431
    Aerospace Manufacturing and the Rework
        Industry for, 51
    Aftermath of Wildfires for, 142
    Agriculture for, 34, 58, 562, 570
    Air Pollution for, 137, 563
    Air Quality for, 125, 172
    Air Quality Management for, 125
    All Forms of Non–Hazardous Solid Waste for, 462
    Alteration of Habitats for, 563
    Aluminum Industry for, 35
    Asbestos for, 61, 172
    Auto Body and Repair Shops for, 52
    Avoid and Mitigate Safety Risks to, 373
    Avoiding and Correcting Cross–Connections and
        Submerged Inlets for, 550
    Avoiding Health Problems Related to Cleaning
        Materials and Disinfectants for, 328
    Avoiding Outbreaks of Disease on Cruise Ships in, 372
    Avoiding Problems in Politics in, 265, 267
    Bedbug Control in, 345
    Biological and Chemical Releases Including Terrorism
        for, 132
    Biological Contaminants for, 61
    Bioterrorism for, xviii, 195, 197, 321, 32
    Building Construction or Renovation for, 292
    Built Environment for, 80, 224
    Carbon Monoxide for, 27, 62
    Carpets for, 62
    Cement Kilns for, 35
    Chemical Industry for, 37, 51
    Chemical Management for, 126
    Chemical Terrorism for, 197
    Childhood Lead Poisoning for, 132
    Childhood Poisoning for, 133
    Children in the Recreational Environment for, 128
    Class I Underground Injection Wells for, 473
    Class II Underground Injection Wells for, 475
    Class III Underground Injection Wells for, 475
    Class IV Underground Injection Wells for, 476

Best Practices (*Continued*)
Class V Underground Injection Wells for, 477
Class VI Underground Injection Wells for, 478
Cleanup of Methamphetamine Contaminated Houses for, 136
Cleaning and Disinfecting Medical and Surgical Instruments and Equipment for, 296
Climate Change for, 564
Collection Systems for Solid Waste in, 447
Collection, Storage, and Disposal of Household Hazardous Waste in, 496
Collection, Storage, and Disposal of School Hazardous Waste in, 497
Combined Storm and Sanitary Sewer for, 564
Composting in, xxiii, 471
Construction and Demolition Waste Landfills for, 468
Control of Bioterrorism and Potential Agents for, 324
Control of Mites in, 349
Control of Production and Use of Agricultural Chemicals in, 483
Control of Production and Use of Basic Chemicals in, 481
Control of Production and Use of Consumer Products in, 485
Control of Production and Use of Pharmaceuticals in, 484
Control of Production and Use of Specialty Chemicals in, 482, 485
Control of Respiratory and Enteric Viruses in Pediatric Care Settings for, 319
Control of Severe Acute Respiratory Syndrome for, 319
Control of the Ebola Virus for, 319
Control of Tuberculosis in, 320
Control of Tularemia as a Biological Weapon for, 324
Controlled Burns for, 38
Controlling Air Emissions in, 313
Controlling Combustion Pollutants for, 94
Controlling Contamination of Surrounding Waters by Graywater Discharge in, 378
Controlling Fleas in, 346
Controlling Mosquitoes in, 349
Controlling Roaches in, 351
Controlling Rodents for, 352
Controlling Smallpox for, 323
Controlling Ticks in, 351
Debris for, 173
Decommissioning a Chemical Manufacturing Facility in, 486
Determining the Size and Nature of Land to Be Used for New Houses for, 423
Diesel Fuel and Emissions for, 49
Disposal of Agricultural and Animal Wastes in, 450
Disposal of Construction and Demolition Wastes in, 454
Disposal of Crude Oil, Coal, and Natural Gas Waste in, 456
Disposal of Hazardous Household Waste in, 96
Disposal of Household Appliances Using Refrigerants for, 467
Disposal of Industrial Wastes in, 452
Disposal of Medical Waste for, 461, 498
Disposal of Solid Waste in Incinerators in, 469
Disposal of Solid Waste in Municipal Landfills in, 466
Drinking Water for, 173, 532, 542
Drinking Water Quality for, 532
Drinking Water Treatment Facilities for, 542
Dry Cleaners for, 53

Best Practices (*Continued*)
Drywall from China for, 95
Electrical Systems and Appliances for, 174
Electricity Production for, 57
Eliminating Formaldehyde for, 96
Employee Safety in, 306, 309
Environmental (Secondhand) Tobacco Smoke for, 134
Environmental Tobacco Smoke for, 62
Erosion for, 564, 572
Explosives Terrorism for, 198
External Combustion Sources for, 40
External Pollution Prevention in, 314
Extraction of Resources for, 565
Extreme Heat for, 141
Extreme Temperatures for, 186
Finished Potable Water Storage Facilities for, 545
Fires for, 41
Floods for, 188
Fly Control in, 347
Food for, 175, 199, 242, 258, 297
Food Preparation, Handling, and Disposal for, 297
Food Product Insect Control in, 347
Food Security and Protection for, 242, 258
Food Terrorism for, 199
Formaldehyde for, 63
Gas Emissions from Mobile Sources for, 57
Gas Stations and Gasoline Distribution Facilities for, 53
Governmental Agencies for, 566
Hand Washing in, 298
Handling and Disposal of Solid and Hazardous Waste and Materials in, 314
Handling Mineral Processing Waste in, 457
Hazardous Materials Use, Waste Storage, and Disposal for, 376
Hazardous Waste Incinerators for, 504
Hazardous Waste Landfills for, 501, 503
Hazardous Waste Piles for, 502
Hazardous Waste Tank Systems for, 501
Health and Safety of Personnel Involved in Emergency/Disaster Cleanups and Repairs for, 180, 203
Heating, Ventilation, Air Conditioning, Fountains, Misting Systems, Humidifiers, and Showers for, 377
Home Health Care for, 329–330
Household Items and Clothing for, 175
Household Products for, 63
Illicit Discharges for, 566
Improving Air Quality for, 375
Improving the Built Environment in Rural Areas for, 90
Improving the Built Environment in Suburban Areas for, 87
Improving the Built Environment in Urban Areas for, 83, 88
Indoor Air Quality for, 60, 125, 173
Industry for, 57
Industry in, 262
Injury Prevention in Occupational Settings for, 232
Insects and Rodents for, 176
Inspection, Maintenance, and Repair of Sewer Laterals in, 404
Installation and Maintenance of Septic Tanks and Distribution Boxes in, 427
Invasive Species for, 567
Iron and Steel Industry for, 42

Best Practices (*Continued*)
Land and Forests for, 56
Land and Superfund Sites for, 507
Land Development and Building Construction for, 42, 567
Land Treatment Units for, 503
Landslides and Mudslides for, 189
Lead for, 29, 64
Lice Control in, 348
Limiting Soil Exposure for, 139
Local Community Recovery from Disastrous Hazardous Materials Transportation Incidents in, 493
Location, Source, and Preliminary Pretreatment for Raw Water for, 539
Mail Screening and Handling Processes for, 327, 342
Managing Livestock and Poultry Manure Contaminants to Avoid Raw Water Contamination in, 537
Medical Waste and Other Hazardous Substances for, 300, 314
Metals for, 568
Mitigate Noise to, 485
Mitigate Odors to, 486
Mitigate or Prevent Infiltration and Inflow into Sewer Pipes to, 404
Mitigation of Occupational Health and Safety Problems in, 510
Mobile Sources for, 33
Molds and Other Fungi for, 176
Mound Systems for, 433
Natural Gas and Crude Oil Extraction Industry for, 44
Natural Sources for, 45
Nitrogen Dioxide for, 64
Nitrogen Oxides for, 29
Non-Point Sources of Contamination for, 569, 572
Nutrients for, 570
Occupational Environment for, 128–129
Occupational Health and Safety for First Responders, Construction Workers, and Volunteers for, 206
Off-Site Hazardous Waste Operations for, 54
Other Healthcare Options for, 333
Oxygen–Demanding Substances for, 568, 571
Ozone for, 28
Packaging of Food in, 256
Painting and Coating of Materials for, 55
Particulate Matter for, 30
Patient Safety in, 306, 309
Perforated Pipe or Tile Field Effluent Removal System for, 429
Personnel and Budgets for, 265
Pest Control through the Use of Integrated Pest Management for, 357
Pesticides for, 64, 353, 571
Pesticides and Disinfectant Residues for, 64
Petroleum Refineries for, 46
Physician's Office or Medical Clinic for, 330
Poultry Processing and, 255
Preparation of Food in, 258
Preschool Environment for, 122
Pressure Systems for, 430
Prevent and Mitigate Disease and Injury in Children's Summer Camps to, 387
Prevent Dog Bites to, 228

Best Practices (*Continued*)
Prevent Fires to, 228
Prevent Injuries to, 135
Prevent or Mitigate Air Quality Problems to, 506
Prevent or Mitigate Damage and/or Destruction of Water and Wastewater Treatment Facilities to, 417
Prevent or Mitigate Seismic Activity Caused by Humans to, 474
Preventing Air or Water Pollution from Cement Kiln Dust in, 455
Preventing Airborne Infections for, 293
Preventing and Mitigating Illness and Injury Associated with Recreational Water Facilities in, 392
Preventing and Mitigating Noise Problems in Residential Areas and Structures for, 98
Preventing and Mitigating the Effects of Chemicals on the Groundwater Supply in, 508
Preventing and Mitigating the Effects of Chemicals on the Surface Water Supply in, 508
Preventing Asbestos Related Problems in, 93
Preventing Botulism in, 323
Preventing Built Environment Poisonings for, 227
Preventing Childhood Asthma for, 121
Preventing Disease Spread by Use of Potable Water Facilities for, 381
Preventing Falls for, 226
Preventing Healthcare–Associated Infections Caused by Medical/Nursing Techniques for, 290
Preventing Infections from Emergency Medical Services System Practices for, 289
Preventing Injuries in Transportation for, 229
Preventing Lead Poisoning in, 97
Preventing Spread of Communicable Disease for, 367
Preventing the Spread of Antibiotic–Resistant Bacteria for, 318
Preventing the Spread of *C. difficile* for, 318
Preventing the Spread of Hospital–Acquired Infections through Laundry for, 298
Prevention and Removal of Mold for, 97
Prevention of Backflow and Backsiphonage for, 380
Printing and Publishing for, 54
Processing Fruits, Vegetables, Juice, and Nuts in, 249
Processing Milk and Dairy in, 254
Processing of Eggs in, 245
Processing of Fish and Shellfish in, 247, 249, 253
Processing of Grains, Soybeans, and Hay in, 251
Processing of Meat and Meat Products in, 253, 255
Production of Food in, 245
Professional Associations for, 263, 269, 274
Protecting Drinking Water Supplies for, 140
Protecting People during Temperature Extremes for, 92
Protecting Workers at Ports and Onboard Ships in, 384
Pulp and Paper Mills, 46
Radiological Terrorism for, 197
Radon Gas for, 65
Recreational Water Facilities Onboard Ship for, 382
Recycling, 463
Reduce Noise to, 301
Reduce or Eliminate Occupational Health and Safety Concerns from Airlines to, 369
Reducing Flow of Water and Contaminants to Collection System in, 405
Reducing Microorganisms on Surfaces in, 302

Best Practices (*Continued*)
Regulatory Process in, 259
Related to Drought, 181
Related to Earthquakes, 183
Repairing Ancillary Structures in, 407
Residential and Commercial Activities for, 58
Resolve Impairment in Housing to, 92
Resolve Problems for Disaster Site Management
    to, 202
Resolving rodent control problems for, 17
Reusing Brownfield Sites in, 94
Sand and Salt Storage for, 571
School Environment for, 125
Sediment for, 572
Sediment Removal from Bodies of Water for, 572
Seepage Pits or Dry Wells for, 430
Septage Treatment/Disposal: Land Application for, 415
Septic Tank Management for, 126
Septic Tank Systems and Public Wastewater Systems
    for, 177
Sewer Inspection and Cleaning for, 410
Sewer Line Replacement in, 404, 411
Shelters for, 178
Shipyards in, 55
Small Chemical Plants for, 55
Soil Analysis for On-Site Sewage Distribution Systems
    for, 425
Solid and Hazardous Waste Removal for, 179
Solid Waste Incinerators for, 47
Solid Waste Landfills for, 47
Solid Waste Management and Recycling for, 126
Solid Waste Storage and Disposal for, 383
Source Reduction/Waste Minimization for, 462
Special Food Management Training Programs
    for, 264
Spills and Run off for, 48
Spills and Runoff from Stored Chemicals and Storage
    Tanks for, 573
Sports and Recreational Facilities for, 231
Stand-Alone Surgical Units for, 331
Stationary Sources for, 32, 51
Storage of Food in, 257
Stormwater Management for, 125
Stoves, Heaters, Fireplaces, and Chimneys for, 65
Surface Impoundments for Farm Waste or Other
    Wastes for, 573
Surface Impoundments Used for Managing Hazardous
    Waste for, 502
Temperature Problems for, 573
The Storage and Disposal of Mine Waste in, 458
The Use of Other Advanced Treatment Systems in, 434
The Use of Pesticides and Antimicrobials in, 356
Total Suspended Solids for, 574
Toxic Chemicals for, 131
Toxic Waste Transfer, Treatment, and Disposal
    Facilities for, 48
Training for Environmental Health Professionals in, 268
Transfer Stations for, 448
Transportation of Food in, 256
Transportation of Hazardous Materials in, 492
Treating Oily Bilge Water for, 379
Treatment of Low–Level Radioactive Waste for, 498
Tsunamis for, 189
Turning Sewage Sludge into Safe Biosolids in, 414

Best Practices (*Continued*)
Underground Injection Wells for, 479
Underground Storage Tanks for, 180
Use and Maintenance of Sand Filters in, 432
Use, Collection, Storage, and Transportation of
    Hazardous Materials for, 487
Use of Combined Sewers in, 408
Use of Disinfectants in Water Treatment Plants
    for, 544
Use of Energy in Water and Wastewater Treatment
    Plants in, 420
Use of Lawnmowers and Other Garden Equipment
    for, 54
Use of Package Treatment Plants in, 419
Use of Water and the Disposal of Waste in, 436
Using Systems Approach with Critical Control Points
    in, 258
Utilizing Private Drinking Water Sources for, 534
Verification of Safety Standards for Imported Foods
    for, 261
Violence-Related Injuries Prevention for, 233
Volatile Organic Compounds for, 65
Volcanoes for, 190
Waste from Pets for, 574
Water Distribution System for, 548
Water Pollution in, 314
Water Problems in, 304
Water Terrorism for, 200
Wet Weather Concerns at Combined Sewer Wastewater
    Treatment Facilities for, 416
Wildfires for, 191
Wildlife or Other Natural Sources of Contamination
    for, 574
Wind-Related Storms for, 193
Wood Furniture Manufacturing for, 55
Bhopal, India, 131, 209
biofilm, 295–296, 412, 535, 546, 548, 550
biological weapons, 8–9, 195, 321, 326
Biomass, 38, 91, 150, 412, 417, 433, 481
Bioreactor landfill, 465
Bioremediation, 476, 478–479
bioretention cell, 569
Biosolids, 409–411, 413–415, 419, 433, 438, 464, 575
Bioterrorism, 3, 9, 194–195, 216, 321, 324–326, 360
Black Death, 345
Blackwater, 377
Blake Medical Center, Bradenton, Florida, 339
Boilers, 22, 38–40, 51, 94, 127, 334–335, 504
Borrelia hermsii, 315
Botulism, 9, 195, 246, 322–323, 325
    Botulism type E, 246
Bradenton, Florida, 339
Bridge Street Corridor, 102
Bridgeport, Connecticut, 339
Bromate, 544
Brownfield, 84, 91, 93–94, 102–103, 139, 507
Brundtland, Gro Harlem, 14
Bubonic plague, 323, 345–346, 366
Budapest, Hungary, 109, 165
Built Environment, 79–83, 224, 329, 445, 530–531, 559
Bunyaviruses, 321
buried water system, 546
Burkholderia, 315, 321
Businesses, 41, 57–58, 185, 299, 354, 493–494, 531

C

C. difficile, 316–318, 332, 336
California, xiv–xv, xx, xxiii, 7, 50, 69, 74, 96, 147–148,
    159–160, 166, 182–183, 210, 213–214, 218,
    220–221, 236–237, 260, 274, 321, 360–364,
    374, 439–440, 523, 526, 553–554, 557, 582
    California Childcare Education Project, 160
    California Department of Public Health, 362–364
    California encephalitis, 321
    California Mosquito Control Programs, xx, 362
    California OSHA, xv, 236
    California Seismic Safety Commission, 210
camping situations, 386
Canada, 7, 24, 39, 74, 76, 105, 109, 130, 154, 156, 164,
    260, 283, 345, 371, 396, 398–399, 494, 527,
    556–557, 574, 577, 583
    Canadian Association of Physicians for the
        Environment, 154, 164
    Canadian Institute for Public Health Inspectors
        Ontario Branch, 154
    Canadian Partnership for Children's Health and
        Environment, 154, 160, 164, 166
cancer, 6–7, 113, 162, 203, 327, 494, 509
Carbon monoxide, 27–29, 62–63, 158, 173–175, 392
Carbon pollution, 24, 68
Carbon Tetrachloride, 50
Carcinogens, 7, 112–113, 129, 355, 480, 507, 565
cardiopulmonary resuscitation (CPR), 184, 337, 392
cardiovascular disease, 6, 22, 48–49, 60, 92, 142
chemical agents, categories of, 196
cellular DNA, 7
Cement Kiln Dust, 36, 455
Centers for Children's Environmental Health and Disease
        Prevention Research, 148, 151
Centers for Disease Control and Prevention (CDC), 143,
        234, 258, 288, 346, 367
Centers for Disease Control and Prevention—
        Environmental Health Services Branch, 151
Centers for Medicare and Medicaid Services, 210
Chemical(s), 33, 159–163, 305, 491, 511–512, 568–574
    Chemical and Biological Releases, 131–132
    chemical industry, 6, 36–37, 51, 70, 170,
        479–481, 508
    chemical manufacturing, 28, 36, 74, 486
    Chemical warfare, 132, 196–197
    chemicals, basic, 36, 480–481, 483, 495
Chernobyl, 170
Chesapeake Bay Watershed, 440
Chicago, 26, 81, 108, 185, 218, 341, 367, 370, 522,
        560, 578
    Great Chicago fire of 1871, 81
Chigger mites, 348
Chikungunya virus, 315, 349
    chikungunya fever, 8
Child labor, 108, 128
    Child labor laws, 108
child mortality rate, 109
Childcare Inspection Checklist, 160
Children's Center Program, 147
    Children's Environment and Health Action Plan for
        Europe, 149
    Children's Environmental Airway Disease
        Center, 148

Children's Environmental Health Center, 146, 148
Children's Environmental Health Network, 145, 154,
    159–160
Children's Health Protection Advisory Committee,
    109, 145
Chlorine, 35, 302–305, 419–420, 435, 538
Chlorine dioxide, 325, 420, 542–544
Chlorites, 544
Chlorofluorocarbons, 23, 56, 68, 126, 375, 465
Chlorpyrifos, 119, 137, 355
Cholera, 15, 107–108, 117, 246, 317, 367, 403
Chromated copper arsenate, 123, 138, 360
Chromium, 48, 50–52, 138, 457, 479, 533, 560
Chronic exposures, 129
Cincinnati, Ohio, 340
    Cincinnati Children's Hospital, Cincinnati, Ohio, 340
    Cincinnati Children's Hospital Medical Center, 148
Cisterns, 174, 533, 569–570
Cities Readiness Initiative, 3
City Centre, Houston, Texas, x, 102
Civil War, 11, 81, 288
Clean Air Act, 11, 23, 33, 51, 66–69, 334, 515
    Clean Air Act standards, 25
Clean Water Act of 1972, 12, 146, 408
Clean Water State Revolving Fund, 438
Clemson University, 103
Climate Change, 24, 117, 531, 564
Climate Change Program, 3
Clinicians, 143, 157
Clinton, President William, 109, 146, 275
Clostridium botulinum, 248, 322
Clostridium difficile, 290, 318
Clostridium perfringens, 316
Coal, xxx, 12, 23–24, 38–40, 456–457, 468, 481
Coalbed Methane Outreach Program, 72
Coalition to End Childhood Lead Poisoning, 155, 159
Coast Guard Regulations, 395
Coastal areas, 5–6, 12, 189, 529
Coastal Zone Management Act of 1972, 12
Coke, 27, 35, 41–42, 48, 50, 504
Cold Stress, 186, 208, 211
Collaborative on Health and the Environment, 155
Collection Systems, 30, 377, 405, 447, 465, 489, 540
Columbia Center for Children's Environmental Health,
    119, 147
Columbus, 26, 76, 102, 164, 260, 275, 277–281, 285, 343,
    362, 364, 400, 441, 524–525
    Columbus Public Health (Ohio) Food Protection
        Program, 277
    Columbus Public Health Division of Environmental
        Health, xx, 362, 364
Combined Sewers, 46, 408–410, 416, 418, 575, 578
Combined Stormwater and Sanitary Sewers, 531
Cometabolism, 479
Commission for Environmental Cooperation, 39, 109,
    130, 494
Community, 12–18, 130, 154–160, 224, 333–334, 514
Community Engagement Initiative, 4
Community Health Aides, 144
Community Mental Health Centers Act, 82
Composite liners, 499
Composting, 58, 314, 436, 471, 519
Comprehensive Environmental Response, Compensation,
    and Liability Act, 12, 146, 395, 513

Comprehensive Preparedness Guide, 214, 219–220
Congress, xxxv, 66–69, 145–146, 270–271, 439, 516
Conservation Movement, 10
Constituents of Air Pollution, 21
Constructed Free Water Surface Wetlands, 413
Construction and Demolition Industry, 531
Construction and Demolition Wastes, 314, 446, 454, 467
Consumer Product Safety Commission, 150–151, 234, 515
contaminated food, 49, 131, 175, 257, 315, 371
Contra Costa counties, 159
control strategy, 71
Corporation for National and Community Service, 211
Corrosivity, 494
Creek Superfund Site, 148
Crimean War, 288
Crimean–Congo hemorrhagic fever, 239, 321
criteria pollutants, 21, 25–27, 31, 41, 46–47
    Criteria Air Pollutants, 22–23, 27, 31, 39, 41–42
cross–connection, 379–380, 386, 398, 421, 547–548, 550–551
Crude Oil, 43–45, 170, 456, 463, 474, 479–481, 491
Cruise ships, 269, 371–376, 378, 380, 385, 393–394, 396–399
    Cruise Vessel Security and Safety Act, 393
Cryptosporidiosis, 86, 108, 403, 450
Cryptosporidium, 199, 315, 317, 380, 390, 393, 541–544
Cryptosporidium parvum, 249, 322, 419, 537
Cyanotoxin Management Plan, 540, 554
Cyclodiene pesticides, 64

D

Dams, 6, 79, 187–188, 456, 459, 554, 563
Day care centers, 163, 230–231, 288
Decentralized Wastewater Management Program, 439
deer fly, 324, 347
Deforestation, 6, 24, 86
Delta Stewardship Council, 554
Dengue fever, 9, 117, 141, 316, 349, 372
Department of Education, 146, 150
Department of Health, Education, and Welfare, 66
Department of Homeland Security, 215
Department of Housing and Urban Development, 89, 91–92, 101, 146, 150, 553
Department of Mental Health, 103
Detroit, Michigan, 26, 86, 157, 367, 529
Diarrheal-type diseases, 109
Diazinon, 119, 137, 360
Diesel exhaust, 7, 33, 49–50, 137, 156
disabilities, 108, 112, 118, 137, 336, 384, 509
Disaster Mitigation Act, 213
Disaster Site Worker Course Number 7600, 212
Disinfectant byproducts, 544–549
Disinfection of Effluent, 419
Dissolved oxygen, 403, 416, 434, 544, 561, 568, 571
District of Columbia Coalition to End Childhood Lead Poisoning, 159
Division of Strategic National Stockpile, 215
Double liners, 500
Drainage wells, 476
drinking water, 126, 140–142, 173–174, 358–361, 515, 529–549, 551–557
Drinking Water Treatment, 538, 540, 542–543, 548

drop shaft, 406–407
drought, 5–6, 24, 189, 210, 459, 571
Dublin, Ohio, x, 102
Duke University Southern Center on Environmentally-Driven Disparities in Birth, 147
Dust mites, 61, 113, 134, 158, 348, 376

E

EAF, 118
Early Care and Learning, 160
Earthquake, 95, 180–185, 205, 210–211, 456, 474, 534
Eastern equine encephalitis, 321, 349
Ebola virus, 319, 324
Ecosystem, xlii, 3–6, 24–25, 190, 365, 403–404, 515–517
Egg Products Inspection Act, 271
ehrlichiosis, 8, 351
Electric Power Plants, 29–31, 38–39, 46, 50, 531
Electronic waste, 496
Emergency Planning and Community Right-to-Know Act, 12, 146, 209, 514
emerging diseases, 5, 8–9, 171, 315
Employee safety, 306, 308–310, 336, 340
Encroachment, 6, 86, 554
Endangered Species Act, 12, 516
endocrine system, 6, 112, 361
    Endocrine disruptors, 6, 112, 137–138, 353, 358, 480
    Endocrine Disruptor Screening Program, 361
ENERGY STAR, 57, 94, 103, 313
Enhanced Fujita Scale (EF-Scale), 193
Entamoeba histolytica, 322, 419, 544
Enterobacteriaceae, 317
Environment and Public Health Committee of the United States Senate, 145
Environmental Cooperation Council, 39
Environmental Emergencies, Disasters, and Terrorism, 531
Environmental Health Aides, 144
Environmental Health and Disease Prevention Research, 147–148, 151
environmental health concerns, 162
Environmental Health Emergency Response Guide, 212
Environmental Health Programs, xlvii–xlviii, 2, 15, 142, 150–151, 156–157, 265–268
Environmental Health Tracking Program, 3
Environmental Justice, 4
Environmental Justice Concerns, 4
environmental pollutants, 6
Environmental Protection Agency (EPA), xxxv, 160, 360, 438, 553, 576, 579
Environmental Public Health Performance Standards, 4
Environmental Specialists (CES), 157
environmental stressors, 9
environmentally attributable fraction, 118
Enzymes, 7, 113
Erosion, 38, 56, 365, 407–408, 502, 564, 574
    Erosion and Sediment, 531, 537, 562, 569, 572
Escherichia coli, 9, 239, 316, 371, 380, 414, 450
    Escherichia coli 0157:H7, 108, 537
    E. coli, 118, 199, 240, 252, 417, 419, 432
    E. coli 0157:H7, 248, 262, 390, 538
ethyl benzene, 7, 44
European Agency for Safety and Health at Work, 337
Evacuation centers, 169–170, 210, 214

Evapotranspiration, 181, 413, 422, 424, 429–430
Executive Orders
    Executive Order 12898, 4
    Executive Order 13045, 109
    Executive Order 13423, xliii, 13, 20
    Executive Order 13514, xliii, 13, 20
    Executive Order 13650, 487
EXTOXNET, 360
Extraction of Resources, 531, 565

F

fact sheets, 70, 160, 195, 197, 360, 572
Fair Labor Standards Act of 1938), 108
Fairfax County, Virginia, 440
    Sewer Maintenance Program, 440
FDA Food Code, 259, 270, 282
FDA Food Safety Modernization Act, 270
Federal Emergency Management Agency, 183, 214–215,
    219–221, 417, 516
Federal Food, Drug, and Cosmetic Act, 11, 271, 358
Federal Hazardous Substances Act, 515
Federal Insecticide, Fungicide, and Rodenticide Act,
    11–12, 146, 271, 335, 357, 516
Federal Land Policy and Management Act of 1976, 12
Federal Meat Inspection Act, 271
Federal Support for State and Local Response Operations
    in Water Contamination Crises, 553
Federal Water Pollution Control Act, 437, 551, 574–575
Fertilizers, 7, 34, 96, 129–130, 251–252, 482–483,
    568–570
Filtration systems, 172, 382, 392, 481, 540
Financial stress, 2
Fish and Wildlife Coordination Act, 515
Fisheries Conservation and Management Act of 1976, 12
fixed film method, 433
flaring, 32, 44, 481
Flint, Michigan, 529, 553, 557
    Flint River, 529
Flooding, 5–6, 59, 192, 262, 304, 417–418, 548
Floods, 95, 141, 187–189, 213–214
Fluorinated gases, 56
Food Allergen Labeling and Consumer Protection
    Act, 271
Food Quality Protection Act, 11, 137, 271, 358, 361
Food Regulatory Program Standards, 258, 260, 276, 278
Food Safety and Inspection Service (FSIS), 258
foodborne diseases, 6, 86, 248, 254, 269, 273
    Foodborne Diseases Active Surveillance Network
        (FoodNet), 273
Forest Service, 516, 581
Formaldehyde, 22, 39, 41, 50, 62–63, 95–96, 134–135
Fourth Ministerial Conference on Environment and
    Health, 109, 149
Fox River Environment and Diet Study, 148
Fracking, 23, 43, 474, 534, 565
Francisella tularensis, 324
Fugitive dust, 21, 30, 41, 457, 489, 506

G

Gary, Indiana, xlv–xlvi, 20, 398, 441, 560, 578
gastroenteritis, 117, 390, 397, 403, 541
Geomembrane or flexible membrane liners, 499

Geosynthetic clay liners, 466, 499
Geothermal reinjection wells, 476
Giardia lamblia, 322, 380, 419, 450, 537
giardiasis, 117, 403, 450
Global Change Research Act, 5
global climate change impacts, 5
Global Methane Initiative, 70–71
Global Warming, 1, 5–6, 8, 56, 191, 445, 477
glycol dehydration unit, 44
Golob, Brian R, 212
Government Accountability Office, 39, 109, 145–146, 474
Government agencies, 69, 155, 169, 487, 566
Gram-negative bacteria, 304
Grand Calumet River, 560, 578–579
Grassy swales, 569
Great Depression, 81–82
Great Lakes, xxviii, 408, 559–560, 567, 577–578, 582–584
    Great Lakes Restoration Initiative, 577–578, 583
    rebuilding of the Great Lakes, 578
Green Infrastructure, 418, 439
Greenhouse gases, 23–24, 56–58, 68, 73, 371, 463, 565
Greenland, 5
Greenville, South Carolina, x, 103
Greenville Mental Health Center, 103
Groundwater, 5–7, 146, 449–451, 476–478, 507–508, 529–530
Groundwater Protection through Density Control, 87

H

H1N1 influenza A virus, 366
H. pylori, 317
habitat, alteration and destruction of, 563
Haloacetic acids, 544
Hamburg, Germany, 408
Hammond, Indiana, 560
hantavirus, 108, 141, 352
Harvard School of Public Health, 148
Hayes, Samuel P, 10
Hazard Analysis and Critical Control Points (HACCP),
    241, 297
Hazard Assessment, 210, 236, 518
Hazardous and Solid Waste Amendments, 146, 473
hazardous household items, 135
Hazardous Materials, 27, 376, 383–384, 445, 477–481,
    499, 530–532
    Hazardous Materials Transportation Uniform Safety
        Act of 1990, 514
    Hazardous Waste Incinerators, 40, 48, 504
    Hazardous Waste Landfills, 501–503, 506–507
    Hazardous Waste Piles, 502
    Hazardous Waste Tank Systems, 501
    Hazardous Waste(s), 14, 126, 387, 445, 502, 516–517, 529
    hazardous waste, categories of, 494
    hazardous materials, classes of, 487
haze prevention program, 67
Hazus-MH, 210, 214
HAZWOPER, 212
Health Protection Advisory Committee, 109, 145
healthcare environment, xxxvii, 229, 287
Healthy Community Design Program, 3
Healthy Home and Community Environments Program, 3
    Healthy Homes and Lead Poisoning Prevention
        Program, 3
    Healthy Homes Project, 158

Healthy Kids, Healthy Schools, 151
Heat stress, 128, 185, 211, 509–510, 512
    heat-related deaths, 24, 185
    heat-related illnesses, 6, 86, 141
heavy metal, 6, 36–37, 41–42, 45, 48, 54–55, 111, 113, 236,
        240, 244–246, 299, 384, 403, 408, 450, 452,
        458, 476–477, 533–534, 560, 563, 578
Helicobacter pylori, 316
Hemorrhagic fever
    Argentine hemorrhagic fever, 324
    Crimean–Congo hemorrhagic fever, 239, 321
    Marburg hemorrhagic fever, 324
    Viral hemorrhagic fevers, 195, 289, 324, 367
Hepatitis, 127–128, 199, 209, 246, 313, 321–322, 371
    hepatitis A and B, 108
    hepatitis C, 8, 205, 289, 296, 308, 316, 334
    hepatitis E, 8
Highway Research Program, 576
HIV, 8, 127–128, 205, 289, 296, 308, 313, 334
Home health care, 329–330
Homelessness, 2, 82, 89, 115, 147
Hospital and Medical Waste, 531
Household Hazardous Waste, 64, 96, 171, 464, 495–496, 579
Household Products, 58, 63, 65, 133, 158, 227, 531
Housing Capacity Study, 87
human genetics, 6
Hunger, 2, 10, 82, 115–116, 165, 198, 482
Hurricane Katrina, 177, 210, 217
hurricane wind scale, Saffir-Simpson, 192
Hydrochlorofluorocarbons, 68, 465
hydrogen sulfide, 21, 189, 384–385, 467–468, 506, 509, 534

I

Ignitability, 494
illegal discharges, 566
Illicit Discharges, 531, 566, 579
immigrants, 81–83, 116, 134
immune system, 7, 22, 61, 147–148, 315–316, 355, 390
Improving Kids Environment, 153
Incinerators, 38, 40–41, 53, 139, 335, 469, 504
Increased temperatures, 5–6
Indiana Harbor and Ship Canal, 560
Indianapolis, Indiana, 26, 102, 153, 164
indoor air quality program, 163
Industrial accidents, 169–170, 211, 214
Industrial wastes, 408, 452, 531, 537
Infant mortality rates, 109
infectious diseases, xxix–xxx, 2, 6, 8–9, 315, 337, 367
infectious hepatitis, 403
Initiative on Children's Environmental Health, 155
injection well, 472–479, 507, 531, 565
Institute for Healthcare Improvement, 338
Institute for Occupational Safety and Health, 129, 152,
        234, 327, 337
Integrated pest management, 159–161, 356–357, 360, 377,
        397, 581
Integrated Public Alert and Warning System, 215
Integrated solid waste management, 519
Internal Combustion Engines, 22, 41
Internal Technical Directive on Reviewing EPA
        Enforcement Cases for Potential
International Convention for the Safety of Life at Sea, 394
International Environmental Education Programme, 14

International Federation of Red Cross and Red Crescent
        Societies, 211
Interstate Highway System, 79, 85, 99
Intervention, 97, 147, 157–158, 235, 258, 276, 305
Invasive Species, 531, 560, 567, 578
irrigation, 5, 59, 124, 251–252, 356, 551, 562

J

Japanese encephalitis, 321
Johns Hopkins Center for Childhood Asthma in
        the Urban Environment, 147
Johnson, President Lyndon B., 18
Joint Commission, 291, 331, 337, 339
Juice Hazard Analysis Critical Control Point
        Regulations, 271

K

King County, Washington, 157
Klebsiella, 316–317
Koren, Dr. Herman "Hank," xxix
Kyasanur Forest disease, 321
Kyoto Protocol, 14, 56

L

La Cross virus, 321
Land Treatment Units, 502–503
Landfill facilities, 465
Landfill gases, 465, 467
Landfill Methane Outreach Program, 72
Large flow applications, 428, 440
Lassa fever, 324, 352
Latino children, 109, 116, 119–120, 158
leachate, 403, 455, 465–468, 499–500, 502–504,
        507–508, 560
    leachate collection system, 465, 500
Lead, 28–29, 97
    Lead poisoning, 28, 97, 114–115, 132, 155–156,
        159–160, 162
    Lead Screening Guidelines, 159
leak detection system, 36, 500–502, 507
Legionella, 26, 304–305, 390, 393
    Legionnaires' disease, 333, 372, 376, 390, 529
Leptospirosis, 352, 403, 450
Lima, Ohio, 339
    Lima Memorial Health System, xix, 339
Los Angeles, California, Sewer Maintenance
        Program, 440
City of Los Angeles
    Emergency Operations Plan, 213
    Hazard Mitigation Plan, xiv, 213
    Sewer Maintenance Program, 440
Low Dissolved Oxygen, 568
Lupus erythematosus, 7
Lyme disease, 8, 86, 351, 363

M

major causes of habitat alteration, 563
major groups of air pollutants, 562
Malaria, 8, 109, 141, 150, 316, 349, 372
Manholes, 406–408, 410, 416, 435, 540

Marburg hemorrhagic fever, 324
Marine Mammal Protection Act of 1972, 12
Marine Protection, Research, and Sanctuaries Act
    (Ocean Dumping Act), 395
Marshfield, Wisconsin, 154
    Marshfield Clinic Research Foundation, 154
materials management, 462
measures of water quality, 561
Medical Waste, 96, 173, 299–300, 334–335, 460–461,
    497–498, 531
Medicare and Medicaid Programs, 210
Membrane Bioreactor, 412, 434
Meningitis, 108, 289, 316, 352
Mercury, 35–36, 50, 96, 119, 150–151, 313–315, 563
    mercury contamination, 117, 134
Metals, 6, 35–36, 205, 315, 408, 476–477, 568–569
metals in gaseous form, 21
Methamphetamine, 135–136
Methane, 23, 34, 43–47, 70–72, 245, 445, 450–451
methyl tert–butyl ether, 7
Mexico, 4, 24, 39, 56, 109, 117, 120, 156, 170, 366, 560
Michigan Network for Children's Environmental Health, 153
microorganisms, 8, 122, 195, 239–240, 321–322, 435,
    547–548
Migrant Clinicians Network, 143, 157
    migrant labor, 89, 109, 119, 163
Migratory Bird Treaty Act, 515
Mineral and fuel recovery wells, 476
Mineral Processing Waste, 457–458
Mining, 6, 12, 71–72, 148, 457–460, 473, 531
    mine drainage, 6, 530, 561, 577
    Mine Waste, 170, 458–460, 561
    Mining and Extraction of Resources Wastes, 531
Ministerial Conference on Environment and Health, 109, 149
Minnesota, xxviii, xlvi, 105, 138, 165–166, 221, 522, 527,
    557, 579, 582
    Minnesota Pollution Control Agency, 579
Mobility, 9, 110, 169, 185, 225, 248, 356
Model Aquatic Health Code, 392, 397
Model Solid Waste Management Program, 520
Mold, 97–98, 175–177, 188, 204, 211, 292–293
monoxide, 21–25, 27–29, 62–63, 173–175, 203–204, 392, 506
    Montreal Protocol, 14, 56, 68
Mosquito(es), 8, 117, 122, 315–317, 349–350, 362, 372
    Aedes, 315–316, 345, 349
    Anopheles, 349
    Culex, 349
mound system, 422–423, 432–433
Mount Sinai Center for Children's Environmental Health
    and Disease Prevention Research, 147
MTBE, 7
Muir, John, 11
Multi-Hazard Mitigation Plan, 213
    Multi-Hazard Mitigation Planning Program, 215
Municipal Solid Waste, 39, 56, 313, 463, 466–471, 494,
    505–507
Mycobacterium abscessus, 316
    Mycobacterium tuberculosis, 316

N

Nashville, Tennessee, 338
National Action Plan, 235
National Air Toxics Assessment Program, 25

National Association of City and County Health
    Officials, 154
National Association of School Nurses, 154
National Asthma Control Program, 3
National Center for Environmental Health, 2, 152,
    211, 396
National Center for Injury Prevention and Control, 234
National Children's Center for Rural and Agricultural
    Health and Safety, 154
National Community Decentralized Wastewater
    Demonstration Project, 439
National Cooperative, 576
National Drought Mitigation Center, 210, 217–218
National Earthquake Hazards Reduction Program, 210, 218
National Emission Standards, 26, 54, 67, 504–505
National Environmental Education Foundation, 152
National Environmental Health Association (NEHA), 155
National Environmental Policy Act, xxxv, 11, 513
National Fire Danger Rating System, 190
National Fire Protection Association, 207, 292
National Flood Insurance Program, 215–216
National Institute for Occupational Safety and Health
    (NIOSH), 129, 234, 327, 337
National Institute of Building Sciences, 183, 210
National Institute of Environmental Health Sciences,
    148, 151
National Library of Medicine Environmental Health and
    Technology, 152
National Marine Sanctuaries Act, 394
National Nosocomial Infection Surveillance System, 288
National Occupational Research Agenda (NORA), 235
National Pesticide Information Center, 71, 360
National Pesticide Program, 361
National Poison Data System, 359
National Pollutant Discharge Elimination System, 413,
    439, 514, 521, 552, 575
National Primary Drinking Water Regulations, 530, 544
National Priorities List, 458, 507, 561
National Research Council of the National Academy of
    Sciences, 24
National Resources Defense Council, 155
National Restaurant Association, 264
National Safety Council, 155
National Science Foundation, 210
National Water Quality Inventory Report, 559
Native American, 89, 101, 148, 195, 209, 227–228, 270
natural disasters, 3, 8, 85, 95, 149, 181, 216
Natural Gas, 30–32, 43–45, 56–58, 70–72, 456, 474, 565
Natural Gas STAR Program, 70, 72
Natural Sources of Pollution, 531, 533
Needlestick Safety and Prevention Act, 334
New Jersey Center for Childhood Neurotoxicology and
    Assessment, 148
New York Comprehensive Emergency Management and
    Continuity of Operations Plan, 214
NIOSH, xxxvi, 129, 152, 155–156, 234, 327, 337
Nipah virus, 86, 239
Nitrification, 412, 419, 450, 539, 544–547, 556
nitrogen oxides, 21–23, 31–36, 45–46, 54, 68–69, 137, 313
Nitrous oxide, 34, 56–58, 128
non-hazardous waste streams, 445
non-point source, 188, 409, 437–438, 532, 560, 563,
    568–569
Non-polio enteroviruses, 316

Noonan, Gary, xlv
Norovirus, 239, 274, 316, 336, 371, 380, 390
Northern Manhattan/South Bronx, New York, 119
Norwalk-like calciviruses, 316
Norway rats, 345
NSF International, 431, 546, 552
Nurses, 143–144, 153–154
Nutrients, 22, 403, 476–477, 483, 532, 568–570, 581–582

O

Obama, President Barack, 68, 276, 487, 578
occupational exposure, 7, 143, 158, 236, 334, 355, 358
   occupational hazards, 128, 333, 369, 511
Occupational Health and Safety Program, 330, 385, 517
Occupational Health Surveillance Program, 236–237
Occupational Safety and Health Administration (OSHA),
   xxxvi, 203, 234, 311, 337, 370
Office of Children's Health Protection, 109
Office of Public Health Preparedness and Response, 215,
   321, 337
Office of the Inspector General, 210
Ohio Department of Health, 423
Oil Pollution Act, 12, 146, 395, 513
On-Site Sewage Disposal, 81, 99–100, 388, 419, 421,
   423–424, 532
On-site water sources, 100
Ontario Association of Supervisors of Public Health
   Inspectors, 154
Open Air-Ways for Schools, 163
Oregon Department of Environmental Quality, 72
   Oregon State University, 360, 364, 523
organisms, categories of, 322
Oriental rat flea, 323
OutbreakNet, 273
Outcomes, 116, 134, 144, 147, 269, 287
Oxygen deficiency, 509–510
Oxygen–Demanding Substances, xxvii, 403, 568, 571
Oxygenated gasoline, 67
Ozonation, 420
Ozone, 5–7, 23–29, 66–69, 121, 420, 542–544
   ozone alert, 26, 54
   ozone depletion, 22, 56

P

package treatment plant, 418–419
Pandemic and All-Hazards Preparedness Reauthorization
   Act, 209
Particulate Matter, 21–25, 28–30, 141, 370, 450, 506
parvovirus B19, 8
pathogens, categories of priority, 195
Pediatric Environmental Health Specialty Units, 156
Pediatrician, 118, 143–144, 153, 161
Penn, William, 80
Percival, Thomas, 288
Permeable pavement, 569–570
permeation and leaching, 547
Pesticide Data Program, 360
Pesticide Registration Improvement Act, 358
Pesticide Worker Safety Program, 4
Pesticides, xxxvii, 8, 100, 137–138, 140, 345–361,
   571–572
Petrochemicals, 480–481

pets, 61, 63–64, 131, 204, 229, 326, 495
   Pets Evacuation and Transportation Standards
      Act, 209
   Pet Waste, 156, 532, 568, 574, 581
Pharmaceuticals, 7, 36, 299–300, 454, 460, 484, 494
Philadelphia, 17–18, 80, 159, 199, 240, 275, 277–281, 285,
      338, 352, 367, 408, 421, 442, 449, 578, 580
   Philadelphia Department of Public Health, 17, 275,
      277–278, 281, 285
   Philadelphia Suburban Water Company, xxviii, 580
   Philadelphia, a Community Childhood Lead Poisoning
      Prevention Project, 159
Phoenix Center, 103
Pittsburgh Childhood Lead Screening Guidelines, 159
Pittsburgh Lead-Safe Coalition, 159
Plague, 195, 321, 323, 326, 345–346, 352, 366–367
$PM_{2.5}$, 25, 29–30, 33
PM_{10}, 25, 29–30, 58, 86, 113
pneumonia, 92, 108–109, 290, 332, 366, 372
Pneumonic plague, 321, 323
point of contact, 7
point of origin, 7, 107, 366, 368, 451
Poisoning, 17–18, 100, 132–133, 155–156, 159–160,
      227–228, 529
Polio entroviruses, 316
political environment, 266
pollution, sources of, 31
Pollutants, 21–27, 92–94, 117, 403
   pollutants found in sewage, 403
Polycyclic organic matter, 50
Population growth, xxxv, 9, 79, 191, 471, 567
Portland Air Toxics Solutions, 72
Potable Water Storage Facilities, 545
Poultry Products Inspection Act, 271
Poverty, 2, 10, 82, 89, 115–116, 119, 149
power outages, 5
Precautionary Principle, 4
President's Council on Sustainable Development, 12, 20
Pressure System, 192, 430
pretreatment for raw water, 538–539
preventative programs, 161
Prevention of Significant Deterioration Program, 67
Prions, 295, 316
priority pathogens, categories of, 195
programs, 8–12, 15–19, 25, 27–28, 33, 51, 66–68,
      70–71, 73, 85, 88, 97, 99, 101, 103, 108,
      115–116, 123–124, 127, 142–145, 147, 150–152,
      154–157, 160–165, 170–171, 181, 199–200,
      207–210, 212–215, 221, 226, 229–230, 232–237,
      244–245, 252–254, 259–260, 262–269,
      271–272, 274–278, 280–282, 285, 299, 307,
      309–310, 312, 314, 325, 331, 334, 337–339,
      350, 354, 357, 359–362, 364–365, 385, 388,
      396–397, 409, 421, 437–440, 446–447, 451,
      453, 459, 461, 463–464, 466, 486, 497, 513,
      515–517, 519, 521, 540, 551, 553–554, 557, 560,
      566, 575–577, 579–581
Project BioShield Act, 209
Promoting Ecological Health for the Whole Child, 155
Protecting Children's Health, 149, 151
Protocol on Substances that Deplete the Ozone Layer, 14
Pseudomonas aeruginosa, 390
Public Health Service Act, 393
Public Works Administration, 81

Pullman, Illinois, 408
PulseNet, 273
Pumps, 44–45, 174, 302, 432–434, 481, 501
Purdom, Dr. Walton P, 17
Pure Food and Drug Act, 11

R

Radioactive materials, 161, 198, 456, 480, 487, 517, 565
    Radioactive Waste, 14, 214, 299, 313, 315, 335,
        498–499
Radiological Emergency Response Related Links, 216
    Radiological Terrorism, 197
Radon, 7, 43, 59, 122–123, 160–161, 225, 533
rain garden, 569
Ravenna, 8
raw water source, 537–540, 545, 552
    raw water, pretreatment for, 538–539
re-emerging diseases, 5, 9, 315
Reactivity, 489–490, 494
Recharge wells, 477
recreational environment, xxxvii, xli, 127–128, 365
Recycling, 27–28, 48, 54, 173, 462–465, 488, 519–523
Red Crescent, 211
Red Cross, 211, 392
reduced crop yields, 5
Registered Environmental Health Specialists, 144–145
Registered Sanitarians, 144
regulatory agencies, goal of the, 259
Resource Conservation and Recovery Act, 12, 146, 334,
        395, 513, 560
resources, 4–5, 9–15, 19, 69, 71, 75–77, 79, 84, 88–90,
        96, 101, 103, 105, 110, 116, 120, 142–144, 146,
        149, 151–152, 155–156, 162, 164–165, 169,
        181–183, 190–191, 195, 200–202, 209–213,
        215, 217, 219, 226, 233–234, 239, 259, 264,
        266–267, 272–277, 310, 315, 321, 334–336,
        338, 343, 359, 363, 369, 385, 392, 394–395,
        399, 438–439, 451, 462–463, 474, 480, 493,
        513–517, 520–523, 525, 527, 531, 537, 552, 559,
        565–566, 575–576, 582
respiratory disease, 24, 62, 86, 129, 316, 509
Restaurant Sanitation Regulations, 270, 278
Ricin toxin, 322
Rio Earth Summit, 14
Riparian buffers, 569–570
Rivers and Harbors Appropriation Act, 11, 574
Road construction, 6, 86, 88, 90
Robert T Stafford Disaster Relief and Emergency
        Assistance Act, 209, 516
Robert Wood Johnson Foundation, 3, 223
    Robert Wood Johnson Wingspread Conference
        Center, 4
Rocky Mountain spotted fever, 351
rodent-borne diseases, 6
Roman fever, 345
Roosevelt, President Theodore, 10
Rotating Biological Contractor Systems, 434
Rotavirus, 108, 317, 319
Rubella, 128, 289, 371–372
Runoff and Non-Point Source Pollution, 532
Runoff from Hazardous Chemicals Spills, 532
rural communities, 88, 91, 148
Rural homelessness, 89

S

S. aureus (MRSA), 288, 317, 336
Sacramento, California, 210
    Sacramento, 74, 210, 218, 221, 260, 274, 363–364, 525,
        554, 557
Safe Drinking Water Act, 12, 146, 334, 359, 515, 551, 575
Safe Water, Food, and Waterborne Illnesses Programs, 3
Saffir-Simpson hurricane wind scale, 192
Salmonella, 120, 248–250, 255, 262, 390, 450, 537–538
    Salmonella typhi, 380
Samuel J. Crumbine Consumer Protection Award, 277, 282
San Francisco, 104, 155, 166, 342, 367, 442, 525, 553
San Joaquin River Delta, 554
Sand and Salt Storage, 532, 571
Sand filters, 422–423, 431–432, 544
SARS outbreak, 315
Scabies mites, 348–349
School environmental health issues, 151
School Hazardous Waste, 496–497
Schools, 79–80, 122–124, 126–127, 154, 161–163,
        165–167, 230–231
scrub typhus fever, 348
sea levels, 5, 24, 564
Seafood Hazard Analysis Critical Control Point
        Regulations, 271
Seattle Public Utilities, 580
Sediment, 6, 45, 376, 413, 541, 545, 567–573
seepage pit, 421, 423, 428, 430–431
Septage, 415, 427, 435
    Septage treatment, 415
septic, 91, 124, 126–127, 323, 413, 539, 575
    Septic systems, 81, 86–87, 140, 422–423, 476, 539, 575
    septic tank systems, 99–100, 137, 177, 427–428, 536,
        552, 572
    SepticSmart Week, 439
service line, 521, 547–548
sewage systems, 99, 174, 386, 388, 404, 421–423, 540
sewage treatment
    primary sewage treatment, 411
    secondary treatment, 377–378, 403, 411–412, 414
    tertiary treatment, 412, 476
Sewage Treatment Plants, 85, 137, 404, 413, 514, 568, 572
Sewage treatment wastes, 455
Sewer Laterals, 403–405
Shigella, 246, 371, 390
    Shigellosis, 9, 108, 403
Sierra Club, 11
Sinclair, Upton, 108
Single liners, 499
slag heaps, 6
Sludge, 47, 411–412, 414–419, 433–434, 455–456, 458, 569
Small System, 435
Smallpox, 107, 131, 195, 321, 323, 325–326, 367
Society for Healthcare Epidemiology of America, 337
Soil amendments, 468, 569–570
Soil Tests, 251, 425
solid waste collection, 176, 445–446
Source reduction, 180, 256, 299, 363, 462, 519–520, 580
Source Review program, 66
South American arenaviruses, 352
Spalding technique, 295
Spring Garden Street Civic Association, 18
St. Thomas Midtown Hospital, 338

St. Vincent's Medical Center, Bridgeport, Connecticut, 339
Standardized Emergency Management Systems, 213
Standards for Criteria Pollutants, 25
Staphylococcal enterotoxin B, 322
    staphylococcal infections, 108
Staphylococci, 246
    Staphylococcus aureus, 9, 127, 288
State and tribal water pollution control programs, 439
State Governmental Agencies, 490, 553
State of Washington Department of Ecology, 580
Stephenson, John B, 146
Stockholm Conference, xxxvi, 13–14
Stormwater, 52–53, 124–125, 314, 439, 560–562, 575–577,
    582–583
    stormwater flow, controlling, 569
    Stormwater Pollution Prevention Plan, 580
    Stormwater Program, 437, 579
Streptococcus, 108, 252, 390
subsurface flow wetland, 413
sulfates, 30–31, 468, 479
    sulfate air pollution, 7
sulfur dioxide, 21–23, 30–33, 45–47, 121, 374, 456
Superfund Amendments and Reauthorization Act,
    334, 513
    Superfund site, 7, 139, 148, 507, 560
Surface Impoundments, 48, 54, 456–458, 502, 507,
    566, 573
Surface Mining Control and Reclamation Act of 1977, 12
Suspended Growth Method, 434
Sustainability, 10–13, 16, 80, 170, 266, 275, 530
Sustainable Communities Program, 439
systems approach in problem solving and resolution, 244

T

tailings, 6, 458–460
    Tampa General Hospital, Tampa, Florida, 340
Tar, 35, 42–43, 50, 148
Temperature, 61, 92, 95–96, 185–187, 241, 404, 573
Terrorism, xxx, 2, 169, 269, 366, 445, 530–532
    Bioterrorism, 3, 9, 194–195, 216, 324–326, 360
    Radiological, 197, 220
    terrorists, 8–9, 131, 206–207, 281, 321–323, 407
Texas Colorado River Floodplain Coalition Multi-
    Jurisdictional Hazard Mitigation Plan, 216
Texas Commission on Environmental Quality, 555
Third World Health Organization International
    Conference on Children's Health and the
    Environment, 150
Thunderstorm, 181, 187, 190, 193, 216, 392
Thymus, 7
tick-borne encephalitis, 321
Toilets to Reduce Water Flow, 436
toluene, 7, 33, 44, 46, 48, 53–55
Tornados, 193
toxic air pollutants, 22–23, 26–27, 33, 39, 45–49, 51, 53–55
Toxic Exposure Surveillance System, 359
Toxic Substances Control Act, 12, 145–146, 335, 515
Toxic Waste Storage and Handling Areas, 532
Toxicity, 27, 36, 55, 112–113, 328, 494–495, 520–521
Toxics Release Inventory, 25–26, 130, 494
TOXNET, 152
Toxoplasma gondii, 317
    T. gondii, 322

transfer stations, 448
Transportation of Hazardous Materials, 487, 491–492, 519
Transportation Research Board, 576
Trust for America's Health, 3, 223
tuberculosis, xxx, 9, 81, 316, 320–321, 366–367, 371–372
Tularemia, 195, 324, 326, 346–347, 351–352
Turbidity, 539, 541–543, 547, 549, 561, 567
Twin Cities Metro Advanced Practice Center, 212, 281
type I diabetes, 7
typhoid fever, 107, 195, 246, 315, 347, 403

U

Ultraviolet light, 111, 115, 321, 412, 420, 543
    Ultraviolet Radiation, 23, 56, 119, 250, 321, 420, 435
underground injection wells, types of, 473
United Nations Conference on Environment and
    Development, 14
United Nations Conference on Sustainable Development,
    xliii, 15
United Nations Environmental Programme, xxxvi, xliii, 14
United Nations Office for the Coordination of
    Humanitarian Affairs, 212
United States Coast Guard, 374, 376, 379, 394, 396, 516, 567
United States Consumer Product Safety Commission,
    150, 234
United States Department of Agriculture, 71, 101, 150,
    258, 360, 449, 581
United States Department of Education, 150
United States Department of Energy, 71, 150
United States Department of Health and Human Services,
    92, 150, 272, 305, 331, 335–336, 553
United States Department of Housing and Urban
    Development, 91–92, 101, 150, 553
United States Department of Labor, 150
United States Department of the Interior, 150
United States Department of Transportation, 492
United States Drought Monitor, 210
United States Environmental Protection Agency, 360, 438,
    521–528, 553, 576
    EPA authority, 575
    Environmental Protection Agency Hotline and Service
        Line Program, 521
United States Geological Service, 183, 581
United States Geological Survey, 211, 517
United States Global Change Research Program, 5
United States Program on Recycling of Tires, 520
United States-Mexico Border 2020 Program, 4
Universal wastes, 494
University Extension Services, 552
University Medical Center at Brackenridge, 338
University of California Davis Center for the Study of
    Environmental Factors in the Etiology of
    Autism, 147
University of California, Berkeley Center for Children's
    Environmental Health Research, 147
University of California, Los Angeles, 148
University of Cincinnati, 148
University of Illinois at Urbana–Champaign, 148
University of Iowa, 148
    University of Iowa Children's Environmental Airway
        Disease Center, 148
University of Medicine and Dentistry of New Jersey, 148
University of Michigan, xxxv, xxxix, 148, 157

University of Nebraska-Lincoln, 217
University of Southern California, 148
University of Washington, 148
unpasteurized milk, 138, 254
Upstate Homeless Coalition of South Carolina, 103
Urban, 6, 80–89, 117–118, 157–158, 162–163, 553, 560
    urban environment, 84–85, 118, 147, 157–158
    urban homelessness, 89
utility waste wells, 477

V

vancomycin–resistant Enterococci (VRI), 317
vector-borne disease, 141
Venezuela equine encephalitis, 321
Vibrio cholerae, 248, 317
Vibrio parahaemolyticus, 246, 390
Vibrio vulnificus, 246, 390
violations of federal EPA regulations, 123
Virginia Graeme Baker Pool and Spa Safety Act, 395
Volatile organic compounds (VOCs), 21, 42–46, 51–52, 137, 313, 450, 506

W

Wagner Steagall Housing Act, 81
Walter S. Mangold Award, xxxi, xlvi, xlviii
War on Poverty Rodent Control Program, 18
Washington State Department of Social and Health Services, 341
Washington, DC., 19–20, 73–77, 104–106, 160, 164–167, 218–221, 237, 282–285, 341–343, 363–364, 367, 398–400, 440–443, 522–528, 556–557, 582–584
waste management program, 446, 450–451, 513, 520–521
Wastewater Collection System Toolbox, 440
wastewater disposal wells, 476
Wastewater pump/lift stations, 406
water contamination, 109, 322, 380, 530, 537, 553, 561
water cycle, 6, 564

water mains, 303, 547, 549
Water pollution, 6, 314, 437, 439, 455, 559, 573–579
waterborne diseases, 109, 117, 175, 195, 371, 450, 550
Waters Program, 4
WaterSense, 438
watershed management, 188, 439, 563
West Nile virus, 141, 211, 321, 362
Western equine encephalitis, 321, 349
Wetland, 6, 413, 439, 575, 578
Wetlands, 45, 84, 413–414, 554, 560, 562, 576
Wildfires, 24, 41, 141–142, 189–193, 213, 532
Wind-Related Storms, 192–193
Wingspread Conference, 4
Wisconsin, 148, 154, 165, 442, 541, 553
Worchester, Massachusetts, x, 103
Workers, xxx, 14–15, 171–172, 174, 200–206, 211–212, 383–385
World Commission on Environment and Development, 14
World Conservation Strategy, 14
World Health Organization, 24, 109, 149–150, 198, 260, 368, 396
World Health Organization Task Force for the Protection of Children's Environmental Health, 109
World Summit on Sustainable Development, 14
World War I, 196–197, 288
World War II, 11, 81, 85, 99–100, 195, 199, 529

X

xylenes, 7, 44, 46

Y

Yellow fever, 8, 195, 317, 345, 349, 367, 372
Yersinia pestis, 323
YWCA, 18

Z

Zoning Board, 101
zoonotic diseases, 8